ECOSYSTEMS OF THE WORLD 15

FORESTED WETLANDS

ECOSYSTEMS OF THE WORLD

Editor in Chief:
David W. Goodall

CSIRO Division of Wildlife and Ecology, Midland, W.A. (Australia)

I. TERRESTRIAL ECOSYSTEMS

    A. Natural Terrestrial Ecosystems
1. Wet Coastal Ecosystems
2. Dry Coastal Ecosystems
3. Polar and Alpine Tundra
4. Mires: Swamp, Bog, Fen and Moor
5. Temperate Deserts and Semi-Deserts
6. Coniferous Forest
7. Temperate Deciduous Forest
8. Natural Grassland
9. Heathlands and Related Shrublands
10. Temperate Broad-Leaved Evergreen Forest
11. Mediterranean-Type Shrublands
12. Hot Desert and Arid Shrubland
13. Tropical Savannas
14. Tropical Rain Forest Ecosystems
15. Forested Wetlands
16. Ecosystems of Disturbed Ground

    B. Managed Terrestrial Ecosystems
17. Managed Grassland
18. Field Crop Ecosystems
19. Tree Crop Ecosystems
20. Greenhouse Ecosystems
21. Bioindustrial Ecosystems

II. AQUATIC ECOSYSTEMS

    A. Inland Aquatic Ecosystems
22. River and Stream Ecosystems
23. Lakes and Reservoirs

    B. Marine Ecosystems
24. Intertidal and Littoral Ecosystems
25. Coral Reefs
26. Estuaries and Enclosed Seas
27. Continental Shelves
28. Ecosystems of the Deep Ocean

    C. Managed Aquatic Ecosystems
29. Managed Aquatic Ecosystems

ECOSYSTEMS OF THE WORLD 15

FORESTED WETLANDS

Edited by

Ariel E. Lugo

Institute of Tropical Forestry
U.S.D.A. Forest Service
Southern Forest Experiment Station
Call Box 25000
Rio Piedras, PR 00928-2500 (U.S.A.)

Mark Brinson

Department of Biology
University of East Carolina
Greenville, NC 27834 (U.S.A.)

and

Sandra Brown

Department of Forestry
University of Illinois
110 Mumford Hall, 1301 W. Gregory
Urbana, IL 61801 (U.S.A.)

ELSEVIER
Amsterdam—Oxford—New York—Tokyo 1990

ELSEVIER SCIENCE PUBLISHERS B.V.
Sara Burgerhartstraat 25
P.O. Box 211, 1000 AE Amsterdam, The Netherlands

Distributors for the United States and Canada:

ELSEVIER SCIENCE PUBLISHING COMPANY INC.
655, Avenue of the Americas
New York, N.Y. 10010, U.S.A.

```
Library of Congress Cataloging-in-Publication Data

Forested wetlands / edited by Ariel E. Lugo, Sandra Brown, and Mark
  Brinson.
      p.   cm. -- (Ecosystems of the world ; 15)
    Includes bibliographies and indexes.
    ISBN 0-444-42812-7 (U.S.) : fl 300.00 (Netherlands : est.)
    1. Wetland ecology.  2. Forest ecology.   I. Lugo, Ariel E.
  II. Brown, Sandra.   III. Brinson, Mark M.   IV. Series.
  QH541.5.M3F67 1989
  574.5'26325--dc19                                          88-39111
                                                                  CIP
```

ISBN 0-444-42812-7

© Elsevier Science Publishers B.V., 1990

All rights reserved. No part of this publication may be reproduced, stored in a retrieval system or transmitted in any form or by any means, electronic, mechanical, photocopying, recording or otherwise, without the prior written permission of the publisher, Elsevier Science Publishers B.V./Physical Sciences & Engineering Division, P.O. Box 330, 1000 AH Amsterdam, The Netherlands.

Special regulations for readers in the USA — This publication has been registered with the Copyright Clearance Center Inc. (CCC), Salem, Massachusetts. Information can be obtained from the CCC about the conditions under which photocopies of parts of this publication may be made in the USA. All other copyright questions, including photocopying outside of the USA, should be referred to the publisher.

No responsibility is assumed by the Publisher for any injury and/or damage to persons or property as a matter of products liability, negligence or otherwise, or from any use or operation of any methods, products, instructions or ideas contained in the material herein.

This book is printed on acid-free paper

Printed in The Netherlands

PREFACE

The impetus to write this book was the enthusiasm of Professor Howard T. Odum who in 1972 organized the Center for Wetlands at the University of Florida and gave two of us the opportunity to work with him in the wetlands of Florida. In 1977, Dr. Odum asked A.E. Lugo to put together a review of available literature on the forested wetlands of the world. With this book, that assignment is finally completed.

Dr. Odum's assignment coincided with the publication of V.J. Chapman's volume, *Wet Coastal Ecosystems*, the first in the series *Ecosystems of the World*. Inquiries showed that the series *Ecosystems of the World* was not planning special coverage of Forested Wetland systems but that an extra volume for the purpose would be welcome. From the beginning, our intention was to prepare a volume that emphasized the structural and functional aspects of forested wetlands. However, we soon realized that many studies were floristically oriented. This volume thus represents a compromise between our initial ecosystem-oriented objective and a more floristic approach to studies of forested wetlands.

The book has two parts. In Part one (Chapters 1–8) we review available information on the structure and function of forested wetlands. This part emphasizes concepts and is strongly biased toward forested wetlands in the Caribbean and the United States of America. This is due partially to our experience with this part of the world, and partially to the fact that much of the ecosystem-oriented research has been conducted in this region. Chapter 8 for example, by Carter, concludes Part one with a pictorial overview of the intensive and technologically advanced research being conducted in the Great Dismal Swamp located between Virginia and North Carolina in the United States.

The second part (Chapters 9–18) presents case studies and descriptions of forested wetlands from those parts of the world where we could find collaborators. We are extremely grateful for the excellent response that we received from these colleagues with interests similar to ours. They provide the international flavor that is typical of volumes in this series. These studies also document the many similarities and differences among forested wetlands growing in different parts of the world. Because of the rudimentary state of understanding of forested wetland function, we emphasize as much as possible functional and structural similarities among wetlands. We believe that, in the absence of specific information, such an approach facilitates management and conservation decisions, and helps focus further research. Chapter 10, by Alvarez deviates from the others by focusing on a wetland type rather than reviewing a region.

The last chapter of the book is a synthesis and an effort to present new paradigms of wetland ecology. Chapter 4 was dedicated to short topical discussions designed to serve two purposes: they clarify concepts and they focus attention on important or unresolved issues. Each topic is a self-contained contribution with supporting data and references.

The book is intended as an introduction to the subject of forested wetlands for anyone interested in these neglected ecosystems. Chapters in Part one also have the objective of stimulating research activity in forested wetlands. Part two has the objective of providing a global overview to the state of knowledge of forested wetlands. We present the volume as a token contribution towards increasing the understanding of natural ecosystems, with the conviction that we must first understand forests in order to manage and conserve them for the benefit of human beings.

Many people and institutions provided support while we worked on this volume. We thank all of them and especially our home institutions and our respective support staffs at the USDA Forest Service's Institute of Tropical Forestry, the Department of Biology at East Carolina University, and the Department of Forestry at the University of Illinois. Earlier, the University of Florida's Center for Wetlands and the US President's Council on Environmental Quality supported our efforts. Helen A. Nunci, Gisel Reyes and Veronica Lugo were especially helpful in the later stages of editing. JoAnne Feheley, librarian at the Institute of Tropical Forestry, was very helpful with library searches.

ARIEL E. LUGO
MARK M. BRINSON
SANDRA BROWN

Rio Piedras, Puerto Rico

LIST OF CONTRIBUTORS

M. ALVAREZ-LOPEZ
University of Puerto Rico
Regional Colleges Administration
Ponce Technological University College
Box 7186
Ponce, PR 00732 (U.S.A.)

P.R. BACON
3 The Woodlands, Brightlingsea
Essex CO7 0RY (Great Britain)

M.M. BRINSON
Department of Biology
East Carolina University
Greenville, NC 27834 (U.S.A.)

S. BROWN
Department of Forestry
University of Illinois
110 Mumford Hall, 1301 W. Gregory
Urbana, IL 61801 (U.S.A.)

E.F. BRUENIG
Institute of World Forestry
University of Hamburg
Leuschnerstrasse 91
D-2050 Hamburg 80 (F.R. Germany)

V. CARTER
U.S. Geological Survey
Reston, VA 22092 (U.S.A.)

M. ISMAIL
University Staff Quarters
16/A Fullar Road
Ramna, Dhaka-2 (Bangladesh)

P.C. KANGAS
Biology Department
Eastern Michigan University
Ypsilanti, MI 48197 (U.S.A.)

A. LOT
Instituto de Biología
Universidad Nacional Autónoma de México
04510 México, D.F. (México)

A.E. LUGO
U.S.D.A. Forest Service
Institute of Tropical Forestry
Call Box 25000
Rio Piedras, PR 00928-2500 (U.S.A.)

R.L. MYERS
Archbold Biological Station
Route 2, Box 180
Lake Placid, FL 33852 (U.S.A.)

A. NOVELO
Instituto de Biología
Universidad Nacional Autónoma de México
04510 México, D.F. (México)

K. PAIJMANS
C.S.I.R.O.
Division of Land Use Research
P.O. Box 1666
Canberra, A.C.T. 2601 (Australia)

R.L. SPECHT
Department of Botany
University of Queensland
St. Lucia, Qld. 4067 (Australia)

A.J. SZCZEPAŃSKI
Institute of Ecology
Polish Academy of Sciences
Wetland Research Laboratory
PL-11-730, Mikotajki (Poland)

J. WIEGERS
Meidoornlaan 24
1185 JX Amstelveen (The Netherlands)

CONTENTS

PREFACE V

LIST OF CONTRIBUTORS VII

Chapter 1. INTRODUCTION
by A.E. Lugo 1

Definition and classification of forested wetlands . 2
Examples of "marginal" forested wetlands . . . 8
References 10

Chapter 2. AN ENERGY THEORY OF LANDSCAPE FOR CLASSIFYING WETLANDS
by P.C. Kangas 15

Introduction 15
Energy sources in ecology 15
Energy signatures 15
Incorporating the spatial expression of energy . . 16
Land classification 17
A wetlands application 20
Conclusion 22
References 23

Chapter 3. LONG-TERM DEVELOPMENT OF FORESTED WETLANDS
by P.C. Kangas 25

Introduction 25
Geology of wetland formation 25
Ecology of wetland formation 28
Conclusions 43
References 44

Chapter 4. CONCEPTS IN WETLAND ECOLOGY
by A.E. Lugo, S. Brown and M.M. Brinson 53

Introduction 53
1. Estimation of wetland areas 53
2. Global role of wetlands 53
3. Organic-carbon export from wetlands . . . 55
4. Forested wetlands and animal life 56
5. Large-scale reduction of forested wetland areas . 58
6. Hydroperiod 58
7. The influence of water on forested wetlands . . 59
8. Xeromorphism in wetland plants for water conservation 61
9. Energy language 63
10. The energy signature 64
11. Zonation and succession 64
12. Wetlands on slopes in Puerto Rico 65
13. Wetland stressors 66
14. Wetland values 68
15. Sediment and peat accumulation 69
16. Definitions of concepts dealing with organic-matter dynamics 70
17. Tolerance of trees to flooding 72
18. Anaerobic processes and detritus processing . 74
19. Response of wetlands to stressors 75
20. Peat loss rates 76
21. The influence of forested wetlands on water and soils 77
References 79

Chapter 5. RIVERINE FORESTS
by M.M. Brinson 87

Introduction 87
Geomorphology 89
Hydrology 92
Community patterns 94
Biomass and production 102
Element distribution and dynamics 108
Geographic variation 113
Impacts and management 129
Conclusions 134
References 134

Chapter 6. FRINGE WETLANDS
by A.E. Lugo 143

Introduction 143
Structure of saltwater fringe forests 144
Dynamics of fringe mangrove forests 152
Freshwater fringe wetlands 165
Human impacts on fringe wetlands 165
References 166

IX

Chapter 7. STRUCTURE AND DYNAMICS OF BASIN FORESTED WETLANDS IN NORTH AMERICA by S. Brown	171

Introduction	171
External factors affecting basin wetlands	172
Distribution and floristic composition	178
Structural characteristics	182
Organic matter dynamics	185
Nutrient storages and dynamics	189
Effects of disturbances on basin wetlands	196
References	196

Chapter 8. THE GREAT DISMAL SWAMP: AN ILLUSTRATED CASE STUDY by V. Carter	201

References	210

Chapter 9. ECOLOGY AND MANAGEMENT OF SWAMP FORESTS IN THE GUIANAS AND CARIBBEAN REGION by P.R. Bacon	213

Introduction	213
The vegetation	213
Swamp structure and development	222
Human influence on swamp vegetation	240
Fauna	241
Management	245
Acknowledgements	248
References	248

Chapter 10. ECOLOGY OF *PTEROCARPUS OFFICINALIS* FORESTED WETLANDS IN PUERTO RICO by M. Alvarez-Lopez	251

Introduction	251
Description of study areas	251
Floristic composition	252
Structural characteristics of *Pterocarpus* forests	256
Ground biomass	258
Phenology	259
Seedling density, germination and regeneration	260
Proposed *Pterocarpus officinalis* seed implantation mechanism	261
Ecotonal zones and *Pterocarpus* climax forests	262
Discussion	264
Acknowledgements	264
References	264

Chapter 11. PALM SWAMPS by R.L. Myers	267

Introduction	267
Palm swamp types	267
Costa Rican palm swamps	273
Discussion	281
Acknowledgements	284
Appendix: Methods used for observations at Tortuguero	284
References	285

Chapter 12. FORESTED WETLANDS OF MEXICO by A. Lot and A. Novelo	287

Introduction	287
High forested wetlands	287
Low forested wetlands	290
Shrub wetlands	295
Discussion	296
Acknowledgements	297
References	297

Chapter 13. OLIGOTROPHIC FORESTED WETLANDS IN BORNEO by E.F. Bruenig	299

Introduction	299
Site conditions	302
Vegetation	309
Utilization and silviculture	330
References	333

Chapter 14. WOODED SWAMPS IN NEW GUINEA by K. Paijmans	335

Introduction	335
Area	336
Classification	336
Swamp forest	336
Swamp woodland	347
Swamp scrub	349
Swamp successions and sequences	351
Uses	354
Acknowledgements	355
References	355

Chapter 15. ENVIRONMENT AND ECOLOGY OF FORESTED WETLANDS OF THE SUNDARBANS OF BANGLADESH by M. Ismail	357

Introduction	357
Regional climate	359
Mangroves of the Sundarbans and Singapore	361
Uses of the mangroves	365
Management	369
Acknowledgements	369
Appendix: Biotic composition of the Sundarbans	370
References	385

CONTENTS

Chapter 16. FORESTED WETLANDS IN
AUSTRALIA
by R.L. Specht 387

Introduction 387
Major tree species 389
Structure of forested wetlands 395
Variation in structure within a climatic region . . 395
Environmental factors: soil fertility 397
Environmental factors: temperature response . . 398
Environmental factors: water balance 400
Fire 402
Regeneration of tree species 403
Community dynamics 404
References 405

Chapter 17. FORESTED WETLANDS IN
WESTERN EUROPE
by J. Wiegers 407

General characteristics of the area surveyed . . . 407
General ecological characteristics 413
Composition, structure and ecology of different types
of wetland forests 416
The animal component 428
Conclusions 432
Acknowledgements 432
References 432

Chapter 18. FORESTED WETLANDS OF
POLAND
by A.J. Szczepański 437

Introduction 437
Alder swamps 437
Forests of other wet sites 440
Succession 442
Soils 443
Litter production 444
Decomposition and nutrient cycling 445
References 446

Chapter 19. SYNTHESIS AND SEARCH FOR
PARADIGMS IN WETLAND
ECOLOGY
by A.E. Lugo, M.M. Brinson and
S. Brown 447

The role of hydrology 447
The influence of nutrients 451
The energy signature approach 457
Elemental cycles 458
References 460

AUTHOR INDEX 461

SYSTEMATIC INDEX 473

GENERAL INDEX 493

Chapter 1

INTRODUCTION

ARIEL E. LUGO

Forested wetlands cover at least 250 million ha (Ch. 4, Sect. 1). They have important roles in global biogeochemical cycles, supporting fresh- and salt-water commercial fisheries, and in providing a place for wildlife of all kinds to flourish (Ch. 4, Sect. 2, 3, 4). Through history people have learned to cultivate wetlands for food (Armillas, 1971), drain them to increase wood or food production (Miller and Maki, 1957; Heikurainen, 1961; Skoropanov, 1968; Campbell and Hughes, 1981), use them to obtain all the life support necessary for survival (Beadle, 1974), or squander the resource through over-exploitation or in search of greater land values (Horwitz, 1978; Ch. 4, Sect. 5).

Whatever the motives for their use, forested wetlands have attracted the attention of people through history, whereas scientific attention towards these ecosystems has lagged. There are few comprehensive works on forested wetlands of the world; the exceptions are Chapman's books on mangrove vegetation (1976, 1977) and a monograph in Russian entitled "Forested Wetlands of the World". Certain aspects of forested wetland ecology, management, or studies of particular forested wetland ecosystems may be well represented in the literature (Table 1.1), but in general forested wetlands have not been studied in the detail typical of upland ecosystems such as boreal or temperate forests (e.g., Larsen, 1980; Persson, 1980; Reichle, 1981).

Using Chapman's review and classification of over 700 titles on mangrove ecosystems (Rollet, 1981, abstracted about 6000 references on mangroves), Lugo, 1984, identified five stages in the development of knowledge and understanding of mangrove forests. These stages were early accounts (325 B.C.–A.D. 136); early descriptions (1230–1731); formalization of descriptive accounts beginning in 1753; ecological studies (still in progress but confined to this century); and the studies of functional attributes beginning in the 1930's. In the 1970's an experimental phase began in Florida with comparative studies of channelized and unmanaged mangrove forests (Carter et al., 1973), and studies of mangrove responses to sewage enrichment (Sell, 1977). Lugo (1984) observed that research related to freshwater forested wetlands in the United States developed in a similar way to that of mangrove wetlands, but with a time lag. Ecosystem function studies in other wetlands, for example, lagged behind those in mangrove ecosystems by about a decade. A major emphasis of this book is to attempt to develop unifying principles and data bases on the structure and function of forested wetlands, with the aim of stimulating scientific study of them.

The lack of systematic studies of forested wetland ecosystems has probably led to an underestimation of their area (particularly in moist and wet tropical environments) and their importance to global and regional biogeochemical cycles. Nowhere is the lack of attention to forested wetlands more evident than in the confused state of the nomenclature used to identify a forested wetland (Table 1.2). The nomenclature for forested wetlands is restrictive because the emphasis has been on species composition and vegetation types which vary geographically, rather than on hydrology and geomorphology which vary much less geographically. As a result, many ecosystem types that should be recognized as forested wetlands are not, leading to underestimates of their importance.

TABLE 1.1

Examples of forested wetland regions or subjects that have been studied in some detail; more examples in other chapters of this book

Type of forest or subject	Geographic location	Source
Riparian and bottomland forests	United States	Johnson and Jones, 1977
		Sands, 1977
		Greeson et al., 1979
		Johnson and McCormick, 1979
		Clark and Benforado, 1981
Pocosins	North Carolina	Richardson, 1981
Peatlands (mires)	World	Moore and Bellamy, 1974
Reclamation of peat bogs	U.S.S.R.	Skoropanov, 1968
Forested wetlands	Far East	Whitmore, 1984
Forested wetlands	Southeastern U.S.	Carter et al., 1973
		Cohen et al., 1984
		Ewel and Odum, 1984
Mangrove forests	World	Chapman, 1976, 1977
Mangrove forests	Australia	Clough, 1982
Effects of flooding	All vegetation	Hook and Crawford, 1978
		Kozlowski, 1984

DEFINITION AND CLASSIFICATION OF FORESTED WETLANDS

Wetlands are areas that are inundated or saturated by surface- or ground-water, at such a frequency and duration that under natural conditions they support organisms adapted to poorly aerated and/or saturated soil (modified from the definition of the U.S. Army Corps of Engineers, 1977). This definition uses two criteria for defining a wetland: saturated or poorly aerated soil, and biotic response to the soil condition. These criteria recognize that the main environmental driving force of a wetland is its hydrologic regime. Sections 6 and 7 of Chapter 4 give reasons why hydroperiod and aeration are the main factors that determine a wetland. The above definition also takes into consideration biotic response in terms of adaptations of organisms. Time is another critical element of any definition of wetland. In this definition time is included in the frequency and duration of soil inundation or saturation. Frequency can span anywhere from days (for example, in mangroves) to centuries (Brink, 1954) or millennia (Buell, 1945), and duration of the hydroperiod can also vary greatly.

A forested wetland is any wetland with a significant component of woody vegetation, regardless of the size of the plants. This lack of limit on tree size is in recognition of the effects of environmental stressors on tree size (Table 1.3). For example, the mangrove *Rhizophora mangle* and the cypress *Taxodium distichum* form mature forest stands at heights that range between 1 and almost 50 m (Brown and Lugo, 1982).

As with any definition, there are advantages and disadvantages in what is proposed above. The main disadvantage is the lack of absolute criteria to identify a wetland or a forested wetland. This problem will be difficult to overcome because most wetlands exist along numerous environmental gradients without clear boundaries between different ecosystems. Environmental variables such as temperature, salinity, or soil fertility often confuse the task of identifying what a wetland is because they modify the ecosystem's response to hydroperiod and soil aeration. For example, leaf sclerophylly is common in wetland plants (see Ch. 4, Sect. 8; Chapter 11; also Bruenig, 1971; Larsen, 1982; Saenger, 1982). This condition both reflects and affects water-nutrient relations. Also, soil salinity reduces the resistance of mangroves to low temperature and inundation (Lugo and Patterson-Zucca, 1977).

On the other hand, the proposed definition of a forested wetland has the advantage of simplicity. Traditional definitions usually require standing water before a forest is recognized as a wetland

INTRODUCTION

TABLE 1.2

Forested wetlands in different parts of the world. The listing is not exhaustive but is compiled to illustrate the diversity in nomenclature used around the world. Terminology used by authors to describe wetlands has been retained when possible. The chapters in this book provide many more examples. The tree species composition of many of these wetlands is given in the species index

Type of wetland[1]	Location	Source
NORTH AMERICA		
Riverside swampland	Hay River, Canada	Hartland-Rowe and Wright, 1975
Fan palm oases	San Adreas fault, California	Vogl and McHargue, 1966
Phreatophyte systems	Rio Grande, New Mexico	Campbell and Dick-Peddie, 1964
Bog forests, moors, and birch bogs	Alberta, Canada	Lewis et al., 1928
Tamarack-dominated swamps, slowly drained swamp, prairie swamps	Northern Lake Michigan to Chicago, Illinois	Coulter, 1904
White cedar swamp	Central Florida	Harper, 1926; Collins et al., 1964
Cypress heads or domes	Central Florida	Monk and Brown, 1965
Deep water swamps, shallow water swamps, peaty freshwater swamps, and minor swamp types	Southern United States	Penfound, 1952
Tupelogum swamp	Alabama	Penfound and Hall, 1939
Bottomland forests	North-central Oklahoma	Rice, 1965
Cypress–gum, cedar, maple–gum, and mixed hardwood swamps	Great Dismal Swamp, Virginia	Wright and Wright, 1932; Whitehead, 1972; Dabel and Day, 1977
Cypress–gum swamp	Selma, Alabama	Hall and Penfound, 1943
Cypress swamps	United States	Mattoon, 1915; Ewel and Odum, 1984
Alder shrub	Finger Lakes, U.S.A.	Tilton and Bernard, 1975
Low floodplain	Central Oklahoma	Ware and Penfound, 1949
Bog forests	Wisconsin–Michigan	DeWitt and Soloway, 1978
Eight types of forested swamps	Okefenokee swamp, Georgia–Florida	Cohen et al., 1984
EUROPE AND RUSSIA		
Riverine forested wetlands	U.S.S.R.	Isakov, 1968
Forested bogs	U.S.S.R.	Yurkevich et al., 1966
Swamp forests	Finland	Heikurainen, 1961, 1967
Forested wetlands	Netherlands	Mörzer Bruyns and Westhoff, 1968
Sitka spruce forests	Northern Ireland	Adams et al., 1970
Mires	England	Ingram, 1967
Basin forested wetlands with marginal wetlands	Wybunbury Moss, Cheshire, England	Green and Pearson, 1968
AFRICA		
Phoenix palm forest	Namanve swamp, Uganda	Eggeling, 1935
Mpanga forest	River Nabu Kongole, East Africa	McCarthy, 1962
Swamp forests of different ages and composition	Western Ghana	Ahn, 1958
Floodplain forest	Bakundu forest reserve, West Africa	Richards, 1962
Mitragyna stipulosa swamp	Inland waters of Africa	Beadle, 1974
AUSTRALIA		
Swamp forests (seasonal tropical and subtropical, 8 types)	All continent	Webb, 1968

(continued)

TABLE 1.2 (*continued*)

Type of wetland[1]	Location	Source
Melaleuca swamp	Western Australia	Beard, 1967
Mangrove forests	Tropical and subtropical coastlines	Clough, 1982
CENTRAL AMERICA AND CARIBBEAN		
Mangrove forests, *Mora*, *Prioria*, and mixed flooded forests	Darien, Panama	Meléndez, 1965
Mangrove forests, freshwater tidal swamp forests, bamboo thickets, and *Bactris* swamps	Honduras	Carr, 1950
Mangrove forests, riverine, and swamp forests	Nicaragua	Taylor, 1961
Raphia swamps	Puerto Viejo, Costa Rica	Anderson and Mori, 1967
Mora, *Pterocarpus* marsh, palm forests and woodlands	Trinidad, West Indies	Beard, 1944, 1946a, b
SOUTH AMERICA		
Perched water table forest	Magdalena Valley, Colombia	Folster et al., 1976
Mangrove forests, *Mora*, Sajal[2], and Guandal[2] forests	Nariño, Colombia	Lamb, 1959
Pterocarpus forests, palm marsh forest	British Guiana	Fanshawe, 1952
Varzea[3] forests	Amazon, Brazil	Williams et al., 1972
Inundation forests	Central Amazonia, Brazil	Fittkau et al., 1975
Igapo[3] forest	Amazon, Brazil	Prance, 1979
Hyperseasonal savannah, swamp, and gallery forests	Cuba and South America	Sarmiento and Monasterio, 1975
Swamp gallery forest	Northeastern Matto Grosso, Brazil	Ratter et al., 1973
Bañados con seibal[4]	Rio La Plata, Argentina	Ringuelet, 1962
ASIA		
Shorea albida peat swamps	Sarawak and Brunei	Anderson, 1958, 1964 Bruenig, 1969, 1970
Alluvial swamps, lowland and hill swamps, and peat swamp forests	Malaya	Wyatt-Smith, 1959, 1964
Fresh and saltwater lowland swamps (six sequences)	Northeastern New Papua	Taylor, 1959
Wetlands	Iran	Firouz, 1974
Myristica swamps	Travancore, India	Moorthy, 1960
Occasionally flooded woodlands	River Vishwamitri, India	Chavan and Sabnis, 1960
Riverine forests	River Indus, West Pakistan	Ansari, 1961
Tropical valley swamps (4 forest formations)	Mothronwala swamp, Dehra Dun, India	Dakshini, 1960, 1965
Spring-fed swamp	Monthronwala swamp, eastern Doon valley, India	Deva and Singh Aswal, 1974
Swamp, riverine, and mangrove forests	West New Britain, Papua New Guinea	Liem, 1976

TABLE 1.2 (*continued*)

¹The plants indicated by vernacular names in this column are as follows:
Alder: *Alnus* spp.
Birch: *Betula* spp.
Cypress: *Taxodium distichum*
Fan palm: *Washingtonia filifera*
Loblolly pine: *Pinus taeda*
Mangrove: *Rhizophora mangle*
Maple: *Acer* spp.
Sitka spruce: *Picea sitchensis*
Tamarack: *Larix laricina*
Tupelogum (or "gum"): *Nyssa* spp.
White cedar or "cedar" (*Chamaecyparis thyoides*)

²Sajal: tall forest dominated by *Campnosperma panamensis*, inundated with fresh water but with some residual salt; Guandal: a tall freshwater swamp, almost a monoculture of *Dialyanthera gordoniaefolia*.
³Varzea, Igapo. These terms are explained in Ch. 5, p. 127.
⁴Bañados con seibal: a community with a sparse tree cover of *Erythrina cristagallii* and a herbaceous understory, periodically flooded.

TABLE 1.3

Examples of studies that document reduction of tree height in forested wetlands

Type of wetland	Factors associated with height reduction	Source
Pocosins	Soil oligotrophy, fire, flooding	Woodwell, 1956; Christensen et al., 1981; Richardson, 1981
Sitka spruce swamps	Low concentration of nutrients in leaves	Adams et al., 1970
	Low nutrient concentrations in leaves and peats	Parker, 1962
Mangrove forests	Soil salinity	Cintrón et al., 1978
	Low temperature	Lugo and Patterson Zucca, 1977
	Latitude	Cintrón et al., 1985
Cypress swamps	Water flow and lack of nutrients	Brown, 1981
Peat swamps	Hydroperiod	Bruenig, 1971

(cf., Beard, 1944; Van Steenis, 1958). Yet saturation of soil, or even a change in the water table, may affect trees and forests as much as flooding. By not considering forests exposed to periodic or occasional soil saturation as wetlands, important opportunities are lost for evaluating the boundary conditions of wetland ecosystems.

The common terminology used to classify forested wetlands (for example, swamp, inundation forest, floodplain forest, riparian forest, etc.) deserves revision to facilitate communication and analysis on a global scale. For example, "swamp" can describe a forested or non-forested wetland, or it can be used independently of hydrology if the soil is well drained most of the year and the flooding occurs when plants are not active. The number of forest classifications are many and are based on a multiplicity of criteria (Table 1.4). Very few efforts have been made to consolidate forested wetlands that are named differently but have similar characteristics. Also, systems with unusual hydroperiods are normally not included in most wetland classifications. For example, should forests that flood once per century be called wetlands? Are montane forests with saturated soil, but without obvious surface flooding, wetlands? Should pine forests with the water table near the surface be termed forested wetlands? Usually these ecosystem types are not included in descriptions of wetlands; however, I will give reasons why they should be.

The purpose of classifying ecosystems is to facilitate their study and management. The study and management of wetlands is primarily concerned with the response of organisms to hydroperiod and poor soil aeration. To better

TABLE 1.4
Examples of forested wetland classification systems

Region covered	Basis for grouping	Number of categories	Prevalent type of wetland	Author
United States	Life form, soil type, water regime and water chemistry	6 major levels 10 regional provinces 5 ecological systems (marine, estuarine, riverine, lacustrine, palustrine) 10 ecological subsystems 8–18 classes based on general vegetation type (forested shrub, submerged, etc.) 12–17 subclasses 8–25 orders	All wetlands including marine	Cowardin et al., 1979
Southern U.S.A.	Floristic description of plant communities in relation to water depth and quality	2 forested/non-forested 4 forested 3 non-forested	Fresh and saltwater	Penfound, 1952
Southeastern U.S.A.	Nutrient and water turnover and % time flooded	Continuum with 5 examples	Fresh and saltwater	Pool et al., 1972
Southwest U.S.A.	Floristic analysis and latitudinal gradients	6 biomes, 9 series, 22 associations	Riparian	Pase and Layser, 1977
Southwest U.S.A.	Plant habit; floristic analysis; and latitudinal gradients	12 riparian communities	Riparian	Brown and Lowe, 1974
Southwest U.S.A.	Latitudinally by plant habit; and floristic analysis	5 main formations	Riparian	Brown et al., 1977
Pacific Northwest, U.S.A.	Geographically in terms of topography, soils, and hydrology; functional roles of the zones	4 sites	Riparian zone	Meehan et al., 1977
Sacramento Valley, California, U.S.A.	Floristic, structural, and habitat analysis	5 types	Riparian	Conard et al., 1977
Inland U.S.A.	Water quality, physiognomy	7 forested wetlands	All types of inland wetlands	Goodwin and Neiring, 1975
Georgia, U.S.A.	Hydroperiod, water quality, geomorphology, species	2 categories by waterflow and hydric systems 39 community types	Coastal and inland	Wharton, 1978
Wisconsin, U.S.A.	Hydrology	4	All state wetlands	Novitzki, 1979
New Mexico, U.S.A.	By the presence of obligate riparian species	1 formation, 3 subformations, 6 series, 19 associations	Riparian	Dick-Peddie and Hubbard, 1977
Florida, U.S.A.	By the water flux and turnover, water source, water quality, and dominant species	5 waterflow types, 24 communities	Freshwater and saltwater	Wharton et al., 1976
North Carolina, U.S.A.	Hydroperiod, fire, and physiognomy	2 community classes and 5 types	Pocosins	Kologiski, 1977

INTRODUCTION

Location	Criteria	Classification	Wetland type	Reference
Lake Agassiz, Minnesota, U.S.A.	Floristic and physiognomic criteria	7 types	Peatlands	Heinselman, 1970
Canada	Physiognomy, surface morphology, vegetation, needs of particular disciplines	4 levels, 8 classes (three are forested) and numerous categories within each class (>30)	All country wetlands	Zoltai et al., 1975
United Kingdom	Soil pH and moisture	8	Moorlands	Pearsall, 1950
Central Europe	Biotic characters	61 peat producing communities	Mires	Tolpa et al., 1967
Finland	By presence of trees and their ground cover	70 swamp types under four general categories	Swamps	Heikurainen, 1961
Poland	Vehicle trafficability		Swamps	Shevchenko, 1973
Arkhangel'sk, U.S.S.R.	Water flow and quality	3 water motion categories, 7 groups of forests, and 17 forest types	Forested wetlands	P'yavchenko, 1957
New World tropics	By geomorphology of the basin	6 types	Saltwater	Lugo and Snedaker, 1974
Tropical America	Hydroperiod and physiognomy	2 formations and 8 communities	Tropical swamp forests	Beard, 1944
Australia	Site fertility and presence of palm		Rain forest vegetation	Webb, 1968
Malaysia	Succession, geography, and species composition	4 successional positions, 17 associations, and 12 types (10 were forested wetlands)	All forested lands	Wyatt-Smith, 1961
Papua New Guinea (wet tropics)	Salinity gradient and depth, duration, and frequency of flooding	8 sequences, 17 forested communities	Tropical swamps	Taylor, 1959

accomplish the goals of research and management it is necessary to determine the complete range of conditions that subject organisms to flooding and poor soil aeration. Traditionally, vegetation has been the unit classified rather than the ecosystem. The result of this approach is that for a given vegetation type literally dozens of communities are identified. For example, over 60 peat-forming communities have been listed (Moore and Bellamy, 1974) and well over 70 swamp forests have been named in Finland (Heikurainen, 1961). Much the same is true for saltwater forests (Chapman, 1976, 1977). Such high numbers of community types make the task of developing general principles or generalizations about the function of these ecosystems difficult. Taylor (1959) suggested that the value of floristic classifications was local and ecologists should use physiognomic classifications such as that of Beard (1944) for global comparisons. I would add that perhaps hydrologic or geomorphologic categories are better units of classification for global comparisons of wetlands.

In the first section of this book we have adopted the strategy of classifying the conditions that control the structure and behavior of forested wetlands by assuming that the physiognomy and floristic composition of the system will reflect the total energy signature of the ecosystem (Ch. 4, Sect. 9). This approach is described in detail by Kangas (Ch. 2), who classifies the embodied energy in the hydroperiod of wetlands as a unifying tool for explaining why wetlands with different species compositions or physiognomy may behave similarly or vice versa. Another approach that may help in understanding the complete diversity of wetlands is to focus attention on how plants or animals respond to soil saturation regardless of frequency or source of water. I propose that if the ecosystem exhibits any quantifiable response to soil saturation, a case can be made that one is dealing with a wetland.

EXAMPLES OF "MARGINAL" FORESTED WETLANDS

In the Río Grande of New Mexico, Campbell and Dick-Peddie (1964) showed that phreatophyte communities could be differentiated by moisture regimes from xerophytic to hydric, and that the species composition of sites responded to changes in the hydric regime. Soils in these communities were influenced by flooding. The case for calling these phreatophyte communities "forested wetlands" rests in the response of its plants to soil saturation caused by groundwater. Similar arguments could be made about rheophyte forests with specialized roots in stream water.

California palm oases provide more insight into phreatophytes and the importance of unusual events in determining wetland characteristics of vegetation (Vogl and McHargue, 1966). Plants in the oases were zoned according to soil moisture. Of 24 oases studied, only one species out of 78, the palm *Washingtonia filifera*, was ubiquitous. Seventeen per cent of the species were phreatophytes. Intense but unusual rains cause catastrophic floods that alter the geomorphology of the oases and vigorously stimulate vegetation regrowth. These events need to occur only once a century to maintain the palm populations in steady state, because the maximum palm age appears to be 200 years. Moreover, changes in water table affected plant survival. Lowering the water table (analogous to drainage) or raising it (analogous to chronic flooding) killed palms. Thus, although fire is also a dominant factor in the control of the physiognomy, species composition, reproduction, and productivity of these palm oases, the ecosystem exhibited periodic responses to hydroperiod and soil aeration typical of wetlands in general. Brink (1954) found that after a 100-yr flood in the Pacific Northwest of the United States, only bog species survived. Plants that had established themselves between catastrophic floods did not survive.

In his classic description of the tropical rain forest, Richards (1964) said: "There is some evidence that aeration as well as the water-supplying power of tropical soils is often a potent ecological factor" (p. 225). Other evidence suggested that root competition for oxygen segregated trees on the basis of their requirement for soil oxygen. No more attention was given to this topic in the book. The study by Odum (1970) in Puerto Rico provided detailed analysis of the interaction between soil saturation, atmospheric saturation deficit, and the water budget of montane tropical forests growing under excessive rainfall. The importance of these studies is that they suggest that wetland conditions in high-rainfall climates may develop almost anywhere, and need not occur over extensive areas. In fact, Odum described anaerobic soils rich in organic matter and the presence of wetland tree species in small pockets within a matrix of "normal" well-drained upland forests. These micro-wetlands are common in most tropical wet and rain forests (*sensu* Holdridge, 1967) and represent another situation where forested wetlands have not been properly assessed. Another similar example is the caatinga vegetation (growing on spodosols) of the Río Negro region in the Venezuelan Amazon (Herrera et al., 1978). These ecosystems are subject to periodic flooding and soil saturation due to high water table, and they export black waters to the Río Negro.

In the Gulf of Mexico coastal plain of the southeastern United States, four pine species are predominant (*Pinus taeda, P. elliottii, P. palustris,* and *P. echinata*). Over large areas of this enormous coastal plain these and other tree species are exposed to wetland conditions which significantly modify forest function, but which are not normally considered when these stands are classified. Studies of pine beetle (*Dendroctonus frontalis*) infestations in Louisiana and Texas led to findings that illustrate the importance of saturated soil conditions on ecosystem function (Lorio and Hodges, 1974). *Pinus taeda* trees growing in low areas are exposed to occasional flooding, acute soil saturation, and higher but more erratic soil moisture conditions than trees of the same species growing on higher ground (mounds). As a result, trees in the low areas had the following characteristics compared to those on mounds: faster but more erratic growth (Lorio and Hodges, 1971); shallow root system with numerous feeder roots but few growing inside the soil; modified root morphology (Lorio et al., 1972); less foliage (Lorio, 1966), lower

TABLE 1.5

Examples of biotic characteristics that recur in forested wetlands

Characteristic	Reason	Source
Presence of palms	Palms have excellent aeration mechanisms	Beard, 1944; Taylor, 1959; Webb, 1968; Frangi and Ponce, 1985
Leaf sclerophylly	May reflect nutrient and/or water limitations (see Ch. 4, Sect. 8)	Bruenig, 1971; Christensen et al., 1981; Saenger, 1982
Gas exchange structures including pneumatophores, lenticels, knees, aerial roots, swelling of base of trees	Required to overcome poor soil aeration	Whitmore, 1984
Sharp zonations	Reflecting gradients induced by hydroperiod or other hydrologic forces (see Ch. 4, Sect. 10)	see Ch. 5
Even-canopied forests	Related to species composition (even age, single species stands) and water relations (see Ch. 4, Sect. 8)	Bruenig, 1971; Brown, 1981 see also Ch. 11
Plank buttresses and stilt roots	Provide stability in muddy or steep conditions	
Surface or aerial roots	Related to poor soil aeration and mineral uptake	Odum, 1970
In moist environments: lower tree species diversity relative to adjacent uplands	Hydroperiod	Monk, 1965, 1966, 1967; Pires, 1978
In arid environments: higher tree species diversity, abundance, and basal area relative to adjacent uplands	Increased soil moisture	see Ch. 5

bark moisture (Lorio and Hodges, 1968a), and altered carbohydrate and nitrogen fractions in the inner bark (Hodges and Lorio, 1969) during periods of drought; reduced production and exudation pressure of oleoresin (Lorio and Hodges, 1968b); higher tree density (Lorio and Bennett, 1974); and greater susceptibility to pine beetle attack (Lorio, 1968).

All the differences discussed above were shown to be caused by rainfall and soil characteristics (Lorio and Hodges, 1971; Lorio, 1977) which in turn led to soil saturation. Trenching experiments were used to confirm the causes (Lorio and Hodges, 1977). In Florida, a similar environment led to reductions in evolution of carbon dioxide from the soil during periods of soil saturation in stands of *P. elliottii* (W.P. Cropper, K. Ewel, and H.L. Goltz, personal communication, 1984). Accordingly, I consider these areas to be forested wetlands because soil saturation is causing a significant ecosystem response even though dominant species composition remains unchanged from that of upland forests. But, because the hydroperiod is so short (weeks to months) and not obvious, these types of stands are not considered in descriptions of the region's wetlands. A listing of some of the characteristics that may be used as indicators of the possible presence of a forested wetland is given in Table 1.5. Periodic flooding or saturated soil may be the cause of these characteristics.

The rest of this book develops in more detail the structural and functional characteristics of forested wetlands from different parts of the world. We use as a unifying concept the grouping of forested wetlands into three types based on hydrology. The three types are riverine (Ch. 5), fringe (Ch. 6), and basin wetlands (Ch. 7). These wetland types are also discussed in Ch. 19. The hydrology of *riverine* wetlands is dominated by river floods and other riverine processes. *Fringe* wetlands grow on oceanic and lake shorelines where water flows are bidirectional. *Basin* wetlands are found in depressions where water accumulates, and may fluctuate in depth depending on the balance of rainfall runoff and evapotranspiration. As the book shows, these contrasting hydrological conditions are accompanied by different water qualities, and together influence wetland structure and function.

REFERENCES

Adams, S.N., Jack, W.H. and Dickson, D.A., 1970. The growth of Sitka spruce on poorly drained soils in northern Ireland. *For., J. Soc. For. G. B.*, 43: 125–133.

Ahn, P., 1958. Regrowth and swamp vegetation in the western forest areas of Ghana. *J. West Afr. Sci. Assoc.*, 4: 163–173.

Anderson, J.A.R., 1958. Observations on the ecology of the peat swamp forests of Sarawak and Brunei. In: *Proc. Symp. Humid Tropics Vegetation, UNESCO, Djakarta*, pp. 141–148.

Anderson, J.A.R., 1964. The structure and development of the peat swamps of Sarawak and Brunei. *J. Trop. Geogr.*, 18: 7–16.

Anderson, R. and Mori, S., 1967. A preliminary investigation of *Raphia* palm swamps, Puerto Viejo, Costa Rica. *Turrialba*, 17: 221–224.

Ansari, T.A., 1961. Riverain forests of Sind — hope or despair. *Emp. For. Rev.*, 40: 228–233.

Armillas, P., 1971. Garden on swamps. *Science*, 174: 653–661.

Beadle, L.C., 1974. *The Inland Waters of Tropical Africa*. Longman, New York, N.Y., 365 pp.

Beard, J.S., 1944. Climax vegetation in tropical America. *Ecology*, 25: 127–158.

Beard, J.S., 1946a. The mora forests of Trinidad, British West Indies. *J. Ecol.*, 33: 172–192.

Beard, J.S., 1946b. *The Natural Vegetation of Trinidad*. Oxford For. Mem., 20, Clarendon Press, Oxford, 152 pp.

Beard, J.S., 1967. Some vegetation types of tropical Australia in relation to those of Africa and America. *J. Ecol.*, 55: 271–290.

Brink, V.C., 1954. Survival of plants under flood in the lower Fraser river valley, British Colombia. *Ecology*, 35: 94–95.

Brown, D.E. and Lowe, C.H., 1974. The Arizona system for natural and potential vegetation — illustrated summary through the fifth digit for the North American southwest. *J. Ariz. Acad. Sci.*, 9 (Suppl. 3): 1–28.

Brown, D.E., Lowe, C.H. and Hausler, J.F., 1977. Southwestern riparian communities: their biotic importance and management in Arizona. In: *Importance, Preservation, and Management of Riparian Habitat: A Symposium*. U.S. Dept. of Agriculture, Forest Service, General Technical Report RM-43, Washington, D.C., pp. 201–211.

Brown, S., 1981. A comparison of the structure, primary productivity, and transpiration of cypress ecosystems in Florida. *Ecol. Monogr.*, 51: 403–427.

Brown, S. and Lugo, A.E., 1982. A comparison of structural and functional characteristics of saltwater and freshwater forested wetlands. In: B. Gopal, R.E. Turner, R.G. Wetzel and D.F. Whigham (Editors), *Wetlands: Ecology and Management*. National Institute of Ecology at Jaipur and International Scientific Publications, Jaipur, pp. 109–130.

Bruenig, E.F., 1969. The classification of forest types in Sarawak. *Malay. For.*, 32: 143–179.

Bruenig, E.F., 1970. Stand structure, physiognomy and environmental factors in some lowland forests in Sarawak. *Trop. Ecol.*, 11: 26–43.

Bruenig, E.F., 1971. On the ecological significance of drought in the equatorial wet evergreen (rain) forests of Sarawak (Borneo). In: J.R. Flenley (Editor), *The Water Relations of*

Malesian Forests. Transactions of the 1st Aberdeen–Hull Symposium on Malesian Ecology, University of Hull. Institute of Southeast Asian Biology, University of Aberdeen, Aberdeen, pp. 66–88.

Buell, M.F., 1945. Late Pleistocene forests of southeastern North Carolina. *Torreya*, 45: 117–118.

Campbell, C.J. and Dick-Peddie, W.A., 1964. Comparison of phreatophyte communities on the Rio Grande in New Mexico. *Ecology*, 45: 492–502.

Campbell, R.G. and Hughes, J.H., 1981. Forest management systems in North Carolina pocosins: Weyerhauser. In: C.J. Richardson (Editor), *Pocosin Wetlands*. Hutchinson Ross Publishing Co., Stroudsburg, Penn., pp. 199–213.

Carr, A.F. Jr., 1950. Outline for a classification of animal habitats in Honduras. *Bull. Am. Mus. Nat. Hist.*, 94: 567–594.

Carter, M.R., Burns, L.A., Cavinder, T.R., Dugger, K.R., Fore, P.L., Hicks, D.B., Revells, H.L. and Cavinder, T.W., 1973. *Ecosystems Analysis of the Big Cypress Swamp and Estuaries*. U.S. Environmental Protection Agency 904/9–74–002, Region IV, Atlanta, Ga.

Chapman, V.J., 1976. *Mangrove Vegetation*. J. Cramer, Leutershausen, 447 pp.

Chapman, V.J. (Editor), 1977. *Wet Coastal Ecosystems*. Elsevier, Amsterdam, 428 pp.

Chavan, A.R. and Sabnis, S.D., 1960. Along the banks of the river Vishwamitri. *Indian For.*, 86: 469–474.

Christensen, N., Burchell, R., Liggett, A. and Simms, E.L., 1981. The structure and development of pocosin vegetation. In: C.J. Richardson (Editor), *Pocosin Wetlands*. Hutchinson Ross Publishing Company, Stroudsburg, Penn., pp. 43–61.

Cintrón, G., Lugo, A.E., Pool, D.J. and Morris, G., 1978. Mangroves of arid environments in Puerto Rico and adjacent islands. *Biotropica*, 10: 110–121.

Cintrón, G., Lugo, A.E. and Martínez, R., 1985. Structural and functional properties of mangrove forests. In: W.G. D'Arcy and M.D. Correa A. (Editors), *The Botany and Natural History of Panama*. Monographs in Systematic Botany, 10. Missouri Botanical Garden, St. Louis, Mo., pp. 53–66.

Clark, J.R. and Benforado, J. (Editors), 1981. *Wetlands of Bottomland Hardwood Forests*. Elsevier, Amsterdam, 401 pp.

Clough, B.F. (Editor), 1982. *Mangrove Ecosystems in Australia*. Australian Institute of Marine Sciences and Australian National University Press, Canberra, 302 pp.

Cohen, A.D., Casagrande, D.J., Andrejko, M.J. and Best, G.R., 1984. *The Okefenokee Swamp: Its Natural History, Geology, and Geochemistry*. Wetland Surveys Inc., Los Alamos, N.M., 709 pp.

Collins, E.A., Monk, C.D. and Spielman, R.H., 1964. Whitecedar stands. *Q.J. Fla. Acad. Sci.*, 27: 107–110.

Conard, S.G., MacDonald, R.L. and Holland, R.F., 1977. Riparian vegetation and flora of the Sacramento valley. In: *Riparian Forests in California: Their Ecology and Conservation*. Institute of Ecology, University of California at Davis, Calif., pp. 47–55.

Coulter, S.M., 1904. An ecological comparison of some typical swamp areas. *Mo. Bot. Garden Rept.*, 15: 39–71.

Cowardin, L.M., Carter, V., Golet, F.C. and LaRoe, E.T., 1979. *Classification of Wetlands and Deep-water Habitats of the United States*. U.S. Fish and Wildlife Service Office of Biological Services (79/31), Washington, D.C., 103 pp.

Dabel, C.V. and Day, F.P., 1977. Structural comparison of four plant communities in the Great Dismal Swamp, Virginia. *Bull. Torrey Bot. Club*, 104: 352–360.

Dakshini, K.M.M., 1960. The vegetation of Mothronwala swamp forest (plant communities of swampy zone). *Indian For.*, 86: 728–733.

Dakshini, K.M.M., 1965. A study of the vegetation of Mothronwala swamp forest. Dehra Dun, India. *J. Indian Bot. Soc.*, 44: 411–428.

Deva, S. and Singh Aswal, B., 1974. Taxonomy and ecology of Mothronwala swamp, a reassessment. *Indian For.*, 100: 12–19.

DeWitt, C.B. and Soloway, E. (Editors), 1978. *Wetlands: Ecology, Values, and Impacts*. Institute for Environmental Studies, University of Wisconsin, Madison, Wisc., 388 pp.

Dick-Peddie, W.A. and Hubbard, J.P., 1977. Classification of riparian vegetation. In: *Importance, Preservation, and Management of Riparian Habitat: A Symposium*. U.S. Dept. of Agriculture, Forest Service, General Technical Report RM-43, Washington, D.C., pp. 85–90.

Eggeling, W.J., 1935. The vegetation of Namanve swamp, Uganda. *J. Ecol.*, 23: 422–435.

Ewel, K.C. and Odum, H.T. (Editors), 1984. *Cypress Swamps*. University Presses of Florida, Gainesville, Fla., 472 pp.

Fanshawe, D.B., 1952. *The Vegetation of British Guiana, a Preliminary Review*. Institute Paper 29, Imperial Forestry Institute, University of Oxford, Oxford, 96 pp.

Firouz, E., 1974. *Environment Iran*. National Society for Conservation of Natural Resources and Human Environment, Tehran, 51 pp.

Fittkau, E.J., Irmler, U., Junk, W.J., Reiss, F. and Schmidt, G.W., 1975. Productivity, biomass, and population dynamics in Amazonian water bodies. In: F.B. Golley and E. Medina (Editors), *Tropical Ecological Ecosystems*. Springer, New York, N.Y., pp. 289–331.

Folster, H., De las Salas, G. and Khanna, P., 1976. A tropical evergreen forest site with perched water table, Magdalena valley, Colombia: biomass and bioelement inventory of primary and secondary vegetation. *Ecol. Plantarum*, 11: 297–320.

Frangi, J.L. and Ponce, M., 1985. The root system of *Prestoea montana* and its ecological significance. *Principes*, 29: 13–19.

Goodwin, R.H. and Neiring, W.A., 1975. *Inland Wetlands of the United States*. Natural History Theme Studies No. 2., National Park Service, Washington, D.C., 550 pp.

Green, B.H. and Pearson, M.C., 1968. The ecology of Wybunbury Moss, Cheshire, 1. The present vegetation and some physical, chemical, and historical factors controlling its nature and distribution. *J. Ecol.*, 56: 245–267.

Greeson, P.E., Clark, J.R. and Clark, J.E. (Editors), 1979. *Wetland Functions and Values: The State of our Understanding*. American Water Resources Association, Minneapolis, Minn., 674 pp.

Hall, T.F. and Penfound, W.T., 1943. Cypress-gum communities in the Blue Girth Swamp near Selma, Alabama. *Ecology*, 24: 208–217.

Harper, R.M., 1926. A middle Florida white cedar swamp. *Torreya*, 26: 81–84.

Hartland-Rowe, R. and Wright, P.B., 1975. Effects of sewage effluent on a swampland stream. *Verhandlungen der Internationale Vereinigung für Theoretische und Angewandte Limnologie*, 19: 1575–1583.

Heikurainen, L., 1961. Swamp forestry research in Finland. *Silva Fenn.*, 108: 5–21.

Heikurainen, L., 1967. Effect of cutting on the groundwater level on drained peatland. In: *Forest Hydrology*. Pergamon Press, New York, N.Y., pp. 345–354.

Heinselman, M.L., 1970. Landscape evolution, peatland types, and the environment in the Lake Agassiz peatlands natural area, Minnesota. *Ecological Monogr.*, 45: 235–261.

Herrera, R., Jordan, C.F., Klinge, H. and Medina, E., 1978. Amazon ecosystems, their structure and functioning with particular emphasis on nutrients. *Interciencia*, 3: 223–232.

Hodges, J.D. and Lorio, P.L., 1969. Carbohydrate and nitrogen fractions of the inner bark of loblolly pines under moisture stress. *Can. J. Bot.*, 47: 1651–1657.

Holdridge, L.R., 1967. *Life Zone Ecology*. Tropical Science Center, San José, Costa Rica, 206 pp.

Hook, D.D. and Crawford, R.M.M., 1978. *Plant Life in Anaerobic Environments*. Ann Arbor Science Publishers, Ann Arbor, Mich., 564 pp.

Horwitz, E.L., 1978. *Our Nation's Wetlands*. U.S. Council on Environmental Quality, U.S. Government Printing Office, Washington, D.C., 70 pp.

Ingram, H.A.P., 1967. Problems of hydrology and plant distribution in mires. *J. Ecol.*, 55: 711–724.

Isakov, Y.A., 1968. The status of waterfowl populations breeding in the U.S.S.R. and wintering in S.W. Asia and Africa. In: *Proceedings of a Technical Meeting on Wetland Conservation*. International Union for Conservation of Nature and Natural Resources, Morges, Switzerland, New Series, 12: 175–186.

Johnson, R.R. and Jones, D.A. (Technical Coordinators), 1977. *Importance, Preservation and Management of Riparian Habitat: A Symposium*. U.S. Dept. of Agriculture, Forest Service, General Technical Report RM-43, Washington, D.C., 217 pp.

Johnson, R.R. and McCormick, J.F. (Technical Coordinators), 1979. *Proceedings of National Symposium on Strategies for Protection and Management of Floodplain Wetlands and Other Riparian Ecosystems*. U.S. Dept. of Agriculture, Forest Service, Washington, D.C., 410 pp.

Kologiski, R.L., 1977. *The Phytosociology of the Green Swamp, North Carolina*. North Carolina Agricultural Experiment Station Technical Bulletin Number 250, Raleigh, N.C., 101 pp.

Kozlowski, T.T., 1984. *Effect of Flooding on Plant Growth*. Academic Press, New York, N.Y., 356 pp.

Lamb, F.B., 1959. The coastal swamp forests of Nariño, Colombia. *Caribb. For.*, 20: 79–89.

Larsen, J.A., 1980. *The Boreal Ecosystem*. Academic Press, New York, N.Y., 500 pp.

Larsen, J.A., 1982. *Ecology of the Northern Lowland Bogs and Conifer Forests*. Academic Press, New York, N.Y., 307 pp.

Lewis, F.J., Dowding, E.S. and Moss, E.H., 1928. The vegetation of Alberta. *J. Ecol.*, 16: 19–70.

Liem, D., 1976. *Report on the Habitat Survey and Habitat Assessment of 76/6*. Research Series 1, Department of Natural Resources, Papua New Guinea, 10 pp.

Lorio, P.L. Jr., 1966. *Phytophthora cinnamomi* and *Pythium* species associated with loblolly pine decline in Louisiana. *Plant Dis. Rep.*, 50: 596–597.

Lorio, P.L. Jr., 1968. Soil and stand conditions related to southern pine beetle activity in Hardin County, Texas. *J. Econ. Entomol.*, 61: 565–566.

Lorio, P.L. Jr., 1977. *Ground-water Levels and Soil Characteristics in a Forested Typic Glossaqualf*. U.S. Dept. of Agriculture, Forest Service, Research Note SO-225, New Orleans, La., 5 pp.

Lorio, P.L. Jr. and Bennett, W.H., 1974. *Recurring Southern Pine Beetle Infestations near Oakdale, Louisiana*. U.S. Dept. of Agriculture, Forest Service, Research Paper SO-95, New Orleans, La., 6 pp.

Lorio, P.L. Jr. and Hodges, J.D., 1968a. Oleoresin exudation pressure and relative water content of inner bark as indicators of moisture stress in loblolly pines. *For. Sci.*, 14: 392–398.

Lorio, P.L. Jr. and Hodges, J.D., 1968b. Microsite effects on oleoresin exudation pressure of large loblolly pines. *Ecology*, 49: 1207–1210.

Lorio, P.L. Jr. and Hodges, J.D., 1971. Microrelief, soil water regime, and loblolly pine growth on a wet, mounded site. *Soil Sci. Soc. Proc.*, 35: 795–800.

Lorio, P.L. Jr. and Hodges, J.D., 1974. Host and site factors in southern pine beetle infestations. In: T.L. Payne, R.N. Coulson and R.C. Thatcher (Editors), *Proceedings, Southern Pine Beetle Symposium*. Texas Agricultural Experiment Station, College Station, Texas, pp. 32–34.

Lorio, P.L. Jr. and Hodges, J.D., 1977. *Tree Water Status Affects Induced Southern Pine Beetle Attack and Brood Production*. U.S. Dept. of Agriculture, Forest Service, Research Paper SO-135, New Orleans, La., 7 pp.

Lorio, P.L. Jr., Howe, V.K. and Martin, C.N., 1972. Loblolly pine rooting varies with microrelief on wet sites. *Ecology*, 53: 1134–1140.

Lugo, A.E., 1984. A review of early forested wetland literature in the United States. In: K.C. Ewel and H.T. Odum (Editors), *Cypress Swamps*. University Presses of Florida, Gainesville, Fla., pp. 7–15.

Lugo, A.E. and Patterson Zucca, C., 1977. The impact of low temperature stress on mangrove structure and growth. *Trop. Ecol.*, 18: 149–161.

Lugo, A.E. and Snedaker, S.C., 1974. The ecology of mangroves. *Annu. Rev. Ecol. Syst.*, 5: 39–64.

Mattoon, W.R., 1915. *The Southern Cypress*. U.S. Dept. of Agriculture, Forest Service Bulletin 272, Washington, D.C., 73 pp.

McCarthy, J., 1962. The colonization of a swamp forest clearing (with special reference to *Mitragyna stipulosa*). *E. Afr. Agric. For. J.*, 25: 22–28.

Meehan, W.R., Swanson, F.J. and Sedell, J.R., 1977. Influences of riparian vegetation on aquatic ecosystems with particular reference to salmonid fishes and their food supply. In: *Importance, Preservation, and Management of Riparian Habitat: A Symposium*. U.S. Dept. of Agriculture, Forest

Service, General Technical Report RM-43, Washington, D.C., pp. 137–145.

Meléndez, E.M., 1965. Algunas características ecológicas de los bosques inundables de Darién, Panamá, con miras a su posible utilización. *Turrialba*, 15: 336–347.

Miller, W.D. and Maki, T.E., 1957. Planting pines in pocosins. *J. For.*, 55: 659–663.

Monk, C.D., 1965. Southern mixed hardwood forest of north central Florida. *Ecol. Monogr.*, 35: 335–354.

Monk, C.D., 1966. An ecological study of hardwood swamps in north central Florida. *Ecology*, 47: 649–654.

Monk, C.D., 1967. Tree species diversity in the eastern deciduous forest with particular reference to north central Florida. *Am. Nat.*, 101: 173–187.

Monk, C.D. and Brown, T.W., 1965. Ecological considerations of cypress heads in north central Florida. *Am. Midl. Nat.*, 74: 126–140.

Moore, P.D. and Bellamy, D.J., 1974. *Peatlands*. Springer-Verlag, New York, N.Y., 221 pp.

Moorthy, K.K., 1960. *Myristica* swamps in the evergreen forests of Travancore. *Indian For.*, 86: 314–315.

Mörzer Bruyns, M.F. and Westhoff, V., 1968. Notes on the economic value of wetlands based on experience in the Netherlands. In: *Proceedings of a Technical Meeting on Wetland Conservation*. International Union for Conservation of Nature and Natural Resources, Morges, Switzerland, New Series, 12: 203–204.

Novitzki, R.P., 1979. Hydrologic characteristics of Wisconsin's wetlands and their influence on floods, streamflow, and sediment. In: P.E. Greeson, J.R. Clark and J.E. Clark (Editors), *Wetland Functions and Values: The State of our Understanding*. American Water Resources Association, Minneapolis, Minn., pp. 377–388.

Odum, H.T., 1970. Rain forest structure and mineral-cycling homeostasis. In: H.T. Odum and R.F. Pigeon (Editors), *A Tropical Rain Forest*. National Technical Information Service, Springfield, Va., Chapter H-1.

Parker, R.E., 1962. Factors limiting tree growth on peat soils. An investigation into the nutrient status of two peatland plantations. *Irish For.*, 19: 60–81.

Pase, C.P. and Layser, E.F., 1977. Classification of riparian habitat in the southwest. In: *Importance, Preservation, and Management of Riparian Habitat: A Symposium*. U.S. Dept. of Agriculture, Forest Service, General Technical Report RM-43, Washington, D.C., pp. 5–9.

Pearsall, W.H., 1950. *Mountains and Moorelands*. Collins Cleartype Press, London, 415 pp.

Penfound, W.T., 1952. Southern swamps and marshes. *Bot. Rev.*, 18: 413–446.

Penfound, W.T. and Hall, T.F., 1939. A phytosociological analysis of tupelo gum forest near Huntsville, Alabama. *Ecology*, 20: 358–364.

Persson, T. (Editor), 1980. *Structure and Function of Northern Coniferous Forests — An Ecosystem Study*. Ecological Bulletins (Stockholm), 32: 609 pp.

Pires, J.M., 1978. The forest ecosystems of the Brazilian Amazon: description, functioning, and research needs. In: *Tropical Forest Ecosystems: a State-of-Knowledge Report*. UNESCO/UNEP/FAO, Paris, pp. 607–627.

Pool, D.J., Searl, L., Kemp, W.M. and Odum, H.T., 1972. *Forested Wetland Ecosystems of the Southern United States*. Center for Wetlands, University of Florida, Gainesville, Fla., 446 pp.

Prance, G.T., 1979. Notes on the vegetation of Amazonia III. The terminology of Amazonian forest types subjected to innundation. *Brittonia*, 31: 26–38.

P'yavchenko, N.I., 1957. *Improvements in Forest Growth on Peat-Bog Soils of the U.S.S.R. Forest Zone and Tundra*. Israel Program for Scientific Translations, Jerusalem.

Ratter, J.A., Richards, P.W., Argent, G. and Gifford, D.R., 1973. Observations on the vegetation of northeastern Matto Grosso, I. The woody vegetation types of the Xavantina — Cachimbo expedition area. *Phil. Trans. R. Soc. Lond.*, 266: 449–492.

Reichle, D.E. (Editor), 1981. *Dynamic Properties of Forested Ecosystems*. International Biological Programme 23, Cambridge University Press, Cambridge, 683 pp.

Rice, E.L., 1965. Bottomland forests of north-central Oklahoma. *Ecology*, 46: 708–713.

Richards, P.W., 1962. Ecological notes on west African vegetation, II. Lowland forest of the southern Bakundu forest reserve. *J. Ecol.*, 50: 123–149.

Richards, P.W., 1964. *The Tropical Rain Forest*. Cambridge University Press, Cambridge, 450 pp.

Richardson, C.J. (Editor), 1981. *Pocosin Wetlands*. Hutchinson Ross Publishing Co., Stroudsburg, Penn., 364 pp.

Ringuelet, R.A., 1962. *Ecología Acuática Continental*. Eudeba Editorial Universitaria de Buenos Aires, 138 pp.

Rollet, B., 1981. *Bibliography on Mangrove Research 1600–1975*. UNESCO, Paris, 479 pp.

Saenger, P., 1982. Morphological, anatomical, and reproductive adaptations of Australian mangroves. In: B.F. Clough (Editor), *Mangrove Ecosystems in Australia*. Australian Institute of Marine Science, Australian University Press, Canberra, pp. 153–191.

Sands, A. (Editor), 1977. *Riparian Forests in California, their Ecology and Conservation*. Institute of Ecology Publication 15. University of California, Davis, Calif., 121 pp.

Sarmiento, G. and Monasterio, M., 1975. A critical consideration of the environmental conditions associated with the occurrence of savanna ecosystems in tropical America. In: F.B. Golley and E. Medina (Editors), *Tropical Ecological Systems*. Springer-Verlag, New York, N.Y., pp. 223–250.

Sell, M.G., 1977. *Modeling the Response of Mangrove Ecosystems to Herbicide Spraying, Hurricanes, Nutrient Enrichment, and Economic Development*. PhD dissertation, University of Florida, Gainesville, Fla., 390 pp.

Shevchenko, L.A., 1973. The possible use of landscape indicators for evaluating transportation through swamps. In: A.G. Chikishev (Editor), *Landscape Indicators*. Consultants Bureau, New York, N.Y., pp. 64–73.

Skoropanov, S.G., 1968. *Reclamation and Cultivation of Peat-Bog Soils*. Translated from Russian by N. Kaner. Israel Program for Scientific Translations, U.S. Dept. of Agriculture and National Science Foundation, Washington, D.C., 234 pp.

Taylor, B.W., 1959. The classification of lowland swamp communities in northeastern Papua. *Ecology*, 40: 703–711.

Taylor, B.W., 1961. An outline of the vegetation of Nicaragua. *J. Ecol.*, 49: 27–54.

Tilton, D.L. and Bernard, J.M., 1975. Primary productivity and biomass distribution in an alder shrub ecosystem. *Am. Midl. Nat.*, 94: 251–256.

Tolpa, S., Jasnowski, M. and Polczynski, A., 1967. System of genetic classification of the peats of Central Europe. *Zesz. Probl. Postepow Nauk Roln.*, 76: 9–99.

U.S. Army Corps of Engineers, 1977. Regulatory program of the Corps of Engineers — general regulatory policies. *Fed. Regist.*, 138 (Tuesday, July 19): 37133–37138.

Van Steenis, C.G.G.J., 1958. Tropical lowland vegetation: the characteristics of its types and their relation to climate. In: *Climate, Vegetation, and Rational Land Utilization in the Humid Tropics.* Vol. 20, Secretariat, Ninth Pacific Science Congress, Bangkok, pp. 25–37.

Vogl, R.J. and McHargue, L.T., 1966. Vegetation of California fan palm oases on the San Andreas fault. *Ecology*, 47: 532–540.

Ware, G.H. and Penfound, W.T., 1949. The vegetation of the lower levels of the floodplain of the South Canadian river in central Oklahoma. *Ecology*, 30. 478–484.

Webb, L.J., 1968. Environmental relationships of the structural types of Australian rain forest vegetation. *Ecology*, 49: 296–311.

Wharton, C.H., 1978. *The Natural Environments of Georgia.* Geologic and Water Resources Division and Georgia Department of Natural Resources, Atlanta, Ga., 227 pp.

Wharton, C.H., Odum, H.T., Ewel, K., Duever, M.J., Lugo, A., Boyt, R., Bartholomew, J., Debellevue, E., Brown, S. and Duever, L., 1976. *Forested Wetlands of Florida, Their Management and Use.* Center for Wetlands, University of Florida, Gainesville, Fla., 421 pp.

Whitehead, D.R., 1972. Development and environmental history of the Dismal Swamp. *Ecol. Monogr.*, 42: 301–315.

Whitmore, T.C., 1984. *Tropical Rain Forests of the Far East.* Oxford University Press, Oxford, 2nd ed., 352 pp.

Williams, W.A., Loomis, R.S. and Alvim, P. de T., 1972. Environments of evergreen rain forests on the lower Rio Negro, Brazil. *Trop. Ecol.*, 13: 65–78.

Woodwell, G.M., 1956. *Phytosociology of Coastal Plain Wetlands of Carolinas.* M.A. Thesis, Duke University, Durham, N.C., 50 pp.

Wright, A.H. and Wright, A.A., 1932. The habitats and composition of the vegetation of Okefenokee swamp, Georgia. *Ecol. Monogr.*, 2: 109–232.

Wyatt-Smith, J., 1959. Peat swamp forests in Malaya. *Malay. For.*, 23: 5–32.

Wyatt-Smith, J., 1961. A note on the fresh-water swamp, lowland and hill forest types of Malaya. *Malay. For.*, 24: 110–121.

Wyatt-Smith, J., 1964. A preliminary vegetation map of Malaya with descriptions of the vegetation types. *J. Trop. Geogr.*, 18: 200–213.

Yurkevich, I.D., Smolyak, L.P. and Garin, B.E., 1966. Content of oxygen in the soil water and of carbon dioxide in the soil, air, and forest bogs. *Sov. Soil Sci.*, 2: 159–167.

Zoltai, S.C., Pollett, F.C., Jeglum, J.K. and Adams, G.D., 1975. Developing a wetland classification for Canada. In: B. Bernier and C.H. Winget (Editors), *Forest Soils and Forest Land Management.* Les Presses de L'Université Laval, Quebec, pp. 497–511.

Chapter 2

AN ENERGY THEORY OF LANDSCAPE FOR CLASSIFYING WETLANDS

PATRICK C. KANGAS

INTRODUCTION

The major premise of this chapter is that environmental energy sources determine ecosystem characteristics. The spatial distribution of energy is considered to determine the spatial form and distribution of ecosystems. The spatial expression of energy sources is then used as the distinguishing feature of a land classification scheme. This classification is applied to forested wetlands in this chapter; but it is general, and applies to all land surfaces.

In classifying wetlands, the role of water as an energy source is obviously important. Different energy values have been assigned to water depending on its use in the system in question (Odum, 1970a). Examples include potential energy as a chemical reactant or as elevated hydrologic head, and kinetic energy as velocity in currents. For the purpose of classifying wetlands, however, the absolute value of water in its different uses is not immediately as important as is the form of its distribution. In this paper, different forms of distribution of water are matched with different wetland forms. Water is used as an index for energy sources.

ENERGY SOURCES IN ECOLOGY

An important conceptual advance in ecology involves the recognition that ecosystems utilize a variety of energy sources that are often termed auxiliary energy sources or energy subsidies. The original energy budget studies in ecology showed only sunlight as the source of energy for building and maintaining ecosystem structure. Organic matter inputs such as detritus were also recognized, where appropriate, but it was not until the 1960's that these inputs received full attention as energy sources (Sibert and Naiman, 1980). Environmental factors such as wind, waves, water currents, dry air, etc. were considered important in influencing ecosystem development and species distributions, but these factors were not viewed as energy sources available to do ecological work. However, Odum (1967, 1968, 1970b) proposed that these environmental factors were actually energy sources and said that "auxiliary energy sources for such processes as mineral cycling, resisting competitive or predatory influences, reproduction, storage, maintenance, etc. are as much energy inputs as the sunlight and organic inflows." The idea was amplified by E.P. Odum (1971) who noted "high rates of production, both in natural and cultured ecosystems, occur when physical factors are favorable and especially when there are energy subsidies from outside the system that reduce the cost of maintenance. Such energy subsidies may take the form of the work of wind and rain in a rain forest, tidal energy in an estuary or the fossil fuel, animal, or human work energy used in the cultivation of a crop."

ENERGY SIGNATURES

The set or combination of different input energies to a system has been called its energy signature (see Ch. 4, Sect. 10). The concept of the energy signature is important because, as already discussed, systems are influenced by energy sources available to them and are limited by their total energy input.

Details on how to determine the energy signature for a system are given in Ch. 4, Sect. 10. After the energy inputs have been expressed in terms of equal quality there are several possible algorithms that can be used to integrate the input energies, the simplest being summation. The result of the calculation is the total weighted energy input to the system of interest. A handbook summarizing the methodology for determining energy signatures is available (Odum et al., 1983). This handbook contains the energy quality ratios, and their method of determination, for 41 environmental energy inputs or energy storages. Included are energy inputs and storages pertinent to wetlands such as: waves, rain, wind, physical and chemical potential energy in stream flow, physical and chemical potential energy in the materials carried by stream flow (e.g., sediments, dissolved substances), potential energy in catastrophes such as floods and cyclones, and potential energy in living biomass.

INCORPORATING THE SPATIAL EXPRESSION OF ENERGY

To produce a more complete explanation of landscape patterns, however, the spatial expression of the separate energies and their combination into an energy signature need to be considered. Fig. 2.1 illustrates four basic spatial distributions that energy inputs can take: sheets, points, fronts, and lines. All of the spatial energy distributions act synoptically over relatively large areas, except for points (see below).

The energy distributions fall into two generic classes based on their angle of input: perpendicular energies (sheet and point), and parallel energies (front and line). Energy sheets are delivered over a broad area perpendicular to the surface. Sunlight and rainfall are examples of sheet energies from above the surface, and a potentiometric surface is an example of a sheet incoming from below. A uniform rock substrate emerging from geologic uplift also qualifies as an energy sheet. The energy point is similar to the sheet in that it comes to the surface perpendicularly, but it is expressed over a smaller area than the sheet. An example of an energy point from below the surface is an artesian spring and, from above, lightning that triggers

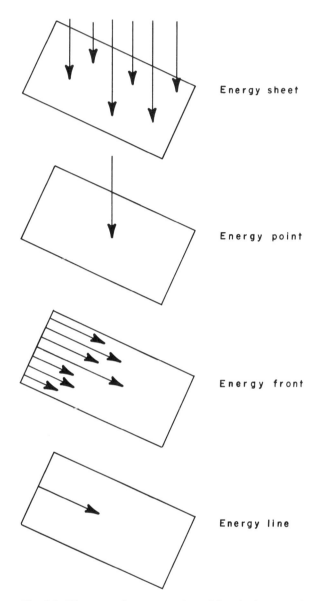

Fig. 2.1. Diagrammatic representation of four basic types of spatial energy distributions. Arrow points show where potential energy is delivered.

fires. A frontal energy is analogous to a weather front. In this case, energy is delivered over a broad area as a moving force parallel to the surface. Examples are runoff flows of water, tides, and air masses. Fronts can be stationary (gradients) or moving. Energy lines, like fronts, move parallel to the land surface, but, like energy points, express their force over a small area. A river is an example of an energy line.

Spatial energy distributions themselves may also have properties of intensity. Within the two generic series, sheets can concentrate to form points, and fronts can concentrate to form lines. Once concentrated, the new energies attract other less concentrated energies from their surroundings, a process which helps maintain the structures that they build. For example, energy lines cause a linear convergence and energy points cause a radial convergence. If the concentrated point or line does not encounter enough energy with which to interact, it will disperse. To illustrate these processes, consider the upper end of a watershed catchment where runoff moving as a front converges to form a concentrated line in the form of a river. The line collects more energy in the form of tributaries as it flows downslope until it reaches the coastal plain. There, the energy gradient decreases and the line reverts to a front as the river forms distributaries of a delta or braided stream.

Concentrated sources can be insulated but often at great cost — by, for instance, a structure to provide containment. An example is a channelized river, whereby the natural river is straightened and confined within artificial levees to protect against flooding and to speed drainage.

LAND CLASSIFICATION

Landscapes can be classified based on the spatial expressions of energy input. Implied is that land forms develop in response to particular spatial energy types. From studying maps, I developed six basic categories of form (called ecosystem forms): background, center, zone, strip, string, and island. These are common shapes of land surface recognizable on vegetation and land-use maps. Different forms are distinguished by their shape, orientation, and, to some extent, relative size. Characteristics of these form categories are shown in Table 2.1. Strip, island, string and zone are generated by the class of frontal energies, while center and background are produced by sheet energies.

Ecosystem forms combine into recurring patterns which are specific to a spatial energy type but can arise under a variety of conditions and developmental mechanisms. Patterns show how particular forms are arranged against backgrounds. These organizational arrangements are displayed in Fig. 2.2 as idealized forms. Attributes of the patterns deal with the relationships between and within forms. Their spacing is of prime importance because this is directly related to the absorption and dissipation of energy. The spacing of individual forms is also important in organizing the larger patterns. If spacing is too close, competition among forms may occur, and if too distant some energy may not be used effectively. Fig. 2.3 illustrates the correspondence between landscape and energy on the patterns from Fig. 2.2 with energy distributions from Fig. 2.1. Energy flow lines are drawn in relation to the pattern of ecosystem forms. It is suggested that the emerging patterns are structures which adapt to, interact with, and amplify spatial energy signatures. Each form category is described below with examples.

Background

Backgrounds develop as a consequence of power sheets. A background is the basic undifferentiated

TABLE 2.1

A classification of ecosystem forms

Form	Shape	Orientation of energy
Generated by frontal energies		
Zone	Rectangular	Perpendicular
String	Linear	Perpendicular
Island	Ovoid, curved, or pointed	Pointed in direction of energy movement
Strip	Linear	Parallel
Generated by sheet energies		
Background	Undifferentiated	No orientation
Center	Circular to irregular	Centrally located

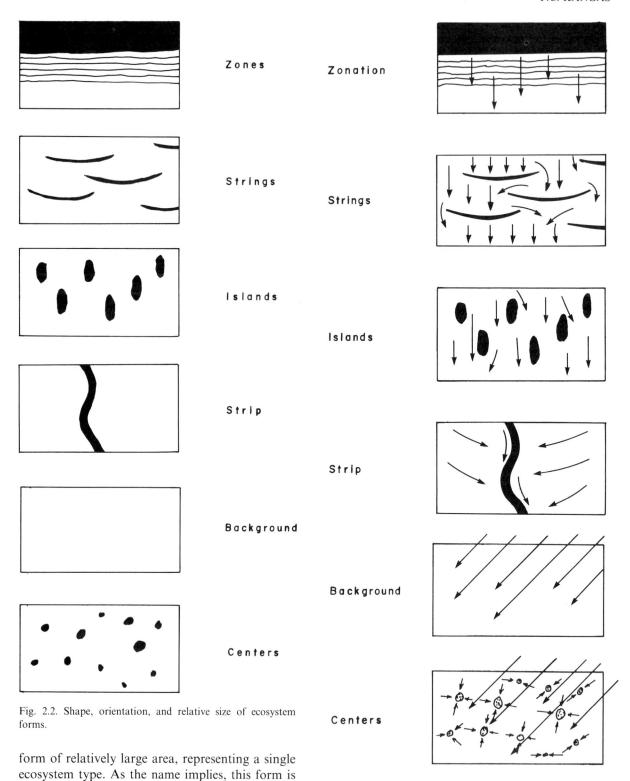

Fig. 2.2. Shape, orientation, and relative size of ecosystem forms.

form of relatively large area, representing a single ecosystem type. As the name implies, this form is the "background" within which other forms are found, though sometimes it will occur alone.

Fig. 2.3. Patterns of ecosystem form from Fig. 2.2 generated by spatial energy distributions in Fig. 2.1.

Although sunlight is a basic sheet energy of all landscapes, the distinctive character of backgrounds may be more reflective of other, higher-quality sheet energies like homogeneous topography or uniform soils, and parent material. At some scales, such as regional-level vegetation maps, backgrounds are the only kinds of systems defined, while the smaller, more complex forms imbedded in the background are not shown. Common wetlands that can be classified as backgrounds include large basins fed mainly by rainfall (e.g. many large bogs in the Boreal and Arctic Zones). Regional wetlands in the southeastern U.S.A., including the Dismal Swamp, the Okeefenokee, and the Big Thicket, may fall into this category with many subsystems imbedded in them.

Center

Under certain conditions, centers can develop within backgrounds, arising directly from point sources, or from the concentration of sheet energies (if of sufficient magnitude) into point sources. Sometimes the convergence of the sheet into points may not be obvious. For example, water holes in an arid system act as centers generated by the convergence of groundwaters. Elevation often is the only feature that distinguishes centers in their background, either as wet depressions against a dry background, or dry mounds against a wet background. The background of pine flatwoods dotted with cypress dome centers found in the Green Swamp and elsewhere in central Florida, U.S.A., is an example of a center pattern. The situation in the northern latitudes of frost-heaved peat mounds against a background of muskeg may be a special case where ice causes the convergence (Lundqvist, 1969). Energy of sufficient intensity can radiate from centers across backgrounds. If a gradient is available, the point will radiate as an energy line, such as a spring-fed river. If lacking a gradient, the point will disperse energy radially over its surroundings.

Zone

Zones develop in response to frontal energies. This ecosystem form occurs in many situations including rocky intertidal shores, beach and barrier island vegetation, coral reefs, various wetlands such as mangroves, salt marshes and littoral macrophytes, and mountain vegetation. Zones can be narrow, or wide like the 160 km long sawgrass plain of the "River of Grass" Everglades (Florida, U.S.A.). At even larger scales, the life zone theories of Merriam (1899) and Holdridge (1967) clearly fit in this series, with biome-type zones generated by broad climatic frontal energies or altitudinal gradients. The spacing between zones may be proportional to the intensity of the energy and the frequency at which it is imposed. If frontal energy is weak, a single zone may develop that is similar to a background generated by sheet energy, but without the requirement of being maintained by a moving physical force.

String and island

Strings and islands are subordinate forms in the frontal series. These are relatively small features within broad zones, being similar to centers in size but usually distinguished by their orientation in the direction of the frontal energy. They may, however, have some central functions in their context within a broad zone.

Strings are transverse features usually of higher elevation than their surroundings. These ecosystem forms are named after the string bogs (*strangmoor*) of the northern latitudes (Drury, 1956; Heinselman, 1963, 1965; Grittinger, 1970; Schenk, 1970; Walter, 1977). Other examples are transverse dunes, cross-valley moraines (Andrews, 1963), certain reefs, levees in floodplains, sand bars and wind breaks or shelter belts. Oyster reefs are particularly good examples, stretching across estuarine channels and river mouths where they capture organic matter in currents. This is a good example of orientation that is adapted to intercepting more current and making available more organic matter. Agricultural terraces are human analogs of nature strings. Terrace engineering, to slow runoff and maintain soils has been formally developed for some time (Ayres, 1936; Archer, 1956; and much earlier) and is still an important technique for slope management (Bostanoglu, 1976). Perhaps strings occurring in nature have similar functional advantages. The rock reefs of the Everglades National Park, U.S.A. (Craighead,

1971) may also be an example of the wetland string.

Islands are elongated forms also, often of higher elevation than the surrounding landscape, and parallel with one another and to the direction of energy flow. Examples are drumlins, most sand dunes, tree islands and some coastal islands and reefs. A remarkable convergence of this form is found in the tree islands of the Everglades in south Florida, U.S.A. (Davis, 1943; Loveless, 1959; Craighead, 1971) and of the peatlands in Boreal regions (Heinselman, 1963, 1970). In both situations the islands are tree-covered against a background of marsh, with streamlined shape and topographic alignment pointing downslope. Islands may be considered as strings that have been rotated 90° in relation to their frontal energy inputs. This rotation occurs as the energy input increases and is presumably a response adapted to exposing more surface area to the passing frontal energy.

Strip

Energy lines through a landscape develop strip forms. An example is the river and floodplain system, where the energy line is the channel flow in the river and its overflow during flooding. Strips are the only other system besides backgrounds that show up on small-scale maps, often exaggerated and out of scale. This may be because they are more energy-intensive relative to surroundings, and thus attract attention. When cutting across large areas, energy lines capture other energies from the surroundings. River piracy (Crosby, 1937; Woodruff, 1977) is an example where a dominant river successfully competes for a subordinate river and captures its catchment. Energy lines and the strips formed by the available energy literally drain excess energy from one sector of the landscape to other areas where it can be used more effectively.

A WETLANDS APPLICATION

To illustrate the land classification scheme presented above, an application is made using Wharton's (1975) descriptions of the natural ecosystems of Georgia, U.S.A. This example was chosen because of the diversity of wetland systems included, and the detailed descriptions given for each.

Wharton defines 38 Georgia systems as hydric, including some not conventionally considered as wetlands (e.g. beaches and oyster reefs). For consistency all will be translated into the energy-form classification by means of a matrix (Table 2.2). Wharton's hydric systems are given as row headings and the spatial energy types such as dominant water flows and land classification units are shown as column headings. Each hydric system is matched with an energy designation and is put in the proper form class with x's. Forested wetlands in the Table include various river swamps (numbers 7–11) and a heterogeneous group of evergreen bogs and ponds (numbers 26–31).

Many of the hydric systems are easily recognized as being formed by one of the four spatial energy types and thus as having a characteristic form. For example, cypress pond systems develop with the concentration of water over a background of flatlands. Other types are harder to classify with this scheme. One reason for the difficulty may be that more than one type of spatial energy distribution may be organizing these systems. A system may have been formed by one kind of source, but when combined with other types none may dominate. The resulting form is a hybrid which has characteristics of several form categories. The tidewater river system is an example which has both a line source from river flow and a frontal source from tides.

The exercise of translating Wharton's inventory suggests a significant amount of redundancy in classes. Twelve kinds of rivers and creeks (numbers 5–14, 21, 22) and five kinds of salt marshes (numbers 15–19) illustrate this point. When classifications are finely broken down based on vegetation type or related characteristics, the zonation pattern is often implied. This is a situation of sequential adaptation of ecosystems to different energy intensities of a common source type. In the salt marsh example, adaptation is commonly attributed to flooding frequency. In rivers, velocity or slope gradient may be the determining feature, with different segments of a river adapted to different conditions in a longitudinal sequence. It may be that different zones within these patterns vary little in structure and function, and could be

TABLE 2.2

Translation of Wharton's (1975) hydric ecosystems of Georgia with the energy classification

Landscape form class						Number	System name	Energy designation			
Background	Center	Zone	String	Island	Strip			Sheet	Point	Front	Line
	x					1	Springs		x		
	x					2	Coastal plains springs		x		
x						3	Aquifers	x			
		x				4	Cliffs and outcrops			x	
					x	5	Mountain river				x
					x	6	Spring-fed stream		x		x
					x	7	Blackwater river and swamp				x
					x	8	Creek swamp				x
					x	9	Piedmont alluvial river				x
					x	10	Coastal plains river				x
					x	11	Coosa River and swamp				x
					x	12	Tidewater river			x	x
					x	13	Backwater streams			x	x
					x	14	River marsh			x	x
		x				15	Cordgrass (*Spartina alterniflora*) marsh			x	
		x				16	Salt grass (*Spartina* spp.) marsh			x	
		x				17	Needle rush (*Juncus roemerianus*) marsh			x	
		x				18	Edge zone marsh			x	
		x				19	Brackish marsh			x	
	x					20	Tide pool			x	
					x	21	Oligohaline creek				x
					x	22	Tidal creek and river			x	x
x						23	Estuary and sound	x		x	
				x		24	Oyster (*Crassostrea*) reef			x	
				x		25	Beach			x	
	x					26	Cypress (*Taxodium* pond)		x		
	x					27	Gum (*Nyssa*) pond		x	x	
				x		28	Carolina bay[1]		x	x	
						29	Bay[1] swamp[2]				
						30	Bog swamp[2]				
						31	Cypress savannah	x			
						32	Herb bog				
						33	Bogs and seeps		x	x	
	x					34	Lime sink		x		
						35	Sag pond[2]				
						36	Marsh pond[2]				
		x			x	37	Levee lake				x
		x			x	38	Beaver pond				x

[1]Bay: *Magnolia* spp., *Persea* spp., *Gordonia* spp.
[2]Unable to translate.

viewed as examples of the same system. The important issue to resolve is the energy limits on thresholds that control zone width and frequency, and thus control distribution and patterns of wetlands.

Another observation on the translation exercise is that few systems have sheet types as their dominant energy source. This may indicate that hydric systems (wetlands) often act as recipients of concentrated flows rather than being broad expanses or dilute backgrounds, as often may be the case with terrestrial ecosystems. Most wetlands have frontal energy sources with water moving along slopes as runoff, seepage, water-level fluctuation, or tide. Line sources are also important, such as flooding from river channels.

CONCLUSION

The reason for classifying a group of objects is to help the human mind organize and understand the diversity of the particular collection of interest. Any set of characteristics can be used to distinguish objects. This is the practice in taxonomy. A variety of schemes has been used in the past to classify wetlands including source of nutrients and water, species composition, acidity, physiognomy, geomorphology, salinity, and hydroperiod. Of special relevance to the classification proposed here is the scheme of Novitzki (1979). The two schemes share some common features. Novitzki's classification deals with two features: surface vs. groundwater as the water source for wetlands, and slopes vs. depressions as geomorphic templates. The latter is related to the distinction made in this chapter between frontal energies, usually generated on slopes, and sheet energies, usually generated on flatlands and concentrated in depressions. Within these features, characteristics such as hydroperiod, depression depth, and vegetation are utilized for refinements. Gosselink and Turner (1978) propose a classification of wetlands along a "hydrodynamic energy gradient", but put little emphasis on spatial form. They recognize six classes of wetlands based on source and velocity of water flow. They also relate net primary productivity, qualitatively, to the energy gradient in water flows for cypress swamps in southeastern U.S.A., as does Odum (1978). Radforth's classification of Canadian muskeg developed over many years (Radforth, 1952, 1962, 1977; Radforth and Bellamy, 1973) is most similar to the system offered here. This is a complicated classification based on "the relationships that exist between the airform pattern, hydrology, and ontogeny of the muskegs". Ten "hydroforms" are recognized with several subsidiary variations.

The preliminary energy classification put forth here is more general than most schemes, since it applies to terrestrial systems as well as wetlands. Ecosystems are classed according to the spatial form of their energy sources. Though not explored in this chapter, further distinctions can be made based on intensity, duration, and frequency of energy input. One value of the energy classification approach is that it uses energy as a common denominator. All physical forces and factors that can be given an energy value and can be assigned to a spatial distribution are used in the classification. In other words, properties of energy should be useful in classifying ecosystems because they drive all systems and are responsible for their structure and functions.

Another advantage of energy classification is that it is holistic and actually approaches the level at which ecosystems themselves operate. The main feature of ecosystems which separates them from other levels of organization, such as communities or populations, is the existence of emergent properties which are attributes of the whole system not predictable from knowledge of the components alone. These are the result of complex interactions between physical and living systems. The ecosystem classification which addresses emergent properties may be the most useful for understanding them at their functional level. It is difficult, however, to work with these properties. The parts are much easier to see than the whole, as is often the situation with classification schemes. An advantage of the classification is that it may allow a view of a class of emergent properties not recognized before. For example, ecologists have studied in detail ecosystem adaptations for assimilating sunlight, such as chlorophyll content, assimilation number, photic zone thickness, and leaf area, some of which are recognized as emergent or summary properties. However, these apply only to one of the energy sources utilized by ecosystems. As noted earlier, the auxiliary energies or energy subsidies may be even more important.

What are some ecosystem adaptations for assimilating auxiliary energies? These may be of a nature very different from those of which we are used to thinking. For example, strategies available for absorbing physical energies in sheetflows of frontal sources may include a variety of sensitive architectural designs of stem density, organic debris dams, reef structure, and other properties. The recognition of different ecosystem forms is a first step towards investigating the ways ecosystems capture all available energy sources. Perhaps the exercise of classifying ecosystems according to energy signatures will help bring up new questions and provide understanding not available from other approaches.

REFERENCES

Andrews, J.T., 1963. Cross-valley moraines of the Rimrock and Isortoq River valleys, Baffin Island, N.W.T., a descriptive analysis. *Geogr. Bull.*, 19: 49–77.

Archer, S.G., 1956. *Soil Conservation.* University of Oklahoma Press, Norman, Okla., 305 pp.

Ayres, Q.C., 1936. *Soil Erosion and its Control.* McGraw-Hill Book Co., New York, N.Y., 365 pp.

Bostanoglu, L., 1976. Restoration and protection of degraded slopes. In: *Conservation in Arid and Semi-arid Zones.* FAO Conservation Guide No. 3. Food and Agriculture Organization, Rome, pp. 105–125.

Craighead, F.C., Sr., 1971. *The Trees of South Florida.* University of Miami Press, Coral Gables, Fla., 212 pp.

Crosby, I.B., 1937. Methods of stream piracy. *J. Geol.*, 45: 465–486.

Davis, J.H., 1943. *The Natural Features of Southern Florida.* Fla. Geol. Surv. Biol. Bull. No. 25. Fla. Geol. Surv., Tallahassee, Fla., 311 pp.

Drury, W.H., Jr., 1956. *Bog Flats and Physiographic Processes in the Upper Kuskokwin River Region, Alaska.* Contrib. Gray Herb., No. 178, Harvard University, Cambridge, Mass., 130 pp.

Gosselink, J.G. and Turner, R.E., 1978. The role of hydrology in freshwater wetland ecosystems. In: R.E. Good, D.F. Whigham, and R.L. Simpson, (Editors), *Freshwater Wetlands.* Academic Press, New York, N.Y., pp. 63–78.

Grittinger, T.F., 1970. String bogs in southern Wisconsin. *Ecology*, 51: 928–930.

Heinselman, M.L., 1963. Forest sites, bog processes, and peatland types in the glacial Lake Agassiz region, Minnesota. *Ecol. Monogr.*, 33: 327–374.

Heinselman, M.L., 1965. String bogs and other patterned organic terrain near Seney, upper Michigan. *Ecology*, 46: 185–188.

Heinselman, M.L., 1970. Landscape evolution, peatland types and the environment in the Lake Agassiz Peatlands Natural Area, Minnesota. *Ecol. Monogr.*, 40: 235–261.

Holdridge, L.R., 1967. *Life Zone Ecology.* Tropical Science Center, San José, Costa Rica, 206 pp.

Loveless, C.M., 1959. A study of the vegetation in the Florida Everglades. *Ecology*, 40: 1–9.

Lundqvist, J., 1969. Earth and ice mounds; a terminological discussion. In: T.L. Pewe (Editor), *The Periglacial Environment.* McGill-Queen's University Press, Montreal, Que., pp. 203–214.

Merriam, C.H., 1899. *Life Zones and Crop Zones of the United States.* U.S. Department of Agriculture Bull. No. 10. Washington, D.C., 79 pp.

Novitzki, R.P., 1979. Hydrologic characteristics of Wisconsin's wetlands and their influence on floods, stream flow, and sediment. In: P.E. Greeson, J.R. Clark, and J.E. Clark (Editors), *Wetland Functions and Values: the State of Our Understanding.* Amer. Water Res. Assoc., Minneapolis, Minn., pp. 377–388.

Odum, E.P., 1971. *Fundamentals of Ecology.* W.B. Saunders Co., Philadelphia, Penn., 574 pp.

Odum, H.T., 1967. Biological circuits and the marine systems of Texas. In: T.A. Olson and F.J. Burgess (Editors), *Pollution and Marine Ecology.* John Wiley and Sons, New York, N.Y., pp. 99–157.

Odum, H.T., 1968. Work circuits and system stress. In: H.E. Young (Editor), *Symposium on Primary Productivity and Mineral Cycling in Natural Ecosystems.* University of Maine Press, Orono, Maine, pp. 81–138.

Odum, H.T., 1970a. Energy values of water resources. In: *Proc. 19th Southern Water Resources and Pollution Control Conf.* Duke University, Durham, N.C., pp. 56–64.

Odum, H.T., 1970b. Summary, an emerging view of the ecological system at El Verde. In: H.T. Odum and R.F. Pigeon (Editors), *A Tropical Rain Forest.* Report TID-24270 (PRNC-138), National Technical Information Service, Springfield, Va., pp. I-191–I-289.

Odum, H.T., 1978. Principles for interfacing wetlands with development. In: M.A. Drew (Editor), *Environmental Quality through Wetlands Utilization.* Coordinating Council on the Restoration of the Kissimmee River, Tallahassee, Fla., pp. 29–56.

Odum, H.T., 1983. *Systems Ecology.* John Wiley and Sons, New York, N.Y., 644 pp.

Odum, H.T., Wang, F.C., Alexander, J., Jr., and Gilliland, M., 1983. Energy analysis of environmental values. In: H.T. Odum, M.J. Lavine, F.C. Wang, M.A. Miller, J. Alexander, Jr., and T. Butler, *A Manual for Using Energy Analysis for Plant Siting.* Report to the Nuclear Regulatory Commission, NUREG/CR-2443, National Technical Information Service, Springfield, Va., 96 pp.

Radforth, N.W., 1952. Suggested classification of muskeg for the engineer. *Eng. J.*, 35: 1199–1210.

Radforth, N.W., 1962. Organic terrain and geomorphology. *Can. Geogr.*, 4: 166–171.

Radforth, N.W., 1977. Muskeg hydrology. In: N.W. Radforth and C.O. Brawner (Editors), *Muskeg and the Northern Environment in Canada.* University of Toronto Press, Toronto, Ont., pp. 130–147.

Radforth, N.W. and Bellamy, D.J., 1973. A pattern of muskeg: a key to continental water. *Can. J. Earth Sci.*, 10: 1420–1430.

Schenk, E., 1970. On the string formation in the aapa moors and raised bogs of Finland. In: *Ecology of the Subarctic Regions.* UNESCO, Paris, pp. 335–342.

Sibert, J.R. and Naiman, R.J., 1980. The role of detritus and the nature of estuarine ecosystems. In: K.R. Tenore and B.C. Coull (Editors), *Marine Benthic Dynamics.* University of South Carolina Press, Columbia, S.C., pp. 311–323.

Walter, H., 1977. The oligotrophic peatlands of western Siberia — the largest Peino-Helobiome in the world. *Vegetatio*, 34: 167–178.

Wharton, C.H., 1975. *The Natural Environments of Georgia.* Georgia Dept. of Natural Resources, Atlanta, Ga., 227 pp.

Woodruff, C.M., Jr., 1977. Stream piracy near the Balcones Fault zone, central Texas. *J. Geol.*, 85: 483–490.

Chapter 3

LONG-TERM DEVELOPMENT OF FORESTED WETLANDS

PATRICK C. KANGAS

INTRODUCTION

Perhaps more than any other kind of ecosystem, forested wetlands (swamps) exhibit the intimate manner in which living and non-living systems are coupled. A swamp is both an ecological system of forest vegetation, microbes, and animals, and a geological system of landform, sediments, and hydrologic regimes. A discussion of wetland development must include both ecological and geological aspects and consider how they relate. In this chapter, concepts and models of wetland development are reviewed.

GEOLOGY OF WETLAND FORMATION

The development of wetlands from a geological perspective has not had an impressively long history of attention on a global basis. Much of this work has been reviewed elsewhere (Gore, 1983). The information below draws heavily on North American studies. It is presented within a framework of causal mechanisms and a survey of chronologies.

Hydrologic mechanisms of forested wetland development

By definition, wetlands are wet, if only temporarily or periodically. The origin and maintenance of wetlands is then tied to local hydrology. A simple hydrologic balance model of a wetland (Fig. 3.1) shows water inputs from rainfall, groundwater seepage, and surface runoff. Storages, not differentiated in this model, occur as water in soil and peat, and as surface water. Outputs of the model are evapotranspiration, infiltration, and surface runoff.

Most wetlands are located in areas where the water table intersects the soil surface, providing a relatively constant source of water. The input is supplied either by groundwater seepage or by surface runoff from surrounding uplands. Wetlands in this category include zones around rivers, lake margins, and along coasts, and seepage wetlands maintained by groundwater discharge (see Forsyth, 1974a, b, for an example of the hydrologic origin of a seepage swamp).

In other situations, wetlands can develop by conserving water outputs or losses. For each outflow special mechanisms exist. Table 3.1 shows the matching of hydrologic output fraction and structural feature responsible for reducing water loss. Although mechanisms presented below are treated separately, several may have been responsible for forming a particular wetland.

Runoff, both as an input and an output, is a function of gravity. In general, wetlands occur on flatlands and on slopes where drainage is impeded due to the nature of soils or lack of hydraulic head (see Ch. 4, Sect. 12). In these situations there is not enough difference in potential energy due to the slope of the land to remove water effectively. Depression landforms also aid in retarding runoff by stabilizing water storage. The origin of these depressions is diverse. Any agent of erosion can be the causal force including stream cutting, past glacial action, wind deflation, and solution in regions of soluble rock. Hutchinson (1957) has reviewed the origins of lake depressions which are similar to wetland situations, and sometimes the precursors of wetlands. On slopes, natural dams can inhibit runoff by blocking drainage and

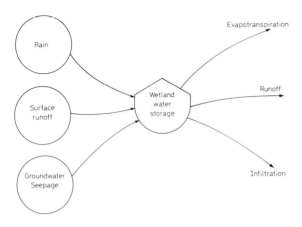

Fig. 3.1. Simple hydrologic balance model of a wetland. The symbols are from energy-circuit language (Odum, 1983; see Ch. 4, Sect. 9).

containing water upslope. Natural dams can arise in a variety of ways including peat dams (Shacklette, 1965; Verry, 1975), sand dunes and barrier islands from times of higher sea level during interglacial periods (Pirkle, 1972), ice mounds of string bogs (Drury, 1956; Washburn, 1973), overwash levees, and beaver dams (to be discussed later). Peat substrates of many kinds of wetlands also slow runoff by absorbing and retaining moisture. Dachnowski-Stokes (1935) found that peat can absorb 6 to 20 times its weight in water depending on the degree of decomposition.

Evapotranspiration presents very different problems. Water is lost to the atmosphere due to gradients of water potential between the atmosphere and the rhizosphere. Transpiration is controlled by sunlight and inputs of dry air which increase the moisture gradient. Possible regulation of evapotranspiration involves microclimate which in turn controls the moisture gradient. Shading by forest and marsh canopies maintains lower temperatures in the surface water and lessens evaporation loss (Linacre, 1976). Further control, at least in wetlands dominated by deciduous species (Odum, 1976; Brown, 1981), may be achieved through leaf loss from the canopy in the dry season, thereby reducing transpiration during the time when water supply is lowest. Tamarack (*Larix laricina*), a common swamp tree in the northern U.S.A. and Canada, is also a deciduous conifer and may play a similar role. The water-budget role of phreatophyte flood plains in the southwestern U.S.A. is an interesting and economically important controversy. Many workers have suggested that through evapotranspiration these forests remove a significant amount of water and that they should be cut to increase water for other uses (see for instance, Davenport et al., 1982). The definitive answer to whether phreatophyte forests transpire more water than they conserve remains unresolved, and optimal management of these forests requires additional considerations other than evapotranspiration (Horton and Campbell, 1974). The highest recorded evapotranspiration, 240 cm yr^{-1}, is from papyrus marshes[1] of the White Nile in the Sudan. Here the river flows through an arid region where the hot, dry air has a large water deficit (Penman, 1970). Evapotranspiration data for forested wetlands reviewed by the National Research Council (1982) ranged from about 60 to 100 cm yr^{-1}. Marshes, which grow where water levels are higher and with a high leaf area index, may generally have higher evapotranspiration losses than swamps. The effect of floating vegetation remains open. Some, like water hyacinth (*Eichhornia crassipes*), may transpire large volumes of water thus speeding succession to dry ground, while others, like duckweed (Lemnaceae), may slow evaporation from surfaces by forming a resistant barrier to water flux.

The main factor retarding infiltration loss is the

TABLE 3.1

Summary of mechanisms for conserving water in wetlands

Hydrologic output	Mechanisms
Surface runoff	Natural dams which block drainage, gentle slopes (reduced relief), absorptive peats and vegetation, and depressions and basins.
Evapotranspiration	Shading of water by vegetation (LAI)[1], deciduous vegetation which drops leaves in the dry season, and floating mats on water surface.
Infiltration	Hardpans which form an impervious confining layer, peat depth, and compaction of peat.

[1]Leaf area index.

---

[1]In this volume, *marsh* is used for wetlands with herbaceous vegetation, whereas *swamps* are dominated by woody species (see also Chapter 9, p. 239). This usage differs from that of Gore (1983).

development of an underlying impervious hardpan, usually a clay stratum. The role of clay pans in forming wetlands has been recognized for some time (Emerson and Smith, 1928; Hesse et al., 1937; Gorham, 1957). Clay strata have been found under many wetlands in the U.S.A. including the Dismal Swamp (Virginia) (Whitehead, 1972), the Okefenokee Swamp (Georgia) (Herrick, 1970), Florida cypress domes (Spangler, 1978), Corkscrew Swamp (Florida) (Duever et al., 1976), and others (Staub and Cohen, 1978, 1979). The origin of clays under wetlands is not completely understood. The clay may wash in with drainage water and accumulate over long periods of time, or it may be the substrate of a lagoonal environment (Van Straaten and Kuenen, 1958) deposited during periods of higher sea levels.

Swamp chronology

The development of wetlands takes variable lengths of time, and for this reason historical chronology of wetlands is often difficult to measure. Methods range from stratigraphic correlation for the longest geologic intervals, to the study of tree rings for shorter intervals. An important method of studying the intermediate history of wetlands is by examination of sediment profiles (Potzger, 1956; Davis, 1973).

Although the earliest known land-plant fossils come from the Silurian age, $400 \cdot 10^6$ yr BP, forests did not develop until the Devonian and early Carboniferous times, $250 \cdot 10^6$ yr BP. The first tree vegetation grew in swamps, making wetlands the "ancestors" of terrestrial ecosystems. These original swamps have been studied in some detail because they formed the great coal deposits which are used in modern times as a fossil fuel. To deposit great volumes of organic matter, these ecosystems had more net production (P) than respiration (R) — that is, $P/R > 1$. Their decline in the Permian age may have been due to the onset of a drier climate. Perhaps a decline of carbon dioxide concentration in the atmosphere due to storage of carbon in sediments may also have contributed (Odum and Lugo, 1970).

The Carboniferous swamps are thought to have been deltaic, growing on the margins of shallow marine seas (Fig. 3.2). The vegetation was composed of now primitive plants related to club mosses, horsetails and ferns, but with dimensions of present-day trees (Thomas, 1978). The two principal genera, *Lepidodendron* and *Sigillaria*, were as much as 40 to 50 m in height with trunks 1 to 4 m in diameter. These trees were woody, but their massive bark layers were thicker than their wood (Phillips et al., 1974). The original swamps had several characteristics of today's swamps (Darrah, 1960). Forests were more or less pure stands of the dominants. Some species had heavily buttressed bases — a feature which has been related to fluctuating water levels and unstable substrates. Aerial roots, similar to those of modern mangroves, have also been found. Leaves of some species were thick and coriaceous with sunken stomata. These are xeromorphic characters which may have been related to occurrence of humic acids in bog waters or to periodic water stress as in modern swamps (see Ch. 4, Sect. 8, and also Ch. 11). Dachnowski (1910, 1911) published early reviews of theories of bog xerophylly and Whitmore (1975) has more recently updated the subject. Fossil tree trunks do not have growth rings suggesting lack of seasonality in temperature and moisture availability. Mycorrhizae, which have been shown to have important ecological roles in modern forests, have been found on fossil tree roots (Wagner and Taylor, 1981). Although there were no terrestrial vertebrates in the swamps, insects such as dragonflies with a wing-span of nearly 1 m were abundant and have never since been equalled in size. Vertebrates came later, and one theory (Smith, 1972) suggests that mammals had their evolutionary origin in wetlands.

The history of recent swamps is tied to Pleistocene climatic activities. The origin of all wetlands in glaciated regions can date no earlier than early post-glacial time, about 12 000 yrs BP. On coastal plains, transgression and regression of the sea, corresponding to expanding and contracting polar ice masses, have set starting dates for wetland development. Parrish and Rykiel (1979) have provided a detailed review of the Pleistocene origin of the Okefenokee Swamp, which they regarded as representative of forested wetlands along the Atlantic and Gulf Coastal Plains of the United States. In unglaciated inland regions, climatic change led to changes in vegetation but not necessarily in geomorphic form. In these cases wetlands may have come back faster during

Fig. 3.2. Carboniferous landscape, combining younger and older taxa. The genera portrayed include:
(Ad) *Adiantites* (As) *Asterocalamites* (C) *Cardiopteris* (Cal) *Calamites* (Cd) *Cordaites*
(Et) *Etapteris* (Lg) *Lyginopteris* (Ln) *Lepidodendron* (N) *Neuropteris* (Ps) *Psygmophyllum*
(Ps) *Psaronius* (Pt) *Pitys* (R) *Rachopteris* (S) *Stauropteris* (Sa) *Stigmaria*
(Sg) *Sigillaria* (Sp) *Sphenophyllum* (U) *Ulodendron* (W) *Walchia*
(By permission from *Plant Life Through the Ages*, by A.C. Seward. Copyright, 1931, by Cambridge University Press, London.)

interglacial periods since only living systems had to be restored, not wetland geomorphology.

Ages of recent wetlands (Table 3.2) are of the same order of magnitude except for the Carolina bays, about which much confusion still exists (Johnson, 1942; Prouty, 1952; Odum, 1952; Wells and Boyce, 1953, 1954; Frey, 1954; Thom, 1970; Wells and Whitford, 1976) and for which earlier dating methods may have over-estimated ages.

ECOLOGY OF WETLAND FORMATION

The ecological mechanism of wetland formation is succession (Tallis, 1983). Ecological succession has been defined (Odum, 1969) as an orderly process of ecosystem development resulting from interactions and modifications between the physical environment and biotic community over time. The process culminates in a stable ecosystem traditionally known as climax. Although the concept of succession is of fundamental importance in ecology, details of its expression are still controversial (Whittaker, 1953; McCormick, 1968; McIntosh and Odum, 1969; Drury and Nisbet, 1973; Horn, 1974). Recent book-length reviews (Knapp, 1974; Golley, 1977; Ewel, 1980; West et al., 1981) provide summaries of some of the controversies. In the following pages the classical ideas of wetland succession are examined, followed by discussion of examples which do not fit with traditional concepts. Simulation models of forested

TABLE 3.2

Age of basal peats from forested wetlands

Ecosystem	Age (yrs B.P.)	References
Muskeg, Alaska, USA	1830	Heilman, 1968
Tropical swamp, Sarawak	4500	Wilford, 1960, in Whitmore, 1975
Okefenokee Swamp, Georgia, USA	4800–5700	Bond, 1970
Lake Agassiz peatland, Minnesota, USA	5000	Heinselman, 1970
Corkscrew Swamp, southwestern Florida, USA	5700	Kropp, 1976
Mangroves, Florida, USA	6500	Egler, 1952
Okefenokee Swamp, Georgia, USA	6500	Cohen, 1973
Peatland, Manitoba, Canada	7900	Reader and Stewart, 1972
Alcovy floodplain, Georgia, USA	8700	Staheli et al., 1974
Great Dismal Swamp, Virginia, USA	9000	Whitehead, 1972
Pine barrens, New Jersey, USA	9000–10 600	Buell, 1970
Carolina bays, North and South Carolina, USA	7000–40 000	Thom, 1970

wetland development are reviewed and a new approach, using embodied energy as a measure of system development, is briefly presented with a sample calculation for a floodplain situation.

The classical model

"Hydrosere" or "hydrarch succession" are the main terms given to wetland succession (Cooper, 1913; Clements, 1916, 1928). These successions begin in open water and progress towards conditions of reduced water depth, due to sedimentation and improved soil aeration. The direction of change in the classic hydrosere is from aquatic towards terrestrial systems. Perhaps because of this directionality, Weber (1908) used the term Verlandung to describe hydrarch succession, the English equivalent of which is "terrestrialization" (Tallis 1973; Malmer 1975). Such wetlands represent a temporary condition for small lakes and ponds which are subject to filling in. Plant communities play important roles in hydrosere development by contributing organic matter to sediments and by creating physical resistance to water movement. This resistance decreases current velocities and causes suspended sediments to be deposited. A typical hydrosere is a simple linear chain succession (Fig. 3.3) — a concept which may be attributed to several authors (Nichols, 1915; Lindeman, 1942). Table 3.3 shows some common species and their depth ranges for the classical hydrosere in central U.S.A. (Weaver and Clements, 1938).

These successions are often related to zonation of plant stands around lakes, streams, and estuaries, with the earliest stages extending out into the deeper water and older stages well back from the margins. In this way, zonation has been said to recapitulate succession and, moreover, zonation patterns have been studied to derive indirect information on successional sequences (e.g., Walker, 1940). Dansereau (1957) discussed the relation between zonation and succession and provided several detailed cross-section diagrams of vegetation zones in tidelands and floodplains for which the direction of succession was indicated. Beard (1944, 1955) also presented cross-sectional diagrams of tropical forested wetland formations according to water level, which imply succession. The relation between hydroperiod and vegetation zonation was described for tropical wetlands by Bellamy (1967). The depth ranges of plant species in Table 3.3 refers to the zonation characteristic of wetlands. More recently, however, the relationship between succession and zonation has been ques-

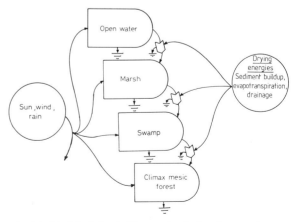

Fig. 3.3. Model of classic Clementsian hydrarch succession.

TABLE 3.3

Summary of classical hydrarch succession for central North America (adapted from Weaver and Clements, 1938); "Swamp" or forested wetland stage is shown as woodland

Successional stage	Common genera at each stage	Depth range
Submerged	*Elodea, Potamogeton, Ceratophyllum, Najas, Utricularia, Chara, Vallisneria*	6 m or less
Floating	*Nymphaea, Potamogeton, Polygonum, Lemna, Eichhornia*	2–3 m
Emergent[1]	*Scirpus, Typha, Phragmites, Sparaganium, Zizania, Sagittaria, Carex, Juncus, Eleocharis, Caltha*	Saturated soil to 1 m
Woodland	*Salix, Cephalanthus, Alnus, Populus*	Soil saturated only in spring and early summer
Climax forest	*Ulmus, Fraxinus, Carya, Salix, Alnus, Celtis, Acer, Quercus*	Moist soil

[1]Combined reed–swamp and sedge–meadow stages of Weaver and Clements.

tioned for mangroves (Lugo, 1980) and for wetlands in general (Egler, 1977) (also see Ch. 4, Sect. 11 and Ch. 7).

Two of the best known examples of hydroseres in the United States are Cedar Creek Bog (Minnesota) and Linsley Pond (Connecticut). The history of Cedar Creek Bog was presented by Lindeman (1941). It is a bog lake in an ice-block depression of an outwash plain formed during the late Wisconsin glaciation. The lake is in an advanced successional stage, with water occupying less than one-tenth of the lake's original depth and area. The lake has been filled with layers of marl, gyttja (organic sediments), silt, and various peats. Bog forests of tamarack (*Larix laricina*) and white cedar (*Thuja occidentalis*) occupy much of the original lake basin peripheral to a floating mat which extends out into the open water. Measurements by Lindeman of inward radial growth of the floating mat indicated a rate of nearly 1 m in 5 years. At this rate, he estimated that the lake would be completely filled-in as a bog forest within 250 years. However, Buell et al. (1968) found that the position of the margin of the mat remained essentially unchanged about 30 years after Lindeman's study. Williams (1971) also considered some of Lindeman's work on Cedar Bog Lake, but offered no reference to bog-filling other than the use of decomposition-rate coefficients in a simulation model.

Linsley Pond further illustrates how eutrophication impinges on the model of hydrarch succession (see, for instance, Hutchinson and Wollack, 1940; Deevey, 1942; Livingstone, 1957). The lake is located in the glaciated region, making its age about 10 000 to 12 000 years. Deevey (1955) has calculated that since its origin, the radius of the hypolimnion has contracted by 50 m and that 15 m of sediments have accumulated at the point of maximum depth. This indicates that 80% of the lake's hypolimnion has been "obliterated". With continued high rates of sedimentation, the indication is that Linsley Pond will eventually be filled in as has happened in Cedar Bog. In fact, a recent study (Brugam, 1978) has shown that human disturbance is increasing rates of eutrophication.

Energy control of basin filling

The classic examples of hydrosere have been reviewed to demonstrate the filling-in of basins as the cause of wetland succession. However, there are different categories of filling which depend on environmental conditions. Table 3.4 summarizes some possible dimensions of hydrosuccession. Along the row headings are energy intensities and along the column headings are aspects of the geometric characteristics of hydric succession. High-energy environments include wave-swept coasts of estuaries and large lakes, and margins of fast-flowing rivers. Because of the high energies, these are erosional rather than depositional en-

TABLE 3.4

Energy control of succession

Energy intensity of environment	Geometric characteristics of succession	
	filling orientation	origin of pioneer stage
High energy	Bottom up	From water to land
Low energy	Top down	From land to water

vironments. Thus, any filling-in of the basin, if it occurs, must be initiated by a benthic system which works from the bottom upwards. Due to the high energies, coasts and margins will be eroding, which precludes initiation of pioneer stages advancing from the shore towards deep water. Pioneer stages can arise from seeds and seedlings carried by currents or from rhizomes. This was noted by Tutin (1941), who goes further to consider seres ending in an underwater climax which he calls "limnoseres". Wetland examples in environments of this type are submerged macrophyte beds and fringe red mangroves (*Rhizophora mangle*).

In low-energy environments, sedimentation occurs because energy required to keep particles in suspension is lacking. In these situations, pelagic systems contribute detritus to the water column below and this leads to filling from the top down. Examples are plankton and floating macrophytes such as water hyacinth (*Eichhornia crassipes*). With lessened erosion due to lower energy along water margins, pioneer stages of succession can invade from the land and resemble terrestrial ecosystems. Examples are white and black mangroves (*Laguncularia racemosa* and *Avicennia germinans*, respectively).

In some environments which, on the average, are characterized by low energies, high-energy events such as floods may occur infrequently (see below, pp. 35–38). However, they may be important in seed dispersal and thus may initiate a typically high-energy succession sequence in a low-energy environment. Some marshes and swamps may have their origins in this way. For example, mangroves may invade newly created substrates formed after hurricanes or extensive flooding.

Mat systems deserve special attention because they occupy a stage in many different hydroseres. Consequently there are many differences in mat structure and community type. Mats usually reflect an advanced stage of eutrophication. These include floating islands (Russell, 1942; Reid, 1952; Sioli, 1968; Trivedy et al., 1978; such communities are also known as "Plav": Pallis, 1916; or "sudd": Richards, 1952; Thompson, 1976), algal mats (Phillips, 1958; Paerl and Ustach, 1982), macrophyte mats (Hunt, 1943; Ultsch, 1973), bog margins (Welch, 1945; Swan and Gill, 1970), *Schwingmoore* (see Gore, 1983), peat batteries or houses (Cypert, 1972), duckweed mats (Lemnaceae — Hillman, 1961) and others. Penfound (1952) discussed the role of mats in relation to succession. At least one view of mat systems is that of an ecosystem-level process which accelerates filling. Fig. 3.4 shows a model of the relations between floating mats and plankton. Although both plant systems contribute to organic sediments and thus to filling of the basin, the mat contributes at a faster rate due to its higher net productivity. The mat reduces wind-driven turbulence which keeps plankton in suspension. Light is also reduced to plankton below the mat (Goulder, 1969). Moreover, by cutting off turbulence, the diffusion of oxygen into the water is decreased, which in turn inhibits microbial decomposition of organic sediments. This speeds filling-in. As an example, Lewis and Bender (1961) documented low oxygen concentrations under a duckweed mat, which they concluded caused fish kills. Not shown on the model are other inhibitions of plankton by macrophytes such as allelopathy and competition for nutrients (Hasler and Jones, 1949; Wetzel, 1975). Another mechanism by which mats accelerate succession was described by Montz and Cherubini (1973), who suggested that alligatorweed (*Alternanthera philoxeroides*) mats aided in germination of bald cypress (*Taxodium distichum*), and therefore its extension into a marsh in Louisiana.

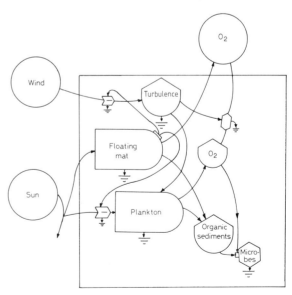

Fig. 3.4. Floating mat model showing the interactions of the mat with plankton. Floating mat and plankton are in parallel, sharing the same energy sources, and thus are in competition.

Another useful classification of wetland sedimentation is the distinction between autogenic and allogenic accretion (Egler, 1977; Tallis, 1983). These terms refer to different sediment sources; autogenic sediments originate from within the system and allogenic sediments originate from outside the system. Models of these two kinds of sedimentation systems are shown in Fig. 3.5. Allogenic sedimentation is shown with a separate source of inorganic sediments, which are advected into the system by currents of floodwaters or tides. Biomass of vegetation decreases the velocity of current leaving the system, thus reducing erosion of sediments. This process follows the Hjulström (1935) relationship which relates sediment-carrying capacity to current velocity. Autogenic sedimentation lacks a significant inorganic sediment source, accretion of organic material taking its place. Brown et al. (1979) related these extremes in a graph showing types of sediment versus current velocities (see Fig. 4.10). Peat was shown as being deposited in stillwater systems, whereas inorganic sediments were deposited in flowing water systems. At intermediate flow rates, mixed deposits were developed.

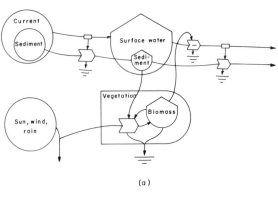

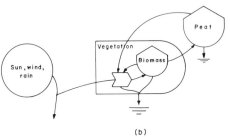

Fig. 3.5. A comparison of sedimentation regimes for wetlands: (a) allogenic sedimentation and (b) autogenic sedimentation.

An important controversy concerning allogenic conditions is the extent to which vegetation causes sedimentation or reacts to it. In some cases vegetation changes are clearly caused by sediment deposition, rather than vegetation development causing sedimentation (Featherly, 1940). However, in other cases the cause-and-effect relationship is not clear. This controversy has been most heated over the role of mangroves in coastal sedimentation. This is an important economic issue because mangroves are being managed for shoreline stabilization (Savage, 1972; Teas, 1977). Reports by researchers often are contradictory. For example, Scoffin (1970) stated that red mangroves "trap and bind sediment for a sufficient time to produce an accumulation higher than in nearby areas without dense mangroves". In contrast, Vann (1959) stated "the mangrove does not precede the appearance of new land along the deltaic seashore, but colonizes flats only after they have risen above low tide level." A great many studies in different parts of the world have been conducted, and the general consensus seems to be that mangroves have a relatively passive role in sedimentation, following the build-up of sediments rather than directly causing it. Excellent case studies are given by Thom (1967; Thom et al., 1975) and reviews on this subject are given by Carlton (1974) and Lugo (1980).

An interesting alternative mangrove system developing on oyster reefs, rather than sediments, in southwest Florida (U.S.A.) has been modeled by Odum (1975). The relationship of mangroves and coral reefs has also been discussed (Welch, 1963; McCoy and Heck, 1976; Stoddart, 1980). These authors described succession to mangroves on reefs whose vertical growth has stopped. Seagrass occurs in these successions as a facultative intermediate stage. Zieman (1972) described the reverse succession from mangrove hammocks to circular beds of seagrass where sea level rose in southern Florida. Examples of shoreline stabilization and succession for freshwater forested wetlands have been given by Shull (1944), McVaugh (1947, 1957), Viereck (1970), Wilson (1970), Graf (1978), Noble (1979) and Blackburn et al. (1982). In certain parts of North America, these successions involve primarily willows (*Salix* spp.) and cottonwoods (*Populus deltoides*) followed by a mixture of hardwoods such as box elder (*Acer negundo*), red maple (*Acer rubrum*), green ash (*Fraxinus pennsylvanica*), and oaks (*Quercus* spp.).

Other detailed examples of hydroseres have been given by Kormondy (1969) and Godwin et al. (1974). Carpenter (1981) quantitatively described basin-filling as a positive feedback system involving littoral vegetation, phytoplankton, and sediments. Recycling or "internal loading" of phosphorus from sediments and littoral plants increases plankton production and, therefore, sedimentation. Greater accretion by plankton increases the colonizable surface area for littoral plants, which expand and continue the cycle. This positive feedback system accelerates eutrophication and succession to more terrestrial vegetation. As a summary, Kerekes (1977) suggested that the ratio of basin volume to shoreline length could be used as an index of basin permanence. The value of the index decreases as basins approach extinction by sedimentation.

Exceptions to the classical model

Although the classical model of hydrarch succession seems to explain many wetland systems, many exceptions occur. Walker (1970) re-examined 20 published pollen diagrams from British hydroseres, including 71 vegetation transitions. Forty-six percent of the transitions fitted the classical model of hydrarch succession as described by Tansley (1939) for that country. Variations occurred in the remaining 54% of the transitions, particularly in the later seral stages. The variations included skipped stages (such as going from reed-swamp to a shrub carr without a fen stage), and in 17% of the transitions, reversed sequences. Systems which failed to comply with the classical model included certain lakes, transition of upland forest to bog (paludification), and cycles maintained by disturbance.

Lakes that do not fill in

There are cases in which filling of lakes never takes place or occurs slowly. These cases speak against the classical concept of the hydrosere. For example, Price (1947) discussed how scouring by wind-driven currents prevents sediment accumulation from occurring in coastal bays. Heinselman (1970) showed a peatland lake which, instead of filling in, had risen intact with the regional buildup of peat.

Slow rates of filling may also give the impression that sediments are not accumulating. Rates at which lakes fill are slow under most conditions (on the order of 10 000 years), but there is a stage in the succession of certain lakes where sedimentation slows even further. This stage has been termed "trophic equilibrium" which is defined as the "dynamic state of continuous complete utilization and regeneration of chemical nutrients in an ecosystem, without loss or gain from the outside, under a periodically constant energy source — such as might be found in a perfectly balanced aquarium or terrarium" (Lindeman, 1942; see Odum and Johnson, 1955, for further discussion of the balanced aquarium concept). Trophic equilibrium follows initiation of eutrophic conditions. In this state, filling is primarily with organic sediments which are decomposed as fast as they sink — that is, $P/R = 1$. Not all lakes can reach trophic equilibrium, and for those that do the stage is ultimately temporary. Drainage lakes have little opportunity to maintain equilibrium since they receive inorganic sediments from surface runoff. Seepage lakes, however, can reach and maintain depths because their inputs (groundwater) carry only dissolved materials (Hutchinson, 1973).

Paludification

Another exception to the classical concept of hydrarch succession is paludification (see, for instance, Drury and Nisbet, 1973; Gore, 1983). Paludification — a concept going back at least as far as Cajander (1913) — is the "swamping" of a landscape, or a kind of retrogressive, or reverse succession with transition from upland forest to wetland (Walker, 1970). This is primarily a northern phenomenon involving bogs and muskeg (moss system with scattered, gnarled trees; Potzger, 1934; Zach, 1950; Drury, 1956; Isaak et al., 1959; Heinselman, 1963, 1970). The process is a lateral overgrowth of mosses and deposition of peat, which aggressively smothers surrounding vegetation.

Two primary causal mechanisms driving paludification are rising water tables and a herbicidal effect due to certain bog conditions. The effect of increased water levels was already mentioned in regard to wetland development. Peat and living mosses act as reservoirs which hold water. The water eventually reduces oxygen diffusion, causing death of trees and decomposers unadapted to anaerobic conditions. In this way, as the bog spreads, surrounding systems are replaced by the advancing front of poorly drained conditions. This is part of an autocatalytic or self-maintaining feedback system. Initially standing water develops due to some interruption of the hydrologic cycle. Because water is standing, it stagnates and its oxygen content drops. The decline of oxygen limits root and soil respiration and decomposition of litterfall and organic sediments. This leads to the preservation of peat which in turn holds more water and amplifies the cycle. As stated by Dansereau and Segadas-Vianna (1952), "the whole bog can be compared to a sponge: the loose texture, the lightness of the materials of the 'soil' and its enormous water-retaining capacity will allow it to grow to great proportions." Extreme examples of paludification by water accumulation, such as raised and climbing bogs (Deevey, 1958; Moore and Bellamy, 1974), occur in the northern maritime climates (cool and moist).

Accompanying increasing water levels in paludification are chemical effects directly due to actions of the bog ecosystem. These include decreasing pH in bog waters, increased concentrations of toxic secondary plant substances, and nutrient stripping by *Sphagnum* spp. and other mosses. The cause and role of acidity have been reviewed by Clymo (1964, 1967) and Walmsley (1977). As an example, Godwin and Turner (1933) document progressive soil acidification accompanying vegetational succession in an English lake hydrosere. One potentially important cause of acidity is through microbial action stimulated by wetland conditions. Sulfur and acidity in mangrove forests are discussed by Hart (1962) and Giglioli and Thornton (1965). Glime et al. (1982) demonstrated the ability of non-*Sphagnum* mosses to lower the pH of water and discussed mechanisms whereby they facilitate succession from alkaline marsh to *Sphagnum* bog.

Many swamp waters are characterized being highly colored and rich in organic compounds. Organic acids contribute to decreasing pH, but also have additional effects. For example, they have been implicated in chelation of iron, thus regulating the availability of phosphorus as a plant nutrient (Shapiro, 1966; Koenings and Hooper, 1976), and in the inhibition of nitrification (Rice and Pancholy, 1973). Direct toxic effects of these substances have long been suspected (Livingston, 1905; Dachnowski, 1909) and have recently been tested (Valiela and Teal, 1980). As an example, Given and Dickinson (1975) suggested that *Sphagnum* moss contains a polyphenol (called sphagnol) which inhibits microbial decay and contributes to the preservation of peat. Cooksey and Cooksey (1978) used growth of benthic diatom species in sediment extracts from a mangrove swamp as a bioassay of the influence of the leachate from decomposition. They found a seasonal pattern of influence with a positive effect during the period of leaf fall, and a negative effect during the rest of the year. They attributed the positive effect or stimulation of growth to the heterotrophic ability of the diatoms to metabolize the soluble organic matter released during early stages of decomposition. Cattails (*Typha* spp.) have been shown to inhibit seed germination with release of autotoxins (McNaughton, 1968). Positive (Prakash and Rashid, 1968) and negative (Jackson and Hecky, 1980) effects of humic substances on phytoplankton production have been found. Fig. 3.6 shows a model of the herbicidal effects of acidity and toxins which contribute to the success of wetlands over alternative terrestrial systems. Rain is separated from the other energy sources as an additional source of acidity to bog waters.

Another chemical impact of paludification is the

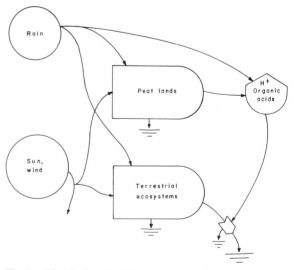

Fig. 3.6. Model of the herbicidal action of acidity in peatlands. Notice the parallel circuit form which indicates competition.

stripping of nutrients from waters passing through peats and mosses, especially *Sphagnum*. This occurs by cation exchange both on cell walls of living mosses and in the peat itself. Bases (for example, calcium) are absorbed from dissolved salts, and H^+ ions are set free. This not only contributes to the decline of pH but also causes an "oligotrophication" of the bog waters (Kurz, 1928; Bell, 1959; Clymo, 1967). Effects of these processes on available nutrient concentrations have been demonstrated by Heilman (1966, 1968) for an advancing bog landscape in Alaska. Concentrations of nitrogen, phosphorus, potassium and magnesium were significantly lower in plant tissues of *Sphagnum* peatlands than in the surrounding upland vegetation. Watt and Heinselman (1965) reported similar results for a spruce bog in northern Minnesota.

An independent demonstration of the inhibiting chemical effect of bog water comes from ecosystem experiments involving the addition of lime to bog systems (Hasler et al., 1951; Waters and Ball, 1957; Waters, 1957; Stross and Hasler, 1960; Stross et al., 1961; O'Toole and Synnott, 1971; Ivarson, 1977), a practice long used in agriculture for raising the pH of acid soils. The purpose of many of these studies was to improve fishery potential, which is low in most bog waters. The immediate effect of the addition of lime was an increase in pH and in plankton productivity which, in most cases, was sustained through at least the following year. The mode of action has been attributed to increasing the availability of phosphorus, because it is insoluble at both low and high pH. Fertilizers that have no neutralizing component do not have the same effects. The influence of calcium concentration is also found in marl lakes, which represent a case of natural "liming". These lakes are often oligotrophic, due to the precipitation of phosphate and other ions as marl is formed (Seischab and Garrisi, 1981; see also Wetzel, 1972, for a detailed discussion of the mechanisms of nutrient limitation in hard-water marl lakes).

The acid effect of paludification is somewhat similar to that of acid rain on ecosystems, and the same procedures for improving bog lakes used in the late 1950's are being "rediscovered" for counteracting the increased acidity. It may be interesting to study how much of the recent increase in acidity in surface waters is due to industrial acid rain, and how much is due to natural paludification (Patrick et al., 1981).

Another unique mechanism of paludification deserves mention. Beavers (*Castor castoroides*) are known to create wetlands with the flooding of terrestrial ecosystems behind their dams (Gates and Woollett, 1926; Lobeck, 1939; Beals, 1965). This mechanism of wetland formation may have been much more important in the Pleistocene, when an extinct genus (*Castoroides*) grew to half the size of a bear, and in Colonial times of North America before beaver populations were drastically reduced by trappers. Beaver distribution in North America covered the entire area above Mexico (Burt and Grossenheider, 1964), which suggests that their potential impact could have been great. Oaks and Whitehead (1979) have even suggested that beavers may have been an important causal force in the creation of the Great Dismal Swamp, along the border of Virginia and North Carolina (U.S.A.), which covers an area of more than 1800 km^2.

Disturbance cycles

The classic model of wetland succession implies constant environmental conditions. However, this assumption is seldom justified as many environmental parameters, such as precipitation, are characterized by much variation. Extreme conditions can disturb the progress of succession towards stable states, arresting it at different stages of development compared to succession without disturbance. The classic model is reversed to an earlier seral stage by disturbance, only again to proceed towards a stable climax which may never be reached. The new state may be considered a cyclical climax if the disturbance is periodic. These periodic disruptions lead to cycles of vegetation instead of the classic unidirectional development of Clementsian succession (Fig. 3.7). Although the disturbance-cycle model is receiving renewed interest (Connell, 1978; White, 1979; Botkin, 1979; Lugo, 1980; Vogl, 1980) it is related to older ideas: Aubréville's mosaic hypothesis (Richards, 1952), Odum's concept of pulse stability (1969), and Watt's study of heather cycles (1947). Disturbance may trigger rejuvenation mechanisms such as replacement of old or senescent individuals in populations and remobilization of accumulated storages in nutrient cycles. In this way the

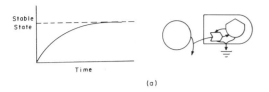

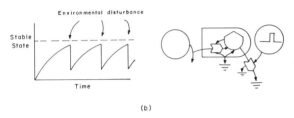

Fig. 3.7. Comparison of time graphs and minimodels of classic hydrarch succession (a) and the disturbance cycle hypothesis (b). The vertical axis of the graphs is biomass but could be other parameters. See Egler (1977) for other possible curves of development.

ecosystem uses the disturbance as a new energy source to do useful work instead of only absorbing the disturbance as a stress.

Fluctuating water levels of wetlands may occasionally result in extremes of high or low water. This is probably the most common type of wetland disturbance. Flooding is only a serious disturbance if inundation is persistent. Those wetland systems with impeded drainage, due to topography or underlying geology, can be severely modified by flooding. A reverse succession is initiated with tree kills, followed by community structure changes towards more aquatic assemblages, depending on the duration of inundation. Many examples of the effects of high water levels can be found in kettle-hole bogs of glaciated regions (Buell et al., 1968; Schwintzer and Williams, 1974; Schwintzer, 1978; Sanger and Gannon, 1979), and in artificial flooding for reservoir construction (see below, p. 38). However, in hydrologically open systems such as floodplains, natural drainage quickly removes flood waters. In these cases flooding is an energy source advecting sediments, detritus, and seeds into wetland systems.

The other disturbance associated with fluctuating water levels is drought and the conditions to which it leads. As a wetland dries, aerobic oxidation of organic sediments can result in subsidence (see, for instance, Sheail and Wells, 1983) which precludes the filling of the depression as required by classic hydrosere succession. Van der Valk and Davis (1978, 1979) have described in much detail the cyclic processes in marshes in Iowa which are driven by drought. The cycle involves a rotation of three vegetation types: submerged, mudflat, and emergent. An energy-circuit translation of this cyclic model (Fig. 3.8) should be compared with that of classic hydrarch succession (Fig. 3.3). Cyclic succession is similar to succession without disturbance except for the reset mechanism of disturbance (see models in Fig. 3.7). Such cycles never allow marshes to succeed to swamps in climates with periodic severe drought. A similar cycle is described by Godfrey and Godfrey (1975) for temperate tidal marshes with storm overwash disturbance. Lugo (1980) presents arguments that mangroves also undergo cyclical succession generated by hurricane wind-throw, frost damage, and a variety of other disturbances.

As in many ecosystems, fire in wetlands often coincides with drought conditions (Kozlowski and Ahlgren, 1974). Although wetlands usually act as firebreaks, there are many reports of the effect of fire on wetlands. Considering North America only, it has been reported for Alaska (Drury, 1956), Alberta (Lewis and Dowding, 1926), Northwest

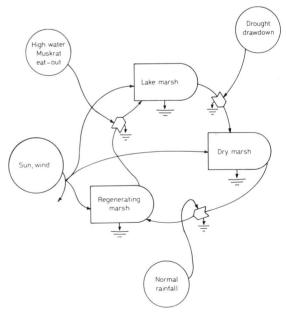

Fig. 3.8. Model of the cyclic marshes of Iowa described by Van der Valk and Davis (1979).

Territories (Black and Bliss, 1978), Wisconsin (Catenhusen, 1950; Curtis, 1959; Vogl, 1964, 1969), Iowa (Davis, 1979), North Carolina (Wells and Whitford, 1976), the Okefenokee Swamp in Georgia (Cypert, 1973; Hamilton, 1984), north-central Florida (Ewel and Mitsch, 1978), and the Everglades in southern Florida (Cohen, 1974). It is also important in Australia (Campbell, 1983) and Africa (Thompson and Hamilton, 1983).

Fires are possible when natural or artificial drainage lowers the water table for a sufficient period of time. Lightning is most often considered to trigger fire once the substrate is dried. The direct effect is to burn off top layers of peat or organic soil and to change, at least temporarily, vegetation structure and composition. Crown fires can occur when the organic debris on the forest floor is too wet to burn. In this case, the fire would probably start in the surrounding uplands and be carried into the wetland by winds. As an example of the effects of fire, Vogl (1964) reported a change from conifer swamp to open *Sphagnum* bog and finally to northern sedge meadow, with increasing prescribed burning. Some wetlands have adapted to recurrent fire. For example, Curtis (1959) suggested that black spruce (*Picea mariana*) may have serotinal cones, in which fire accelerates the release of seeds. At least one forested wetland type, the Atlantic white cedar (*Chamaecyparis thyoides*) forest of the eastern coastal plain of the United States, may even be a fire climax in the sense that the fire maintains the wetland condition by favoring reproduction of white cedar (Buell and Cain, 1943; Little, 1979). Once fire ceases, however, succession returns towards pre-fire conditions as described by the disturbance-cycle graph in Fig. 3.7.

An interesting side effect occurs when the pre-fire system is dominated by *Sphagnum* or other acid-producing vegetation, as is the situation in many bogs and black-water swamps. If the fire is sufficiently intense to remove the *Sphagnum* mat, the pH of the system increases to more alkaline conditions. This chemical change may in part explain the vegetation changes that follow fire, because it was noted earlier that acidity may act as an herbicide excluding certain species from wetlands. Loveless (1959) gives an example where pH, calcium, and magnesium increase with the burning of tree islands in southern Florida.

Perhaps because of the significance of fluctuating water levels in the development of wetlands, many species have become adapted to these conditions. This is particularly striking in the reproduction strategies of wetland trees. Flooding is used by many tree species as an energy source for seed dispersal. Prime examples are the mangrove genera *Rhizophora* and *Avicennia*, which are viviparous, the seeds germinating while still attached to the parent tree. When the seedlings fall, they are dispersed by tidal currents. This kind of mechanism of seed dispersal is important in reforesting bare areas in the forest mosaic (Egler, 1948). Rabinowitz (1978a, b) explains the zonation of mangroves with a hypothesis of tidal sorting of the propagules according to size and by differential ability of propagules to establish in deep water.

Another unique seed-dispersal mechanism is described by Gottsberger (1978) and Goulding (1980). They suggest that, in the Amazon's floodplain forests, a large number of fruit-eating fish aid in dispersal of the tree seeds. The relationship is actually symbiotic, the fish getting food from the trees and the trees getting seed dispersal work from the fish (Fig. 3.9).

Germination is even more closely tied with hydroperiod. Two dominant northern swamp trees, tamarack and black spruce (*Picea mariana*), and a dominant southern swamp tree, cypress, require dry conditions for germination to occur (Kurz, 1933; Curtis, 1959). Standing water is usually fatal to the seedlings, which makes it imperative that seeds remain dormant until lower water levels occur. This reproductive strategy has also been suggested for a tropical swamp tree, *Pterocarpus officinalis* (Ch. 10, pp. 261–262) and is probably widespread. Pulses of germination following disturbance may explain the origin of the large, pure even-aged stands characteristic of many

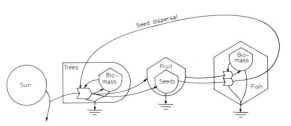

Fig. 3.9. Model of Goulding's (1980) system for tree seed dispersal in Amazon floodplain forests.

kinds of forested wetlands. Casual references are made to the floating abilities of many wetland tree seeds, but the hypothesis that they float more effectively than seeds from upland species appears not to have been tested yet.

Although drought, fire and flooding are the most common kinds of wetland disturbances, other extreme conditions, such as wind-throw, peatslips, and disruption by animal populations, can initiate cycles of succession (see Ch. 4, Sect. 4 and 11). Wind-throw leads to a gap-phase type of cycle over small areas, which is characteristic of many forest ecosystems (Whitmore, 1978; Runkle, 1981). An open question is whether wetland trees are more susceptible to wind-throw because they are usually shallow-rooted and the substrate in which they root is less stable than upland soils. Peat-slip is a special disturbance which occurs in blanket bogs and moorland where wetlands grow on hillslopes. Under these conditions, if peat accumulation becomes great enough, the bogs literally "burst" (Wace, 1961) and slip downslope, leaving bare spots within the larger bog landscape. Other examples of natural peat erosion are given by Tallis (1965, 1983) and Crisp (1966).

Animal populations can also cause disturbance to wetland ecosystems when their densities are high. Most significant may be muskrats which are known to "eat-out" marshes by feeding and incorporating vegetation, including the root systems, into their lodges (O'Neil, 1949; Danell, 1977; Van der Valk and Davis, 1979). Moose can cause major disruption in that they "may retard or set back pond hydrosere by cutting up the marginal *Carex* mats and dispersing them with their hooves" (Daubenmire, 1968). A similar example is given by Walker (1972) for feral pigs in a New Guinea swamp. The role of alligator holes in maintaining standing water refuges during droughts in southern Florida is well known (Craighead, 1968; Kushlan, 1974), but the mounds built by the alligators, as nests for their eggs, may also be important as colonization sites initiating the transition from sawgrass (*Cladium jamaicense*) marsh to the forest vegetation of tree islands (Craighead, 1968). An interesting controversy developed over the role of a wood-boring isopod, *Sphaeroma terebrans*, on the prop roots of red mangroves in southern Florida. Rehm and Humm (1973) and Rehm (1976) suggest that the isopod was destroying the mangrove forests, whereas Snedaker (1974) and Simberloff et al. (1978) provided arguments for positive effects of the isopod on mangroves. This situation illustrates the complexities of the effects of animals on wetland development.

Another type of disturbance is due to the actions of human populations. Lumbering, burning, and grazing by domestic animals lead to changes in species composition, but may maintain wetland conditions. Examples are given by Sampson (1930) for a temperate area and by Flenley (1979) for the Old World tropics. Often these changes are retrogressive, with transitions from forested swamps to herbaceous marshes. Changes in hydroperiod brought about by human actions can be more serious. Drainage for agriculture or construction causes shifts from wetland to terrestrial conditions. Burns (1978) showed changes in productivity upon draining cypress swamps of southwestern Florida. LeBarron and Neetzel (1942) noted that the construction of drainage canals in the upper peninsula of Michigan caused a switch from coniferous bog forest to a more advanced stage of succession dominated by red maple (*Acer rubrum*) and alder (*Alnus* sp.). Flooding from the construction of reservoirs causes a shift from wetland to completely aquatic conditions. Water levels are not only raised but also stabilized. Harms et al. (1980) documented tree kills due to the impoundment of Lake Ocklawaha in north-central Florida. They found that after three years of inundation, all trees had died in those parts of the reservoir flooded to a depth in excess of 1.3 m. Impacts by humans, then, can either accelerate or reverse hydrarch succession.

Cycles of longer intervals have caused "recurrence horizons" in certain raised bogs (Gore, 1983). These bog profiles have the form of horizons of dark, humified peat containing much *Calluna* (heath) pollen alternating with bright orange-colored, less humified *Sphagnum* peat. The well-humified recurrence horizons are recognizable as crusty or harder sections laid down under drier climates and indicate adverse conditions for bog growth. The *Sphagnum*-dominated horizons are attributed to periods of higher precipitation with improved growth. This type of stratigraphy was used in dating or timing Pleistocene climatic change before ^{14}C dating techniques were developed (Godwin, 1954). Recent, extensive reviews on the subject have been published by Barber (1981) and Godwin (1981).

Simulation models of forested wetland succession

The models included in this paper are descriptive, for the purpose of depicting systems and processes. However, they could all be simulated because the symbols and pathways of the energy-circuit language have direct mathematical translations (Odum, 1972, 1983; Hall et al., 1977). A survey of existing simulation models of forested wetland succession is described next.

Shugart and West (1980) have provided a general review of forest succession models, emphasizing terrestrial examples. They grouped succession models into three categories: tree models, gap models, and forest models. Tree models use the individual tree as the unit of simulation. They include growth and probabilities of replacement of an individual tree of one species by another individual of the same or of a different species. Gap models also simulate attributes of individual trees, but over a small area (i.e. gap size due to disturbance) and considering interspecific interactions. Forest models simulate the forest as a whole, aggregating individual and sometimes species roles into state variables. These models primarily consider transitions between community types or successional stages, and patterns of development of forest biomass over time.

Characteristics of 10 succession models of forested wetlands are summarized in Table 3.5. Except for one tree model (Franz and Bazzaz, 1977) and one gap model (Phipps, 1979), all of these models are probably best categorized as forest models according to Shugart and West's scheme. Most use differential equations to describe vegetation transitions, with biomass as a measure of importance. Abiotic state variables, such as nutrients and non-living organic matter (detritus or peat), which influence the development of the biotic state variables, are included in most models. The objective of the majority of the models was to investigate the response of the wetland to some kind of disturbance, either human management or impact (harvest, impoundment, etc.) or a natural event (fire, hurricane, etc.). Succession models of non-forested wetlands (marshes) are described by Bayley and Odum (1976), Friedman and DeWitt (1977), Zieman and Odum (1977), Browder (1978), Jones and Gore (1978), Randerson (1979), and Van der Valk (1981).

The models themselves are hypotheses about the systems they portray. All of the models described in Table 3.5 have been verified or validated. However, the primary value of the models at present may not be in the accuracy of their predictions, but rather in their ability to summarize a number of experiments and data into explicit statements of system behavior. The variety of different types of models suggest that there are many ways of viewing the development of forested wetlands.

Succession as accumulations of information and energy

According to Margalef (1968) "the process of succession is equivalent to a process of accumulating information." Margalef (1963, 1968) has developed a terminology of maturity to describe the relative degree of organization of ecosystems as an alternative to seral stages of vegetation. The P/B ratio (P = primary production, B = biomass) is most often used as a measure of maturity. This ratio indicates the relative amount of energy needed to maintain ecosystem biomass. Conceptually the P/B ratio is a turnover rate, the reciprocal being turnover time. It suggests the extent to which ecosystems can "accumulate history" (Margalef, 1961). In this sense the amount of information stored by the ecosystem is directly related to the turnover time. The longer the turnover time, the more information that can be stored before the system "turns over". Turnover time, as measured in this way, may also be related to the period of the disturbance cycle discussed earlier. Examples of turnover times of forested wetlands are given in Table 3.6. Values range from about 7 years for northern bog forest to a high of about 23 years for an Illinois floodplain forest. These data suggest that the Illinois floodplain is more mature than the others, with the potential to accumulate more information and to be more organized. For comparison, the turnover time for a marsh is included. Since temperate-zone marshes die back each year, their turnover times will approach one year. A value for peat storage in an English

TABLE 3.5

Characteristics of simulation models of forested wetland development

Ecosystem	Type of model	Simulation	Reference
Swiss peat bog	Differential equation	Succession of bog and fen vegetation as a function of slope and nutrient status	Wildi, 1978
English peat bog	Differential equation	Development of depth profiles of peat bulk density, mass and volume	Clymo, 1978
English peat bog	Differential equation (compartment model)	Development of peat storage with harvesting and fertilization	Gore and Olson, 1967; refined by Jones and Gore, 1978
English hydroseres	Transition probability tables	Empirical frequencies of transitions between wetland vegetation types	Walker, 1970
Okefenokee Swamp, Georgia	Fractals (see Mandelbrot, 1977)	Analysis of spatial patterns of vegetation with diffusion functions to predict successional status	Hastings et al., 1980
Florida cypress dome	Differential equation (energy circuit language)	Oscillatory patterns of biomass with timber harvest and fire	Mitsch, 1975
Illinois floodplain	Normal probability functions	Tree species distributions and community similarity before and after reservoir impoundment	Franz and Bazzaz, 1977
Arkansas floodplain	Computer program subroutines (conversion of the JABOWA model of Botkin et al., 1972)	Tree growth as a function of water-level manipulation and logging	Phipps, 1979
Iowa shoreline vegetation	Differential equation (conversion of Bledsoe and Van Dyne's 1971 compartment model)	Oscillatory succession of vegetation types with an historical sequence of floods	Austin et al., 1979
Florida mangroves	Differential equation (energy circuit language)	Development of biomass and detritus under different metabolic rates and regrowth after hurricane stress	Lugo et al., 1978

moorland is also included, which is more than 20 times longer than any ecosystem. Other tables of P/B ratios for wetlands are given by Reader and Stewart (1972) and Lugo (1980).

Another way to quantify the successional development of an ecosystem is with a new concept of embodied energy. The total energy accumulated in the development of a structure, such as the biomass and organization of an ecosystem, is termed its embodied energy (Odum, 1983; see also Ch. 4, Sect. 10). The embodied energy of a structure is its energy input expressed in units of equivalent quality, integrated over the time needed to develop the structure. Embodied energy is somewhat similar to the economic notion of capital cost. A stepwise procedure for calculating embodied energy of a given structure is as follows (Odum et al., 1983).

(1) Make a model of the structure showing important inputs, outputs, feedbacks, and storages.

(2) Evaluate the model by quantifying all pathways and storages, first in units appropriate to the pathway or storage (i.e. rainfall in mm, wind speed in $km\ h^{-1}$, sediment in $g\ m^{-2}$), second in units of actual energy (joules or calories of heat), and finally in units of equivalent energy quality.

(3) Calculate the total energy input to the structure per unit time (most simply done by adding energies in equivalent units).

TABLE 3.6

Turnover times (above ground B/P ratio) using net production for some forested wetland ecosystems, with comparisons with a marsh and peat storage

Ecosystem	Turnover time (yrs)	Reference
Illinois floodplain forest	23	Johnson and Bell, 1976
Florida floodplain forest	17	Brown, 1981
Manitoba muskeg	10	Reader and Stewart, 1972
Florida mangroves	9	Snedaker and Lugo, 1973; Lugo and Snedaker, 1974
Michigan bog forest	7	Ulrich, 1980
New Jersey marsh	0.9	Jervis, 1969
English moorland peat	526	Gore and Olson, 1967

(4) Multiply the energy input rate from step 3 by the time period required to build the structure to find the total energy embodied in it.

Several assumptions are involved in calculating embodied energy with the procedure described above. The total energy signature is assumed to be embodied in the structure (see Ch. 4, Sect. 10). Some energies, however, may pass the structure without being completely expended. Are these embodied in the structure? Development time is taken to be the length of time from initiation of development of the structure to the present. This definition assumes that development is continuous, whereas certain systems may reach steady states which may not be considered to develop further. Cyclic systems also present problems of defining development time. Use of addition as the algorithm for interacting energies of the signature may be too simplistic. Other algorithms are considered by Regan (1977), Swaney (1978) and Odum (1983). A detailed description and review of this method is given by Odum et al. (1983).

A preliminary calculation for a temperate floodplain system is developed below. Fig. 3.10 shows a model of some of the main energy sources (the energy signature) of the floodplain. The model was evaluated using data representative of southeastern Michigan (U.S.A.) and, where available, for the Huron River which drains about 2300 km^2 into western Lake Erie. Derivations of the actual

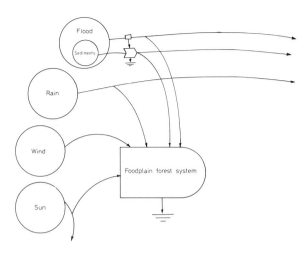

Fig. 3.10. Model of energy sources for a floodplain. Calculations of an energy signature appear in Table 3.7.

energy values for each source are given in Table 3.7 along with the energy transformation ratios which convert energies into equivalent quality units. Here, solar calories are used for embodied energy, as shown by the product of actual energies multiplied by the transformation ratios, in the final column in Table 3.7. The total energy input from these sources (the energy signature) is $44.8 \cdot 10^6$ solar cal. m^{-2} yr^{-1}.

To find the embodied energy in the floodplain, the energy signature must be multiplied by the length of time needed to build the structure. The floodplain, as all ecosystems, is a combination of living ecosystem and non-living landform. These are separated for comparison. The length of time needed to develop the living ecosystem can be taken as the successional time to climax forest. Several floodplain forest values are given in Table 3.8. The average, about 500 years is used here. The difference may be noted between this value and turnover time in Table 3.6, which is for turnover of an existing system rather than the time it takes to develop that system through a successional series of systems. The length of development for the non-living landform of the floodplain is taken as the time since the last glacial stage covered southeastern Michigan, or about 10 000 years. The calculation of embodied energy is summarized in Table 3.9. The energy signature, multiplied by development time, yields embodied energy. Obviously, the landform portion of the floodplain takes longer to develop and thus has a greater embodied energy.

TABLE 3.7

Floodplain energy signature[1]

Source	Actual energy (J m^{-2} yr^{-1})	Transformation ratio (solar J J^{-1})	Embodied energy (solar J m^{-2} yr^{-1})
Sun[2]	$4.2 \cdot 10^9$	1.0	$4.2 \cdot 10^9$
Wind[3]	$4.2 \cdot 10^6$	56.7	$0.24 \cdot 10^9$
Rain			
chemical potential[4]	$10.0 \cdot 10^3$	$6.9 \cdot 10^3$	$0.07 \cdot 10^9$
kinetic impact[5]	$22.6 \cdot 10^3$	$2.4 \cdot 10^5$	$5.4 \cdot 10^9$
Flood head[6]	$19.7 \cdot 10^3$	$4.0 \cdot 10^5$	$8.0 \cdot 10^9$
Organic sediments[7]	$368.5 \cdot 10^3$	$4.6 \cdot 10^5$	$170 \cdot 10^9$
Total	$4.2 \cdot 10^9$	–	$188 \cdot 10^9$

[1] Editor's note: Energy units have been converted to joules to maintain uniformity. However, not all kinds of energy can be converted with 100% efficiency to mechanical energy; thus, the conversion ignores differences in energy quality (Odum, 1983).
[2] Sunlight = $1.0 \cdot 10^6$ cal. m^{-2} yr^{-1} (Odum et al., 1983).
[3] Wind = $1/2\ \rho\ V^2\ C\ (1/d) = 1.0 \cdot 10^3$ cal. m^{-2} yr^{-1}, where: ρ = density of air = $1.2 \cdot 10^{-3}$ g cm^{-3}; V = wind velocity = 478.3 cm sec^{-1} (National Oceanographic and Atmospheric Administration, 1974); C = eddy diffusion coefficient = 10^4 cm^2 sec^{-1} (Kemp, 1977); d = height of boundary layer = 10^4 cm.
[4] Rain (chemical potential) = $nRT \ln C_2/C_1 M$ = 2.4 cal. m^{-2} yr^{-1}, where: n = 1 mole/18 g of H$_2$O; R = gas constant = 1.99 cal K^{-1} mole^{-1}; T = absolute temperature = 281°C; C_2 = water content of rain = 999 986 ppm (Odum et al., 1983); C_1 = water content of receiving water = 999 900 ppm; M = mass of rain per yr = 78 cm yr^{-1} (National Oceanographic and Atmospheric Administration, 1974) or = $7.8 \cdot 10^5$ g m^{-2} yr^{-1}.
[5] Rain (kinetic impact) = $1/2M\ V^2$ = 5.4 cal. m^{-2} yr^{-1}, where: M = mass of rain per yr (see footnote 4); V = velocity of raindrops = 762 cm sec^{-1} for an average drop diameter (Odum et al., 1983).
[6] Flood head = $\rho\ DGHF\ (1/A)$ = 4.68 cal m^{-2} yr^{-1}, where: ρ = density of water = 1 g cm^{-3}; D = flood discharge = $8.5 \cdot 10^7$ cm^3 sec^{-1} (Corps of Engineers, 1976) for a 20 year flood; G = gravity = 980 cm sec^{-2}; H = elevation drop or head = 7625 cm (Say and Jansson, 1976); F = duration of flood per year (5 days flood^{-1}) (flood 20 yr^{-1}) ($8.64 \cdot 10^4$ sec day^{-1} = $2.16 \cdot 10^4$ sec yr^{-1}); A = area of floodplain = $6.96 \cdot 10^7$ m^2 (Corps of Engineers, 1976).
[7] Organic inflows of floodwaters = 88 cal. m^{-2} yr^{-1}. Wolman and Leopold (1957) give an average deposition rate in major floods of about 3.1 cm. Assuming a recurrence interval of 20 yr the annual rate of deposition is: (3.1 cm)/(20 yr) = 0.16 cm yr^{-1}. In terms of mass of sediments this is $2.2 \cdot 10^3$ g m^{-2} yr^{-1} which assumes a bulk density of 1.4 g cm^{-3}.
  Assume an organic content of 1% (Brown et al., 1979) and a heat content of 4.0 cal. g^{-1} (Odum, 1971).

Also shown in Table 3.9 are the actual energy values of the two structures of the floodplain. For the ecosystem this is the energy in the biomass itself, and for the landform it is the gravitational potential generated by sediment deposition. The embodied energy can be divided by the actual energy content of the structure to yield transformation ratios, which are shown in the last column of Table 3.9. This is a unit measure of energy quality or value.

Other calculations of embodied energy in ecosystems and landforms are given by Kangas (1983). These represent first attempts to estimate embodied energy accumulations rather than only energy in biomass in areas of landscape. The chief interest in these values, for a discussion of succession and development of forested wetlands, is in comparison of the living and non-living portions of the wetlands. In the sample calculation given here, the landform has about 20 times more energy embodied in its structure (on an area basis) than the forest ecosystem, and the quality factor (transformation ratio) is a million times larger. A principal utility of calculations of embodied energy comes from its

TABLE 3.8

Floodplain forest characteristics

Location	Dominant species	Age of climax forest (yrs)
Lower Mississippi river, USA (Shelford, 1954)	*Liriodendron tulipifera* *Tilia heterophylla* *Quercus* sp. *Acer saccharinum*	600
Olympic mountains, Washington, USA (Fonda, 1974)	*Picea sitchensis* *Tsuga heterophylla*	750
North-central Texas, USA (Nixon, 1975)	*Ulmus alata* *Quercus stellata* *Prunus mexicana*	100[1]

[1]Proposed by Nixon, oldest study site was a *Celtis–Juniperus* community, 47 years old.

better values for development times and energy inputs become available.

CONCLUSIONS

The summary view which emerges from this survey is that aspects of ecosystem and landform develop together in a wetland landscape. Fig. 3.11 illustrates this main feature in a simple model. Development is shown as a production process with ecosystems and landforms as the storages of landscape. Both storages share the production process and influence its rate through feedback. Although this form of interaction between the storages is indirect, ecosystems sometimes contribute directly to landform — for example in the deposition of peat and in the stabilizing structure of heavy root mats. The landform has a higher

role in evaluating environmental impacts and economic investments. Comparisons of embodied energy and other parameters of energy analysis for alternative actions can be a basis for objective and quantitative decision-making. Wang et al. (1980) used embodied energy as a criterium for power-plant siting by comparing a forested floodplain and agricultural land as alternative locations. Kangas (1980) calculated embodied energies for the channel volume, substrate rocks, and animal community of a stream ecosystem, and discussed their use in evaluating human impacts such as channelization and organic pollution. It should be recognized that this approach is still under development and that the results of calculations may change as

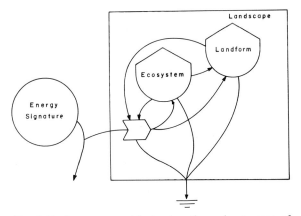

Fig. 3.11. Summary model showing the main storages of landscape as ecosystem and landform.

TABLE 3.9

Embodied energy[1] values for the floodplain example

Storage	Energy signature (solar J m^{-2} yr^{-1})	Development time (yrs)	Embodied energy (solar J m^{-2})	Actual energy[2] (J m^{-2})	Transformation ratio (solar J J^{-1})
Ecosystem	$1.88 \cdot 10^{11}$	500	$9.4 \cdot 10^{13}$	$3.3 \cdot 10^7$	$2.9 \cdot 10^5$
Landform	$1.88 \cdot 10^{11}$	10 000	$188 \cdot 10^{13}$	$6.7 \cdot 10^3$	$2.8 \cdot 10^{11}$

[1]Editor's note: Energy units have been converted to joules to maintain uniformity; see Table 3.7.
[2]Actual energy calculations: Ecosystem actual energy = BE, where: B = biomass of forest = 20 kg m^{-2} (Brown et al., 1979); E = energy content of biomass = 4 cal. g^{-1} (Odum, 1971).
Landform actual energy = MGH, where: M = mass = volume × density; G = gravity = 980 cm sec^{-2}; H = average height = 50 cm (assumed); density = 1.36 g cm^{-3}; volume = area × depth; area = $6.96 \cdot 10^7$ m^2 (United States Geological Survey, 1979); depth = 1 m (assumed).

embodied energy per unit area and per unit of actual energy (Table 3.9), thus giving it a role in the landscape as an amplifying feedback on the ecosystem storage. In wetlands, where water is needed by both ecosystem and landform processes, this relation takes the form of landforms orchestrating hydrology while ecosystems provide the fine-tuning.

Either form of growth shown in Fig. 3.7 may characterize the storages of landscape depending on the program of forcing functions of the energy sources. This accounts for both the classic steady-state model of hydrarch succession and the oscillating disturbance-cycle model. Human action can change the pattern of development of either the geological or ecological storages by accelerating or reversing succession, by changing a steady-state climax to a cyclic climax or vice versa, or by complete replacement and substitution of systems. The hope is that with better understanding of wetland systems as being the product of long and complex development processes, better management of wetland resources will be possible.

The simple model given above also allows identification of values of a landscape. True value of a swamp is in both the ecosystem and the landform. If the trees are cut down, or the water drained away, the swamp still remains in the storage of landform. In land policy, priorities should be given to storages with the highest quality, which in most cases will probably be the swamp landform (see Ch. 4, Sect. 14).

The important distinction between ecosystems and landforms is their time constants. Although there is a symbiosis between geology and ecology of landscape (Fig. 3.11), time periods often distinguish the two subsystems. In discussions of ecosystems and landforms as interacting systems, the importance of their different time scales has been emphasized (Drury and Nisbet, 1971; Swanson, 1979). Turnover of ecosystems is often much faster than turnover of landform. In the floodplain example, the forest ecosystem would turn over 20 times faster than the floodplain landform. However, in other examples discussed here, the time periods may not be so different — as, for instance, in coastal wetlands where sedimentation is great. Here, ecological succession quickly tracks geomorphic events, suggesting that these ecosystems are more closely coupled in time with their landforms. Time periods of each type of wetland system need to be evaluated. If the turnover time of the landform is close to the turnover time of the ecosystem, then a small change in one may result in a larger change in the other. Consider the example of the effect of forest growth on erosion rates. The logging of a small amount of forest on a mountain slope will result in loss of a greater amount of sediment than logging that same amount of forest on flatlands. One explanation for this is that the turnover time of the mountain slope is closer to that of its forest than the turnover time of the flatland sediment compared with its forest. The study and comparison of time periods for storages of landscape can then form the basis for land-use management and planning.

REFERENCES

Aubréville, A., 1938. La forêt coloniale: les forêts de l'Afrique occidentale française. *Ann. Acad. Sci. Colon., Paris*, 9: 1–245.

Austin, T.A., Riddle, W.F. and Landers, R.Q., Jr., 1979. Mathematical modeling of vegetative impacts from fluctuating flood pools. *Water Resour. Bull.*, 15: 1265–1280.

Barber, K.R., 1981. *Peat Stratigraphy and Climatic Change*. Balkema, Rotterdam, 219 pp.

Bayley, S. and Odum, H.T., 1976. Simulation of interrelations of the Everglades marsh, peat, fire, water and phosphorus. *Ecol. Model.*, 2: 169–188.

Beals, E.W., 1965. An anomalous white cedar–black spruce swamp in northern Wisconsin. *Am. Midl. Nat.*, 74: 244.

Beard, J.S., 1944. Climax vegetation in tropical America. *Ecology*, 25: 127–158.

Beard, J.S., 1955. The classification of tropical American vegetation-types. *Ecology*, 36: 89–100.

Bell, P.R., 1959. The ability of sphagnum to absorb cations differentially from dilute solutions resembling natural waters. *J. Ecol.*, 47: 351–355.

Bellamy, D.R., 1967. Succession and depth time scale in ephemeral swamp ecosystem. *Trop. Ecol.*, 8: 67–75.

Black, R.A. and Bliss, L.C., 1978. Recovery sequence of *Picea mariana*/*Vaccinum uliginosum* forests after burning near Inuvik, Northwest Territories, Canada. *Can. J. Bot.*, 56: 2020–2030.

Blackburn, W.H., Knight, R.W. and Schuster, J.L., 1982. Saltcedar influence on sedimentation in the Brazos River. *J. Soil Water Conserv.*, 37: 298–301.

Bledsoe, L.J. and Van Dyne, G.M., 1971. A compartmental model simulation of secondary succession. In: B.C. Patten (Editor), *Systems Analysis and Simulation in Ecology*, Vol. 1. Academic Press, New York, N.Y., pp. 479–511.

Bond, T., 1970. Radiocarbon dates of peat from the Okefenokee Swamp, Georgia. *Southeast. Geol.*, 11: 199–201.

Botkin, D.B., 1979. A grandfather clock down the staircase:

stability and disturbance in natural ecosystems. In: R.H. Waring (Editor), *Forests: Fresh Perspectives from Ecosystem Analysis*. Oregon State Univ. Press, Corvallis, Ore., pp. 1–10.

Botkin, D.B., Janak, J.F. and Wallis, J.R., 1972. Some ecological consequences of a computer model of forest growth. *J. Ecol.*, 60: 849–872.

Browder, J.A., 1978. A modeling study of water, wetlands and wood storks. *Wading Birds Res. Rept. 7*. National Audubon Soc., Washington, D.C., pp. 325–346.

Brown, S., 1981. A comparison of the structure, primary productivity and transpiration of cypress ecosystems in Florida. *Ecol. Monogr.*, 51: 403–427.

Brown, S., Brinson, M.M. and Lugo, A.E., 1979. Structure and function of riparian wetlands. In: *Strategies for Protection and Management of Floodplain Wetlands and Other Riparian Ecosystems*. U.S. Department of Agriculture, For. Serv. Gen. Tech. Rept. WO-12. U.S. Department of Agriculture Forest Service, Washington, D.C., pp. 17–31.

Brugam, R.B., 1978. Human disturbance and the historical development of Linsley Pond. *Ecology*, 59: 19–36.

Buell, M.F., 1970. Time of origin of New Jersey Pine Barrens bogs. *Bull. Torrey Bot. Club*, 97: 105–108.

Buell, M.F. and Cain, R.L., 1943. The successional role of southern white cedar, *Chamaecyparis thyoides*, in southeastern North Carolina. *Ecology*, 24: 85–93.

Buell, M.F., Buell, H.F. and Reiners, W.A., 1968. Radial mat growth on Cedar Creek Bog, Minnesota. *Ecology*, 49: 1198–1199.

Burns, L.A., 1978. *Productivity, Biomass and Water Relations in a Florida Cypress Forest*. Ph.D. Diss., Univ. North Carolina, Chapel Hill, N.C., 170 pp.

Burt, W.H. and Grossenheider, R.P., 1964. *A Field Guide to the Mammals*. Houghton Mifflin Co., Boston, Mass., 284 pp.

Cajander, A.K., 1913. Studien über die Moore Finnlands. *Acta For. Fenn.*, 2(3): 1–208.

Campbell, E.O., 1983. Mires of Australasia. In: A.J.P. Gore (Editor), *Mires: Swamp, Bog, Fen and Moor*. Ecosystems of the World 4B, Elsevier, Amsterdam, pp. 153–180.

Carlton, J.M., 1974. Land-building and stabilization by mangroves. *Environ. Conserv.*, 1: 285–294.

Carpenter, S.R., 1981. Submersed vegetation: an internal factor in lake ecosystem succession. *Am. Nat.*, 118: 372–383.

Catenhusen, J., 1950. Secondary successions on the peat lands of glacial Lake Wisconsin. *Trans. Wisc. Acad. Sci.*, 40: 29–48.

Clements, F.E., 1916. Plant succession: an analysis of the development of vegetation. *Carnegie Inst. Wash. Publ.*, 242: 1–512.

Clements, F.E., 1928. *Plant Succession and Indicators: a Definitive Edition of Plant Succession and Plant Indicators*. Wilson Publ., New York, N.Y., 453 pp.

Clymo, R.S., 1964. The origin of acidity in sphagnum bogs. *Bryologist*, 67: 427–431.

Clymo, R.S., 1967. Control of cation concentrations, and in particular of pH, in sphagnum dominated communities. In: H.L. Golterman and R.S. Clymo (Editors), *Chemical Environment in the Aquatic Habitat*. N.V. Noord-Hollandsche Uitgevers Maatschappij, Amsterdam, pp. 273–284.

Clymo, R.S., 1978. A model of peat bog growth. In: O.W. Heal and D.F. Perkins (Editors), *Production Ecology of British Moors and Montane Grasslands*. Springer-Verlag, New York, N.Y., pp. 187–223.

Cohen, A.D., 1973. Possible influences of subpeat topography and sediment type upon the development of the Okefenokee Swamp–Marsh complex of Georgia. *Southeast. Geol.*, 15: 141–151.

Cohen, A.D., 1974. Evidence of fires in the ancient Everglades and coastal swamps of southern Florida. In: P.J. Gleason (Editor), *Environments of South Florida: Present and Past*. Mem. 2, Miami Geol. Soc., Miami, Fla. pp. 213–218.

Connell, J.H., 1978. Diversity in tropical rain forests and coral reefs. *Science*, 199: 1302–1309.

Cooksey, K.E. and Cooksey, B., 1978. Growth-influencing substances in sediment extracts from a subtropical wetland: investigation using a diatom bioassay. *J. Phycol.*, 14: 347–352.

Cooper, W.S., 1913. The climax forest of Isle Royale, Lake Superior, and its development. *Bot. Gaz.*, 55: 1–44, 115–140, 189–235.

Corps of Engineers, 1976. *Special Flood Hazard Information Report, Huron River, Ann Arbor-Ypsilanti, Michigan and Vicinity*. Detroit District, U.S. Army Corps of Engineers, Detroit, Mich., 26 pp.

Craighead, F.C., 1968. The role of the alligator in shaping plant communities and maintaining wildlife in the southern Everglades. *Fl. Nat.*, 41: 2–7, 69–74.

Crisp, D.T., 1966. Input and output of minerals for an area of Pennine moorland: the importance of precipitation, drainage, peat erosion and animals. *J. Appl. Ecol.*, 3: 327–348.

Curtis, J.T., 1959. *The Vegetation of Wisconsin*. Univ. Wisconsin Press, Madison, Wisc., 657 pp.

Cypert, E., 1972. The origin of houses in the Okefenokee prairies. *Am. Midl. Nat.*, 87: 448–458.

Cypert, E., 1973. Plant succession on burned areas in Okefenokee Swamp following the fires of 1954 and 1955. *Proc. Ann. Tall Timbers Fire Ecol. Conf., Tallahassee, Fla.*, pp. 199–217.

Dachnowski, A., 1909. Bog toxins and their effect upon soils. *Bot. Gaz.*, 47: 389–405.

Dachnowski, A., 1910. Physiologically arid habitats and drought-resistance in plants. *Bot. Gaz.*, 49: 325–340.

Dachnowski, A., 1911. The vegetation of Cranberry Island (Ohio) and its relations to the substratum, temperature and evaporation. *Bot. Gaz.*, 52: 1–33, 126–149.

Dachnowski-Stokes, A.P., 1935. Peat land as a conserver of rainfall and water supplies. *Ecology*, 16: 173–177.

Danell, K., 1977. Short-term plant successions following the colonization of a northern Swedish lake by the muskrat, *Ondatra zibethica*. *J. Appl. Ecol.*, 14: 933–948.

Dansereau, P., 1957. *Biogeography, an Ecological Perspective*. Ronald Press, New York, N.Y., 394 pp.

Dansereau, P. and Segadas-Vianna, F., 1952. Ecological study of the peat bogs of eastern North America, I. Structure and evolution of vegetation. *Can. J. Bot.*, 30: 490–520.

Darrah, W.C., 1960. *Principles of Paleobotany*. Ronald Press, New York, N.Y., 295 pp.

Daubenmire, R., 1968. *Plant Communities*. Harper and Row Publ., New York, N.Y., 300 pp.

Davenport, D.C., Martin, P.E. and Hagan, R.M., 1982. Evapotranspiration from riparian vegetation: conserving water by reducing salt cedar transpiration. *J. Soil Water Conserv.*, 37: 237–239.

Davis, A., 1979. Wetland succession, fire and the pollen record: a midwestern example. *Am. Midl. Nat.*, 102: 86–94.

Davis, M.B., 1973. Ecological history of wetlands. In: T. Helfgott, M.W. Lefor and W.C. Kennard (Editors), *Wetlands Conference*. Inst. Water Res., Univ. Connecticut, Storrs, Conn., pp. 43–124.

Deevey, E.S., Jr., 1942. Studies on Connecticut lake sediments, III. The biostratonomy of Linsley Pond. *Am. J. Sci.*, 240: 233–264, 313–338.

Deevey, E.S., Jr., 1955. The obliteration of the hypolimnion. *Mem. Ist. Ital. Idrobiol., Suppl.*, 8: 9–38.

Deevey, E.S., Jr., 1958. Bogs. *Sci. Am.*, 199(4): 114–124.

Drury, W.H., Jr., 1956. *Bog Flats and Physiographic Processes in the Upper Kuskokwim River Region, Alaska*. Contrib. Gray Herbarium, Harvard Univ., Cambridge, Mass., 178 pp.

Drury, W.H. and Nisbet, I.C.T., 1971. Interrelations between developmental models in geomorphology, plant ecology and animal ecology. *Gen. Syst.*, 16: 57–68.

Drury, W.H. and Nisbet, I.C.T., 1973. Succession. *J. Arnold Arbor., Harv. Univ.*, 54: 331–368.

Duever, M.J., Carlson, J.E., Gunderson, L.A. and Duever, L., 1976. Ecosystem analysis at Corkscrew Swamp. In: H.T. Odum and K.C. Ewel (Editors), *Cypress Wetlands for Water Management, Recycling and Conservation*. Third Ann. Rept., Center for Wetlands, Univ. Florida, Gainesville, Fla., Report PB-273-097, National Technical Information Service, Springfield, Va., pp. 707–737.

Egler, F.E., 1948. The dispersal and establishment of red mangrove, *Rhizophora*, in Florida. *Caribb. For.*, 9: 299–320.

Egler, F.E., 1952. Southeast saline Everglades vegetation, Florida, and its management. *Vegetatio*, 3: 213–265.

Egler, F.E., 1977. *The Nature of Vegetation, its Management and Mismanagement*. Connecticut Conservation Association, Bridgewater, Conn., 527 pp.

Emerson, F.V. and Smith, J.E., 1928. *Agricultural Geology*. John Wiley and Sons, New York, N.Y., 377 pp.

Ewel, J. (Editor), 1980. Tropical succession. *Biotropica*, Suppl., 12: 1–95.

Ewel, K.C. and Mitsch, W.J., 1978. The effects of fire on species composition in cypress dome ecosystems. *Fla. Sci.*, 41: 25–31.

Featherly, H.I., 1940. Silting and forest succession on Deep Fork in southwestern Creek County, Oklahoma. *Proc. Okla. Acad. Sci.*, 21: 63–65.

Flenley, J.R., 1979. *The Equatorial Rain Forest: a Geological History*. Butterworths, London, 162 pp.

Fonda, R.W., 1974. Forest succession in relation to river terrace development in Olympic National Park, Washington. *Ecology*, 55: 927–942.

Forsyth, J.L., 1974a. Geologic conditions essential for the perpetuation of Cedar Bog, Champaign County, Ohio. *Ohio J. Sci.*, 74: 116–125.

Forsyth, J.L., 1974b. The hydrologic conditions that maintain Cedar Bog. In: C.C. King and C.M. Fredrick (Editors), *Cedar Bog Symposium*. Ohio Biol. Surv. Inform. Circ. No. 4, Columbus, Ohio, pp. 12–14.

Franz, E.H. and Bazzaz, F.A., 1977. Simulation of vegetation response to modified hydrologic regimes: a probabilistic model based on niche differentiation in a floodplain forest. *Ecology*, 58: 176–183.

Frey, D.G., 1954. Evidence for the recent enlargement of the "bay" lakes of North Carolina. *Ecology*, 35: 78–88.

Friedman, R.M. and DeWitt, C.B., 1977. Spatial modeling of lake-edge wetland development. In: C.B. DeWitt and E. Soloway (Editors), *Wetlands Ecology, Values, and Impacts*. Inst. for Environmental Studies, Univ. Wisconsin, Madison, Wisc., pp. 232–241.

Gates, F.C. and Woollett, E.C., 1926. The effect of inundation above a beaver dam upon upland vegetation. *Torreya*, 26: 45–50.

Giglioli, M.E.C. and Thornton, I., 1965. The mangrove swamps of Keneba, Lower Gambia River Basin, II. Sulphur and pH in the profiles of swamp soils. *J. Appl. Ecol.*, 2: 257–269.

Given, P.H. and Dickinson, C.H., 1975. Biochemistry and microbiology of peats. In: E.A. Paul and D. McLaren (Editors), *Soil Biochemistry, 3*, Marcel Dekker, New York, N.Y., pp. 123–212.

Glime, J.M., Wetzel, R.G. and Kennedy, B.J., 1982. The effects of bryophytes on succession from alkaline marsh to sphagnum bog. *Am. Midl. Nat.*, 108: 209–223.

Godfrey, P.J. and Godfrey, M.M., 1975. Some estuarine consequences of barrier island stabilization. In: L.E. Cronin (Editor), *Estuarine Research*. Vol. II, Academic Press, New York, N.Y., pp. 485–516.

Godwin, H., 1954. Recurrence-surfaces. *Dan. Geol. Unders., II. Raekke*. Nr. 80.

Godwin, H., 1981. *The Archives of the Peat Bogs*. Cambridge Univ. Press, Cambridge, Mass., 229 pp.

Godwin, H. and Turner, J.S., 1933. Soil acidity in relation to vegetational succession in Calthrope Broad, Norfolk. *J. Ecol.*, 21: 235–262.

Godwin, H., Clowes, D.R. and Huntley, B., 1974. Studies in the ecology of Wicken Fen, V. Development of fen carr, *J. Ecol.*, 62: 197–214.

Golley, F. (Editor), 1977. *Ecological Succession*. Benchmark Papers in Ecology 5. Dowden, Hutchinson and Ross, Stroudsburg, Penn., 373 pp.

Gore, A.J.P. (Editor), 1983. *Mires: Swamp, Bog, Fen and Moor*. Ecosystems of the World 4A and B, Elsevier, Amsterdam, 440 and 479 pp.

Gore, A.J.P. and Olson, J.S., 1967. Preliminary models for accumulation of organic matter in an Eriophorum/Calluna ecosystem. *Aquilo, Ser. Bot.*, 6: 297–313.

Gorham, E., 1957. The development of peat lands. *Q. Rev. Biol.*, 32: 145–166.

Gottsberger, G., 1978. Seed dispersal by fish in the inundated regions of Humaita, Amazonia. *Biotropica*, 10: 170–183.

Goulder, R., 1969. Interactions between the rates of production of a freshwater macrophyte and phytoplankton in a pond. *Oikos*, 20: 300–309.

Goulding, M., 1980. *The Fishes and the Forest: Explorations in Amazonian Natural History*. Univ. of California Press, Berkeley, Calif., 280 pp.

Graf, W.L., 1978. Fluvial adjustments to the spread of tamarisk in the Colorado Plateau region. *Geol. Soc. Am. Bull.*, 89: 1491–1501.

Hall, C.A.S., Day, J.W., Jr. and Odum, H.T., 1977. A circuit language for energy and matter. In: C.A.S. Hall and J.W. Day, Jr. (Editors), *Ecosystem Modeling in Theory and Practice*. John Wiley and Sons, New York, N.Y., pp. 38–48.

Hamilton, D.B., 1984. Plant succession and the influence of disturbance in Okefenokee Swamp. In: A.D. Cohen, D.J. Casagrande, M.J. Andrejko and G.R. Best (Editors), *The Okefenokee Swamp: its Natural History, Geology, and Geochemistry*. Wetland Surveys, Los Alamos, N.M., pp. 86–111.

Harms, W.R., Schreuder, H.T., Hook, D.D., Brown, C.L. and Shropshire, F.W., 1980. The effects of flooding on the swamp forest in Lake Ocklawaha, Florida. *Ecology*, 61: 1412–1421.

Hart, M.G.R., 1962. Observations on the source of acid in empoldered mangrove soils, I. Formation of elemental sulphur. *Plant Soil*, 17: 87–98.

Hasler, A.D. and Jones, E., 1949. Demonstration of the antagonistic action of large aquatic plants on algae and rotifers. *Ecology*, 30: 359–364.

Hasler, A.D., Brynildson, O.M. and Helm, W.T., 1951. Improving conditions for fish in brown-water bog lakes by alkalization. *J. Wildl. Manage.*, 15: 347–352.

Hastings, H.M., Monticciolo, R., Pekelney, R. and van Kannon, D., 1980. *Averaging, Control and Persistence, and Vegetation in the Okefenokee Swamp*. Publ. No. 4, Ecosystems Modelling Group, Hofstra Univ., New York, N.Y., 23 pp.

Heilman, P.E., 1966. Change in distribution and availability of nitrogen with forest succession on north slopes in interior Alaska. *Ecology*, 47: 825–831.

Heilman, P.E., 1968. Relationship of availability of phosphorus and cations to forest succession and bog formation in interior Alaska. *Ecology*, 49: 331–336.

Heinselman, M.L., 1963. Forest sites, bog processes and peatland types in the glacial Lake Aggassiz region, Minnesota. *Ecol. Monogr.*, 33: 327–374.

Heinselman, M.L., 1970. Landscape evolution, peatland types, and the environment in the Lake Agassiz peatland natural area, Minnesota. *Ecol. Monogr.*, 40: 235–261.

Herrick, S.M., 1970. New data regarding the Okefenokee Swamp in southeastern Georgia. *Geol. Soc. Am. Abstr. with Programs for 1970 Annual Meetings*, 2: 215.

Hesse, R., Allee, W.C. and Schmidt, K.P., 1937. *Ecological Animal Geography*. John Wiley and Sons, New York, N.Y., 597 pp.

Hillman, W.S., 1961. The Lemnaceae, or duckweeds. *Bot. Rev.*, 27: 221–287.

Hjulström, F., 1935. Studies of the morphological activity of rivers as illustrated by the river Fyria. *Bull. Geol. Inst. Univ. Uppsala*, 25: 221–527.

Horn, H.S., 1974. The ecology of secondary succession. *Ann. Rev. Ecol. Syst.*, 5: 25–37.

Horton, J.S. and Campbell, C.J., 1974. *Management of Phreatophyte and Riparian Vegetation for Maximum Multiple Use Values*. U.S. Department of Agriculture For. Serv. Res. Pap. RM-117, Forest Service, Fort Collins, Colo., 23 pp.

Hunt, K.W., 1943. Floating mats on a southeastern coastal plain reservoir. *Bull. Torr. Bot. Club*, 70: 481–488.

Hutchinson, G.E., 1957. *A Treatise in Limnology*. Vol. 1, John Wiley and Sons, New York, N.Y., 978 pp.

Hutchinson, G.E., 1973. Eutrophication. *Am. Sci.*, 61: 269–279.

Hutchinson, G.E. and Wollack, A., 1940. Studies on Connecticut lake sediments, II. Chemical analyses of a core from Linsley Pond, North Branford. *Am. J. Sci.*, 238: 493–517.

Isaak, D., Marshall, W.H. and Buell, M.F., 1959. A record of reverse plant succession in a tamarack bog. *Ecology*, 40: 317–320.

Ivarson, K.C., 1977. Changes in decomposition rate, microbial population and carbohydrate content of an acid peat bog after liming and reclamation. *Can. J. Soil Sci.*, 57: 129–138.

Jackson, T.A. and Hecky, R.E., 1980. Depression of primary productivity by humic matter in lake and reservoir waters of the boreal forest zone. *Can. J. Fish. Aquatic Sci.*, 37: 2300–2317.

Jervis, R.A., 1969. Primary production in the freshwater marsh ecosystem of Troy Meadows, New Jersey. *Bull. Torrey Bot. Club*, 96: 209–231.

Johnson, D., 1942. *The Origin of the Carolina Bays*. Columbia Univ. Press, New York, N.Y., 341 pp.

Johnson, F.L. and Bell, D.T., 1976. Plant biomass and net primary production along a flood-frequency gradient in the streamside forest. *Castanea*, 41: 156–165.

Jones, H.E. and Gore, A.J.P., 1978. A simulation of production and decay in blanket bog. In: O.W. Heal and D.F. Perkins (Editors), *Production Ecology of British Moors and Montane Grasslands*. Springer-Verlag, New York, N.Y., pp. 160–186.

Kangas, P., 1980. *Energy Analysis of Storages in a Stream Ecosystem*. Unpublished Manuscript of a paper presented at Conf. on Lotic Ecosystems, Savannah River Ecology Lab., Augusta, Georgia, available from author.

Kangas, P., 1983. *Energy Analysis of Landforms, Succession and Reclamation*. Ph.D. Diss., Univ. of Florida, Gainesville, Fla., 186 pp.

Kemp, W.M., 1977. *Energy Analysis and Ecological Evaluation of a Coastal Power Plant*. Ph.D. Diss., Univ. of Florida, Gainesville, Fla., 560 pp.

Kerekes, J., 1977. The index of lake basin permanence. *Int. Rev. Gesamten Hydrobiol.*, 62: 291–293.

Knapp, R. (Editor), 1974. *Vegetation Dynamics. Handbook of Vegetation Science, 8*. W. Junk Publ., The Hague, 364 pp.

Koenings, J.P. and Hooper, F.F., 1976. The influence of colloidal organic matter on iron-phosphorus cycling in an acid bog lake. *Limnol. Oceanogr.*, 21: 684–696.

Kormondy, E.J., 1969. Comparative ecology of sandspit ponds. *Am. Midl. Nat.*, 82: 28–61.

Kozlowski, T.T. and Ahlgren, C.E. (Editors), 1974. *Fire and Ecosystems*. Academic Press, New York, N.Y., 542 pp.

Kropp, W., 1976. Geochronology of Corkscrew Swamp Sanctuary. In: H.T. Odum and K.C. Ewel (Editors), *Cypress Wetlands for Water Management, Recycling and Conservation*. Third Ann. Rept., Center for Wetlands, Univ. Florida, Gainesville, Fla. Report PB 273-097,

National Technical Information Service, Springfield, Va., pp. 772–785.

Kurz, H., 1928. Influence of sphagnum and other mosses on bog reactions. *Ecology*, 9: 56–69.

Kurz, H., 1933. Cypress domes. *Fl. Geol. Surv., Annu. Rep.*, 23–24: 54–56.

Kushlan, J.A., 1974. Observations on the role of the American alligator, *Alligator mississippiensis*, in the southern Florida, USA, wetlands. *Copeia*, 1974: 993–996.

Le Barron, R.K. and Neetzel, J.R., 1942. Drainage of forested swamps. *Ecology*, 23: 457–465.

Lewis, F.J. and Dowding, E.S., 1926. The vegetation and retrogressive changes of peat areas ("muskegs") in central Alberta. *J. Ecol.*, 14: 317–341.

Lewis, W.M. and Bender, M., 1961. Effect of a cover of duckweeds and the alga *Pithophora* upon the dissolved O_2 and free CO_2 of small ponds. *Ecology*, 42: 602–604.

Linacre, E.T., 1976. Swamp. In: J.L. Monteith (Editor), *Vegetation and the Atmosphere*. Vol. 2, Academic Press, London, pp. 329–347.

Lindeman, R.L., 1941. The developmental history of Cedar Creek Bog, Minnesota. *Am. Midl. Nat.*, 25: 101–112.

Lindeman, R.L., 1942. The trophic–dynamic aspect of ecology. *Ecology*, 23: 399–418.

Little, S., 1979. Fire and plant succession in the New Jersey Pine Barrens. In: R.T.T. Forman (Editor), *Pine Barrens: Ecosystem and Landscape*. Academic Press, New York, N.Y., pp. 297–314.

Livingston, B.E., 1905. Physiological properties of bog water. *Bot. Gaz.*, 39: 348–355.

Livingstone, D.A., 1957. On the sigmoid growth in the history of Linsley Pond. *Am. J. Sci.*, 255: 364–373.

Lobeck, A.K., 1939. *Geomorphology*. McGraw-Hill Book Co., New York, N.Y., 731 pp.

Loveless, C.M., 1959. A study of the vegetation in the Florida Everglades. *Ecology*, 40: 1–9.

Lugo, A.E., 1980. Mangrove ecosystems: successional or steady state? In: J.J. Ewel (Editor), *Tropical Succession. Biotropica, Suppl.*, 12: 65–72.

Lugo, A.E. and Snedaker, S.C., 1974. The ecology of mangroves. *Annu. Rev. Ecol. Syst.*, 5: 39–64.

Lugo, A.E., Sell, M. and Snedaker, S.C., 1978. Mangrove ecosystem analysis. In: B.C. Patten (Editor), *Systems Analysis and Simulation in Ecology*. Academic Press, New York, N.Y., pp. 133–145.

Malmer, N., 1975. Development of bog mires. In: A.D. Hasler (Editor), *Coupling of Land and Water Systems*. Springer-Verlag, New York, N.Y., pp. 85–92.

Mandelbrot, B.B., 1977. *Fractrals: Form, Chance and Dimension*. Freeman Press, San Francisco, Calif., 346 pp.

Margalef, R., 1961. Communication of structure in planktonic populations. *Limnol. Oceanogr.*, 6: 124–128.

Margalef, R., 1963. On certain unifying principles in ecology. *Am. Nat.*, 97: 357–374.

Margalef, R., 1968. *Perspectives in Ecological Theory*. Univ. of Chicago Press, Chicago, Ill., 111 pp.

McCormick, I., 1968. Succession. *Via*, 1: 1–16.

McCoy, E.D. and Heck, K.L., Jr., 1976. Biogeography of corals, seagrasses and mangroves: an alternative to the centre of origin concept. *Syst. Zool.*, 25: 201–210.

McIntosh, R.P. and Odum, E.P., 1969. Ecological succession. *Science*, 166: 403.

McNaughton, S.J., 1968. Autotoxic feedback in the regulation of *Typha* populations. *Ecology*, 49: 367–369.

McVaugh, R., 1947. Establishment of vegetation on sand-flats along the Hudson River, New York. *Ecology*, 28: 189–193.

McVaugh, R., 1957. Establishment of vegetation on sand flats along the Hudson River, New York, II. The period 1945–1955. *Ecology*, 38: 23–28.

Mitsch, W.J., 1975. Simulation of possible effects of sewage application, fire and harvesting on a cypress dome. In: H.T. Odum and K.C. Ewel (Editors), *Cypress Wetlands for Water Management, Recycling and Conservation*. 2nd Ann. Rept., Center for Wetlands, Univ. of Florida, Gainesville, Fla., pp. 276–304.

Montz, G.N. and Cherubini, A., 1973. An ecological study of a baldcypress swamp in St. Charles Parish, Louisiana. *Castanea*, 38: 378–386.

Moore, P.D. and Bellamy, D.J., 1974. *Peatlands*. Springer-Verlag, New York, N.Y., 221 pp.

National Oceanographic and Atmospheric Administration, 1974. *Climates of the States*. Water Information Center, Inc., Port Washington, N.Y., 975 pp.

National Research Council, 1982. *Ecological Aspects of Development in the Humid Tropics*. National Academy Press, Washington, D.C., 297 pp.

Nichols, G.E., 1915. The vegetation of Connecticut, IV. Plant societies in lowlands. *Bull. Torrey Bot. Club*, 42: 169–217.

Nixon, E.S., 1975. Successional stages in a hardwood bottomland forest near Dallas, Texas. *Southwest. Nat.*, 20: 323–331.

Noble, M.G., 1979. The origin of *Populus deltoides* and *Salix interior* zones on point bars along the Minnesota River. *Am. Midl. Nat.*, 102: 59–67.

Oaks, R.Q., Jr. and Whitehead, D.R., 1979. Geologic setting and origin of the Dismal Swamp, southeastern Virginia and northeastern North Carolina. In: P.W. Kirk, Jr. (Editor), *The Great Dismal Swamp*. Univ. Press of Virginia, Charlottesville, Va., pp. 1–24.

Odum, E.P., 1969. The strategy of ecosystem development. *Science*, 164: 262–270.

Odum, E.P., 1971. *Fundamentals of Ecology*. W.B. Saunders Co., Philadelphia, Penn., 574 pp.

Odum, H.T., 1952. The Carolina bays and a Pleistocene weather map. *Am. J. Sci.*, 25: 263–270.

Odum, H.T., 1972. An energy circuit language for ecological and social systems: its physical basis. In: B.C. Patten (Editor), *Systems Analysis and Simulation in Ecology*. Vol. II, Academic Press, New York, N.Y., pp. 139–211.

Odum, H.T., 1975. Mangrove models. In: S.C. Snedaker and L.A. Burns (Editors), *Evaluation of South Florida Mangrove–Estuarine Systems by Ecosystem Models*. Rept. to U.S. Dept. of the Interior. Contract No. 14–1T10004–425, Washington, D.C., pp. III-1–III-5.

Odum, H.T., 1976. Introductory narrative. In: H.T. Odum and K.C. Ewel (Editors), *Cypress Wetlands for Water Management, Recycling and Conservation*. Center for Wetlands, Univ. of Florida, Gainesville, Florida Rep. PB-273-097, National Technical Information Service, Springfield, Va., pp. 1–26.

Odum, H.T., 1983. *Systems Ecology*. John Wiley and Sons, New York, N.Y., 644 pp.
Odum, H.T. and Johnson, J., 1955. Silver Springs and the balanced aquarium controversy. *Sci. Counselor*, 15: 128–130.
Odum, H.T. and Lugo, A.E., 1970. Metabolism of forest floor microcosms. In: H.T. Odum and R.F. Pigeon (Editors), *A Tropical Rainforest*. TID-24270 (PRNC-138), National Technical Information Service, Springfield, Va., pp. 135–156.
Odum, H.T., Wang, F.C., Alexander, J., Jr. and Gilliland, M., 1983. Energy analysis of environmental values. In: H.T. Odum, M.J. Lavine, F.C. Wang, M.A. Miller, J. Alexander, Jr. and T. Butler (Editors), *A Manual for Using Energy Analysis for Plant Siting*. Report to the Nuclear Regulatory Commission, NUREG/CR-2443, National Technical Information Service, Springfield, Va., 96 pp.
O'Neil, T., 1949. *The Muskrat in the Louisiana Coastal Marshes*. Louisiana Dept. of Wild Life and Fisheries, New Orleans, La., 152 pp.
O'Toole, M.A. and Synnott, D.M., 1971. The bryophyte succession on blanket peat following calcium carbonate, nitrogen, phosphorus and potassium fertilizers. *J. Ecol.*, 59: 121–126.
Paerl, H.W. and Ustach, J.R., 1982. Blue-green algal scums: an explanation for their occurrence during freshwater blooms. *Limnol. Oceanogr.*, 27: 212–217.
Pallis, M., 1916. The structure and history of Plav; the floating fen of the delta of the Danube. *J. Linn. Soc. Lond.*, 43: 233–290.
Parrish, F.K. and Rykiel, E.J., 1979. Okefenokee Swamp origin, review and reconsideration. *J. Elisha Mitchell Sci. Soc.*, 95: 17–31.
Patrick, R., Binetti, V.P. and Halterman, S.G., 1981. Acid lakes from natural and anthropogenic causes. *Science*, 211: 446–448.
Penfound, W.T., 1952. Southern swamps and marshes. *Bot. Rev.*, 18: 413–446.
Penman, H.L., 1970. The water cycle. *Sci. Am.*, 223(3): 99–108.
Phillips, R.C., 1958. Floating communities of algae in a North Carolina pond. *Ecology*, 39: 765–766.
Phillips, T.L., Peppers, R.A., Aucin, M.J. and Laughnan, P.F., 1974. Fossil plants and coal: patterns of change in Pennsylvanian coal swamps of the Illinois basin. *Science*, 184: 1367–1369.
Phipps, R.L., 1979. Simulation of wetlands forest vegetation dynamics. *Ecol. Model.*, 7: 257–288.
Pirkle, W.A., 1972. *Trail Ridge, a relic shoreline of Florida and Georgia*. Ph.D. Diss., Univ. of North Carolina, Chapel Hill, N.C., 90 pp.
Potzger, J.E., 1934. A notable case of bog formation. *Am. Midl. Nat.*, 15: 567–580.
Potzger, J.E., 1956. Pollen profiles as indicators in the history of lake filling and bog formation. *Ecology*, 37: 476–483.
Prakash, A. and Rashid, M.A., 1968. Influence of humic substances on the growth of marine phytoplankton: dinoflagellates. *Limnol. Oceanogr.*, 13: 598–606.
Price, W.A., 1947. Equilibrium of form and forces in tidal basins of coast of Texas and Louisiana. *Bull. Am. Assoc. Pet. Geol.*, 31: 1619–1663.
Prouty, W.F., 1952. Carolina bays and their origin. *Geol. Soc. Am. Bull.*, 63: 167–224.
Rabinowitz, D., 1978a. Early growth of mangrove seedlings in Panama, and an hypothesis concerning the relationship of dispersal and zonation. *J. Biogeogr.*, 5: 113–133.
Rabinowitz, D., 1978b. Dispersal properties of mangrove propagules. *Biotropica*, 10: 47–57.
Randerson, P.T., 1979. A simulation model of salt-marsh development and plant ecology. In: B. Knights and A.J. Phillips (Editors), *Estuarine and Coastal Land Reclamation and Water Storage*. Saxon House Publ., Teakfield, pp. 48–67.
Reader, R.J. and Stewart, J.M., 1972. The relationship between net primary production and accumulation for a peatland in southeast Manitoba. *Ecology*, 53: 1024–1037.
Regan, E.J., Jr., 1977. *The Natural Energy Basis for Soils and Urban Growth in Florida*. M.S. Thesis, Univ. of Florida, Gainesville, Fla., 176 pp.
Rehm, A.E., 1976. The effects of the wood-boring isopod *Sphaeroma terebrans* on the mangrove communities of Florida. *Environ. Conserv.*, 3: 47–57.
Rehm, A.E. and Humm, H.J., 1973. *Sphaeroma terebrans*: a threat to the mangroves of southwestern Florida. *Science*, 182: 173–174.
Reid, G.K., 1952. Some considerations and problems in the ecology of floating islands. *Q. J. Fla. Acad. Sci.*, 15: 63–66.
Rice, E.L. and Pancholy, S.K., 1973. Inhibition of nitrification of climax ecosystems, II. Additional evidence and possible role of tannins. *Am. J. Bot.*, 60: 691–702.
Richards, P.W., 1952. *The Tropical Rain Forest*. Cambridge Univ. Press, New York, N.Y., 450 pp.
Runkle, J.R., 1981. Gap regeneration in some old-growth forests of the eastern United States. *Ecology*, 62: 1041–1051.
Russell, R.J., 1942. Floatant. *Geogr. Rev.*, 32: 74–98.
Sampson, H.C., 1930. Successions in the swamp forest formation in northern Ohio. *Ohio J. Sci.*, 30: 340–357.
Sanger, R. and Gannon, J.E., 1979. Vegetation succession in Smith's Bog Cheboygan County, Michigan. *Mich. Bot.*, 18: 59–69.
Savage, T., 1972. *Florida Mangroves as Shoreline Stabilizers*. Fla. Dept. of National Resources, Marine Research Lab., St. Petersburg, Fla., Prof. Pap. Ser., No. 19, 46 pp.
Say, E.W. and Jansson, O., 1976. *The Huron River and its Watershed*. Huron River Watershed Council, Ann Arbor, Mich. 34 pp.
Schwintzer, C.R., 1978. Nutrients and water levels in a small Michigan bog with high tree mortality. *Am. Midl. Nat.*, 100: 441–451.
Schwintzer, C.R. and Williams, G., 1974. Vegetation changes in a small Michigan bog from 1917 to 1972. *Am. Midl. Nat.*, 92: 447–459.
Scoffin, T.P., 1970. The trapping and binding of subtidal carbonate sediments by marine vegetation in Bimini Lagoon, Bahamas. *J. Sediment. Petrol.*, 40: 249–273.
Seischab, F.K. and Garrisi, P., 1981. The Byron-Bergen swamp, Genesee County, NY, as it relates to the classification of wetlands. *Bull. Ecol. Soc. Am.*, 62: 95 (abstr.).
Shacklette, H.T., 1965. A leafy liverwort hydrosere on Yakobi Island, Alaska. *Ecology*, 46: 377–378.

Shapiro, J., 1966. The relation of humic color to iron in natural waters. *Int. Ver. Theor. Angew. Limnol., Verh.*, 16: 477–484.

Sheail, J. and Wells, T.C.E., 1983. The Fenlands of Huntingdonshire, England: a case study in catastrophic change. In: A.J.P. Gore (Editor), *Mires: Swamp, Bog, Fen and Moor. Regional Studies*. Ecosystems of the World 4B. Elsevier, Amsterdam, pp. 375–395.

Shelford, V.E., 1954. Some lower Mississippi Valley flood plain biotic communities, their age and elevation. *Ecology*, 35: 126–142.

Shugart, H.H., Jr. and West, D.C., 1980. Forest succession models. *Bioscience*, 30: 308–313.

Shull, C.A., 1944. Observations of general vegetational changes on a river island in the Mississippi River. *Am. Midl. Nat.*, 32: 771–776.

Simberloff, D., Brown, B.J. and Lowrie, S., 1978. Isopod and insect root borers may benefit Florida mangroves. *Science*, 201: 630–632.

Sioli, H., 1968. Principal biotopes of primary production in the waters of Amazonia. In: R. Misra and B. Gopal (Editors), *Proc. Symp. Recent Advances in Tropical Ecology*. Part II, Int. Soc. for Tropical Ecology, Varanasi, pp. 591–600.

Smith, H.M., 1972. The palustral origin of mammals. *Biologist*, 54: 49–52.

Snedaker, S.C., 1974. Mangroves, isopods, and the ecosystem. *Science*, 183: 1036–1037.

Snedaker, S. and Lugo, A., 1973. *The Role of Mangrove Ecosystems in the Maintenance of Environmental Quality and a High Productivity of Desirable Fisheries*. U.S. Bureau of Sports Fisheries and Wildlife, Final Rept., Contract No. 14–16–008–606, Atlanta, Ga., 404 pp.

Spangler, D.P., 1978. Hydrogeologic aspects of cypress domes in north central Florida — a summary. In: H.T. Odum and K.C. Ewel (Editors), *Cypress Wetlands for Water Management, Recycling and Conservation*. 4th Ann. Rept., Center for Wetlands, Univ. of Florida, Gainesville, Fla. Report PB-282-159, National Technical Information Service, Springfield, Va., pp. 89–91.

Staheli, A.C., Ogren, D.E. and Wharton, C.H., 1974. Age of swamps in the Alcovy River drainage basin. *Southeast. Geol.*, 16: 103–106.

Staub, J.R. and Cohen, A.D., 1978. Kaolinite-enrichment beneath coals; a modern analog, Snuggedy Swamp, South Carolina. *J. Sediment. Petrol.*, 48: 203–210.

Staub, J.R. and Cohen, A.D., 1979. The Snuggedy Swamp of South Carolina: a back-barrier coal-forming environment. *J. Sediment. Petrol.*, 49: 133–144.

Stoddart, D.R., 1980. Mangroves as successional stages, inner reefs of the northern Great Barrier Reef. *J. Biogeogr.*, 7: 269–284.

Stross, R.G. and Hasler, A.D., 1960. Some lime-induced changes in lake metabolism. *Limnol. Oceanogr.*, 5: 265–272.

Stross, R.G., Neess, J.C. and Hasler, A.D., 1961. Turnover time and production of planktonic cructacea in limed and reference portions of a bog lake. *Ecology*, 42: 237–245.

Swan, J.M.A. and Gill, A.M., 1970. The origins, spread and consolidation of a floating bog in Harvard Pond, Petersham, Massachusetts. *Ecology*, 51: 829–840.

Swaney, D.P., 1978. *Energy Analysis of Climatic Inputs to Agriculture*. M.S. Thesis, Univ. of Florida, Gainesville, Fla., 198 pp.

Swanson, F.J., 1979. Geomorphology and ecosystems. In: R.H. Waring (Editor), *Forests: Fresh Perspectives from Ecosystem Analysis*. Oregon State Univ. Press, Corvallis, Ore., pp. 159–170.

Tallis, J.H., 1965. Studies on southern Pennine peats, IV. Evidence of recent erosion. *J. Ecol.*, 53: 509–520.

Tallis, J.H., 1973. The terrestrialization of lake basins in north Cheshire, with special reference to the development of a "schwingmoor" strucutre. *J. Ecol.*, 61: 537–567.

Tallis, J.H., 1983. Changes in wetland communities. In: A.J.P. Gore (Editor), *Mires: Swamp, Bog, Fen and Moor*. Ecosystems of the World 4A, Elsevier, Amsterdam, pp. 311–347.

Tansley, A.G., 1939. *The British Islands and Their Vegetation*. Cambridge Univ. Press, Cambridge, 930 pp.

Teas, H.J., 1977. Ecology and restoration of mangrove shorelines in Florida. *Environ. Conserv.*, 4: 51–58.

Thom, B.G., 1967. Mangrove ecology and deltaic geomorphology, Tabasco, Mexico. *J. Ecol.*, 55: 301–343.

Thom, B.G., 1970. Carolina bays in Horry and Marion Counties, South Carolina. *Geol. Soc. Am. Bull.*, 81: 783–814.

Thom, B.G., Wright, L.D. and Coleman, J.M., 1975. Mangrove ecology and deltaic-estuarine geomorphology Cambridge Gulf–Ord River, western Australia. *J. Ecol.*, 63: 203–232.

Thomas, B.A., 1978. Carboniferous Lepidodendraceae and Lepidocappaceae. *Bot. Rev.*, 44: 321–364.

Thompson, K., 1976. Swamp development in the head waters of the White Nile. In: J. Rzoska (Editor), *The Nile, Biology of an Ancient River*. W. Junk Publ., The Hague, pp. 177–196.

Thompson, K. and Hamilton, A.C., 1983. Peatlands and swamps of the African continent. In: A.J.P. Gore (Editor), *Mires: Swamp, Bog, Fen and Moor*. Ecosystems of the World 4B, Elsevier, Amsterdam, pp. 331–373.

Trivedy, R.K., Sharma, K.P., Goel, P.K. and Gopal, B., 1978. Some ecological observations on floating islands. *Hydrobiologia*, 60: 187–190.

Tutin, T.G., 1941. The hydrosere and current concepts of the climax. *J. Ecol.*, 29: 268–279.

Ulrich, K., 1980. Net primary productivity of a mature southern Michigan bog. *Mich. Acad.*, 12: 289–295.

Ultsch, G.R., 1973. The influence of water hyacinths (*Eichhornia crassipes*) on the microenvironment of aquatic communities. *Arch. Hydrobiol.*, 72: 143–160.

United States Geological Survey, 1979. *Flood Prone Map — Huron River, Southeastern Michigan*. Geol. Surv., Lansing, Mich.

Valiela, I. and Teal, J.M., 1980. *Secondary Plant Substances in Detritus and Their Effects on Detritus Feeders and Decay in Salt Marsh Ecosystems*. Am. Soc. of Limnology and Oceanography, Second Winter Meeting, Grafton, Wisconsin, p. 49 (Abstr.).

Van der Valk, A.G., 1981. Succession in wetlands: a Gleasonian approach. *Ecology*, 62: 688–696.

Van der Valk, A.G. and Davis, C.B., 1978. The role of seed banks in the vegetation dynamics of prairie glacial marshes. *Ecology*, 59: 322–335.

Van der Valk, A.G. and Davis, C.B., 1979. A reconsideration of the recent vegetational history of a prairie marsh, Eagle Lake, Iowa, from its seed bank. *Aquatic Bot.*, 6: 29–51.

Van Straaten, L.M.J.U. and Kuenen, Ph.H., 1958. Tidal action as a cause of clay accumulation. *J. Sediment. Petrol.*, 28: 405–409.

Vann, J., 1959. Landform-vegetation relationships in the Atrato Delta. *Ann. Assoc. Am. Geogr.*, 49: 345–360.

Verry, E.S., 1975. Streamflow chemistry and nutrient yields from upland-peatland watersheds in Minnesota. *Ecology*, 56: 1149–1157.

Viereck, L.A., 1970. Forest succession and soil development adjacent to the Chena River in interior Alaska. *Arctic Alpine Res.*, 2: 1–26.

Vogl, R.J., 1964. The effects of fire on a muskeg in northern Wisconsin. *J. Wildl. Manage.*, 28: 317–329.

Vogl, R.J., 1969. One hundred and thirty years of plant succession in a southeastern Wisconsin lowland. *Ecology*, 50: 248–255.

Vogl, R.J., 1980. The ecological factors that produce perturbation dependent ecosystems. In: J. Cairns, Jr. (Editor), *The Recovery Process in Damaged Ecosystems*. Ann Arbor Sci. Publ., Ann Arbor, Mich., pp. 63–94.

Wace, N.M., 1961. The vegetation of Gough Island. *Ecol. Monogr.*, 31: 337–367.

Wagner, C.A. and Taylor, T.N., 1981. Evidence for endomycorrhizae in Pennsylvanian age plant fossils. *Science*, 212: 562–563.

Walker, D., 1970. Direction and rate in some British postglacial hydroseres. In: D. Walker and R. West (Editors), *The Vegetational History of the British Isles*. Cambridge Univ. Press, Cambridge, pp. 117–139.

Walker, D., 1972. Vegetation of the Lake Ipea region, New Guinea highlands, II. Kayamanda Swamp. *J. Ecol.*, 60: 479–504.

Walker, R.E., 1940. Biotic succession in a coastal salt marsh. *Proc. Okla. Acad. Sci.*, 20: 95–97.

Walmsley, M.E., 1977. Physical and chemical properties of peat. In: N.W. Radforth and C.O. Brawner (Editors), *Muskeg and the Northern Environment in Canada*. Univ. of Toronto Press, Toronto, Ont., pp. 82–129.

Wang, F.C., Odum, H.T. and Kangas, P.C., 1980. Energy analysis for environmental impact assessment. *J. Water Resour. Planning Manage. Div., Am. Soc. Civ. Eng.*, 106: 451–466.

Washburn, A.L., 1973. *Periglacial Processes and Environments*. St. Martin's Press, New York, N.Y., 320 pp.

Waters, T.F., 1957. The effects of lime application to acid bog lakes in northern Michigan. *Trans. Am. Fish. Soc.*, 86: 329–344.

Waters, T.F. and Ball, R.C., 1957. Lime application to a softwater, unproductive lake in Northern Michigan. *J. Wildl. Manage.*, 21: 385–392.

Watt, A.S., 1947. Pattern and process in the plant community. *J. Ecol.*, 35: 1–22.

Watt, R.F. and Heinselman, M.L., 1965. Foliar nitrogen and phosphorus level related to site quality in a northern Minnesota spruce bog. *Ecology*, 46: 357–360.

Weaver, J.E. and Clements, F.E., 1938. *Plant Ecology*. McGraw-Hill Book Co., New York, N.Y., 601 pp.

Weber, C.A., 1908. Aufbau und Vegetation der Moore Norddeutschlands. *Englers Bot. Jahrb.*, 40 (Suppl): 19–34.

Welch, B.L., 1963. From coral reef to tropical island via *Thalassia* and mangrove. *Va. J. Sci.*, 14: 213–214.

Welch, P.S., 1945. Some limnological features of water impounded in a northern bog-lake mat. *Trans. Am. Microsc. Soc.*, 65: 183–195.

Wells, B.W. and Boyce, S.G., 1953. Carolina bays: additional data on their origin, age and history. *J. Elisha Mitchell Sci. Soc.*, 69: 119–141.

Wells, B.W. and Boyce, S.G., 1954. Carolina bay lakes: the bog margin problem. *Ecology*, 35: 584.

Wells, B.W. and Whitford, L.A., 1976. History of stream-head swamp forests, pocosins, and savannahs in the Southeast. *J. Elisha Mitchell Sci. Soc.*, 92: 148–150.

West, D., Shugart, H.H., Jr. and Botkin, D. (Editors), 1981. *Forest Succession*. Springer-Verlag, New York, N.Y., 517 pp.

Wetzel, R.G., 1972. The role of carbon in hard-water marl lakes. In: G.E. Likens (Editor), *Nutrients and Eutrophication*. Am. Soc. Limnol. Oceanogr. Spec. Symp., 1, pp. 84–97.

Wetzel, R.G., 1975. *Limnology*. W.B. Saunders Co., Philadelphia, Penn., 243 pp.

White, P.S., 1979. Pattern, process and natural disturbance in vegetation. *Bot. Rev.*, 45: 229–300.

Whitehead, D.R., 1972. Developmental and environmental history of the Dismal Swamp. *Ecol. Monogr.*, 42: 301–315.

Whitmore, T.C., 1975. *Tropical Rain Forests of the Far East*. Clarendon Press, Oxford, 282 pp.

Whitmore, T.C., 1978. Gaps in the forest canopy. In: P.B. Tomlinson and M.H. Zimmerman (Editors), *Tropical Trees as Living Systems*. Cambridge Univ. Press, Cambridge, pp. 639–655.

Whittaker, R.H., 1953. A consideration of climax theory: the climax as a population and pattern. *Ecol. Monogr.*, 23: 41–78.

Wildi, D., 1978. Simulating the development of peat bogs. *Vegetatio*, 37: 1–18.

Wilford, G.E., 1960. Radiocarbon age determinations of Quaternary sediments in Brunei and north east Sarawak. *Br. Borneo Geol. Surv. Annu. Rep.*, 1959.

Williams, R.B., 1971. Computer simulation of energy flow in Cedar Bog Lake, Minnesota based on the classical studies of Lindeman. In: B.C. Patten (Editor), *Systems Analysis and Simulation in Ecology*, Vol. I, Academic Press, New York, N.Y., pp. 544–582.

Wilson, R.E., 1970. Succession in stands of *Populus deltoides* along the Missouri River in southeastern South Dakota. *Am. Midl. Nat.*, 83: 330–342.

Wolman, M.G. and Leopold, L.B., 1957. River flood plains: some observations on their formation. *U.S. Geol. Surv. Prof. Pap.*, 282-C, pp. 87–107.

Zach, L.W., 1950. A northern climax, forest or muskeg? *Ecology*, 31: 304–306.

Zieman, J.C., Jr., 1972. Origin of circular beds of *Thalassia* in South Biscayne Bay, Florida and their relationship to mangrove hammocks. *Bull. Mar. Sci.*, 22: 559–575.

Zieman, J.C., Jr. and Odum, W.E., 1977. Modeling of ecological succession and production in estuarine marshes. *Dredged Material Research Program Tech. Rept. D-77–35*, U.S. Army Engineer Waterways Experiment Station, Vicksburg, Miss., 278 pp.

Chapter 4

CONCEPTS IN WETLAND ECOLOGY

ARIEL E. LUGO, SANDRA BROWN and MARK M. BRINSON

INTRODUCTION

While reviewing the literature for this book, we were faced with the complex job of integrating a great diversity of topics. It soon became obvious that some of the topics were relevant to most types of wetlands. For this reason we decided to treat these topics individually rather than repetitively throughout the book, which led to the 21 topics in wetland ecology in this chapter. This approach also gave us the opportunity to treat each topic in more detail than would have been possible if they were presented as parts of one or several chapters.

Each of the sections that follow are self-contained literature reviews of the subject. We made an effort to assemble data and figures to illustrate the concepts. Further research is needed to test some of the more speculative statements. Book chapters dealing with any of these topics are cross-referenced to the relevant section in this chapter. Such cross-referencing integrates the volume, provides background, and reduces redundancy. This chapter in turn serves as a short introduction to important concepts of wetland ecology.

1. ESTIMATE OF WETLAND AREAS

The area of all wetlands (non-forested and forested) is estimated to be between 200 and 530×10^6 ha (Table 4.1), or about 3% of the land surface. It is not known how much of this area is forested; Matthews and Fung (1987) estimate 60% of the area is forested. Another estimate of wetland areas is based on the assumption that histosols and histic and humic gleysols are mostly capable of supporting wetland vegetation. The global area of these soils was estimated at 537 to 759×10^6 ha, of which 41 to 58% are histosols (Armentano, 1979). Even though there were large errors associated with the soil approach and not all histosols and gleysols are wetlands, the conservative estimate (537×10^6 ha) is almost twice most of the estimates of area of wetland ecosystems in Table 4.1 and similar to the estimate by Matthews and Fung (1987; based on a variety of sources including vegetation, soil, and fractional inundation maps). This suggests that large-scale conversions of wetland areas to other land uses have occurred in the past.

It appears that the tropical region contains considerably more wetland area than the temperate and boreal region combined (Table 4.1). In addition, most of the tropical wetlands appear to be forested. Using the mid-point of the range of values given in Table 4.1, forested wetlands account for approximately 60% of the total tropical wetland area.

Estimates of salt water forested wetlands (mangroves) vary two-fold (Table 4.1). However, the estimate by Saenger et al. (1983) of approximately 17×10^6 ha is probably the best one to date.

2. GLOBAL ROLE OF WETLANDS

Wetland areas are important in many global processes. Here we discuss their role in the global carbon cycle, fisheries production, and biogeochemical cycles. Although this discussion relates to all wetlands, forested ones account for a large proportion of the total area, particularly in the tropics (see Sect. 1, this chapter) and thus contribute significantly to these processes.

TABLE 4.1

Estimates of wetland areas of the world (10^6 ha)

	Forested wetlands			All wetlands	Source
	fresh-water	salt-water	total		
Global:				200	Whittaker and Likens, 1973
				230	Ajtay et al., 1979
				250[1]	Olson et al., 1983
			316	530	Matthews and Fung, 1987
Temperate and boreal:				50	Ajtay et al., 1979
				90	Olson et al., 1983
Tropical:		30		180	Ajtay et al., 1979
		20–30		160	Olson et al., 1983
		15.47			Lanly, 1982
		16.9			Saenger et al., 1983
Tropical Africa:			24.5–49.2[2]		Armentano and Lawler, 1983
	47.4[2,3]	10.2[2]	57.6[4]		Lanly, 1982
Tropical Asia:			15.6–30.9[2]		Armentano and Lawler, 1983
Tropical America:	29.7[5]	4.6[5]	34.3[5]	73.9	Lanly, 1982
	42.3	3.7	46.0		Hueck, 1978 in FAO, 1981
Caribbean Basin:			5.1–10.3[2]		Armentano and Lawler, 1983

[1] Conservative estimate because additional wetlands are included in other land use categories used by authors.
[2] Area estimates were obtained from the global vegetation map of Olson et al. (1983) by counting the number of map cells (~ 3080 km^2) denoted as mangroves or tropical swamp woods. The area was estimated as a range between one-half and the total area of cells to account for the possibility that wetlands occupied from 50–100% of a given cell (Armentano and Lawler, 1983).
[3] May be an overestimate because 40% of the area was classified as a mosaic of swamp forest and wet Guineo-Congolian lowland rain forest.
[4] Areas obtained from a summary table, given in Lanly (1982), of the UNESCO/AETFAT vegetation map of Africa.
[5] Areas obtained from a summary table, given in Lanly (1982), of the UNESCO/CITV vegetation map at South America.

Wetlands occupy only a small fraction of the world's land area ($\sim 3\%$, see Sect. 1, this chapter) but because their soil organic carbon content is generally high, they are relatively more important to the global carbon cycle than their area suggests. The average organic carbon content of the soil profile (1 m deep) of wetland areas is 72 kg m^{-2} (Schlesinger, 1984), the highest of all ecosystem types. Using this value for the carbon content of wetland soils and a wetland area of 250×10^6 ha (Olson et al., 1983), the carbon pool of wetland soils is about 180×10^9 t, or 12% of the world's total soil organic carbon. In contrast, the carbon content of wetland vegetation is less than 10×10^9 t (Olson et al., 1983). It is the large soil organic carbon pool that makes wetlands potentially important in the carbon cycle.

Undisturbed wetland areas may accumulate organic carbon in their soils and vegetation, thus acting as sinks of atmospheric carbon through fixation by plants via photosynthesis. It has been estimated that undrained histosols (associated with wetlands areas) sequester about 0.14×10^9 t C yr^{-1} and gleysols (also associated with wetlands) account for an additional, but unknown, amount (Armentano, 1980).

Wetland soils change from a carbon sink to a carbon source, however, when wetlands are drained for agriculture. Drainage causes the highly organic soils to subside, with biochemical oxidation being an important component of this process (Armentano, 1980). Draining the world's histosols and gleysols in recent years has produced an estimated carbon output to the atmosphere of between 0.03×10^9 and 0.37×10^9 t C yr^{-1} (Armentano, 1980).

During the past decade or so, reclamation of tropical histosols (and gleysols to a minor extent) for agricultural uses has created a source of atmospheric carbon estimated to be between

0.07×10^9 and 0.18×10^9 t C yr^{-1} (Armentano et al., 1983). In many parts of the tropics, mangrove wetlands are being converted to fishponds. Preliminary estimates suggest that 0.23×10^6 ha yr^{-1} are being converted giving an annual release of carbon to the atmosphere of about 0.007×10^9 t yr^{-1} (Armentano et al., 1983). This is an insignificant amount of carbon, but the effect of this conversion on fisheries resources could be far-reaching (see below). Widespread drainage has also significantly altered the carbon balance of wetlands in the temperate zone (Armentano and Menges, 1986).

It is possible that, at present, the carbon sources due to draining wetland areas are balanced by the carbon sinks of undisturbed wetland areas. Reliable estimates of wetland areas are not available and all information indicates that the best estimates are probably underestimates. These "missing" wetlands are probably in inaccessible areas and undisturbed, and are more likely to be accumulating carbon in soil and wood. In addition, as demonstrated elsewhere (Sect. 3, this chapter), waters draining wetland areas export significant amounts of organic carbon (0.08×10^9 t C yr^{-1}), thus acting as an additional sink of atmospheric carbon from the terrestrial biota. The future role of the world's wetland areas in the global carbon cycle will depend to a large degree on the success of drainage programs that increase their utilization for agriculture and timber management.

Floodplain wetlands are also areas where organic carbon is sequestered. Likens et al. (1981) have estimated that as much as 0.19×10^9 t C yr^{-1} are deposited in the world's floodplains, though more than half of this may be oxidized. Altogether, wetland areas could be a sink for as much as 0.4×10^9 t ha^{-1} of atmospheric carbon, or about 9 to 22% of the range of carbon fluxes between the terrestrial biota and atmosphere estimated for 1980 (Houghton et al., 1983).

Natural wetlands also contribute to the global carbon cycle through their emission of methane, another "greenhouse" gas. Methane is a result of anaerobic decomposition that often occurs in wetlands. A recent estimate of the magnitude of this input to the atmosphere is about 110×10^{12} g yr^{-1} (Matthews and Fung, 1987).

Wetlands, particularly coastal ones, provide important nursery areas and food sources for commercial fisheries. For example, Turner (1977) found that the annual catch of inshore shrimp in the Louisiana area (southern U.S.A.) was significantly correlated with the area of wetland vegetation in the estuary. On a global scale, he also found that the average shrimp yield of intertidal wetlands varied between 10 and 200 kg ha^{-1} yr^{-1}, and that this yield per unit area was significantly related to latitude. High yields per unit area were found for the tropical latitudes, with yield per unit area declining exponentially at higher latitudes to about 35°. The generally high productivity of tropical mangrove wetlands and the habitat that they provide could account for the high yield of shrimp per unit area at lower latitudes.

In light of the above discussion, losses of wetlands are expected to result in losses of fisheries. Projects that are designed to convert wetland area to other land uses should take heed of these findings. Turner (1977) used the above results to determine the impact of the common practice in many tropical countries of converting mangrove areas to fishponds. His analysis suggested that converting a hectare of mangrove forest to a fishpond would result in a net *loss* of about 480 kg of fish, not a very favorable trade-off. His calculation did not consider the other values that mangrove forests provide (see Sect. 14, this chapter) and thus underestimated the total impact of these conversion projects.

Wetlands are important components in the biogeochemical cycles of nitrogen, sulfur, and oxygen. Deevey (1970) suggested that the most important organisms of wetlands are the reducing bacteria. These organisms close the biogeochemical cycles of nitrogen, sulfur, oxygen and carbon.

3. ORGANIC-CARBON EXPORT FROM WETLANDS

Waters draining wetlands have higher rates of organic carbon export per unit area than any other major biogeographic region of the world (Table 4.2). Saltwater forested wetlands (mangroves) have the highest rates of export. The mean export rate of total organic carbon (TOC) for temperate freshwater wetlands is about three times higher than its upland counterpart and similar to the export rate from tropical forests (Table 4.2).

TABLE 4.2

Export of total organic carbon (TOC) from catchments in the major biogeographic regions of the world[1]

Region	TOC			Area[2] (10^{12} m^2)	% of total area	TOC export (10^{14} g C yr^{-1})	% of total export
	No. of sites	mean (g C m^{-2} yr^{-1})	(S.E.)				
Tundra and taiga	25	2.3	(0.4)	29.5	33	0.68	16
Temperate[3]	69	3.2	(0.2)	39.1	44	1.25	29
Tropical forests[4]	22	8.7	(2.3)	18.7	21	1.63	37
Wetlands:							
marshes (salt)	7	95	(32)	0.3[5]	0.3	0.29	7
marshes and forests (fresh)	17	11.6	(2.3)	1.6[6]	1.8	0.18	4
mangroves	5	171	(86)	0.2	0.2	0.34	8
Total				89.4[7]		4.37	

[1]The data base for this table can be obtained from Ariel Lugo at the Institute of Tropical Forestry.
[2]From Olson et al. (1983).
[3]Includes all forests, scrub and woodlands, second growth woods and field complexes, and grasslands.
[4]Includes all humid and dry forests and woodlands, montane complexes, and secondary growth woods, and field complexes.
[5]Assumed that the coastal complex is composed mainly of marshes from which TOC data are derived.
[6]Does not include bogs or mires of cool or cold climates; these ecosystem types are included with tundra and taiga.
[7]Represents 60% of the land area; missing are tropical savannas, deserts, ice covered areas, and residential, commercial and agricultural land.

Most of the carbon exported from wetlands is dissolved organic carbon (DOC). The ratio of DOC to particulate organic carbon (POC) varies between 3 and 6 for freshwater and saltwater wetlands. The exception is for fringing mangroves, which have a DOC/POC ratio of about 0.2 due to the unusually high rate of POC export.

There appears to be no clear relationship between runoff and TOC export for temperate forested wetlands (Fig. 4.1a), although the higher runoff values (>90 cm yr^{-1}) clearly correspond with the highest rates of TOC export. This relationship may be significant for upland tropical forests (Fig. 4.1b). In contrast, upland temperate forests appear to have an upper limit of TOC export, independent of runoff, at about 6 g m^{-2} yr^{-1}.

The export of TOC from all wetlands accounts for about 20% of the world's total continental export (Table 4.2) and coastal wetlands account for most of the wetland percentage. This high proportion of the world's carbon export from an extremely small percentage of the land area (2.3%, Table 4.2) suggests that wetlands are far more important in this global process than their areal extent would indicate.

4. FORESTED WETLANDS AND ANIMAL LIFE

Animals constitute a small but important fraction of the total biomass in forested wetlands. For example, in a Malaysian forested wetland, animal biomass was 30 g m^{-2}, or less than 0.1% of the total organic matter in the stand. In spite of their small biomass, animal communities recycled a significant amount of calcium in the system (Furtado and Mori, 1982). The role of insect and vertebrate consumers in wetlands and the effect of hydroperiod on consumption by animals was reviewed by Brinson et al. (1981a) while Beadle (1974) reviewed animal adaptations to scarcity of oxygen in tropical wetland environments. The diversity of animal life in mangroves was reviewed by MacNae (1968). Other reviews emphasize the importance of wetlands for avian wildlife, particularly waterfowl (e.g. Viosca, 1928; Hopkins, 1947; Isakov, 1968; Firouz, 1974; Anderson et al., 1977; Johnson et al., 1977; Brinson et al., 1981b). Riverine forests in both humid and arid environments support more avian species than do adjacent uplands of the same geographic region (Fig. 4.2).

Recent work has emphasized the role of forested wetlands in supporting animal life in other ecosys-

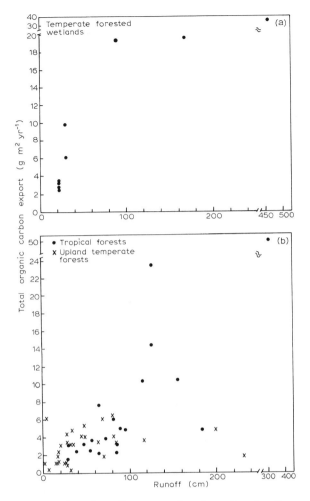

Fig. 4.1. Effect of runoff on export of total organic carbon from (a) temperate forested wetlands and (b) upland tropical and temperate forests. (Data are from many sources; a copy of the data can be obtained from Ariel E. Lugo.)

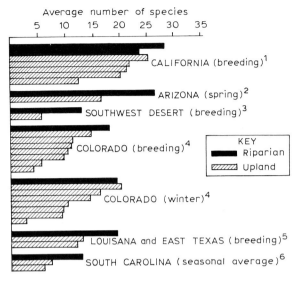

Fig. 4.2. Number of bird species in riparian and upland vegetation types. After Brinson et al. (1981b). Sources: *1* — Gaines, 1974; *2* — Stevens et al., 1977; *3* — Austin, 1970; *4* — Bottorff, 1974; *5* — Dickson, 1978; *6* — Hair et al., 1979.

tems coupled hydrologically to the wetland. The thorough and fascinating account by Goulding (1980) of the dependence of riverine fish on the forested wetlands of the Amazon and for other rivers by Welcomme (1979) are examples. Forest fruits, seeds and leaves, as well as insects and other foods are the base of the food webs that support 75% of the riverine fishery of the Amazon. The food webs that depend on freshwater forested wetlands have been described in detail for the Tasek Bera forest in Malaysia where most of the food sources were shown to originate from the forest (Furtado and Mori, 1982). In temperate systems, Meehan et al. (1977) documented the dependence of salmonoid fisheries on the riparian forested wetlands. Turner (1977) and Turner and May (1977) used statistical correlations to show that global shrimp production is dependent on the area of wetlands draining into estuarine and oceanic waters. Stomach analyses alone (Odum, 1970) or coupled with electron microscopy (Yee Mun, 1984) have been used to show dependence on mangroves of commercial marine fisheries of the Gulf coast of the United States, and of sesarmid crabs in Malaysia. Over two-thirds of all commercial fish and shellfish species in the Gulf, Atlantic, and Pacific fisheries in U.S.A. waters depend on wetlands for food or shelter (Horwitz, 1978).

The studies quoted above underscore the open nature of the biogeochemical cycles of forested wetlands and their importance to consumer metabolism in coupled ecosystems. Carter et al. (1973) and Twilley (1982) estimated that over 50% of the carbon consumed in estuarine waters originated in forested wetlands. They also demonstrated responses in heterotrophic metabolism of receiving ecosystems. Consumers in aquatic ecosystems not only utilize the amount and quality of organic foods exported from forested wetlands, but also are adapted to the timing or periodicity of such inputs (Twilley, 1982). For these reasons, the drainage or channelization of wetlands is usually associated with dramatic reductions in the abundance of animal populations. These reductions have been documented in the United States in terms of fish (Viosca, 1928; Wharton, 1971; Starrett, 1972; Lugo, 1973; Horwitz, 1978; Street and McCless, 1981) and many other groups of organisms (Horwitz, 1978; Monschein, 1981). However, some wildlife species increase in abundance, at least temporarily, when a forested wetland is stressed. Examples are woodpeckers, which take advantage of the large number of dead trees produced in thermally stressed wetlands (Straney et al., 1974).

Because of the periodicity in the hydrology of forested wetland, many studies have shown the importance of hydroperiod to animal population behavior (e.g. Wasawo and Visser, 1959; Banage, 1966; Heatwole and Sexton, 1966; Gentry et al., 1971; Fittkau et al., 1975; Scheller, 1979; Swift et al., 1984). Animals have been assigned important roles in the ecology of forested wetlands. Some of these involve enrichment of soil fertility (Wasawo and Visser, 1959; Walker, 1972), acceleration of nutrient cycles of the wetland (Young, 1967; Fittkau et al., 1975), changing the saturated hydraulic conductivity of the soil (Denning et al., 1978), altering vegetation zonation (Walker, 1972), or making seedling establishment possible through mound-building (McCarthy, 1962).

5. LARGE-SCALE REDUCTION OF FORESTED WETLAND AREAS

Throughout the globe, vast areas of forested wetlands are being converted to other uses (Sect. 2, this chapter). In the contiguous United States for example, about 2.4×10^6 ha of forested wetlands were converted to other land uses, mainly agriculture, during the period from 1950 to 1970 (Frayer et al., 1983). However, reliable statistics on these conversions at a global scale are not available. For illustrative purposes only, we present examples of these large-scale conversions of forested wetland to draw attention to those human activities that in some cases threaten the integrity of these forests (see Table 4.3).

Apparently, those forested wetlands in developing regions that survive the conversion are usually those subjected to the longest hydroperiods or growing in close proximity to channels. These are often the most stressful environments where structural development may be low.

6. HYDROPERIOD

The single most important regulator in wetlands is usually hydroperiod — the duration, frequency, depth, and season of flooding. Hydroperiod is especially critical in determining species composition in the plant community. For example, the early life history stages of trees are particularly sensitive to hydroperiod. Few seedlings can tolerate flooding under nondormant conditions and few seeds are capable of germinating while submerged. In later stages of forest development, trees are less sensitive to seasonal and annual variations in hydroperiod (Harms et al., 1980). However, permanent changes, such as those induced by human activities, are likely to cause drastic changes in

TABLE 4.3

Areas of rapid conversion of forested wetland

Type of wetland	Where it occurs	Human activity
Cativo swamps (*Prioria copaifera*)	Central America	Selectively cut for timber
Mangroves	Globally	Various: aquaculture, tin mining, coastal development
Pterocarpus swamps	Islands in the Caribbean	Agriculture
Floodplain forests	United States Argentina	Agriculture Hydroelectric development
Várzea forests	Brazilian Amazon	Rice production
Bog forests	Northern latitudes	Peat and wood harvests

species composition. Because trees have a longer life history than herbs, changes in a community brought about by species replacement occur more slowly in forested wetlands than in marshes. However, extensive tree mortality caused by subtle changes in hydroperiod may occur rapidly (within weeks to months) and are common (Jiménez et al., 1985). Frequently a longer hydroperiod will result in tree mortality and replacement of forest by marsh. The zonation of communities in geomorphologically stable wetlands is a function of hydroperiod rather than succession (see Ch. 9, and Sect. 10 of this chapter; also Lugo, 1980, for mangroves).

Fig. 4.3 illustrates idealized hydrographs for the three wetland categories used in this book — riverine (stream-swamp and alluvial swamp), basin (pocosin or Carolina bay) and fringe (irregularly flooded tidal swamp). Riverine wetlands usually have relatively short periods of flooding associated with the peak hydrograph of a stream after precipitation events or during spring snow-melt. Basin wetlands have longer hydroperiods, and water level fluctuations do not occur as rapidly as for riverine wetlands. Fringe wetlands may flood regularly in coastal areas with astronomic tides or irregularly where water levels are controlled by wind. Each of the variables that determine hydroperiod differs in these examples.

A wetland's hydroperiod is dependent on climate, topography, channel slope (for riverine situations), catchment area, soils, and geology of the region. For riverine wetlands, the duration and depth components are largely a function of catchment size. Floodplains of large drainage basins have deeper and longer flood periods than floodplains of lower order streams (Jackson, 1976; Richards, 1982). Precipitation and evapotranspiration control hydroperiod in basin wetlands because lateral inflows and outflows are usually small relative to atmospheric exchanges. Fringe wetlands normally have frequent flooding of short duration. For example, coastal mangrove fringes may have semidiurnal or diurnal flooding due to tides, while flooding in lakeside forests may depend upon wind patterns that drive seiches.

Flood stages may be classified according to recurrence intervals or mean return periods. Thus the terms "10-yr flood", "50-yr flood", etc., refer to the average interval of time before a flood of that magnitude will be equalled or exceeded. Because of the relatively flat topography of many wetlands, small floods that cover much of the system will occupy nearly as much surface area as much larger floods. However, the amount of water stored in the wetland will be nearly proportional to stage height.

Hydroperiod can function as both a subsidy and a stress for wetlands. In riverine wetlands, flooding generally results in an influx and deposition of sediments and associated nutrients. Also, floods create currents that aid in mobilization and export of toxic compounds that may otherwise accumulate in sediments. On the other hand, extended flooding during the growing season may cause death in some species adapted to flooding only during the dormant season (Lugo and Brown, 1984). This is seldom observed under natural conditions because selective mortality generally occurs in the younger, more sensitive life-history stages of trees.

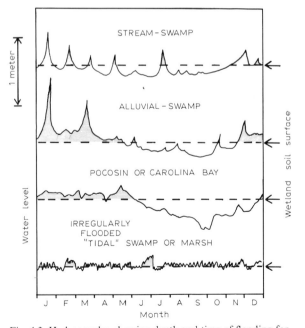

Fig. 4.3. Hydrographs, showing depth and time of flooding for four wetlands in the coastal plain of North Carolina. From Brinson (1985). Reprinted with permission of Van Nostrand Reinhold Company.

7. THE INFLUENCE OF WATER ON FORESTED WETLANDS

The flow of water through an ecosystem has many effects on its plants, animals, and soil. Water is required for transpiration, cell turgidity and metabolism, nutrient mobility, and animal life. In wetlands, water also floods and saturates soil, a condition that has many consequences to the function of the ecosystem. This topic outlines these and other effects that water has on forested wetland ecosystems. Sect. 21 of this chapter summarizes the effects of wetland ecosystems on water.

Excessive amounts of water cause damage to ecological systems by killing or damaging trees and other plants; causing soil erosion, landslides, excessive sedimentation, and other geomorphological changes (Brink, 1954); and causing asphyxiation of organisms if the flooding is prolonged. Plants respond to flooding by closing stomata; by reducing photosynthesis, translocation of carbohydrates, and absorption of minerals; by altering their hormonal balance; and by reducing growth (Kozlowski, 1984). In humid tropical environments high rainfall leaches nutrients from plant and soil surfaces, and is responsible for pulsed and high rates of export of materials from the ecosystem (Shure and Gottschalk, 1978; Frangi and Lugo, 1985). In some instances, inputs of materials in hydrologic fluxes may compensate for losses that occur by runoff (Frangi and Lugo, 1985).

Water movement is very important for a forested wetland because it ventilates and renews the immediate environment of surfaces that come in contact with the moving water. Water movement creates a favorable exchange gradient through which diffusion of gases or removal of toxic materials may be facilitated. Aeration always improves with water movement (Yurkevich et al., 1966). The direction of flow is a critical factor in the organization of plant and animal communities. Lateral flow influences the dispersal of plant propagules from one sector of the forest to another. Vertical movement transports ions and gases among soil horizons and is especially important in controlling the distribution of roots.

The quality of water in wetlands depends upon its source, for example, rain, snow-melt, groundwater, or surface runoff (Bay, 1967; Moore and Bellamy, 1974; Haapala et al., 1975) and its rate of movement through soil horizons. The movement of water through the soil depends on its conductivity. For example, peats have filtration coefficients of only $0.0014–0.00083$ cm day^{-1} while sands may be as high as $251–259$ cm day^{-1} (Skoropanov, 1968). These different rates of filtration result in different residence times of water in the soil. Therefore, the substrate may modify the chemical potential and composition of water by affecting its residence time in different soil horizons. This residence time, if long, provides an opportunity for autogenic control of water quality by vegetation and animals in the ecosystem (Sect. 6, this chapter).

The distribution of water in a given soil horizon will determine what portion of the root mat will be irrigated, what the nature of its nutrition will be, and the intensity of aeration (Lahde, 1966; Ingram, 1967). Additionally, the level of the water table will determine root development (Boggie, 1972). As water moves through a community it forms "water tracks" where vegetation responds differentially. Water tracks are usually more eutrophic and better aerated than the rest of the wetland (Ingram, 1967; Siegel and Glaser, 1987).

The above discussion helps explain the increase in species richness with increasing water motion in northern Minnesota peatlands, which was observed by Heinselman (1970). His data and other similar data were analyzed by Gosselink and Turner (1978) who showed that wetlands become more productive in terms of organic matter turnover as hydrologic energy increases. Similar analyses were presented by Brown et al. (1979) for other freshwater forested wetlands.

Construction of drainage canals in wetland areas results in zones of increasing tree size and production with decreasing distance from the canal (Fig. 4.4). We propose that this increase in structure and production of the forest is partly a result of the increase in water movement (which affects soil aeration, for instance) due to the canal rather than due only to a lowering of the water table. The effect of the gradient of water movement is also important in promoting zonations in other wetland forests (Sect. 11, this chapter).

Soil saturation is the main reason why wetland soils become anoxic (Sect. 18, this chapter), but plants may gain some advantages from anoxic soils. One of these is the capacity of anoxic soils to form complexes with metals which are toxic to vegetation at high concentrations. The toxicity of the metal is neutralized through chelation pro-

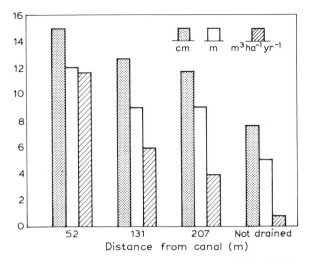

Fig. 4.4. Change in average diameter at breast height (dbh in cm), tree height (m), and wood volume yield (m³ ha⁻¹) as a function of distance from a drainage canal. Data are for a 17-yr-old *Pinus taeda* plantation planted at 1.83 m × 1.83 m spacing, trees ≥ 1.6 cm dbh, and wood volume is under bark (Miller and Maki, 1957).

cesses or root immobilization and, in effect, plants are isolated from potentially deleterious effects (Dykeman and Sousa, 1966).

Shortage of water is also stressful to wetlands. Tree dieback, particularly in the upper branches of taller trees (Broadfoot and Toole, 1957), rapid defoliation or increased rate of leaf fall (Furtado and Mori, 1982), and leaf wilting are examples of wetland plant responses to drought. Xeromorphic plant characteristics and leaf sclerophylly have attracted the attention of wetland ecologists as unusual adaptations in ecosystems growing in flooded soils. Several studies have suggested that these adaptations are responses to erratic water supply and/or nutrient shortages (cf. Sect. 8 of this chapter for further discussion).

8. XEROMORPHISM IN WETLAND PLANTS FOR WATER CONSERVATION

It is often assumed that forested wetlands seldom have limitations imposed by a water shortage and thus lack adaptations for controlling water loss. However, in this topic we will demonstrate that some wetland forest types appear to experience water deficits at a frequency that is high enough to select for species that have developed xerophytic characteristics. The adaptive significance at the species level is tolerance to water stress and at the ecosystem level it is control of water loss.

Actual transpiration from plant communities results from a balance between atmospheric conditions (such as water vapor pressure deficit, wind, temperature, etc.) and availability of soil moisture. When soil moisture is relatively unlimited, transpiration is controlled by the prevailing atmospheric conditions. If, on the other hand, soil moisture is relatively unavailable, transpiration is controlled by the plants through adaptations. The environmental conditions at a given site (wetland or upland) select for species best adapted to those conditions.

In riverine forests water is seldom limiting to ecosystem processes because of the close proximity of tree roots to the water table which is replenished by inflowing water. Therefore, there is little to no selection for plant characteristics associated with water stress. In basin wetlands, which depend primarily on precipitation, water stress may occur because it is more directly coupled with local vagaries of precipitation (such as drought). Although water stress may not occur often, it is the severe environmental condition, rather than the average, that determines the characteristics of plants. Thus water stress becomes a selective factor in basin forests where water tables are not supplemented by inflows.

The few data on transpiration rates of forested wetlands suggest that transpiration is lower in basin wetlands than in riverine wetlands, even when standing water is present. Brown (1981) obtained average transpiration rates of 3.1 to 3.8 mm day⁻¹ for small cypress (*Taxodium distichum* var. *nutans*) dome ecosystems and 0.9 mm day⁻¹ for a scrub cypress ecosystem. Both ecosystems are basin wetlands located in Florida, U.S.A., and the measurements were made between June and September. These transpiration rates are well below the 5.2 mm day⁻¹ measured for a bald cypress (*Taxodium distichum*) riverine ecosystem in the same vicinity as the domes. When transpiration data are collected under different atmospheric conditions, as in this study, a useful index for comparison is the ratio of transpiration to standard pan evaporation (transpiration-to-pan ratio). Transpiration-to-pan ratios for the scrub cypress

(0.19) and cypress domes (0.66) were lower than those for the cypress riverine forest (0.95) (Brown, 1981). Heimburg (1984) measured evapotranspiration (ET) rates from the same cypress domes and obtained ET-to-pan ratios of 0.8 in the spring and fall, 0.6 in the summer wet season, and less than 0.1 in the winter months when the trees drop their leaves. Thus water is conserved in basin wetlands relative to riverine ones. Similar measurements are unavailable for northern basin forests.

There are several features of cypress basin wetlands that may account for their low rates of water loss. Cypress trees have needle-shaped leaves with thick cuticle layers and deeply sunken stomata. The leaves are vertically oriented on the branches thus minimizing heating loads and maximizing cooling by convection rather than by evaporation of water (Brown, 1981). Cypress canopies have a high albedo (Capehart et al., 1977) and light-green to yellow appearance when observed from low altitudes in spite of their low leaf area index and high chlorophyll concentrations (Brown, 1981). The idea that wetland plants have adaptations for tolerating water stress appears to be a contradiction because many wetland plants are also well adapted to flooded conditions. However, droughts can occur in the growing season. For example, the area where Brown did her study received little rainfall for a period of several weeks during one growing season and as a result many trees in the cypress domes had leaves that were turning brown and falling. When more favorable conditions returned later in the growing season a new crop of leaves was produced.

Similar xeromorphic characteristics have been observed for plants in tropical basin wetland forests of Malaysia (see also Ch. 13). Whitmore (1975) proposed that these characteristics are a response to both potential water deficits and low soil fertility. Bruenig (1971), on the other hand, strongly suggested that soil-water deficits were the key factor because water available for growth could become potentially limiting during periods of low rainfall and the reduction of transpiration to conserve water loss is a necessary adaptation for survival.

Many of the leaf characteristics described here for cypress basin wetlands and tropical wetlands of Malaysia have been described for plants growing in other basin-type forested wetlands (for example, pocosins of southeastern U.S.A. and peat bogs of northern U.S.A.). In fact, the apparent paradox of the presence of plants possessing leaves with xeromorphic characteristics (particularly sclerophylly) growing in wetlands, where it is assumed that water is unlimited, has intrigued scientists for decades. In an attempt to explain this apparent paradox numerous theories have been presented, including "physiological drought", presence of toxins in bog water, anoxic soils, poorly developed and shallow root system, low temperature of bog water, water-retaining nature of peats, and soil nutrient deficiencies.

The earliest theory, by Schimper (1898), proposed that water uptake was inhibited due to the presence of humic acids or salts in bog water thus creating a "physiological drought". This theory was not supported by Livingston (1904), who found no evidence of high concentrations of dissolved substances in bog waters. However, Livingston (1905) and Dachnowski (1908) demonstrated that bog water and bog soil extract contained substances toxic to some plants.

Measurements of transpiration rates of bog plants (per unit of leaf area) with sclerophyllous leaves (Gates, 1914; Caughey, 1945; Small, 1972; Schlesinger and Chabot, 1977; Brown, 1981; Brown et al., 1984) do not support the theory of "physiological drought" either. Transpiration rates for these bog plants were no different from either those for bog plants without the sclerophyllous features or those for upland plants. It should be noted, however, that water was never limiting to the plants in these experiments.

The root systems of bog plants tend to be poorly developed and confined to shallow depths (Dachnowski, 1908; Burns, 1911; Caughey, 1945; Philipson and Coutts, 1978). Burns (1911) found that the root system in a bog in Michigan (U.S.A.) penetrated less than 45 cm — even the roots of large trees. Apparently, the properties of bog water and soil are a causative factor because plants grown in bog water or bog soil extract produced stunted roots and tops, whereas those grown in treated bog water (aerated or neutralized) did not develop these characteristics (Dachnowski, 1908).

Low soil temperatures, particularly between fall and spring, were implicated by Gates (1914) to contribute to the presence of xerophytes in wetlands. Because plants possessing xeromorphic leaf characteristics are also generally evergreen, they "need" water-conserving adaptations in order to survive periods when transpiration can occur but water uptake is inhibited because of very low soil temperature. However, the effect of low soil temperatures does not explain why plants with xeromorphic characteristics are present in southern bogs where the winter climate is milder (Caughey, 1945; Schlesinger and Chabot, 1977).

Although peat often appears to be wet, its water retaining nature may result in low water availability to the plants. For example, wilting occurs when the soil moisture is 1.5 to 7.8% for sands and loams compared to 50% for peat because of its low density (Burns, 1911). In addition, Boelter (1966) demonstrated that, while the porosity of many peats is high (80 to 100%), specific yields and hydraulic conductivities of peats vary considerably and are not linearly related. Mossy peats have a specific yield and hydraulic conductivity of 819 $l\,m^{-3}$ and 36.6 m day^{-1}, respectively, while well-decomposed peats have a specific yield of 264 $l\,m^{-3}$ and hydraulic conductivity of 0.01 m day^{-1} (Boelter, 1966). Other factors affecting the availability of water in peaty soils are irreversible drying and the small amount of capillary rise of water (Boelter, 1966).

Soil nutrient deficiencies have also been proposed to account for the presence of sclerophyllous plants in bogs (Loveless, 1962; Small, 1972, 1973; Schlesinger and Chabot, 1977). Phosphorus deficiencies are common to both wet (bogs) and dry habitats, thus explaining the presence of sclerophyllous plants in both environments (Loveless, 1962). Small (1973) suggested that deficiencies in both nitrogen and phosphorus were the cause of sclerophylly in bog plants. Schlesinger and Chabot (1977) concluded that evergreen, sclerophyllous leaves give a competitive advantage to plants possessing them because of greater efficiency of mineral use and higher annual productivity.

The theory that plants with xeromorphic characteristics (or sclerophylly) are a response to soil nutrient deficiencies has merit because it can explain the presence of these plants in both northern and southern wetlands. However, this line of reasoning should not exclude the idea that these characteristics are also a response to water deficits. The low productivity of basin-type forested wetlands (Ch. 7) in response to nutrient-poor soils implies that the plants which these wetlands support do not have the resources nor the genetic plasticity for rapid development of structure. These constraints prevent plants from allocating energy for rapid root growth when water tables drop below the shallow rhizosphere during dry periods. The only strategy that remains for these plants, then, is to tolerate water stress by developing xeromorphic leaves. Thus water and nutrient deficiencies can result in convergence toward xeromorphism, a strategy of conservatism for the use of limited resources. In nutrient-poor wetlands that also experience water deficits, the same xeromorphic characteristics are adaptive strategies for both stresses.

9. ENERGY LANGUAGE

Several chapters in this book use an energy language to represent wetland ecosystem structures and functions. This language was developed by H.T. Odum and is useful for presenting concepts, organizing knowledge, planning research, formulating hypotheses, and simulating with computers. The language follows the laws of energy and has been devised in such a way that each symbol has a precise mathematical equivalent. Thus, a model that uses the symbols correctly is also a mathematical representation. The qualitative meaning of each symbol used in this book is given in Fig. 4.5 and the reader is referred to Odum (1983) for comprehensive overview of the language and its uses.

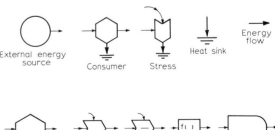
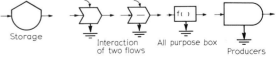

Fig. 4.5. Description of energy language symbols.

10. THE ENERGY SIGNATURE

The sum of all incoming energy flows to an ecosystem and the pattern of their delivery is called the "energy signature" (Odum, 1983). Fig. 4.6a illustrates the energy signature of a coastal marsh wetland subjected to thermal stress. However, these energy inputs are not comparable because they are of different types (solar, fossil fuel, wave, etc.). Use of the concept of "energy quality" is the way by which energies of different types can be compared, because "energy quality" recognizes that different energy sources have different abilities to do work — that is, different qualities (Odum, 1983). Some forms of energy — for example, solar energy — are dilute, whereas others, like fossil fuels, are concentrated. Therefore, a unit of solar energy cannot do the same amount of work as a unit of fossil fuel because to concentrate a dilute energy source requires additional energy. Odum et al. (1977) developed factors, called energy quality ratios, which enable energies of different types to be expressed on the same quality basis, either coal or solar equivalents. As an example, 2000 units of solar energy need to be concentrated to do the same amount of work as one unit of fossil fuel energy (as coal); thus the energy quality ratio for converting sunlight energy to coal equivalents is the reciprocal of 2000 (0.0005 solar/coal equivalent). Other energy quality ratios are given by Odum et al. (1977).

To determine the energy signature of an ecosystem, each component is first evaluated in heat equivalents (Fig. 4.6b), and then in coal equivalents by multiplying the heat value by the corresponding energy quality ratio (Fig. 4.6c). By using heat equivalents, heat stress from heated effluent and sunlight energy appeared to be the dominant energy sources for the marsh. However, when energy quality was taken into consideration, there is more similarity among sources, and wave and tidal energies gain importance.

The change in the relative value of each energy source (particularly waves and tides), when the contribution of each component of the signature is corrected for quality, provides insight into why most wetlands are sensitive to some environmental changes and not to others. In general, hydrologic fluxes are primary energy sources to wetlands because these fluxes perform many vital services upon which the biota of the wetland are highly dependent.

It is suggested that studies or analyses of forested wetland ecosystems should consider their complete energy signature as a vital first step to understanding their functional and structural response under changing environments, stressors, or management.

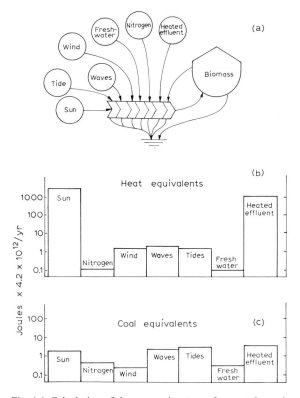

Fig. 4.6. Calculation of the energy signature of a coastal marsh ecosystem under the influence of thermal pollution. The symbols in (a) are described in Section 9, this chapter. Note that when all the energies are expressed on the basis of equal quality (coal equivalents) (c), their relative importance changes in comparison to calculations based on heat equivalents (b). (From Kemp, 1977.)

11. ZONATION AND SUCCESSION

Zonation of species in a community is a spatial phenomenon while succession of species is a temporal process. In forested wetlands, zonations are common and have been extensively described (for example, Chapman, 1976). Forested wetlands are also very dynamic in the time dimension and numerous studies describe temporal changes (cf. Ch. 3). In some instances species zonations have been equated to community succession (for example, Wright and Wright, 1932; Eggeling, 1935; Stewart and Merrell, 1937). For example, studies on changes in peats are one source of evidence for the idea of recapitulation of succession by zonation (Davis, 1940; Anderson, 1960). However, this may not be a general rule for ecosystems that are open to hydrologic fluxes controlled by forces outside of the wetland (allogenic).

A study of 300 years succession in a floodplain forest in the boreal region of Alaska showed that the control of succession changed from mostly allogenic forces during the first 40 years, to a combination of autogenic (internal) and allogenic control during the next 60 to 100 years, and mostly to internal control during the mature stages (Van Cleve and Viereck, 1981). Much of the successional theory is based on the assumption of autogenic control of environmental change (for example, Odum, 1969). However, Odum described "pulsed" succession in wetlands as a separate case. These types of successions have recently been given more attention (West et al., 1981). In saltwater forested wetlands where tidal forces are predominant, zonations generally do not recapitulate successions. Allogenic successions predominate and many series involve autosuccession or cyclical succession (Fig. 4.7; see also Lugo, 1980 and Snedaker, 1982). In freshwater basin wetlands there is more opportunity for the operation of autogenic factors, but autogenic control may decrease with increasing hydrologic energy. This is reflected in the accumulation of soil organic matter (Fig. 4.10). Penfound (1952) described allogenically controlled cyclic successions in peaty forested wetlands. Fire and aggradation were the two factors responsible for the reversal in direction of succession. Autogenic succession is also arrested by fire in stillwater Carolina bays (Buell, 1946) and pocosins (Christensen et al., 1981).

Evidence is available to suggest that even in basin-type forested wetlands allogenic factors may have a predominant role in determining zonations and the direction of succession. For example, the seedlings of *Mitragyna stipulosa* grow preferentially on the top of mounds formed by microrelief of the swamp (McCarthy, 1962). This leads to a segregation of species according to elevation within the wetland. Frey (1954) found evidence of erosional events alternating with periods of accumulation in fringe-type and basin-type cypress wetlands. Ingram (1967) questioned autogenic succession in mires and showed that hydrological conditions, independent of their effect on aeration, were the forcing factors of zonation and succession through their differential effect on nutrient availability. Groundwater fluctuations and water quality were clearly associated with the zonation of species in mires (basins) where vegetation zones were not serally related (Green and Pearson, 1968). Changes in drainage and soil moisture were also considered the primary determinants of successional relationships in the Mothronwala swamp in India (Dakshini, 1965). Finally, observations since 1917 in a Michigan (U.S.A.) bog demonstrated that seral changes that appeared to be autogenic successions were in fact associated with periods of low water (Schwintzer, 1978). Small water level changes affected all vegetation zones.

12. WETLANDS ON SLOPES IN PUERTO RICO

The northern and northeastern slopes of the mountains of Puerto Rico intercept the path of the tropical trade winds as they cool adiabatically after a long fetch over the Atlantic Ocean. Rainfall on these slopes increases with elevation reaching maximum annual values of about 5000 mm on the high windward peaks (Brown et al., 1983). Fig. 4.8 illustrates the wetland forests (and an herbaceous bog) that are located on the steep slopes and river valleys of the Luquillo Mountains and on other mountains of the island.

Frangi (1983) explained the presence of forested wetlands on these slopes on the basis of excessive rainfall and poorly drained, clayey soils. In spite of the steepness of the slopes, soil saturation occurs year round. Among the characteristics typical of wetland forests Frangi (1983) found the following in these wetlands on slopes: low species richness; abundance of palms; low leaf area indices; high stem densities; abundance of adventitious roots, lenticels, and specialized structures such as pneumatorrhizae (Frangi and Ponce, 1985); shallow distribution of roots; scattered pools of standing water; anaerobic soils; high concentration of organic matter in the soil profile; and high

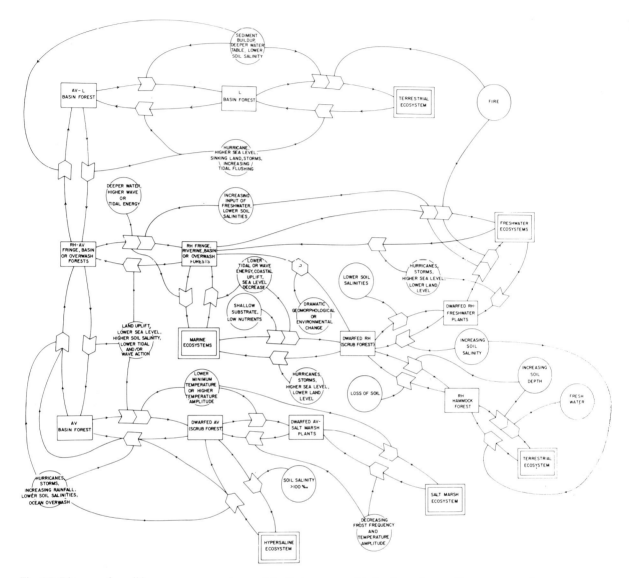

Fig. 4.7. Diagram of possible mangrove successions and the factors responsible for the pathways. Boxes represent the mangrove sere (RH is *Rhizophora*, L is *Laguncularia*, and AV is *Avicennia*). Large arrows indicate the direction of succession, and they connect with the physical factor(s) (circles) believed to be responsible for the pathway. Boxes with double walls represent non-mangrove ecosystems. All successional pathways are reversible. More details are given by Lugo (1980).

concentrations of organic carbon in runoff waters and high rates of carbon export.

Some of the wetlands are background ones (cf. Ch. 2) under the influence of sheet water flow such as the Colorado forest and palm forests on steep slopes. Others are lines such as the floodplain palm wetlands or *Pterocarpus* swamps. Others are points such as the herbaceous bogs, while some are strings such as the cloud forests. Viewed at a larger scale, the forested wetlands are arranged as zones along an altitudinal gradient of water availability.

13. WETLAND STRESSORS

Selye (1956) defined a stressor as any factor or situation that forces a system to mobilize its resources and spend increased amounts of energy to maintain homeostasis. He defined stress as the

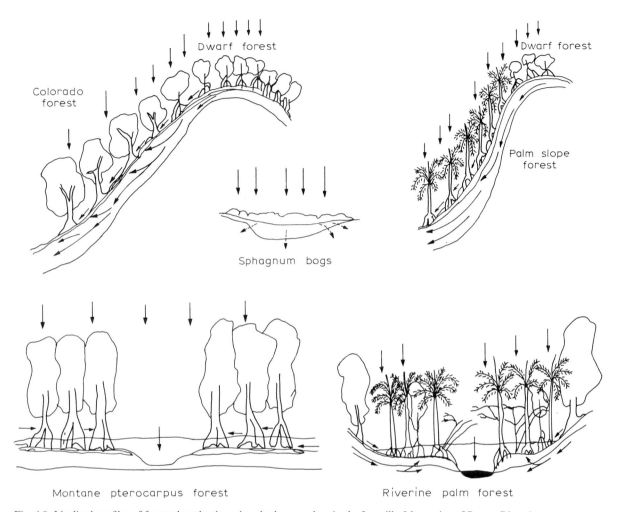

Fig. 4.8. Idealized profiles of forested wetlands and an herbaceous bog in the Luquillo Mountains of Puerto Rico. Arrows represent direction of water flow.

response of the system to the stressor. Odum (1967) suggested that all stresses are energy drains because they involve the diversion of potential energy flows that otherwise could be used to do useful work in a system. Lugo (1978) reviewed the issue of stress and the ecosystem and suggested that, when all conditions are equal, ecosystem response to a stressor depends on the point of attack of the stressor on the system (Fig. 4.9). If the stressor interferes with the primary energy sources and/or initial energy transformation processes of the system, recovery is slow. Stressors that act on plants, or factors which directly affect plant photosynthesis (nutrients, sunlight, or water), are examples of this type of stressor. However, if the stressor acted on the higher-quality energy flows of the system without directly affecting input energies, recovery would be more rapid. Stressors that act on animals or ecosystem respiration processes are examples of this group. Lugo classified stressors into five types: type 1, those that alter the nature of the main energy source; type 2, those that divert a portion of the main energy source before it is incorporated into the system; type 3, those that remove potential energy before it is stored but after it is transformed by plant photosynthesis; type 4, those that remove storages (this type of stressor may be subdivided into type 4a, those that remove the storage of limiting factors, and type 4b–4d, those that remove the structure of the ecosystem); and type 5, stressors that increase the rate of respiration.

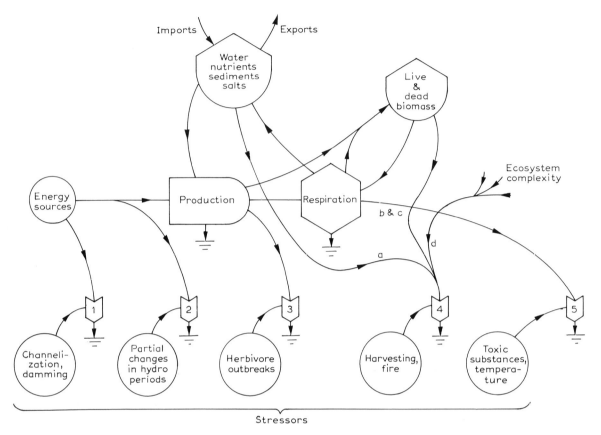

Fig. 4.9. Diagram illustrating the point of attack of stressors on wetland ecosystems, modified from Lugo (1978). An example is given for each type of stressor. Symbols are described in Fig. 4.5. All heterotrophs are represented by the symbol labelled "respiration".

Examples of stressors that affect forested wetlands (mangrove) are given in Table 4.4 (Lugo et al., 1981).

14. WETLAND VALUES

Wetlands, including forested areas, have many values; however, not all wetlands have all these values. Some of those frequently present are (from Lugo and Brinson, 1978):

(1) *Water*: storage of flood waters; conservation of water during dry periods; and desalination of salty water.

(2) *Organic productivity*: high primary productivity; high secondary productivity (e.g., commercial and sport fisheries); high export of organic foods to other ecosystems; and high wood production.

(3) *Biogeochemical values*: high capacity to recycle nutrients; high storage of organic matter and sink for carbon dioxide; net oxygen production; closure of many biogeochemical cycles by reduction of nitrogen, carbon, sulphur, iron, etc., in anaerobic muds; and sequestration of heavy metals, radioactive isotopes, and other poisonous chemicals in anaerobic muds.

(4) *Geomorphological values*: high potential for erosion control; protection of coastlines against storms, tides, and winds; and high potential to build land.

(5) *Biotic values*: fish nurseries, bird rookeries, and refuges and corridors for terrestrial animals; and gene banks for many plant and animal species.

(6) *Other values*: natural laboratories for teaching and research; location for recreation and relaxation; rich organic soils for agriculture, aquaculture, or as fuels; location for solid waste disposal; historical legacy in sediments of climate,

CONCEPTS IN WETLAND ECOLOGY

TABLE 4.4

Natural and human-induced stressors of mangrove ecosystems. Numbers in parentheses identify them according to the classification by Lugo (1978)

Stressor	Causal force or factor	Primary point of attack on ecosystem
Natural		
Low temperature	Atmospheric fronts	Leaves (4b, 5, 2, 1)
High temperature amplitude	Latitudinal gradient	Structural complexity (4d, 5, 1)
Hypersalinity	Aridity of climate or lack of terrestrial inputs	Structural complexity (4d, 5)
Low fertility	Edaphic factors or lack of terrestrial inputs	Structural complexity (4b, 4d)
Siltation (acute)	Hurricanes or floods	Root gas exchange, nutrient, and water uptake (4a, 5)
Windthrows	High winds	Structural complexity (4b, 4d)
High wave energy	Water motion	Structural complexity (4a, 4b, 4c)
Hydrogen sulfide accumulation	Water stagnation	Root and sediment respiration (5)
Chronic flooding	Hurricane or large flood	Root and sediment gas exchange (4a, 5, 1)
Change in drainage pattern	Hurricane or geologic change	Structural complexity (1, 2, 5)
Herbivore outbreaks	Unknown environmental change	Plants (3, 4b)
Human induced		
High temperature and high temperature amplitude	Power-plant cooling	Structural complexity, leaves, seedlings (2, 4d, 5)
Oil coating	Accidental spills	Gas exchange of oil-covered surfaces, leaves (2, 5)
Fire	Human set	Structural complexity (4a, 4b, 4c, 4d)
Chronic drainage	Channelization, road construction, dikes	Structural complexity and all ecosystem processes (1)
Excess harvesting	Human need	Plants (4b)
Herbicides	War or land-clearing practices	Leaves and buds (2, 4b)
Filling or siltation (chronic)	Construction activities	Structural complexity and gas exchange (1, 5)
Heavy-metal runoff	Runoff from urban systems	Stored in leaves and sediments (5)
Chronic flooding	Diking, road construction	Gas exchange of roots and sediments (4a, 5, 1)

hydrology, vegetation, and disturbance; importance as natural heritage, particularly when they become scarce; and representative of personal intangible values.

15. SEDIMENT AND PEAT ACCUMULATION

Most wetlands are depositional environments. However, many wetland sediments are a mixture of inorganic and organic materials from both fluvial and biogenic sources. In flowing-water wetlands such as riverine floodplains, and particularly deltas, inorganic sediments preferentially accumulate near the channel in a levee-building process, while backswamp areas (depressions on the floodplain which remain flooded but are far removed from the main channel) may become progressively wetter and more isolated from the river. In contrast, basin wetlands often accumulate peat and have no appreciable source of inorganic sediments. Fluvial deposition and organic matter accumulation appear to be a function of water flow (Fig. 4.10).

We have assembled some deposition rates (Table 4.5) for predominantly inorganic sediments in floodplains and for peat deposits of bogs, fens, and stream floodplains. In floodplains, sediment accumulation was measured during short intervals (single flood events to several years). Data are biased toward demonstrating accumulation, and no rates of floodplain erosion could be found even though it is commonly observed. The ^{14}C dating method used for most peat accumulation rates constrains measurements to much longer intervals (centuries to millenia). These "averaged" rates of accumulation undoubtedly encompass episodes of both peat loss (drought and fire) and optimal, rapid accumulation. Conditions that initiate oxidation and subsidence (such as drainage, cultivation; see Sect. 20, this chapter) are better understood than conditions that promote accumulation. For example, rising sea level, blocked drainage, wetter climate, beaver impoundments, and activities of Neolithic man have been suggested as factors causing paludification (Moore, 1975). Optimal rates of accumulation may be restricted to a narrow range of hydrologic and related physicochemical conditions which allow high rates of biomass production while they simultaneously or intermittently restrict decomposition. Several of the peat accumulation rates in Table 4.5 are due to *Sphagnum* accumulation in raised bogs which may not be considered forested wetlands.

Vertical measurements of accumulation, without accompanying data on bulk density, carbon concentration, and compaction coefficients, are not comparable with each other, nor are they interpretable vis-à-vis flux units used in describing carbon dynamics of ecosystems.

16. DEFINITIONS OF CONCEPTS DEALING WITH ORGANIC-MATTER DYNAMICS

One of the difficulties in interpreting the literature on organic-matter dynamics is that the terminology is confusing. Additional confusion about productivity values derives from the array of differing assumptions, methods of collection, and computation of results. To avoid ambiguity, we define the terminology used in this book (from Brinson et al., 1981a).

(1) *Primary productivity* is the process by which plants convert solar energy into chemical energy.

(2) *Net daytime primary productivity* (P_N) is net daytime photosynthesis (per unit of leaf area) during the daylight hours, measured by carbon dioxide uptake, multiplied by the leaf area index of the stand.

(3) *Night-time respiration* (R_D) is the respiration rate of leaves (per unit of leaf area) during the night-time hours, measured by the production of carbon dioxide, multiplied by the leaf area index of the stand.

(4) *Gross primary productivity* (P_G) is the sum of

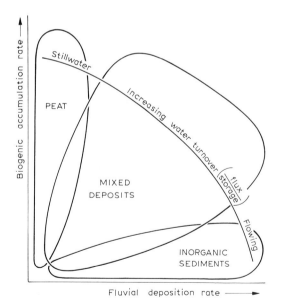

Fig. 4.10. Composition of wetland soils, postulated as a function of biogenic and fluvial deposition rates. Water turnover is shown as increasing from peat deposits (still water) to inorganic alluvial sediments (flowing water). (After Brown et al., 1979.)

TABLE 4.5

Accumulation of sediment and peat

Locality	Method, event or period of measurement	Accretion rate	Source
Floodplains			
Missouri River near Bismarck, North Dakota	Largest flood on record; 1952	8–10 cm	Johnson et al., 1976
Lowlands between Bismarck and Mandan, North Dakota		180 cm	
Cimmarron R., SW Kansas	About 12 yr since destructive flood using tree age	5.1 cm yr^{-1}	Schumm and Lichty, 1963
Cache R., Illinois	Measured with sediment traps	0.8 cm yr^{-1}	Mitsch et al., 1979a
Upper Mississippi R., Iowa	Deposition in floodplain lake	1.7 cm yr^{-1}	Eckblad et al., 1977
Kankakee R., Illinois	Total sedimentation during spring flood of which 80% was inorganic	590 ± 121 g m^{-2}	Mitsch et al., 1979b
Ohio R., Ohio	Mean deposition during 100 yr flood, Jan–Feb 1937	0.24 cm	Wolman and Leopold, 1957
Connecticut R.	March 1936	3.47 cm	Wolman and Leopold, 1957
	Sept. 1935	2.23 cm	
Kansas R., Kansas	July 1951	2.97 cm	Wolman and Leopold, 1957
Rio Grande, New Mexico	Mean aggradation for 16-yr period between Albuquerque and Socorro	1.5 cm yr^{-1}	Thompson, 1957
Alexandra R., Alberta	Fed by glacial meltwater; average aggradation during past 2500 yr	0.3 cm yr^{-1}	Smith, 1976
MacKenzie R., N.W.T.	Mean for sand deposition along point bar for 2 months during each of two summers	1.3–1.9 cm	Gill, 1972
Peat bogs, fens, and stream margins		mm yr^{-1}	
Itasca State Park, Minnesota	4–8 yr accumulation above cotyledon scars and wire marker	14–15	Leisman, 1953, 1957
Okefenokee Swamp Georgia	Six cores, ^{14}C of basal samples	0.48 ± 0.05 (S.E.)	Cohen, 1973
Everglades, Florida	^{14}C of basal peat profile	0.76–0.84	Stephens, 1956; McDowell et al., 1969
	Ratio of SiO$_2$ in plant to that of peat	1.4	Davis, 1946
Dismal Swamp, Virginia and North Carolina	^{14}C of peat; 3580 yr	1.0	Whitehead, 1972
Rockyhock Bay, North Carolina	^{14}C of peat above lake sediment	0.22	Whitehead, 1981
Shippea Hill, Cambridgeshire, England	^{14}C stratigraphy	0.24	Clark and Godwin, 1962
Western Netherlands and northern Germany	^{14}C of riverine peatland: *Betula* zone	0.6	Pals et al. (1980) in Wiegers, this volume, Ch. 17
	Salix–Betula zone	1.3	
	^{14}C of peat: riverine *Alnus* woodland	0.9–1.1	Van der Woude (1981) in Wiegers, this volume, Ch. 17
	ombrogenous *Sphagnum* bog	0.12–0.84	Overbeck (1975) in Wiegers, this volume, Ch. 17
Various sites, England	Fen peat	0.7–1.0	Walker, 1970
	Bog peat	0.46–1.03	
Argentine Isles and Signy Island, Antarctic	Upward growth rates	0.9–1.3	Fenton, 1980
	^{14}C of basal peat	0.96–1.4	
Grampian region, Scotland	Pollen analysis distinguishing four zones	0.1–1.5	Durno, 1961
Puget Sound area bogs in Washington	^{14}C of peat	0.62	Rigg and Gould, 1957
Draved Mose, Denmark	^{14}C stratigraphy	0.16–0.80 ($\bar{x}=0.44$)	Aaby and Tauber, 1974
Wales and Northern England	^{14}C stratigraphy	0.39–1.06	Hibbert and Switsur, 1976
Shrub carr in Wisconsin	^{137}Cs accumulation: alluvial zone	13	Johnston et al., 1984
	non alluvial zone	5	

P_N and an estimate for daytime respiration. Daytime respiration was estimated as $R_D \times$ (number of hours daylight/number of hours dark). This assumes that daytime respiration rate is the same as the night-time rate and it does not account for the effects of higher temperatures on respiration nor for photorespiration.

(5) *Net 24-hour primary productivity* (P_{N24}) is defined as the difference between P_G and 24-h plant respiration (R_{T24}). R_{T24} is equal to the sum of 24-h respiration of leaves [(R_D/number of hours dark) $\times$ 24], 24-h respiration of stem and branches (respiration rate per unit area of woody tissue $\times$ surface area index of woody tissue), and 24-h respiration of roots. In practice it is difficult to separate root respiration from soil respiration.

(6) *Net biomass production* was defined by Newbould (1967) as the production of flowers + fruits + leaves + current twigs + branches + stems + roots during the time interval t_1-t_2 (usually one year for forested wetlands). The production of flowers, fruits, and leaves for deciduous or evergreen species is usually obtained from litterfall collections or from allometric relationships. Branch and stem biomass production is obtained from allometric relationships, either from increment coring of trees or measurements of diameter over 1-yr time intervals. No corrections for tree mortality were made for estimates reported in this volume. Several methods of estimating root biomass production were proposed by Newbould; however, the data available for forested wetlands use the following relationship: (aboveground production)/(aboveground biomass) = (belowground production)/(belowground biomass).

(7) *Organic matter export* is the quantity of organic carbon removed from the ecosystem by water transport. Separation of dissolved organic carbon from particulate fractions is normally made with 0.4 to 0.8 μm pore filters.

(8) *Decomposition* includes nearly all changes in organic matter that has undergone senescence and death. Decay refers to losses due to respiration and assimilation of organic matter by microbes and detritus feeders. Leaching is the loss of "soluble" materials from organic matter and includes both inorganic and organic constituents mobilized either by passive physical processes or secondarily from decomposer activity. For litter-bag decomposition studies, unknown losses of incompletely decayed fragments occur (the amount depending largely on the mesh size of the bag) and are included in decomposition estimates. Immobilization is the net accumulation of inorganic ions over that which is lost by mineralization and leaching during decomposition. This is a consequence of selective nutrient uptake by microbes during their growth and results in a concomitant depletion of the organic matter that serves as their energy supply.

17. TOLERANCE OF TREES TO FLOODING

The capacity of wetland trees to thrive in waterlogged, anaerobic soil has been attributed to metabolic and morphologic adaptations of roots and to morphologic adaptations of stems. These morphological, anatomical, and physiological adaptations either facilitate the transport of oxygen to roots, permit the plant to function under low oxygen tensions, or both. Reviews on the subject can be found in Hook and Crawford (1978), Armstrong (1981), Drew (1983), and Kozlowski (1984).

At the plant level many morphological and anatomical adaptations have been described. Adventitious roots in both tropical and non-tropical trees are included in the general category of pneumorhizae (Longman and Jenik, 1974). These include lateral knee-roots, serial knee-roots, root knees, peg roots (pneumatophores), stilted peg roots, pneumathodes, and others. These roots invariably have lenticels and aerenchymous tissue which provide a permeable pathway for oxygen diffusion from the atmosphere (and outward diffusion of carbon dioxide) which appears to compensate for the lack of oxygen in flooded soils (Armstrong, 1964). Morphological plasticity of many wetland plants to flooding is shown through changes in root morphology (Hook et al., 1970; Boggie, 1972), increased permeability of the cambium (Hook and Brown, 1972), and the development of adventitious roots (Ellmore et al., 1983).

Physiological responses of plants to flooding include increased stomatal resistance to diffusion (Kozlowski, 1983), cessation of root growth in the absence of oxygen (Leyton and Rousseau, 1958), altered metabolic pathways in roots (Crawford,

1967; Crawford and McManmon, 1968) and inhibited translocation of nutrients (McKee et al., 1984).

The relative flood tolerance of trees has been related to the number of morphological and physiological adaptations that a particular species possesses (Hook and Scholtens, 1978). Experimental confirmation of these adaptations corresponds well with the relative flood tolerance under field conditions (Fig. 4.11).

The experimental work by Keeley (1979) on three populations of Nyssa sylvatica has helped to resolve much of the confusion on strategies to flooding. His populations were distributed along a gradient from long hydroperiod (swamp plants), to intermediate hydroperiod (floodplain plants), and finally to well drained (upland plants). He concluded that swamp plants adapt to flooding by reducing biomass allocation to roots (and hence oxygen demand), and by increasing oxygen transport to roots. Increased oxygen transport is achieved by morphological adaptation through (a) a rapid but temporary response of producing succulent, adventitious roots, and (b) long-term development of lenticels and a pervious cambium for oxygen transport from stems. This acclimation explains enhanced oxygen diffusion in seedlings of lodgepole pine (Pinus contorta) that grew under waterlogged conditions compared to those that grew under drained conditions (Philipson and Coutts, 1978).

Although Keeley (1979) indicated that temporary roots may be more tolerant to ethanol produced by anaerobic fermentation, no support was found for a malate-producing pathway to alleviate ethanol accumulation as hypothesized by McManmon and Crawford (1971). Jackson and Drew (1984) failed to produce evidence for the significance of malate from an extensive review of the literature. Oxidation in the rhizosphere prevents accumulation of toxins (H_2S, Fe, Mn) characteristic of anaerobic soils (Sect. 18, this chapter) rather than requiring physiological adaptations to them. Keeley further concluded that the adaptations for enhanced oxygen transport may create water control problems because stem transpiration is also enhanced. This puts wetland-adapted species at a competitive disadvantage during periods of water stress.

At the ecosystem level, we postulate that hydroperiod may be responsible for several structural patterns. Basin wetlands tend to have little water flow and moderate water level fluctuations in comparison with riverine and fringing wetlands. Therefore, basin wetlands may be constrained to: (a) lower root/shoot biomass ratios which reduce demand for oxygen by below-ground parts; (b) lower species diversity, limited to the few with special morphological and physiological adaptations; and (c) higher stem density which serves to reduce distance for oxygen transport to roots and/or increase surface area for gas exchange. In general, root biomass in wetlands is restricted to the upper, partially aerated layer of soil (Boggie, 1972; Montague and Day, 1980). However, varia-

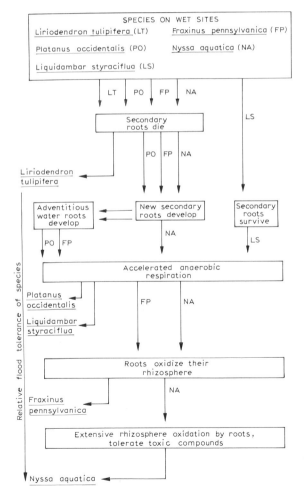

Fig. 4.11. The relationship of specific physiological and morphological adaptations of tree species to their relative flood tolerance. (After Hook and Scholtens, 1978.)

tions in aeration are induced by microsite elevation. These variations are reflected in a tree's ability to survive water or nutrient stress in low depressions because of poor root growth (Lorio et al., 1972).

18. ANAEROBIC PROCESSES AND DETRITUS PROCESSING

The extent to which anaerobic microbial processes predominate over aerobic ones depends largely on hydroperiod. For example, if a sediment has high oxygen demand due to high concentrations of biologically labile organic carbon, anoxic conditions will be maintained until conditions are altered by water flow and mixing, exposure to the atmosphere during a drydown period, or some other event that transports oxygen to the sediments. Because of the high resistance of saturated and flooded soils to oxygen diffusion, microorganisms in wetland sediments depend upon alternate electron acceptors for respiration. It is common for only the top 1 cm or less of flooded sediments to be oxic.

Redox potential (oxidation-reduction state) usually decreases with increasing depth in sediments to a minimum of about -400 mV. Thus, respiration of oxygen is restricted to a small region at and just below the surface. At progressively greater depths and lower redox potentials, a sequence of electron acceptors may become involved in microbial respiration (Fig. 4.12). These gradients have been demonstrated in muds that are homogenized in the laboratory and allowed to develop vertical redox zonations (Reddy and Patrick, 1975). The gradients occur naturally in marine (ZoBell, 1946) and freshwater (Mortimer, 1941, 1942; Ponnamperuma, 1972) sediments. However, redox readings seldom correspond ideally with the reduction of specific electron acceptors because aqueous oxidation-reduction reactions are generally not at equilibrium (Lindberg and Runnells, 1984). For example, sulfate reduction and methanogenesis (CO_2 reduction) have been shown to co-occur in sediment samples (e.g., Nedwell, 1982). Anaerobic microsites may develop in sediments that otherwise give readings of relatively high redox potential and are seemingly homogeneous. Alternatively, transport of oxygen through the plant to roots and the

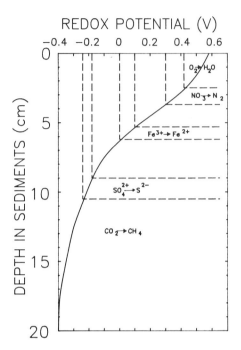

Fig. 4.12. Microbially mediated redox processes in relation to redox potential and depth in sediment. See Munch and Ottow (1983) for evidence of microbially mediated reduction of ferric oxides.

radial diffusion of oxygen to the surrounding sediments (Armstrong, 1964) would create oxidized microsites in an otherwise reducing environment. Burrowing activities of crayfish and other sediment-dwelling organisms contribute to redox heterogeneity.

One of the fundamental differences between freshwater and saltwater wetlands (i.e., those derived from marine salts) is the presence of sulfate in the latter. The concentration of sulfate in seawater is 0.92 g l^{-1}, whereas nitrate, often an important electron acceptor in anoxic freshwater environments, generally is of the order of µg l^{-1}. Even though the turnover of nitrate may be quite high (Sørensen et al., 1979), the potential for anaerobic decomposition is much higher in saltwater wetlands than in freshwater ones because of the abundance of sulfate. A consequence of the use of sulfate as a respiratory electron acceptor is the production of hydrogen sulfide. Plants, animals, and microbes of saltwater wetlands must have adaptations to survive high concentrations of this toxic gas. Sulfate-reducing microbes, which are incapable of utilizing complex organic compounds such as cellulose, also remove low-molecular-weight end-products of fermentation, thus maintaining favorable conditions for the activity of fermentative organisms.

Anaerobic processes may be especially significant in detritus food webs where sulfur cycling plays a large role. It may be argued that the higher energy yield of respiration with oxygen (42 kJ g^{-1} organic carbon) compared to respiration with sulfate (11 kJ g^{-1} organic carbon) would imply greater contribution of aerobic processes in detritus food webs. However, inorganic end products of sulfate respiration, such as sulfide, contain energy available for chemoautotrophic biomass production through fixation of carbon dioxide when aerating mechanisms discussed above make oxygen available (Howarth and Teal, 1980). Perhaps the greater number of steps involved in anaerobic respiration and chemoautotrophic microbial production is more important than the greater energy yield from organic carbon when respired by oxygen. Further, the more diversified and perhaps less efficient pathways may actually favor the production of microbial biomass in detritus food webs. These complex pathways may make pulses of plant detritus production (autumn litter-fall, fall senescence and die-off of aquatic macrophytes, blooms of phytoplankton) more available for a longer period to consumers, and thus result in more stable energy flow. Peterson et al. (1980) have suggested that the microbial biomass derived from the oxidation of reduced sulfur compounds in salt marshes may be an important connection between plant primary production and estuarine food webs. Similar pathways undoubtedly exist for mangrove ecosystems (Ch. 6). The relative importance of aerobic vs. anaerobic pathways in microbial production and transfer to higher trophic levels is just beginning to be studied.

Associated with the normally longer time necessary for anaerobic decomposition are other microbial processes which improve the quality of detritus as a food source (Fenchel and Jørgensen, 1977). Decreasing C/N ratios of organic detritus are often interpreted as higher protein concentration, and thus greater nutritive value to detritivores. Atomic C/N ratios decrease during decomposition from initially high values in wetland detritus [32 for *Nyssa aquatica* leaves; Brinson (1977); 59 for *Rhizophora mangle* leaves; and 30 for *Avicennia germinans* leaves; Twilley et al. (1987)] to relatively low values in less than a year (13, 28 and 15, respectively). Both microbial immobilization of ambient nitrogen and decreasing carbon content by respiration are involved. However, Rice (1982) cautions that organic nitrogen concentrations are not always indicative of protein, and that the ratio of nitrogen to actual protein varies widely. Much of the organic nitrogen may be in condensation products of phenolic and carbohydrate groups with amino acids. Thus, much of the accumulated nitrogen may actually be non-labile humic nitrogen rather than microbial protein. During the initial stages of decomposition, the large mass of detritus normally available to an ecosystem for microbial colonization would favor the importance of the protein enrichment hypothesis. Toward the terminal states of decomposition, non-labile humic nitrogen formation likely predominates, but the mass remaining would be small by comparison. At this point, humic compounds become involved in functions somewhat remote from food webs — that is, their roles in chelation, ion exchange, and sediment structure.

19. RESPONSE OF WETLANDS TO STRESSORS

The response of wetlands to stressors depends on the point of attack of the stressor and its intensity (Lugo, 1978), as well as the type of wetland being stressed. [See Sect. 13, this chapter; and also Lugo et al. (1981) and Lugo and Brown, (1984).]

In humid environments, such stressors as (a) diking, damming, channelization, and levee and road construction have a more severe impact on wetlands than (b) fires, harvesting, or moderate grazing because stressors in group (a) affect the energy signature of the wetland whereas stressors in group (b) affect ecosystem structure.

Mangrove wetlands growing in arid climates (mostly fringe forests) are less susceptible to alterations in runoff from the land than those that grow in humid climates (mostly riverine and basins) where the ecosystem is more dependent on freshwater runoff to regulate soil salinity. On the other hand, hypersalinity is a severe natural stress to arid coastal mangrove forests and its effects are aggravated by human actions that reduce seawater flushing.

Wetlands located in naturally stressed environ-

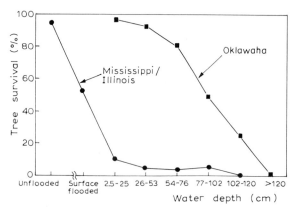

Fig. 4.13. The effect of water depth on tree survivorship in the Oklawaha and Illinois and Mississippi rivers. Trees in the Illinois and Mississippi rivers had been flooded for 5 yr while trees in the Oklawaha had been flooded for 7 yr. (From Lugo and Brown, 1984.)

ments appear to tolerate less additional stress than those located in more favorable conditions. For example, those growing at higher latitudes are less tolerant of chronic flooding than comparable wetlands growing at lower latitudes where climate is more favorable for growth (Fig. 4.13). When mangrove wetlands are exposed to periodic frost, their salinity tolerance diminishes. Alternatively, mangroves that grow under higher-salinity regimens tolerate less frost stress than mangroves that grow at lower salinities (Lugo and Patterson-Zucca, 1977). The combination of increased hydroperiod from sea level rise and salinity intrusions during drought years and hurricanes can initiate tree mortality in coastal freshwater forests (Brinson et al., 1985; DeLaune et al., 1987).

The most severe stressors of forested wetlands are related to factors that diminish aeration of the rooting zone. For example, forested wetlands are very sensitive to chronic flooding (Lugo and Brown, 1984), to oil coating of roots and sediments (Lugo et al., 1981), or to changes in hydroperiods (Carter et al., 1973). A specific effect of these stressors on wetland trees is the reduction of leaf size (Fig. 4.14).

20. PEAT LOSS RATES

Peat deposits are susceptible to oxidation and subsidence when they are drained and cultivated. At least three processes are involved:

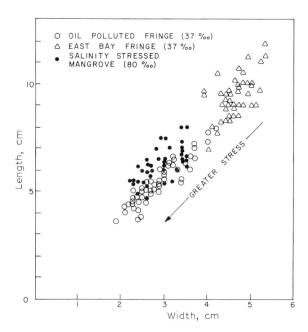

Fig. 4.14. Relationship between leaf length and leaf width of red mangrove leaves in the Bahia Sucia region in Puerto Rico (latitude 18°N). The relationship between leaf length and width remained constant but the size of the leaves decreased with increasing stress caused by increased soil salinity or oil pollution. (From Lugo et al., 1981.).

(1) *Lowering of water table*. Some subsidence may occur because of loss of buoyancy and "shrinkage" of peat from drying.

(2) *Oxidation of organic matter*. The ratio of aerobic to anaerobic microbial respiration increases.

(3) *Clearing practices and cultivation*. Removal of the litter layer, the rhizosphere, and decaying logs immediately reduces elevation of landscape surface. Wind erosion, fire, and tilling all accelerate losses, and exposure of the surface by removing cover may increase microbial respiration through heating.

Depth to water table is roughly proportional to the desaturated zone where aerobic metabolism can occur. Stephens (1956) developed the relationship between year-round depth to water tables (y, in cm) and subsidence rate (x, in cm yr^{-1}) for Florida peats, U.S.A. (30°N latitude) where $x = 0.0677y - 0.419$. The coefficients for the equation for Indiana peats, U.S.A. (40°N latitude) resulted in somewhat lower rates, suggesting temperature dependency of the relationship. However, initial rates for peats in Florida and England are surprisingly similar and the effects of renewed pumping in England (arrows in Fig. 4.15) clearly show that depth to water table is a major control. In a simulation model that encompassed a large number of variables, Browder and Volk (1978) confirmed the importance of temperature and depth to water table.

The short-term losses of peat (Table 4.6) tend to be one to two orders of magnitude greater than long-term accumulation rates listed in Sect. 15, this chapter. In spite of problems in comparing short- and long-term rates, peat formation is a process that is unique to wetlands. Large-scale clearing and draining have initiated a vast experiment that has reversed a trend of accumulation over several millennia. The role of wetland vegetation in water balance and conservation, in shading soil from direct solar heating, and in producing above-ground and below-ground organic matter are all stabilizing features that are interrupted when a wetland is converted to agricultural use.

21. THE INFLUENCE OF FORESTED WETLANDS ON WATER AND SOILS

Because forested wetlands are interface or ecotonal ecosystems, water and sediments may be exchanged with ecosystems upstream and downstream from the wetland. Evidence presented below shows that the quality and pattern of water discharge from wetland ecosystems are modified from that of inflowing water, and that soil conditions within the wetland are different from those of soils a few meters outside. Other sections of this chapter (7 and 17) discuss the impact of soil and water on trees. These sections demonstrate that hydroperiod is the most critical environmental factor in the wetland's energy signature and that it is very difficult to separate cause–effect relations among the many interactions that typify wetland ecosystems.

In comparison with upland ecosystems, basin-type forested wetlands may conserve water by reducing evapotranspiration rates (Sect. 8, this chapter). Basin forests also affect runoff by reducing peak flows, retarding the time to peak flow, and increasing the magnitude and length of base flow conditions (Daniel, 1981; Gilliam and Skaggs, 1981). Conversion of wetlands to agricul-

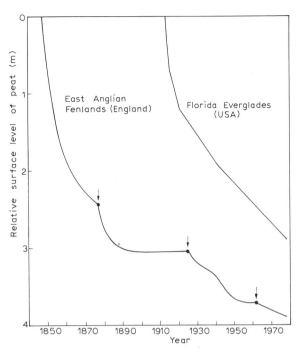

Fig. 4.15. Plots of surface level of peat at East Anglian Fenlands (after Hutchinson, 1980) and Florida Everglades (various sources). Arrows indicate initiation of new pumping regime.

ture by drainage, for example, destroys these functions (Fig. 4.16). The storage capacity of the flooded area determines how much peak flow will be retarded. For example, a storm hydrograph depends upon the level of the water table at the beginning of the event (Bay, 1969; Daniel, 1981); if the water table is high at the time of the storm, water runs off more rapidly than it would with a lower water table.

On a regional level, the area of wetlands in catchments, as well as their storage capacity, determines the overall effect wetlands have on stream and flood flows (Novitski, 1979; Brown, 1984). An analysis of the flood characteristics of streams in Wisconsin (U.S.A.) suggested that the relative flood flow was curvilinearly related to the relative area of wetlands and lakes (the analysis grouped forested and non-forested wetlands, as well as lakes, as one parameter) within the catchments (Fig. 4.17; Novitski, 1979). Although the analysis was based on a theoretical reduction in wetland and lake area, the analysis suggested that with only 40% of the catchment area composed of wetlands and lakes, flood flows were reduced to 20% of those for catchments without wetlands.

TABLE 4.6

Peat loss rates

Locality	Type of wetland	Disturbance		Rate (cm yr^{-1})	Source
Jalankebun, Selangor, Malaysia	Forested peat	Coffee crop; frequent disking.		8.5	Coulter, 1957
		Pineapple crop		3.3	
East Anglian Fenlands, England	Fen	Drainage and cultivation	initial max. recent max. mean for 128 yr	22.0 1.1 3.05	Hutchinson, 1980
Washington and Hyde Counties, North Carolina	Mixed hardwood on blanket peat	Drainage, fire, and cultivation		3.5	Dolman and Buol, 1967
Everglades, Florida	Herbaceous	Drainage cultivation, and fire		2.5–3.8	Stephens, 1956, 1969
San Joaquin Delta, California	*Scirpus* with *Salix* fringe	Drainage, cultivation, wind erosion, and fire		8.1	Weir, 1950
Southwestern Quebec, Canada	Drained histosol, probably fen	Drained and protected		1.59	Parent et al., 1982
		Drained and unprotected		4.53	

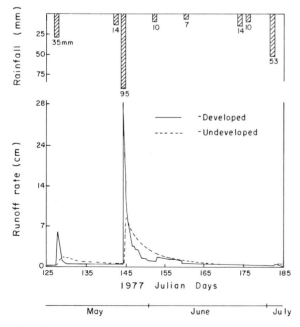

Fig. 4.16. Changes in hydroperiod in areas with forested wetlands (undeveloped) and without them (developed). Data are for pocosin wetlands in North Carolina, U.S.A. (From Gilliam and Skaggs, 1981.)

Flood flows were reduced by 50% with as little as 5% of the catchment in wetlands (Fig. 4.17). A similar analysis was done by Brown (1984) for the Green Swamp region (an area of about 220 000 ha) of central Florida (U.S.A.) of which about 30% was occupied by wetlands (>80% forested). A reduction of the wetland area by about half resulted in an approximately 30% increase in flood flows.

By increasing the residence time of water within wetlands, storm hydrographs are modified relative to what they would be without wetlands. This characteristic of wetlands has led the United States Army Corps of Engineers to purchase wetlands for

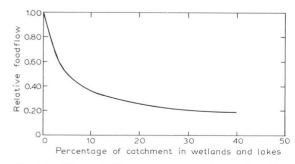

Fig. 4.17. Effects of different percentages of wetland and lake area within catchments of Wisconsin, U.S.A., on relative flood flows. (From Novitski, 1979.)

flood control rather than constructing more expensive and less efficient engineering structures (Horwitz, 1978). Water budget studies show that although wetlands change the hydrograph, they do not change the total amount of water discharged (Daniel, 1981; Gilliam and Skaggs, 1981).

A longer residence time of water within wetlands favors modification of water and soil quality by microbial, plant, and animal activity. These effects are usually measured by comparing water quality before and after it passes through a wetland or by comparing soil characteristics inside and outside of the wetland. Another approach involves the study of paired catchments that have been either treated experimentally or which support different land uses.

Water and nutrient balance studies in riverine forested wetlands have shown that isolation or uncoupling of wetlands from the stream channel leads to higher concentrations of phosphorus and nitrogen in the stream water (Carter et al., 1973; Kuenzler et al., 1977). The export of these materials was more erratic without the wetland, and the quality of the organic material differed from that when the stream and floodplain forest were coupled. Without the forested wetland, waters became eutrophic and algal production was partly responsible for the change in quality of organic material. The excess phosphorus and nitrogen in the stream isolated from the forested wetland was considered a measure of the biotic activity of wetlands when water was allowed to flood them.

Budget studies using catchment approaches document the role of forested wetlands in maintaining water quality. Cumulative losses of nitrate-nitrogen, Kjeldahl-nitrogen, and phosphorus were lower from an undeveloped wetland than from a developed site (Gilliam and Skaggs, 1981). In the same study, pH, dissolved oxygen, biochemical oxygen demand, total organic carbon, chlorine, sodium, potassium, calcium, magnesium and copper were lower in drainage waters from the undisturbed wetland than in those from drained and agriculturally developed wetlands.

Where floodplain forests are negligible, water quality may be controlled by streamside riparian forests. For example, Peterjohn and Correll (1984) found that significant quantities of carbon, nitrogen, and phosphorus were removed from runoff waters after just 50 m of transit over the riparian forest. A comparison of an upland cropped area with a vegetated floodplain in the southeastern U.S.A. (Yates and Sheridan, 1983) showed that reductions in nitrate-nitrogen, nitrite-nitrogen, and orthophosphate-phosphorus between the cropped area and floodplain outlets exceeded reductions that would be caused by mere dilution effects. The excess nutrient runoff from the cropped area was retained, utilized, and/or transformed by the floodplain community. Other studies of riparian zones have demonstrated similar results (McColl, 1978; Lowrance et al., 1984).

The cumulative effects of riverine forests are reflected in the soils. Wetland soils may accumulate heavy metals (Gardner et al., 1978) and certain radioisotopes (Hay and Ragsdale, 1978). At least three factors are responsible for this: the depositional nature of flooded areas, the redox conditions in soils due to hydroperiod (Sect. 18, this chapter), and differential uptake by trees (Golley et al., 1975). In forested wetlands in Panama, Golley et al. found a higher concentration of heavy metals in tree tissue than was present in tissue of non-wetland forests. Usually, root tissue will concentrate heavy metals and radioactive cesium (Shure and Gottschalk, 1978). These investigators also showed that floodplain forests removed radioactive cesium from rainfall and fallout, and sequestered the material in soil, from where a small portion cycled through the ecosystem.

REFERENCES

Aaby, B. and Tauber, H., 1974. Rates of peat formation in relation to the degree of humification and local environment, as shown by studies of a raised bog in Denmark. *Boreas*, 4: 1–17.

Ajtay, G.L., Ketner, P. and Duvigneaud, P., 1979. Terrestrial primary production and phytomass. In: B. Bolin, E.T. Degens, S. Kempe and P. Ketner (Editors), *The Global Carbon Cycle, SCOPE 13*, John Wiley and Sons, New York, N.Y., pp. 129–182.

Anderson, B.W., Higgins, A.E. and Ohmart, R.D., 1977. Avian use of salt cedar communities in the lower Colorado river valley. In: *Importance, Preservation, and Management of Riparian Habitat: A Symposium*. U.S. Department of Agriculture Forest Service, Gen. Tech. Rep. RM-43, pp. 128–136.

Anderson, J.A.R., 1960. The structure and development of the peat swamps of Sarawak and Brunei. *J. Trop. Geogr.*, 18: 7–16.

Armentano, T.V. (Editor), 1979. *The Role of Organic Soils in the World Carbon Cycle — Problem Definition and Research Needs*. The Institute of Ecology, Indianapolis, Ind., 37 pp.

Armentano, T.V., 1980. Drainage of organic soils as a factor in the world carbon cycle. *BioScience*, 30: 825–830.

Armentano, T.V. and Lawler, D.J., 1983. *Analysis of the Data Base on Tropical Wetland Area Estimates*. Holcomb Research Institute, Butler University, Indianapolis, Ind., 33 pp.

Armentano, T.V. and Menges, E.S., 1986. Patterns of change in the carbon balance of organic soil-wetlands of the temperate zone. *J. Ecol.*, 74: 755–774.

Armentano, T.V., De la Cruz, A., Duever, M., Loucks, O.L., Meijer, W., Mulholland, P.J., Tate, R.L. and Whigham, D., 1983. *Recent Changes in the Global Carbon Balance of Tropical Organic Soils*. Holcomb Research Institute, Butler University, Indianapolis, Ind., 57 pp.

Armstrong, W., 1964. Oxygen diffusion from the roots of some British bog plants. *Nature*, 204(4960): 801–802.

Armstrong, W., 1981. The water relations of heathlands: general physiological effects of waterlogging. In: R.L. Specht (Editor), *Heathlands and Related Shrublands of the World, B. Analytical Studies*. Elsevier, Amsterdam, pp. 111–121.

Austin, G.T., 1970. Breeding birds of desert riparian habitat in southern Nevada. *Condor*, 72: 431–436.

Banage, W.B., 1966. Survival of a swamp nematode (*Dorylaimus* sp.) under anaerobic conditions. *Oikos*, 17: 113–120.

Bay, R.R., 1967. Factors influencing soil–moisture relationships in undrained forest bogs. In: *Forest Hydrology*. Pergamon Press, New York, N.Y., pp. 335–343.

Bay, R.R., 1969. Runoff from small peatland watersheds. *J. Hydrol.*, 9: 90–102.

Beadle, L.C., 1974. *The Inland Waters of Tropical Africa*. Longmann Group Limited, London, 365 pp.

Boelter, D.H., 1966. Hydrologic characteristics of organic soils in Lake States watersheds. *J. Soil Water Conserv.*, 21: 50–53.

Boggie, R., 1972. Effect of water-table height on root development of *Pinus contorta* on peat in Scotland. *Oikos*, 23: 304–312.

Bottorff, R.L., 1974. Cottonwood habitat for birds in Colorado. *Am. Birds*, 28: 975–979.

Brink, V.C., 1954. Survival of plants under flood in the lower Fraser river valley, British Columbia. *Ecology*, 35: 94–95.

Brinson, M.M., 1977. Decomposition and nutrient exchange of litter in an alluvial swamp forest. *Ecology*, 58: 601–609.

Brinson, M.M., 1985. Management potential for nutrient removal in forested wetlands. In: P.J. Godfrey, E.R. Kaynor, S. Pelezarski and J. Benforado (Editors), *Ecological Considerations of Wetland Treatment of Municipal Wastewater*. Van Nostrand Reinhold Co., New York, N.Y. pp. 405–414.

Brinson, M.M., Lugo, A.E. and Brown, S., 1981a. Primary productivity, decomposition, and consumer activity in freshwater wetlands. *Annu. Rev. Ecol. Syst.*, 12: 123–161.

Brinson, M.M., Swift, B.L., Plantico, R.C. and Barclay, J.S., 1981b. *Riparian Ecosystems: Their Ecology and Status*. U.S. Fish and Wildlife Service, Office of Biological Services — 81/17, Washington, D.C., 155 pp.

Brinson, M.M., Bradshaw, H.D. and Jones, M.N., 1985. Transitions in forested wetlands along gradients of salinity and hydroperiod. *J. Elisha Mitchell Sci. Soc.*, 101: 76–94.

Broadfoot, W.M. and Toole, E.R., 1957. *Drought Effects on Southern Hardwoods*. U.S. Department of Agriculture Forest Service, Southern Forest Experiment Station Research Note 111, New Orleans, La.

Browder, J.A. and Volk, B.G., 1978. Systems model of carbon transformations in soil subsidence. *Ecol. Modelling*, 5: 269–292.

Brown, S., 1981. A comparison of the structure, primary productivity, and transpiration of cypress ecosystems in Florida. *Ecol. Monogr.*, 51: 403–427.

Brown, S., 1984. The role of wetlands in the Green Swamp. In: K.C. Ewel and H.T. Odum (Editors), *Cypress Swamps*. University Presses of Florida, Gainesville, Fla., pp. 405–415.

Brown, S., Brinson, M.M. and Lugo, A.E., 1979. Structure and function of riparian wetlands. In: *Strategies for Protection and Management of Floodplain Wetlands and Other Riparian Ecosystems*. U.S. Department of Agriculture, Forest Service, Gen. Tech. Rep. WO-12, Washington, D.C., pp. 17–31.

Brown, S., Lugo, A.E., Silander, S. and Liegel, L., 1983. *Research History and Opportunities in the Luquillo Mountains*. U.S. Department of Agriculture, Forest Service, Southern Forest Experiment Station, Gen. Tech. Rep. SO-44, 128 pp.

Brown, S.L., Cowles, S.W. and Odum, H.T., 1984. Metabolism and transpiration of cypress domes in north-central Florida. In: K.C. Ewel and H.T. Odum (Editors), *Cypress Swamps*. University Presses of Florida, Gainesville, Fla., pp. 145–163.

Bruenig, E.F., 1971. On the ecological significance of drought in the equatorial wet evergreen (rain) forests of Sarawak (Borneo). In: J.R. Flenley (Editor), *The Water Relations of Malesian Forests*. Transactions of the First Aberdeen–Hull Symposium on Malesian Ecology, Hull, England. Institute of Southeast Asian Biology, Univ. Aberdeen, Aberdeen, pp. 66–88.

Buell, M.F., 1946. Jerome bog, a peat-filled "Carolina bay". *Bull. Torrey Bot. Club*, 73(1): 24–33.

Burns, G.P., 1911. A botanical survey of the Huron River Valley. 8. Edaphic conditions in peat bogs of southern Michigan. *Bot. Gaz.*, 52: 105–125.

Capehart, B.L., Ewel, J.J., Sedlik, B.R. and Myers, R.L., 1977. Remote sensing of *Melaleuca*. *Photogramm. Eng. Remote Sensing*, 43: 198–206.

Carter, M.R., Burns, L.A., Cavinder, T.R., Dugger, K.R., Fore, P.L., Hicks, D.B., Revells, H.L. and Schmidt, T.W., 1973. *Ecosystem Analysis of the Big Cypress Swamp and Estuaries*. U.S. Environmental Protection Agency 904/9–74–002, Atlanta, Ga., 478 pp.

Caughey, M.C., 1945. Water relations of pocosin or bog shrubs. *Plant Physiol.*, 20: 671–689.

Chapman, V.J., 1976. *Mangrove Vegetation*. J. Cramer, 447 pp.

Christensen, N., Burchell, R.B., Liggett, A. and Simms, E.L., 1981. The structure and development of pocosin vegetation. In: C.J. Richardson (Editor), *Pocosin Wetlands*. Hutchinson Ross Publishing Co., Stroudsburg, Pa., pp. 43–61.

Clark, J.G.D. and Godwin, H., 1962. The Neolithic in the Cambridgeshire Fens. *Antiquity*, 36: 10–23.

Cohen, A.D., 1973. Possible influence of subpeat topography and sediment type upon the development of the Okefenokee swamp–marsh complex of Georgia. *Southeast. Geol.*, 15: 141–151.

Coulter, J.K., 1957. Development of the peat soils of Malaya. *Malay. Agric. J.*, 40: 188–199.

Crawford, R.M.M., 1967. Alcohol dehydrogenase activity in relation to flooding tolerance in roots. *J. Exp. Bot.*, 18: 458–464.

Crawford, R.M.M. and McManmon, M., 1968. Inductive responses of alcohol and malic dehydrogenases in relation to flooding tolerance in roots. *J. Exp. Bot.*, 19: 435–441.

Dachnowski, A., 1908. The toxic property of bog water and bog soil. *Bot. Gaz.*, 46: 130–143.

Dakshini, K.M.M., 1965. A study of the vegetation of Mothronwala swamp forest, Dehra Dun, India. *J. Ind. Bot. Soc.*, 44(4): 411–428.

Daniel, C.C., 1981. Hydrology, geology, and soils of pocosins: a comparison of natural and altered systems. In: C.J. Richardson (Editor), *Pocosin Wetlands*. Hutchinson Ross Publishing Company, Stroudsburg, Pa., pp. 69–108.

Davis, J.H., 1940. The ecology and geologic role of mangroves in Florida. *Pap. Tortugas Lab.*, 32: 302–412. Publication of the Carnegie Institution No. 517.

Davis, J.H., 1946. The peat deposits of Florida, their occurrence, development, and uses. *Fla. Geol. Surv. Bull.* No. 30, Tallahassee, Fla., 247 pp.

Deevey, E.S., 1970. In defense of mud. *Bull. Ecol. Soc. Am.*, 51(1): 4–8.

DeLaune, R.D., Pezeshki, S.R. and Patrick, W.H., Jr. 1987. Response of coastal plants to increase in submergence and salinity. *J. Coastal Res.*, 3: 535–546.

Denning, J.L., Hole, F.D. and Bouma, J., 1978. Effects of *Formica cenerea* on a wetland soil on West Blue Mound, Wisconsin. In: C.B. DeWitt and E. Soloway (Editors), *Wetlands: Ecology, Values, and Impacts*. Institute for Environmental Studies, University of Wisconsin, Madison, Wisc., pp. 276–287.

Dickson, J.G., 1978. Forest bird communities of the bottomland hardwoods. In: *Proceedings of the Workshop on Management of Southern Forests for Nongame Birds*. U.S. Department of Agriculture Forest Service, Gen. Tech. Rep. SE-14, Atlanta, Ga., pp. 66–73.

Dolman, J.D. and Buol, S.W., 1967. *A study of organic soils (histosols) in the tidewater region of North Carolina*. Tech. Bull. No. 181, North Carolina Agricultural Experiment Station, Raleigh, N.C., 52 pp.

Drew, M.C., 1983. Plant injury and adaptation to oxygen deficiency in the root environment: a review. *Plant Soil*, 75: 179–199.

Durno, S.E., 1961. Evidence regarding the rate of peat growth. *J. Ecol.*, 49: 347–351.

Dykeman, W.R. and Sousa, A.S., 1966. Natural mechanisms of copper tolerance in a swamp forest. *Can. J. Bot.*, 44: 871–878.

Eckblad, J.W., Peterson, N.L. and Ostlie, K., 1977. The morphometry, benthos and sedimentation rates of a floodplain lake in pool 9 of the upper Mississippi River. *Am. Midl. Nat.*, 97: 433–443.

Eggeling, W.J., 1935. The vegetation of Namanve swamp, Uganda. *J. Ecol.*, 23: 422–435.

Ellmore, G.S., Lee, S.C. and Nickerson, N.H., 1983. Plasticity expressed by root ground tissues of *Rhizophora mangle* L. (red mangrove). *Rhodora*, 85: 397–403.

Fenchel, T.M. and Jørgensen, B.B., 1977. Detritus food chains of aquatic ecosystems. In: M. Alexander (Editor), *Advances in Microbial Ecology*, 1: 1–56.

Fenton, J.H.C., 1980. The rate of peat accumulation in Antarctic moss banks. *J. Ecol.*, 68: 211–228.

Firouz, E., 1974. *Environment Iran*. The National Society for the Conservation of Natural Resources and Human Environment, Tehran, 51 pp.

Fittkau, E.J., Irmler, U., Junk, W.J., Reiss, F. and Schmidt, G.W., 1975. Productivity, biomass, and population dynamics in Amazonian water bodies. In: F.B. Golley and E. Medina (Editors), *Tropical Ecological Systems*. Springer-Verlag, New York, N.Y., pp. 289–311.

Food and Agriculture Organization (FAO), 1981. *Los Recursos Forestales de la América Tropical*. UN 32/6.1301-78-04, Informe Técnico 1, FAO, Rome, 343 pp.

Frangi, J., 1983. Las tierras pantanosas de la montaña Puertorriqueña. In: A.E. Lugo (Editor), *Los Bosques de Puerto Rico*. U.S. Department of Agriculture Forest Service, Institute of Tropical Forestry, and Puerto Rico Department of Natural Resources, San Juan, Puerto Rico, pp. 233–247.

Frangi, J.L. and Lugo, A.E., 1985. Ecosystem dynamics of a subtropical floodplain forest. *Ecol. Monogr.*, 55(3): 351–369.

Frangi, J. and Ponce, M., 1985. The root system of *Prestoea montana* and its ecological significance. *Principes*, 29: 13–19.

Frayer, W.E., Monahan, T.J., Bowden, D.C. and Graybill, F.A., 1983. *Status and Trends of Wetlands and Deepwater Habitats in the Conterminous United States, 1950's to 1970's*. Department of Forest and Wood Sciences, Colorado State University, Fort Collins, Colo., 32 pp.

Frey, D.G., 1954. Evidence for the recent enlargement of the "bay" lakes of North Carolina. *Ecology*, 35: 78–88.

Furtado, J.I. and Mori, S. (Editors), 1982. *Tasek Bera*. Dr. W. Junk Publishers, The Hague, 413 pp.

Gaines, D.A., 1974. A new look at the nesting riparian avifauna of the Sacramento Valley, California. *West. Birds*, 5: 61–80.

Gardner, L.R., Chen, H.S. and Settlemyre, J.L., 1978. Comparison of trace metals in South Carolina floodplain and marsh sediments. In: D.C. Adriano and I.L. Brisbin, Jr. (Editors), *Environmental Chemistry and Cycling Processes*. U.S. Department of Energy Symposium Series 45, CONF-760429, National Technical Information Service, Springfield, Va., pp. 446–461.

Gates, F.C., 1914. Winter as a factor in the xerophylly of certain evergreen ericads. *Bot. Gaz.*, 57: 445–489.

Gentry, J.B., Golley, F.B. and Smith, M.H., 1971. Yearly fluctuations in small mammal populations in a southeastern United States hardwood forest. *Acta Theriol.*, 15(12): 179–190.

Gill, D., 1972. The point bar environment in the Mackenzie River Delta. *Can. J. Earth Sci.*, 9: 1382–1393.

Gilliam, J.W. and Skaggs, R.W., 1981. Drainage and agricultural development: effects on drainage waters. In: C.J. Richardson (Editor), *Pocosin Wetlands*. Hutchinson Ross Publishing Company, Stroudsburg, Pa., pp. 109–124.

Golley, F.B., McGinnis, J.T., Clements, R.G., Child, G.I. and Duever, M.J., 1975. *Mineral Cycling in a Tropical Moist Forest Ecosystem*. Univ. of Georgia Press, Athens, Ga., 248 pp.

Gosselink, J.G. and Turner, R.E., 1978. The role of hydrology in freshwater wetland ecosystems. In: R.E. Good, D.F. Whigham and R.L. Simpson (Editors), *Freshwater Wetlands: Ecological Processes and Management Potential*. Academic Press, New York, N.Y., pp. 63–78.

Goulding, M., 1980. *The Fishes and the Forest*. University of California Press, Berkeley, Calif., 280 pp.

Green, B.H. and Pearson, M.C., 1968. The ecology of Wybunbury moss, Cheshire. I. The present vegetation and some physical, chemical and historical factors controlling its nature and distribution. *J. Ecol.*, 56: 245–267.

Haapala, H., Sepponen, P. and Meskus, E., 1975. Effects of spring floods on water acidity in the Kiiminkijoki area, Finland. *Oikos*, 26: 26–31.

Hair, J.D., Hepp, G.T., Luckett, L.M., Reese, K.P. and Woodward, D.K., 1979. Beaver pond ecosystems and their relationships to multi-use natural resource management. In: *Strategies for Protection and Management of Floodplain Wetlands and Other Riparian Ecosystems*. U.S. Department of Agriculture Forest Service, Gen. Tech. Rep. WO-12, Washington, D.C., pp. 80–92.

Harms, W.R., Schreuder, H.T., Hook, D.D., Brown, C.L. and Shropshire, F.W., 1980. The effects of flooding on the swamp forest in Lake Oklawaha, Florida. *Ecology*, 61: 1412–1421.

Hay, D. and Ragsdale, H.L., 1978. Patterns of cesium-137 distribution across two disparate floodplains. In: D.C. Adriano and I.L. Brisbin, Jr. (Editors), *Environmental Chemistry and Cycling Processes*. U.S. Department of Energy Symposium Series 45, CONF-760429, National Technical Information Service, Springfield, Va., pp. 462–478.

Heatwole, H. and Sexton, O.J., 1966. Herpetofaunal comparisons between two climatic zones in Panama. *Am. Midl. Nat.*, 75: 45–60.

Heimburg, K., 1984. Hydrology of north-central Florida cypress domes. In: K.C. Ewel and H.T. Odum (Editors), *Cypress Swamps*. University Presses of Florida, Gainesville, Fla., pp. 72–82.

Heinselman, M.L., 1970. Landscape evolution: peatland types and the environment in the Lake Agassiz Peatlands Natural Area, Minnesota. *Ecol. Monogr.*, 40: 235–261.

Hibbert, F.A. and Switsur, V.R., 1976. Radiocarbon dating of Flandrian pollen zones in Wales and northern England. *New Phytol.*, 77: 793–807.

Hook, D.D. and Brown, C.L., 1972. Permeability of the cambium to air in trees adapted to wet habitats. *Bot. Gaz.*, 133: 304–310.

Hook, D.D. and Crawford, R.M.M. (Editors), 1978. *Plant Life in Anaerobic Environments*. Ann Arbor Science, Ann Arbor, Mich., 564 pp.

Hook, D.D. and Scholtens, J.R., 1978. Adaptations and flood tolerance of tree species. In: D.D. Hook and R.M.M. Crawford (Editors), *Plant Life in Anaerobic Environments*. Ann Arbor Science, Ann Arbor, Mich., pp. 299–331.

Hook, D.D., Brown, C.L. and Kormanik, P.P., 1970. Lenticel and water root development of swamp tupelo under various flooding conditions. *Bot. Gaz.*, 131: 217–224.

Hopkins, J.M., 1947. Forty-five years with the Okefenokee swamp. *Bull. Ga. Soc. Nat.*, 4: 1–69.

Horwitz, E.L., 1978. *Our Nation's Wetlands*. U.S. Council on Environmental Quality, U.S. Government Printing Office, Washington, D.C., 70 pp.

Houghton, R.A., Hobbie, J.E., Melillo, J.M., Moore, B., Peterson, B.J., Shaver, G.R. and Woodwell, G.M., 1983. Changes in the carbon content of terrestrial biota and soils between 1860 and 1980: a net release of CO^2 to the atmosphere. *Ecol. Monogr.*, 53: 235–262.

Howarth, R.W. and Teal, J.M., 1980. Energy flow in a salt marsh ecosystem: the role of reduced inorganic sulphur compounds. *Am. Nat.*, 116: 862–872.

Hutchinson, J.N., 1980. The record of peat wastage in the East Anglian Fenlands at Holme Post, 1849–1978 A.D. *J. Ecol.*, 68: 229–249.

Ingram, H.A.P., 1967. Problems of hydrology and plant distribution in mires. *J. Ecol.*, 55: 711–724.

Isakov, Y.U., 1968. The status of waterfowl population breeding in the USSR and wintering in S.W. Asia and Africa. In: *Proceedings of a Technical Meeting on Wetland Conservation*. IUCN Publications, New Ser. (12), Morges, pp. 175–186.

Jackson, M.B. and Drew, M.C., 1984. Effects of flooding on growth and metabolism of herbaceous plants. In: T.T. Kozlowski (Editor), *Flooding and Plant Growth*. Academic Press, Orlando, Fla., pp. 47–128.

Jackson, N.M., Jr., 1976. Magnitude and frequency of floods in North Carolina. U.S. Geological Survey Resources Investigations, 76–17. Raleigh, N.C., 26 pp.

Jiménez, J.A., Lugo, A.E. and Cintrón, G., 1985. Tree mortality in mangrove forests. *Biotropica*, 17(3): 177–185.

Johnson, R.R., Haight, L.T. and Simpson, J.M., 1977. Endangered species vs. endangered habitats: a concept. In: *Importance, Preservation, and Management of Riparian Habitat: A Symposium*. U.S. Department of Agriculture Forest Service, Gen. Tech. Rep. RM-43, Fort Collins, Colo., pp. 68–78.

Johnson, W.C., Burgess, R.L. and Keammerer, W.R., 1976. Forest overstory vegetation on the Missouri River floodplain in North Dakota. *Ecol. Monogr.*, 46: 59–84.

Johnston, C.A., Bubenzer, G.D., Lee, G.B., Madison, F.W. and McHenry, J.R., 1984. Nutrient trapping by sediment deposition in a seasonally flooded lakeside wetland. *J. Environ. Qual.*, 13: 283–290.

Keeley, J.E., 1979. Population differentiation along a flood frequency gradient: physiological adaptations to flooding in *Nyssa sylvatica*. *Ecol. Monogr.*, 49: 89–108.

Kemp, W.M., 1977. *Energy Analysis and Ecological Evaluation* of a Coastal Power Plant. PhD Dissertation, Univ. Florida, Gainesville, Fla., 560 pp.

Kozlowski, T.T., 1983. Plant responses to flooding of soil. BioScience, 34: 162–166.

Kozlowski, T.T. (Editor), 1984. Flooding and Plant Growth. Academic Press, Orlando, Fla., 356 pp.

Kuenzler, E.J., Mulholland, P.J., Ruley, L.A. and Sniffen, R.P., 1977. Water quality in North Carolina coastal plain streams and effects of channelization. Rep. No. 127, Water Resources Research Institute, University of North Carolina, Raleigh, N.C., 160 pp.

Lahde, E., 1966. Vertical distribution of biological activity in peat of some virgin and drained swamp types. Acta For. Fenn., 81: 1–15.

Lanly, J.P., 1982. Tropical forest resources. FAO Forestry Paper 30, Food and Agriculture Organization, Rome, 106 pp.

Leisman, G.A., 1953. The rate of organic matter accumulation on the sedge mat zones of bogs in the Itasca State Park region of Minnesota. Ecology, 34: 81–101.

Leisman, G.A., 1957. Further data on the rate of organic matter accumulation in bogs. Ecology, 38: 361.

Leyton, L. and Rousseau, L.Z., 1958. Root growth of tree seedlings in relation to aeration. In: K.V. Thimann (Editor), Physiology of Forest Trees. The Ronald Press Co., New York, N.Y., Chapter 23.

Likens, G.E., Mackenzie, F.T., Richey, J.E., Sedell, J.R. and Turekian, K.T., 1981. Flux of Organic Carbon by Rivers to the Oceans. CONF-8009140, National Technical Information Service, Springfield, Va., 397 pp.

Lindberg, R.D. and Runnells, D.D., 1984. Ground water redox reactions: analysis of equilibrium state applied to Eh measurements and geochemical modeling. Science, 225: 925–927.

Livingston, B.E., 1904. Physical properties of bog water. Bot. Gaz., 37: 383–385.

Livingston, B.E., 1905. Physiological properties of bog water. Bot. Gaz., 39: 348–355.

Longman, K.A. and Jenik, J., 1974. Tropical Forest and its Environment. Longman Group Limited, London, 196 pp.

Lorio, P.L., Jr., Howe, V.K. and Martin, C.N., 1972. Loblolly pine rooting varies with microrelief on wet sites. Ecology, 53: 1134–1140.

Loveless, A.R., 1962. Further evidence to support a nutritional interpretation of sclerophylly. Ann. Bot. (New Ser.), 26: 551–561.

Lowrance, R., Todd, R., Fail, J., Hendrickson, O., Jr., Leonard, R. and Asmussen, L., 1984. Riparian forests as nutrient filters in agricultural watersheds. Bioscience, 34: 374–377.

Lugo, A.E., 1973. Ecological management of south Florida range ecosystems for maximum environmental quality. In: Range Resources of the Southeastern United States. American Society of Agronomy and Crop Society of America, Madison, Wisc., pp. 57–68.

Lugo, A.E., 1978. Stress and ecosystems. In: J.H. Thorp and J.W. Gibbons (Editors), Energy and Environmental Stress in Aquatic Ecosystems. U.S. Department of Energy Symposium Series. CONF-771114, National Technical Information Services, Springfield, Va., pp. 62–101.

Lugo, A.E., 1980. Mangrove ecosystems: successional or steady state? Biotropica, 12 (Suppl., 2): 65–72.

Lugo, A.E. and Brinson, M.M., 1978. Calculation of the value of saltwater wetlands. In: P.E. Greeson, J.R. Clark and J.E. Clark (Editors), Wetland Functions and Values: the State of Our Understanding. American Water Resources Association, Minneapolis, Minn., pp. 120–130.

Lugo, A.E. and Brown, S., 1984. The Oklawaha River forested wetlands and their response to chronic flooding. In: K.C. Ewel and H.T. Odum (Editors), Cypress Swamps. University of Florida Presses, Gainesville, Fla., pp. 356–373.

Lugo, A.E. and Patterson-Zucca, C., 1977. The impact of low-temperature stress on mangrove structure and growth. Trop. Ecol., 18: 149–161.

Lugo, A.E., Cintrón, G. and Goenaga, C., 1981. Mangrove ecosystems under stress. In: G.W. Barrett and R. Rosenberg (Editors), Stress Effects on Natural Ecosystems. John Wiley and Sons, New York, N.Y., pp. 129–153.

MacNae, W., 1968. A general account of the fauna and flora of mangrove swamps and forests in the Indo-West Pacific region. Adv. Mar. Biol., 6: 73–270.

McCarthy, J., 1962. The colonization of a swamp forest clearing (with special reference to Mitragyna stipulosa). East Afr. Agric. For. J., 25(1): 22–28.

McColl, R.H.S., 1978. Chemical runoff from pasture: the influence of fertilizer and riparian zones. N.Z. J. Mar. Freshwater Res., 12: 371–380.

McDowell, L.L., Stephens, J.C. and Stewart, E.H., 1969. Radiocarbon chronology of the Florida Everglades peat. Soil Sci. Soc. Am. Proc., 33: 743–745.

McKee, W.H., Jr., Hook, D.D., DeBell, D.S. and Askew, J.L., 1984. Growth and nutrient status of loblolly pine seedlings in relation to flooding and phosphorus. Soil Sci. Soc. Am. J., 48: 1438–1442.

McManmon, M. and Crawford, R.M.M., 1971. A metabolic theory of flooding tolerance: the significance of enzyme distribution and behaviour. New Phytol., 70: 299–306.

Matthews, E. and Fung. I., 1987. Methane emission from natural wetlands: global distribution, area, and environmental characteristics of sources. Global Biochem. Cycles, 1: 61–86.

Meehan, W.R., Swanson, F.J. and Sedell, J.R., 1977. Influences of riparian vegetation on aquatic ecosystems with particular references to salmonid fishes and their food supply. In: Importance, Preservation, and Management of Riparian Habitat: A Symposium. U.S. Department of Agriculture Forest Service, Gen. Tech. Rep. RM-43, Fort Collins, Colo., pp. 137–145.

Miller, W.D. and Maki, T.E., 1957. Planting pines in pocosins. J. For., 55(9): 659–663.

Mitsch, W.J., Dorge, C.L. and Wiemhoff, J.R., 1979a. Ecosystem dynamics and a phosphorus budget of an alluvial cypress swamp in southern Illinois. Ecology, 60: 1116–1124.

Mitsch, W.J., Rust, W., Behnke, A. and Lai, L., 1979b. Environmental Observations of a Riparian Ecosystem during Flood Season. Water Resources Center, University of Illinois, Urbana-Champaign, Ill.

Monschein, T.D., 1981. Values of pocosins to game and fish species in North Carolina. In: C.J. Richardson (Editor), Pocosin Wetlands. Hutchinson Ross Publishing Co., Stroudsburg, Pa., pp. 155–170.

Montague, K.A. and Day, F.P., Jr. 1980. Belowground biomass of four plant communities of the Great Dismal Swamp, Virginia. *Am. Midland Nat.*, 103: 83–87.

Moore, P.D., 1975. Origin of blanket mires. *Nature*, 256: 267–269.

Moore, P.D. and Bellamy, D.J., 1974. *Peatlands*. Springer-Verlag, New York, N.Y., 221 pp.

Mortimer, C.H., 1941. The exchange of dissolved substances between mud and water in lakes. *J. Ecol.*, 29: 280–329.

Mortimer, C.H., 1942. The exchange of dissolved substances between mud and water in lakes. *J. Ecol.*, 30: 147–201.

Munch, J.C. and Ottow, J.C.G., 1983. Reductive transformation mechanism of ferric oxides in hydromorphic soils. *Ecol. Bull.* (Stockholm), 35: 384–394.

Nedwell, D.B., 1982. The cycling of sulphur in marine and freshwater sediments. In: D.B. Nedwell and C.M. Brown (Editors), *Sediment Microbiology*. Academic Press, London, pp. 73–106.

Newbould, P.J., 1967. *Methods for Estimating the Primary Production of Forests*. IBP Handbook No. 2. Blackwell Book Co., New York, N.Y., 62 pp.

Novitski, R.P., 1979. Hydrologic characteristics of Wisconsin's wetlands and their influence on floods, stream flow, and sediment. In: P.E. Greeson, J.R. Clark and J.E. Clark (Editors), *Wetland Functions and Values: the State of our Understanding*. American Water Resources Association, Minneapolis, Minn., pp. 377–388.

Odum, E.P., 1969. The strategy of ecosystem development. *Science*, 164: 262–270.

Odum, H.T., 1967. Work circuits and systems stress. In: H.E. Young (Editor), *Symposium on Primary Productivity and Mineral Cycling in Natural Ecosystems*. University of Maine Press, Orono, Me., pp. 81–138.

Odum, H.T., 1983. *Systems Ecology*. John Wiley and Sons, New York, N.Y., 644 pp.

Odum, H.T., Kemp, W., Sell, M., Boynton, W. and Lehman, M., 1977. Energy analysis and the coupling of man and estuaries. *Environ. Manage.*, 1(4): 297–315.

Odum, W.E., 1970. *Pathways of Energy Flow in a South Florida Estuary*. PhD Dissertation, University of Miami, Miami, Fla., 162 pp.

Olson, J.S., Watts, J.A. and Allison, L.J., 1983. *Carbon in Live Vegetation of Major World Ecosystems*. DOE/NBB-0037, National Technical Information Service, Springfield, Va., 152 pp. and map.

Parent, L.E., Millette, J.A. and Mehuys, G.R., 1982. Subsidence and erosion of a histosol. *Soil Sci. Soc. Am. J.*, 46: 404–408.

Penfound, W.T., 1952. Southern swamps and marshes. *Bot. Rev.*, 18: 413–446.

Peterjohn, W.T. and Correll, D.L., 1984. Nutrient dynamics in an agricultural watershed: observations on the role of a riparian forest. *Ecology*, 65(5): 1466–1475.

Peterson, B.J., Howarth, R.W., Litshultz, F. and Ashendorf, D., 1980. Salt marsh detritus: an alternative interpretation of stable carbon isotope ratios and the fate of *Spartina alterniflora*. *Oikos*, 34: 173–177.

Philipson, J.J. and Coutts, M.P., 1978. The tolerance of tree roots to waterlogging. III. Oxygen transport in lodgepole pine and Sitka spruce roots of primary structure. *New Phytol.*, 80: 341–349.

Ponnamperuma, F.N., 1972. The chemistry of submerged soils. *Adv. Agron.*, 22: 29–96.

Reddy, K.R. and Patrick, W.H., Jr., 1975. Effect of alternate aerobic and anaerobic conditions on redox potential, organic matter decomposition and nitrogen loss in a flooded soil. *Soil Biol. Biochem.*, 7: 87–94.

Rice, D.L., 1982. The detritus nitrogen problem: new observations and perspectives from organic geochemistry. *Mar. Ecol. Prog. Ser.*, 9: 153–162.

Richards, K., 1982. *Rivers: Form and Process in Alluvial Channels*. Methuen and Co., London, 358 pp.

Rigg, G.B. and Gould, H.R., 1957. Age of Glacier Peak eruption and chronology of post-glacial peat deposits in Washington and surrounding areas. *Am. J. Sci.*, 155: 341–363.

Saenger, P., Hegerl, E. and Davie, J.D.S. (Editors), 1983. Global status of mangrove ecosystems. *Environmentalist*, 3, suppl., 88 pp.

Scheller, U., 1979. *Hanseniella arborea* n. sp., a migrating symphylan from an Amazonian blackwater inundation forest (myriapoda, Symphyla, Scutigerellidae). *Acta Amazónica*, 9(3): 603–607.

Schimper, A.F.W., 1898. *Plant Geography upon a Physiological Basis*. (English translation, 1903.) Clarendon Press, Oxford, 839 pp.

Schlesinger, W.H., 1984. Soil organic matter: a source of atmospheric CO_2? In: G.M. Woodwell (Editor), *The Role of the Terrestrial Vegetation in the Global Carbon Cycle: Methods for Appraising Changes*. John Wiley and Sons, New York, N.Y., pp. 111–127.

Schlesinger, W.H. and Chabot, B.F., 1977. The use of water and minerals by evergreen and deciduous shrubs in Okefenokee Swamp. *Bot. Gaz.*, 138: 490–497.

Schumm, S.A. and Lichty, R.W., 1963. *Channel Widening and Floodplain Construction along Cimarron River in Southwestern Kansas*. U.S. Geol. Surv. Prof. Pap. 352-D, Washington, D.C.

Schwintzer, C.R., 1978. Vegetation changes and water levels in a small Michigan bog. In: C.B. DeWitt and E. Soloway (Editors), *Wetlands: Ecology, Values, and Impacts*. Institute for Environmental Studies, University of Wisconsin, Madison, Wisc., pp. 326–336.

Selye, H., 1956. *The Stress of Life*. McGraw-Hill Book Co., New York, N.Y., 324 pp.

Shure, D.J. and Gottschalk, M.R., 1978. Radiocesium transfer through aerial pathways in a South Carolina floodplain forest. In: D.C. Adriano and I.L. Brisbin, Jr. (Editors), *Environmental Chemistry and Cycling Processes*. U.S. Department of Energy Symposium Series 45, CONF-760429, National Technical Information Service, Springfield, Va., pp. 709–724.

Siegel, D.I. and Glaser, P.H., 1987. Groundwater flow in a bog-fen complex, Lost River Peatland, northern Minnesota. *J. Ecol.*, 74: 743–754.

Skoropanov, S.G., 1968. *Reclamation and Cultivation of Peat-bog Soils*. Translated from Russian by N. Kramer. Israel Program for Scientific Translations, 234 pp.

Small, E., 1972. Water relations of plants in raised sphagnum peat bogs. *Ecology*, 53: 726–728.
Small, E., 1973. Xeromorphy in plants as a possible basis for migration between arid and nutritionally-deficient environments. *Bot. Notiser*, 126: 534–539.
Smith, D.G., 1976. Effect of vegetation on lateral migration of anastomosed channels of a glacier meltwater river. *Bull. Geol. Soc. Am.*, 87: 857–860.
Snedaker, S.C., 1982. Mangrove species zonation: why? In: D.N. Sen and K.S. Rajpurohit (Editors), *Tasks for Vegetation Science*, Volume 2. Dr. W. Junk Publishers, The Hague, pp. 111–125.
Sørensen, J., Jørgensen, B.B. and Revsbech, N.P., 1979. A comparison of oxygen, nitrate, and sulfate respiration in coastal marine sediments. *Microb. Ecol.*, 5: 105–115.
Starrett, W.C., 1972. Man and the Illinois river. In: P.T. Oglesby, C.A. Carlson and J.A. McCann (Editors), *River Ecology and Man*. Academic Press, New York, N.Y., pp. 131–169.
Stephens, J.C., 1956. Subsidence of organic soils in the Florida Everglades. *Soil Sci. Soc. Am. Proc.*, 20: 77–80.
Stephens, J.C., 1969. Peat and muck drainage problems. *J. Irrigation Drainage Div. Proc. Am. Soc. Civ. Eng.*, 95(IR2): 285–305.
Stevens, L.E., Brown, B.T., Simpson, J.M. and Johnson, R.R., 1977. The importance of riparian habitat to migrating birds. In: *Importance, Preservation and Management of Riparian Habitat: A Symposium*. U.S. Department of Agriculture, Forest Service, Gen. Tech. Rep. RM-43, Fort Collins, Colo., pp. 156–164.
Stewart, P.A. and Merrell, W.D., 1937. The Bergen swamp: an ecological study. *Proc. Rochester Acad. Sci.*, 8: 209–262.
Straney, D.O., Briese, L.A. and Smith, M.H., 1974. Bird diversity and thermal stress in a cypress swamp. In: J.W. Gibbons and R.R. Sharitz (Editors), *Thermal Ecology*. CONF-730505, NTIS, Springfield, Va., pp. 572–578.
Street, M.W. and McCless, J.D., 1981. North Carolina's coastal fishing industry and the influence of coastal alterations. In: C.J. Richardson (Editor), *Pocosin Wetlands*. Hutchinson Ross Publishing Co., Stroudsburg, Pa., pp. 238–251.
Swift, B.L., Larson, J.S. and DeGraaf, R.M., 1984. Relationship of breeding bird density and diversity to habitat variables in forested wetlands. *Wilson Bull.*, 96(1): 48–59.
Thompson, J.C., 1957. Conditions on irrigated sections in the middle Rio Grande in New Mexico. In: P.C. Duisberg (Editor), *Problems of the Upper Rio Grande: An Arid Zone River*. Publ. No. 1., U.S. Commission for Arid Resource Improvement and Development, American Association for the Advancement of Science, Washington, D.C.
Turner, R.E., 1977. Intertidal vegetation and commercial yield of penaeid shrimp. *Trans. Am. Fish. Soc.*, 106: 411–416.
Turner, R.E. and May, N., 1977. An alternative evaluation of the fishery value of tropical and subtropical wetlands. *Actas del Cuarto Simposio Internacional de Ecologia Tropical*. Universidad de Panamá, Panamá, 3: 837–852.
Twilley, R.R., 1982. *Litter Dynamics and Organic Carbon Exchange in Black Mangrove* (Avicennia germinans) *Basin Forests in a Southwest Florida Estuary*. PhD Dissertation, University of Florida, Gainesville, Fla., 260 pp.
Twilley, R.R., Lugo, A.E. and Patterson-Zucca, C., 1987. Litter production and turnover in basin mangrove forests in southwest Florida. *Ecology*, 67(3): 670–683.
Van Cleve, K. and Viereck, L.A., 1981. Forest succession in relation to nutrient cycling in the boreal forest of Alaska. In: D.C. West, H.H. Shugart and D.B. Botkin (Editors), *Forest Succession, Concepts and Application*. Springer-Verlag, New York, N.Y., Chapter 13.
Viosca, P., 1928. Louisiana wetlands and the value of their wildlife and fisheries resources. *Ecology*, 9: 216–229.
Walker, D., 1970. Direction and rate in some British postglacial hydroseres. In: D. Walker and R.G. West (Editors), *Studies in the Vegetational History of the British Isles*. Cambridge University Press, Cambridge, pp. 117–139.
Walker, D., 1972. Vegetation of the Lake Ipea region, New Guinea highlands. II. Kayamanda swamp. *J. Ecol.*, 60: 479–499.
Wasawo, D.P.S. and Visser, S.A., 1959. Swamp worms and tussock mounds in the swamps of Teso, Uganda. *East Afr. Agric. J.*, 25: 86–90.
Weir, W.W., 1950. Subsidence of peat lands of the Sacramento–San Joaquin Delta, California. *Hilgardia*, 20: 37–56.
Welcomme, R.L., 1979. *Fisheries Ecology of Floodplain Rivers*. Longmann, New York, N.Y., 317 pp.
West, D.C., Shugart, H.H. and Botkin, D.B. (Editors), 1981. *Forest Succession, Concepts and Application*. Springer-Verlag, New York, N.Y., 517 pp.
Wharton, C.H., 1971. *Hearing on Dredging, Modification, and Channelization of Rivers and Streams*. Congress of the United States, Committee on Government Operations, Sub-Committee on Conservation and Natural Resources, Washington, D.C., 265 pp.
Whitehead, D.R., 1972. Developmental and environmental history of the Dismal Swamp. *Ecol. Monogr.*, 42: 301–315.
Whitehead, D.R., 1981. Late-Pleistocene vegetational changes in northeastern North Carolina. *Ecol. Monogr.*, 51: 451–471.
Whitmore, T.C., 1975. *Tropical Rainforests of the Far East*. Oxford University Press, Oxford, 282 pp.
Whittaker, R.H. and Likens, G.E., 1973. Carbon in the biota. In: G.M. Woodwell and E.V. Pecan (Editors), *Carbon and the Biosphere*. CONF-720510, National Technical Information Service, Springfield, Va., pp. 281–302.
Wolman, M.G. and Leopold, L.B., 1957. *River Flood Plains: Some Observations on their Formation*. U.S. Geol. Surv. Prof. Pap. 282-C, Washington D.C., 107 pp.
Wright, A.H. and Wright, A.A., 1932. The habitats and composition of the vegetation of Okefenokee swamp, Georgia. *Ecol. Monogr.*, 2: 109–232.
Yates, P. and Sheridan, J.M., 1983. Estimating the effectiveness of vegetated floodplains/wetlands as nitrate-nitrite and orthophosphorus filters. *Agric. Ecosystems Environm.*, 9: 303–314.
Yee Mun, L., 1984. *Energetics of Leaf Litter Production and its Pathway through the Sesarmid Crabs in a Mangrove Ecosystem*. Masters Thesis, University Sains Malaysia, Penang, 140 pp.
Young, F.N., 1967. A possible recycling mechanism in tropical forests. *Ecology* 48(3): 506.
Yurkevich, I.D., Smolyak, L.P. and Garin, B.E., 1966. Content of oxygen in the soil water and of carbon dioxide in the soil air of forest bogs. *Sov. Soil Sci.*, 2: 159–167.
ZoBell, C.E., 1946. Studies on redox potential of marine sediments. *Bull. Am. Assoc. Pet. Geol.*, 30: 477–513.

Chapter 5

RIVERINE FORESTS

MARK M. BRINSON

INTRODUCTION

The floodplains of large rivers give the impression of supporting very high amounts of biomass, and having dynamic nutrient cycling and high primary productivity. This is especially true of the forests that have developed in the humid tropics and warm, humid temperate zones. The impressive stature of the trees in these forests and, at times, their high stem density are convincing evidence of the large amount of biomass that they support. The dynamic nature is suggested by their fluctuating water levels driven by the seasonal floodwaters from the river. Whereas upland forests may attain the stature of riverine forests, uplands lack episodes of flooding and thus appear to be more static.

Riverine forests are those that owe their dynamics, structure, and composition to river processes of inundation, transport of sediments, or the abrasive and erosive forces of water and ice movement. The typical riverine forest of a large river begins at the natural levee next to the channel where coarse-grained deposits result in soils that drain quickly. These levees support the "gallery" or streamside forests which develop high leaf area indices because of the edge effect created by the stream channel. Toward the floodplain interior, surface elevation decreases and the soil becomes more poorly drained because the groundwater table is closer to the soil surface. Many floodplains have ridge-and-swale topography throughout, where relict point bars[1], levees, and channels were abandoned by the meandering stream. When this undulating topography occurs along a transect from stream channel to upland, community types likewise alternate between topographic lows with species adapted to long hydroperiods and topographic highs with species also found in mesic uplands. Hydroperiod in places may be too long and water depth too great for trees to become established. This is where shallow lakes or herbaceous wetland communities are located. Narrower floodplains that lack the topographic variation just described often show a gradual transition from the longest hydroperiod and most hydric plant communities at the stream-side to infrequently flooded forests at the boundary between the floodplain and upland ecosystem. In some cases where a floodplain is poorly developed, streamside forests differ from uplands merely by having greater moisture availability from groundwater seepage. However, this seepage is derived ultimately from geologic riverine processes through intersection of the water table by the stream channel, thus providing the setting for forests to develop. In fact, it is often the infrequent, catastrophic floods that are most influential in changing channel and floodplain morphometry, through massive erosion and deposition of sediments. The more frequent floods do not have the power to develop the complex structural features of most floodplains. However, some rivers are more "youthful" than others, so that floods of lower power cause greater changes in landform of floodplains in comparison with more mature and stable river systems.

The "power line" designation developed by Kangas for riverine forests (Chapter 2) can be developed further by considering the power and frequency of inundation and the way in which flood events organize the ecosystem. Flood power

---

[1] A point bar is formed by the deposition of sediment on the inner side of a curve in a river — see Fig. 5.3 below.

and frequency of inundation are inversely proportional and exist in a continuum from high-power, low-frequency floods that affect the whole floodplain to low-power, high-frequency floods that influence only lower elevations. It is the high-power, low-frequency type that determine patterns of the large geomorphic features that persist on the order of hundreds to thousands of years. These features include oxbow lakes and relic levees that result in the commonly observed ridge and swale topography. Medium-power, intermediate-frequency floods determine patterns of ecosystem structure that have lifetimes of tens to hundreds of years. Zonation of tree species associations are influenced at this scale because of similar generation times of many tree species. The low-power, high-frequency floods that occur annually determine short-term patterns such as seed germination and seedling survival, and activities of mammals, amphibians, fish, and reptiles. The generation times of both landforms and organisms are roughly proportional to the size of these features (Fig. 5.1). Thus large features and large flood events have long-term effects on ecosystem pattern and operate at the highest level of hierarchical control according to the power line concept. However, when viewed from a smaller scale, a floodplain contains a mixture of forest types. Some types receive flowthrough of floodwaters and sediment deposition, and thus have attributes of true riverine forests. Others are more isolated from the influence of the river and assume characteristics and species composition of basin forests.

Emphasis is placed on the wide variety of riverine forests that occur in the U.S.A. and Canada, ranging from floodplains in the Southeast, to arid climates in the Southwest, and even to streamside vegetation in the Arctic of Alaska. First the geomorphic and hydrologic settings will be discussed, because they are what make riverine forests unique. In this context, nutrient-rich alluvial soils and seasonal flooding are generally two conditions that, by occurring together, support high rates of production and demand adaptation to anaerobic conditions in the rhizosphere. This will be followed by an examination of the structure of the vegetation, with summaries of data on biomass production and nutrient cycling. The abundant data on structure from many geographic locations show that riverine forests range from

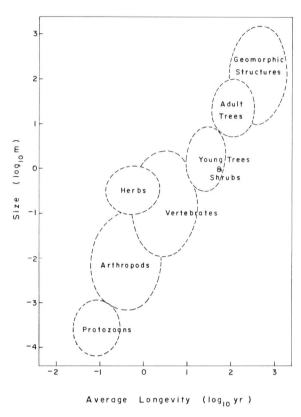

Fig. 5.1. Relationship between size and generation time of components in riverine forests.

those with closed canopies, abundant lianas, and large basal areas, to those whose sparsely occurring trees make them appear more as savannas. These latter ecosystems may be influenced as much by water deficits as by flooding, yet they are unquestionably influenced by riverine processes. Next, the geographic variation in riverine forests in the U.S.A. and Canada will be described. In spite of being categorized as a functional ecosystem type as a result of the unique features that they share, there are many differences in species composition and structure. Riverine forests in the tropics and subtropics are treated very briefly because other chapters are devoted exclusively to these zones. Finally, the alterations by humans that these ecosystems have received are described. These "experiments" (seldom with controls) yield convincing results on how dependent the forests are on riverine processes. By understanding the mechanisms of this dependency, there is an opportunity for constructively managing and protecting these ecosystems.

Other chapters in this volume describe riverine

forests in areas outside the U.S.A. and Canada. Riverine forests are described for the Caribbean (Ch. 10), Australia (Ch. 16), New Guinea (Ch. 14), Borneo (Ch. 13), Poland (Ch. 18) and Western Europe (Ch. 17).

GEOMORPHOLOGY

Floodplains upon which riverine forests develop are constantly undergoing change. A river valley from its headwaters to its mouth may be aggrading, degrading, or maintaining a steady state. Even when in a steady state or "poised" condition, where the downstream movement of floodplain deposits is balanced by transport of alluvium from upstream, stream channels will continue to meander laterally and downvalley so that morphologic features of the floodplain continue to change. This includes the development of oxbow lakes, ridge-and-swale topography, and other familiar floodplain features. Under non-steady-state conditions, the alluvial valley may be aggrading or degrading. The formation of the Mississippi alluvial valley (U.S.A.) has been reconstructed by Fisk and coworkers (Fisk, 1944, 1952; Fisk and McFarlan, 1955) and provides an example of the complexity of this process. The valley has undergone at least five periods of valley cutting and alluvial deposition that correspond with sea-level changes during the Quaternary. When sea level was low, erosion of an extensive valley system occurred across the Gulf Coastal Plain. As ice sheets retreated, sea level rose, and the entrenced valley system became alluviated during interglacial stages. Coarse material was introduced from steep tributaries which built alluvial cones of gravel and sands. When these materials reached the Mississippi, they were transported seaward and deposited over wide areas as aggradation occurred. With a thickening of the basal portion of the alluvium, stream gradients became reduced and deposited alluvium became progressively finer as both the quantity and grain size of the source material diminished. As sea level stabilized, the braided channel was replaced by a single meandering one through a combination of diminished load, smaller particle size, and deeper scouring. The details of change in riverine forests of the valley is not known. However, considering the array of environments in which riverine forests exist today, it is certain that forests have always been a part of the floodplain surface.

Smaller streams have been shown to undergo similar but less dramatic phases of downcutting and alluvial fill (Fig. 5.2). Factors that cause these shifts can be the result of change of base level as just described, tectonic activity, or change in climate and land use. Burnett and Schumm (1983) reported that tectonic uplifting has changed the valley floor of several streams in Louisiana and Mississippi. Above the axis of uplift, a reduction in valley slope has caused a reduction in channel slope and depth, thus resulting in more frequent flooding over the banks. In the region of uplift, channel slope and depth increased, thus causing the floodplain to become elevated relative to the stream. This incision has yielded increased sediment transport downstream from the uplift, and is expressed in the development of locally braided reaches and a new floodplain. Forests adapted to a particular hydrologic regime undoubtedly respond with subtle, long-term changes in species composition and dominance. Floodplain in tectonically active Andes has been proposed as a speciation mechanism for forest and aquatic communities of upper Amazonia (Räsänen et al., 1987).

Particularly for smaller floodplains, colluvium, or material transported from valley sides, can be a source of material for floodplain deposits. In narrow portions of floodplains, this material may be the dominant substrate for floodplain forests. Approximately one-fifth of the cross-sectional area of the alluvium of Beaverdam Run, Pennsylvania, consists of colluvium (Lattman, 1960). The remainder consists of channel fill, lag deposits (boulders), lateral accretion, and vertical accretion, including peaty material.

Flooding from channel overflow provides the opportunity for vertical accretion by depositing sediments derived from upstream (Allen, 1965). This sediment contributes to the notoriously high fertility of riverine forest soils. Even if upland soils that provide this source are relatively infertile, continual inputs of these deposits and their concentration in the relatively smaller floodplain surface should augment the nutrient capital of riverine forests relative to that of uplands, which depend solely on weathering and meteorologic inputs as sources of nutrients. The rate of deposition over the river banks may be associated,

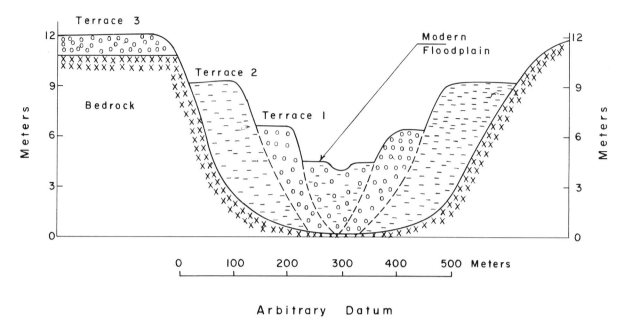

Fig. 5.2. Generalized cross-section of the valley of Fivemile Creek, Wyoming (U.S.A.) showing modern floodplain and remaining terraces from past periods of aggradation. Modern floodplain and Terrace 1 are mainly sand with gravel lenses and stringers; Terrace 2 is predominately very fine sand and gravel in contact with bedrock; Terrace 3 is mostly gravel overlying Cretaceous and Tertiary sedimentary rocks. No soil profile development in Terrace 1 and modern floodplain indicates youth of deposits. Adapted from Hadley (1960).

in part, with hydroperiod and the amount of suspended sediment load of the stream. While suspended sediment load varies in proportion to the erodibility of the catchment (Patric et al., 1984), local effects of hydroperiod are a function of floodplain topography and flood frequency. The recurrence intervals for bank-full flows for 19 streams in the U.S.A. summarized by Wolman and Leopold (1957) ranged from 1.07 to 4.0 years. However, this frequency is lower than that of floodplain inundation because of breaches in levees and backwater flooding. In fact, many floodplain areas are inundated several times a year. For sites in the bottomland forests of the White River basin in Arkansas, flooding occurs annually and persists for as much as 40% of the year (Bedinger, 1979).

Floodplains differ greatly in rate of deposition. Observations on the rate of vertical accretion in riverine forests range from a few millimeters per year to over a meter during a single flood episode (Ch. 4, Sect. 15). It cannot be determined from these values whether or not the standing stock of alluvium is increasing or decreasing, because few authors report rates of floodplain erosion. Beaverdam Run, Pennsylvania, which switched to an aggrading regime perhaps 200 years ago due to deforestation, consists of vertical accretion in at least the upper 2 m (Lattman, 1960). The floodplain of Cimarron River in southwestern Kansas has been undergoing accretion, most of which is vertical, at the rate of 2.1 cm yr^{-1} since a major flood widened the floodplain and destroyed channel meandering (Schumm and Lichty, 1963).

Some typical floodplain features are illustrated schematically for a section of the Mississippi River, Louisiana in Fig. 5.3. These include (a) natural levees adjacent to the channel which contain coarser material deposited during flood overflow; (b) meander scrolls and ridge-and-swale topography created by abandoned point bar deposits as the channel migrated laterally and downslope; (c) oxbow lakes which are relict meander bends that have been cut off; and (d) point bars on the inside curve of river bends where deposition is rapid. Levees support riverside gallery forests that may flood frequently, but the coarse deposits normally result in rapid drainage when water levels drop. Oxbow depressions and lakes are the most hydric of floodplain communities and support species adapted to constant flooding and anaerobic soils.

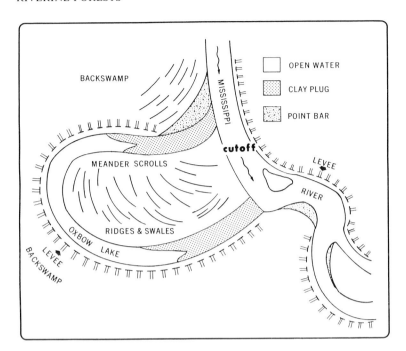

Fig. 5.3. Typical floodplain topographic features, illustrated diagrammatically, of the Mississippi River near False River, Louisiana (U.S.A.). Adapted from Fisk (1952).

Point bars, because of rapid deposition rates, are usually invaded by colonizing species which initiate successional development of floodplain forests. If the rate of meander movement occurs on a time scale similar to that of succession, it is possible in some cases to calculate the rate of lateral channel movement from the gradient of tree age in a transect perpendicular to the inside of the meander curve (Everitt, 1968). On the basis of successional development in a section of the Missouri River, South Dakota, it has been demonstrated that the youngest communities are located near the center of the floodplain, while the oldest ones are located at the edge (Johnson et al., 1976). Rivers in the southeastern Atlantic states of the U.S.A. appear to be migrating southward, as indicated by their proximity to bluffs on the south side and by the presence of broad floodplain on the north side of rivers. In extremely broad floodplains, such as the lower Mississippi River, large areas of the floodplain have not been occupied by the river channel in recent times.

Thermo-erosional processes are important agents in stream morphology and bank erosion in regions of permafrost (Church, 1977). Outhet (1974) classified bank types in the Mackenzie River delta (Canada) according to shape and erosional rates (Fig. 5.4). River channels in permafrost erode the bank on the outside of meanders as elsewhere; however, the development of thermo-erosional undercuts or "niches" and the presence of structural weaknesses (ice wedges and other forms of ground ice) result in large-scale sloughing to a somewhat greater extent than occurs in temperate environments. Although nearshore currents and thermal exchange are usually responsible for niche development, erosion by wave action may be significant where long open-water fetch exists. Continuous removal by high current velocities during summer is why type 1 banks (see Fig. 5.4) have higher rates of erosion than other types. Type 2 banks are a result of intermittent removal of material by variations in channel discharge or variation in wind velocity and direction. Type 3 banks result from soil flow where ice-rich faces retreat continuously through summer. Destruction of cut bank levees is accompanied by deposition along their back-slopes. Thus, the form of the levee is maintained without local destruction (Gill, 1972b). Only where thermo-erosional niches are active (type 1) is the floodplain destroyed without compensating alluviation.

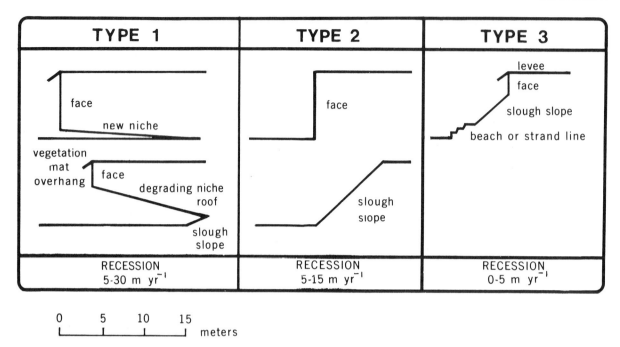

Fig. 5.4. Diagrammatic side view of bank shapes in permafrost environments and rates of erosion. Modified from Outhet (1974).

As one proceeds upstream from higher- to lower-order streams, the amount of material available for alluvial deposition decreases due to diminishing size of the catchment (Allen, 1965). Even where floodplains are negligible, streamside vegetation is often distinguishable from surrounding upland communities. Riverine forest vegetation may be simply a consequence of moister conditions where water tables approach the rooting depth of trees, thus resulting in more mesic conditions than the upland environment. Leopold et al. (1964) have suggested that in humid environments the upper limit of floodplain development may be the point at which flow in the channel changes from perennial to ephemeral — that is, where groundwater supply is insufficient to sustain flow between storm periods. In arid climates where intermittent streams are common because of protracted drought and high evapotranspiration, this criterion may not apply. The vicinity of headward gully erosion and gully wall collapse (Leopold and Miller, 1956) may represent the upper limit of floodplains. However, distinctive riverine forests continue upstream from the upper limit of floodplain development in both arid and humid climates.

HYDROLOGY

Riverine forests vary considerably from stream to stream, and even within sectors of a single stream. Because they all experience flooding and have topographic features of fluvial origin, the influence of hydrology on ecosystem structure and function will depend on the particular nature of the hydroperiod. Climate, topography, channel slope, soils and geology are all factors that may influence depth of flooding (see Ch. 4, Sect. 6). Many of these factors are involved in the gradient beginning in the eastern slope of the Appalachian Mountains (U.S.A.), continuing through the Piedmont province, and terminating in the coastal plain along the south Atlantic seaboard (Coble, 1979). In the mountainous headwater streams, catchments are small and slopes are steep. This results in constricted V-shaped valleys, and the shallow soils of drainage basins have limited storage capacity. Orographic rains result in greater precipitation than occurs at lower altitudes. Consequently hydrographic peaks are sharp and frequent, particularly toward the end of the winter season and into the spring, when evapotranspiration is low and soil water storage reaches annual highs. Riverine forests in this region consist of narrow bands of streamside river birch (*Betula nigra*). In the rolling topography of the Piedmont, flood peaks are the highest among the three physiographic provinces (Coble, 1979) and tend to occur when frontal weather systems stabilize over the region and provide abundant precipitation. Flash floods are less likely than in the mountains, partly because of large catchments and greater storage capacity of the deeply weathered soils. Coastal-plain rivers that originate in the Piedmont and mountains tend toward a prolonged hydrographic pulse as a result of integrating the upstream peaks. Floodplains in low elevations of the broad alluvial valleys may remain flooded for months at a time. For the Ouachita River, Arkansas, flood duration ranges from 10% to 18% of the year for sites having drainage areas from 13 000 to 18 000 km^2 and from 5% to 7% for sites having drainage areas less than 780 km^2 (Bedinger, 1979). If other factors could be held constant, communities with more flood-tolerant species should be more prevalent on floodplains of larger rivers than on small ones.

A fundamental issue is the degree to which the hydroperiod of floodplains influences the hydrograph of rivers. Where floodplains are narrow and alluvium mostly lacking there is little capacity for the floodplain to store water overflowing from the stream channel. On the other hand, where the alluvium is deep it may serve as both a source of water for the channel at low river stage and a recipient of water from the channel at high river stage. These reversals in hydraulic gradient between stream and floodplain have been documented for the lower Missouri River (Grannemann and Sharp, 1979). Other factors that control groundwater flows and levels are (Grannemann and Sharp, 1979):

(a) Distance from the river channel. The piezometric pressure changes more slowly farther from the river than close to it.

(b) Time elapsed since the river has risen or fallen. Provided the river stage does not overtop the levee system, a sustained flood peak will contribute more water to the groundwater system than a higher flood of shorter duration.

(c) Geometry of river meander and valley walls. Near sharp river meanders where an area of floodplain is bounded on more than one side or where floodplain segments are narrow, river stage and groundwater levels are more tightly coupled.

(d) Composition of alluvium. Thick clay strata and clay plugs reduce transmissivity and create longer time lags in the response of groundwater level to river stage than do sand or silt.

(e) Tributary creeks flowing into the floodplain. These may cause permanent groundwater highs and promote downvalley flow where they are oriented parallel to the receiving river.

In addition to exchanges with the river and drainage from higher elevations, water-table fluctuations in the floodplain respond to evapotranspiration of vegetation. At middle elevations between stream and uplands of the upper Sangamon River, Illinois, water loss by evapotranspiration during certain summer periods may exceed infiltration from the stream and from uplands (Bell and Johnson, 1974). However, quantification of evapotranspiration relative to inflows and outflows of streams is restricted to arid regions where groundwater is subjected to competitive demand by phreatophytes and by withdrawals for consumptive human use and irrigation. Gatewood et al. (1950) estimated that of the losses from the Gila River, Arizona, 12.3% was due to evapotranspiration from floodplain vegetation and 2.5% was due to evaporation from river surface and wet sand bars. Greatest groundwater use occurred by evapotranspiration during the warm months when stream flows were lowest, and greatest groundwater recharge coincided with early winter months when stream flows were highest.

Topographic features of floodplains may impound water and cause flooding as a result of precipitation, independent of stream discharge. This is particularly common in oxbows, backswamp depressions, and swales between relict levees where drainage patterns to the stream channel are poorly developed. Where complex topographic features are lacking and the floodplain slopes gently from the river channel to uplands, flooding frequency and depth are inversely proportional to floodplain elevation. Forest communities respond to this gradient with corresponding patterns of zonation. However, vegetation does not respond strictly to elevation because of confounding variables that affect texture, internal drainage, and aeration of soil (Robertson et al., 1978). In floodplains of small streams in Virginia, flooding frequency, depth, and duration do not follow the same pattern as soil moisture between flooding episodes (Parsons and Ware, 1982). In floodplains of large streams, minor drainages that intersect floodplains may dominate local drainage patterns and affect the distribution of woody species more than the major channel (Buchholz, 1981).

In cold climates, spring snowmelt and the presence of ice floes have a pronounced effect on floodplain hydroperiod and vegetation. Five hydrologic periods are recognized in the annual cycle of rivers in interior and northern Alaska (MacDonald and Lewis, 1973). The longest period is when the river is frozen, beginning as early as October and lasting into May. Rising temperatures in May melt snow, and flow is initially on top of the winter ice cover during the pre-breakup phase. The breakup phase may last only a few days and may be accompanied by ice jamming, depending on local conditions such as river level during initial freezing and whether the stage rises sufficiently to allow ice to float freely downstream. Damage by ice undoubtedly alters the development of streamside forests. A post-breakup flood normally coincides with peak snowmelt. Finally the summer flow phase is established by mid-June when the general trend is of decreasing discharge except for occasional summer storms that may cause rapid rises in river stage. The abrasive force of ice floes may be particularly destructive to vegetation when ice movement accompanies flooding. In Alaska and other areas of high latitude, the paucity of large woody vegetation may be due, in part, to ice stress and, in part, to massive outburst floods from glacier-dammed lakes (Post and Mayo, 1971). However, *Populus balsamifera* has been reported to survive burial to 2.4 m depth and later scour that exhumed the trees (Brice, 1971). Damaged and partially buried trees in riverine forests can be used to reconstruct past flood events (Sigafoos, 1964; Phipps, 1983).

COMMUNITY PATTERNS

The pattern of vegetation most often observed in riverine forests is that species composition changes along a gradient of flooding frequency. This change is the result of the responses of species to flooding and factors associated with soil aeration (Gill, 1970; Clark and Benforado, 1981). At the community level, however, there may be structural and compositional patterns that are repeated in geographically separated regions which deserve closer examination. The purpose of this section is to identify a few patterns and determine if they can be attributed to the hydrologic and geologic forces operating in floodplains.

Structure

Structural characteristics of riverine forests are listed in Tables 5.1 and 5.2. Of the data available for freshwater riverine forests, it appears that forested wetlands of the southeastern U.S.A. and the humid tropics tend toward higher stem density and basal area than those in more arid regions (e.g. Colorado) and more northern latitudes (e.g. New Jersey and North Dakota). The stature of the southeastern riverine forests is visually impressive, especially if they have not been harvested in the past 75 years or so. However, values vary widely even within this group, in part due to stand age. Even in the arid southwestern U.S.A., some of the stands attain basal areas greater than 50 m^2 ha^{-1}.

Basal areas of freshwater riverine forests are plotted with those for upland ecosystems in Fig. 5.5. The precipitation scale is used merely to provide a frame of reference rather than to suggest that structure is under control of precipitation. The most obvious difference appears in the low precipitation regimes where floodplains support forest communities, but uplands either do not or basal areas are lower. However, even in regions where upland forest is present, the basal area of riverine forests is usually as high as or higher than that of upland forests (Schmelz and Lindsey, 1965; Robertson et al., 1978; Adams and Anderson, 1980).

Understory development

Frequent references have been made to the lower density and reduced species numbers in the understory of riverine forests, particularly in the wettest portions of floodplains. The paucity of shrub species in the Southeastern Forest Region (Hall and Penfound, 1943; Robertson et al., 1978; Marks and Harcombe, 1981) and in mature cottonwood (*Populus* spp.) stands in arid regions (Campbell and Dick-Peddie, 1964) appears to be

TABLE 5.1

Structural characteristics of freshwater riverine forests arranged in order of decreasing north latitude. Values for trees ≥2.5 cm diameter at breast height except as noted in first column

Location and type	No. of species	Density (no. ha^{-1})	Basal area (m^2 ha^{-1})	Source
Montana (49°N) *Picea engelmannii* (≥1.0 cm)	3	715	34.6	Lee, 1983
Washington (48°N)				Lee, 1983
Populus trichocarpa (≥1.0 cm)	8	494	37.9	
Alnus[1]	3	793	38.8	Fonda, 1974
Czechoslovakia (48°N)				Vyskot, 1976a
Quercus robur				
38 yr-old	—	1724	25.0	
49 yr-old[1]	—	1104	27.9	
54 yr-old	—	884	—	
North Dakota (47°N), Missouri R., range of 34 stands (≥10.2 cm)	9	574 (93–1105)	28.8 (15.8–56.9)	Johnson et al., 1976
South Dakota (41°N), Missouri R. *Populus deltoides* >1 m tall	4	350	25.2	Wilson, 1970
Colorado (38°–41°N),				Lindauer, 1983
Arkansas River[1]	3	71	13.1	
S. Platte R.[1]	2	126	18.0	
New Jersey (41°N), mixed hardwood				
High area	24	771	20.9	Frye and Quinn, 1979
Low area	16	768	16.5	
Terrace[1]	11	720	21.9	Buell and Wistendahl, 1955
Inner floodplain[1]	12	670	28.6	
Outer floodplain[1]	14	800	25.7	
Island site[1]	9	490	27.1	
Illinois (40°N) *Acer*[1]	6	423	32.1	Brown and Peterson, 1983
Indiana and Illinois (38°–42°N) (≥10 cm)				Schmelz and Lindsey, 1965
Taxodium	—	272	52.3	
Celtis–Tilia	—	393	38.3	
Q. schumardii–Liquidambar	—	284	45.7	
Ulmus–Liquidambar	—	284	28.9	
Celtis–Ulmus–Platanus	—	250	27.8	
Acer–Celtis	—	284	26.4	
High Wabash	—	469	25.3	

(*continued*)

TABLE 5.1 (*continued*)

Location and type	No. of species	Density (no. ha^{-1})	Basal area (m^2 ha^{-1})	Source
Indiana and Illinois (*continued*)				
High Tippecanoe	—	259	24.1	
Q. palustris	—	336	23.0	
Q. stellata	—	326	21.6	
Composite ($\bar{X} \pm$ SE)	—	316 ± 22	31.3 ± 3.3	
Virginia (39°N), mixed hardwood				Hupp, 1982
Site B[1]	8	—	27.1	
Site D[1]	6	—	30.4	
Site E[1]	15	—	40.9	
New Jersey (39°N), mixed hardwood				Ehrenfeld and Gulick, 1981
Muskingum[1]	5	1370	27.3	
Sooy[1]	5	910	33.4	
Hampton[1]	3	1200	37.2	
Clark[1]	2	1570	19.5	
Wesickaman[1]	6	1400	28.9	
Illinois (39°N), mean of six sites ± SE (≥ 10 cm)	8	614 ± 22	37.4 ± 2.7	Crites and Ebinger, 1969
Kansas (39°N), *Quercus–Celtis* (≥ 10 cm), average of 18 stands ± SE	6	390 ± 48	28.1 ± 2.7	Abrams, 1986
California (38–39°N) (> 10 cm)				Conard et al., 1977
Quercus lobata	7	125	18.4	
Populus fremontii	7	252	38.6	
Illinois (37°N) (≥ 6.6 cm) *Taxodium–Nyssa*				Robertson et al., 1978
Old growth				
mesic	—	459	33.5	
transition	—	423	35.7	
wet	—	372	41.9	
Secondary				
mesic	—	602	31.9	
wet	—	535	33.7	
Illinois (37°N) (≥ 8.9 cm at 2.1 m) *Taxodium–Nyssa*				Anderson and White, 1970
Old-growth	—	325	62.7	
Cut-over	—	449	55.5	
N. Carolina (36°N) *Nyssa aquatica*[1]	4	2730	69.0	Brinson et al., 1980
N. Carolina (35°N) mixed hardwood[1]	16	705	47.8	Mulholland, 1979

(*continued*)

TABLE 5.1 (continued)

Location and type	No. of species	Density (no. ha^{-1})	Basal area (m^2 ha^{-1})	Source
Oklahoma (36°N) ($\geqslant$7.6 cm)				Rice, 1965
Western ($n=3$)	11.3	533	19.8	
Central ($n=3$)	16.7	494	22.7	
Eastern ($n=4$)	21.8	548	21.4	
S. Carolina (33°N) ($\geqslant$4.5 cm)				Good and Whipple, 1982
Nyssa–Taxodium	13	703	77.6	
Nyssa–Liquidambar	19	921	38.4	
Nyssa–Magnolia	20	1395	33.1	
Liquidambar–Nyssa	17	797	37.3	
Liquidambar–Acer	18	962	39.8	
Nyssa–Taxodium	5	1484	77.4	
New Mexico (33°N) ($\geqslant$5.1 cm)				Freeman and Dick-Peddie, 1970
E. Slope Black Range				
1853 m elev.	7	—	66.4	
2006 m	8	—	39.8	
2158 m	13	—	34.5	
2310 m	11	—	20.7	
2463 m	6	—	1.3	
W. Slope Black Range				
1853 m elev.	6	—	33.2	
2006 m	9	—	29.2	
2158 m	11	—	62.8	
2310 m	9	—	46.9	
2463 m	7	—	11.2	
Sacramento Slope				
2006 m elev.	5	—	10.6	
2158 m	6	—	18.6	
2310 m	10	—	13.3	
2463 m	8	—	16.6	
Alabama (32°N)				Hall and Penfound, 1943
Nyssa–Taxodium[1]	4	1423	92.3	
Nyssa–Taxodium[1]	2	1082	77.1	
Georgia (31°N)				
Nyssa sylvatica ($\geqslant$14 cm)				
Swamp, 37 plots	—	988	51.7	Applequist, 1959
Riverbottom, three plots	—	1055	48.4	
Nyssa aquatica				
Pure, 9 plots	—	1006	57.9	
Mixed, 8 plots	—	815	59.9	
Florida (30°N) *Taxodium*[1]	—	1644	32.5	Brown, 1981

(continued)

TABLE 5.1 (*continued*)

Location and type	No. of species	Density (no. ha^{-1})	Basal area (m² ha^{-1})	Source
Louisiana (30°N)				Conner and Day, 1976
mixed hardwood[1]	23	1710	24.3	
Taxodium–Nyssa[1]	9	1235	56.2	
Louisiana, Pearl River (30°N)				White, 1983
Mixed hardwood (east)[1]	23	904	14	
Mixed hardwood (west)[1]	22	965	12	
Nyssa–Taxodium[1]	11	1588	55	
Louisiana (30°N)				
N. sylvatica[1]	4	746	77.7	Hall and Penfound, 1939b
Nyssa–Taxodium[1]	4	1018	46.6	Hall and Penfound, 1939a
Florida (30°N)				Leitman et al., 1983
Apalachicola R. (⩾7.5 cm)				
Liquidambar– Celtis	31	1340	28.5	
Carya–Fraxinus	25	1360	32.8	
Nyssa aquatica– N. ogeeche	25	2210	53.6	
N. aquatica– N. sylvatica var. *biflora*	8	2050	66.1	
Nyssa–Taxodium	6	1120	59.2	
Central Texas (30°N) (⩾1 cm at ground level) (Average of six stands ± SE)	21	1396 ± 337	67.5 ± 6.4	Ford and Van Auken, 1982
Central Texas (29°N) (⩾1 cm) (Average of four stands ± SE)	14	1531 ± 148	26.0 ± 1.3	Bush and Van Auken, 1984
Florida (26°N)				Duever et al., 1984
Corkscrew Swamp *Taxodium* (⩾3.8 cm)				
Large forest	7	856	80.2	
Small forest	8	2032	42.9	
Puerto Rico (18°N) Floodplain palm forest (⩾1 cm)	27	3059	42.4	Frangi and Lugo, 1985
Puerto Rico (17–19°N)				Alvarez-Lopez, this volume, Ch. 10
Pterocarpus				
Montane[1]	13	1080	55.0	
Coastal[1]	2	950	27.7	
Coastal[1]	1	1770	43.0	
Costa Rica (11°N) (⩾10 cm)				Holdridge et al., 1971
Prioria copaifera	3.6	290	54.9	

(*continued*)

TABLE 5.1 (continued)

Location and type	No. of species	Density (no. ha^{-1})	Basal area (m^2 ha^{-1})	Source
Costa Rica (10°N) (⩾10 cm)				
Parkinsonia oculeata	2.0	133	3.0	
Erythrina glauca	4.2	335	70.5	
Costa Rica (9°N) (⩾10 cm)				
Mora oleifera	4.5	235	35.0	
Terminalia lucida	25.0	540	65.0	
Anacardium excelsum	24.0	440	60.7	
Panama (9°N) (⩾10 cm)				Mayo Meléndez, 1965
Mora oleifera	7	297	51.5	
Prioria copaifera	7	473	49.3	
Panama (8°N) Mixed forest (⩾10 cm)	—	3792	59.6	Golley et al., 1975
Amazonia (3°S) Seasonal *várzea*				Klinge et al., 1988
Western plots	38	1100	33	
Eastern plots				
[1]Average	8.3	1076	37.8	
[1]Range	1–23	71–2730	12.0–92.3	
[1]Number of Sites	30	29	32	

[1]Data thus marked are used in calculating the average and range at the foot of the table.

related in part to complete canopy closure, which implies that light may be limiting. A significant negative correlation between tree basal area and sapling density led Marks and Harcombe (1981) to suggest light limitation. Conner et al. (1981) said that canopy closure limits understory development, but also increases competition among overstory trees resulting in slower increases in basal area. Alternatively, the unusually high shrub development in riverine forests of the New Jersey Pine Barrens was attributed to poor development of the *Acer rubrum* canopy (Ehrenfeld and Gulick, 1981). However, none of these studies actually measured understory development as a function of light penetration. Barnes (1978) found no clear relationship between subcanopy light conditions and elevation, or between light conditions and distance from the bank of the Chippewa River, Wisconsin. Competition among understory species has been postulated in only one study (Robertson et al., 1978) as an explanation for the lower numbers of species.

An alternative explanation for poor understory development is the unfavorable growth conditions due to flooding. Along a streamside elevation gradient in Illinois, 80% of the understory species were represented in elevations free from flooding (Bell, 1974). The same study showed a sharp increase in species numbers above the elevation corresponding with a flood recurrence once per century, and a fairly constant species composition at even higher elevations. In a Puerto Rican palm forest, palm seedlings reached maximum densities where flooding occurs infrequently, and bromeliads were entirely absent from the forest floor in contrast to adjacent upland forests (Frangi and

TABLE 5.2

Structural characteristics of saltwater riverine forests. Values for trees $\geqslant 2.5$ cm diameter at breast height except for Panama. Dominant genus is *Rhizophora*

Location and type	No. of species	Density (no. ha^{-1})	Basal area (m^2 ha^{-1})	Stand height (m)	Complexity index[1]	Source
Florida (26°N) Ten Thousand Islands	2	4000	38.5	9.0	27.7	Pool et al., 1977
Mexico (23°N)						Pool et al., 1977
Isla La Palma	3	2360	60.8	17.0	73.2	
Roblitos	2	2240	29.6	8.0	10.6	
Río de las Cañas	3	1790	57.8	16.0	49.7	
Puerto Rico (18°N)						Pool et al., 1977
Vacia Talega	3	1890	20.9	13.0	15.4	
Puerto Rico (18°N)						Martínez et al., 1979
Río Cocal	1	400	11.5	7.5	0.4	
Río Cocal	2	4670	20.0	13.0	24.3	
Río Espíritu Santo	2	1460	37.0	16.0	17.3	
Río Espíritu Santo	2	680	17.5	12.3	2.9	
Río Espíritu Santo	2	3230	32.5	16.5	34.6	
Río Espíritu Santo	3	700	14.5	14.3	4.4	
Río Antón Ruíz	3	3000	25.5	19.8	45.4	
Río Antón Ruíz	3	1140	21.5	21.8	16.0	
Río Antón Ruíz	3	4070	15.0	18.0	33.0	
Puerto Rico (18°N)						Mosquera, 1979
Río Espíritu Santo	2	2340	28.0	—	—	
Río Espíritu Santo	2	1610	34.7	—	—	
Río Espíritu Santo	2	1910	45.0	—	—	
Río Espíritu Santo	2	1660	51.3	—	—	
Costa Rica (10°N)						Pool et al., 1977
Moín	4	1370	96.4	16.0	84.5	
Boca Barranca	3	1100	32.9	9.5	10.3	
Panama (9°N) ($\geqslant 10$ cm dbh)	8	712	12.2	41.0	—	Golley et al., 1975
Panama (9°N) ($\geqslant 10$ cm)	—	313	25.1	34	—	Mayo Meléndez, 1965
Average[2]	2.6	2131	33.2	14.2	28.1	
Range[2] ($n=16$)	1–4	400–4670	11.5–96.4	7.5–21.8	0.4–84.5	

[1] The Complexity Index, due to Holdridge et al. (1971) is the product of the four preceding columns multiplied by 10^{-5}. The number of species is the mean for a sample area of 0.1 ha.
[2] Does not include Mosquera (1979), Golley et al. (1975), or Mayo Meléndez (1965).

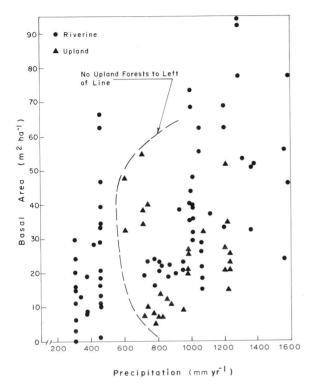

Fig. 5.5. Basal area of riverine forests (circles) and upland forests (triangles) in the U.S.A. plotted along a scale of precipitation. To left of dashed line, precipitation is too low to support upland forest, but riverine forest persists. Basal areas for riverine forests from Table 5.1; those for upland forests from Hough (1936), Eggler (1938), Stearns (1951), Rice (1965), Whittaker et al. (1974), Gilman (1976), and McEvoy et al. (1980). Lower limit for diameter at breast height varies with data source.

Lugo, 1985). Bryophyte cover decreased and *Selaginella* cover increased with elevation on the floodplain. An increase in leaf litter with elevation, apparently due to decreasing transport by flood waters, may have been partly responsible for the bryophyte decrease. In a Florida cypress strand, ferns and shrubs grow epiphytically around the bases of wetland trees and on stumps and fallen logs. Root clusters of these plants are actually elevated above the forest floor. Both density of ferns and the elevation of root clusters decrease in lower elevations of the strand where flooding is greatest. However, optimum fern growth appears to occur where maximum flooding depth is 15 cm (Lugo et al., 1984). Fallen logs support a somewhat greater number of herbaceous species than tree bases in a *Taxodium–Nyssa* forest in southern Illinois (Anderson and White, 1970). Several shrub species and *Acer rubrum* depended on both these substrates for establishment. These data argue strongly for control by flooding frequency and depth.

Rather than understory development being under the control of either flooding or light limitation separately, it is more likely that both factors are important. Menges and Waller (1983) argued that, in the heavily shaded areas of lower floodplains in Wisconsin, the combination of low light and frequent flooding precludes adaptation of many herbaceous species. Using the strategy categories of Grime (1979), they stated that "competitors" respond to decreasing flood frequency and thus are more abundant in the upper floodplain. Toward increasing flooding frequencies, "stress tolerators" and "ruderals" share ground, perhaps because the former respond to flooding as a stress and the latter as a disturbance. In addition, microtopographic features of the floodplain surface that affect patterns of herbaceous vegetation are related to tree location (Hardin and Wistendahl, 1983). Upstream from tree bases, litter accumulates, but, on the downstream side, the soil surface is scoured and forms depressions.

In some cases, shrubs dominate and persist in riverine wetlands because they compete successfully with potential tree species. For example, shrubs may suppress tree development in gaps created by dead canopy individuals (McBride, 1973; Huenneke and Sharitz, 1986), where salinity has caused tree mortality (Penfound and Hathaway, 1938; White, 1983; Brinson et al., 1985), or in arid floodplains where saltcedar (*Tamarix pentandra*) invades (Campbell and Dick-Peddie, 1964). Even a rich herbaceous flora may inhibit succession toward woody species in thermally stressed swamps (Sharitz et al., 1974). In the New Jersey Pine Barrens, Ehrenfeld and Gulick (1981) found the highest shrub biomass on the wettest sites, where the openness of the *Acer rubrum* canopy was given as a reason for high shrub development.

Species abundance of the tree stratum

The number of tree species ranges from two in the arid South Platte River in Colorado to as high as 31 in floodplain stands in Florida (Table 5.1).

Stands with few species — a single dominant or up to four codominants — are found in either the longest hydroperiods or in arid regions like Colorado and South Dakota. The small number of species adapted to the longest hydroperiods would explain the low species richness in some of the stands listed. In arid regions, drought tolerance and phreatophytic habit may be restrictive. Even conditions associated with low pH may exclude species that potentially could dominate the community (McClelland and Ungar, 1970).

Almost without exception there is an increase in the number of tree species from the most frequently flooded to the less frequently flooded, higher elevations of the floodplain. For example, permanently flooded sites of riverine forests in the southeastern U.S.A. are occupied by *Nyssa aquatica* and *Taxodium distichum*, and often only one of these species is dominant. *Acer saccharinum* often grows in monospecific stands in the most frequently flooded areas further north (Hosner and Minckler, 1963; Johnson and Bell, 1976; Franz and Bazzaz, 1977). This high species dominance may persist indefinitely in these stands as long as the channel is stable and flooding frequency remains constant (Adams and Anderson, 1980). Where channels are not so stable, colonization by pioneer species such as *Salix* spp. and *Populus* spp. on newly formed point bars forms temporary communities that will be succeeded by shade- and flood-tolerant assemblages. Thus it appears that Menges and Waller's (1983) adaptations of Grime's strategy categories to floodplain herbs would hold for the tree stratum as well. *Taxodium distichum* and *Nyssa aquatica* could be classified as stress tolerators that occupy stable habitats. *Salix* spp. and *Populus* spp. would be classified as ruderals capable of rapid establishment of unstable habitats, but would also possess characteristics of stress tolerance.

Spatial patterns

If microtopographic features of the floodplain surface due to scouring and deposition are responsible for the distribution of herbaceous species (Hardin and Wistendahl, 1983), it seems reasonable that the establishment of tree and shrub seedlings would be under similar controls. In fact there appears to be some substrate preference in the establishment of vascular plant species (Huenneke and Sharitz, 1986). Differential germination and mortality occurs on muck, logs, cypress knees, etc. Beyond these young life stages, sapling growth and eventual occupation of the canopy depends on whether the species is shade-tolerant. "Groves" of shade-intolerant *Platanus occidentalis* and *Sassafras albidum* in an otherwise mature floodplain forest were interpreted by Robertson et al. (1978) as evidence that gap-phase regeneration was the method for replacement of canopy trees.

Tests for dispersion of individuals ($\geqslant 4.5$ cm diameter) in bottomland forests of South Carolina revealed that uniform dispersion patterns were common when all species were considered together, and that this uniformity in dispersion seemed to be associated with uniformity in tree diameter (Good and Whipple, 1982). However, when individual species populations were analyzed separately, aggregation or clumping seemed to be a more common mode of dispersal. There was a decrease in clumping with increasing size, apparently because competition and differential mortality lead to more uniform dispersal patterns. Gap-phase replacement might encourage the small size classes to be clumped. Two species that remained clumped in the large size classes were *Liriodendron tulipifera* and *Taxodium distichum*. These are both long-lived species which are intolerant of shading.

As Hupp and Osterkamp (1985) explained, distribution patterns of vegetation are predictable only when both hydrogeomorphic conditions and adaptations of tree species are taken into account. For example, survival after destructive floods requires the ability to tolerate inundation and to resprout, while less severe flooding requires only the ability to withstand inundation. Sediment grain size and depth to zone of saturation may be of secondary importance in the distribution of riverine forest species, at least along streams with enough energy to have relatively strong and direct control of their fluvial landforms.

BIOMASS AND PRODUCTION

Standing stocks of aboveground biomass in riverine forests tend to fall between 100 and 300 t ha^{-1} (Table 5.3) with the exception of one stand in

TABLE 5.3

Biomass of riverine forests (t ha^{-1})

Location	Aboveground		Below-ground	Source
	leaves	total		
FRESHWATER				
Czechoslovakia (48°N)	3.5	314	46	Vyskot, 1976b
Quercus robur				
Illinois (40°N)		290	—	Johnson and Bell, 1976
Acer	—			
New Jersey (39°N)				Ehrenfeld and Gulick, 1981
Mixed hardwood				
Muskingum	—	130	—	
Sooy	—	189	—	
Hampton	—	178	—	
Clark	—	79	—	
Wesickaman	—	176	—	
North Carolina (35°N)				
Mixed hardwood	—	276	27	Mulholland, 1979;
Nyssa	—	—	23	Brinson et al., 1981a
South Carolina (33°N)				Muzika et al., 1987
Stream recovering	3.6²	25.5	—	
Stream undisturbed	5.4²	347.8	—	
Riverine recovering	4.6²	25.8	—	
Riverine undisturbed	4.7²	383.2	—	
Louisiana (30°N)				Conner and Day, 1976
Mixed hardwood	—	165	—	
Taxodium–Nyssa	—	372	—	
Florida				
Taxodium (30°N)	6.6	284	—	Brown, 1981
Taxodium (30°N)	5.7	286	50	Nessel, 1978
Taxodium (26°N)	3.5	192	84	Burns, 1978
Taxodium (26°N)				Duever et al., 1984
Large¹	4.8	608	32.3	
Small¹	6.0	240	39.8	
Puerto Rico (18°N)				
Palm swamp	2.8	223.5	69.2	Frangi and Lugo, 1985
Panama (9°N)				
Mixed hardwood	12.1	118.9	12	Golley et al., 1975
Average freshwater (n)	5.6(8)	242(17)	43(9)	
SALTWATER				
Florida (26°N)	3.8	98	—	Lugo and Snedaker, 1974
	9.5	174	—	
Panama (8°N)	3.6	279	190	Golley et al., 1975
Amazonia (3°S)				Klinge et al., 1988
Seasonal *várzea*				
Bio 1	—	97.52	—	
Bio 2	—	257.6	—	
Australia (34°S)	—	128	154	Briggs, 1977
Average saltwater (n)	5.6(3)	170(4)	172(2)	

¹Average of two plots.
²Leaf and fruit annual litter fall.

Florida that exceeded 600 t ha^{-1}. Leaves represent between 1% and 10% of the total. Belowground biomass varies greatly, partly as a result of different sampling methods, and most values are much lower than aboveground biomass (Table 5.3). Data for total aboveground biomass production in riverine forests ranges from 6.48 to 21.36 t ha^{-1} yr^{-1} (Table 5.4). Litter fall accounts for an average of 47% of the production in those studies that measured both litter fall and wood growth. No latitudinal patterns emerge from the data set. Stand age probably accounts for much of the variation among sites.

Data for primary productivity and respiration from freshwater riverine forests are much too scanty to elucidate patterns. Gross primary productivity (26 g C m^{-2} day^{-1}) and community respiration (25 g C m^{-2} day^{-1}) of a riverine *Taxodium* forest in Florida are among the highest reported for any ecosystem (Brown, 1981). The respiration (2 g C m^{-2} day^{-1}) of the forest floor (soil and water) was a small part of total community respiration, but about twice that found in the floodplain in North Carolina (Mulholland, 1979). Most of the carbon dioxide evolution in the North Carolina forest occurred when the forest floor was unflooded. Daily gross primary productivity of saltwater riverine forests in Florida generally is about half of that for *Taxodium* just mentioned (Lugo and Snedaker, 1974). Gross primary productivity and respiration tend to increase when salinities decrease during freshwater runoff. Methane evolution appears to be small when compared to the flux rates of carbon dioxide. High rates of methane evolution are associated with organic content, thickness of the organic layer, and nutrient enrichment (Harriss and Sebacher, 1981). Thus, in still-water peat deposits of basin wetlands a greater proportion of carbon loss is in the form of methane than in riverine forests.

Information from the oak–gum–cypress (*Quercus–Nyssa–Taxodium*) forest type in the southeastern U.S.A. provides a picture of changes in biomass allocation with age (Fig. 5.6). These values should not be confused with biomass production rates (see Ch. 4, Sect. 16). When stands are approximately 20 years old, there is a sharp transition from increasing to decreasing stem density. Up to 20 years, much of the accumulation of biomass and volume is due to recruitment into the ≥ 2.5 cm dbh size class. Between 20 and 40 years mortality is high and indices of production (slope of curves for total stem-wood volume and basal area) begin to level off from maximum values as biomass is allocated among fewer individuals. Steady state appears to be achieved after 70 years.

Most evidence supports the notion that riverine forests have relatively high rates of biomass production in comparison with other forest ecosystems in similar geographic locations. The values for leaf and fruit fall, and for wood biomass production, suggest that there are no strong limiting factors or stressors that result in notably low values. In a few cases there is a suggestion that carbon allocation to litter fall may be at the expense of wood biomass production (Table 5.4). In comparing a drained and undrained *Taxodium* forest, Burns (1978) reported similar litter fall, but much lower wood biomass production in the drained forest. Similarly, the Illinois *Acer* forest had litter fall comparable to locations further south, but the wood production was much lower, possibly a consequence of ice storm damage (Brown and Peterson, 1983). If carbon allocation to leaves is given priority, as seems reasonable, its usefulness as an index of total biomass production may be questioned, especially in ecosystems undergoing changes toward environmental conditions less favorable for biomass production. This is supported by the fact that the ratio of litter production to wood production is not constant as usually assumed (Brown and Lugo, 1982). Rates of belowground biomass production have received little attention in riverine forests.

One of the difficulties in separating factors controlling biomass production at the ecosystem level is that the hydroperiod and water flow control so many factors, such as nutrient delivery, soil redox status, and physical damage by floods. Nevertheless, it is useful to examine the control mechanisms in order to gain insight into the environmental conditions to which species are adapted. Most of the data are based on observations of changes in tree growth under changing field conditions and on response of seedlings in greenhouse experiments. Controlled field experiments have not been conducted.

TABLE 5.4

Aboveground biomass production of riverine forests (t ha^{-1} yr^{-1})

Location and type	Litter fall	Wood biomass production	Total biomass production	Source
FRESHWATER				
Czechoslovakia (48°N)	3.48[1]	17.88	21.36	Vyskot, 1976b
Quercus robur				
Illinois (40°N)	4.91	1.77	6.68	Brown and Peterson, 1983
Acer				
Illinois (37°N)	3.48	3.30	6.78	Mitsch et al., 1977
Nyssa–Taxodium				
North Carolina (36°N)	5.14	7.86	13.00	Mulholland, 1979
Mixed hardwood				
North Carolina (36°N)	6.13	—	—	Brinson et al., 1980
Nyssa aquatica				
South Carolina (33°N)				Muzika et al., 1987
Stream recovering	4.4	2.8	7.2	
Stream undisturbed	6.1	13.4	19.6	
Riverine recovering	5.9	2.8	8.8	
Riverine undisturbed	5.5	2.9	8.5	
Florida (30°N)	5.21	10.86	16.07	Brown, 1981
Taxodium				
Louisiana (30°N)				
Mixed hardwood	5.74	8.00	13.74	Conner and Day, 1976
Nyssa–Taxodium	6.20	5.00	11.20	
Natural flooding	4.17	7.49	11.66	Conner et al., 1981
Permanent flooding	3.28	5.59	8.87	
Controlled flooding	5.49	12.31	17.80	
Florida (30°N)				Elder and Cairns, 1982
Apalachicola R.				
Liquidambar–Celtis	4.73[2]	—	—	
Carya–Fraxinus	5.13[2]	—	—	
Nyssa aquatica–N. sylvatica	4.81[2]	—	—	
Nyssa–Taxodium	4.56[2]	—	—	
Composite (including wood)	7.95	—	—	
Florida (30°N)	6.50	6.40	12.90	Nessel, 1978
Enriched				
Florida (26°N)				Burns, 1978
Undrained	3.45	7.72	11.17	
Drained	3.15	3.70	6.85	
Florida (26°N)				Duever et al., 1984
Large cypress (*Taxodium*)	7.00	1.96	8.96	
Small cypress	7.24	8.18	15.42	
Puerto Rico (17–19°N)				Alvarez-Lopez, this volume
Pterocarpus	14.1	—	—	
Prestoea montana	17.0	3.0	20.0	Frangi and Lugo, 1985
Panama (8°N)				Golley et al., 1975
Mixed hardwood	11.61	—	—	
Malaysia (3°N)				Furtado et al., 1980
Eugenia swamp	9.15	—	—	
Average freshwater[3] (n = 16)	5.72	6.94	12.65	

(continued)

TABLE 5.4 (continued)

Location and type	Litter fall	Wood biomass production	Total biomass production	Source
SALTWATER				
Florida (26°N)				
Ten Thousand Is.	10.66, 11.73	—	—	Snedaker and Brown, 1981
Chokolsee Bay	11.75, 11.83	—	—	Sell, 1977
Gordon River	14.43, 9.09	—	—	Sell, 1977
Puerto Rico (18°N)				Pool et al., 1975
Vacia Talega	14.45	—	—	
Colombia				
Cienaga Grande de Santa Marta (11°N)	11.97	—	—	Hernández et al., 1980
El Encanto River, Guapí (3°N)	14.7	—	—	Hernández and Mullen, 1979
Malaysia				
Sungai Merbok Estuary, Kedah (6°N)	11.96	—	—	Ong et al., 1980
Mantang, Port Weld, Perak[4]	16.0	—	10.2[5]	Ong et al., 1979
Average saltwater (n=8)	13.0	—	—	

[1] Leaf biomass.
[2] Leaf fall only; does not include fruit or wood.
[3] Averages for studies that reported both litter fall and wood biomass production.
[4] Average for three plots >20 years old.
[5] Annual increment over 28 years based on biomass harvests.

Soil moisture and hydroperiod

Riverine forests occur within a moisture gradient that is limited at the wet end by water currents that are too strong or water that is too deep for trees to grow, and at the dry end by upland communities that have reduced moisture requirements. At the wet end of the gradient, the stressor is either unstable or anaerobic soils, and at the dry end the stressor is low water potential. Between these extremes, species appear to be distributed in a pattern that reflects the long-term regime of soil moisture and hydroperiod. Deviations from this regime allow observations on the effects of soil moisture and flooding on biomass production (Ch. 4, Sect. 7).

To determine if reduced flooding would affect tree growth on the Missouri River floodplain, Johnson et al. (1976) compared radial wood growth during 15-year periods prior to and following flood control by reservoirs. Significant decreases were shown for three tree species, *Fraxinus* spp., *Acer negundo*, and *Ulmus americana*, which germinate under shaded forest conditions. Simulation of actual evapotranspiration rates showed that, when water surpluses from flooding were absent, low fall and winter precipitation in the region was insufficient to bring moisture in the surface soil to field capacity by the start of the growing season. Even in the more humid climate of Mississippi, bottomland hardwood vegetation shows accelerated growth when spring flood waters are artificially impounded until about June (Broadfoot, 1960). In southeastern Arkansas, perhaps three-fourths of the annual radial growth occurs between late April and late June (Phipps, 1979). However, similar water management in a riverine forest at higher latitude reduced growth rates of dominant tree species (Malecki et al., 1983).

In a comparison of forests with changed flooding regimes in Louisiana, Conner et al. (1981) found that less frequent flooding resulted in higher biomass production (17.8 t ha^{-1} yr^{-1}) and permanent flooding resulted in lower biomass production (8.87 t ha^{-1} yr^{-1}) than in the natural swamp (11.66 t ha^{-1} yr^{-1}). Mitsch et al. (1979a), in the Cache River floodplain in Illinois, reported an

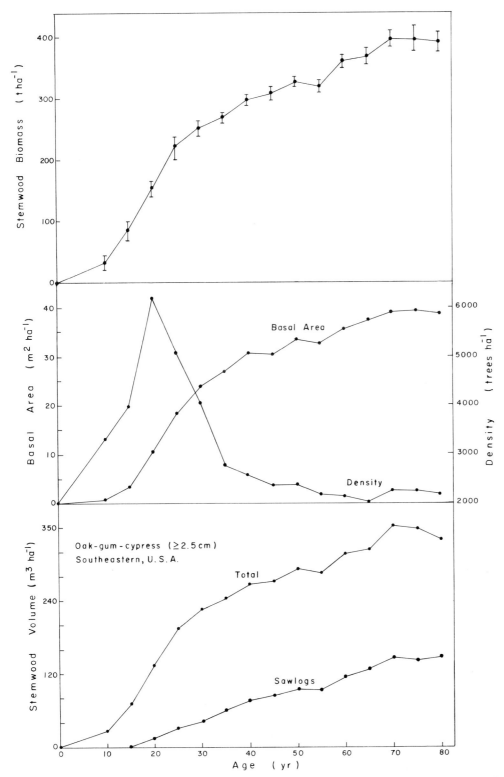

Fig. 5.6. Trends in stand characteristics for the oak–gum–cypress forest type in the southeastern U.S.A. Only trees >2.5 cm dbh are included. Volume excludes bark, and also twigs less than 1.25 cm in diameter, and biomass excludes foliage and twigs. Adapted from McClure and Knight (1984).

increase in basal area growth rate as a function of average river discharge, but slower growth rates when water levels were raised by beaver dams.

Seedlings of many flood-adapted species of trees show suboptimal growth rates either under extreme saturated and stagnant conditions or under low moisture content resulting in water stress. Seedlings grow faster in moving water than in stagnant water, even when depths of inundation are the same (Harms, 1973), apparently in response to greater aeration under flowing conditions. Root growth appears to be depressed more than that of shoots under more reduced soil conditions (Dickson and Broyer, 1972). Seedlings of the most flood-tolerant species grew best under saturated soil conditions (Ch. 4, Sect. 17) and were more susceptible to water stress at low soil moisture than species less tolerant of flooding (Hosner et al., 1965).

Nutrient limitation

Nutrient limitation has been demonstrated occasionally for riverine forests of humid climates in relationship to wastewater additions (Nessel, 1978), but has never been reported in arid regions where water stress occurs commonly. In greenhouse experiments, urea gave better growth than nitrate additions under most conditions of soil moisture status (Dickson and Broyer, 1972). Foliar enrichment of nitrogen occurred in understory trees that received ammonium amendments to the soil, but no foliar response was noted for phosphorus additions (Brinson et al., 1984). Sediments would be expected to have more abundant supplies of available phosphorus and ammonium in soil water under anaerobic than under aerobic conditions. Riverine forests that receive annual inputs of fine sediments during flooding are less likely to be limited by nutrients than either basin wetlands, which are more closed to geologic and hydrologic inputs, or riverine forests that grow on sandy, nutrient-impoverished soils, such as the igapó of the Amazon. A cypress floodplain forest, which received higher rates of annual phosphorus loading than three other cypress forest types, ranked highest in ecosystem metabolism, biomass production, and leaf biomass (Brown, 1981).

Flooding stress and nutrient supply may interact in complex ways for a given species. McKee et al. (1984) showed that phosphorus additions to 2-yr-old *Pinus taeda* seedlings increased total biomass under continuous flooding, but gave no growth response in treatments flooded during the dormant season. The explanation may be that phosphorus translocation is reduced under continuous flooding. *Pinus taeda* is unlikely to compete successfully with more flood-tolerant species in wetlands, so phosphorus amendments in practice would be a means to alleviate the need for drainage in mesophytes, but not hydrophytes, some of which respond poorly to phosphorus amendments (Hook et al., 1983). However, the results contribute to an understanding of interaction between nutrient uptake and effects of hydroperiod.

Other factors

Other agents acting singly or in concert with those described above can affect biomass production in riverine forests. Physical damage by ice, strong water currents, and erosion may essentially remove biomass and thus temporarily reduce primary production. Ice damage is primarily responsible for the low stature of streamside woody vegetation along Arctic streams (Bliss and Cantlon, 1957), while in the arid southwestern U.S.A. canyons support little woody vegetation because of flood damage (Zimmerman, 1969). Fire is second only to inundation frequency as a perturbation affecting riverine vegetation in northern Montana (Lee, 1983). Defoliation by insects (Conner and Day, 1976) represents a temporary loss which not only must be replaced by additional allocations of carbon, but curtails the period during which fully developed leaf area is available for photosynthesis. Disturbance, stand topography, and soil characteristics may be equally important in determining the species composition of woody and herbaceous vegetation in forested wetlands in southeastern Wisconsin (Dunn and Stearns, 1987).

ELEMENT DISTRIBUTION AND DYNAMICS

Elemental cycling in riverine forests is influenced by hydrology in two ways. One is the development of anaerobic conditions in the sediments under waterlogged conditions associated with seasonal flooding. Oxygen availability influences metabolic pathways of microbial communities and, for several elements, determines whether they exist in a predominantly soluble or particulate phase. Anaerobic processes are discussed in Sect. 18 of Ch. 4, but specific examples for nitrogen and phosphorus are given below. The other influence of hydrology is that elements are transported laterally in riverine forests. This occurs when dissolved and particulate forms are carried into the wetland by floodwaters originating upstream, or are exported downstream. In addition, groundwater flows into floodplain and streamside forests transports dissolved elements from the catchment in a manner that makes these elements available for uptake by plants. Most studies of elemental cycling have been conducted in floodplains of the Southeastern Forest Region, where long hydroperiods and broad floodplains contribute to the potential significance of these systems in stabilizing nutrient concentrations of streams.

In forested ecosystems, the distribution of elements among ecosystem components, and annual changes in elemental content of these components, tend to be proportional to the distribution and changes in biomass. High or low standing stocks of elements generally correspond with high or low standing stocks of organic matter. In riverine forests the rank, from highest to lowest amounts of phosphorus, is usually (a) soil (total P to approximately 25 cm depth), (b) aboveground wood, (c) belowground wood, (d) canopy leaves, (e) litter layer, and (f) surface water (Table 5.5). Canopy leaves and other non-perennial structures such as flowers and fruits tend to have high phosphorus concentrations relative to other biomass components, particularly woody ones, but the quantity per unit area is lower. Sediments represent a large proportion of the phosphorus capital of the ecosystem, although only a small proportion may be in a form available for plants. Herbaceous vegetation in floodplains plays a lesser role in nutrient dynamics than in uplands (Peterson and Rolfe, 1982c).

Major flows of elements that are most frequently measured are nutrient return from the canopy as litter fall (Table 5.6), decomposition of litter, accumulation of wood, and sedimentation. Rates of nutrient return from the canopy to the forest floor for temperate-zone riverine wetlands are higher than those for upland ecosystems and basin wetlands of similar latitudes (Brinson et al., 1980). This suggests that fluvial processes are important in maintaining the relative high fertility of riverine forests. Where floodplain soils are relatively nutrient poor, such as the igapó forests of the Amazon, nutrient resorption from leaves prior to abscission may be important in conservation of elements (Adis et al., 1979). Considerable retranslocation by palm leaves was noted in a

TABLE 5.5

Distribution of phosphorus in riverine forests (kg P ha^{-1})

Location and Forest type	Leaves	Aboveground wood	Lateral roots	Surface water	Litter layer	Soil (total)	Source
Illinois (37°N) *Nyssa*	12.2	50.9	28.2	1.8	—	119[1]	Mitsch et al., 1979a
North Carolina (37°N) Mixed hardwood	12	54.5	15.2	0.1	4.5	337[2]	Yarbro, 1979; Brinson et al., 1981a
Florida (30°N) *Taxodium*	12.6	35.2	—	1.9	—	466[2]	Brown, 1978
Florida (30°N) *Taxodium* (sewage-enriched)	4	36	62	8.0	21	902[2]	Nessel, 1978
Puerto Rico (18°N) Palm forest	29.2	51.4	12.5	—	1.4	705[3]	Frangi and Lugo, 1985
Panama (8°N) Mixed forest	16.9	1163.8	13.4	—	12.7	17.3[4]	Golley et al., 1975

[1] To 24-cm depth; [2] to 20-cm depth; [3] to 100-cm depth; [4] to 30-cm depth.

TABLE 5.6

Organic carbon (OC) and nutrients in leaf and fruit fall for riverine forests

Location and forest type	kg ha^{-1} yr^{-1}							Sources
	OC	N	P	S	K	Ca	Mg	
Illinois (40°N) *Acer saccharinum* forest	2210	82.6	8.1	—	22.2	86.8	14.3	Peterson and Rolfe, 1982a
North Carolina (36°N) *Nyssa aquatica* forest	2254	60.1	4.6	6.4	19.2	37.3	15.6	Brinson et al., 1980
Puerto Rico (18°N) Palm forest	3370	—	2.2	—	—	—	—	Frangi and Lugo, 1985
Malaysia (3°N) *Eugenia* spp. swamp	4575	85.0	1.9	—	30.6	129.2	37.2	Furtado et al., 1980
Manaus, Amazonas (3°S) Igapó forest	3316	86.7	1.5	—	17.0	29.9	7.38	Adis et al., 1979

tropical lower montane floodplain forest (Frangi and Lugo, 1985).

Annual phosphorus uptake by stem wood appears to respond to changes in phosphorus supply. Rate of accumulation in stem wood increased approximately threefold when nutrient-rich sewage effluent was released into a cypress (*Taxodium*) strand in Florida (Nessel, 1978). Compared with other cypress forests that had lower fluvial inputs, a floodplain forest in Florida had greater stem wood production as measured by annual basal area increment (Brown, 1981). However, because phosphorus concentrations in wood are extremely low, annual accumulation in wood tends to be quite low when compared with recycling in litter fall.

Release of nutrients by decomposing leaf litter in riverine forests is usually rapid enough for there to be little or no accumulation from year to year (Table 5.7). Decomposition rates vary greatly, in part due to the quality of the material (Brinson et al., 1981b). Peat accumulation in floodplains occurs only in areas with long hydroperiod, usually where topographic lows intersect the water table of the floodplain. These areas of poor flushing and infrequent aeration consequently acquire characteristics of basin wetlands. In better flushed areas, floodwaters alter litter through transport, concentration, sorting, physical destruction, siltation, and increased moisture regimes (Bell and Sipp, 1975) and by creating debris piles that accumulate on the upstream side of the bases of trees (Hardin and Wistendahl, 1983). However, where water currents on floodplains are muted, decomposing litter may immobilize nitrogen and phosphorus for several months after litter fall, thus providing a mechanism for nutrient conservation during the dormant season when losses might be expected to be greatest (Fig. 5.7).

Sedimentation rates on floodplains vary greatly (Ch. 4, Sect. 15), and have been documented for phosphorus in a few cases (Table 5.8). The phosphorus deposition rates approach or exceed some of the litter-fall fluxes mentioned earlier (Table 5.6), but the proportion of the total phosphorus that is available for plant uptake has not been determined. Where deposition rates are large, as in the case of marshes to the mouth of the Mississippi River (Baumann et al., 1984; DeLaune et al., 1987a), sedimentation represents permanent accumulation due to burial. Destruction of forested wetlands in the Mississippi River deltaic plain is anticipated because vertical accretion rates are insufficient to conpensate for land subsidence and sea level rise (DeLaune et al., 1987b).

When floodwaters come in contact with the sediments and litter layer of riverine forests, the slow movement of these water masses provides an opportunity for biological and physicochemical processes to alter elements dissolved in the water. Immobilization and release of nitrogen and phosphorus in decomposing litter (Fig. 5.7) is but one

TABLE 5.7
Decomposition rates of litter (in bags) in riverine forests

Forest type	Duration of measurement (weeks)	Litter type	Site	mm mesh	Decay rate, kg yr^{-1}	Half life of loss, years[1]	Source
Beaver pond, Alberta	75	*Salix* sp., leaves	Shallow water	3.5	0.98	0.71	Hodkinson, 1975
	75	*Juncus tracyi*, leaves		3.5	0.41	1.69	
	75	*Pinus contorta*, leaves		3.5	0.21	3.30	
	75	*Deschampsia caespitosa*, leaves		3.5	0.67	1.03	
Mixed floodplain forest, Michigan	50	*Fraxinus nigra*, leaves	Forest floor	0.05	1.08	0.64	Merritt and Lawson, 1979
				0.5	1.69	0.41	
				8.0	4.95	0.14	
Floodplain, Illinois	60	*Acer saccharinum*, leaves	Forest floor	7	2.57	0.27	Peterson and Rolfe, 1982a
Alluvial swamp, North Carolina	48	*Nyssa aquatica*, leaves	Forest floor	1.6	1.87	0.37	Brinson, 1977
	48	*Nyssa aquatica*, twigs		1.6	0.28	2.48	
Cypress (*Taxodium*) strand, Florida	52	Site litter, leaves	Forest floor	0.8	0.86	0.81	Burns, 1978
				1.6	1.39	0.50	
			Debris pile	0.8	0.75	0.92	
				1.6	0.69	1.00	
	52	Site litter, woody	Forest floor	0.8	0.45	1.54	
				1.6	0.52	1.33	
			Debris pile	0.8	0.87	0.80	
				1.6	0.39	1.78	
Cypress strand, Florida	51	Site litter	Flooded 0% time	1.6	0.47	1.47	Duever et al., 1975
			Flooded 50% time	1.6	0.23	3.01	
			Flooded 61% time	1.6	0.30	2.31	
Cypress strand, Florida	52	*Taxodium ascendens*, leaves	Wet site	1.6	0.55	1.26	Nessel, 1978
			Dry site	1.6	0.46	1.51	
	52	*Nyssa sylvatica*, leaves	Wet site	1.6	0.85	0.82	
			Dry site	1.6	0.76	0.91	
	52	*Acer rubrum*, leaves	Wet site	1.6	0.51	1.36	
			Dry site	1.6	0.73	0.95	
Palm swamp, Puerto Rico	37	Dicotyledonous leaves	Forest floor	1.0	0.83	0.84	Frangi and Lugo, 1985
	37	Palm leaves	Forest floor	1.0	0.55	1.27	
	37	Palm leaves	Hanging in tree	1.0	1.35	0.52	
Inundation Forests, Central Amazonia	14	Site litter, leaves	Igapó forest	15	0.48	1.44	Irmler and Furch, 1980
			Várzea forest	15	1.08	0.64	
All values (n)					32	32	
Average					0.93	1.19	
Range					0.21–4.95	0.14–3.30	

[1] Half life is the time required for disappearance of one half of the dry weight, according to the exponential decay formula $X/X_0 = e^{-kt}$ where X_0 is the dry weight initially present, X the dry weight remaining at the end of the measurement period, t, in years, and k is the annual decay coefficient. Half time is calculated as $0.693/k$.

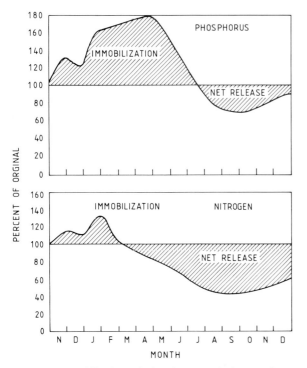

Fig. 5.7. Immobilization of phosphorus and nitrogen from decaying leaf litter in an alluvial swamp. Adapted from Brinson (1977).

example. An anaerobic zone near the surface profoundly affects the pathways of nitrogen by creating low redox conditions for denitrification. The source of nitrate may be external, as in the case of transport by flood waters (Kitchens et al., 1975), or internal, as in the pathways identified by Patrick and Tusneem (1972). These pathways include regeneration of ammonium from organic matter in the deeper sediments, diffusion of ammonium toward the surface, its nitrification in the oxygenated layer, and finally diffusion of nitrate back to low redox conditions. The capacity of highly organic, anoxic sediments for denitrification appears to be limited only by nitrate supply rate (Brinson et al., 1984). Vegetation growing in anaerobic sediments of some floodplains depends almost exclusively on ammonium for its inorganic nitrogen supply, because the high potential for denitrification limits nitrate supplies and the low nitrification rates conserve ammonium.

In view of the high potential for denitrification in many wetland sediments, the apparently high fertility of riverine forests, as suggested by rapid rates of recycling by litter fall (Table 5.6) seems paradoxical. Brinson et al. (1983) argued that

TABLE 5.8

Sedimentation rates of total phosphorus in the floodplains of riverine forests

Locality	Sedimentation rate	Source
Stream levee, northeastern Wisconsin	8.2 g P m^{-2} yr^{-1} during 25-year period	Johnston et al., 1984
Cache River, Illinois	3.6 g P m^{-2} contributed by flood of 1.13 year recurrence interval	Mitsch et al., 1979a
Sangamon River, Illinois	0.12 g P m^{-2} during unusually deep and long flood in spring 1979	Peterson and Rolfe, 1982b
Kankakee River, Illinois	1.357 g P m^{-2} contributed by unusually large spring flood lasting 62–80 days	Mitsch et al., 1979b
Creeping Swamp, North Carolina	0.17 g P m^{-2} yr^{-1} sedimentation on floodplain floor from stream overflow	Yarbro, 1979
Creeping Swamp, North Carolina	0.315–0.730 g P m^{-2} yr^{-1} based on input–output budget of floodplain (most was filterable reactive phosphorus)	Yarbro, 1979
Prairie Creek, Florida	3.25 g P m^{-2} yr^{-1} as sedimentation from river overflow	Brown, 1978

oxygen supplies for rapid nitrate regeneration by nitrification are limiting under flooded conditions and, furthermore, a number of competing pathways for ammonium favor nitrogen conservation by the ecosystem. These include cation exchange, immobilization (Qualls, 1984), uptake by roots, seasonal uptake by algae and aquatic macrophytes, and perhaps, as for other forests, the production by the vegetation of allelopathic inhibitors to nitrifying micro-organisms (Lodhi, 1978; Thibault et al., 1982). However, when the water table drops below ground level during the growing season, exposure of the sediments to the atmosphere stimulates nitrification and increases soil nitrate concentrations (Brinson et al., 1983).

Phosphorus dissolved in floodwaters also interacts strongly with floodplain sediments. The mechanisms for phosphate uptake by the forest floor include uptake by filamentous algae (Yarbro, 1983), accumulation in leaf litter, presumably by microbial immobilization, and abiotic sorption by the sediments (Brinson et al., 1983). In contrast to nitrogen, which has an atmospheric sink through denitrification to N_2 and N_2O, there are no similar pathways of escape from the ecosystem for phosphorus. Thus, phosphorus uptake from floodwaters is limited by the capacity of floodplain sediments to store this element. Phosphorus deposited in the particulate form may later be released to flood waters by upward diffusion from higher concentrations in the interstitial water. The net effect of all of these cycles in large floodplain systems may be to execute transformations among species of the same element rather than to serve either as a large source or sink for a particular element (Elder, 1985).

Due to hydrologic differences, stream-side forests of small catchments with restricted floodplains appear to process dissolved elements differently from the examples given above. Small catchments generally have short hydroperiods and small floodplains, so the opportunity for interaction of floodwaters with floodplain surfaces is minimal. Rather, groundwater seepage through the riparian zone toward the stream channel transports dissolved elements that come in contact with riverine forest. Peterjohn and Correll (1984) have shown reductions in concentration of nitrogen and phosphorus in both runoff and groundwater in the riparian zone in the course of movement from upland fields to streams. Nutrients that showed substantial reductions in riparian forests were nitrogen, phosphorus, calcium, magnesium, and potassium in Georgia (Lowrance et al., 1984) and in New Zealand (McColl, 1978). As both these sites received drainage from agricultural fields, the high removal rates may have been in part due to high loading rates. A substantial component of removal for most nutrients was storage in aboveground biomass production, suggesting that the forests were not mature. In North Carolina watersheds, riparian areas between field edge and perennial stream were a sink for 50% of the phosphorus leaving agricultural areas (Cooper and Gilliam, 1987).

GEOGRAPHIC VARIATION

In spite of the utility of considering processes and dynamics of riverine forests without regard to geographic variation, one must depart from this approach when dealing with species composition and physiognomy. Within a given geomorphic context, four factors play major roles in influencing species composition: hydroperiod, climate, salinity, and biogeographic location.

The isopleth for 25 mm runoff annually shows good agreement in the central U.S.A. with the separation between wet and dry climatic zones (Fig. 5.8). Where runoff is less than 25 mm yr^{-1}, intermittent streams are more common in relatively large drainage basins than is the case for more humid climates. As a result, riverine forest vegetation is subjected to both water deficiency and water excess. Corresponding changes in species composition would be expected. In floodplains of arid regions, plants that can tolerate periods of drought by extending roots to the water table (phreatophytes) and also withstand flooding are most likely to survive. Water limitation in riverine forests in humid climates is seldom experienced, and adaptations for water excess become important for survival in areas with long hydroperiod. Species richness consequently is higher in riverine forests of humid climates because adaptation to water potential stress or a phreatophytic habit are unnecessary. Thus, climate and hydrology are closely associated factors in explaining gross differences in adaptations necessary for riverine

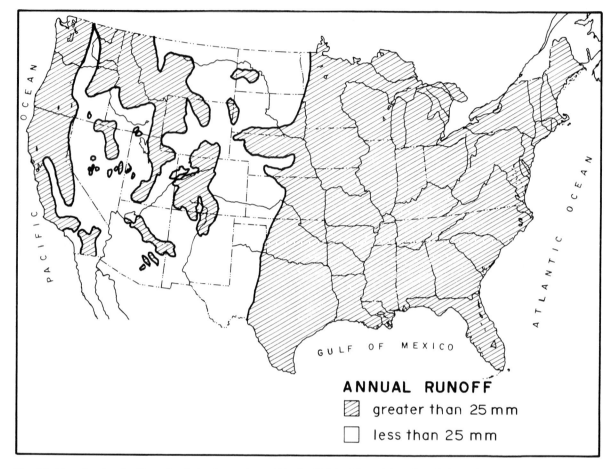

Fig. 5.8. Map showing the 25 mm isopleth of annual runoff in the U.S.A. Adapted from Langbein et al. (1949).

environments. Temperature and length of growing season also influence the distribution of floodplain species, as demonstrated for *Taxodium distichum* in Maryland, U.S.A. (Brush et al., 1980). Lugo and Brown (1984) have argued that freezing temperatures and shorter growing season are responsible for the smaller number of wetland species at higher latitudes.

Salinity requires physiologic adaptations which are associated with the taxonomic and structural uniqueness of mangroves. In higher latitudes where no tree species have adapted to high salinities and freezing temperatures, herbaceous salt marsh replaces mangroves. Freshwater swamp communities that receive intrusions of brackish water undergo structural changes as a result of water stress and tree mortality (Penfound and Hathaway, 1938; White, 1983; Brinson et al., 1985).

Biogeographic location is closely associated with climate, and explains why riverine and upland forests often share species. It is not surprising that similar species and genera recur in riverine forests of many geographic regions; it merely confirms the importance of common edaphic conditions and physiologic stresses that plants of riverine ecosystems share regardless of climatic differences.

In the sections that follow, I describe riverine forests of North America that include temperate, boreal, and tundra climates, and provide some general information on riverine forests of the tropics and subtropics. For convenience of discussion, the community descriptions of North America are organized according to broad biogeographical units in the U.S.A. and Canada. The regions are shown in Fig. 5.9 merely to take advantage of biogeographic differences in commu-

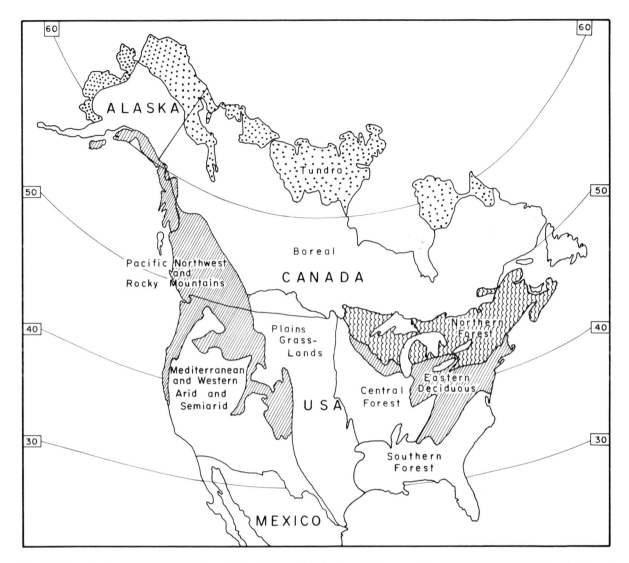

Fig. 5.9. Forest regions in the United States and Canada for which riverine plant communities are described. Terminology after Rowe (1972) and Bailey (1978).

nity composition among regions, rather than to suggest that species in riverine forests demonstrate fidelity for a particular region. The Pacific Northwest and Rocky Mountain region extends northward into Canada and Alaska and would include the subalpine, montane, coastal, and Columbia forest regions described by Rowe (1972). The Northern Forest Region of the U.S.A. extends northward to include the Great Lakes–Saint Lawrence forest region of Rowe (1972). The greater part of Canada, however, is occupied by Boreal forest region which forms a continuous belt from Newfoundland and the Labrador coast westward to the Rocky Mountains and northwestward into Alaska. North of this zone, predominantly coniferous forest in uplands eventually gives way to Arctic tundra. In an earlier volume of this series, Hofstetter (1983) described wetlands of the United States, and Zoltai and Pollett (1983) did the same for Canada. However, neither of these chapters dealt with riverine forests in the detail described below, because of emphasis on covering the large diversity of wetland types found in North America. Information in this chapter on the tropics and subtropics will not be separated by geographic regions.

U.S.A. and Canada

Southern forest region

The geographic distribution of *Taxodium distichum* corresponds approximately with the distribution of southern riverine forests (Mattoon, 1915). However, this species is not always an important component of major floodplains because it tends to be restricted to the wettest and most deeply flooded conditions. Some of the most extensive floodplain areas are along the lower Mississippi River as well as its large tributaries such as the Arkansas, Red, Ouachita, Yazoo, and St. Francis rivers. Vegetation that characterizes these forests has been described for areas as far west as eastern Texas (Chamless and Nixon, 1975; Nixon et al., 1977, 1983) and central Texas (Ford and Van Auken, 1982). Some of the larger rivers draining in a southerly direction into the Gulf of Mexico are the Pearl, Tombigbee, Alabama, Pascagoula, Chattahooche, Apalachicola, and Sewanee rivers. Those draining to the southern and middle Atlantic coast in a southeasterly direction include the Altamaha, Ogeechee, Santee, Pee Dee, Cape Fear, Neuse, Tar, and Roanoke rivers. Rivers as far north as Virginia and Maryland that drain into the Chesapeake Bay support southern riverine forest (Glascock and Ware, 1979; Parsons and Ware, 1982).

Riverine forests on major and minor streams of the Southeast may have a large array of community types. Vegetation varies from communities adapted to extremely long hydroperiods, such as the association of *Nyssa aquatica* and *Taxodium distichum*, to communities dominated by *Quercus* spp. and located relatively high on the floodplain, some of which may not flood annually (Fig. 5.10). If the stream channel has undergone recent reorientation, newly formed point bars and levee deposits may support nearly pure stands of *Salix* spp. and mixtures of this and *Populus heterophylla*, *Betula nigra*, and possibly scattered *Acer saccharinum*. If the river channel remains stable, the species

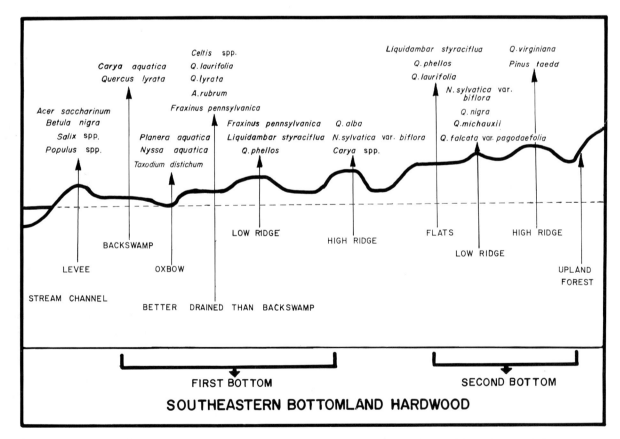

Fig. 5.10. Idealized profile of species associations in southeastern bottomland hardwood forests. After Wharton (1978).

may change to those normally found at higher elevations because the coarsely textured soils drain rapidly after saturation.

Areas in deeper depressions that have long hydroperiods, such as swales and oxbows, will develop *N. aquatica, T. distichum*, and frequently *Planera aquatica* and *Fraxinus caroliniana*; where sufficient sunlight penetrates, *Cephalanthus occidentalis* may occur. Communities where *Quercus lyrata* and *Carya aquatica* occur are usually among the next most poorly drained sites. With even shorter hydroperiod, *Acer rubrum, Celtis laevigata, C. occidentalis, Fraxinus pennsylvanica, Quercus laurifolia* and *Ulmus americana* may be common. Low ridges in the low elevations may be dominated by *Liquidambar styraciflua* while higher ridges that have quite short hydroperiods may be occupied by *Nyssa sylvatica* var. *biflora, Carya* spp., and *Quercus alba*.

The flats of the high floodplain elevations are likely to have poorer internal drainage than the high ridges of the lower elevations. As a result, the species composition may appear similar to that of the low ridges of the lowest area of the floodplain. Where *Quercus falcata* var. *pagodaefolia, Q. michauxii*, and *Q. nigra* occur, hydroperiods are among the shortest or drainage the best among all bottomland sites. *Pinus taeda* and *Quercus virginiana* are usually confined to the highest "islands" in floodplain topography.

The zonation of these communities and fidelity of a species to the zones described above are seldom as ideal as that depicted in Fig. 5.10. Species indicated for different zones may actually coexist. Dominance types have been developed for each of four forest zones that receive inundation (Clark and Benforado, 1981; Wharton et al., 1982). Many old stands have undergone selective timber removal and younger stands undoubtedly will undergo further changes in species composition. Further, the composition of a stand may reflect the stochastic events of hydroperiod and mast production that were present during the critical phases of seed germination and seedling establishment.

So few undisturbed bottomland hardwood stands now exist that cyclic changes in ancient stands are difficult to deduce. Forested wetlands have been lost at rates five times higher than non-wetland forests (Abernethy and Turner, 1987). In the Congaree Swamp of South Carolina, where eleven distinct communities can be delineated, Gaddy et al. (1975) suggested that shade-tolerant hardwoods such as *Quercus laurifolia* eventually overtop the *Liquidambar styraciflua* and other hardwoods for protracted periods of time. Tree fall was offered as a mechanism to create canopy openings so that a mosaic pattern is maintained.

Point bar deposition and other new land forms are initially stocked with *Populus* and *Salix*. These are succeeded by *Acer negundo, A. saccharinum, Fraxinus* spp., and *Ulmus* spp., a community that may persist indefinitely in southern Illinois (Hosner and Minckler, 1963). For more poorly drained sites of the same region, secondary succession has been observed to be initiated by *Cephalanthus occidentalis, Foresteria acuminata, Fraxinus caroliniana, F. pennsylvanica, Nyssa aquatica, Populus* spp., *Taxodium distichum* and *Salix* spp. According to Hosner and Minckler (1963), further fluvial deposition or other events that lead to improved drainage will result in replacement of this community by species found on successively better drained sites (Fig. 5.10). Swamps dominated by *T. distichum* tend to be replaced with hardwood species after logging (Conner et al., 1986).

In narrow bottoms of small streams where the alluvial soils may be moderately well drained, *Nyssa aquatica* and *Taxodium distichum* generally are absent. The mixture of tree species in small stream bottoms in Alabama includes those from the large bottomlands discussed above, from moist coves, and from mesic uplands (Golden, 1979). After agricultural abandonment there is a distinct trend toward dominance by light-seeded hardwoods (*Acer rubrum, Liriodendron tulipifera*, and *Liquidambar styraciflua*) from the reproduction of mature individuals in uncut strips left over from incomplete clearing for agriculture. In Virginia, vegetational ordination revealed two broad classes: (a) stands mainly dominated by *Acer rubrum, Fraxinus pennsylvanica*, and *Ulmus americana*, although two were dominated by *Taxodium distichum*, and (b) a group of communities in which *Betula nigra, Carpinus caroliniana, Fagus grandifolia, Ilex opaca, Liquidambar styraciflua, Liriodendron tulipifera, Quercus michauxii, Q. nigra, Q. phellos*, and *Pinus taeda* assumed high importance (Glascock and Ware, 1979; Parsons and Ware, 1982). The absence of large homogeneous terrain in narrow stream bottoms probably precludes the development of more readily classified community types such as are found in larger floodplains. Disturbance by logging in most of these stands makes it difficult to project the stable state of species composition that they might have over the long term.

Central forest region

Riverine forests of this region have strong affinities with those described for adjacent regions. For example, the studies by Hosner and Minckler (1963) in southern Illinois have already been mentioned in reference to succession in forests of the Southern Forest Region. Robertson et al. (1978) showed that the southern floodplain forest extends up the Mississippi valley and further northward up the Ohio and Wabash rivers. *Acer saccharinum*, a dominant in the canopy, and *Fagus grandifolia*, which is not regenerating, are species which have more northern affinities. Lindsey et al. (1961) conducted studies on the Wabash River which are equally applicable to the Eastern Deciduous Forest Region and these are referred to below. The western part of the Central Forest Region approaches areas where floodplain forests in the Plains Grassland Region have been studied.

The admixture in the Central Forest Region of floral components from surrounding regions makes sweeping generalizations about species composition and succession in riverine forests unwarranted.

Vegetation along the Sangamon River in central Illinois illustrates the rapid transition in species composition, biomass, and annual biomass accumulation from floodplain to upland forest (Table 5.9). In the floodplain, *A. saccharinum* is clearly dominant, *Celtis occidentalis* and *Quercus imbricaria* are codominants in the transition zone, and *Q. alba* dominates the upland community. Individuals of nearly all species are normally distributed around a mean floodplain elevation — a situation analogous to the population continuum concept of Whittaker (Franz and Bazzaz, 1977). Total tree biomass and estimated net biomass accumulation were greatest in the floodplain followed by the upland and transition stands. Dutch elm disease (*Ceratocystis ulmi*) and phloem necrosis have contributed to low biomass of the transition zone by eliminating all large trees of *Ulmus americana* which probably dominated the zone prior to 1959 (Johnson and Bell, 1976). In southeastern Iowa, McBride (1973) studied the

TABLE 5.9

Distribution of biomass among species (%), aboveground biomass, and net biomass production of trees along a flooding frequency gradient for a stream in Illinois. From Johnson and Bell (1976)

Species	Percent of total biomass		
	Floodplain	Transition	Upland
Acer saccharinum	73.6	15.7	—
Gleditsia triacanthos	10.9	—	—
Fraxinus pennsylvanica	9.4	—	—
Platanus occidentalis	3.6	—	—
Euonymus atropurpureus	—	9.3	—
Quercus imbricaria	—	22.3	—
Carya cordiformis	—	2.2	—
Prunus serotina	—	4.7	—
Ulmus rubra	—	6.2	—
Ulmus americana	—	5.8	3.3
Quercus velutina	—	—	6.2
Quercus alba	—	—	84.9
Aboveground biomass (t ha^{-1})	289	135	227
Estimated net biomass production (t ha^{-1}yr^{-1})	11.5	7.0	10.0
Probability of flooding in any one year	3–25%	0.5–3%	0.5%

impact of loss of *U. americana* on *Acer saccharinum–Populus* communities which occur on the floodplains of large rivers and *Celtis–Ulmus* communities along the banks of smaller streams. *Celtis* spp. and *Prunus serotina*, both potential overstory species, were the most important species regenerating under dead elm canopies, while *Carpinus caroliniana* was most important under the living forest canopy of the *A. saccharinum–Populus* spp. type. McBride predicted future dominance by *A. saccharinum* and *Celtis* spp. in this community. However, for the *Celtis–Ulmus* type, *Celtis* may be replacing *Ulmus*. Dense growths of shrubs in areas previously occupied by *Ulmus* suggest that forest development may be retarded by competition. In central New York, multiple rather than single tree gaps left by dead *Ulmus* individuals are necessary for dense shrub cover to become established (Huenneke, 1983).

Eastern deciduous forest region
Riverine forests in this region range from those located along small to moderate sized streams draining the Appalachian Mountains to rivers that are relatively large by the time they pass through the region. Some of the larger rivers are the upper Mississippi, those of the upper Ohio River valley, Susquehanna, Shenandoah, and Delaware.

Floodplains[1] on the Wabash and Tippecanoe rivers in Indiana, which could as easily be included in the Central Forest Region, are dominated by *Populus* spp., *Salix nigra*, and *Ulmus americana* on "first bottoms" or lowest floodplain elevations (Lindsey et al., 1961; Schmelz and Lindsey, 1965). "Second bottoms" which are infrequently flooded were occupied by *Acer saccharum, Aesculus glabra, Cercis canadensis, Fagus grandifolia,* and *Ulmus americana* as well as sixteen other species with individuals exceeding 10 cm diameter at breast height.

In stands on the floodplain of the Raritan River, New Jersey, Buell and Wistendahl (1955) mentioned fourteen woody species. On the inner floodplain where erosion produced a series of ridges and poorly drained sloughs, *Acer saccharinum* was the dominant tree, followed by *A. rubrum, Fraxinus americana,* and *Ulmus americana*. In less frequently flooded and less severely scoured areas, *Fagus grandifolia* and *Liriodendron tulipifera* trees were abundant, along with *A. saccharinum*.

In a follow-up study of the Raritan floodplain 20 years later, Frye and Quinn (1979) described successional trends in the same area studied by Buell and Wistendahl (1955) and compared succession on floodplains with that of uplands. Forest development occurred more rapidly on high areas of the floodplain than on uplands. This was expressed both by greater number of species and by a decrease in single species dominance in the floodplain. Although both habitats passed through a thicket stage, it appeared earlier on the floodplain and passed through it more rapidly, thus resulting in a structurally more complex forest. However, in the lowest area of the floodplain, higher flooding frequency apparently results in physical damage, and especially affects seedlings, saplings, shrubs, and lianas. Not only was there less cover by shrubs and lianas on the lower floodplain, but five species were absent. On the wettest sites on the Connecticut River floodplain, however, herbaceous species replace woody vegetation at most frequently flooded sites (Metzler and Damman, 1985). On the upland sites, Frye and Quinn attributed lower structural development and species richness to water stress during drought.

Successional development on barren sites created by stream migration on Wissahickon Creek in southeastern Pennsylvania began with *Acer saccharinum* and *Platanus occidentalis* after the initial herbaceous weed cover (Sollers, 1973). This was replaced by a community dominated by *Acer rubrum, Fraxinus americana, Juglans nigra, Lindera benzoin,* and *Ulmus americana*. With improved drainage, *Quercus–Fraxinus* stands will eventually develop. Highest bottoms, or areas which are inundated only by the most severe floods, are dominated by typical upland species (Lindsey et al., 1961). Thus, well-drained portions of riverine forests may be expected to reflect as much variation as do upland forests throughout this physiographically diverse region. This was evident in a tributary of the Shenandoah River in northern Virginia where the floodplain supports typical mesic upland species (Hupp, 1982). Flood-adapted species are restricted to below bank-full stage where the recurrence interval of flooding is

---

[1] Floodplains, because of their lower elevation than adjacent uplands, are often called "bottoms" or "bottomlands" in the United States literature.

1.9 years. Along gentle channel gradients of this stream, dominants are *Acer rubrum*, *F. pennsylvanica*, and *Platanus* spp. Where the channel gradient steepens, and where flood damage and flow velocity are higher, a comparatively wider active channel shelf supports shrubs (*Alnus serrulata*, *Cornus amomum*, *Hamamelis virginiana*, *Ilex decidua*, *Physocarpus opulifolius*, *Salix nigra*).

Two community types can be distinguished on the narrow floodplains of the Little Tennessee River system in the Appalachian Mountains of western North Carolina: those strongly dominated by *Betula nigra* and those in which *B. nigra* is absent (Wolfe and Pittillo, 1977). Dominants in the latter communities are *Acer rubrum*, *Liriodendron tulipifera*, *Prunus serotina*, and *Robinia pseudo-acacia*. The major environmental variables separating the two types are higher proportions of silt, clay, and organic matter where *B. nigra* stands occur. Wolfe and Pittillo (1977) attribute the difference to greater moisture requirements of *B. nigra* which would not be met by the sandier soils of the other stand type.

Northern forest region

Riverine forests in this region have received little study, possibly because attention has been directed to extensive peat bogs located in the western portion (Hofstetter, 1983). In comparison with other regions, rivers tend to be small because many either drain headwater catchments of the Mississippi River or terminate in the Great Lakes after flowing a short distance. The Hudson and Connecticut rivers in New England are exceptions. Vegetation developing on mineral-rich soil and in areas of distinct water flow is different from that occurring on surrounding low-lying shrub and sphagnum bog. Heinselman (1970) describes these as rich swamp forest; they typically include high densities of *Thuja occidentalis*, which may be overtopped by *Fraxinus nigra*, *Larix laricina* or *Picea glauca*. Except where *T. occidentalis* is dense, *Alnus rugosa* forms a shrub layer. *A. rugosa* and *F. nigra* usually disappear in transition from marginal fen to poorer swamp where water flow is more sluggish, water is less rich in minerals, and peat is deeper and contains less inorganic material. Where more apparent floodplain features exist and there is little peat accumulation, *Ulmus americana* may play a larger role, although *F. nigra* is still important (Janssen, 1967).

Species composition of riverine forests along the Susquehanna, Chemung, and Delaware rivers in the Appalachian Uplands of New York appear to be influenced by five characteristic floodplain features (Morris, 1977; Morris et al., 1979). These are: (a) flood basins with poorly drained silts and high organic matter content, dominated by *Acer saccharinum*, *Populus* spp., *Prunus serotina*, and *Salix* spp.; (b) point bars and stream confluence areas with well drained silts which lack *Salix*, but have *Fraxinus* and *Platanus* in addition to the species listed for (a); (c) frequently and destructively flooded point bars and confluence areas of sand and silt mixtures which support *Acer saccharinum*, *A. saccharum*, *Robinia pseudo-acacia* and *Ulmus americana*; (d) less frequently flooded stable point bars and confluence areas of coarsely textured sands, which support *Carya* spp., in addition to the two *Acer* species; and (e) seldom flooded Pleistocene terraces where *Acer rubrum*, *Pinus* spp., *Prunus serotina*, and *Quercus* spp. dominate. Many of the same species were mentioned for the floodplain of the Chippewa River, Wisconsin (Barnes, 1978, 1985) and on islands of the Saint Lawrence River downstream from Montreal, Canada (Tessier et al., 1981). Huenneke (1982) demonstrated that seasonality of water levels and soil texture are the major environmental variables separating types of forested wetland vegetation in central New York. Mixing of boreal relict vegetation with temperate swamp species of the region contributes to high species richness (Paratley and Fahey, 1986).

Plains grassland region

Riverine forests of this region become conspicuous features of the landscape as upland forest gives way to savanna in this transition from humid to arid climate. Floodplains of this region differ floristically from those further east by a general absence of *Quercus* spp. Some of the major rivers that cross this region are the Missouri, Platte, upper Arkansas, and the Canadian rivers.

Transitions due to moisture are particularly well illustrated in Oklahoma where Bruner (1931) distinguished between riverine forests of the eastern, central, and western parts of the state. Species that occurred in more than one of these zones showed decreasing height from east to west (Fig. 5.11). In the east, continuous flow in even small streams supports forests rich in species of trees, shrubs, vines, and herbs. *Betula nigra*, *Liquidambar styraciflua*, *Nyssa sylvatica*, *Platanus occidentalis*, and *Taxodium distichum* are common. Dominants of central Oklahoma floodplains, such as *Celtis* spp., *Gleditsia triacanthos*, *Juglans nigra*, *Robinia pseudo-acacia*, and *Ulmus* spp., occur also in the east and augment the species diversity there. In the arid west, trees are usually rather widely spaced, and neither *Salix* nor *Populus* reach the stature that they attain eastward. *Acer negundo* and *Ulmus* spp. are usually found only in valleys or near streams where the water supply is constant. With only a 2° change in longitude but a 240 mm change in annual precipitation in central Oklahoma, species richness in floodplain communities increased from 11 in the west to 23 in the east (Rice, 1965).

In the Missouri River floodplain of North Dakota where floodplain width varies from about 1 to 11 km, three forest types can be distinguished (Fig. 5.12) (Keammerer et al., 1975; Johnson et al., 1976). On the lowest and most frequently flooded areas, few species other than those of *Populus* and *Salix* occur, and trees are small (6 to 12 m tall). At higher elevations, forests consist of older trees of *Populus*, whose tall open canopies overtop *Acer negundo* and *Quercus macrocarpa*. At the highest elevations, floodplains are dominated by *A. negundo*, *Fraxinus pennsylvanica*, *Q. macrocarpa* and *Ulmus americana*. Canopies are relatively closed and lack the tall shrub and sapling layer characteristic of *Populus* forests. Wikum and Wali (1974) also noted the poor development of shrubs in a *A. negundo–U. americana* floodplain on a tributary of the Red River in North Dakota.

In the absence of rejuvenation by flooding due to upstream impoundment in 1954, Johnson et al. (1976, 1982) stated that *Populus–Salix* forests would eventually disappear because seedbed requirements for regeneration were lacking. *Ulmus americana* would decline because of rapidly declining growth rates related to altered moisture levels, almost complete absence of seedling survivorship, and mortality due to Dutch elm disease. A slower decline in *Acer negundo* would be accompanied by an increase in *Fraxinus pennsylvanica*. Overall, they predicted decreases in species and landscape diversity.

At the western edge of the Plains Grasslands in eastern Colorado, Lindauer (1983) identified four community types of woody vegetation on floodplains of the Arkansas and South Platte rivers. These included a community where *Populus sargentii* is dominant, areas which remain moist throughout most of the year dominated by *Salix* spp., a mixed community including *P. sargentii*, *Salix amygdaloides*, *S. exiqua*, *S. interior*, *Tamarix pentandra* (= *T. chinensis*), and other tree species, and a *T. pentandra* community on the Arkansas River where it covered approximately one-third of the floodplain. Treeless areas on the floodplains were attributed to insufficient moisture.

Mediterranean and western arid forest region

Some of the major drainages of this region are the San Joaquin, Sacramento, Salt-Gila, and Rio Grande–Pecos rivers. Along these rivers and their tributary streams, riparian vegetation provides a striking contrast to the drought stressed semidesert and chaparral of uplands. Species of riverine

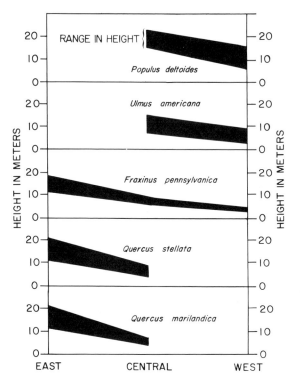

Fig. 5.11. Changes in height and species composition of floodplain forest stands along an east to west gradient in Oklahoma. Adapted from Bruner (1931).

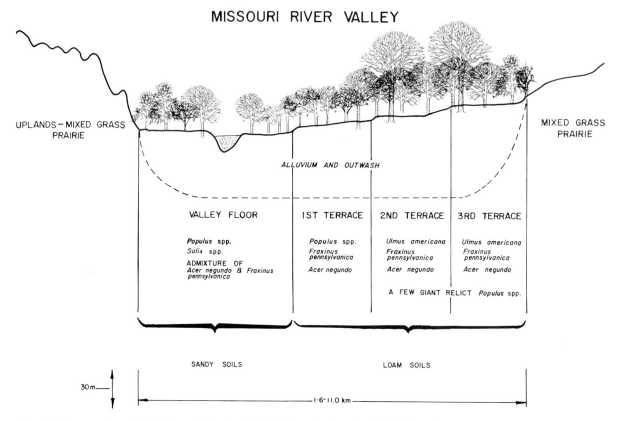

Fig. 5.12. Cross-section of the Missouri River in North Dakota showing the distribution of important tree species. Reprinted by permission from Keammerer et al. (1975). Copyright 1975 by the *Canadian Field-Naturalist*.

forests include those that are confined to moist areas as well as those that can survive under the drier upland conditions (Campbell and Green, 1968). Differentiation between valley floor and upland vegetation increases with increasing drainage areas (Zimmerman, 1969). Headwaters of intermittent streams have available little more water than well drained upland slopes. There are also dramatic changes in riparian vegetation with increasing elevation.

Along the Rio Grande between El Paso, Texas and Albuquerque, New Mexico, a distance of 480 km, five vegetation classes were distinguished by Campbell and Dick-Peddie (1964). These form a continuum from south to north with gradual and almost imperceptible changes between dominant and subdominant species (Fig. 5.13).

Class 1. In the most xeric class of riverine vegetation, *Prosopis pubescens* dominates and the cover or density is determined by age of the stand and moisture availability.

Class 2. Where moisture is greater and flooding may occur during the growing season, *Tamarix pentandra* becomes a competitor with *P. pubescens*. In areas with a high water table and occasional flooding during the growing season, *T. pentandra* thrives to the exclusion of *P. pubescens*.

Class 3. In these dense covers of *T. pentandra*, few shrubs and grasses occur, unlike classes 1 and 2. This class predominates in the southern sector of the river and in disturbed areas further north.

Class 4. Populus fremontii stands attain great height relative to other floodplain species. *Elaeagnus angustifolia*, *T. pentandra*, and *Salix gooddingii* may become codominants. *Prosopis juniflora* occurs occasionally in the northern localities.

Class 5. These are stands with a dense overstory of *Populus fremontii* and a separate understory of *E. angustifolia* and *S. gooddingii*. *Tamarix pentandra* is found only in disturbed areas.

In southern New Mexico (1800–2560 m elevation) *P. fremontii* and *Juglans major* dominated

RIVERINE FORESTS

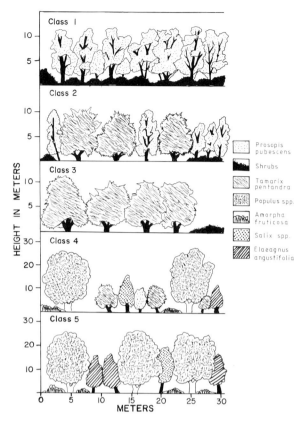

Fig. 5.13. Profiles of five vegetation types along the Rio Grande from El Paso to Albuquerque. Reprinted by permission from Campbell and Dick-Peddie (1964). Copyright 1964 by the Ecological Society of America.

riparian communities at lower elevations and *Acer negundo* and *Alnus oblongifolia* at higher elevations (Medina, 1986). These stands had been protected from livestock grazing since 1937.

The introduction of *Tamarix pentandra* and *Elaeagnus angustifolia* in the last 50 to 60 years has changed succession and ultimate dominants in some communities. *T. pentandra* is in more than 50% of the floodplain plant communities of the lower Gila River, Arizona (Haase, 1972). However, Irvine and West (1979) have presented evidence that stream flow regulation, which reduces flood frequency and severity, is responsible for widespread colonization by *T. pentandra*.

Elsewhere, Freeman and Dick-Peddie (1970) noted a trend toward shrub dominance at lower and upper elevations in southern New Mexico, while trees dominate intermediate elevations. This supports Zimmerman's (1969) observations of increasing upland–riverine differentiation with larger drainage area. However, exceptions exist where streams flow from moist mountain environments and actually decrease discharge due to evapotranspiration as they pass through more arid climates of lower altitude (Kondolf et al., 1987). Species such as *Pseudotsuga menziesii* and *Pinus ponderosa* are restricted to the highest elevations studied (1400 m) because of their intolerance of high temperature at lower elevations (Campbell and Green, 1968). Where adequate moisture is available for woody vegetation, species such as *Celtis reticulata*, *Fraxinus velutina*, *Juglans major*, and *Platanus wrightii* may be excluded due to high ratios of monovalent to divalent cations, a condition to which *Tamarix pentandra* seems well adapted. *Populus fremontii* and *Salix gooddingii* appear somewhat tolerant of this salt status (Zimmerman, 1969).

A floodplain in the Sacramento Valley, California, in the Mediterranean Region (Fig. 5.14), shows a gradient in species composition and canopy height. Forest vegetation develops only in areas that have not been frequently flooded or that have not undergone recent lateral erosion. Dominant species are *Acer negundo*, *Alnus rhombifolia*, *Fraxinus latifolia*, *Juglans hindsii*, *Platanus racemosa*, *Populus fremontii*, *Quercus lobata* and *Salix* spp. However, future generations of *Populus fremontii* are dependent on the open, moist sand bars which have resulted in stream instability (Conard et al., 1977). Over 90% of riparian forests in California's Central Valley have been lost (Warner, 1984).

Pacific Northwest and Rocky Mountain Region

Because of the rugged relief of much of this region, stream gradients are steep and channel degradation probably predominates. Riverine forest frequently consists of narrow, interrupted bands along small streams or as uninterrupted zones in broad river valleys (Walters et al., 1980). In mesic sites along streams, gradients of riverine vegetation are probably a result rather of stand age as dictated by time since the last disturbance than of the limiting effects of flooding. Distinct streamside communities are a result either of new land being exposed by destructive floods or of the higher groundwater table along streams (Fonda, 1974). A typical gradient, beginning at stream-side

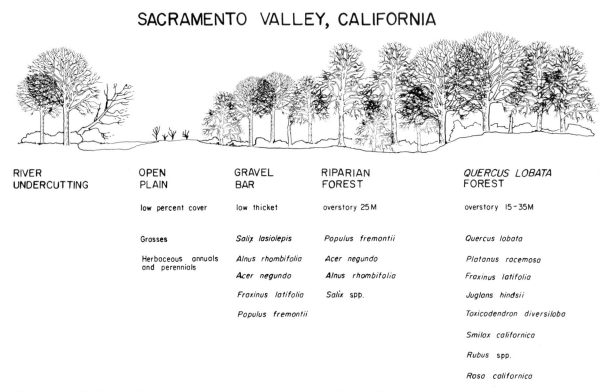

Fig. 5.14. Profile of vegetation along major rivers in the Sacramento Valley, California. Adapted from Conard et al. (1977).

in the *Tsuga heterophylla* zone of the Olympic Mountains, Washington, was (a) gravel bars dominated by *Salix scouleriana*; (b) elevated flats dominated by *Alnus rubra*, which gives way to *Acer macrophyllum*, *Picea sitchensis*, and *Populus trichocarpa* with time; and (c) second terraces occupied typically by *P. sitchensis* and *T. heterophylla*. This is similar to the gradient of riverine forests along the McKenzie River, Oregon (Fig. 5.15). Flooding may occur annually on the lowermost floodplain.

Some species occupy riverine forests only at higher elevations. For example, *Thuja plicata* and *Tsuga heterophylla* are restricted generally to less than 550 m, but will reach altitudes of 600 m along waterways. In the coastal region of northern California, *Sequoia sempervirens* takes the place occupied by *Abies amabilis*, *Picea sitchensis*, and *Tsuga heterophylla* in the floodplains of Oregon and Washington. Not only is *S. sempervirens* adapted to survive rapid sedimentation by producing additional roots, but it is fire-tolerant as well.

In the Rocky Mountains, species on wet sites include *Alnus tenuifolia*, *Populus angustifolia*, *P. balsamifera*, *P. tremuloides*, *Rubus* spp., and *Salix* spp. At lower elevations, *Picea pungens* may replace the wet-site species when drainage is improved and disturbance is lacking. Upon reaching even lower elevations the vegetation changes southward to the drier riverine forests of the Western Arid Region and eastward to that described for the Plains Grassland Region.

Canada and Alaska

Sufficient descriptive work has been done in high-latitude riverine forests to identify species assemblages and to characterize the zonation of woody vegetation on floodplains. There are few data on ecosystem structure and dynamics, however, that allow quantitative comparisons. Work on riverine forests may have been neglected in part because of the overwhelming abundance of basin peat wetlands (Brown, this volume, Ch. 7; Zoltai and Pollett, 1983), which, in the Hudson Bay lowland alone, cover 85% of the region (Sims et al., 1982). The other contributing factor may be the lack of merchantable timber in the shrubby floodplains of the tundra in comparison with the

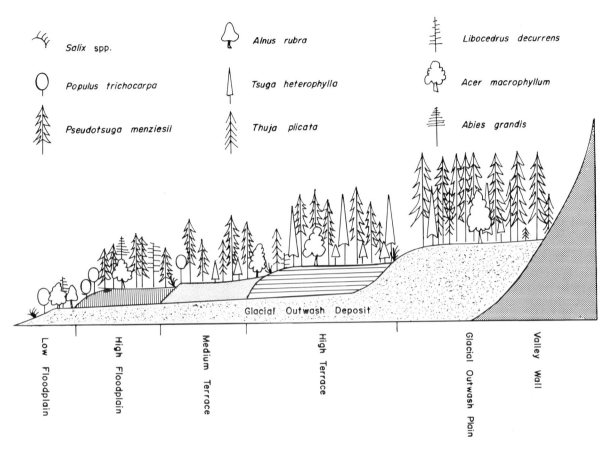

Fig. 5.15. Cross-section of floodplain and terrace communities of the McKenzie River, Oregon. Reprinted by permission from Hawk and Zobel (1974). Copyright 1974 by the Washington State University Press.

economically more important forest of boreal upland regions.

On the arctic slope north of the Brooks Range where permafrost prevails, *Alnus–Salix* communities along streams are in striking contrast to the shorter tussock–heath tundra and sedge–grass marsh that surrounds them (Bliss and Cantlon, 1957; Spetzman, 1959). Sage (1974) described three riverine forest types in arctic Alaska. On alluvial deposits that form gravel and silt bars and islands in braided streams, usually no vegetation develops, but in areas not regularly submerged *Equisetum* will develop, as will occasional dwarf *Salix*. Along streams of small drainages, shrub communities of up to 1 m height are composed of *Alnus crispa*, *Betula nana*, *Salix pulchra*, and *S. lanata*. A less common community is restricted to streams and drainage canals in the foothills region, and is described as tall shrub (>1 m tall), dominated by *Salix alaxensis*.

In regions where *Picea mariana* forests replace the tussock–heath tundra, more elevated portions of the floodplain support stands of *Populus balsamifera*, which are eventually replaced by *Picea glauca*. Fig. 5.16 illustrates an idealized profile for riverine vegetation of the Mackenzie River, NorthWest Territory, Canada (Gill, 1972a). In the *Salix alaxensis* zone, other species of the genus (for example *S. arbusculoides*, *S. glauca*, *S. pulchra*) and *Alnus crispa* may assume importance with increasing stand age. *Salix* spp. are eventually eliminated by a combination of short life span, herbivory by hares, and shade intolerance (Walker et al., 1986). *Picea glauca* appears to

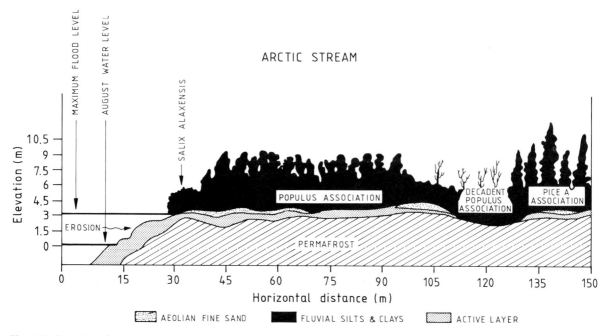

Fig. 5.16. Zonation of plant communities along an arctic stream. Reprinted by permission from Gill (1972a). Copyright 1972 by the *Canadian Journal of Earth Science*.

assume dominance only after longer periods without disturbance from flooding. *Picea mariana* occurs only at the uppermost floodplain elevations as described by Drury (1956) for the upper Kuskokwim River region just northwest of the Alaska Range.

In the extensive peatlands near James Bay, Ontario, Sims et al. (1982) found stands of *P. mariana* often over 4 m high as elongated features paralleling watercourses. Jeglum and Cowell (1982) found that *Abies balsamea* was occasionally important in this region. However, an even more frequently flooded community along streams included *Alnus rugosa*, *Betula pumila* var. *glandulifera*, *Cornus stolonifera*, and *Salix planifolia*. Near the southern boundary of the boreal forest in Ontario, *Fraxinus nigra* dominates wetlands that have the least organic soils and the most eutrophic conditions (Kenkel, 1987). Further west in the boreal forest, levees of the Saskatchewan River delta support woody species different from those of the surrounding bog and fen (Dirschl and Coupland, 1972). *Alnus rugosa* and *Salix discolor* dominated low levees while *Betula papyrifera*, *Picea glauca*, *Populus balsamifera*, and *Ulmus americana* were associates on the higher levees.

Europe

Riverine forests in Europe have been extensively damaged. Because they are considered to be an oustanding part of European natural heritage, much of the descriptive literature has been synthesized (Yon and Tendron, 1981) to aid in the process of protecting them. Other chapters in this volume (17 and 18) provide information on riverine forests in Europe.

Tropics and subtropics

Other chapters in this volume and others in this series treat forested wetlands in the tropics and subtropics by geographic region, but, except for the chapter on Australia (Ch. 16), and that of Junk (1983) covering the Amazon wetlands in another volume, none deals exclusively with the riverine type. In this brief section, I will emphasize how the variables of water quality (including salinity), hydroperiod, and regional rainfall serve to influence species variety and physiognomy of riverine forests. Unlike the previous section, it is necessary to bring salinity into the picture for the tropics and subtropics, because of the presence of mangrove forests in estuaries. Because of obvious biogeographic variation in taxa throughout these zones, there will be no attempt to describe species composition of common or typical communities in the detail that has been done in the preceding section. Brunig (Ch. 13) provides this level of detail in his discussion of Borneo.

The array of riverine forest types present in the Amazon basin (see Junk, 1983) illustrates how hydroperiod and salinity can be used to distinguish several riverine forest types. In an attempt to standardize the terminology applied to the largest system of riverine forests in the world, Prance (1979) has provided a key that separates types based first on hydroperiod and second on water type:

Flooded by regular annual cycles of rivers: white water: seasonal várzea; black or clear water: seasonal igapó.

Flooded by tides: salt water: mangrove; fresh water: tidal várzea.

Implicit in his water types are geomorphic variables in which clear-water and black-water rivers are low in ionic content, high in organic carbon (black water only), and low in suspended sediments in comparison with white-water rivers. Permanently inundated forests have been omitted from Prance's scheme as they are clearly of the basin type. Some of the distinguishing features of the vegetation of these riverine forests are described.

Seasonal *várzea* forest is the major riverine forest type of the Amazon and is flooded annually by white-water rivers. Therefore, *várzea* floodplain soils consist primarily of clays and silts of alluvial origin. *Várzea* forests are characterized by high biomass, abundant lianas, many large and buttressed trees with pneumatophores, and a well-developed shrub (Williams et al., 1972) and herbaceous stratum. There may be many species associations in this vast region, but they have not been adequately distinguished.

Seasonal *igapó* forests occur on the floodplains of black- and clear-water acid rivers which usually have sandy soil. Water stress and nutrient limitation may together be responsible for the lower species diversity (with many endemics), stunted stature, and sclerophyllous leaves (Ch. 4, Sect. 8). Where sandy soils are replaced by richer soil in upper Amazonia (Colombia and Peru) and in the Amazon delta, the physiognomy of the forest approaches that of seasonal *várzea*. In transition zones where black or clear acid waters mix with the white waters, the change in species composition is marked by a peak in dominance of the palm, *Eupera simoni* (Irmler, 1977).

Floodplain forests are restricted to small creeks in the upper reaches of Amazonia. These forests receive only short and irregular periods of inundation in response to local rainfall and flash flooding. At higher elevations above the stream where flooding is of short duration, many upland species occur. Otherwise, physiognomy and species composition approaches that of seasonal *várzea*.

Tidal *várzea* forests encompass considerable variation in species composition. Prance observed that studies had not yet been conducted by which further divisions might be distinguished. For example, areas flooded only by spring tides are more similar to seasonal *várzea*, while sites subjected to twice-daily tides tend toward pure stands of palms, *Raphia taedigera* and *Manicaria saccifera*. Some of these palm swamps may develop under basin conditions (see also Myers, Ch. 11, on distinctions among palm-swamp types). Both seasonal and tidal *várzea* forests, however, are similarly species-poor in comparison with neighbouring *terra firma* forests (Klinge et al., 1988).

Finally, the presence of salt water in combination with twice-daily tides supports *mangrove* forests with species characteristic of the Western Hemisphere (red mangrove, *Rhizophora* spp.; "siriuba" or black mangrove, *Avicennia germinans*; white mangrove, *Laguncularia racemosa*). Red mangroves dominate riverine forests, while the other species are more common in both basin and fringe forests of the delta. *Rhizophora racemosa* extends as far upstream as the tides. Riverine mangroves are those that receive runoff from upland drainage and are normally in oligohaline portions of estuaries. Pool et al. (1977) suggested that differences in height of mangrove stands may be due to differences in size of drainage areas and the nutrient content of runoff waters. In fact, a relationship between potential nutrient loading from runoff and litter production may exist under optimal salinity regimes (Pool et al., 1975). Fanshawe (1954) described a floristic gradient upstream from the mangrove zone in Guyana in which changes in species composition are likely due to changes in both salinity and hydroperiod.

It appears that some of the general statements concerning high biomass for riverine forests compared with uplands hold also for the humid tropics. A freshwater riverine forest in Panama dominated by *Prioria copaifera* had the largest basal area, greatest maximum tree height, and by far the highest biomass compared with upland forests and a riverine mangrove forest also in Panama (Table 5.10). The freshwater riverine forest also had substantially more litter than upland types, but understory development was not sparse as observed for temperate riverine forests. Riverine mangroves, however, differed from all other forest types by low stem density and large, widely spaced trees, and by the absence of palm species, vines, and a developed understory. A substantial portion of both aboveground biomass and basal area appears to be shifted toward prop roots, which were included in the root compartment. High litter values indicate that the dense prop-root environment is an effective trap for materials transported by twice-daily tides as described by Wolanski and Ridd (1986).

The forest types described for Amazonas and Panama do not include those that occur in arid climates, and, with the exception of seasonal igapó, are not subjected to long periods of water stress. Like the riverine forests of the Mediterranean and Arid Southwest regions of the U.S.A., arid biomes of the tropics and subtropics normally support forests in river valleys. Holdridge et al. (1971) described such a forest dominated by *Parkinsonia oculeata* in Costa Rica, in which stand height was only 7 m and structural indices were lower than those of riverine forests in areas of higher rainfall. As Specht (Ch. 16) points out, most floodplain trees of arid Australia are exposed to some water deficit annually, and increasing aridity results in a transition from closed canopy to open woodland.

Several studies in tropical Africa provide additional comparisons between adjacent riverine and upland forests, and, at the same time, illustrate the effects of increasing aridity on the physiognomy of riverine forests. In the relatively moist climate of southern Nigeria (precipitation = 2080 mm yr^{-1}), the uneven distribution of tall trees and the intermixture of impenetrable shrub thickets of a stream swamp was attributed to variations in floodplain topography (Richards, 1939). Lowest and wettest elevations were dominated by shrubs where the canopy was thin. Buttressing was not as common as in the nearby upland forest, but stilt roots were common in the swamp. Shade-tolerant herbaceous species were scarce, but woody vines

TABLE 5.10

Comparison of four Panamanian forests. From Golley et al. (1975)

	Riverine		Upland	
	Freshwater	Mangrove	Tropical moist[1]	Premontane wet
Density of palm trees[2] (no. ha^{-1})	1524	0	568	3592
Density of other trees[2] (no. ha^{-1})	3792	712	6048	6904
Density of vines (no. ha^{-1})	12 332	0	11 424	15 520
Basal area ($m^2\ ha^{-1}$)	14.91	3.39	11.32	8.27
Tree height (m)	49	41	40	31
Biomass (t ha^{-1})				
Aboveground	1176.58	279.33	367.96	271.37
Understory	1.24	0.02	1.71	0.91
Roots	12.19	189.76	9.85	12.71
Litter	14.15	102.11	2.91	4.82

[1] Wet season site, Rio Lara.
[2] Trees with diameter at breast height ≥ 10 cm.

were much more abundant in the riverine forest. Only 29% of the 38 tree species present in the swamp were restricted to the riverine forest, and there was a clear dominance by *Mitragyna stipulosa* and *Spondianthus ugandensis*. Regeneration of *M. stipulosa* after clearing has been attributed to germination success on mounds created by the worm *Alma* sp. (McCarthy, 1962).

In the slightly drier Budongo forest region of Uganda (precipitation = 1500 mm yr^{-1}), Eggeling (1947) described a stream-side forest dominated by *M. stipulosa*, similar to that described for Nigeria, but with *Pseudospondias microcarpa* as the subdominant. In comparison with upland forest types nearby, the swamp was the only one in which the same species was dominant in all size classes of the overstory and understory. Eggeling mentioned several life forms (palms, rattans, bamboo, pandan) that were completely absent in all upland forest types.

For an even drier climate in Kenya (precipitation = 600 mm yr^{-1}), short (<8 m) thorn woodlands of *Acacia* spp. prevailed in upland sites (Bogdan, 1958). The floodplain was dominated by the much taller (18 m) *Acacia usambarensis*, but even this forest is considerably shorter than the riverine forests (>30 m height) in the wetter climates just described.

These sites in Africa would support the concept of greater species dominance in riverine than upland forests, but buttressing, a possible adaptation to unstable swamp substrate, is less common in these wetlands than in uplands. With increasing aridity, canopy height in riverine forests decreases, but less so than in uplands.

Thompson and Hamilton (1983) have provided an overview of wetlands on the African continent, mostly non-forested. Floodplain systems have been described in reference to riverine fisheries by Welcomme (1979).

IMPACTS AND MANAGEMENT

This section illustrates, by example, how alterations of rivers and their drainage basins affect riverine forests. Riverine forests are extremely susceptible to changes in hydrology and fluvial processes. Few other land forms are as dynamic as floodplains, where the channel adjusts its capacity to natural episodes of large, infrequent floods and variations in sediment load. Although topographic relief is muted in comparison to many upland landforms, the presence of surface water and flooding from natural hydrologic events impose strong control over the microenvironments to which the vegetation is adapted.

Table 5.11 summarizes alterations that are common in streams, the structure and flows that are affected, and the generic category of alteration. The consequences of these alterations are predictable in a general way and, at times, the consequences can be anticipated with high certainty. In general, those alterations that interact with primary sources of energy and material flow will have the greatest probability of altering form and function of riverine forests (Ch. 4, Sect. 13 and 19). Primary energy sources in riverine forests include sunlight, water flow and hydroperiod, and accompanying sediment gains and losses.

Stream channelization

One of the purposes of stream channelization is to improve downstream conveyance of water. This is usually achieved by deepening, widening, and straightening the channel. It represents initially a disruptive geomorphic change that would never occur under natural conditions. Of the normal sources of energy to the floodplain surface such as water flow and sediment exchange, all are eliminated except sunlight.

The geomorphic consequences of reducing sinuosity is to increase channel gradient, which results in sharper pulses in flow and a concentration of water flow over a shorter time. Because the kinetic energy of water flow is concentrated in the stream channel rather than being dissipated across the floodplain, gullying may be initiated, with transport of sediments downstream. Channelization in low-order streams usually entails removal of stream-side vegetation, which precludes influxes of litter. To the extent that better drainage from channelization lowers water tables, any floodplain vegetation remaining is likely to experience water stress. However, stream-side forests are often removed in the channelization process, and land clearing for agriculture commonly follows improved drainage.

TABLE 5.11

Examples of riverine forest alteration and their relationship to categories of alteration shown in Ch. 4, Sect. 13. Alterations are listed in approximate order of decreasing severity of their impacts on riverine forests, and in decreasing time required for recovery following cessation of perturbations

Intrusions and alterations	Ecosystem component affected		Category of alteration
	Structure	Function	
Stream channelization	Channel depth increased	Decreases in floodplain-channel exchanges of water, nutrients, and organisms	Water delivery and geomorphology
	Channel gradient increased and sinuosity decreased	Sharper pulses in flow, increased effectiveness of material transport, loss of sinuosity	Water delivery and geomorphology
Containment of streamflow and channel constriction	Restricted floodplain storage	Increased channel scour and greater deposition in narrowed floodplain	Water delivery and geomorphology
Impoundments and diversions:			
Upstream in flooded area	Biomass and water depth	Primary productivity, nutrient cycling, upstream–downstream exchange of organisms	Water delivery and geomorphology
Downstream	Channel depth increased	Sediment supply decreased, scour continues	Water delivery
Introduction of toxic compounds:			
Herbicides	Plant biomass	Primary productivity, trophic structure, and nutrient cycling	Physiological stress
Insecticides	Animal biomass	Trophic structure	Physiological stress
Heavy metals	Plant and animal biomass	Primary productivity, trophic structure, and nutrient cycling	Physiological stress
Timber harvest followed by agriculture	Standing stocks of plant biomass and nutrients. Stream bank deterioration	Decreased primary productivity, increased nutrient export, and increased sediment supply and transport	Biomass removal and geomorphology
Grazing by livestock	Plant age structure	Primary productivity and biomass accumulation	Biomass removal
	Stream bank deterioration	Increased sediment supply and transport	Geomorphology
Timber harvest followed by silviculture	Standing stocks of plant biomass and nutrients	Temporarily decreased primary productivity	Biomass removal
Hunting and fishing	Standing stocks of animal biomass	Grazing and predation	Biomass removal

Stream channel containment and constriction

When levees are built to contain floodwaters in the central portion of the floodplain, the surface area for distribution and deposition of sediments is reduced. This causes more rapid deposition between levees and more rapidly obliterates topographic features of the floodplain. This is occurring on a large scale along the Mississippi River (Belt, 1975) and its distributary, the Atchafalaya River in Louisiana (Van Beek, 1979). At the same time that accretion accelerates and channels become elevated above their original level, floodplains outside the levees are deprived of sediments, just as channelization reduces exchanges between stream channel and floodplain.

Dikes and jetties also contribute to containment of stream flow and channel constriction which stabilizes channel position, narrows channel width, and results in swifter currents. An increase in riverine forest may actually result because broad, braided streams may be converted into relatively narrow and swift channels. For example, in a 830 km reach of the Missouri River, the surface area of the stream was reduced by half (24 600 ha between 1879 and 1972; Funk and Robinson, 1974). Islands, sandbars, snags, and marshes have been virtually eliminated (Fig. 5.17). Construction of dikes and revetments have been responsible for the surface area lost, but levees, mainstream dams, and tributary reservoirs also contributed to changes in channel configuration. Much of the recently accreted floodplain has been put into cultivation. The resulting narrower, swifter, and deeper channel has reduced aquatic habitat diversity, species richness of fish, and commercial catches of fish.

Impoundments and diversions

Upstream from impoundments in reservoirs, the change from floodplain vegetation to deep aquatic habitat is so obvious and extreme that it does not warrant further consideration. In upstream areas where floodplain forests are converted from a seasonal to a chronic flooding regime by impoundments, survival of trees is lower in colder climates than in warmer ones (Lugo and Brown, 1984). Downstream from impoundments, water delivery patterns are altered and sediment supplies reduced. Williams and Wolman (1984) observed that sediment concentrations and suspended loads were lower than those of pre-dam conditions for hundreds of kilometers downstream in rivers of the U.S.A. Large reservoirs commonly trap more than 99% of the sediment entering the reservoir. Other documented changes in streams regulated by reservoir storage are changes in water chemistry (Hannan, 1979; Krenkel et al., 1979), effects on channel morphology (Simons, 1979), and changes in water temperature (Fraley, 1979). Conversion of land use to agriculture is frequently the result of stream-flow regulation by impoundments. The Platte Rivers in Nebraska and Colorado have undergone a reduction in width by 80 to 95%

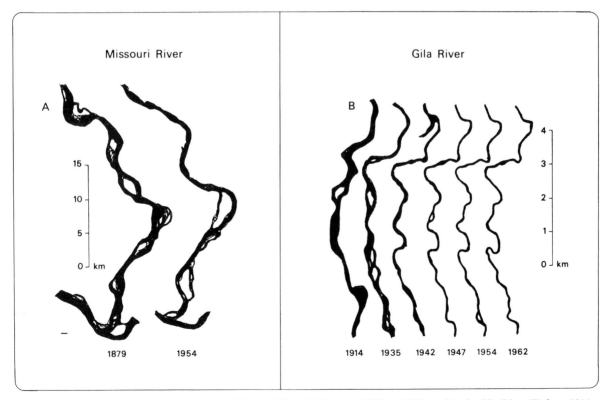

Fig. 5.17. Changes in channel morphology of the Missouri River (A) between 1879 and 1954, and in the Gila River (B) from 1914 to 1962. After Funk and Robinson (1974) and Turner (1974).

during the past 100 years (Williams, 1978). The amount of floodplain vegetation has increased considerably at the expense of aquatic surface area and sparsely vegetated islands. Nadler (1978) attributed this trend to more stable flow regimes. Irrigation water, which is withdrawn from the river and reduces its sediment load, raises floodplain water tables in irrigated regions and produces more uniform stream flow. Riverine forests become more dense and may invade channels during drought years. The result has been a transition from relatively straight, wide, and intermittent streams to narrow and swift channels with more sinuous configuration. Perhaps for similar reasons, the Gila River, Arizona has undergone a gradual narrowing since 1914 (Fig. 5.17). Establishment of more riverine vegetation has also resulted from changes in the hydrologic regime of the Colorado River in the Grand Canyon (Turner and Karpiscak, 1980). Before Glen Canyon Dam was built, seasonal variations in discharge were large and daily variations low (Fig. 5.18). A reversal in this hydroperiod after dam operation in 1963 has resulted in the establishment of vegetation along the river, especially exotic species such as saltcedar (*Tamarix pentandra*) and Russian olive (*Elaeagnus angustifolia*). However, most of the examples surveyed by Williams and Wolman (1984) were cases where channel width has increased as a result of lateral erosion of the steep banks created by channel degradation.

Even if the floodplain is not subjected to changes in land use, the decrease in sediment supply below the impoundment will result in channel scouring (Galay, 1983), greatly reduce flooding frequency, and may eliminate sediment delivery to the floodplain. Usually this occurs in the first decade after dam closure, and may degrade the bed up to 7.5 m (Williams and Wolman, 1984). More subtle effects of reduced moisture supplies to riverine forests have already been described for the Missouri River in the Plains Grassland Region (Johnson et al., 1976, 1982). Because fish yield is positively correlated with the surface area of floodplain inundated (Welcomme, 1979), reduction in peak flows downstream from reservoirs reduces a potential source of food. Other consequences are reduced sediment supply to estuaries, and upstream salinity intrusion (Lagler, 1969; National Research Council (U.S.A.), 1982).

Introduction of toxic compounds

Herbicides, insecticides, toxic metals, acid rain, and other toxic compounds introduced to riverine forests either directly or indirectly from the stream by flooding over the banks can be regarded as a physiological stress to either the vegetation or the microbial communities that regulate elemental cycling. If water delivery and geomorphic changes do not occur, the primary energy sources are maintained and recovery may be possible if disruptions are not renewed. In fact, organic-rich sediments of riverine wetlands may immobilize heavy metals and retain pesticides long enough for them to be detoxified (Pionke and Chesters, 1973). Little is known of the capacity of floodplains for processing these residues of spills, leaks, and the appearance of synthetic chemicals in runoff. Although this function potentially helps to maintain water quality of streams, accidental or intentional discharges into riverine ecosystems are likely to have widespread and uncontrollable effects because of the strong hydrological dispersal powers of river systems.

Livestock grazing

Grazing by livestock removes plant biomass, alters the age structure of plant populations, and may change species composition of riverine forests. Although these changes are not restricted to

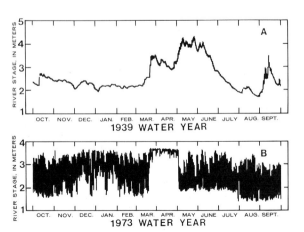

Fig. 5.18. Daily variation in river stage for the Colorado River at Lees Ferry during water years 1939 and 1973. From Turner and Karpiscak (1980).

riverine forests, cattle spend more time in streamside forests than they do in adjacent uplands in arid regions (Martin, 1979). Tree reproduction is affected most by heavy browsing on young plants (Dahlem, 1979). Without recruitment of young trees, riverine forests develop an unstable age structure and become biased toward large, older trees (Warner, 1984).

Intense grazing and browsing may secondarily result in increased runoff from uplands and reduction in stability of stream channels. Restoration may require construction of check dams, in addition to reductions in or elimination of grazing, for the recovery of plant communities. Rangeland restoration in Colorado demonstrated that streams may be transformed from intermittent to perennial flow when restoration resulted in retention of alluvial fill and re-establishment of stream-side vegetation (Heede, 1977). This appears to conflict with the notion that stream-side vegetation reduces water yield. However, annual water yields were not determined, and increase in water storage capacity of newly acquired alluvial fill appears to be responsible for the perennial flow pattern.

Timber harvest

Many riverine forests are attractive for removal of wood products because of high stocking density and rapid rates of growth. Forest management practices range from selective removal of mature trees to clear-cutting and the transformation of natural stands to intensive silviculture. The capacity of floodplain forests to recover from harvests depends on the extent to which propagules of native species are available for reproduction, provided that drainage patterns and hydroperiod are not altered. Clear-cutting causes temporary decreases in evapotranspiration, primary productivity, and probably the capacity to recycle nutrients. Selective cutting would have negligible effects on these processes but may greatly affect species composition. Animal communities are most severely disrupted by clear-cutting, but even selective cutting can alter certain populations. For example, *Ficus sycomorus* is the single most important food for arboreal primates on the Tana River, Kenya, and is also heavily exploited for dugout canoes (Marsh, 1980).

Related to timber harvesting is the removal of phreatophytic vegetation in floodplains of arid climates. The practice has been used in the southwestern U.S.A. as a measure to conserve water for consumptive use, mainly in irrigating farmland. However, because earlier studies of evapotranspiration did not take into account that these species react to high wind-speeds and temperature by stomatal closure even when water is freely available, early estimates of water use by phreatophytes were excessive (Van Hylckama, 1980). Culler et al. (1982) reported that dense stands of phreatophytes in Arizona evapotranspire 1420 mm yr^{-1} in comparison with 630 mm yr^{-1} in areas with no phreatophytes. Biomass removal reduced evapotranspiration to that of areas originally devoid of phreatophytes, supporting the observation of Davenport et al. (1982) that evapotranspiration by saltcedar is dependent on stand density. However, the reduction is temporary because of natural recolonization. Frequent clearing would be necessary to maintain this artificial condition.

Secondary practices of flood control and drainage are more seriously damaging to riverine floodplains than biomass removal. When upland and bottomland forests are converted to agricultural crop production, changes in flood regime and exposure to toxic compounds normally occur.

Hunting and fishing

Removal of animal biomass from riverine forest would have a negligible effect in altering primary energy sources that control overall ecosystem function. An exception is beaver populations which have the capacity to alter the hydrology of floodplains and have large effects on plant community composition. Activity of beavers is usually inconsistent with management of riverine forests for timber production. It has been demonstrated that the fisheries of certain tropical rivers are dependent upon floodplain inundation. Seasonality of reproduction and feeding is closely keyed to hydrologic events (Ch. 4, Sect. 4) in floodplains of Africa (Welcomme, 1979) and the Amazon, Orinoco, and other South American rivers (Bayley, 1981; Goulding, 1981; Smith, 1981; Hamilton and Lewis, 1987). However, there are examples of wildlife management practices that alter hydroperiod for selected game species.

CONCLUSIONS

Riverine forests owe their uniqueness to the special hydrologic conditions and geomorphic dynamics characteristic of floodplains. They vary in structure from communities of massive trees which tolerate deep flooding in the humid tropics, to shrub thickets in the Arctic tundra which sprout after periodic scour by ice and gravel. Arid climates support floodplain forests in a treeless landscape. Riverine forests even occur in oligohaline portions of estuaries in the tropics and subtropics. Within a given geomorphic context, the climate, hydroperiod, salinity, and biogeographic location are the four principal factors that influence species composition.

Historical patterns of river channel movement, induced by infrequent catastrophic floods, are responsible for the gross topographic features found in riverine forests. Within this topographic pattern, more frequent, lower power floods determine the environmental conditions for tree reproduction, animal activity, and other short term phenomena.

Although the depth of flooding in a riverine forest is determined by climate, topography, channel slope, soils, and lithology, it is the size of the drainage basin that controls the flood-duration component of hydroperiod, if all other factors are held constant. Therefore, flood-tolerant species tend to be more prevalent on floodplains of large rivers than small ones. On a smaller scale, the arrangement of plant communities within a given riverine forest will be a reflection of complex hydrologic patterns that are dictated by local drainage.

Flooding frequency alone seldom adequately explains the species mixture at a given site because of the influence of soil texture and drainage on moisture regimes. Different stressors impose limits at the wet and dry ends of the moisture continuum. At the wet end, anaerobic soils, deep water, and strong currents limit forest development. At the dry end, the absence of supplementary water from the river marks the transition to upland ecosystems. Species richness of trees is limited at the wet end; only those few species that are adapted to anaerobic conditions and frequent disturbance survive. A phreatophytic habit is a necessity for survival of forest species in floodplains of arid climates.

In general, riverine forests have greater basal area, biomass, and biomass production rates than uplands of the same location, a pattern which is most striking in arid climates. Factors implicated in regulating biomass production include deviations from normal levels of soil moisture and hydroperiod, nutrient limitation, and physical damage by flooding, fire, and herbivores. However, the sedimentary regime of most riverine forests makes nutrient limitation less likely than in other forests. Cycling of elements is generally rapid, in part, because alternating soil moisture levels are induced by fluctuating water levels. Consequently, decomposition is rapid and there is little tendency toward accumulation of soil organic matter. However, mechanisms of nutrient conservation operate as they do in other ecosystems.

As a result of their position in the landscape, riverine forests play a critical role, disproportionately large for their size, in buffering potential impacts on water quality of rivers from disturbances in upland ecosystems. The fact that riverine forests are so dependent on imported materials (i.e. water, nutrients, and sediments) means that they are also vulnerable to alteration when deprived of these materials.

REFERENCES

Abernethy, Y. and Turner, R.E., 1987. U.S. forested wetlands: 1940–1980. *Bioscience*, 37: 721–727.

Abrams, M.D., 1986. Historical development of gallery forests in northeast Kansas. *Vegetatio*, 65: 29–37.

Adams, D.E. and Anderson, R.C., 1980. Species response to a moisture gradient in central Illinois forests. *Am. J. Bot.*, 67: 381–392.

Adis, J., Furch, K. and Irmler, U., 1979. Litter production of a Central-Amazonian black water inundation forest. *Trop. Ecol.*, 20: 236–245.

Allen, J.R.L., 1965. A review of the origin and characteristics of recent alluvial sediments. *Sedimentology*, 5: 89–191.

Anderson, R.C. and White, J., 1970. A cypress swamp outlier in southern Illinois. *Trans. Ill. Acad. Sci.*, 63: 6–13.

Applequist, M.B., 1959. *A Study of Soil and Site Factors Affecting the Growth and Development of Swamp Blackgum and Tupelo Gum Stands in Southeastern Georgia*. Ph.D. Dissertation, Duke University, Durham, N.C., 181 pp.

Bailey, R.G., 1978. *Description of the Ecoregions of the United States*. USDA Forest Service, Intermountain Region, Ogden, Utah, 77 pp.

Barnes, W.J., 1978. The distribution of floodplain herbs as influenced by annual flood elevation. *Wisc. Acad. Sci., Arts Lett.*, 66: 254–266.

Barnes, W.J., 1985. Population dynamics of woody plants on a river island. *Can. J. Bot.*, 63: 647–655.

Baumann, R.H., Day, J.W., Jr. and Miller, C.A., 1984. Mississippi deltaic wetland survival: Sedimentation versus coastal submergence. *Science*, 224: 1093–1095.

Bayley, P.B., 1981. Fish yield from the Amazon in Brazil: Comparison with African river yields and management possibilities. *Trans. Am. Fish. Soc.*, 110: 351–359.

Bedinger, M.S., 1979. *Forests and Flooding with Special References to the White River and Ouachita River Basins, Arkansas*. U.S. Geol. Surv. Water Resour. Invest., Open-file Rep. 79–68, U.S. Gov. Printing Office, Washington D.C.

Bell, D.T., 1974. Studies on the ecology of a streamside forest: composition and distribution of vegetation beneath the tree canopy. *Bull. Torrey Bot. Club*, 101: 14–20.

Bell, D.T. and Johnson, F.L., 1974. Ground-water level in the flood plain and adjacent uplands of the Sangamon River. *Trans. Ill. State Acad. Sci.*, 67: 376–383.

Bell, D.T. and Sipp, S.K., 1975. The litter stratum in the streamside forest ecosystem. *Oikos*, 26: 391–397.

Belt, C.B., Jr., 1975. The 1973 flood and man's constriction of the Mississippi River. *Science*, 189: 681–684.

Bliss, L.C. and Cantlon, J.E., 1957. Succession on river alluvium in northern Alaska. *Am. Midl. Nat.*, 58: 452–469.

Bogdan, A.V., 1958. Some edaphic vegetational types at Kiboko, Kenya. *J. Ecol.*, 46: 115–126.

Brice, J., 1971. *Measurement of Lateral Erosion at Proposed River Crossing Sites of the Alaska Pipeline*. U.S. Geological Survey, Water Resource Division, Alaska District, 39 pp.

Briggs, S.V., 1977. Estimates of biomass in a temperate mangrove community. *Aust. J. Ecol.*, 2: 369–373.

Brinson, M.M., 1977. Decomposition and nutrient exchange of litter in an alluvial swamp forest. *Ecology*, 58: 601–609.

Brinson, M.M., Bradshaw, H.D., Holmes, R.N. and Elkins, J.B., Jr., 1980. Litterfall, stemflow, and throughfall nutrient fluxes in an alluvial swamp forest. *Ecology*, 61: 827–835.

Brinson, M.M., Bradshaw, H.D. and Kane, E.S., 1981a. *Nitrogen Cycling and Assimilative Capacity of Nitrogen and Phosphorus by Riverine Wetland Forests*. Water Resources Research Institute, University of North Carolina, Rep. No. 167, Raleigh, N.C., 84 pp.

Brinson, M.M., Lugo, A.E. and Brown, S., 1981b. Primary productivity, consumer activity, and decomposition in freshwater wetlands. *Annu. Rev. Ecol. Syst.*, 12: 123–161.

Brinson, M.M., Bradshaw, H.D. and Holmes, R.N., 1983. Significance of floodplain sediments in nutrient exchange between a stream and its floodplain. In: T.D. Fontaine and S.M. Bartell (Editors), *Dynamics of Lotic Ecosystems*. Ann Arbor Science, Ann Arbor, Mich., pp. 199–221.

Brinson, M.M., Bradshaw, H.D. and Kane, E.S., 1984. Nutrient assimilative capacity of an alluvial floodplain swamp. *J. Appl. Ecol.*, 21: 1041–1058.

Brinson, M.M., Bradshaw, H.D. and Jones, M.N., 1985. Transitions in forested wetlands in North Carolina along gradients of salinity and hydroperiod. *J. Elisha Mitchell Sci. Soc.*, 102: 76–94.

Broadfoot, W.M., 1960. Soil water shortages and a means of alleviating resulting influences on southern hardwoods. In: *Eighth Annual Forestry Symposium Proceedings*, Louisiana State University, Baton Rouge, La., pp. 115–119.

Brown, S., 1978. *A Comparison of Cypress Ecosystems and Their Role in the Florida Landscape*. Ph.D. Dissertation, University of Florida, Gainesville, Fla., 569 pp.

Brown, S., 1981. A comparison of the structure, primary productivity, and transpiration of cypress ecosystems in Florida. *Ecol. Monogr.*, 51: 403–427.

Brown, S. and Lugo, A.E., 1982. A comparison of structural and functional characteristics of saltwater and freshwater forested wetlands. In: B. Gopal, R.E. Turner, R.G. Wetzel and D.F. Whigham (Editors), *Wetlands: Ecology and Management*. National Institute of Ecology and International Scientific Publications, Jaipur, pp. 109–130.

Brown, S. and Peterson, D.L., 1983. Structural characteristics and biomass productivity of two Illinois bottomland forests. *Am. Midl. Nat.*, 110: 107–117.

Bruner, W.E., 1931. The vegetation of Oklahoma. *Ecol. Monogr.*, 1: 99–188.

Brush, G.A., Lenk, C. and Smith, J., 1980. The natural forests of Maryland: an explanation of the vegetation map of Maryland. *Ecol. Monogr.*, 50: 77–92.

Buchholz, K., 1981. Effects of minor drainages on woody species distributions in a successional floodplain forest. *Can. J. For. Res.*, 11: 671–676.

Buell, M.F. and Wistendahl, W.A., 1955. Flood plain forests of the Raritan River. *Bull. Torrey Bot. Club*, 82: 463–472.

Burnett, A.W. and Schumm, S.A., 1983. Alluvial-river response to neotectonic deformation in Louisiana and Mississippi. *Science*, 222: 49–50.

Burns, L.A., 1978. *Productivity, Biomass, and Water Relations in a Florida Cypress Forest*. Ph.D. Dissertation, University of North Carolina, Chapel Hill, N.C., 170 pp.

Bush, J.K. and Van Auken, O.W., 1984. Woody-species composition of the upper San Antonio River gallery forest. *Texas J. Sci.*, 36: 139–148.

Campbell, C.J. and Dick-Peddie, W.A., 1964. Comparison of phreatophyte communities on the Rio Grande in New Mexico. *Ecology*, 45: 492–502.

Campbell, C.J. and Green, W., 1968. Perpetual succession of stream-channel vegetation in a semiarid region. *J. Ariz. Acad. Sci.*, 5: 86–98.

Chamless, L.F. and Nixon, E.S., 1975. Woody vegetation–soil relations in a bottomland forest of east Texas. *Texas J. Sci.*, 26: 407–416.

Church, M., 1977. River studies in northern Canada: reading the record from river morphology. *Geosci. Can.*, 4: 4–12.

Clark, J.R. and Benforado, J. (Editors), 1981. *Wetlands of Bottomland Hardwood Forests*. Elsevier, New York, N.Y., 401 pp.

Coble, R.W., 1979. *A Technique for Estimating Heights Reached by the 100-year Flood in Unregulated, Nontidal Streams in North Carolina*. U.S. Geol. Surv. Water-Resour. Invest., Open-file Rep. 79–69, U.S. Gov. Printing Office, Washington, D.C.

Conard, S.G., MacDonald, R.L. and Holland, R.F., 1977. Riparian vegetation and flora of the Sacramento Valley. In: A. Sands (Editor), *Riparian Forests in California: Their Ecology and Conservation*. Publ. No. 15, Institute of Ecology, Davis, Calif., pp. 47–55.

Conner, W.H. and Day, J.W., Jr., 1976. Productivity and composition of a baldcypress–water tupelo site and bottomland hardwood site in a Louisiana swamp. *Am. J. Bot.*, 63: 1354–1364.

Conner, W.H., Gosselink, J.G. and Parrondo, R.T., 1981. Comparison of the vegetation of three Louisiana swamp sites with different flooding regimes. *Am. J. Bot.*, 63: 320–331.

Conner, W.H., Toliver, J.R. and Sklar, F.H., 1986. Natural regeneration of baldcypress (*Taxodium distichum* (L.) Rich.) in a Louisiana swamp. *For. Ecol. Manage.*, 14: 305–317.

Cooper, J.R. and Gilliam, J.W., 1987. Phosphorus redistribution from cultivated fields into riparian areas. *Soil Sci. Soc. Am. J.*, 51: 1600–1604.

Crites, R.W. and Ebinger, J.E., 1969. Vegetation survey of floodplain forests in east-central Illinois. *Trans. Ill. Acad. Sci.*, 62: 316–330.

Culler, R.C., Hanson, R.L., Myrick, R.M., Turner, R.M. and Kipple, F.P., 1982. *Evapotranspiration before and after clearing phreatophytes, Gila River flood plain, Graham County, Arizona.* U.S. Geol. Surv. Prof. Pap. 655-P, U.S. Gov. Printing Office, Washington, D.C., 51 pp.

Dahlem, E.A., 1979. The Mahogany Creek watershed — with and without grazing. In: O.B. Cope (Editor), *Grazing and Riparian/Stream Ecosystems*. Trout Unlimited, Denver, Colo., pp. 31–34.

Davenport, D.C., Martin, P.E. and Hagan, R.M., 1982. Evapotranspiration from riparian vegetation: water relations and irrecoverable losses for saltcedar. *J. Soil Water Conserv.*, 37: 233–236.

DeLaune, R.D., Smith, C.J., Patrick, W.H., Jr. and Roberts, H.H., 1987a. Rejuvenated marsh and bay-bottom accretion on the rapidly subsiding coastal plain of the U.S. Gulf Coast: a second-order effect of the emerging Atchafalaya delta. *Estuarine, Coastal Shelf Sci.*, 25: 381–389.

DeLaune, R.D., Patrick, W.H., Jr. and Pezeshki, S.R., 1987b. Forseeable flooding and death of coastal wetland forests. *Environ. Conserv.*, 14: 129–133

Dickson, R.E. and Broyer, T.C., 1972. Effects of aeration, water supply, and nitrogen source on growth and development of tupelo gum and bald cypress. *Ecology*, 53: 626–634.

Dirschl, H.J. and Coupland, R.T., 1972. Vegetation patterns and site relationships in the Saskatchewan River delta. *Can. J. Bot.*, 50: 647–675.

Drury, W.H., Jr., 1956. Bog flats and physiographic process in the upper Kuskokwim River region, Alaska. *Contributions Gray Herbarium, Harvard Univ.* 1128, 130 pp.

Duever, M.J., Carlson, J.E. and Riopelle, L.A., 1975. Ecosystem analysis at Corkscrew Swamp. In: H.T. Odum and K.C. Ewel (Editors), *Cypress Wetlands for Water Management, Recycling and Conservation*. 2nd Annu. Rep., University of Florida, Gainesville, Fla., pp. 627–725.

Duever, M.J., Carlson, J.E. and Riopelle, L.A., 1984. Corkscrew Swamp: a virgin cypress strand. In: K.C. Ewel and H.T. Odum (Editors), *Cypress Swamps*. University Presses of Florida, Gainesville, Fla., pp. 334–348.

Dunn, C.P. and Stearns, F., 1987. Relationship of vegetation layers to soils in southeast Wisconsin forested wetlands. *Am. Midl. Nat.*, 118: 366–374.

Eggeling, W.J., 1947. Observations on the ecology of the Budongo rain forest, Uganda. *J. Ecol.*, 34: 20–87.

Eggler, W.A., 1938. The maple–basswood forest type in Washburn County, Wisconsin. *Ecology*, 19: 243–263.

Ehrenfeld, J.G. and Gulick, M., 1981. Structure and dynamics of hardwood swamps in the New Jersey Pine Barrens: contrasting patterns in trees and shrubs. *Am. J. Bot.*, 68: 471–481.

Elder, J.F., 1985. Nitrogen and phosphorus speciation and flux in a large Florida river wetland system. *Water Resour. Res.*, 21: 724–732.

Elder, J.F. and Cairns, D.J., 1982. *Production and Decomposition of Forest Litter Fall on the Apalachicola River Flood Plain, Florida.* U.S. Geological Survey Water-Supply Paper 2196-B. U.S. Gov. Printing Office, Washington, D.C.

Everitt, B.L., 1968. Use of the cottonwood in an investigation of the recent history of a flood plain. *Am. J. Sci.*, 266: 417–439.

Fanshawe, D.B., 1954. Riparian vegetation of British Guiana. *J. Ecol.*, 42: 289–295.

Fisk, H.N., 1944. *Geological Investigation of the Alluvial Valley of the Lower Mississippi River.* Mississippi River Commission, Vicksburg, Miss., 78 pp.

Fisk, H.N., 1952. *Geological Investigation of the Atchafalaya Basin and the Problem of the Mississippi River Diversion. Volume 1.* Miss. River Commission, Vicksburg, Miss., 145 pp.

Fisk, H.N. and McFarlan, E., Jr., 1955. Late Quaternary deltaic deposits of the Mississippi River. *Geol. Soc. Am. Spec. Pap.*, 62: 279–302.

Fonda, R.W., 1974. Forest succession in relation to river terrace development in Olympic National Park, Washington. *Ecology*, 55: 927–942.

Ford, A.L. and Van Auken, O.W., 1982. The distribution of woody species in the Guadalupe River floodplain forest in the Edwards Plateau of Texas. *Southwest. Nat.*, 27: 383–392.

Fraley, J.J., 1979. Effects of elevated stream temperatures below a shallow reservoir on a cold water macroinvertebrate fauna. In: J.V. Ward and J.A. Stanford (Editors), *The Ecology of Regulated Streams*. Plenum Press, New York, N.Y., pp. 257–272.

Frangi, J.L. and Lugo, A.E., 1985. Ecosystem dynamics of a subtropical floodplain forest. *Ecol. Monogr.*, 55: 351–369.

Franz, E.H. and Bazzaz, F.A., 1977. Simulation of vegetation response to modified hydrologic regimes: a probablistic model based on niche differentiation in a floodplain forest. *Ecology*, 58: 176–183.

Freeman, C.E. and Dick-Peddie, W.A., 1970. Woody riparian vegetation in the Black and Sacramento Mountain Ranges, southern New Mexico. *Southwest. Nat.*, 15: 145–164.

Frye, R.J., II and Quinn, J.A., 1979. Forest development in relation to topography and soils on a floodplain of the Raritan River, New Jersey. *Bull. Torrey Bot. Club*, 106: 334–345.

Funk, J.L. and Robinson, J.W., 1974. *Changes in the Channel of the Lower Missouri River and Effects on Fish and Wildlife.* Aquatic Series No. 11, Missouri Dept. of Conservation, Jefferson City, Mo.

Furtado, J.I., Verghese, S., Liew, K.S. and Lee, T.H., 1980. Litter production in a freshwater swamp forest, Tasek Bera, Malaysia. In: J.I. Furtado (Editor), *Tropical Ecology and Development*. International Society of Tropical Ecology, Kuala Lumpur, pp. 815–822.

Gaddy, L.L., Kohlsaat, T.S., Laurent, E.A. and Stansell, K.B., 1975. *A Vegetation Analysis of Preserve Alternatives Involving the Beidler Tract of the Congaree Swamp*. South Carolina Wildlife & Marine Resources Dept., Columbia, S.C.

Galay, V.J., 1983. Causes of river bed degradation. *Water Resour. Res.*, 19: 1057–1090.

Gatewood, J.S., Robinson, J.W., Colby, B.R., Hem, J.D. and Halpenny, L.C., 1950. *Use of Water by Bottomland Vegetation in Lower Stafford Valley, Arizona*. U.S. Geol. Surv. Water-Supply Pap. 1103. U.S. Gov. Printing Office, Washington, D.C.

Gill, C.J., 1970. The flooding tolerance of woody species — a review. *For. Abstr.*, 31: 671–688.

Gill, D., 1972a. The point bar environment in the Mackenzie River Delta. *Can. J. Earth Sci.*, 9: 1382–1393.

Gill, D., 1972b. Modification of levee morphology by erosion in the Mackenzie River Delta, Northwest Territories. In: R.J. Price and D.E. Sugden (Editors), *Polar Geomorphology*. Institute of British Geographers Spec. Publ. No. 4, pp. 123–138.

Gilman, B.A., 1976. *Wetland Plant Communities along the Eastern Shoreline of Lake Ontario*. M.S. Thesis, SUNY College of Environmental Science and Forestry, Syracuse, N.Y.

Glascock, S. and Ware, S., 1979. Forests of small stream bottoms in the peninsula of Virginia. *Va. J. Sci.*, 30: 17–21.

Golden, M.S., 1979. Forest vegetation of the lower Alabama Piedmont. *Ecology*, 60: 770–782.

Golley, F.B., McGinnis, J.T., Clements, R.G., Child, G.I. and Duever, M.J., 1975. *Mineral Cycling in a Tropical Moist Forest Ecosystem*. Univ. Georgia Press, Athens, Ga., 248 pp.

Good, B.J. and Whipple, S.A., 1982. Tree spatial patterns: South Carolina bottomland and swamp forests. *Bull. Torrey Bot. Club*, 109: 529–536.

Goulding, M., 1981. *Man and Fisheries on an Amazon Frontier*. Dr. W. Junk Publ., The Hague, 137 pp.

Grannemann, N.G. and Sharp, J.M., Jr., 1979. Alluvial hydrogeology of the lower Missouri River valley. *J. Hydrol.*, 40: 85–99.

Grime, J.P., 1979. *Plant Strategies and Vegetation Processes*. Wiley, New York, N.Y.

Haase, E.F., 1972. Survey of floodplain vegetation along the lower Gila River in southwestern Arizona. *J. Ariz. Acad. Sci.*, 7: 66–81.

Hadley, R.F., 1960. *Recent Sedimentation and Erosional History of Fivemile Creek, Fremont County, Wyoming*. U.S. Geol. Surv. Prof. Pap. 352-A, U.S. Gov. Printing Office, Washington, D.C.

Hall, T.F. and Penfound, W.T., 1939a. A phytosociological study of cypress–gum swamp in southeastern Louisiana. *Am. Midl. Nat.*, 21: 378–395.

Hall, T.F. and Penfound, W.T., 1939b. A phytosociological study of *Nyssa biflora* consocies in southeastern Louisiana. *Am. Midl. Nat.*, 22: 369–375.

Hall, T.F. and Penfound, W.T., 1943. Cypress–gum communities in the Blue Girth Swamp near Selma, Alabama. *Ecology*, 24: 208–217.

Hamilton, S.K. and Lewis, W.M., Jr., 1987. Causes of seasonality in the chemistry of a lake on the Orinoco River floodplain, Venezuela. *Limnol. Oceanogr.*, 36: 1277–1290.

Hannan, H.H., 1979. Chemical modifications in reservoir-regulated streams. In: J.V. Ward and J.A. Stanford (Editors), *The Ecology of Regulated Streams*. Plenum Press, New York, N.Y., pp. 75–94.

Hardin, E.D. and Wistendahl, W.A., 1983. The effects of floodplain trees on herbaceous vegetation patterns, microtopography and litter. *Bull. Torrey Bot. Club*, 110: 23–30.

Harms, W.R., 1973. Some effects of soil type and water regime on growth of tupelo seedlings. *Ecology*, 54: 188–193.

Harriss, R.C. and Sebacher, D.I., 1981. Methane flux in forested freshwater swamps of the southeastern United States. *Geophys. Res. Lett.*, 8: 1002–1004.

Hawk, G.M. and Zobel, D.B., 1974. Forest succession on alluvial landforms of the McKenzie River valley, Oregon. *Northwest Sci.*, 43: 245–265.

Heede, B.H., 1977. *Case Study of Watershed Rehabilitation Project: Alkali Creek, Colorado*. USDA For. Serv. Res. Pap. RM-189, U.S. Gov. Printing Office, Washington, D.C.

Heinselman, M.L., 1970. Landscape evolution, peatland types, and the environment in the Lake Agassiz Peatlands Natural Area, Minnesota. *Ecol. Monogr.*, 40: 235–261.

Hernández, R.A. and Mullen, K.P., 1979. Observaciones preliminares sobre la productividad primaria neta en un ecosistema de manglar-estuario (Guapi–Colombia). *Memorias del II Simposio Latinoamericano sobre Oceanografía Biológica*. Imprenta de la Universidad de Oriente, Cumana, pp. 89–98.

Hernández, C.J., Van Hildebrand, P. and Alvarez, L., 1980. Problematica del manejo de manglares con especial referencia al sector occidental de la Ciénaga Grande de Santa Marta, Magdalena, Colombia. In: *Estudio Científico e Impacto Humano en el Ecosistema de Manglares*. Oficina Regional de Ciencia y Tecnología de la UNESCO, Montevideo, pp. 364–386.

Hodkinson, I.D., 1975. Dry weight loss and chemical changes in vascular plant litter of terrestrial origin, occurring in a beaver pond ecosystem. *J. Ecol.*, 63: 131–142.

Hofstetter, R.H., 1983. Wetlands in the United States. In: A.J.P. Gore (Editor), *Mires: Swamp, Bog, Fen and Moor: Regional Studies*. Ecosystems of the World 4B. Elsevier, Amsterdam, pp. 201–244.

Holdridge, L.R., Grenke, W.C., Hatheway, W.H., Liang, T. and Tosi, J.A., Jr., 1971. *Forest Environments in Tropical Life Zones*. Pergamon Press, New York, N.Y., 747 pp.

Hook, D.D., DeBell, D.S., McKee, W.H., Jr. and Askew, J.L., 1983. Responses of loblolly pine (mesophyte) and swamp tupelo (hydrophyte) seedlings to soil flooding and phosphorus. *Plant Soil*, 71: 387–394.

Hosner, J.F. and Minckler, L.S., 1963. Bottomland hardwood forests of southern Illinois — regeneration and succession. *Ecology*, 44: 29–41.

Hosner, J.F., Leaf, A.L., Dickson, R.E. and Hart, J.B., Jr., 1965. Effects of varying soil moisture upon the nutrient uptake of four bottomland tree species. *Soil Sci. Soc. Am. Proc.*, 29: 313–316.

Hough, A.F., 1936. A climax forest community on East Tionesta Creek in northwestern Pennsylvania. *Ecology*, 17: 9–28.

Huenneke, L.F., 1982. Wetland forests of Tompkins County, New York. *Bull. Torrey Bot. Club*, 109: 51–63.

Huenneke, L.F., 1983. Understory response to gaps caused by death of *Ulmus americana* in central New York. *Bull. Torrey Bot. Club*, 110: 170–175.

Huenneke, L.F. and Sharitz, R.R., 1986. Microsite abundance and distribution of woody seedlings in a South Carolina cypress–tupelo swamp. *Am. Midl. Nat.*, 115: 328–335.

Hupp, C.R., 1982. Stream-grade variation and riparian-forest ecology along Passage Creek, Virginia. *Bull. Torrey Bot. Club*, 109: 488–499.

Hupp, C.R. and Osterkamp, W.R., 1985. Bottomland vegetation distribution along Passage Creek, Virginia, in relation to fluvial landforms. *Ecology*, 66: 670–681.

Irmler, U., 1977. Inundation forest types in the vicinity of Manaus. In: P. Muller (Editor), *Ecosystem Research in South America, Biogeographica*, Volume 8. Dr. W. Junk Publishers, The Hague, pp. 17–29.

Irmler, U. and Furch, K., 1980. Weight, energy, and nutrient changes during the decomposition of leaves in the emersion phase of Central-Amazonian inundation forests. *Pedobiologia*, 20: 118–130.

Irvine, J.R. and West, N.E., 1979. Riparian tree species distribution and succession along the lower Escalante River, Utah. *Southwest. Nat.*, 24: 331–346.

Janssen, C.R., 1967. A floristic study of forests and bog vegetation, northwestern Minnesota. *Ecology*, 48: 751–765.

Jeglum, J.K. and Cowell, D.W., 1982. Wetland ecosystems near Kinoje Lakes, southern interior Hudson Bay lowland. *Nat. Can.*, 109: 621–635.

Johnson, F.L. and Bell, D.T., 1976. Plant biomass and net primary production along a flood-frequency gradient in the streamside forest. *Castanea*, 41: 156–165.

Johnson, W.C., Burgess, R.L. and Keammerer, W.R., 1976. Forest overstory vegetation on the Missouri River floodplain in North Dakota. *Ecol. Monogr.*, 46: 59–84.

Johnson, W.C., Reily, P.W., Andrews, L.S., McLellan, J.F. and Brophy, J.A., 1982. Altered Hydrology of the Missouri River and Its Effects on Floodplain Forest Ecosystems. Virginia Water Resources Research Center, Virginia Polytechnic and State University, Bull. 139, Blacksburg, Va., 83 pp.

Johnston, C.A., Bubenzer, G.D., Lee, G.B., Madison, F.W. and McHenry, J.R., 1984. Nutrient trapping by sediment deposition in a seasonally flooded lakeside wetland. *J. Environ. Qual.*, 13: 283–290.

Junk, W.J., 1983. Ecology of swamps on the Middle Amazon. In: A.J.P. Gore (Editor), *Mires: Swamp, Bog, Fen and Moor: Regional Studies*. Ecosystems of the World 4B. Elsevier, Amsterdam, pp. 269–294.

Keammerer, W.R., Johnson, W.C. and Burgess, R.L., 1975. Floristic analysis of the Missouri River bottomland forests in North Dakota. *Can. Field-Nat.*, 89: 5–19.

Kenkel, N.C., 1987. Trends and interrelationships in boreal wetland vegetation. *Can. J. Bot.*, 65: 12–22.

Kitchens, W.M., Jr., Dean, J.M., Stevenson, L.H. and Cooper, J.H., 1975. The Santee Swamp as a nutrient sink. In: F.G. Howell, J.B. Gentry and M.H. Smith (Editors), *Mineral Cycling in Southeastern Ecosystems*. ERDA Conf-740513, NTIS, Springfield, Va., pp. 349–366.

Klinge, H., Adis, J. and Revilla C., J. 1988. Studies of the vegetation of seasonal *várzea* forest, Ilha de Marchantaria/Lower Solimões River, Amazon region of Brazil. *Vegetatio*. In press.

Kondolf, G.M., Webb, J.W., Sale, M.J. and Felando, T., 1987. Basic hydrologic studies for assessing impacts of flow diversions on riparian vegetation: examples from streams of the eastern Sierra Nevada, California, USA. *Environ. Manage.*, 11: 757–769.

Krenkel, P.A., Lee, G.F. and Jones, R.A., 1979. Effects of TVA impoundments on downstream water quality and biota. In: J.V. Ward and J.A. Stanford (Editors), *The Ecology of Regulated Streams*. Plenum Press, New York, N.Y., pp. 289–306.

Lagler, K.F. (Editor), 1969. *Man-made Lakes: Planning and Development*. Food and Agriculture Organization, Rome, 71 pp.

Langbein, W.B. et al., 1949. *Annual Runoff in the United States*. U.S. Geological Survey Circular 52. U.S. Gov. Printing Office, Washington, D.C.

Lattman, L.H., 1960. Cross section of a floodplain in a moist region of moderate relief. *J. Sediment. Petrol.*, 30: 275–282.

Lee, L.C., 1983. *The Floodplain and Wetland Vegetation of Two Pacific Northwest River Ecosystems*. Ph.D. Dissertation, University of Washington, Seattle, Wash., 245 pp.

Leitman, H.M., Sohm, J.E. and Franklin, M.A., 1983. *Wetland Hydrology and Tree Distribution of the Apalachicola River Flood Plain, Florida*. U.S. Geological Survey Water Supply Paper, 2186-A, Washington, D.C., 52 pp.

Leopold, L.B. and Miller, J.P., 1956. *Ephemeral Streams — Hydraulic Factors and Their Relation to the Drainage Net*. U.S. Geol. Surv. Prof. Pap. 282A, Washington, D.C.

Leopold, L.B., Wolman, M.G. and Miller, J.P., 1964. *Fluvial Processes in Geomorphology*. W.H. Freeman and Co., San Francisco, Calif.

Lindauer, I.E., 1983. A comparison of the plant communities of the South Platte and Arkansas River drainages in eastern Colorado. *Southwest. Nat.*, 28: 249–259.

Lindsey, A.A., Petty, R.O., Sterlin, D.K. and Van Asdall, W., 1961. Vegetation and environment along the Wabash and Tippecanoe Rivers. *Ecol. Monogr.*, 31: 105–156.

Lodhi, M.A.K., 1978. Comparative inhibition of nitrifiers and nitrification in a forest community as a result of the allelopathic nature of various tree species. *Am. J. Bot.*, 65: 1135–1137.

Lowrance, R., Todd, R., Fail, J., Hendrickson, O., Jr., Leonard, R. and Asmussen, L., 1984. Riparian forests as nutrient filters in agricultural watersheds. *Bioscience*, 34: 374–377.

Lugo, A.E. and Brown, S., 1984. The Oklawaha River forested wetlands and their response to chronic flooding. In: K.C. Ewel and H.T. Odum (Editors), *Cypress Swamps*. University Presses of Florida, Gainesville, Fla., pp. 365–373.

Lugo, A.E. and Snedaker, S.C., 1974. The ecology of mangroves. *Annu. Rev. Ecol. Syst.*, 5: 39–64.

Lugo, A.E., Nessel, J.K. and Hanlon, T.M., 1984. Root distribution in a north-central Florida cypress strand. In: K.C. Ewel and H.T. Odum (Editors), *Cypress Swamps*. University Presses of Florida, Gainesville, Fla., pp. 279–285.

MacDonald, B.C. and Lewis, C.P., 1973. *Geomorphic and Sedimentologic Processes of Rivers and Coast, Yukon Coastal Plain*. Geological Survey of Canada, Cat. No. R72-12173. Information Canada, Ottawa, Ont.

McBride, J., 1973. Natural replacement of disease-killed elms. *Am. Midl. Nat.*, 90: 300–306.

McCarthy, J., 1962. The colonization of a swamp forest clearing (with special reference to *Mitragyna stipulosa*). *East Afr. Agric. For. J.*, 28: 22–28.

McClelland, M.K. and Ungar, I.A., 1970. The influence of edaphic factors on *Betula nigra* L. distribution in southeastern Ohio. *Castanea*, 35: 99–117.

McClure, J.P. and Knight, H.A., 1984. *Empirical Yields of Timber and Forest Biomass in the Southeast*. U.S. Department of Agriculture, Forest Service, Southern Forest Experiment Station Res. Pap. SE-245, Asheville, N.C., 75 pp.

McColl, R.H.S., 1978. Chemical runoff from pasture: the influence of fertiliser and riparian zones. *N. Z. J. Marine Freshwater Res.*, 12: 371–380.

McEvoy, T.J., Sharik, T.L. and Smith, D.W., 1980. Vegetative structure of an Appalachian oak forest in southwestern Virginia. *Am. Midl. Nat.*, 103: 96–105.

McKee, W.H., Jr., Hook, D.D., DeBell, D.S. and Askew, J.L., 1984. Growth and nutrient status of loblolly pine seedlings in relation to flooding and phosphorus. *J. Soil Sci. Soc. Am.*, 48: 1438–1442.

Malecki, R.A., Lassoie, J.R., Rieger, E. and Seamans, T., 1983. Effects of long-term artificial flooding on a northern bottomland hardwood forest community. *For. Sci.*, 29: 535–544.

Marks, P.L. and Harcombe, P.A., 1981. Forest vegetation of the Big Thicket, southeast Texas. *Ecol. Monogr.*, 51: 287–305.

Marsh, C., 1980. Primates and economic development on the Tana River, Kenya: The monkey in the works. In: J.I. Furtado (Editor), *Tropical Ecology and Development*. Int. Soc. Trop. Ecol., Kuala Lumpur, pp. 373–376.

Martin, S.C., 1979. Evaluating the impacts of cattle grazing on riparian habitats in the national forests of Arizona and New Mexico. In: O.B. Cope (Editor), *Grazing and Riparian/Stream Ecosystems*. Trout Unlimited, Denver, Colo., pp. 35–38.

Martínez, R., Cintrón, G. and Encarnación, L.A., 1979. *Mangroves in Puerto Rico: A Structural Inventory*. Final report of the Office of Coastal Zone Management, NOAA. Department of Natural Resources, Area of Scientific Research, Government of Puerto Rico, San Juan, P.R.

Mattoon, W.R., 1915. *The Southern Cypress*. USDA Bull. 272, Washington, D.C., 74 pp.

Mayo Meléndez, E., 1965. Algunas características ecológicas de los bosques inundables de Darién, Panamá, con miras a su posible utilización. *Turrialba*, 15: 336–347.

Medina, A.L., 1986. Riparian plant communities of the Fort Bayard watershed in southwestern New Mexico. *Southwest. Nat.*, 31: 345–359.

Menges, E.S. and Waller, D.M., 1983. Plant strategies in relation to elevation and light in floodplain herbs. *Am. Nat.*, 122: 454–473.

Merritt, R.W. and Lawson, D.L., 1979. Leaf litter processing in floodplain and stream communities. In: R.R. Johnson and J.F. McCormick (Technical Coordinators), *Strategies for Protection and Management of Floodplain Wetlands and Other Riparian Ecosystems*. Gen. Tech. Rep. WO-12, USDA Forest Service. U.S. Gov. Printing Office, Washington, D.C., pp. 93–105.

Metzler, K.J. and Damman, A.W.H., 1985. Vegetation patterns in the Connecticut River floodplain in relation to frequency and duration of flooding. *Nat. Can.*, 112: 535–547.

Mitsch, W.J., Dorge, C.L. and Wiemhoff, J.R., 1977. *Forested Wetlands for Water Resource Management in Southern Illinois*. Res. Rep. No. 132. Water Resources Center, Univ. Illinois, Urbana, Ill.

Mitsch, W.J., Dorge, C.L. and Wiemhoff, J.R., 1979a. Ecosystem dynamics and a phosphorus budget of an alluvial cypress swamp in southern Illinois. *Ecology*, 60: 1116–1124.

Mitsch, W.J., Rust, W., Behnke, A. and Lai, L., 1979b. *Environmental Observations of a Riparian Ecosystem during Flood Season*. Res. Rep. No. 142, Water Resources Center, Univ. Illinois, Urbana, Ill.

Morris, L.A., 1977. *Evaluation, Classification and Management of the Floodplain Forests of South Central New York*. M.S. Thesis, SUNY College Environmental Science and Forestry, Syracuse, N.Y., 181 pp.

Morris, L.A., Mollitor, A.V., Johnson, K.J. and Leaf, A.L., 1979. Forest management and floodplain sites in the northeastern United States. In: R.R. Johnson and J.F. McCormick (Technical Coordinators), *Strategies for Protection and Management of Floodplain Wetlands and Other Riparian Ecosystems*. Gen. Tech. Rep. WO-12, USDA Forest Service. U.S. Gov. Printing Office, Washington, D.C., pp. 236–242.

Mosquera, R.A., 1979. *Química del agua intersticial y varios parámetros estructurales cuantificados en un manglar riverino, en el Río Espíritu Santo, Río Grande, Puerto Rico*. Center for Energy and Environmental Research, University of Puerto Rico and U.S. Dept. of Energy, Publ. No. 18, San Juan, P.R., 118 pp.

Mulholland, P.J., 1979. *Organic Carbon Cycling in a Swamp-Stream Ecosystem and Export by Streams in Eastern North Carolina*. Ph.D. Dissertation, Univ. of North Carolina, Chapel Hill, N.C., 152 pp.

Muzika, R.M., Gladden, J.B. and Haddock, J.D., 1987. Structural and functional aspects of succession in southeastern floodplain forests following major disturbance. *Am. Midl. Nat.*, 117: 1–9.

Nadler, C.T., Jr., 1978. *River Metamorphosis of the South Platte and Arkansas Rivers, Colorado*. M.S. Thesis, Colorado State Univ., Fort Collins, Colo., 151 pp.

National Research Council (U.S.A.), 1982. *Ecological Aspects of Development in the Humid Tropics*. Committee on Selected Biological Problems in the Humid Tropics. National Academy Press, Washington, D.C.

Nessel, J., 1978. *Distribution and Dynamics of Organic Matter and Phosphorus in a Sewage Enriched Cypress Stand*. M.S. Thesis, Univ. of Florida, Gainesville, Fla., 159 pp.

Nixon, E.S., Willett, R.L. and Cox, P.W., 1977. Woody vegetation of a virgin forest in an eastern Texas river bottom. *Castanea*, 42: 227–236.

Nixon, E.S., Ehrhart, R.L., Jasper, S.A., Neck, J.S. and Ward, J.R., 1983. Woody, streamside vegetation of Prairie Creek in east Texas. *Texas J. Sci.*, 35: 205–213.

Ong, J.E., Gong, W.K., Wong, C.H. and Dhanarajan, G., 1979. Productivity of a managed mangrove forest in West Malaysia. In: *Proceedings of the International Conference on "Trends in Applied Biology in South East Asia", Penang, Malaysia, 11–14 October 1979.* pp. 274–284.

Ong, J.E., Gong, W.K. and Wong, C.H., 1980. *Ecological Survey of the Sungei Merbok Estuarine Mangrove Ecosystem.* School of Biological Sciences, Universiti Sains Malaysia, Penang, 83 pp.

Outhet, D.N., 1974. Progress report on bank erosion studies in the Mackenzie River Delta, N.W.T. In: *Hydrologic Aspects of Northern Pipeline Development.* Environmental-Social Program, Northern Pipeline. Cat. No. R57–1/1974. Information Canada, Ottawa, Ont., pp. 298–345.

Paratley, R.D. and Fahey, T.J., 1986. Vegetation–environment relations in a conifer swamp in central New York. *Bull. Torrey Bot. Club*, 113: 357–371.

Parsons, S.E. and Ware, S., 1982. Edaphic factors and vegetation in Virginia Coastal Plain swamps. *Bull. Torrey Bot. Club*, 109: 365–370.

Patric, J.H., Evans, J.O. and Helvey, J.D., 1984. Summary of sediment yield data from forested land in the United States. *J. For.*, 82: 101–104.

Patrick, W.H., Jr. and Tusneem, M.E., 1972. Nitrogen loss from flooded soil. *Ecology*, 53: 735–737.

Penfound, W.T. and Hathaway, E.S., 1938. Plant communities in the marshlands of southeastern Louisiana. *Ecol. Monogr.*, 8: 1–56.

Peterjohn, W.T. and Correll, D.L., 1984. Nutrient dynamics in an agricultural watershed: observations on the role of a riparian forest. *Ecology*, 65: 1466–1475.

Peterson, D.L. and Rolfe, G.L., 1982a. Nutrient dynamics and decomposition of litterfall in floodplain and upland forests of central Illinois. *For. Sci.*, 28: 667–681.

Peterson, D.L. and Rolfe, G.L., 1982b. Seasonal variation in nutrients of floodplain and upland forest soils of central Illinois. *J. Soil Sci. Soc. Am.*, 46: 1310–1315.

Peterson, D.L. and Rolfe, G.L., 1982c. Nutrient dynamics of herbaceous vegetation in upland and floodplain forest communities. *Am. Midl. Nat.*, 107: 325–339.

Phipps, R.L., 1979. Simulation of wetlands forest vegetation dynamics. *Ecol. Modelling*, 7: 257–288.

Phipps, R.L., 1983. Streamflow of the Occoquan River in Virginia as reconstructed from tree-ring series. *Water Resour. Bull.*, 19: 735–743.

Pionke, H.B. and Chesters, G., 1973. Pesticide–sediment–water interactions. *J. Environ. Qual.*, 2: 29–45.

Pool, D.J., Lugo, A.E. and Snedaker, S.C., 1975. Litter production in mangrove forests of southern Florida and Puerto Rico. In: G.E. Walsh, S.C. Snedaker and H. Teas (Editors), *Proceedings International Symposium on Biology and Management of Mangroves.* IFAS, University of Florida, Gainesville, Fla.

Pool, D.J., Snedaker, S.C. and Lugo, A.E., 1977. Structure of mangrove forests in Florida, Puerto Rico, Mexico, and Costa Rica. *Biotropica*, 9: 195–212.

Post, A. and Mayo, L.R., 1971. *Glacier Dammed Lakes and Outburst Floods in Alaska.* U.S. Geological Survey Hydrologic Investigation, Atlas HA-455. U.S. Gov. Printing Office, Washington, D.C., 10 pp. + maps.

Prance, G.T., 1979. Notes on the vegetation of Amazonia, III. The terminology of Amazonian forest types subject to inundation. *Brittonia*, 31: 26–38.

Qualls, R.G., 1984. The role of leaf litter nitrogen immobilization in the nitrogen budget of a swamp stream. *J. Environ. Qual.*, 13: 640–644.

Räsänen, M.E., Salo, J.S. and Kalliola, R., 1987. Fluvial perturbance in the western Amazon basin: regulation by long-term sub-Andean tectonics. *Science*, 238: 1398–1401.

Rice, E.L., 1965. Bottomland forests of north-central Oklahoma. *Ecology*, 46: 708–713.

Richards, P.W., 1939. Ecological studies on the rain forest of southern Nigeria. I. The structure and floristic composition of the primary forest. *J. Ecol.*, 27: 1–53.

Robertson, P.A., Weaver, G.T. and Cavanaugh, J.A., 1978. Vegetation and tree species patterns near the northern terminus of the southern floodplain forest. *Ecol. Monogr.*, 48: 249–267.

Rowe, J.S., 1972. *Forest Regions of Canada.* Canadian Forest Service Publication 1300. Information Canada, Ottawa, Ont., 172 pp.

Sage, B.L., 1974. Ecological distribution of birds in the Atigum and Sagavanirktor River Valleys, Arctic Alaska. *Can. Field-Nat.*, 88: 281–291.

Schmelz, D.V. and Lindsey, A.A., 1965. Size-class structure of old-growth forests in Indiana. *For. Sci.*, 11: 258–264.

Schumm, S.A. and Lichty, R.W., 1963. *Channel Widening and Floodplain Construction along Cimarron River in Southwestern Kansas.* U.S. Geol. Surv. Prof. Pap. 352-D, U.S. Gov. Printing Office, Washington, D.C.

Sell, M.G., 1977. *Modeling the Response of Mangrove Ecosystems to Herbicide Spraying, Hurricanes, Nutrient Enrichment and Economic Development.* Ph.D. Dissertation. University of Florida, Gainesville, Fla., 390 pp.

Sharitz, R.R., Irwin, J.E. and Christy, E.J., 1974. Vegetation of swamps receiving reactor effluents. *Oikos*, 25: 7–13.

Sigafoos, R.S., 1964. *Botanical Evidence of Floods and Flood-Plain Deposition.* U.S. Geol. Surv. Prof. Pap. 485-A, Washington, D.C., 35 pp.

Simons, D.B., 1979. Effects of stream regulation on channel morphology. In: J.V. Ward and J.A. Stanford (Editors), *The Ecology of Regulated Streams.* Plenum Press, New York, N.Y., pp. 95–111.

Sims, R.A., Cowell, D.W. and Wickware, G.M., 1982. Use of vegetational physiognomy in classifying treed peatlands near southern James Bay, Ontario. *Nat. Can.*, 109: 611–619.

Smith, N.J.H., 1981. *Man, Fishes, and the Amazon.* Columbia University Press, New York, N.Y., 180 pp.

Snedaker, S.C. and Brown, M.S., 1981. *Water Quality and Mangrove Ecosystem Dynamics.* Final Report to United States Environmental Protection Agency, EPA-600/4-81-022, Gulf Breeze, Fla.

Sollers, S.C., 1973. Substrate conditions, community structure,

and succession in a portion of the floodplain of Wissahickon Creek. *Bartonia*, 42: 24–42.
Spetzman, L.A., 1959. *Vegetation of the Arctic Slope of Alaska*. Geol. Surv. Prof. Pap. 302-B, U.S. Gov. Printing Office, Washington, D.C. pp. 19–58.
Stearns, F., 1951. The composition of the sugar maple–hemlock–yellow birch association in northern Wisconsin. *Ecology*, 32: 245–265.
Tessier, C., Maire, A. and Aubin, A., 1981. Etude de la végétation des zones riveraines de l'archipel des Cent-îles du fleuve Saint-Laurent, Quebec. *Can. J. Bot.*, 59: 1526–1536.
Thibault, J.-R., Fortin, J.-A. and Smirnoff, W.A., 1982. In vitro allelopathic inhibition of nitrification by balsam poplar and balsam fir. *Am. J. Bot.*, 69: 676–679.
Thompson, K. and Hamilton, A.C., 1983. Peatlands and swamps of the African continent. In: A.J.P. Gore (Editor), *Mires: Swamp, Bog, Fen and Moor: Regional Studies*. Ecosystems of the World 4B. Elsevier, Amsterdam, pp. 331–373.
Turner, R.M., 1974. *Quantitative and Historical Evidence of Vegetation Changes along the Upper Gila River, Arizona*. U.S. Geol. Surv. Prof. Pap. 655-H, U.S. Gov. Printing Office, Washington, D.C.
Turner, R.M. and Karpiscak, M.M., 1980. *Recent Vegetation Changes along the Colorado River between Glen Canyon Dam and Lake Mead, Arizona*. U.S. Geol. Surv. Prof. Pap. 1132, U.S. Gov. Printing Office, Washington, D.C.
Van Beek, J.L., 1979. *Hydraulics of the Atchafalaya Basin Main Channel System*. EPA-600/4-79-036, U.S. Environmental Protection Agency, Washington, D.C.
Van Hylckama, T.E.A., 1980. *Weather and Evapotranspiration Studies in a Saltcedar Thicket, Arizona*. U.S. Geol. Surv. Prof. Pap. 491-F, U.S. Gov. Printing Office, Washington, D.C., 78 pp.
Vyskot, M., 1976a. Biomass production of the tree layer in a floodplain forest near Lednice. In: H.E. Young (Editor), *Oslo Biomass Studies*. University of Maine, Orono, Me., pp. 175–202.
Vyskot, M., 1976b. Floodplain forest in biomass. In: H.E. Young (Editor), *Oslo Biomass Studies*. University of Maine, Orono, Me., pp. 203–229.
Walker, L.R., Zasada, J.C. and Chapin, F.S., III. 1986. The role of life history processes in primary succession on an Alaskan floodplain. *Ecology*, 67: 1243–1253.
Walters, M.S., Teskey, R.O. and Hinckley, T.M., 1980. *Impact of Water Level Changes on Woody Riparian and Wetland Communities: Vol. VIII. Pacific Northwest and Rocky Mountain Regions*. USDI Fish and Wildlife Service OBS-78/94, Washington, D.C.
Warner, R.E., 1984. Structural, floristic, and condition inventory of Central Valley riparian systems. In: R.E. Warner and K.M. Hendrix (Editors), *California Riparian Systems: Ecology, Conservation and Productive Management*. University of California Press, Berkeley, Calif., pp 356–274.
Welcomme, R.L., 1979. *Fisheries Ecology of Floodplain Rivers*. Longman, New York, N.Y., 317 pp.
Wharton, C.H., 1978. *The Natural Environments of Georgia*. Georgia Dept. Natural Resources, Atlanta, Ga., 227 pp.
Wharton, C.H., Kitchens, W.M., Pendleton, E.C. and Sipe, T.W., 1982. *The Ecology of Bottomland Hardwood Swamps of the Southeast: a Community Profile*. FWS/OBS-81/37, USDI Fish and Wildlife Service, Biological Services Program, Washington, D.C., 133 pp.
White, D.A., 1983. Plant communities of the lower Pearl River basin, Louisiana. *Am. Midl. Nat.*, 110: 381–396.
Whittaker, R.H., Bormann, F.H., Likens, G.E. and Siccama, T.G., 1974. The Hubbard Brook ecosystem study: Forest biomass and production. *Ecol. Monogr.*, 44: 233–254.
Wikum, D.A. and Wali, M.K., 1974. Analysis of a North Dakota gallery forest: vegetation in relation to topographic and soil gradient. *Ecol. Monogr.*, 44: 441–464.
Williams, G.P., 1978. Historical perspective of the Platte Rivers in Nebraska and Colorado. In: W.D. Graul and S.J. Bissell (Technical Coordinators), *Lowland River and Stream Habitat in Colorado: A Symposium*. Colorado Div. Wildl., Denver, Colo., pp. 11–41.
Williams, G.P. and Wolman, M.G., 1984. *Downstream Effects of Dams on Alluvial Rivers*. U.S. Geol. Surv. Prof. Pap. 1286, U.S. Gov. Printing Office, Washington, D.C., 83 pp.
Williams, W.A., Loomis, R.S. and Alvim, P. de T., 1972. Environments of evergreen rain forests on the lower Rio Negro, Brazil. *Trop. Ecol.*, 13: 65–78.
Wilson, R.E., 1970. Succession in stands of *Populus deltoides* along the Missouri River in southeastern South Dakota. *Am. Midl. Nat.*, 83: 330–342.
Wolanski, E. and Ridd, P., 1986. Tidal mixing and trapping in mangrove swamps. *Estuarine, Coastal Shelf Sci.*, 23: 759–771.
Wolfe, C.B., Jr. and Pittillo, J.D., 1977. Some ecological factors influencing the distribution of *Betula nigra* L. in western North Carolina. *Castanea*, 42: 18–30.
Wolman, M.G. and Leopold, L.B., 1957. *River Flood Plains: Some Observations on their Formation*. U.S. Geol. Surv. Prof. Pap. 282-C, U.S. Gov. Printing Office, Washington, D.C. 107 pp.
Yarbro, L.A., 1979. *Phosphorus Cycling in the Creeping Swamp Floodplain Ecosystem and Exports from the Creeping Swamp Watershed*. Ph.D. Dissertation, Univ. of North Carolina, Chapel Hill, N.C., 231 pp.
Yarbro, L.A., 1983. The influence of hydrologic variations on phosphorus cycling and retention in a swamp stream ecosystem. In: T.D. Fontaine, III and S.M. Bartell (Editors), *Dynamics of Lotic Ecosystems*. Ann Arbor Science, Ann Arbor, Mich., pp. 223–245.
Yon, D. and Tendron, G., 1981. *Alluvial Forests of Europe*. European Committee for the Conservation of Nature and Natural Resources, Council of Europe, National Museum of Natural History, Paris, 65 pp.
Zimmerman, R.C., 1969. *Plant Ecology of an Arid Basin: Tres Alamos–Redington area, southeastern Arizona*. U.S. Geol. Surv. Prof. Pap. 485-D, U.S. Gov. Printing Office, Washington, D.C., 47 pp.
Zoltai, S.C. and Pollett, F.C., 1983. Wetlands in Canada: their classification, distribution and use. In: A.J.P. Gore (Editor), *Mires: Swamp, Bog, Fen and Moor: Regional Studies*. Ecosystems of the World 4B. Elsevier, Amsterdam, pp. 245–268.

Chapter 6

FRINGE WETLANDS

ARIEL E. LUGO

INTRODUCTION

Fringe wetlands are characterized by the dominance of few species, a clear species' zonation (although they can be monospecific), synchrony of ecological processes with episodic events, and simplicity in the structure of vegetation. In this chapter I will discuss what is known about the structure and ecosystem dynamics of fringe forested wetlands, with emphasis on saltwater wetlands because they have been studied more than freshwater ones. My discussion centers on Caribbean and Florida mangroves.

Fringe wetlands are found on the water edge of oceans, inland estuaries, and lakes. Water motion in the fringe is bi-directional and perpendicular to the forest (a power front, *sensu* Kangas, Ch. 2). Water motion in oceanic and estuarine fringes is due mostly to tidal energy. In lakes, water moves in and out of the fringe under the influence of wind, waves, or seiches. Some fringe forests are occasionally flushed by terrestrial runoff or aquifer discharge. These unidirectional events have profound effects on the function of fringe forests because they reduce salinity, remove organic matter, and transport terrigenous materials into the forest. In contrast, fringe forests located on small offshore islands or steep coastal shores are isolated from terrestrial runoff or aquifer discharge, and their hydroperiod is controlled by tides and waves only.

Any discussion of the characteristics of fringe forested wetlands must address the question of why these ecosystems are structured into vegetation bands or zones. Most of the literature on fringe wetland vegetation deals with descriptions of zonations; unfortunately equal effort has not been placed on the study of the physical environment responsible for the zonations (Snedaker, 1982). However, Watson (1928) and Chapman (1944) showed that the frequency of tidal inundation correlated with species zonation. Literature reviews suggest that such ecosystem parameters as vegetation structure, tree growth, primary productivity, and organic matter in sediments respond proportionally to hydrologic energy (Gosselink and Turner, 1978; Brown et al., 1979; Brinson et al., 1981).

Because hydrologic energy appears to be of fundamental importance to the fringe forest, wherever possible I will discuss the characteristics of fringe forests in the context of this factor. The criterion that I use to evaluate the intensity of tidal or wave energy is the position and elevation of the fringe relative to the main body of water it fringes. Fringes growing in exposed locations under intensive wave or tidal energy will be termed high-energy fringes. The width of the fringe under these conditions can be extensive because there is enough kinetic energy to flush a larger area, and the physical and chemical environment of soil and water remain fairly constant in time. Those growing in more protected areas, such as calm lakes or shallow estuaries, or at higher elevations above mean water level, will be termed low-energy fringes. This type of fringe is characterized by steep gradients in the soil and water environment. Species are segregated into zones according to these gradients and the fringe itself is narrow in comparison to those influenced by high energy environments. For example, fringes in a windward zone were 3.5 times wider than fringes in the leeward zone of mangrove islands on the south coast of Puerto Rico (Cintrón et al., 1978). Many examples fall within the two extremes, while others cannot be grouped because of lack of information.

STRUCTURE OF SALTWATER FRINGE FORESTS

Zonation

In most new world mangroves, species of *Rhizophora* dominate the water edge of the fringe. These species grow best in oligohaline environments and are well adapted to strong wind and tidal forces converging on the fringe. *Rhizophora* spp. always grow where water is in motion. The best-studied adaptations of *R. mangle* are its extensive prop-root system, its thick xeromorphic leaves, and its large viviparous seedlings (compare Chapman, 1976, 1977 for reviews of this literature). The extensive prop-root system is critical for providing support in muddy, unstable substrates and constant water turbulence. Saenger (1982) calculated that the root-to-shoot ratio of *Rhizophora* spp. ranged from 0.68 to >1.7 compared to 0.2 for upland tree species. He interpreted these high ratios as adaptations to unstable soils.

The thick leaves of *Rhizophora* spp. and its many other xeromorphic characteristics (for example, most stomata on the lower epidermis, thick cuticles, and large number of vessels with small pores: Saenger, 1982) provide tolerance to frequent abrasion by off-shore winds that are typical of its habitat. These leaf characteristics also have important implications for gas exchange and nutrient cycling as discussed below. Vivipary and large size of seedlings provide the species with strong regenerative and dispersal mechanisms. These seedlings usually fall in deep turbulent water or on soft muds that are frequently inundated. Under either circumstance the propagules are able to establish themselves successfully (for example, Banus and Kolehmainen, 1975; Rabinowitz, 1975, 1980; Saenger, 1982).

In many old world mangrove stands it is common to find species of *Avicennia* on the water edge and species of *Rhizophora* in inland locations (the opposite pattern found in most new world mangroves). The shift in the zonation of species may be due to a shift in the conditions of the fringe (H.T. Odum, pers. commun. 1976). *Avicennia* species are euryhaline adapted to fluctuating and often extreme environments, while *Rhizophora* species are stenohaline species which prefer environments of constant salinity. Fringe environments with large tidal amplitude, and high sedimentation and freshwater runoff from terrigenous sources (euryhaline) are the ones that support *Avicennia* fringes. In these instances the more stable conditions occur inland under the influence of high tides which push seawater to the higher elevations where *Rhizophora* spp. grow (compare Watson, 1928). At low tides the *Rhizophora* zone escapes the salinity change that occurs in the lower *Avicennia* zone. When the fringe is oligohaline, *Rhizophora* or an analogous old world mangrove species will occupy the fringe.

Other mangrove species may occupy the outer zone of a fringe depending on the conditions that prevail and the species found in the region. For example, Cintrón et al. (1978) and Martínez et al. (1979) described *Laguncularia racemosa* fringes under two apparently different conditions: (1) high wave energy and seawater salinity; and (2) high salinity (>60‰) but almost still-water conditions. The common denominator to both examples was a sharp change in topography which allowed the species to "escape" from the direct influence of high wave energy or high salinity. *L. racemosa* grows well in the elevated portions of mangrove environments. Therefore, when conditions in the fringe are too extreme for other species, *L. racemosa* may occupy the fringe if there is enough topographic gradient to allow its growth without significant exposure to stressors such as high soil salinity or constant wave activity.

Conocarpus erectus is another mangrove species that can occupy a fringe when there is considerable wave action. Usually, this species grows at higher land elevations than *L. racemosa*, and it is not unusual to find fringes of *C. erectus* on steep rocky terrain of high energy coastlines. *Avicennia germinans* occupies fringes in Florida where periodic frost kills *Rhizophora mangle* (Lugo and Patterson Zucca, 1977). Many other examples of dominance of fringe forests by different species in different parts of the world are reviewed in Chapman (1976, 1977). In this chapter I will use *R. mangle* to illustrate adaptation to the water edge of the fringe, and *A. germinans* to illustrate adaptations to landward and basin conditions.

The effects of hydrologic energy on the zonation

and structure of fringe forest vegetation can best be studied in offshore mangrove islands on arid coastlines. These locations are ideal for these studies because any effects of terrestrial runoff are eliminated. Also, the islands are usually small and abundant, allowing replication as well as providing examples of fringes subjected to different intensities of hydrologic conditions. Cintrón et al. (1978, 1980) and Martínez et al. (1979) on the south coast of Puerto Rico described three tiers of offshore islands plus the mainland shoreline, and seven types of fringe forests differentiated by the intensity of wave and tidal energy (Fig. 6.1).

Mangrove growth is absent in the outer tier of islands due to exposure to conditions of strong wave energy (No. 1 and 2, Fig. 6.1). Occasionally, fringes of *L. racemosa* occur if the substrate is sufficiently elevated above the water to allow its establishment. Trees under these conditions are stunted and gnarled, and storms periodically set back the colonization of these fringes.

A wider mangrove fringe forms in the second tier of islands. At these locations wave energy has been reduced enough to allow the formation of overwash islands (*sensu* Lugo and Snedaker, 1974) which are basically a monospecific fringe forest (No. 4 in Fig. 6.1). Hydrologic energy is strong enough to flush all the island and maintain conditions very close to those of the ocean (Fig. 6.2a). If there is enough substrate area for expansion of the fringe, a point is reached when all the fringe cannot be overwashed. This condition

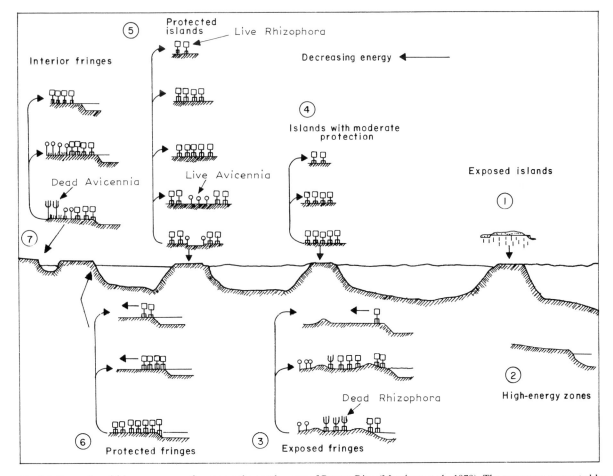

Fig. 6.1. Examples of fringe mangrove forests on the south coast of Puerto Rico (Martínez et al., 1979). The sequences connected by arrows illustrate the progression of succession in a particular type of fringe. Succession starts on the upper diagram of each sequence. Most colonizations begin from the ocean and progress inland. Arrows illustrate the reversal of succession by storms or other episodic events. Each diagram illustrates zonations at specific moments during succession.

(a) ENRIQUE (Parguera)

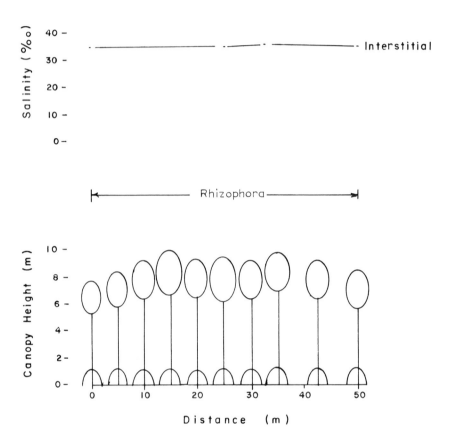

Fig. 6.2a. For explanation see p. 150.

also develops in fringes exposed to little wave action (No. 5 and 6, Fig. 6.1). In either case, increased resistance to water flow by mangrove prop-roots or by higher topographic relief causes a reduction in the frequency of flushing towards the internal portion of the fringe. As a result, gradients of aeration, soil salinity, water temperature, and frequency and depth of flooding develop throughout the fringe. Associated with these gradients are changes in the stature of the forest (No. 6 in Fig. 6.1) and its species composition (No. 5 in Fig. 6.1). Species zonations become more pronounced with time since colonization. Because these forests lack terrestrial runoff and grow in arid environments, zones of tree mortality and internal lagoons develop in the areas with the least hydrologic energy (Cintrón et al., 1978). The incipient effects of reduced water circulation are illustrated in Fig. 6.2b, where the onset of tree mortality was observed in the central portions of an overwash island located behind the island in Fig. 6.2a.

Mangrove fringes on the mainland behind the offshore islands are well protected, and hydrologic energy is not high enough to support wide fringe forests. Instead, vegetation zones are narrow, and extensive areas of tree mortality develop in the rear of the forest, as do salt flats (No. 7 in Fig. 6.1; Fig. 6.2c and d). However, the stature of vegetation in the narrower *Rhizophora mangle* fringe is much greater than that of the higher-energy fringes offshore (Fig. 6.2a). It is possible that the lower kinetic energy of ocean water causes less leaching of nutrients and export of organic materials from

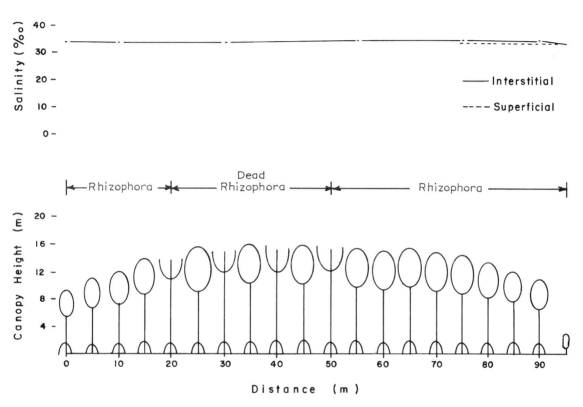

Fig. 6.2b. For explanation see p. 150.

protected than from exposed fringes. This in turn increases the opportunity for more recycling and accumulation of biomass and materials in the protected fringe than in high-energy fringes. Alternatively, trees in high-energy fringes may be younger due to more frequent die-back induced by storm waves.

In moist climates with flat topography, the fringe receives freshwater runoff from the uplands. Salt flats and zones of tree mortality do not develop on the back of these fringes as they do in the arid-climate fringes (compare Figs. 6.2e and Fig. 6.3) because the greater freshwater input leaches salts and prolongs the hydroperiod. These conditions favor the invasion of the landward portions of the fringe by species with lower tolerance to salinity. The complexity of fringe vegetation increases as a result of the species enrichment (Table 6.1). In some instances oligohaline basin mangrove forests or even freshwater basin forests develop behind the fringe forest in moist and wet climates (Chs. 9, 10, 11, 14 and 15).

Because the fringe forest is zoned and each zone could (and frequently is) treated as a community, care is needed when delimiting the boundaries of the fringe ecosystem. The boundaries of the fringe can easily be confused with boundaries of its zones. The limit of the fringe forest coincides with the area influenced by the two-way tidal motion. Usually this limit occurs at a topographic high (a berm) or where tidal waters dissipate into a basin

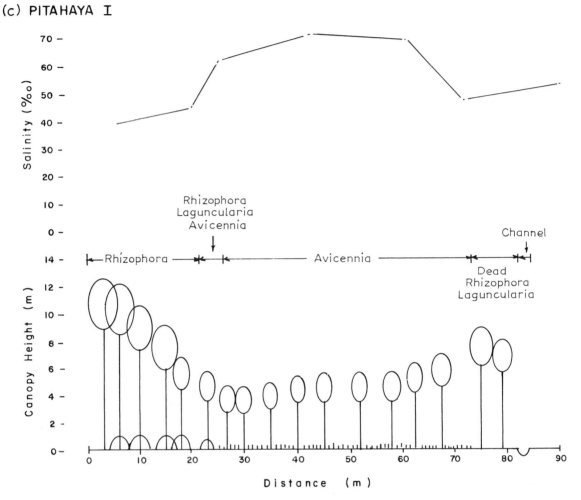

Fig. 6.2c. For explanation see p. 150.

wetland. Species changes usually accompany the hydrologic transition.

Structural indices

Given the impact of hydrologic energy on the species composition and zonation of fringe forests, it is possible to interpret the structural characteristics of fringe vegetation as residual effects of hydrologic conditions in a given zone. Fringe forests in the Caribbean are characterized by low species richness (1–3 tree species in 0.1 ha), low basal area (range from 1 to 35 m^2 ha^{-1} for trees with dbh $\geqslant$ 10 cm), high stem density (up to 2200 trees ha^{-1} for stems with dbh $\geqslant$ 10 cm), and trees that seldom exceed 25 m in height (Tables 6.1 and 6.2).

Stand structural indices increase considerably if measurements of trees $\geqslant$ 2.5 cm dbh are included. This is due to the large density of small stems in mangrove fringes. The average tree diameter in these stands ranges from 4 to 25 cm (Table 6.2). The low tree height is probably associated with exposure to winds and unstable substrate which cannot support large trees. Trees of low-energy fringes are usually taller along the water edge of the ocean or internal canals. In high-energy fringes tree height increases away from the water's edge.

Soil salinity and fertility also have a role in determining tree height in the fringe. As a rule, tree height decreases with increasing soil salinity and oligotrophy. Latitude affects many structural characteristics of mangroves (Cintrón et al., 1985).

FRINGE WETLANDS

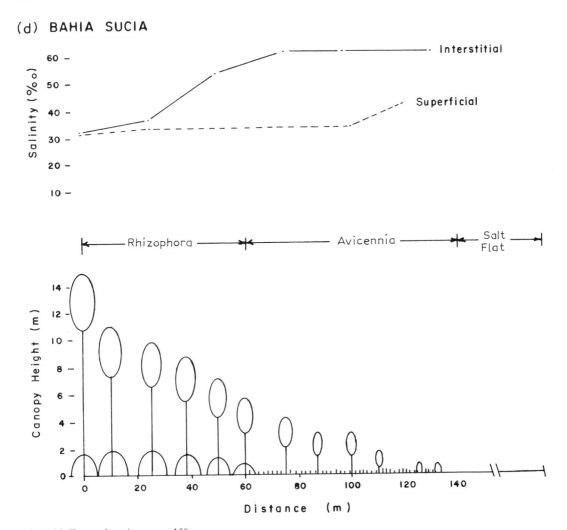

Fig. 6.2d. For explanation see p. 150.

In general, stature increases towards the moist lower latitudes. However, the available data for fringe forests are biased to higher latitudes and it is difficult to evaluate the role of latitude.

The low species richness in fringes reflects the pattern of monospecific zones. These zones coincide with zones of environmental stress in the system. For example, each species appears to be adapted to a given range of soil salinity. As gradients of soil salinity become broader (as they do in moist environments), more species can grow in the fringe (Table 6.1). Low basal area and high stem density reflect the rapid turnover of trees in the fringe environment. Tree mortality in the fringe is high. It is common to find pockets of mortality, usually associated with changes in water circulation (see p. 154). As a result, fringe stands support young trees of low basal area, and high density (Table 6.2). The average diameter of trees is highest near the water edge and decreases with distance from the water. Age and rate of tree growth account for this difference. However, faster growth rates in this zone are only observed in low-energy fringes.

Aboveground biomass of fringe forests is low (range from 8 to 160 t ha^{-1}, Table 6.3). The value from Thailand is similar to those from Florida, suggesting little latitudinal effect. However, because the ages of stands are not known, it is difficult to assess the effect of latitude on stand biomass. Roots comprised a high proportion of mass in the stands where this compartment has

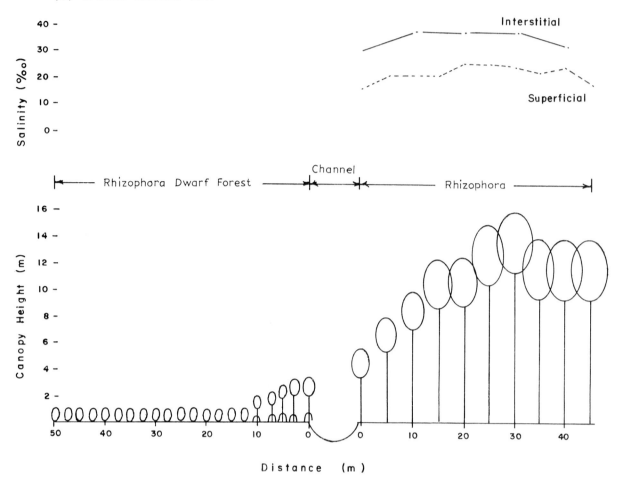

Fig. 6.2. Profiles of vegetation and salinity in fringe mangrove forests on the south and north coasts of Puerto Rico (Martínez et al., 1979). Wave and tidal energy decrease from (a) to (d), all located on the arid south coast. The forest in (e) was located on the moist north coast. Soil salinities are similar to seawater in (a) and (b), although in (b) the vegetation shows signs of reduced aeration and soil salinities are slightly higher than seawater. Profiles (c) and (d) illustrate protected fringes with sharp zonations and corresponding gradients of soil salinity. The profile in the moist coastline was protected but had lower salinities due to greater runoff. This forest was located across a channel supporting a basin dwarf forest. Notice the variations in tree height with salinity or location within the fringe forest.

been evaluated (Table 6.3). This agrees with the high root-to-shoot ratios reported by Saenger (1982) for Australian mangroves.

Ground litter and flotsam may accumulate in fringe forests. These materials tend to concentrate in large piles where waves break or where tidal waters lose their initial velocity. In many instances organic materials from *Thalassia* beds and other subtidal ecosystems accumulate in piles inside the fringe. In the flushed areas of fringe forests, litter accumulations are small. The actual amount of litter on the forest floor is highly variable because of constant transport and deposition by water movement (Table 6.3). In some cases litter mass approaches aboveground biomass.

Seedlings of mangrove species are also zoned and may pile up in the forward sectors of the fringe. Red mangrove seedlings usually grow only in areas affected by daily tides (Fig. 6.4). The smaller seedlings of *Avicennia* grow best in the less frequently inundated portions of the forest.

Fringe forests are characterized by low leaf area

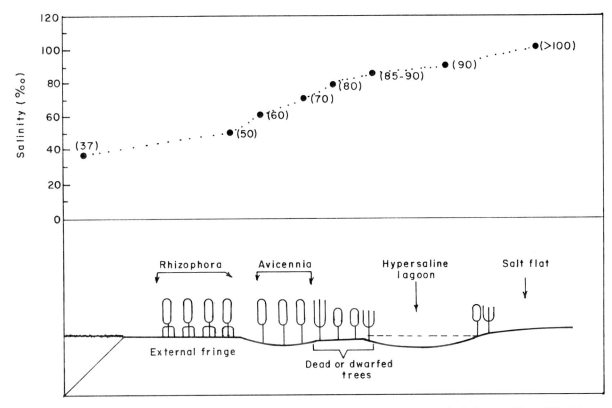

Fig. 6.3. Typical fringe forest on an arid coastline. Notice that *Rhizophora mangle* grows under oligohaline conditions while *Avicennia germinans* is located in the zone of rapid change in salinity (Cintrón et al., 1980).

TABLE 6.1

Structural characteristics and complexity index[1] of mangrove zones arranged according to decreasing wave and tidal energy but increasing influence of terrestrial runoff (Martínez et al., 1979), SE is standard error of the mean

Physiographic type		No. species	Basal area (m² ha⁻¹)		Density (ha⁻¹)		Mean dbh (cm)		Height (m)	Complexity index[1]	
			⩾2.5[2]	⩾10[2]	⩾2.5[2]	⩾10[2]	⩾2.5[2]	⩾10[2]		⩾2.5[2]	⩾10[2]
Offshore fringes, high energy	mean	1.60	21.7	13.7	4620	930	13.1	14.2	9.2	12.5	3.8
	SE	0.40	4.7	5.6	2040	380	1.8	1.3	2.6	4.6	2.7
Low-energy fringes	mean	2.20	22.0	16.0	6960	1010	10.9	15.2	10.3	18.8	4.4
	SE	0.63	9.3	2.4	2110	140	0.9	0.9	0.9	5.2	1.0
Landward on arid fringes	mean	1.00	6.5	4.8	3010	410	6.1	11.6	3.7	0.7	0.1
	SE	0	1.3	0.8	430	—	1.0	0.4	0.5	0.1	—
Landward on moist fringes	mean	3.13	20.8	20.8	2800	740	14.0	—	13.1	26.0	4.7
	SE	0.13	2.6	2.6	710	120	4.1	—	0.7	11.2	2.1

[1]See footnote to Table 5.2 (p. 100).
[2]Diameter at breast height.

TABLE 6.2

Structural characteristics of fringe mangrove forests (data are from Martínez et al., 1979, unless otherwise indicated)

Stand location	Latitude (N)	No. of tree species in 0.1 ha	Mean dbh (cm)	Number of trees (ha^{-1})		Basal area (m^2 ha^{-1})		Stand height (m)	Complexity index	
				≥2.5[1]	≥10[1]	≥2.5[1]	≥10[1]		≥2.5[1]	≥10[1]
Florida										
Rookery Bay[2]	25°55′	3	7.7	3400	—	15.9	—	6.5	10.5	—
Ten Thousand Islands[3]	25°50′	2	10.3	1800	600	15.0	11.0	6.3	3.4	0.8
		2	10.3	2800	800	23.5	12.4	7.3	9.6	1.4
Mexico										
Isla Roscell[3]	22°30′	3	15.5	1480	910	28.5	26.2	8.0	10.1	5.7
Laguna Terminos[4]	18°40′	3	6.3	7510	—	23.3	—	—	—	—
Puerto Rico										
Ceiba[3]	18°15′	2	6.1	5690	260	16.7	3.4	8.5	16.2	0.2
Bahía Medio Mundo	18°15′	1	5.7	16 760	550	43.0	5.0	9.8	70.6	0.3
		1	12.2	2800	1890	33.0	28.0	16.0	14.8	8.5
Ensenada Honda	18°14′	1	3.2	7490	—	6.0	—	5.7	2.6	—
		3	4.7	11 830	—	20.5	—	7.2	52.4	—
Laguna Joyuda	18°08′	1	24.9	440	440	21.5	21.5	22.0	2.1	2.1
		1	11.2	1780	1150	17.5	15.5	13.8	4.3	2.5
		2	18.9	1180	1180	33.0	33.0	20.0	15.6	15.6
		1	8.4	1820	110	10.0	3.0	8.3	1.5	<0.1
Laguna Las Salinas	17°58′	2	9.3	3860	1600	26.0	16.5	9.0	18.0	4.8
Bahía de Guánica	17°58′	2	10.8	2510	1130	23.0	16.5	10.8	12.5	4.0
Caballo Blanco	17°58′	1	14.2	630	330	10.0	8.0	14.8	0.9	0.4
		1	13.1	1400	1040	19.0	17.5	14.3	3.8	2.6
Cayo Enrique	17°58′	1	7.2	6590	640	27.0	6.0	9.2	16.4	0.4
Bahía Sucia	17°58′	1	14.1	1930	1500	30.0	28.0	15.0	8.7	6.3
			7.3	5200	330	21.5	3.0	7.0	7.8	<0.1
Punta Pitahaya	17°57′	1	8.2	2830	500	15.0	8.0	8.8	3.7	0.4
Bahía Salinas	17°57′	1	14.5	1750	1750	29.0	29.0	12.0	6.1	6.1
		2	11.3	2080	1390	21.0	18.0	10.8	9.4	5.4
Punta Gorda[3]	17°57′	1	7.0	1780	260	6.9	3.1	7.0	0.9	0
Aguirre[3]	17°57′	3	8.9	3670	110	22.6	13.9	12.0	29.9	0.6
Punta Pitahaya	17°57′	1	11.1	2770	1700	27.0	23.0	12.2	9.1	4.8
		2	5.4	7980	100	18.5	1.0	6.0	17.7	<0.1
		1	12.8	2940	2140	38.0	35.0	13.8	15.4	10.3
Bahía de Jobos[3]	17°57′	3	4.1	11 650	260	15.5	3.2	4.8	26.0	0
Cayos Caribe	17°55′	2	13.1	2210	1910	30.0	28.0	9.0	11.9	9.6
		2	10.2	2550	1020	21.0	14.0	9.0	9.6	2.6
Costa Rica										
Santa Rosa[3]	10°45′	2	16.8	1050	800	23.2	22.2	10.0	4.9	3.6

[1] dbh in cm; [2] Twilley, 1982; [3] Pool et al., 1977; [4] J.H. Day, Jr., pers. commun., 1983.

indices. Values range from 3.5 and 4.4 (Golley et al., 1962; Pool et al., 1975) in the areas facing the water to 1.7 and 1.8 (Carter et al., 1973) in the landward portions of the fringe. On the water edge light strikes the forest laterally, increasing leaf area. Leaves cover the side of the forest to the high water mark and one can estimate the tidal amplitude in a fringe forest by the gap created between low water and the lower edge of leaf cover.

DYNAMICS OF FRINGE MANGROVE FORESTS

Succession

Fringe forests are constantly undergoing change. These changes are induced either by catastrophic or periodic events external to the mangrove or by the maturation of the mangrove fringe. Both mechanisms promote succession and change in

TABLE 6.3

Biomass of fringe mangrove ecosystems

Ecosystem type and location	Latitude (N)	Age (years)	Biomass (t ha^{-1})			
			aboveground	roots	total vegetation	litter
Fringe						
Florida[1]	25°–26°	—	129.6	—	—	17.3
Florida[1]	25°–26°	—	119.6	—	—	14.0
Florida[1]	25°–26°	—	117.5	—	—	60.2
Florida[1]	25°–26°	—	152.9	—	—	98.4
Florida[1]	25°–26°	40–50	49.0	—	—	—
Florida[1]	25°–26°	5	8.1	14.1	22.2	0.3
Puerto Rico[2]	18°	—	62.8	50.0	112.8	—
Thailand[3]	8°	15	159.1	—	—	0.3

[1]Lugo and Snedaker (1975).
[2]Golley et al. (1962).
[3]Christensen (1978).

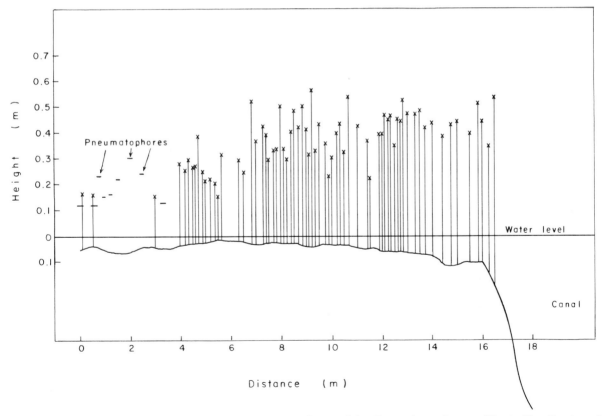

Fig. 6.4. Distribution and height of seedlings in a fringe mangrove forest at Jobos Bay on the south coast of Puerto Rico. Courtesy of Nydia Dávila.

plant zonation. In general, autosuccessions predominate under extreme environmental conditions, and species change cyclically under more moderate conditions (Ch. 4, Sect. 10). However, in spite of constant change, mangrove ecosystems are steady-state components of tropical coastlines (Lugo, 1980).

The salient feature of colonization patterns and succession in high-energy fringes is that the advance of trees is from the ocean to the land (Figs. 6.1 and 6.5). Trees at the water edge are older than those inland. This alone may explain the presence of large diameter trees at the water's edge. In contrast, very low-energy fringes, such as those in the west coast of Florida, are characterized by a colonization of tree species from the land to the ocean (Davis, 1940). These contrasting patterns of succession exemplify the opportunistic nature of mangrove succession (Thom, 1982).

Mangroves are opportunistic colonizers in the variable coastal environment. When the outside of the fringe is exposed to high energy, wave and tidal motion transport propagules towards areas that can be colonized. Colonization begins at the water edge, progressing inland under the protection of the first band of colonizer trees. As this band of trees expands inland, the internal environment changes and other species can grow on the leeward side. In the classic fringe succession described by Davis (1940), wave and tidal energy was low and terrestrial runoff abundant. Extensive mangrove areas behind the fringe provided propagules. Under these conditions trees advance from the land to the ocean, particularly if there is a source of sediments to build up the land before colonization (Thom, 1967, 1975, 1982).

Succession in mangrove fringes usually starts after massive tree mortalities (Jiménez et al., 1985). These mortalities are caused by the synchronous maturation of trees or by destructive episodic

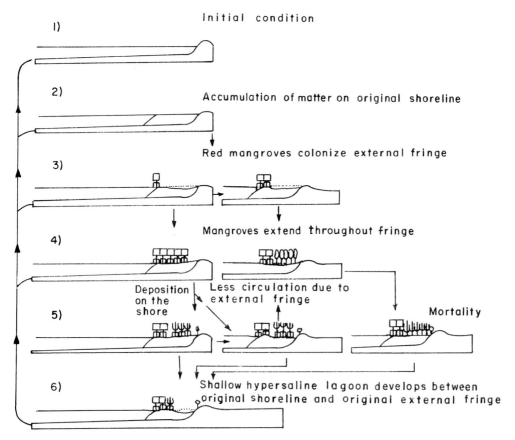

Fig. 6.5. Mangrove succession in high-energy arid fringe mangrove forests. Three alternatives are given in stage 5 of the high-energy succession (Cintrón et al., 1980).

events (Fig. 6.6). Because episodic events promote development of even-age stands, they are also indirectly responsible for mortalities caused by the synchronous maturation of trees. In the example on Fig. 6.6 a storm deposited a new berm on a fringe forest. This resulted in some tree mortality due to deposition of sand on the fringe, and additional tree mortality inside the stand due to a reduction of water circulation and soil aeration. The resulting circular area of dead trees either formed a lagoon or was invaded by trees of lower stature than those in the original stand.

The successional events described in Figs. 6.1, 6.5, and 6.6 were documented using sequential aerial photographs (Cintrón et al., 1980). These showed that, once the external mangrove fringe was established, its width did not change, and the expansion of the mangrove basin behind the fringe was at the expense of the internal hypersaline lagoons (Fig. 6.7). These sequences suggest that

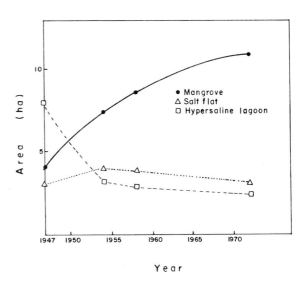

Fig. 6.7. Change in the area of mangroves and adjacent ecosystems over a 38-year period after passage of a hurricane in 1932. From Cintrón et al. (1980).

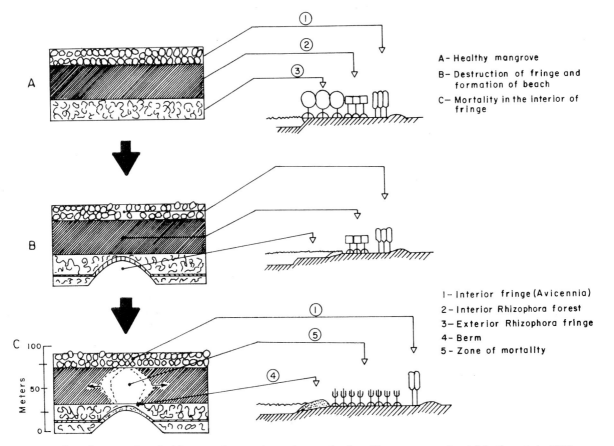

Fig. 6.6. Effect of a storm-deposited berm on the zonation and succession in a fringe mangrove forest (Martínez et al., 1979).

the width of the external fringe is a function of the hydrologic energy available to maintain adequate water circulation. Trees will not expand beyond the limit imposed by hydrologic energy.

The hydrologic environment affects many other processes in the fringe. Among the documented functional responses of mangroves to hydrologic turnover are the following: seedling establishment, growth, and mortality; litter production, decomposition, and turnover; photosynthesis, respiration, and transpiration rates of trees; tree growth and primary productivity; and nutrient cycling. In general all of these processes occur at faster rates or at greater efficiency in the water edge than in inland zones of the fringe.

Seedling establishment, growth, and mortality

Mangrove seedlings are viviparous and have a wide range of sizes. They tend to be larger in those species that grow at the water edge than in those that grow in the landward portions of the forest (Rabinowitz, 1975, 1980). Once a seedling falls from the parent tree, dispersal through the maze of surface roots of established forests would be made difficult by large size. Size is important to dispersal, tidal sorting within the mangrove zones, and establishment. Larger seedlings have larger stores of food which are needed to sustain metabolism during dispersal, before the roots of the plant are established in a new substrate, and until net daytime photosynthesis exceeds night-time respiration. Diurnal measures of daytime net photosynthesis (P) and night-time respiration (R) show more respiration than photosynthesis in recently established seedlings ($P/R = 0.1$: Lugo et al., 1975).

Because wave and tidal action is reduced within the fringe, it acts as a trap for large quantities of flotsam, litter, and propagules as described above. For example, Lugo and Snedaker (1975) found that seedling biomass at the high tide mark of a low-energy fringe in southern Florida changed from 1.1 g m^{-2} to 267 g m^{-2} after such an event. The growth of these seedlings is usually negligible compared to the growth of seedlings established elsewhere in the forest. The reason may be competition for space, damage to seedlings during or after transport, predation by animals, or the poor growth conditions in this part of the forest. There is a frequent recurrence of these events at the high water mark, and more material is frequently deposited over seedlings.

The rates of establishment and mortality of seedlings in fringe forests are often quite high (Table 6.4). In a Florida fringe, mortality of first-year seedlings was 13% and 8% during a 124-day period and 34% for established seedlings of unknown age. The number of new seedlings arriving during this period was 28% of the existing population (Lugo and Snedaker, 1975). Lugo and Snedaker (1975) found that seedlings on branches facing the water suffered greater mortality than those on branches facing inland. Apparently, wind, desiccation, and predation cause more stress to exposed seedlings. The size and weight of *Rhizophora mangle* seedlings were smaller in stressed than in non-stressed fringes (Martínez et al., 1979).

Under canopy cover the rate of seedling growth is faster by the water edge than at inland locations

TABLE 6.4

Production, establishment, and mortality rates for propagules of Central Queensland mangroves (data from Saenger, 1982)

Genus	No. of propagules establishing per 30 m of shoreline during 4 years	No. of established propagules per adult of same species	Mortality of established propagules during first year (%)	Mean mortality of adults during 1 year (%)
Rhizophora	276	1.64	71.7	2.98
Aegialitis	3	1.50	0	0
Avicennia	199	1.47	22.1	5.97
Lumnitzera	9	1.00	0	2.78
Aegiceras	27	0.18	14.8	1.51
Ceriops	52	0.13	36.5	1.01

(Lugo and Snedaker, 1975). This may be simply a response to more light intensity reaching the forest floor in the fringe. The seedling population profile in Fig. 6.4 shows the reduction in seedling height with distance from the water edge.

It is common to find two size classes of seedlings in mangrove fringes. One represents the recent arrivals to the population (averaging about 30 cm tall and with one or two pairs of leaves) and the other is the stagnated population at about twice that height. When the forest canopy is disturbed, light incident on the forest floor increases and explosive growth follows. Impenetrable thickets of fast growing trees result.

If the canopy is not disturbed, most surviving seedlings will grow to about 0.5 to 1 m and stagnate (Lugo and Snedaker, 1975). This stagnation may be caused by light limitation or a physiological change when seedlings mature into saplings. Such a change may make saplings less tolerant of the environment or perhaps more demanding for resources. Evidence for the physiological shift is suggested by the results of McMillan (1971, 1975), who found seedlings of *Avicennia germinans* to be more tolerant of salinity and low temperature than adult trees.

Photosynthesis, respiration, and transpiration

The physiological aspects of gas exchange in mangroves were recently reviewed by Clough et al. (1982). In general, the behavior of mangroves reflects the xeromorphic nature of leaves and whether they are in shade or open conditions (Table 6.5). Leaf xeromorphism is reflected in the low rates of transpiration measured in fringe forests. Shade leaves transpired more than sun leaves probably due to leaf orientation (Saenger, 1982). However, in the United States, transpiration rates were higher towards the water where *Rhizo-*

TABLE 6.5

Gas exchange rates of mangrove species and compartments in a fringe forest in southern Florida[1] (Lugo and Snedaker, 1975; the standard error of the mean is in parentheses)

Species and compartment	n	Net daytime photosynthesis (g C m^{-2} day^{-1})		Night-time respiration (g C m^{-2} day^{-1})		P/R		Transpiration (g m^{-2} day^{-1})	
Rhizophora mangle									
Sun leaves	8	1.125	(0.30)	0.439	(0.16)	3.3	(0.7)	633	(169)
Shade leaves	3	0.755	(0.11)	0.461	(0.10)	1.9	(0.6)	1744	(964)
Stems	3	−2.500[2]	(0.90)	1.123	(0.38)			154	
Branches	2	0.001		0.007				9	(1)
Prop roots	3	0.042	(0.33)	0.081	(0.05)			25	
Seedlings	5	0.954	(0.70)	1.397	(0.58)	1.7	(0.9)		
Avicennia germinans									
Sun leaves	9	1.145	(0.54)	0.468	(0.17)	3.1	(0.6)	228	(85)
Shade leaves	1	0.723		2.31		0.3		715	
Stems	3	−1.463[2]	(0.52)	0.727	(0.25)			134	
Pneumatophores	6	0.051	(0.04)	0.26	(0.10)			163	(42)
Laguncularia racemosa									
Sun leaves	5	0.511	(0.24)	0.139	(0.04)	3.8	(1.2)	261	(77)
Shade leaves	1	0.524		0.348		1.5		812	
Stems	1	−0.831[2]		0.400					
Conocarpus erectus									
Sun leaves	1	0.239		0.208		1.2		332	
Salicornia sp.	1	1.83		0.04		45.8			
Sesuvium portulacastrum	1	2.49						145	
Soil	1	0.04		0.49				25	

[1]All values are means of determinations over 24 h, and are calculated, in the case of leaves, on the basis of a single leaf surface.
[2]Daytime respiration.

phora mangle is dominant compared to inland where *Avicennia germinans* predominates. This was not anticipated, because leaves of *R. mangle* are more xeromorphic than those of *A. germinans*. Differences in soil salinity may explain this trend. As soil salinity increases inland, the "cost" associated with desalinization of water (Clough et al., 1982) may contribute to a reduction in plant transpiration.

The rate of net daytime photosynthesis of a given mangrove species varies according to the zone in which the species grows. A species may exhibit less daytime net photosynthesis and transpiration, and higher night-time respiration, when growing out of its normal zone than it does while growing in its usual position in the zone. For example, *R. mangle* exhibited faster net photosynthesis rates near the water than any other species, but lower rates when growing inland (Fig. 6.8). Studies in progress in a mangrove fringe at Jobos Bay, Puerto Rico, confirm these results. Rates of gross primary productivity (GPP) in three mangrove species of south Florida exhibited different responses to increases in chlorinity (Fig. 6.9). Gross primary productivity was positively correlated with chlorinity for *Avicennia germinans* and *Laguncularia racemosa* but negatively correlated for *Rhizophora mangle*. This relationship coincides with the position of these species in the zonation. *Rhizophora mangle* grows better at salinities below those of sea-water, while the other two species predominate in zones of higher salinity.

Species differences in the rates of net daytime photosynthesis, night-time respiration, and 24-h-transpiration are also illustrated in Table 6.5. Rates are lowest in the species that grow in the rarely inundated berm (*Conocarpus erectus*, *Salicornia* sp., *Sesuvium portulacastrum*) and highest at the water edge (*Rhizophora mangle*).

Gas exchange rates of mangrove fringe forests change vertically as well. Sun leaves in the canopy exhibit maximum gain of carbon, with P/R values between three and four (Table 6.5). At night, shade leaves exhibit high respiration rates. The main trunk surfaces lose significant amounts of carbon dioxide to the atmosphere, while branches lose little. Pneumatophores of *Avicennia germinans* were found to lose more carbon dioxide to the atmosphere per unit ground area than prop roots of *Rhizophora mangle*. The opposite was true when the respective stems were compared. In addition to

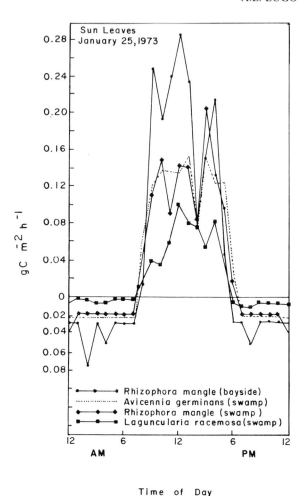

Fig. 6.8. Daytime net photosynthesis and night-time respiration in sun leaves of several mangrove species growing in a fringe mangrove forest in Florida (Lugo et al., 1975). Measurements on all four systems were made simultaneously. Bayside represents trees on the water edge and swamp represents those inland. Values are based on leaf area.

exchanging carbon dioxide with the atmosphere, pneumatophores, prop-roots, and soil surface exhibited a small amount of net daytime photosynthesis due to algae growing on them. This photosynthetic activity is important to sediments and roots because it supplements the oxygen supplied by tidal transport and diffusion.

Roots, stems, and soil together accounted for 25% of total ecosystem respiration in a fringe forest, and 30% in a basin forest in southern Florida (Lugo et al., 1975). These results were used to explain the vulnerability of mangrove ecosystems to chronic flooding or to stressors, such as oil,

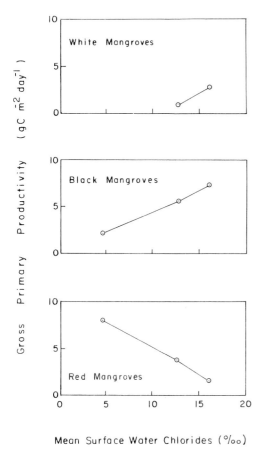

Fig. 6.9. Response of gross primary productivity of three mangrove species to chlorinity (Hicks and Burns, 1975). White mangroves: *Laguncularia racemosa*; black mangroves: *Avicennia germinans*; red mangroves: *Rhizophora mangle*.

that affect these gas-exchange locations (Lugo et al., 1981).

Seedlings of *Rhizophora mangle* had an average P/R ratio of 1.7 (Table 6.5) because most were in active growth rather than in the stagnation stage (discussed above). Other plants in the forest had high rates of daytime net photosynthesis, but one, *Sesuvium portulacastrum*, had a typical crassulacean acid metabolism (CAM) pattern of carbon dioxide exchange. Succulent plants are typical of the high portions (the berm) of the forest, where inundation seldom occurs and soil conditions are xeric.

Primary productivity and tree growth

The review of Lugo and Snedaker (1974) on this subject is still pertinent as few if any new primary productivity studies have been made at the ecosystem level. Most of the recent work on mangrove primary productivity uses the sum of litter fall and biomass increment. Biomass increments are estimated from allometric relations using tree diameter data (for example Ong et al., 1980a, 1980b, 1981). These studies are excellent for assessing net biomass accumulation by trees, but they miss two important components of the primary productivity of mangroves — photosynthesis by other plants, and ecosystem respiration. For example, the net photosynthesis of seedlings and periphyton in a fringe forest in southern Florida was 32% of the net primary productivity of the system. This is too high a proportion to be ignored. In the absence of these fluxes, the rates of ecosystem productivity are underestimated (Odum, 1964).

The first study of primary productivity of fringe mangrove forests (Golley et al., 1962) reported a 24-h respiration rate of 18.2 g m^{-2} day^{-1}, a gross primary productivity rate of 16.4 g m^{-2} day^{-1}, and no net primary productivity. In Florida, Lugo et al. (1975) found corresponding values of 3.8, 12.4, and 8.8. The difference in estimates was probably due to the length of the measurement period. Golley et al. (1962) used sporadic measures of gas exchange during the month of May, while Lugo et al. (1975) conducted 24-h measurements over a period of several months. The gross primary productivity reported by Lugo et al. (1975), extrapolated to a year, results in an estimate of annual gross primary production of 45 t ha^{-1}, a value below the average for six tropical forests (68.6 t ha^{-1} yr^{-1}; Brown and Lugo, 1982). The factors that regulate the productivity of these stands were summarized by Carter et al. (1973) as follows:

(1) *Tidal factors*. (a) Water currents transport oxygen to the root system; (b) exchange of the soil-water solution with the overlying water mass removes toxic sulfides and reduces soil-water salinity; (c) tidal flushing and surface water particulate load determine the rate of sediment deposition or erosion of stands; (d) vertical motion of groundwater transports nutrients regenerated by detrital food webs into the mangrove root zone.

(2) *Water chemistry factors*. (a) Total salt content governs transpiration rates through the water potential gradient between soil solution and plant vascular system; (b) a high macro-nutrient

content of the soil solution may enable the maintenance of high productivity in mangrove ecosystems by offsetting the stress of low transpiration rates caused by high salinity; (c) allochthonous macro-nutrients contained in wet-season surface runoff may dominate macro-nutrient budgets of mangrove ecosystems.

In the observations of Carter et al. (1973), the metabolism of the mangrove ecosystems responded to increasing chlorinity by increasing respiration and gross primary productivity, and decreasing net primary productivity (Fig. 6.10). These measurements were taken over a narrow range of values, but they reflect the respiratory cost of increased salinity in the mangrove environment. As this cost increases, net production available for growth and development of structure decreases. This helps to explain why community stature decreases with increasing salinity. When a mangrove fringe loses the subsidy of hydrologic energy, salinity increases and trees are stressed. To overcome stress, respiration increases and net primary production decreases. If gross primary production remains constant or decreases (compare Carter et al., 1973; Lugo and Snedaker, 1974; Hicks and Burns, 1975), community structure decreases to a level comensurate with available net energy yield. The metabolic cost of increased

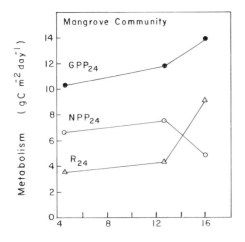

Fig. 6.10. Changes in mangrove ecosystem metabolism with chlorinity (Carter et al., 1973). GPP_{24}: gross primary productivity over 24 h; NPP_{24}: net primary productivity over 24 h; R_{24}: respiration over 24 h.

salinity has been discussed by Poljakoff-Mayber and Wainwright (1984).

The growth of individual trees is usually faster at the water edge of low-energy fringes than in the landward portions of the forest. On an area basis, Lugo and Snedaker (1975) found faster rates of tree growth in the water edge than the landward *Avicennia germinans* zone, and slower rates in the transitional zone between them. The estimated wood growth in this fringe forest was 6.4 t ha^{-1} yr^{-1} or almost the same as leaf fall. Clearly rates of wood production and net primary productivity are high in fringe mangrove forests.

Litter production, decomposition, and turnover

The production of litter in fringe mangroves is fairly constant throughout the world. Pool et al. (1975) found an average rate of about 2.7 g m^{-2} day^{-1} in the Caribbean and Florida, while Bunt (1982) reported an average of 2.3 g m^{-2} day^{-1} in Australia. Results from Asia (Ong et al., 1980a, 1980b, 1981) lie within the same range. However, the production of litter in high-energy fringes like those in Fig. 6.1 (No. 2) has not been measured. I hypothesize that under these conditions litter fall is lower. A lower rate of litter fall conserves nutrients in high-energy environments (see next section).

In low-energy fringe forests, litter fall increases towards the water edge (Lugo and Snedaker, 1975; Twilley, 1982; Twilley et al., 1986).

The proportion of leaves in total litter fall ranges between 60% and 90% (Pool et al., 1975; Bunt, 1982). Leaf fall occurs throughout the year, but peaks are coincident with the rainy season, low temperature, and extensive drought (Bunt, 1982; Lugo and Snedaker, 1975; Pool et al., 1975). Leaf fall may be a mechanism for removing salts from mangroves, and thus have a role in osmoregulation (Saenger, 1982).

The decomposition of leaf litter in mangroves is fast (Table 6.6). The half-life of red mangrove leaves is less than a year and averages 170 days. Rates of decomposition decrease towards the water edge, and are also species-specific (Twilley, 1982; Table 6.7). Leaves of *Rhizophora mangle* decompose slower than those of *Avicennia germinans*. This difference has been attributed to the chemical quality of the leaves (Twilley, 1982). Those of *R. mangle* are more xeromorphic and

TABLE 6.6

Litter degradation rates in mangrove forests

Forest/litter type	Location	Decay rate k (yr^{-1})	Half-life (days)	Source
Fringe				
Rhizophora mangle	Florida	8.39	30	Heald, 1971
	Florida	2.55	99	Heald, 1971
	Florida	1.46	173	Heald, 1971
	Florida	1.10	231	Heald, 1971
	Florida	0.85	346	Lugo and Snedaker, 1975
	Puerto Rico	2.55	99	Golley et al., 1962
	Mexico	1.42	178	J.H. Day, Jr., pers. commun. 1983
	Brazil	1.23	206	Schaeffer-Novelli and Cintrón, 1980
Mean			170 ± 34	
Laguncularia racemosa	Brazil	3.08	82	Schaeffer-Novelli and Cintrón, 1980

have a higher carbon to nitrogen ratio (C/N) than *A. germinans* leaves. Twilley reported an inverse linear relation between C/N and decay rate (k) in mangrove leaves.

Canopy leaves turn over at a rate of about 1 to 1.4 yr^{-1} in fringe forests, but the turnover of leaf litter is much higher (Pool et al., 1975). The position of the fringe relative to hydrologic factors influences the turnover of its litter compartment. The amount of litter storage can either increase or decrease, depending on tidal and wave energy. At the water edge rapid litter turnover is due to export rather than decomposition of falling materials. Very little decomposition occurs in place before litter is transported by water. However, the process is complicated by the variation in hydrologic energy at the fringe. Exposed high-energy fringes (No. 2, Fig. 6.1) are devoid of litter because all the leaf fall is exported by the daily inundation. Other fringe forests may trap this litter. As a result, some fringe forests may have a low amount of ground litter while others have large amounts (Table 6.3).

Inland in the fringe, the influence of hydrologic forces is reduced, the quality of leaf litter changes, and rates of leaf litter decomposition are high. Decomposition becomes more important than whole-leaf export in determining litter turnover rates. Because leaves of inland species are less xeromorphic and have a lower C/N ratio than those on the water edge, they suffer more damage and fragmentation due to grazers. This litter is also more susceptible to leaching (Twilley, 1982). The result is that a high proportion of litter fall is exported seasonally by tidal or runoff waters, either in solution or as small particulate detritus (Ch. 4, Sect. 3).

Nutrient cycling

The conditions for nutrient cycling in fringe forests are regulated by hydroperiod through two

TABLE 6.7

Decomposition of *Rhizophora mangle* and *Avicennia germinans* leaves at two distances from open water (Twilley, 1982)

Time (days)	*Rhizophora mangle*		*Avicennia germinans*	
	30 m	150 m	30 m	150 m
Carbon (mg g^{-1})				
10	465	453	473	447
148	411	420	387	452
Nitrogen (mg g^{-1})				
10	6.8	6.9	13.5	13.3
148	12.4	9.7	21.8	19.2
C/N ratio				
10	68	66	35	33
148	33	43	17	23
Decomposition rate (% day^{-1})				
148	0.365	0.236	0.601	0.534

mechanisms, one with and the other without biotic control. Water movement affects the amount and turnover of litter on the forest floor, and transports dissolved nutrients to and from the fringe forest. Plants have little control over these conditions. Hydroperiod also affects the rate of litter decomposition and the chemistry of the soil. The availability of nutrients from these reservoirs is environmentally controlled, but plant adaptations can influence the efficiency of nutrient uptake or use. For example, the high root biomass of mangrove fringe forests (Table 6.3) must obviously play an important role in capturing nutrients. Plants can also regulate the amount of nutrients returned to the forest floor in litter fall by retranslocation before abscission.

Information required to clarify how fringe forests cope with nutrient cycling in their environment is scarce (for example, Rodin and Bazilevich, 1967; Clough and Attiwill, 1975; Golley et al., 1975; Ong et al., 1980a, 1980b; Snedaker and Brown, 1981; Bunt, 1982; Twilley, 1982). In the discussion that follows I use data from these sources and some of my unpublished data to describe some emerging patterns. I will use *Rhizophora* sp. as an indicator of how fringe species may respond to conditions on the water edge. The summary chapter (Ch. 19) discusses these data in the context of wetland type and in comparison with other species.

Leaves of *R. mangle* have high concentrations of nutrients except for phosphorus (Table 6.8). The phosphorous concentration is low in comparison to upland species (Rodin and Bazilevich, 1967) or deciduous trees (Medina, 1980). However, nitrogen and phosphorus concentrations in Table 6.8 are slightly higher than the average values reported by Medina for evergreen plants — that is, 0.96 and 0.06% respectively. The other nutrient concentrations reported for these mangrove leaves are within the range for upland species (Rodin and Bazilevich, 1967). The concentrations of nutrients in wood were lower than in leaves except for calcium. Calcium concentrations were in the same order of magnitude in wood as in leaves. The high phosphorus values in *Rhizophora* wood from Brazil and Panama may be due to a highly fertile site.

Nutrient concentrations of *R. mangle* leaves changed with age (Fig. 6.11). After leaf expansion, phosphorus concentrations decreased rapidly while

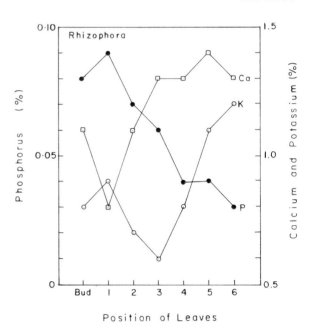

Fig. 6.11. Changes in chemical composition of *Rhizophora mangle* leaves with position in the stem (Snedaker and Brown, 1981).

calcium concentrations increased. The pattern for potassium is similar to that for calcium, except for a delay by about two age classes. Other elements studied by Snedaker and Brown (1981) did not change with age. They attributed the accumulation in calcium to increased cell-wall formation and the decrease of phosphorus to retranslocation. However, such a decrease could also be explained by dilution caused by accumulation of dry weight during leaf expansion.

Within a well-flushed monospecific fringe forest that I am studying in Jobos Bay, Puerto Rico, there is little change in the characteristics of *R. mangle* leaves collected in horizontal and vertical transects. However, there were significant differences between leaves of seedlings and adult trees (Table 6.8). The leaves of seedlings had a greater specific area, higher C/N and lower N/P ratio. These differences were probably due to lower nitrogen concentrations in seedling leaves. Phosphorus and potassium concentrations in seedling leaves were slightly higher than those of adult leaves. It is possible that viviparous seedlings show remnants of the high nutrient concentrations that typify most fruits.

To describe the within-stand efficiency of nutri-

TABLE 6.8

Characteristics of *Rhizophora* spp. leaves and wood. Standard error of the mean is in parentheses

Location and species	Specific area of leaves (cm² g⁻¹)	Composition (% of dry matter)					N/P	Source
		Ash	N	P	K	Ca		
Puerto Rico								
Rhizophora mangle								
Fringe, Jobos								
Seedlings (n=5)	58 (3)	15 (0.3)	1.1 (0.1)	0.085 (0.002)	0.66 (0.13)		13 (1)	Lugo, unpublished[1]
Adult trees (n=10)	51 (0.8)	14 (0.7)	1.5 (0.02)	0.082 (0.001)	0.53 (0.06)		18 (0.3)	Lugo, unpublished[1]
Nine sites								
Leaves				0.11				Snedaker, 1975
Wood				0.02				Snedaker, 1975
Vacia Talega	57							Pool et al., 1977
Florida								
Rhizophora mangle								
Fringe	53							Pool et al., 1977
	48							Pool et al., 1977
Eleven sites								
Leaves			1.3	0.05	1.6	1.7	26	Snedaker, 1975
Wood			0.5	0.02	0.4	2.7	25	Snedaker, 1975
Panama								
Rhizophora mangle								
Riverine								
Leaves				0.09	0.8	1.2		Golley et al., 1975
Wood				0.09	0.3	1.3		Golley et al., 1975
Brazil								
Rhizophora mangle								
Leaves			1.0	0.22	0.6	2.2	5	Rodin and Bazilevich, 1967
Wood			0.7	0.23	0.4	1.2	3	Rodin and Bazilevich, 1967
Rhizophora sp.								
Leaves				0.15	0.9	1.4		Lamberti, 1969
China								
Rhizophora mucronata								
Leaves			1.3	0.08	0.8	1.3	16	Rodin and Bazilevich, 1967

[1] In collaboration with H. Serrano and I. Monefeldt.

ent cycling by leaf fall (Vitousek, 1984), data on mass and nutrient return by *Rhizophora* spp. were reviewed (Table 6.9). The ratio of mass fall to nutrients in mass was used by Vitousek as a measure of within-stand recycling efficiency. The higher the ratios, the more efficient is the process because it is circulating more carbon per unit of nutrient.

The few data available for calcium return in leaf litter fall show that *Rhizophora* spp. behave as do upland forests reported by Vitousek (1984). None of the ratios exceed the threshold of 200 suggested by Vitousek as the value that separated the most calcium-efficient systems. Efficiency ratios for nitrogen recycling are higher than the value of 130 used by Vitousek to identify nitrogen-efficient upland ecosystems. This suggests a high within-stand nitrogen-recycling efficiency.

The efficiency ratios for phosphorus recycling in *Rhizophora* spp. varies widely, from low values in Malaysian forests to intermediate values for Australian forests and very high values in Florida (Table 6.9). These differences accentuate the functional plasticity of mangrove species and

TABLE 6.9

Litter turnover and efficiency of nutrient cycles in *Rhizophora* spp. stands

Location and ecosystem	Leaf fall (t ha^{-1} yr^{-1})	Litter turnover (yr^{-1})	Return of nutrients (kg ha^{-1} yr^{-1})			Efficiency ratio[1]			Source
			Ca	N	P	Ca	N	P	
Florida									
Monospecific basin									
Rhizophora mangle	0.33			2			165		Twilley, 1982
Monospecific basin									
Rhizophora mangle	0.61			4			153		Twilley, 1982
Mixed basin									
Rhizophora mangle	3.15			19			60		Twilley, 1982
Hammock									
Rhizophora mangle	0.27		45	11	0.37	59	241	7162	Snedaker and Brown, 1981
Dwarf (*Rhizophora mangle*)	0.12		17	6	0.08	70	189	14 875	Snedaker and Brown, 1981
	0.20		27	10	0.13	70	188	14 769	Snedaker and Brown, 1981
Puerto Rico									
Fringe (*Rhizophora mangle*)	9.45			67			141		Levine, 1981
Malaysia									
Rhizophora apiculata									
5 years	6.18	5	91	69	4.9	68	90	1261	Ong et al., 1980a; 1982
10 years	8.09	9	125	64	4.9	65	126	1651	Ong et al., 1980a; 1982
15 years	8.02	4	98	52	—	82	154	—	Ong et al., 1980a; 1982
20 years	8.08	5	102	40	—	79	202	—	Ong et al., 1980a; 1982
30 years	8.44		118	86	9.0	72	98	938	Ong et al., 1980a; 1982
Virgin	5.76	4	65	31	6.3	89	186	914	Ong et al., 1980a; 1982
Rhizophora–Bruguiera	8.0	24	77	40	4.9	103	200	1633	Ong et al., 1980b
Panama									
Riverine (*Rhizophora mangle*)			87	6					Golley et al., 1975
Australia									
Rhizophora apiculata	6.00			29	2.2		204	2778	Bunt, 1982
Rhizophora stylosa	5.59			19	2.6		294	2175	Bunt, 1982
Rhizophora lamarckii	4.28			15	1.2		278	3567	Bunt, 1982

[1] Dry mass/nutrient mass (Vitousek, 1984).

cannot be attributed to geographical location. Rather, they may reflect site differences. Apparently, *Rhizophora* spp. can modify the efficiency of phosphorus use according to site conditions (see Twilley et al., 1986, for a description of a similar phenomenon in *Avicennia germinans*). Very little else can be inferred from these data relative to the behaviour of the species in the fringe environment. From the above, it appears that mangrove species growing in fringes have developed efficient mechanisms for recycling of nitrogen and perhaps phosphorus.

Other biotic adaptations that conserve nutrients in fringe mangroves include nutrient storage in biomass, and nutrient uptake by plants and microbial populations during the process of decomposition. For example, the root compartment is an important biotic nutrient reservoir in mangrove fringes. A fringe forest of *Avicennia marina* in Australia stored 68% of the total phosphorus and 71% of the total nitrogen, but only 62% of the mass of organic matter in vegetation (Clough and Attiwill, 1975). Studies of nutrient uptake by trees are not available, but uptake must be significant given the high rates of primary productivity discussed above. Twilley (1982) documented net nutrient uptake by micro-organisms decomposing *Rhizophora* and *Avicennia* leaves.

Biotic adaptations to conserve and recycle nutrients certainly mitigate many of the hydrologic forces that transport organic matter (and the nutrients bound therein) from the fringe to open waters. However, data are not available to construct nutrient balances, and it is thus difficult to distinguish the portion of the nutrient capital of a fringe forest that recycles within the fringe from the portion that has to be supplied from outside sources or from soil storage. Similarly, there is no information on how nutrient cycles operate in the various types of fringes described in Figs. 6.1–6.3.

FRESHWATER FRINGE WETLANDS

Most of the work on freshwater fringe wetlands is beyond the scope of this chapter because it has been conducted in marshes or papyrus swamps rather than forested wetlands (for example, Lind and Visser, 1962; Prentki et al., 1978; Thompson and Hamilton, 1983). These studies document the role of the fringe in sequestering nutrients and the close relation between vegetation, hydrology, soils, and topography. I found few descriptions of high- and low-energy freshwater fringe forests. In Canada, for example, Zoltai et al. (1975) described lakeside swamps which I interpret to be high-energy freshwater forested wetland fringes. Bellamy and Rieley (1967) and Yurkevich et al. (1966) studied what appear to be low-energy fringe forests developing on the margins of small bodies of open water. In some instances, trees may grow in the landward zones of complex lake fringes dominated by non-woody vegetation. One such example in Uganda, dominated by the date palm (*Phoenix dactylifera*), was planted with *Eucalyptus* spp. in an effort to increase wood production (Eggeling, 1935).

A high-energy forested wetland fringe in the Rideau Lakes region of Ontario was similar to high-energy saline fringes in terms of its homogeneous environmental conditions throughout the community (Beschel and Webber, 1962). Distinct communities did not emerge from the 161 species of tracheophytes, bryophytes, and ground lichens that they found. However, they did observe a gradual substitution of tree species from the lake to the upland, at which point a sharp transition was observed (Fig. 6.12). As is typical of saltwater fringes, physiological responses precede species changes (for example reduction of tree growth with increasing water depth; Frey, 1954). In low-energy fringe wetlands, soil carbon dioxide increases, oxygen decreases, and temperature decreases under homogeneous vegetation but with increasing distance from open water (Yurkevich et al., 1966).

Similar kinds of vegetation gradients form along lake fringes managed by humans. For example, Silker (1948) reported the results of plantings in the fringes of lakes of the Tennessee Valley Authority in the United States. After about 12 years of growth, eight species had segregated along subtle gradients of hydroperiod observed in two habitats.

Succession in freshwater fringe wetlands

The succession of freshwater fringe wetland vegetation is usually controlled by allogenic factors (discussed above for saltwater fringes). Such external control of succession is greater in the high-energy fringe than in the low-energy fringe. But under low-energy conditions, the proximity of the water table may be critical in determining the direction that succession will take (Ingram, 1967). Autogenic control of succession in low-energy fringes has been documented by Bellamy and Rieley (1967), who showed that 7 years of autogenic peat accumulation led to a succession of species in a low-energy bog fringe.

In high-energy fringes the overwhelming effects of strong winds or rapid water exchange are more obvious. When the wetland is located at the limits of tolerance of its species, changes in one factor may swiftly tilt succession. This occurs, for example, when soil salinity increases at the ecotone between mangroves and freshwater *Pterocarpus* or *Mora* forests (Chs. 9 and 10, see also Beard, 1946), or with the chronic inundation of any fringe wetland (Lugo, 1978).

HUMAN IMPACTS ON FRINGE WETLANDS

Humans cause two kinds of impacts on fringe forested wetlands. One involves direct interventions with the wetland. Examples are harvesting of trees or wood products such as tannins from *Rhizophora* spp., extraction of sand, peat, or organisms such as shellfish from the roots of

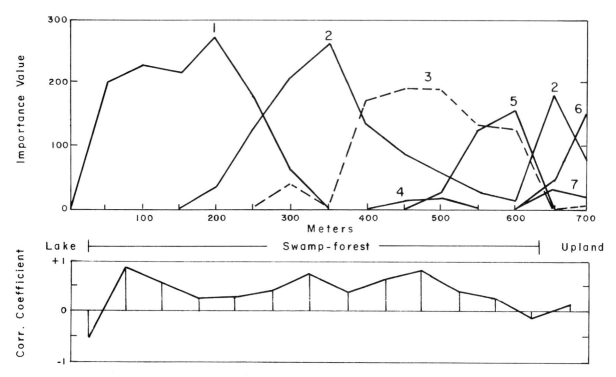

Fig. 6.12. Transect of a swamp forest near Portland, Ontario (Beschel and Webber, 1962). Importance values (the sum of relative density %, relative frequency % and relative basal area %) for the sampled stations of (1) *Larix laricina*, (2) *Thuja occidentalis*, (3) *Ulmus americana*, (4) *Fraxinus nigra*, (5) *Acer saccharinum*, (6) *Acer saccharum*, (7) *Fraxinus americana*. Correlation coefficients between the importance values for a given species at two adjoining stations are given below.

Rhizophora spp. Provided that the intensity of intervention is low, these human impacts can be overcome because they only affect the structure of the forest rather than the conditions that support its growth (Ch. 4, Sect. 13, 19). Therefore the forest can recover after intervention ceases. Obviously, extensive clearing without management, or the use of fringe-forest areas for urban or other types of development cannot be overcome.

A second type of human impact on fringe forests involves changes in hydrology and water quality. These kinds of impacts can be very damaging to fringe forests. For example, a change in water circulation can alter succession (Fig. 6.6), and leads to tree mortality and change in forest stature. Humans can induce such types of hydrologic change in lake and oceanic fringes by dredging in the vicinity of the fringe, or through the construction of structures that change local water currents or tidal conditions. Stabilization of lakes through impoundment usually kills fringe forests by causing chronic inundation, which also prevents regeneration.

Changes in the quality of water which affect fringe forests include oil pollution and eutrophication. However, eutrophication could increase rather than decrease the productivity of the fringe forest because it removes nutrient limitations and stimulates plant growth (Sell, 1977).

REFERENCES

Banus, M.D. and Kolehmainen, S.E., 1975. Floating, rooting and growth of red mangrove (*Rhizophora mangle* L.) seedlings: effect on expansion of mangroves in southwestern Puerto Rico. In: G. Walsh, S.C. Snedaker and H. Teas (Editors), *Proceedings of International Symposium on Biology and Management of Mangroves*. Institute of Food and Agricultural Sciences, University of Florida, Gainesville, Fla., pp. 370–384.

Beard, J.S., 1946. The mora forests of Trinidad, British West Indies. *J. Ecol.*, 33: 173–192.

Bellamy, D.J. and Rieley, J., 1967. Some ecological statistics of a "miniature bog". *Oikos*, 18: 33–40.

Beschel, R.E. and Webber, P.J., 1962. Gradient analysis in swamp forests. *Nature*, 194(4824): 207–209.

Brinson, M.M., Lugo, A.E. and Brown, S., 1981. Primary productivity, decomposition, and consumer activity in freshwater wetlands. *Annu. Rev. Ecol. Syst.*, 12: 123–161.

Brown, S. and Lugo, A.E., 1982. The storage and production of organic matter in tropical forests and their role in the global carbon cycle. *Biotropica*, 14: 161–187.

Brown, S., Brinson, M.M. and Lugo, A.E., 1979. Structure and function of riparian wetlands. In: R.R. Johnson and J.F. McCormick (Technical Coordinators), *Proceedings of the National Symposium on Strategies for Protection and Management of Floodplain Wetlands and Other Riparian Ecosystems*. U.S. Department of Agriculture Forest Service, Gen. Tech. Rep. WO-12, Washington, D.C., pp. 17–31.

Bunt, J.S., 1982. Studies of mangrove litter fall in tropical Australia. In: B.F. Clough (Editor), *Mangrove Ecosystems in Australia*. Australian Institute of Marine Science and Australian National University Press, Canberra, A.C.T., pp. 223–238.

Carter, M.R., Burns, L.A., Cavinder, T.R., Dugger, K.R., Fore, P.L., Hicks, D.B., Revells, H.L. and Schmidt, T.W., 1973. *Ecosystems Analysis of the Big Cypress Swamp and Estuaries*. U.S. Environmental Protection Agency 904/9–74–002, Region IV, Atlanta, Ga., 478 pp.

Chapman, V.J., 1944. 1939 Cambridge University expedition to Jamaica. *J. Linnean Soc. Bot.* (London), 52: 407–433.

Chapman, V.J. 1976. *Mangrove Vegetation*. J. Cramer, Leutershausen, 447 pp.

Chapman, V.J.(Editor), 1977. *Wet Coastal Vegetation*. Ecosystems of the World 1. Elsevier, Amsterdam, 428 pp.

Christensen, B., 1978. Biomass and primary production of *Rhizophora apiculata* Bl. in a mangrove in southern Thailand. *Aquat. Bot.*, 4: 43–52.

Cintrón, G., Lugo, A.E., Pool, D.J. and G. Morris, 1978. Mangroves of arid environments in Puerto Rico and adjacent islands. *Biotropica*, 10: 110–121.

Cintrón, G., Goenaga, C. and Lugo, A.E., 1980. Observaciones sobre el desarollo del manglar en costas aridas. In: *Memorias del Seminario Sobre el Estudio Cientifico e Impacto Humano en el Ecosistema de Manglares*. Oficina Regional de Ciencia y Tecnologia de la UNESCO Para America Latina y el Caribe, Montevideo, 405 pp.

Cintrón, G., Lugo, A.E. and Martínez, R., 1985. Structural and functional properties of mangrove forests. In: W.G. D'Arcy and M.D. Correa A. (Editors), *The Botany and Natural History of Panama*. Monographs in Systematic Botany 10. Missouri Botanical Garden, St. Louis, Mo. pp. 53–66.

Clough, B.F. and Attiwill, P.M., 1975. Nutrient cycling in a community of *Avicennia marina* in a temperate region of Australia. In: G. Walsh, S.C. Snedaker and H. Teas (Editors), *Proceedings of International Symposium on Biology and Management of Mangroves*. Institute of Food and Agricultural Sciences, University of Florida, Gainesville, Fla., pp. 137–146.

Clough, B.F., Andrews, T.J. and Cowan, I.R., 1982. Physiological processes in mangroves. In: B.F. Clough (Editor), *Mangrove Ecosystems in Australia*. Australian Institute of Marine Science and Australian National University Press, Canberra, A.C.T., pp. 193–210.

Davis, J.H., 1940. The ecology and geologic role of mangroves in Florida. *Pap. Tortugas Lab.*, 32: 303–412.

Eggeling, W.J., 1935. The vegetation of Namanve swamp, Uganda. *J. Ecol.*, 23: 422–435.

Frey, D.G., 1954. Evidence for the recent enlargements of the "bay" lakes of North Carolina. *Ecology*, 35: 78–88.

Golley, F.B., Odum, H.T. and Wilson, R.F., 1962. The structure and metabolism of a Puerto Rican red mangrove forest in May. *Ecology*, 43: 9–19.

Golley, F.B., McGinnis, J.T., Clements, R.G., Child, G.I. and Duever, M.J., 1975. *Mineral Cycling in a Tropical Moist Forest Ecosystem*. University of Georgia Press, Athens, Ga., 248 pp.

Gosselink, J.G. and Turner, R.E., 1978. The role of hydrology in freshwater wetland ecosystems. In: R.E. Good, D.F. Whigham and R.L. Simpson (Editors), *Freshwater Wetlands Ecological Processes and Management Potential*. Academic Press, New York, N.Y., pp. 63–78.

Heald, E., 1971. *The Production of Organic Detritus in a South Florida Estuary*. University of Miami Sea Grant Tech. Bull. 6, Miami, Fl.

Hicks, D.B. and Burns, L.A., 1975. Mangrove metabolic response to alterations of natural freshwater drainage to southwestern Florida estuaries. In: G. Walsh, S.C. Snedaker and H. Teas (Editors), *Proceedings of International Symposium on Biology and Management of Mangroves*. Institute of Food and Agricultural Sciences, University of Florida, Gainesville, Fla., pp. 238–255.

Ingram, H.A.P., 1967. Problems of hydrology and plant distribution in mires. *J. Ecol.*, 55: 711–724.

Jiménez, J.A., Lugo, A.E. and Cintrón, G., 1985. Tree mortality in mangrove forests. *Biotropica*, 17(3): 177–185.

Lamberti, A., 1969. *Contribuicao a conhecimento da Ecologia das Plantas do Manguezal de Itanhaém*. University of Sao Paulo, Sao Paulo, 218 pp.

Levine, E.A., 1981. *Nitrogen Cycling by the Red Mangrove, Rhizophora mangle L. in Joyuda Lagoon, on the West Coast of Puerto Rico*. Masters Thesis, University of Puerto Rico at Mayaguez, Mayaguez, P.R., 103 pp.

Lind, E.M. and Visser, S.A., 1962. A study of a swamp at the north end of Lake Victoria. *J. Ecol.*, 50: 599–613.

Lugo, A.E., 1978. Stress and ecosystems. In: J.H. Thorp and J.W. Gibbons (Editors), *Energy and Environmental Stress in Aquatic Systems*. U.S. Department of Energy Symposium Series (CONF-77114) National Technical Information Services, Springfield, Va., pp. 62–101.

Lugo, A.E., 1980. Mangrove ecosystems: successional or steady state. *Biotropica*, 12(supplement): 65–73.

Lugo, A.E. and Patterson Zucca, C., 1977. The impact of low temperature stress on mangrove structure and growth. *Trop. Ecol.*, 18: 149–161.

Lugo, A.E. and Snedaker, S.C., 1974. The ecology of mangroves. *Annu. Rev. Ecol. Syst.*, 5: 39–64.

Lugo, A.E. and Snedaker, S.C., 1975. Properties of a mangrove forest in southern Florida. In: G. Walsh, S.C. Snedaker and H. Teas (Editors), *Proceedings of International Symposium on Biology and Management of Mangroves*. Institute of Food and Agricultural Sciences, University of Florida, Gainesville, Fla., pp. 170–212.

Lugo, A.E., Evink, G., Brinson, M.M., Broce, A. and

Snedaker, S.C., 1975. Diurnal rates of photosynthesis, respiration, and transpiration in mangrove forests of Florida. In: F.B. Golley and E. Medina (Editors), *Tropical Ecological Systems*. Springer-Verlag, New York, N.Y., pp. 335–350.

Lugo, A.E., Cintrón, G. and Goenaga, C., 1981. Mangrove ecosystems under stress. In: G.W. Barrett and R. Rosenberg (Editors), *Stress Effects on Natural Ecosystems*. John Wiley and Sons, New York, N.Y., pp. 129–153.

McMillan, C., 1971. Environmental factors affecting seedling establishment of the black mangrove on the central Texas coast. *Ecology*, 52: 927–930.

McMillan, C., 1975. Adaptive differentiation to chilling in mangroves. In: G. Walsh, S.C. Snedaker and H. Teas (Editors), *Proceedings of International Symposium on Biology and Management of Mangroves*. Institute of Food and Agricultural Sciences, University of Florida, Gainesville, Fla., pp. 62–68.

Martínez, R., Cintrón, G. and Encarnación, L., 1979. *Mangroves in Puerto Rico: a Structural Inventory*. Puerto Rico Department of Natural Resources, San Juan, P.R., 149 pp.

Medina, E., 1980. Ecology of tropical American savannas: an ecophysiological approach. In: D.R. Morris (Editor), *Human Ecology in Savanna Environments*. Academic Press, New York, N.Y., pp. 297–319.

Odum, H.T., 1964. A symposium on net production of terrestrial communities (Book Review). *Ecology*, 45: 415–416.

Ong, J.E., Gong, W.K. and Wong, C.H., 1980a. *Studies on Organic Productivity and Mineral Cycling in a Mangrove Forest*. School of Biological Sciences, University Sans Malaysia, Penang, 48 pp.

Ong, J.E., Gong, W.K. and Wong, C.H., 1980b. *Ecological Survey of the Sungei Merbok Estuarine Mangrove Ecosystem*. School of Biological Sciences, University Sains Malaysia, Penang, 83 pp.

Ong, J.E., Gong, W.K. and Wong, C.H., 1981. *Ecological Monitoring of the Sungai Merbok Estuarine Mangrove Ecosystem*. School of Biological Sciences, University Sains Malaysia, Penang, 49 pp.

Ong, J.E., Gong, W.K. and Wong, C.H., 1982. *Studies on Nutrient Levels in Standing Biomass, Litter and Slash in a Mangrove Forest*. School of Biological Sciences, University Sains Malaysia, Penang, 44 pp.

Poljakoff-Mayber, A. and Gale, J. (Editors), 1975. *Plants in Saline Environments*. Springer-Verlag, New York, N.Y., 213 pp.

Pool, D.J., Lugo, A.E. and Snedaker, S.C., 1975. Litter production in mangrove forests of southern Florida and Puerto Rico. In: G.E. Walsh, S.C. Snedaker and H. Teas (Editors), *Proceedings of the International Symposium on Biology and Management of Mangroves*. Institute of Food and Agricultural Sciences, University of Florida, Gainesville, Fla., pp. 213–237.

Pool, D.J., Snedaker, S.C. and Lugo, A.E., 1977. Structure of mangrove forests in Florida, Puerto Rico, Mexico, and Costa Rica. *Biotropica*, 9: 195–212.

Prentki, R.T., Gustafson, T.D. and Adams, M.S., 1978. Nutrient movements in lakeshore marshes. In: R.E. Good, D.F. Whigham and R.L. Simpson (Editors), *Freshwater Wetlands*. Academic Press, New York, N.Y., pp. 169–194.

Rabinowitz, D., 1975. Planting experiments in mangrove swamps of Panama. In: G. Walsh, S.C. Snedaker and H. Teas (Editors), *Proceedings of International Symposium on Biology and Management of Mangroves*. Institute of Food and Agricultural Sciences, University of Florida, Gainesville, Fla., pp. 385–393.

Rabinowitz, D., 1980. Dispersal properties of mangrove propagules. *Biotropica*, 10: 47–57.

Rodin, L.E. and Bazilevich, N.I., 1967. *Production and Mineral Cycling in Terrestrial Vegetation*. Oliver and Boyd, London, 288 pp.

Saenger, P., 1982. Morphological, anatomical and reproductive adaptations of Australian mangroves. In: B.F. Clough (Editor), *Mangrove Ecosystems in Australia*. Australian Institute of Marine Science and Australian National University Press, Canberra, A.C.T., pp. 153–192.

Schaeffer-Novelli, Y. and Cintrón, G., 1980. *Estimados de produccion de hojas, hojarasca, y tasas de descomposicion foliar en un manglar de Cananeia, Sao Paulo, Brazil*. Department of Natural Resources, San Juan, P.R.

Sell, M.G., 1977. *Modeling the Response of Mangrove Ecosystems to Herbicide Spraying, Hurricanes, Nutrient Enrichment, and Economic Development*. PhD Dissertation, University of Florida, Gainesville, Fla., 390 pp.

Silker, T.H., 1948. Planting of water-tolerant trees along margins of fluctuating-level reservoirs. *Iowa State Coll. J. Sci.*, 22: 431–447.

Snedaker, S.C., 1975. *Reports on grant R-803340–02*. Institute of Food and Agricultural Sciences, University of Florida, Gainesville, Fla.

Snedaker, S.C., 1982. Mangrove species zonation: why? In: D.N. Sen and K.S. Rajpurohit (Editors), *Tasks for Vegetation Science*. Volume 2. Dr. W. Junk Publishers, The Hague, pp. 111–125.

Snedaker, S.C. and Brown, M.S., 1981. *Water Quality and Mangrove Ecosystem Dynamics*. U.S. Environmental Protection Agency 600/4–81–022, Environmental Research Laboratory Gulf Breeze, Fla., 80 pp.

Thom, B.G., 1967. Mangrove ecology and deltaic geomorphology, Tabasco, Mexico. *J. Ecol.*, 55: 301–343.

Thom, B.G., 1975. Mangrove ecology from a geomorphic viewpoint. In: G. Walsh, S.C. Snedaker and H. Teas (Editors), *Proceedings of International Symposium on Biology and Management of Mangroves*. Institute of Food and Agricultural Sciences, University of Florida, Gainesville, Fla., pp. 469–481.

Thom, B.G., 1982. Mangrove ecology — a geomorphological perspective. In: B.F. Clough (Editor), *Mangrove ecosystems in Australia*. Australian Institute of Marine Sciences and Australian National University Press, Canberra, A.C.T., pp. 3–18.

Thompson, K. and Hamilton, A.C., 1983. Peatlands and swamps of the African continent. In: A.J.P. Gore (Editor), *Mires: Swamp, Bog, Fen and Moor: Regional Studies*. Ecosystems of the World 4B. Elsevier, Amsterdam, pp. 331–373.

Twilley, R.R., 1982. *Litter Dynamics and Organic Carbon Exchange in Black Mangrove* (Avicennia germinans) *Basin*

Forests in a Southwest Florida Estuary. PhD Dissertation, University of Florida, Gainesville, Fl., 260 pp.

Twilley, R.R., Lugo, A.E. and Patterson-Zucca, C., 1986. Litter production and turnover in basin mangrove forests in southern Florida. *Ecology*, 67: 670–683.

Vitousek, P.M., 1984. Litterfall, nutrient cycling, and nutrient limitation in tropical forests. *Ecology*, 65: 285–298.

Wainwright, S.J., 1984. Adaptations of plants to flooding with salt water. In: T.T. Kozlowski (Editor), *Flooding and Plant Growth*. Academic Press, New York, N.Y., pp. 295–343.

Watson, J.G., 1928. Mangrove forests of the Malay peninsula. *Malay. For. Rec.*, 6: 1–275.

Yurkevich, I.D., Smolyak, L.P. and Garin, B.E., 1966. Content of oxygen in the soil water and of carbon dioxide in the soil air of forest bogs. *Sov. Soil Sci.*, 2: 159–167.

Zoltai, S.C., Pollett, F.C., Jeglum, J.K. and Adams, G.D., 1975. Developing a wetland classification for Canada. In: B. Bernier and C.H. Winget (Editors), *Forest Soils and Forest Land Management*. Les Presses de L'Université Laval, Quebec, Que., pp. 497–511.

Chapter 7

STRUCTURE AND DYNAMICS OF BASIN FORESTED WETLANDS IN NORTH AMERICA

SANDRA BROWN

INTRODUCTION

Freshwater basin wetlands are found in depressions of various depths, generally in areas where precipitation exceeds evapotranspiration or where the depression intersects the water table creating groundwater seeps or springs. Forested basins are those that contain woody vegetation with the potential for reaching tree stature; they do not include woody shrub wetlands. In North America (U.S.A. and Canada) these areas are mainly in the central and eastern region (see Fig. 5.8). Typical basin wetland forests of this region include coniferous peat bogs and swamps in high latitudes; pocosins of the southeastern coastal plain of the U.S.A.; and mixed deciduous hardwood, evergreen hardwood, and coniferous forested wetlands of the eastern and southern U.S.A. Saltwater basin forested wetlands, or mangroves, are found along subtropical and tropical coastlines, including the coast from Florida to Texas in the continental U.S.A.

Freshwater basin wetlands receive inputs of water mainly from rainfall, overland flow from drainage basins that are small or have gentle relief, and in some cases from groundwater seepage. Inputs of water to saltwater wetlands are from freshwater runoff, rainfall, and tidal exchanges of seawater. The common hydrological feature of basin wetlands is that floodwaters tend to be stagnant, or move very slowly as sheet flow over gentle topographic gradients. This similarity among basins is due to the prevalent energy signature (Ch. 4, Sect. 9): energy points (over small areas) and energy sheets (over broad areas) (*sensu* Kangas, Ch. 2); that is, major inputs of water are perpendicular to the land surface. As a result, vertical rather than lateral fluxes of water predominate, in sharp contrast to strong lateral flows in riverine forests (Ch. 5) and two-way water flow in fringe forests (Ch. 6). Water flows downward during precipitation and upward to replace evapotranspiration losses (Hemond, 1980). The restricted lateral water movement has important implications for the conditions to which plants must adapt. For example, water levels may fluctuate widely throughout the year, dissolved oxygen may be depleted for extended periods, reduced and toxic substances may accumulate, and redox potentials tend to be low. In saltwater basin forests, slowly moving water may concentrate salts in the soil, particularly in dry seasons when freshwater inputs are reduced and evapotranspiration is high.

Because of the general lack of a strong lateral component of water movement, basin wetlands are rarely flushed completely. Thus export of organic materials is low, and instead basins accumulate organic matter (Ch. 4, Sect. 15). As a result, peat deposits or highly organic sediments develop — a fact which, in turn, affects water chemistry and nutrient availability. The typically low pH of organic substrates coupled with anoxic conditions influences root functions and microbial activity.

Temperature and fire are other factors affecting basin forested wetlands. In northern latitudes, surface waters and substrates may remain frozen for many months of the year. Lack of water movement is likely to result in longer periods of frozen conditions as compared to riverine or fringe wetlands. Even after the thaw, temperatures of the organic substrates may still be low well into the growing season, and thus effectively shorten its length by slowing temperature-dependent physiological processes such as root respiration and water uptake. Temperature also affects species range — for instance, mangroves and other tropical species are limited to frost-free areas. Fires are a common feature in many freshwater wetlands, including those of northern and southern latitudes. Depending upon their severity and frequency, fires may influence nutrient availability, species composition, forest regeneration, and beta diversity of the wetland landscape.

The combination of restricted water movement, accumulation of organic materials, temperature, and fire produces the most stressful wetland environment compared to fringes and riverine wetlands. It is probably for these reasons that many basin wetlands are dominated by coniferous and sclerophyllous plants that have the ability to adapt to these conditions (see Ch. 4, Sect. 8).

Basin wetlands may cover large areas (hundreds of hectares) such as those in the northern latitudes or along subtropical or tropical coastlines, may be part of large wetland areas that also contain other wetland types such as those in the Okefenokee Swamp (Georgia) or Great Dismal Swamp (Virginia and North Carolina), or may cover small areas (a few hectares), as is common of the coniferous or evergreen hardwood basin wetlands of the southeastern U.S.A. There are about 1.7 million ha of all types of freshwater basin forests in the southeastern U.S.A., and about 2 million ha of coniferous peat bogs and swamps in the north-central U.S.A. (Johnson, 1979).

The purpose of this chapter is to summarize the pertinent information on the distribution, floristic composition, structure and dynamics of basin forested wetlands of North America. Their geomorphology is discussed in Ch. 3 by Kangas. Major emphasis will be on freshwater wetlands, but data for saltwater wetlands (mangroves), mainly from Florida and tropical America, will be included where appropriate. The first section of this chapter describes the external factors affecting basin wetlands or the important components of a wetland's energy signature. This is followed by a description of the distribution and floristic composition of representative basin wetlands. Sections on structural characteristics, organic matter dynamics, and nutrient cycling comprise the bulk of quantitative information. The chapter concludes with a section on effects of disturbance.

EXTERNAL FACTORS AFFECTING BASIN WETLANDS

The hydrologic regime of basin forested wetlands, particularly the frequency, depth and duration of surface flooding, has a strong influence on the structure and function of these ecosystems. It is surprising, however, how few studies of wetlands report hydrologic characteristics. Another factor that strongly influences basin wetlands, particularly vegetation patterns, is the chemical composition of surface waters (for example, Heinselman, 1970) and soil (for example, Monk and Brown, 1965). Fires commonly occur in basin wetlands, where they have a marked influence on ecosystem processes. This section briefly describes some representative hydrologic budgets, typical patterns of flood frequency and depth, chemical characteristics of surface waters, physical and chemical characteristics of the soil, and the effects of fire.

Hydrologic budgets

Although of limited duration (1–3 years), the examples of hydrologic budgets in Fig. 7.1 clearly demonstrate that precipitation is the major source of water to freshwater basin wetlands. As expected, tidal exchanges (about 430 tides yr^{-1}: Twilley, 1982), dominate inputs and outputs of the saltwater system. Evapotranspiration losses account for 57% and 70% of precipitation for the bog and pocosin, respectively, whereas the proportion is considerably higher for the cypress dome[1] and mangrove forest (119% and 89%, respectively). The apparently higher evapotranspiration losses of these last two forests includes evaporation of intercepted water, a significant quantity (300 mm and 110 mm for the cypress dome and mangroves, respectively). Losses by evapotranspiration from vegetation alone are 91% of precipitation for the cypress dome and 80% for the mangroves.

The hydrologic budgets of the four wetlands in Fig. 7.1 may be divided into two general classes based on peat depth. The bog and pocosin wetlands, often underlain by very deep peats (up to several meters thick), have virtually no recharge to deeper aquifers and no surface inflows, and 17 to 41% of their inputs are lost as surface and

---

[1]See explanation on p. 181.

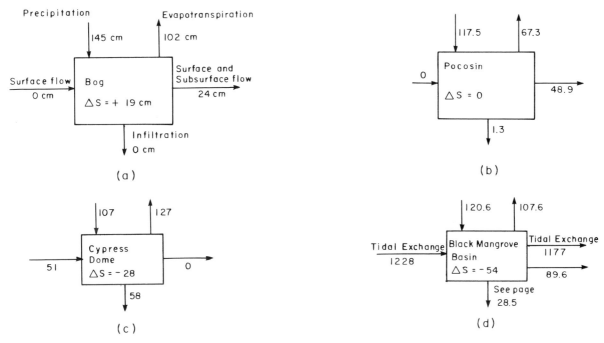

Fig. 7.1. Hydrologic budgets for four forested basin wetlands. ΔS = change in water storage. Data are from (a) Hemond (1980), (b) Richardson (1983, the average of 3 years), (c) Heimburg (1984), and (d) Twilley (1982). Refer to (a) for description of the budget components.

subsurface outflow, primarily in spring, with little to no change in storage. The cypress dome and mangrove, usually with shallow peats, receive surface inflows, experience losses through infiltration or seepage, and exhibit water deficits over the annual cycle which are made up by surface inflows. Thus these last two wetland forests have the potential for being richer in nutrients than the bog and pocosin forests, which must rely on precipitation alone for nutrient inputs. It is highly likely that these differences in hydrologic budgets explain the differences in the structure and function among these systems that are described later in this chapter.

Physical and chemical characteristics of surface water and soil

Forested wetlands of northern latitudes can be divided into two kinds based on nutrient and pH values (Jaworski and Raphael, 1979; Gore, 1983). Minerotrophic wetlands, influenced by groundwater, are mineral-rich with nearly neutral pH. Ombrotrophic wetlands are nourished by precipitation and dust and, as such, tend to be nutrient-poor and low in pH. In the regional terminology, fens are minerotrophic, bogs are ombrotrophic, and swamps have characteristics of both (Jaworski and Raphael, 1979). This system was further modified by Novitzki (1979) to include surface topography (depression or slope). His system identified four wetland types of which two can be considered as basin forested wetlands. The first (ombrotrophic) occurs when precipitation and sometimes overland flow collects in surface depressions, flooding is infrequent, and soils are organic. Water remains in these depressions long enough for water-tolerant plants to become established, because the bottom of the depression retards leakage. The second (minerotrophic) occurs when depressions intersect the groundwater and there is little to no surface drainage away from the wetland. Different plant associations develop depending upon the depth of the basin, source of water, and fluctuations in groundwater levels (Fig. 7.2). Deep basins with stable groundwater levels favor forested bogs or fens. Fens are favored when plant roots have access to mineral-rich groundwater; bogs are favored when plant roots are restricted to the peat layers. I suggest that this

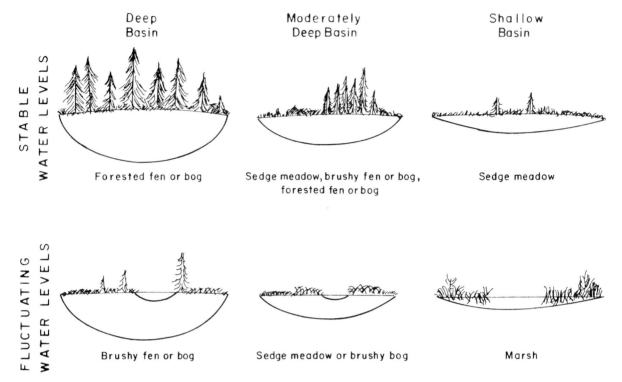

Fig. 7.2. Relationship between groundwater level fluctuations, basin depth, and plant communities in groundwater depression wetlands of Wisconsin (U.S.A.). (From Novitzki, 1979.)

classification system is applicable to both northern and southern freshwater basin wetlands (see bottom).

Examples of northern basin forested wetlands are those of Manitoba, Canada (50°N: Reader and Stewart, 1972), the Lake Agassiz Peatlands Natural Area, Minnesota (48°N: Heinselman, 1970), Cedar Creek Natural History Area, Minnesota (45°35'N: Reiners, 1972; Tilton, 1977), Wisconsin (Novitzki, 1979), and Thoreau's Bog, Massachusetts (42°28'N: Hemond, 1980). Minerotrophic basin wetlands have water tables frequently at, near, or above the ground surface; surface waters close to neutral (pH 6–7.5) and high in calcium and magnesium (Table 7.1; see also Heinselman, 1970; Tilton, 1977). Peat depth may vary from 0.3 to 1.8 m. The minerotrophic wetland often intergrades into one that has influences from both minerotrophic and ombrotrophic sources and, where, in consequence, surface waters are also frequently at or above the ground surface and intermediate in pH and mineral concentrations (Table 7.1; see also Heinselman, 1970; Tilton,

1977). Peat layers are generally thicker than those in strictly minerotrophic wetlands (Heinselman, 1970; Reiners, 1972). The level of the water table of ombrotrophic bogs is generally stable, and may be below the ground surface (Heinselman, 1970; Hemond, 1980) or above (Tilton, 1977) during most of the growing season. Because bogs tend to have thick peat layers (from 1 m to as much as 9 m: Heinselman, 1970), waters are very acid and low in minerals (Table 7.1).

In the eastern and southeastern region of the United States are a variety of wetlands that have surface water and soil characteristics similar to those of the northern region (Monk and Brown, 1965; Monk, 1966; Schlesinger, 1978; Brown, 1981; Christensen et al., 1981; Richardson et al., 1981; Gomez and Day, 1982; Brown et al., 1984b). One difference, however, is that surface waters in some types of southern basin forests are deeper than those in the north. For example, in wetlands dominated by cypress (*Taxodium*), a very flood-tolerant species, hydroperiods are long [from almost continuously (Schlesinger, 1978; Brown,

TABLE 7.1

Characteristics of surface waters of freshwater basin forested wetlands

Location, latitude, and forest type[1]	Nutrient concentration (mg l^{-1})							Conductivity (25°C) (μS cm^{-1})	pH	Source
	total N	NO$_3$-N	NH$_4$-N	total P	Ca	K	Mg			
Minnesota, 48°N										
Rich swamp forest (Northern white cedar)	—	—	—	—	20.1–25.2	—	0.5–8.1	84–119[2]	6.0–6.5	Heinselman, 1970
Poor swamp forest (Tamarack)	—	—	—	—	3.2–5.8	—	1.3–3.4	24–61[2]	3.9–5.7	Heinselman, 1970
Black spruce–moss bog	—	—	—	—	5.3–6.2	—	0.5–1.4	40–60[2]	3.2–3.5	Heinselman, 1970
Sphagnum–black spruce bog	—	—	—	—	1.6–2.6	—	0.1–0.2	35–53[2]	3.1–3.8	Heinselman, 1970
Minnesota, 47.5°N										
Black spruce bogs	0.69	0.20	0.45	0.19	2.4	1.3	0.97	51	3.6	Verry, 1975
Black spruce fen	0.33	0.10	0.15	0.09	16.6	1.1	2.88	125	6.5	Verry, 1975
Michigan, 45.5°N										
Bog forests	—	0.008	0.23	0.032[3]	3.2	0.5	0.7	58	4.1	Schwintzer, 1978
North Carolina, 36°N										
Pocosins	1.3	0.05	—	0.05–0.08	—	—	—	50–180	2.2–6.6	Daniel, 1981
Florida, 29.5°N										
Cypress dome	1.4	0.08	0.14	0.07	2.87	0.34	1.37	60	4.5	Dierberg and Brezonik, 1984
Florida, 26°N										
Scrub cypress	0.92	0.10	0.07	0.01	—	—	—	—		Brown et al., 1984b

[1]Scientific names of the plants mentioned in this column are: black spruce, *Picea mariana*; northern white cedar, *Thuja occidentalis*; cypress, *Taxodium* spp.; tamarack, *Larix laricina*.
[2]Conductivity measured at 20°C.
[3]Orthophosphate only.

1981) to about 6 months (Gomez and Day, 1982)] and floodwaters are deep (15–100 cm: Schlesinger, 1978; Brown, 1981). Surface waters are acid and low in nutrients (Table 7.1). Peat layers may be as thick as 1 m (Schlesinger, 1978; Spangler, 1984). Ombrotrophic southern wetlands, such as pocosins and bayheads (Monk, 1966) develop mainly on thick peat deposits (as much as several meters deep; Richardson, 1983). Their hydroperiod is long, but surface waters are stable and shallow (up to 15 cm: Monk, 1966; Christensen et al., 1981), acid, and nutrient-poor (Table 7.1; see also Monk, 1966; Daniel, 1981). Thick peats, however, are not always required to produce basin wetlands that are, in many respects, very similar to ombrotrophic bogs. For example, the scrub cypress forest in southern Florida and some pocosins grow on very sandy, nutrient-poor soils that lack organic matter, and their surface waters are shallow, acid and very poor in nutrients (Brown, 1981; Richardson et al., 1981).

The significant feature of surface waters and soils of saltwater basin forests is due to the influence of tidal exchanges of seawater. Characteristics of surface waters and soils of saltwater basin forests, dominated by black mangroves (*Avicennia*), are summarized by Twilley (1982) and Jiménez and Lugo (1985). Tidal exchanges in mangrove forests mainly control depth and duration of flooding. Surface waters are usually a few centimeters deep, except during periods of tidal exchange when they may increase to 15 to 20 cm deep (Twilley, 1982). Duration of flooding may be continuous, or only during tidal exchanges, which vary from 152 to 432 tides yr^{-1} (Jiménez and Lugo, 1985). Furthermore, the number of tidal exchanges may vary seasonally — in Florida, for instance, from less than ten per month during the winter months from November to April to more than twenty per month in September and October (Twilley, 1982). Soils of black mangrove forests vary greatly in texture (Jiménez and Lugo, 1985). When present, peat varies in depth from 0.5 to 2 m. Soils are usually saline; dependent on climate and proximity to the sea, salinities may range from zero to 100‰ (Jiménez and Lugo, 1985).

Data on concentrations of nutrients in the soils of freshwater basin wetlands (Table 7.2) confirm the general patterns described for surface waters. Basin wetland soils are acidic and low in nutrients.

In general, it appears that depth of the peat is an important factor influencing water depth and water quality — the deeper the peat the greater tendency for waters to be shallow, acidic, and poor in nutrients. However, similar characteristics in surface waters can be produced in wetlands growing on very sandy soils. Species composition is clearly strongly influenced by these factors (see below, p. 178).

Fire

Many basin wetland types of North America are affected by fire, including northern bog forests (for example, Lewis et al., 1928; Heilman, 1966), Atlantic white cedar (*Chamaecyparis thyoides*) forests of the Great Dismal Swamp (McKinley and Day, 1979), pocosins (Christensen et al., 1981), wetland forests of the Okefenokee Swamp (Cypert, 1961; Schlesinger, 1978), cypress domes (Ewel and Mitsch, 1978), and scrub cypress (Craighead, 1971). Fires affect these wetlands in several ways depending upon their frequency and severity, the amount of peat, and the soil moisture conditions. Extremely hot fires, occurring after severe droughts, burn deeply into the peat and destroy the forest stand (Lewis et al., 1928; Cypert, 1961; Ewel and Mitsch, 1978). Destruction of the peat substrate usually prevents re-establishment of the forest, because the basins are deeper and flooding is too great. Such severe fires are more than likely responsible for the formation of many of the lakes in the Okefenokee Swamp (Cypert, 1961) and the treeless, deep pools sometimes present in cypress domes (Ewel and Mitsch, 1978). In the scrub cypress ecosystem fires are common but the trees are rarely damaged or killed, because insufficient litter accumulates and there are no peat deposits (Craighead, 1971).

Moderate fires may benefit basin wetlands by increasing the rate of mineralization of litter and peat surface (Lewis et al., 1928; Heilman, 1966), by reducing competition from hardwoods and pines in cypress forests (Ewel and Mitsch, 1978; Schlesinger, 1978), by promoting successful establishment of some forested wetland species and maintaining species richness (McKinley and Day, 1979; Christensen et al., 1981), and by adding diversity to the wetland landscape (Christensen et al., 1981). In moderate fires, cypresses are more fire

TABLE 7.2
Nutrient concentrations of soils in freshwater basin forested wetlands

Location, latitude, and forest type[1]	Depth (cm)	pH	Concentration C (%)	N (%)	P (%)	Ca (μeq g^{-1})	Mg (μeq g^{-1})	K (μeq g^{-1})	CEC[2] (μeq g^{-1})	Source
Alaska, 65°N: Black spruce–*Sphagnum*	0–16	3.9–4.1	—	0.53–0.68	—	120–160	48–79	49.9–51.2	850–1000	Heilman, 1968
	16–34	3.7–4.1	—	0.65–1.13	—	60–175	35–67	11.8–27.1	850–1000	
Sphagnum–black spruce	0–16	3.6–4.0	—	0.43–0.59	—	155	53–64	39–42	1000–1070	
	16–30	3.9–4.2	—	0.44–0.92	—	230	49–71	8.2–14	1070–1100	
Minnesota, 45°N: Northern white cedar swamp	0–70	—	—	1.87	0.10	12.2	1.3	—	—	Reiners and Reiners, 1970
Marginal fen	0–25	—	—	1.03	0.08	0.5	1.0	—	—	Schlesinger, 1978
Georgia, 31°N: Cypress swamp	0–91.5	—	—	1.85	—	0.7	0.4	0.1	—	Coultas and Duever, 1984
Florida, 26–29°N: Cypress domes	0–15	3.3	55	1.57	—	6	23	8	880	
	15–76	3.2	49	1.54	—	2	13	5	1450	
	76–112	3.3	48	0.88	—	1	4	1	1160	
	0–56	4.1	—	4.69	—	1	2	1	310	
	56–76	4.4	—	0.56	—	1	1	1	34	
	76–127	4.4	—	0.77	—	7	17	1	160	
	0–20	—	—	—	1.35–3.02	—	—	—	—	Brown, 1981
Florida, 29°N: Evergreen hardwood swamps	15–30	3.9	—	—	0.008	0.3	0.3	0.1	61	Monk, 1966

[1] For scientific names of species mentioned see Table 7.1.
[2] Cation exchange capacity.

resistant, particularly large trees, than pines or hardwoods (Ewel and Mitsch, 1978). Thus, fire reduces the relative importance of hardwoods and enhances cypress dominance. Fire is also an important thinning agent in cypress swamps of the Okefenokee, and tends to maintain biomass at constant levels (Schlesinger, 1978). The life histories of many wetland forest plants depend on fire. For example, fire is necessary for maintaining Atlantic white cedar forests because of the light requirement of the seedlings (McKinley and Day, 1979). *Pinus serotina* (pond pine) is a particularly fire-resistant plant of the pocosins, and because of its moderately serotinous cones it is successful in establishing itself after a fire (Christensen et al., 1981). Landscape diversity is increased after fires because of the uneven burn pattern, which produces lightly burned areas and others where the fire burns deep into the peat (Christensen et al., 1981). This diversity in the wetland landscape has the potential for maintaining species diversity.

DISTRIBUTION AND FLORISTIC COMPOSITION

In this section the major types of basin forested wetlands and their plant associations are briefly described by region. Further information on the distribution and floristics of these wetland types is given in another book in this series (Gore, 1983). The emphasis in this section is on the basin wetlands that are described elsewhere in this chapter in relation to their surface water, soils, and ecosystem structure and function. Further detailed descriptions of the composition of basin wetlands of North America are found in Ch. 3 by Kangas, and in the publications of Richardson (1981), Larsen (1982), Cohen et al. (1984), and Ewel and Odum (1984).

The most northerly wetlands cited elsewhere in this chapter are a bog forest and muskeg located in Manitoba, Canada (50°N: Reader and Stewart, 1972). The bog forest was composed of a canopy of mature *Picea mariana* (black spruce) trees, a shrub layer dominated by *Ledum groenlandicum* (Labrador tea), with a carpet of *Pleurozium schreberi* underneath (Reader and Stewart, 1972). The relative frequency[1] of these three species totalled 75%, the remaining 25% being made up of seven other species. Adjacent to the bog forest was an area referred to as a muskeg in which black spruce trees were more widely spaced. This wetland also contained ericaceous shrubs such as *Chamaedaphne calyculata* (leatherleaf), *Ledum groenlandicum* and *Oxycoccus quadripetalus*. *Sphagnum* and *Polytrichum* species were common on the forest floor. The muskeg had more species than the bog forest (16 and 10 respectively). Further descriptions of bog vegetation of Canada have been given by Lewis et al. (1928), Dansereau and Segadas-Vianna (1952) and Zoltai and Pollet (1983).

The glaciated region of the United States, including Minnesota, Michigan and Wisconsin, contains large areas of basin forested wetlands, both minerotrophic and ombrotrophic. Many studies have been done on a variety of wetland types in this region (for example, Conway, 1949; Heinselman, 1970; Reiners and Reiners, 1970; Reiners, 1972; Parker and Schneider, 1974; Tilton, 1977). Heinselman's (1970) study of the Lake Agassiz Peatlands of northern Minnesota (48°N) recognized seven vegetation types, of which four are basin forested wetlands. The species-rich swamp forest (Heinselman, 1970) was dominated by *Thuja occidentalis* (northern white cedar), but black spruce, *Larix laricina* (tamarack), and *Fraxinus nigra* (black ash) often overtopped the cedar. *Abies balsamea* (balsam fir) and *Betula papyrifera* (paper birch) were also present but at low abundances. *Alnus rugosa* (speckled alder) dominated the shrub layer; herbs, grasses, ferns, and mosses were present in the ground layer. Transition from the species-rich swamp forest to species-poor swamp forest corresponded to a shift in dominance from northern white cedar to *Larix laricina*, the stands of which were almost pure and stunted (Heinselman, 1970). A dense understory of *Betula pumila* (bog birch) was common. The low shrub layer was dominated by ericaceous plants, such as *Andromeda glaucophylla* and *Chamaedaphne calyculata*. Much of the ground was covered by *Sphagnum* mosses.

Two types of black spruce wetlands were found in the Lake Agassiz Peatlands (Heinselman, 1970). The black spruce – *Pleurozium schreberi* forest was a mature, tall, dense stand of black spruce trees

---

[1]Frequency (f_i) is here defined as the proportion of the total observation units (in this case 50 points) in which the i th species was recorded, while relative frequency is $100 f_i / \sum_i f_i$.

with a continuous carpet of *Pleurozium schreberi*. *Ledum groenlandicum* and *Vaccinium* spp. were common in the sparse shrub layer. On very nutrient-poor sites, a *Sphagnum* – black spruce – *Chamaedaphne calyculata* bog was found. This is the typical open, stunted muskeg forest widespread in the northern Lake States (U.S.A.) and Canada (Heinselman, 1970; Reader and Stewart, 1972). These wetlands are dominated by black spruce, although *Larix laricina* is sometimes found. In addition to *Chamaedaphne calyculata*, *Kalmia polifolia* and *Ledum groenlandicum* are also common in the shrub layer.

Of the four types of wetlands described by Heinselman (1970), the species-rich swamp forest had the highest total number of plant species (71), followed by the species-poor swamp forest (51), black spruce – *Pleurozium schreberi* forest (25), and *Sphagnum*–black spruce–*Chamaedaphne calyculata* forest (20). This order was the same for all life forms.

In east-central Minnesota (45.5°N, Cedar Creek Natural History Area), Reiners and Reiners (1970) and Reiners (1972) studied two forested wetlands: a marginal fen and a cedar swamp. The marginal fen, a term coined by Conway (1949), had a high water table and was under the influence of relatively base-rich waters draining from the uplands. This wetland was dominated by both deciduous broadleaf species such as *Fraxinus nigra* (importance value[1] [IV300] = 72), *Acer rubrum* (red maple; IV = 33), *Ulmus americana* (American elm; IV = 29), and *Quercus ellipsoidalis* (northern pin oak; IV = 25), and the coniferous northern white cedar (IV = 51). The understory shrub layer was composed mainly of *Alnus rugosa* and *Corylus cornuta*, while the herb layer contained mainly *Athyrium filix-femina*, *Onoclea sensibilis* and *Osmunda cinnamonea*. The cedar swamp was strongly dominated by northern white cedar (IV = 127), with *Betula papyrifera* (IV = 42), *Fraxinus nigra* (IV = 32), *Ulmus americana* (IV = 28), and *Betula lutea* (yellow birch; IV = 17) as subdominants. Shrubs and herbs were very sparse in the cedar swamp (Reiners and Reiners, 1970). The marginal fen was richer in species than the cedar swamp (total number of plant species in the fen was 59 versus 43 in the swamp) because the former was at the ecotone between the oak upland forest and lowland cedar swamp and had species from both types of forest (Reiners, 1972). Furthermore, the fen had more understory species than the cedar swamp. This greater richness in the understory of the fen was attributed to the higher occurrence of windthrows there than in the swamp, because the substrate was a soft, wet muck compared to the firmer peat substrate of the swamp.

Wetlands of the Cedar Creek Natural History Area (Minnesota) were also studied by Tilton (1977). The bog forest was similar in species composition to those described above, composed of black spruce and *Larix laricina* trees, with an understory of *Chamaedaphne calyculata* and *Ledum groenlandicum*. In contrast to the fen studied by Reiners (1972), which was dominated by broadleaf deciduous species and northern white cedar, the fen studied by Tilton was dominated by *Larix laricina* with a sparse understory of *Alnus* spp. Ericaceous shrubs and *Sphagnum* mosses were absent. This variability in species composition between two fens from the same area agrees with the statement by Conway (1949) that fens cannot be described in general terms. Conway (1949) believed that this was partly due to their marginal position, which makes them vulnerable to human disturbance. For example, *Fraxinus nigra* trees are often logged from fens. The susceptibility of fens to disturbance may explain the dominance of *Larix laricina* in the fen described by Tilton (1977). *Larix laricina* is a successful early colonizer of northern wetland forests because of its high light requirement and good growth in constantly moist soils (Conway, 1949; Duncan, 1954).

Broad expanses of *Alnus* swamps are found in the upper peninsula of Michigan (Parker and Schneider, 1974). Dominant tree species in these wetlands were *Alnus rugosa* and *Fraxinus nigra*; other tree species present, 12 in all, included *Abies balsamea*, *Larix laricina*, *Populus balsamifera*, and *Ulmus americana*. The main shrub species was *Cornus stolonifera*. The herbaceous layer was dominated by 57 species predominantly from the genera *Aster*, *Carex*, *Glyceria*, *Impatiens*, and

---

[1] Importance value (IV, or IV300) is defined as relative frequency + relative dominance + relative density where relative frequency is defined as in the footnote on p. 178, relative dominance is $b_i/\sum_i b_i$ where b_i is the basal area of the ith species, and relative density is defined as $d_i/\sum_i d_i$ where d_i is the number of individuals of the ith species.

Solidago. The total number of plant species from the wetland area was 80 (Parker and Schneider, 1974).

In the northeastern United States, forested basin wetlands are widely scattered. Floristically they are similar in composition to those of the north-central United States (Hemond, 1980). A study of a bog in Massachusetts (42°N) by Hemond (1980) described three concentric vegetation zones: the open, floating mat of *Sphagnum*; a transition zone of tall, dense shrubs; and a forested swamp. The floating mat of *Sphagnum*, composed of two species (*S. magellanicum* and *S. rubellum*) formed a matrix upon which other vegetation grew. The tree species consisted of black spruce, *Acer rubrum*, *Larix laricina*, and *Pinus strobus* (white pine). The conifers tended to grow in mixed clumps, representing a small percentage of the plant cover. The trees were often gnarled, and seldom exceeded 2 m in height. The shrub layer was composed mainly of *Chamaedaphne calyculata*, but *Andromeda glaucophylla* and *Kalmia augustifolia* were also present. *Vaccinium macrocarpon* and *V. oxycoccos* were scattered on the bog. *Drosera rotundifolia*, *Eriophorum* spp., and *Sarracenia purpurea* were present in the herbaceous layer. The forested swamp zone was dominated by *Acer rubrum*, with a relatively open shrub layer of *Clethra alnifolia* (sweet pepperbush), *Kalmia augustifolia*, and *V. corymbosum*.

The Great Dismal Swamp in Virginia and North Carolina (about 37°N) is a large (85,000 ha) forested basin wetland that contains several wetland associations based on different hydroperiods, peat depths, and disturbance history (Day and Dabel, 1978; also see Ch. 8). Four types of forested wetlands have been studied in the Dismal Swamp; two were dominated by pre-disturbance vegetation — cypress and Atlantic white cedar communities — and two were dominated by post-disturbance vegetation, *Acer–Nyssa* (maple–gum) and mixed hardwood communities (Day and Dabel, 1978). The cypress community, the most extensively flooded one, was dominated by *Taxodium distichum* (bald cypress) with co-dominants of *Acer rubrum* and *Nyssa sylvatica* var. *biflora* (black gum). These three species accounted for 83% of the aboveground biomass (Dabel and Day, 1977). Other species present in this community were *Nyssa aquatica* (tupelo gum), *Liquidambar styraciflua* (sweet gum), *Fagus grandifolia* (American beech), *Fraxinus caroliniana* (Carolina ash), and *Quercus laurifolia* (laurel oak).

The next most flooded forest (fluctuating water levels for 6 months of the year) studied by Dabel and Day (1977), the maple–gum community, was dominated by *Acer rubrum*, *Nyssa aquatica*, and *Nyssa sylvatica* var. *biflora* (accounting for 96% of the aboveground biomass). Other tree species of minor importance included *Fraxinus caroliniana*, *Ostrya virginiana* (American hophornbeam), *Prunus serotina* (black cherry), and *Quercus laurifolia*. *Clethra alnifolia* and *Vaccinium corymbosum* were present in the shrub layer. The area covered by the maple – gum forest used to include cypress, which was harvested several decades ago and has not regenerated. Dabel and Day (1977) considered that this was due to reduced frequency of flooding in the area, the conditions being more conducive to the successful establishment of maple and gum trees.

Forested wetlands dominated by Atlantic white cedar (*Chamaecyparis thyoides*) grow on less frequently flooded sites in the Dismal Swamp (Dabel and Day, 1977). In addition to Atlantic white cedar, these forests are dominated by *Acer rubrum* and *Nyssa sylvatica* var. *biflora*, these three species accounting for about 97% of the aboveground biomass. Individuals of *Persea borbonia* (red bay) and *Magnolia virginiana* (sweet bay) were also present. The understory in the cedar forest, the densest of the four communities studied by Dabel and Day (1977), was dominated by *Clethra alnifolia* and *Lyonia lucida* (fetterbush). Seedlings of Atlantic white cedar were absent in the understory, and as mature cedar trees were dying they were being replaced by red maple and black gum (Dabel and Day, 1977).

The least flooded community in the Dismal Swamp was dominated by *Liquidambar styraciflua*, *Quercus alba* (white oak), *Q. laurifolia*, and *Nyssa sylvatica* var. *biflora*. Other associated species included: *Acer rubrum*, *Carpinus caroliniana* (American hornbeam), *Fagus grandifolia*, *Ilex opaca* (American holly), *Liriodendron tulipifera* (yellow-poplar), *Magnolia virginiana*, *Oxydendrum arboreum* (sourwood), *Quercus michauxii* (swamp chestnut oak), *Q. nigra* (black oak), and *Taxodium distichum*. The dense understory was dominated by *Arundinaria gigantea*. This community was the least disturbed of the four, and had developed four vegetation layers — canopy, subcanopy, shrub and herb strata (Dabel and Day, 1977).

Species richness of trees was generally low, the highest number of species occurring in the rarely flooded mixed hardwood community (14), the lower values in the other three flooded communities (7–8). Dabel and Day (1977) attributed the higher species richness of the mixed hardwood site to the infrequent flooding and absence of recent disturbance. Although all four community types had different flooding regimes, two species (*Acer rubrum* and *Nyssa sylvatica* var. *biflora*) were present in all. Apparently these two species are able to adapt to a wide range of flooding conditions.

Located on the Atlantic coastal plain, from Virginia to northern Florida, are areas of freshwater wetlands commonly referred to as pocosins. Once covering about 1.2 million ha, most of which was in North Carolina, by 1980 they were reduced to only 31% of the original area (Richardson, 1983). Pocosins are evergreen shrub bogs that can be divided into two kinds based on the physiognomy of the vegetation: short pocosins, where the vegetation is dominated by shrubs less than 6 m tall, and tall pocosins, where the vegetation is taller and also contains emergent trees (Richardson, 1983). Common tree species are *Acer rubrum*, *Gordonia lasianthus* (loblolly bay), *Magnolia virginiana*, *Persea borbonia*, and *Pinus serotina*. In a tall pocosin many of these broadleaf species are part of the shrub layer, underneath a *Pinus serotina* canopy. Atlantic white cedar and bald cypress may also occur, although less frequently because of past harvesting (Richardson, 1983). The shrub and vine layer is composed generally of *Cyrilla racemiflora*, *Ilex glabra*, *Lyonia lucida*, *Smilax laurifolia* and *Zenobia pulverulenta* (Christensen et al., 1981). Peat mosses (*Sphagnum* spp.) occur in pocosins but their cover is dependent upon the density of the shrub layer. In tall pocosins they are usually restricted to light gaps and small disturbed areas (Christensen et al., 1981).

Various wetland types, both forested and non-forested, collectively compose what is known as the Okefenokee Swamp, an area of 165,000 ha in southeastern Georgia (Schlesinger, 1978). Four major kinds of forested wetlands have been identified for this area (McCaffrey and Hamilton, 1984).

(1) A needle-leaved evergreen wetland (pine wetland) forms when *Pinus elliottii* (slash pine) extends from the uplands into wet depressions.

(2) Broad-leaved evergreen wetland or bay forest is dominated by *Gordonia lasianthus*, *Ilex cassine*, *I. coriacea*, *Magnolia virginiana*, and *Persea borbonia*. Small patches of shrubs such as *Clethra alnifolia*, *Cyrilla racemiflora*, *Ilex glabra*, and *Lyonia lucida* are found in the understory. *Sphagnum* moss is a common ground cover.

(3) Needle-leaved deciduous wetlands are dominated by *Taxodium distichum* var. *nutans* (pond cypress). It is this forest type for which data on structure and dynamics are available (Schlesinger, 1978; Best, 1984). The canopy of this forest is composed of pond cypress, and the subcanopy of *Gordonia lasianthus*, *Ilex cassine*, *Magnolia virginiana*, *Nyssa sylvatica* var. *biflora*, and *Persea borbonia*. Shrubs include *Clethra alnifolia*, *Ilex cassine*, *Itea virginica*, *Leucothoe racemosa*, *Lyonia lucida*, *Rhus radicans*, and *Vaccinium* spp. (Schlesinger, 1978; Best, 1984). *Tillandsia usneoides* (Spanish moss) is an abundant epiphyte (Schlesinger, 1978), and *Sphagnum* is the ground cover in most areas (McCaffrey and Hamilton, 1984).

(4) Broad-leaved deciduous wetland forests are dominated by *Nyssa sylvatica* var. *biflora* with scattered *Pinus serotina*. The subcanopy, when it occurs, contains broad-leaved evergreens and *Acer rubrum*, and the understory is generally open with a ground cover of *Sphagnum* moss (McCaffrey and Hamilton, 1984).

Combinations of these four forest types also occur in the Okefenokee Swamp, such as mixtures of bay and cypress (dominated by bays but also containing >25% of pond cypress in the canopy) and cypress–black gum and cypress–bay mixtures (containing >50% cypress) (McCaffrey and Hamilton, 1984).

Throughout the flatwoods of *Pinus elliottii* in the southern Atlantic and Gulf coastal plain of the United States is a unique forested wetland referred to as cypress heads or cypress domes (Monk and Brown, 1965). From a side view, these forests appear to have a smooth, symmetrical, domed shape (hence their name) because the tallest trees grow in the center of the depression with tree height decreasing toward the edge (see Fig. 18.1). Other characteristics of the wetland decrease from the center of the depression toward the edge — water depths and concentrations of organic matter, clay, and minerals in the soil (Monk and Brown, 1965; Brown, 1978; Marois and Ewel, 1983). There are several hypotheses as to why trees in the center of the cypress domes are larger than those towards the edge. Among them are differences in age (Kurz, 1933), rises in sea level (Vernon, 1947), and the presence of more favorable conditions for tree growth in the center of the dome (Kurz and Wagner, 1953). Results from my research (Brown, 1981) invalidated the first two hypotheses and supported that of Kurz and Wagner (1953). Cypress domes are fairly uniform floristically (Monk and Brown, 1965; Brown, 1981). *Taxodium distichum* var. *nutans* is the dominant canopy tree and *Nyssa sylvatica* var. *biflora* the usual dominant subcanopy tree. Other tree species sometimes present are *Acer rubrum*, *Magnolia virginiana*, and *Persea borbonia*; *Pinus elliottii* is often present towards the edge of the dome where the hydroperiod is shorter. In the shrub layer *Ilex glabra*, *Itea virginica*, *Lyonia lucida*, *Myrica cerifera*, and *Vaccinium* sp. are common. Ferns, such as *Osmunda cinnamonea* and *Woodwardia virginica*, and grasses such as *Panicum* spp. are often present in the herb layer. In the deeper, central zone of the dome the shrubs and herbs generally grow near the bases of large cypress trees where there are accumulations of organic material and the water is shallower. A total of 11 tree species and 26 shrub and herb species were found growing in 15 cypress domes in north-central Florida (Monk and Brown, 1965).

The bay forested wetland of the Okefenokee Swamp (type (2), p. 181) are also found in Florida where they are referred to as bayheads (Monk, 1966) and are composed of similar species. Other species which may also be locally present are *Acer rubrum*, *Liquidambar styraciflua*, *Nyssa sylvatica* var. *biflora*, *Pinus elliottii*, and *Quercus nigra* (Monk, 1966). The understory species include *Cephalanthus occidentalis* and *Ilex glabra*.

In southern Florida are large expanses of scrub cypress (sometimes referred to as dwarf cypress) wetlands. This community has been described as a "scrubby, stunted cypress growing in marsh-like, seasonally wet prairies" (Davis, 1943). The length of the hydroperiod and depth of soil has a marked effect on the physiognomy of the scrub cypress forest. The shorter the hydroperiod and the shallower the soils the scrubbier the forest becomes (Brown et al., 1984b). *Taxodium distichum* var. *nutans* is the dominant tree, forming a very open canopy (Brown et al., 1984b). Bromeliads such as *Tillandsia fasciculata* are abundant on the cypress trunks. The understory consists of *Ilex cassine* and *Myrica cerifera*. The herb layer tends to be sparse, with *Panicum hemitomon* dominating. During periods of inundation, thick mats of *Utricularia* spp. are present. In areas affected by drainage projects, *Pinus elliottii* is often found invading. When the hydroperiod is reduced due to higher elevations and shallower soils, the scrub cypress forest becomes even shorter and less dense (Brown et al., 1984b).

Saltwater basin forests — mangrove forests — are found along tropical and subtropical coastlines, often inland from the fringe forest (see Ch. 6). Basin mangrove forests of tropical America and Florida mostly consist of pure stands of *Avicennia germinans* (black mangrove). Mixed stands of *A. germinans* and *Laguncularia racemosa* (white mangrove) and/or *Rhizophora mangle* (red mangrove) are also found in saltwater basins (Twilley et al., 1986). Mixtures with red mangroves occur where there is more lateral movement of water. Many saltwater basin forests have berms, varying in size and continuity, that separate the forest from the estuary (Twilley, 1982). In a site in Florida, the berm was sandy and supported *Yucca* spp. and *Conocarpus erectus* (buttonwood) (Twilley, 1982). In the same site, halophytes such as *Salicornia* spp. and *Sesuvium portulacastrum* may grow on higher ground within the forest. Twilley (1982) also found colonization of pneumatophores by algae such as *Bostrychia* spp. and *Cladophora repens*.

STRUCTURAL CHARACTERISTICS

Available data on the structural characteristics of fresh and saltwater basin wetlands are summarized in Table 7.3. Although some of the differences could be attributed to differences in ages of the stands, most of the stands in Table 7.3 are relatively mature. Species richness of trees tends to be low, stem densities are high, and basal areas vary over a wide range. Stand height tends to

TABLE 7.3

Structural indices of freshwater and saltwater basin forested wetlands

Location, latitude and forest type[1]	Number of tree species	Stem density[2] (number ha^{-1})	Basal area[2] (m^2 ha^{-1})	Canopy height (m)	Source
FRESHWATER					
Manitoba, Canada, 50°N					
Picea mariana bog	1	7820	—	—	Reader and Stewart, 1972
Muskeg	1	1970	—	—	Reader and Stewart, 1972
Minnesota, 47.5°N					
Picea mariana bog	1	—	9.5	13	Bay, 1967
Groundwater bog	4	—	11.3	15–18	Bay, 1967
Picea mariana bog	1	—	19.0	14	Verry, 1975
Michigan, 46.5°N					
Alnus swamp	3–8	1640 ($\geqslant$1 cm dbh)	—	5	Parker and Schneider, 1974
Michigan, 45.5°N					
Bog forest[3]	5	3894	32.1	—	Schwintzer, 1977
Minnesota, 45°N					
Northern white cedar swamp	9	2755	42.2	17	Reiners, 1972
Marginal Fen	14	3348	25.1	—	Reiners, 1972
Virginia, 37°N[4]					
Cypress	8	1560	59.3	—	Dabel and Day, 1977
Maple – Gum	9	2080	39.0	—	Dabel and Day, 1977
Atlantic white cedar	7	2000	55.8	—	Dabel and Day, 1977
Mixed hardwoods	14	1440	33.2	—	Dabel and Day, 1977
Georgia, 31°N					
Cypress swamp	4–9	1891	70.7	18	Schlesinger, 1978
Cypress swamp	5	1739	56.3	30	Best, 1984
Florida, 29.5°N					
Cypress domes					
— secondary (4 sites)	5	2735	51.6	15	Brown, 1981
— mature	6	3951	70.8	20	Brown, 1981
Florida, 26°N					
Scrub cypress (2 sites)	3	2496	21.9	6	Brown et al., 1984b
Mean	5.3	2753	39.9	15	
SALTWATER					
Mangroves (9 sites) (Florida, 26°N)	2–3	5130	21.3	6.5	Cintrón et al., 1985
Mangroves (2 sites) (Brazil, 25°S)	2–3	3550	23.2	4.5	Cintrón et al., 1985
Mangroves (Mexico, 23°N)	2	3120	15.2	9.0	Pool et al., 1977
Mangroves (5 sites) (Puerto Rico, 18°N)	2–3	2468	21.2	13.9	Pool et al., 1977
Mangroves (14 sites) (Puerto Rico, 18°N)	1–3	3017	15.2	9.4	Cintrón et al., 1985
Mean	2	3580	18.5	9.0	

[1] The scientific names of plants mentioned in this column are: alder, *Alnus* spp.; Atlantic white cedar, *Chamaecyparis thyoides*; black spruce, *Picea mariana*; cypress, *Taxodium* spp.; gum, *Nyssa* spp.; maple, *Acer* spp.; northern white cedar, *Thuja occidentalis*.
[2] For trees with dbh $\geqslant$ 2.5 cm, unless noted otherwise.
[3] Based on data for 1969 (before massive tree mortality).
[4] These four sites are arranged in order of decreasing flood frequency and depth.

decrease with increasing latitude for both freshwater and saltwater wetlands. Some differences are exhibited between structural indices of freshwater and saltwater wetlands. Stem densities of mangroves tend to be higher and trees, on the average, shorter than in freshwater wetlands (Table 7.3).

With the exception of the scrub cypress wetland in Florida, total biomass is higher in southern wetlands than in northern ones (Table 7.4). (No biomass

TABLE 7.4

Biomass of freshwater basin forested wetlands of North America

Location, latitude and forest type[1]	Tree biomass (t ha^{-1})				Understory biomass (t ha^{-1})	Source
	aboveground	belowground	total	leaves		
Manitoba, Canada, 50°N						
Bog forest	43.2	22.8	66.0	—	3.35	Reader and Stewart, 1972
Muskeg	4.0	16.4	20.4	—	5.94	Reader and Stewart, 1972
Michigan, 46.5°N						
Alnus swamp						
— poorly drained	51.4	—	—	3.54	1.62	Parker and Schneider, 1975
— better drained	28.9	—	—	2.96	2.23	Parker and Schneider, 1975
Minnesota, 45°N						
Northern white cedar swamp	159.4	—	—	7.76	0.54	Reiners, 1972
Marginal fen	98.1	—	—	3.86	1.88	Reiners, 1972
Virginia, 37°N[2]						
Cypress	345.2	15.3	360.5	5.99	0.14	Day and Dabel, 1978
Maple–Gum	193.9	12.2	206.1	5.81	0.68	Day and Dabel, 1978
Atlantic white cedar	218.5	18.0	236.5	10.83[3]	1.88	Day and Dabel, 1978
Mixed hardwood	194.7	31.0	225.7	5.23	0.99	Day and Dabel, 1978
Georgia, 31°N						
Cypress swamp	301.0	—	—	2.31	5.38	Schlesinger, 1978
Florida, 29.5°N						
Cypress domes						
— secondary (4 sites)	222.3	—	—	4.57	—	Brown, 1981
— mature	266.0	—	—	4.65	—	Brown, 1981
Florida, 26°N						
Scrub cypress (2 sites)	76.8	7.8	84.6	2.65	0.44	Brown et al., 1984b

[1]For scientific names of plants in this column, see Table 7.3.
[2]Arranged in order of decreasing flood frequency and depth.
[3]Contains small twigs which could not be separated from the leaves.

data for saltwater wetlands are available.) It is also clear that the understory contributes a larger proportion of the total biomass in the northern forests. The low density of trees allows greater development of the understory.

The proportion of aboveground biomass of trees in the leaves varies between 1 and 5% for low-latitude forests and 4 and 10% in higher latitudes. The higher proportion of leaf biomass in the northern forests is consistent with observations in other forests growing in stressful environments or with short growing seasons.

Although there are few data on belowground biomass, two patterns emerge from the data. The ratio of aboveground to belowground biomass for four wetland types from Virginia (Dabel and Day, 1978) increases with increasing frequency and depth of flooding (from 6.3 for the mixed hardwood forest to 22.6 for the cypress forest). In all cases more than 63% of the root biomass was located in the top 29 cm of soil and 58 to 62% of the roots were smaller than 1 cm (Dabel and Day, 1978).

The second pattern that emerges is that forests at high latitudes have a high proportion of biomass in the roots; the two Manitoba sites in Table 7.4 have aboveground to belowground ratios of 2.0 and 0.6 for vascular plants (shrubs and trees; calculated from additional information in Reader and Stewart, 1972). Apparently in the northern wetland forests a large proportion of biomass is allocated to both roots and leaves, important plant parts for nutrient uptake and energy fixation, at the expense of woody tissue.

Interesting structural features of cypress trees are the aerial roots, commonly referred to as "knees". Densities of cypress knees range from 2500 to 3400 ha^{-1} in cypress domes to as few as

900 ha^{-1} in a scrub cypress forest (Brown, 1981). Black gum trees (*Nyssa sylvatica* var. *biflora*) also develop inverted U-shaped aerial roots containing obvious lenticels, although there are fewer of them than cypress knees. There has been much discussion in the literature regarding their function (Mattoon, 1915; Kurz and Demaree, 1934; Kramer et al., 1952; Brown, 1981; Odum, 1984). Cypress knees occur mostly where the hydroperiod is long, and their heights often correspond to the mean high water level (Mattoon, 1915). It is generally assumed that the cypress knees and aerial roots of black gums are used for gas exchange between the anaerobic substrate and atmosphere. Kramer et al. (1952) concluded from field experiments that little gas exchange occurred between cypress knees and roots. Results from my work (Brown et al., 1984a) showed that knees give off several times more carbon dioxide than an equal area of bole surface, but it is still unknown if they also transport oxygen to the roots. Odum (1984) hypothesized that oxygen is drawn in through the aerial portions of the roots by the transpiration stream, that metabolism is completed below ground, and that carbon dioxide is released through trunks and leaves. Cypress trees and, to a lesser extent, black gum trees are also two of the few non-tropical tree species that produce well-developed buttresses, the height of which is also related to the height of the mean water level.

Black and red mangroves also have aerial root structures for facilitating gas exchange. In black mangroves (*Avicennia* spp.), these are spongy, finger-like structures called pneumatophores, and in red mangroves (*Rhizophora* spp.) they are the lenticels on the prop roots. Like cypress knees, the height of the pneumatophores is directly related to the depth of the surface water (Lugo, 1981). Although the actual number of pneumatophores is highly variable in black mangrove stands, average densities as high as 672 m^{-2} have been reported (Twilley, 1982). Scholander et al. (1955) have demonstrated that these structures transport oxygen through the roots into the anaerobic sediments. As with cypress knees, the carbon dioxide release rates from pneumatophores and prop roots were many times higher than from an equivalent area of trunk surface (Lugo et al., 1975).

ORGANIC MATTER DYNAMICS

Primary productivity and respiration

Measurements of primary productivity and respiration, using gas exchange methods (Ch. 4, Sect. 16), are available for very few basin wetlands (Table 7.5). It appears that gross primary productivity (GPP) was higher for freshwater forests than for saltwater ones; no generalization can be made about respiration (R). Gross primary productivity and respiration of the cypress dome was significantly higher than for the scrub cypress, but the net primary productivity (NPP) of the two forests was

TABLE 7.5

Rates of primary productivity and respiration (measured by CO_2 gas exchange methods) of basin forested wetlands. All values are in units of g OM m^{-2} day^{-1}

Rate	Freshwater[1]		Saltwater[2]	
	cypress dome	scrub cypress	mangrove	scrub mangrove
Gross primary productivity (GPP)	25.3	6.2	18.0	2.8
Total 24-h plant respiration (R)	21.9	2.9	12.4	4.0
Net primary productivity (NPP)	3.4	3.3	5.6	0

[1] Brown (1981) for sites dominated by cypress (*Taxodium distichum* var. *nutans*) in Florida (U.S.A.).
[2] From Lugo and Snedaker (1975) for sites in Florida (U.S.A.). The scrub mangrove was dominated by *Rhizophora mangle*, but because it was far removed from the sea and waters were stagnant and shallow it was classified as a basin forest.

almost equal. However, when the scrub cypress was not flooded, respiration rates were high (4.1 g OM m^{-2} day^{-1}) and NPP was about half the average rate given in Table 7.5 (Brown, 1981). The saltwater forests exhibited a similar trend to that of the freshwater forests with respect to GPP and R, but the NPP of the mangrove forest was higher than that for the scrub mangrove.

That rates of GPP, NPP, and R in cypress domes are influenced by nutrient availability in the substrate is illustrated by the response of these rates to nutrient enrichment. Application of secondarily treated sewage effluent for about 15 months to another cypress dome resulted in higher rates of GPP (35.3 g OM m^{-2} day^{-1}), R (30.4 g OM m^{-2} day^{-1}), and NPP (4.9 g OM m^{-2} day^{-1}) than in the untreated dome (Brown et al., 1984a). The increase in all these rates due to the addition of sewage effluent was a result of an increase in leaf area index of the cypress forest rather than an increase in the metabolism of leaves per unit area (Brown et al., 1984a).

The data in Table 7.5 are based on average daily rates during the height of the growing season. Collection of such data over an annual cycle may produce different trends. Cypress trees are deciduous and remain leafless for about 4 months of the year, whereas mangroves are evergreen and are capable of photosynthesizing throughout the year. If the rates in Table 7.5 were extrapolated to a year, GPP of the freshwater forests would appear lower and perhaps similar to that of the saltwater ones. Net primary productivity of the mangrove forest would become considerably higher than that of the cypress dome, because of the longer season of the former.

Biomass production

Aboveground biomass production of freshwater forests is highly variable (Table 7.6) and is likely to be related, in part, to nutrient availability of the site. The low production of the bog forests in Manitoba is generally attributed to low fertility of the deep peats, due to slow decomposition and nutrient recycling rates (Reader and Stewart,

TABLE 7.6

Aboveground biomass production of basin forested wetlands

Location, latitude and forest type[1]	Biomass production (t ha^{-1} yr^{-1})				Source
	total[2]	leaves and reproductive parts	wood	under-story	
Manitoba, Canada 50°N					
Bog forest	3.03	1.19	—	0.63	Reader and Stewart, 1972
Muskeg	0.72	0.35	—	2.54	Reader and Stewart, 1972
Michigan, 45.5°					
Alnus swamp					
— poorly drained	5.40	3.49	1.91	1.01	Parker and Schneider, 1975
— better drained	4.59	2.86	1.73	1.17	Parker and Schneider, 1975
Minnesota, 45°N					
Northern white cedar swamp	10.14	4.93	5.21	0.18	Reiners, 1972
Marginal fen	6.51	3.17	3.34	0.55	Reiners, 1972
Georgia, 31°N					
Cypress swamp	5.95	2.65	3.30	1.06	Schlesinger, 1978
Florida, 29.5°N					
Cypress domes					
— secondary (3 sites)	10.05	5.46	4.59	—	Brown, 1981
— mature	10.29	4.88	5.41	—	Brown, 1981
Florida, 26°N					
Scrub cypress	2.95	2.47	0.48	—	Brown et al., 1984b

[1]Alder: *Alnus* sp.; cypress: *Taxodium distichum* var. *nutans*; northern white cedar: *Thuja occidentalis*.
[2]Total for trees; it does not contain understory.

1972). The low production of the scrub cypress forest is also due to low soil fertility, but in this case the highly leached sandy soils are practically devoid of organic matter and extremely low in phosphorus (Brown, 1981). Fens are generally considered to be richer in nutrients than swamps, but the biomass production of the marginal fen in Minnesota is lower than the nearby white cedar swamp. The high biomass production of the cypress domes suggests that these sites are richer in nutrients than the other sites in Table 7.6 (see also p. 173 above).

In general, biomass production is about equally divided between leaves and wood (Table 7.6), except for the stunted alder (Parker and Schneider, 1975) and scrub cypress wetlands (Brown et al., 1984b). Variation in understory production is most likely a function of differences in stem density of the forest (see Table 7.3) and corresponding canopy development (for example, leaf area index).

It appears that differences in biomass production can be better described by differences in standing biomass of the wetlands. There is a statistically significant linear relationship between aboveground biomass production and aboveground standing stock of biomass over the range of values available (Fig. 7.3). The slope of the relationship represents turnover time for biomass, and suggests that wetland forests turn over about once in 25 years, regardless of latitude. The outlier point is for a cypress swamp in Georgia (31°N) which appears to require more than 25 years for complete turnover.

No biomass production data are available for saltwater wetlands, because little information exists on the rates of wood production in basin mangrove forests. Lugo and Snedaker (1975) estimated an average rate of wood production for a basin forest in Florida to be 6.4 t ha^{-1} yr^{-1}, similar in value to leaf production.

Litter fall, litter standing crop, and decomposition

Litter fall rates, like those for biomass production, are also variable (Table 7.7). However, excluding the scrub forests, the mean rate of litter fall for saltwater basin forests is higher than for freshwater forests. Scrub cypress exceeds scrub mangrove in litter fall.

Saltwater forests of mixed mangrove species appear to produce almost twice as much litter as monospecific stands. Monospecific mangrove stands typically grow in areas of higher soil salinity than do mixed stands. High soil salinities place a stress on mangroves, and likely reduce the amount of photosynthate allocated to litter production. For example, at soil salinities of 55 to 65‰ litter production in monospecific stands was approximately 4 t ha^{-1} yr^{-1}; at lower soil salinities (about 40‰) litter fall of a mixed mangrove forest was about double that of the monospecific stands (Lugo et al., 1980).

There is little variation in the standing crop of fine litter (non-woody) in freshwater wetlands. In the Dismal Swamp and wetlands of Minnesota, standing crops are similar to litter fall rates (Table 7.7). In contrast, the mixed mangrove stands have more litter than the monospecific stands, and in both cases the standing crop is about half the value of annual litter fall (Table 7.7). Higher decomposition rates of mangrove litter, and periodic tidal flushing which exports the litter, are possible explanations (Twilley et al., 1986).

Few studies have measured the amount of coarse litter (dead wood) in wetlands. Of note is the study by Day (1979), who recorded the following amounts of dead wood (in t ha^{-1}) in four forest types in Virginia: 45.4 for a cypress forest, 26.8 for

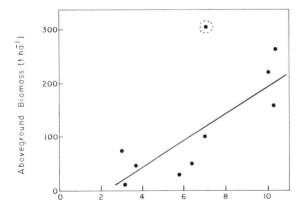

Fig. 7.3. Relationship between aboveground biomass and aboveground biomass production for freshwater basin wetlands. Data are from Tables 7.4 and 7.6. The equation describing the curve is $Y = -61.89 + 25.7X$ ($r^2 = 0.74$, $P = 0.01$). The circled point appears to be an outlier and was not used in the regression.

TABLE 7.7

Litter fall, litter standing crop, and decomposition rates of basin forested wetlands

Location, latitude and forest type[1]	Litter standing crop (t ha^{-1})	Litter fall (t ha^{-1} yr^{-1})		Decomposition constant (k) (yr^{-1})	Source
		leaves and reproductive parts	total		
FRESHWATER					
Minnesota, 45°N					
Northern white cedar swamp	4.89	—	4.99	—	Reiners, 1972
Marginal fen	5.16	—	4.41	—	Reiners, 1972
Virginia, 37°N[2]					
Cypress	4.80	5.47	6.78	0.55	Day, 1982; Gomez and Day, 1982
Maple–Gum	5.11	5.53	6.58	0.59	Gomez and Day, 1982
Atlantic white cedar	4.03	5.55	7.57	0.39	Gomez and Day, 1982
Mixed hardwoods	5.48	5.15	6.52	0.38	Gomez and Day, 1982
Georgia, 31°N					
Cypress swamp	—	2.66	3.28	—	Schlesinger, 1978
Florida, 29.5°N					
Cypress domes					
— secondary (3 sites)	—	5.21	5.85	—	Brown, 1981
— mature	—	4.15	4.88	1.87	Brown, 1981; Dierberg and Ewel, 1984
— average of 4 domes	—	—	—	1.19	Deghi et al., 1980
Florida, 26°N					
Scrub cypress	4.41	—	2.47	0.25	Brown et al., 1984b
Mean	4.91	—	5.33	0.75	
SALTWATER					
Florida, 26°N					
Scrub mangroves (3 sites)	—	—	1.86	1.5–2.3	Twilley et al., 1986
Black mangroves	1.3[3]	—	4.84[4]	3.9–6.0	Twilley et al., 1986
Mixed mangroves (3 sites)	4.21	—	7.90	—	Twilley et al., 1986
Puerto Rico, 18°N					
Mixed mangroves	—	—	9.70	—	Pool et al., 1975
Mean (excluding scrub mangrove)	3.05	—	6.60	—	

[1]Scientific names of species mentioned in this column: Atlantic white cedar: *Chamaecyparis thyoides*; black mangrove: *Avicennia* sp.; cypress: *Taxodium distichum* var. *nutans*; gum: *Nyssa* sp.; maple: *Acer* sp.; northern white cedar: *Thuja occidentalis*.
[2]Arranged in order of decreasing flood frequency and depth.
[3]Average of 2 sites.
[4]Average of 4 sites.

a maple–gum forest, 50.2 for an Atlantic white cedar forest, and 8.4 for a mixed hardwood forest. By far the largest fraction of this dead wood was >2 cm in diameter. Schlesinger (1978) found 49 t ha^{-1} of standing dead and 2.7 t ha^{-1} of fallen dead wood in a cypress swamp in Georgia. Clearly woody debris is a large component of the organic matter pool of some forested wetlands, and as such it may be important in these systems. The coniferous forests of the Pacific Northwest (U.S.A.) contain large quantities of woody debris, serving as a nutritional substrate and habitat for many micro-organisms, invertebrates, and vertebrates; as a nursery for several tree seedlings; and as an important long-term reservoir in nutrient cycles (Triska and Cromack, 1980). The relative importance of these roles of woody debris in wetland forests needs to be determined.

Decomposition of litter is slower in freshwater wetlands than in saltwater wetlands (Table 7.7).

This difference is explained in part by more dynamic flushing in the saltwater forests, which maintains more optimal conditions for microbial action, increases leaching losses and fragmentation, and provides a medium for export.

The influence of the hydrologic regime on short-term decomposition rates for freshwater wetlands can be inferred from the data in Table 7.7. The decomposition coefficient decreases with decreasing frequency of inundation for the four wetland forests in Virginia (Day, 1982). This trend was confirmed by Deghi et al. (1982), who found that cypress litter had decomposed almost twice as fast in wet sites than in dry sites at the end of one year. Day (1982) believed that the characteristics of the litter were the major determinants of its decomposition rate, because the quality of the litter changed due to the effect of changing hydrologic regime on species composition. For example, he found that the decomposition coefficient was highly correlated with the initial phosphorus concentration of the litter. This was supported by Dierberg and Ewel (1984), who found that decomposition rates of cypress litter were faster in a sewage-enriched site than in a control site. They attributed the increased rate to increased nutrient concentrations in the litter, because the frequency of submergence of the two sites was similar. The factors affecting decomposition of litter in basin wetlands are complex, due to the interactions of litter quality, frequency of submergence, and quality of surface water. Determining the relative importance of these factors in influencing decomposition rates is an important area of research.

Although the decomposition rates in Table 7.7 are not particularly low, it is clear that organic matter accumulates in most basin wetlands, often to significantly higher values than the mass of live vegetation. For example, Schlesinger (1978) reported that the mass of peat to 91.5 cm (formed from cypress vegetation) was 915 t ha^{-1}, or about three times the mass of the cypress forest vegetation. In Minnesota, the mass of the humus layer under a marginal fen (25 cm deep) and under a northern white cedar swamp (70 cm deep) was respectively about three times and six times the aboveground biomass of the forests (Reiners, 1972; also see Table 7.4). Peat accumulation is a slow process (see Ch. 4, Sect. 15), and it is difficult to understand the process from short-term decomposition studies. Furthermore, it is highly likely that much of the peat is formed from belowground biomass, the dynamics of which are poorly understood.

NUTRIENT STORAGES AND DYNAMICS

Standing stocks of nutrients in vegetation

Concentrations of nutrient elements in the foliage of basin wetland trees appear to vary by species and by site (Table 7.8). Differences in site fertility and species composition explain most of the patterns in the data, particularly for concentrations of nitrogen and phosphorus. For example, the nitrogen in black spruce leaves in Alaska is below 1%, a value proposed by Heilman (1966) as indicating deficiency. The three sites of Heilman (1968) are ranked in order of decreasing site quality, from shallow to deep peat. Nitrogen and phosphorus concentrations in the black spruce foliage of the poorest site were both lower than in the other two sites; concentrations of magnesium and potassium decreased along the site-quality gradient; no significant trend was observed for calcium. The three sites of Tilton (1977; Table 7.8) also follow this pattern. A similar trend was observed for black spruce foliage from a variety of sites in Minnesota (Watt and Heinselman, 1965; only the average concentrations for black spruce vegetation are given in Table 7.8 because most data in their paper were presented as graphs). In this study foliage was collected from stands along transects of increasing peat thickness, and thus decreasing site index, which extended from channels where the water was flowing (water tracks) and the peat was shallow (20–30 cm deep) to a poor muskeg where the peat was almost 2 m deep. There were highly significant linear relations between foliar concentrations of nitrogen and phosphorus and site index, attributed to greater availability of these elements in stands with greater water movement in the soil and shallower peats. The inverse relationship with site index found for potassium was explained by a dilution effect due to more rapid growth by the foliage on richer sites.

Concentrations of elements in foliage vary seasonally, but the pattern is different for different elements. Patterns for nitrogen and phosphorus for

TABLE 7.8

Concentration of nutrient elements in foliage of basin forested wetlands

Location, latitude and forest type[1]	Concentrations (% dry mass)					Source
	N	P	Ca	Mg	K	
Alaska, 65°N						
North slope bogs[2]						Heilman, 1968
Black spruce–moss	0.873	0.123	0.303	0.447	0.122	
Black spruce–*Sphagnum*	0.873	0.142	0.328	0.382	0.105	
Sphagnum–black spruce	0.752	0.116	0.339	0.358	0.101	
Minnesota, 48°N						
Black spruce bog	0.83	0.130	0.459	0.420	0.119	Watt and Heinselman, 1965
Ottawa, Canada, 45.5°N						
Peat bogs						Small, 1972
Evergreen species[3]	1.22	0.077	0.52	—	—	
Black spruce	1.02	0.086	0.59	—	—	
Deciduous species[3]	1.19	0.097	0.70	—	—	
Larix laricina	1.28	0.080	0.68	—	—	
Minnesota, 45.5°N						
Larix laricina forests[4]						Tilton, 1977
Bog	1.24	0.17	0.62	0.37	0.09	
Conifer swamp	1.06	0.25	0.47	0.42	0.14	
Fen	1.95	0.26	0.67	0.52	0.19	
Georgia, 31°N						
Cypress swamp	1.43	0.098	0.49	1.35	0.30	Schlesinger, 1978
Florida, 29.5°N						
Cypress domes						Straub and Post, 1978
Cypress	1.48	0.084	—	—	—	
Hardwoods	1.965	0.063				
Florida, 26°N						
Scrub cypress	—	0.052	—	—	—	Brown, 1978

[1] For scientific names of species in this column, see Table 7.1.
[2] Nutrient concentrations for all these sites is for foliage of black spruce trees.
[3] Average values for tree and understory species.
[4] Values for *Larix* sp. leaves; sites arranged in increasing order of fertility.

leaves and branches of cypress and black gum trees growing in a cypress dome in Florida are shown in Fig. 7.4. The concentrations of both elements in leaves are lower at the end of the growing season, just before leaf abscission, than at the beginning suggesting that these two elements are retranslocated to stems. Concomitant increases in nitrogen and phosphorus for branches confirm this nutrient-conserving mechanism. Seasonal patterns of nitrogen concentrations in the foliage of *Larix laricina* trees from three types of bogs in Minnesota (Tilton, 1977) were in agreement with those for cypress and black gum foliage. However, Tilton found no significant difference between the phosphorus concentration of fully developed leaves and the concentration just prior to abscission. Apparently more studies are needed before generalizations about phosphorus movement can be made.

Evidence for retranslocation of nutrients can also be obtained from comparisons of concentrations of nutrients in fresh foliage and leaf litter (Schlesinger, 1978). Such comparisons confirm the trend exhibited in Fig. 7.4 — that is, nitrogen and phosphorus are retranslocated by cypress (Deghi, 1977; Schlesinger, 1978). Small (1972) found that northern bog species retranslocated a greater proportion of nitrogen from leaves before abscission than non-bog species. However, there was no evidence for retranslocation of phosphorus in either bog or non-bog species. Retranslocation of nitrogen has also been shown to occur in man-

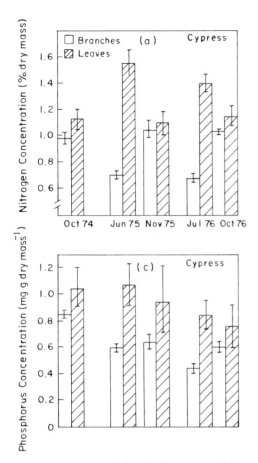
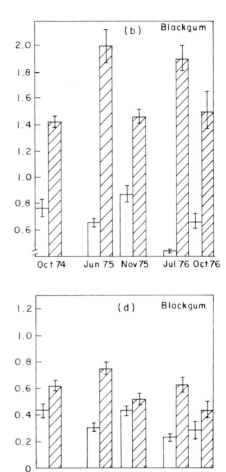

Fig. 7.4. Seasonal variation in nitrogen (a and b) and phosphorus (c and d) concentrations of branch and leaf tissues for cypress (*Taxodium distichum* var. *nutans*) (a) and (c) and black gum (*Nyssa sylvatica* var. *biflora*) (b) and (d). Samples are from a cypress dome in Florida (U.S.A.). (Data from Straub and Post, 1978.) Sampling dates for phosphorus are the same as those for nitrogen.

groves (*Avicennia* and *Rhizophora*) (Twilley et al., 1986).

Nutrient concentrations in the foliage of basin wetland plants is often low, which suggests that there may be efficient use of the nutrients. Based on the potential amount of photosynthate that northern bog and non-bog species could produce per unit of foliar nitrogen, Small (1972) found that the efficiency of nitrogen use was greater in evergreen than in deciduous species in bogs and greater in deciduous species from bogs than from other environments. For example, bog evergreens were found potentially to be able to produce 235% more photosynthate per unit of foliar nitrogen than bog deciduous species. Small (1972) hypothesized that the highly efficient use of nitrogen by bog plants was an adaptive strategy for overcoming the problems of acquiring nitrogen from the very nutrient-deficient bog substrates. This hypothesis was supported by Schlesinger and Chabot (1977) for plants growing in a low latitude cypress forest (see also Ch. 4, Sect. 8).

Few studies have measured nutrient concentrations of all plant components and estimated nutrient pools in the vegetation of basin forests. Data for three types of cypress forests are shown in Table 7.9. As is typical of most forests, nutrient concentrations follow the following pattern: foliage > branches > bark > bole wood. However, because most of the biomass of a forest is in the bole wood and branches, the largest proportion of the standing crop of a given element

TABLE 7.9

Concentration of nutrient elements in plant tissue and standing crop of nutrients in aboveground biomass of freshwater basin forested wetlands

Location, latitude and forest type[1]	Concentrations (% dry mass)					Standing crop (g m^{-2})					Source
	N	P	Ca	Mg	K	N	P	Ca	Mg	K	
Georgia, 31°N											
Cypress swamp											Schlesinger, 1978
Bole wood	0.24	0.009	0.10	0.03	0.05	60.8	2.28	25.3	7.6	12.7	
Bole bark	0.76	0.040	0.75	0.05	0.13	27.9	1.47	27.5	1.8	4.8	
Branches	0.56	0.054	1.06	0.08	0.32	4.6	0.44	8.6	0.7	2.6	
Leaves	1.43	0.098	1.35	0.30	0.49	3.3	0.23	3.1	0.7	1.1	
Evergreen shrubs	1.20	0.052	0.61	0.24	0.67	—	—	—	—	—	
Deciduous shrubs	1.43	0.069	0.95	0.39	0.67	2.2	0.13	1.9	0.3	1.2	
Shrub stems	0.35	0.02	0.29	0.05	0.20	—	—	—	—	—	
Herbs	1.39	0.05	0.20	0.12	1.35	0.7	0.05	0.1	0	0.6	
Total						99.6	4.60	66.6	11.1	23.0	
Florida, 29.5°N											
Cypress domes											Brown, 1978; Straub
Bole wood and bark											
— cypress	0.30	0.003	—	—	—	56.6[2]	0.64				
— hardwood	0.31	0.006	—	—	—	14.4[2]	0.28				
Branches											
— cypress	0.69[1]	0.044	—	—	—	15.6[2]	1.00				
— hardwood	0.54[1]	0.023	—	—	—	1.9[2]	0.08				
Leaves											
— cypress	1.48[1]	0.084	—	—	—	3.9[2]	0.22				
— hardwoods	1.95[1]	0.063	—	—	—	3.9[2]	0.13				
Total						96.3	2.35				
Florida, 26°N											
Scrub cypress											Brown, 1978; Brown et al., 1984b
Bole wood and bark	—	0.004	—	—	—	—	0.13				
Branches	—	0.03	—	—	—	—	0.05				
Leaves	—	0.052	—	—	—	—	0.08				
Roots ≤ 1 mm diameter (to 35 cm)	—	0.050	—	—	—	—	0.20				
Roots > 1 mm diameter (to 35 cm)	—	0.018	—	—	—	—	0.07				
Total (without roots)							0.26				
(with roots)							0.53				

[1]Average of June–July values given in Straub and Post (1978).
[2]Calculated from biomass values in Brown (1978) and concentration data.

in these cypress forests is in this component (69–94%). Foliage accounts for about 5 to 15% of the nutrient pools in the cypress swamp and dome, compared to about 30% for the scrub cypress. With the exception of the scrub cypress forest in Florida, the standing stocks of nutrients in the vegetation are lower than those in the soil (Table 7.10).

The standing stock of nitrogen is similar for the cypress swamp in Georgia and cypress dome in Florida. However, the standing stock of phosphorus for the former is almost twice that of the latter, even though their biomasses are similar (see Table 7.4). The major cause for this large difference is the low phosphorus concentration in cypress wood of the Florida forest.

TABLE 7.10

Nutrient content of soils of freshwater basin forested wetlands

Location, latitude, and forest type[1]	Depth (cm)	Nutrient content (g m^{-2})					Source
		N (total)	P (total)	Ca	K	Mg	
Alaska, 65°N							
Black spruce–*Sphagnum* bogs	41–51	91–210	11–19	150–388	51–120	149–712	Heilman, 1968
Sphagnum–black spruce bogs	38–51	187–269	9.6–61	58–261	31–66	35–323	Heilman, 1968
Minnesota, 45°N							
Northern white cedar swamp	70	2184	120	2864	—	178	Reiners and Reiners, 1970
Marginal fen	25	845	69	815	—	98	Reiners and Reiners, 1970
Georgia, 31°N							
Cypress swamp	91.5	1693	—	115	33	44	Schlesinger, 1978
Florida, 29.5°N							
Cypress dome	20	—	9–18	—	—	—	Brown, 1981
Florida, 26°N							
Scrub cypress	20	—	0.23	—	—	—	Brown, 1981

[1]Scientific names of species, see Table 7.1

Although the biomass of the scrub forest is about one-fourth that of the two other forests in Table 7.9, its standing stock of phosphorus is an order of magnitude lower. Furthermore this forest has as much phosphorus in the small roots as in the aboveground components, even though roots accounted for only about 10% of the aboveground biomass (Table 7.4). The scrub cypress forest grows on very infertile sandy soils with a very low total phosphorus content (0.009%) and practically devoid of organic matter (Brown, 1981). The relatively high root biomass and high phosphorus storage in the roots may indicate that roots are very important for nutrient uptake and cycling. In contrast to peatland ecosystems, the standing stock of phosphorus in vegetation of the scrub cypress forest is almost twice as high as that in the soil (0.23 g m^{-2}; Table 7.10). This concentration of phosphorus in the vegetation is also observed in other forests growing on poor inorganic soils, particularly in the wet tropics, which retain most of the nutrients in the vegetation as a possible mechanism to avoid losses by leaching.

Data on the nutrient content of soils (Table 7.10) vary, in part due to differences in depth of the samples. It is clear, however, that wetland soils contain a large pool of total nutrients, mainly due to their accumulation in peats. Few of these nutrients are available to the plants. Most, however, are removed from biogeochemical circulation (Schlesinger, 1978) until deep peat fires or extraction for fuel take place.

Nutrient cycling and cycling efficiency

An index of the efficiency of within-stand nutrient cycling has been proposed by Vitousek (1982), by calculating the ratio of litter fall (dry matter) to the nutrient content in the litter (nutrient cycling efficiency index). Vitousek found that the nutrient cycling efficiency index for nitrogen increases exponentially with decreasing rates of nitrogen return in litter to the forest floor, and that forests with an efficient use of nitrogen had nutrient cycling efficiency indices above 120. Similarly, he suggested that efficient use of phosphorus and calcium occurred when the cycling efficiency indices were above 2000 and 120 respectively. Nutrient deposition in litter and the corresponding nutrient cycling efficiency indices for basin wetlands are given in Table 7.11. Three of the wetland forests in Table 7.11 are efficient in their cycling of nitrogen (the Northern white cedar swamp in Minnesota, the cypress swamp in Georgia, and mixed mangrove forest in Florida), but none is particularly efficient in its cycling of

TABLE 7.11

Nutrient deposition in litter fall and nutrient cycling efficiency index[1] (in parentheses) of basin forested wetlands

Location, latitude and forest type[2]	Litter fall (g m^{-2} yr^{-1})	Nutrient deposition (g m^{-2} yr^{-1})					Source
		N	P	Ca	Mg	K	
Minnesota, 45°N							
Northern white cedar swamp	488.1	4.2 (116)	0.62 (787)	9.06 (54)	1.20 (407)	—	Reiners and Reiners, 1970
Marginal fen	411.5	4.5 (91)	0.65 (633)	8.70 (47)	1.15 (358)	—	Reiners and Reiners, 1970
Virginia, 37°N[3]							
Cypress	210.9	2.66 (79)	0.18 (1172)	2.58 (82)	0.63 (335)	0.66 (320)	Gomez and Day, 1982
Maple-Gum	278.0	3.35 (83)	0.19 (1463)	3.26 (85)	0.84 (331)	0.95 (293)	Gomez and Day, 1982
Atlantic white cedar	172.9	2.19 (79)	0.13 (1330)	2.60 (67)	0.47 (368)	0.33 (524)	Gomez and Day, 1982
Mixed hardwood	174.6	2.08 (84)	0.12 (1455)	2.30 (76)	0.49 (356)	0.43 (406)	Gomez and Day, 1982
Georgia, 31°N							
Cypress swamp	328.5	2.71 (121)	0.13 (2527)	3.93 (84)	0.64 (513)	0.34 (966)	Schlesinger, 1978
Florida, 29.5°N							
Cypress domes	461.0	—	0.22[4] (2134)	—	—	—	Deghi et al., 1980
Florida, 26°N							
Mixed mangroves (red/black)	751	6.3 (120)	—	—	—	—	Twilley et al., 1986
Black mangroves	469	5.3 (89)	—	—	—	—	Twilley et al., 1986
	538	7.0 (77)	—	—	—	—	

[1] Nutrient cycling efficiency index = litter fall (g dry matter m^{-2} yr^{-1})/nutrient content of litter fall (g m^{-2} yr^{-1}) (Vitousek, 1982).
[2] For scientific names of species, see Table 7.1.
[3] All data are for the period of peak litter fall (October–November). Sites are in decreasing order of flooding frequency and duration.
[4] Average of 2 years.

calcium. The cypress swamps in Georgia and Florida are efficient in their use of phosphorus as shown by litter composition (Table 7.11), possibly in response to the low availability of phosphorus in their soils. In fact the cycling efficiency indices for all nutrients are highest for the Georgia cypress swamp, a nutrient-poor site (Schlesinger, 1978). The high nutrient-use efficiencies could be due to more organic matter fixed per unit of nutrient and/or retranslocation of nutrients before leaf abscission (Vitousek, 1982), both of which are adaptations to nutrient-poor environments.

If nutrient availability in wetlands is increased, cycling efficiency indices should decrease. Addition of sewage effluent to a cypress dome in Florida resulted in higher rates of litter fall and phosphorus deposition (Deghi et al., 1980), and a nutrient cycling efficiency index of 1280. This is about half the value for a control dome (Table 7.11), indicating that efficiency of phosphorus cycling decreased in response to greater nutrient availability.

The accumulation of nitrogen and phosphorus in decomposing litter is a commonly observed phenomenon, as these elements are often immobilized by microbes from other external sources. Gains of nutrients are likely to occur when the initial nutrient concentrations in the litter are low enough to limit decomposer activity (Day, 1982). Three studies in basin wetlands, however, demonstrate that this phenomenon does not always occur. Day (1982) and Yates and Day (1983) measured the changes in nitrogen, phosphorus, potassium, calcium and magnesium in decaying litter in four wetland forests with different flooding regimes in the Great Dismal Swamp (Virginia). Their results indicated that the amount of phosphorus and nitrogen remaining tended to decrease initially during the first 3 months followed by an increase (phosphorus) or constant value (nitrogen) in all sites, regardless of litter type and flooding regime. The amount of potassium, calcium and magnesium remaining in the litter declined for all sites as a result of net mineralization of these elements.

A study by Deghi et al. (1980) of the phosphorus dynamics of decomposing cypress litter in cypress domes of Florida produced different results. About 50% of the original phosphorus was lost in the first 4 months of the study, followed by a period of no change in content even though mass continued to decline. Although the phosphorus concentration in the cypress litter was less than that of the wetlands studied by Day (1982), less immobilization of phosphorus occurred. Apparently other factors were influencing the ability of decomposers to immobilize phosphorus; for instance, surface waters were very low in nutrients (Table 7.1) and leaching losses of phosphorus from the cypress canopy were low (Deghi, 1977).

The percentage of nitrogen increased constantly (during the 148 days of the study) in decomposing leaves of black and red mangroves in a basin saltwater forest in Florida; the C/N ratio decreased from 47 to 20 and from 98 to 38 respectively (Twilley et al., 1986). These differences in the rates of nitrogen immobilization are partially explained by the differences in initial concentrations in the litter; leaves of black mangroves had almost twice the initial nitrogen concentration of those of red mangroves (Twilley et al., 1986).

Models of nutrient cycling are useful tools for understanding the relationships between storages and flows of nutrients. Some examples of such models for phosphorus and calcium for cypress ecosystems of the southeastern United States are shown in Fig. 7.5; sufficient data for developing models for other basin wetland forests are lacking. Uptake of nutrients was calculated by assuming that part of the nutrient requirement for NPP is met by the reuse of nutrients from retranslocation from foliage. Retranslocation was estimated from the difference between the nutrient content of foliar standing stocks and that of litter fall. Retranslocation of phosphorus accounts for a significant fraction of the requirement for NPP (about 43%); calcium on the other hand is not readily retranslocated, and it accounts for only 5% of the requirement for NPP. Uptake and return of phosphorus and calcium for the two cypress swamps are low compared to temperate upland forests (Schlesinger, 1978). Although the two cypress forests are different structurally and have very different rates of net biomass accumulation (Table 7.6), their rates of phosphorus uptake, reuse of phosphorus by retranslocation from foliage, and

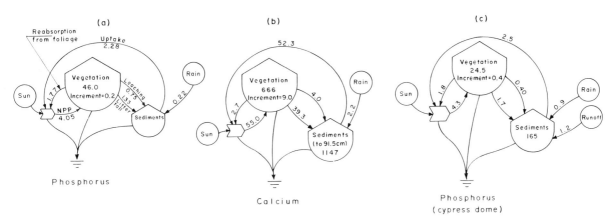

Fig. 7.5. Examples of nutrient budgets for cypress wetlands. Units are in kg ha^{-1} (storages) and kg ha^{-1} yr^{-1} (flows). (a) and (b) from Schlesinger (1978) and (c) from Brown (1978).

return to sediments are very similar. The ratio between return and uptake for calcium (0.83) in the Georgia swamp is slightly lower than for phosphorus (0.91), due partly to the higher increment or net accumulation of this element in net biomass production. These ratios for phosphorus and calcium are typical of temperate upland forests. The difference between uptake and return is more than adequately compensated for by the inputs from rainfall for phosphorus, though not for calcium.

These cypress swamps do not retain a large fraction of the uptake as a consequence of biomass accumulation — only 9 to 17%. As cypress trees are deciduous they are limited in their ability to conserve nutrients, particularly as a large fraction of their annual NPP and nutrient uptake is in leaves (Schlesinger, 1978). In northern latitudes, where evergreen black spruce dominates the very nutrient-poor bogs, it might be expected that a larger fraction of the nutrient uptake would be retained by the plants; however, as no information is available for these sites this hypothesis cannot be tested.

EFFECTS OF DISTURBANCES ON BASIN WETLANDS

Basin wetlands are often subjected to a variety of disturbances (or stressors), both natural and human-induced, with varying degrees of impact depending upon the intensity and on the part of the ecosystem to which the stressor is applied (see Ch. 4, Sects. 13 and 19). Examples of stressors in basin wetlands are water impoundment, water diversion, thermal stress from hot water, sedimentation, addition of toxic substances, addition of waste-water, oil spills (mostly mangroves), and harvesting (Brown and Lugo, 1982).

Altering the hydrologic regime has an impact on wetlands because hydroperiod is a primary factor controlling their structure and function (Ch. 4, Sect. 7). Increased flooding of basin wetlands generally results in a decline in tree vigor or in tree mortality (for example, Schwintzer and Williams, 1974; Schwintzer, 1977), whereas reduction in flooding by diversions often results in changes in species composition and dominance (for example, Marois and Ewel, 1983). In peatland areas, draining wetlands for agriculture or to improve forest growth may result in peat loss (Ch. 4, Sect. 20) or change the regional hydrologic cycle (Ch. 4, Sect. 21). Altering the hydrologic regime of saltwater wetlands may also affect the salt budget (Brown and Lugo, 1982). In saltwater wetlands, changes in the flooding regime are often accompanied by increases in soil salinity which results in changes in species dominance and structural characteristics. For example, impounding a basin forest in Florida produced a complete change in species composition, lower stand height, lower basal area, and a higher accumulation of litter on the forest floor (Lugo et al., 1980). Therefore, chronic alterations to the hydrologic regime have the most severe impact on a basin wetland and the probability of recovery is low.

Additions of toxic substances, such as oil, have long-lasting consequences on basins because of their lack of flushing by water, and they remain in the system. An oil spill off the coast of Puerto Rico interfered with gas exchange of a basin mangrove forest, and as a result all the trees died (Lugo et al., 1981). In contrast, a fringing forest which is flushed regularly experienced only partial defoliation.

Not all disturbances have negative effects on basin wetlands. Because they tend to be nutrient-poor, addition of sewage effluent, rich in nitrogen and phosphorus, to a basin wetland may result in faster rates of organic matter accumulation and nutrient cycling, and changes in structural components. Effects of sewage effluent on a cypress dome in Florida was the subject of a multi-disciplinary investigation, the results of which were summarized by Ewel and Odum (1984). The response to the effluent of productivity rates, nutrient cycling, producer and consumer diversity, and distribution of micro-organisms were the major focus of the study. Rapid increases in biomass production (for example, two-fold increase in wood production and litter fall) and turnover of the forest occurred, and the soil compartment, with a slow turnover rate, stored more phosphorus (Brown, 1981).

REFERENCES

Bay, R.R., 1967. Ground water and vegetation in two peat bogs in northern Minnesota. *Ecology*, 48: 308–310.

Best, G.R., 1984. An old-growth cypress stand in Okefenokee Swamp. In: A.D. Cohen, D.J. Casagrande, M.J. Andrejko and G.R. Best (Editors), *The Okefenokee Swamp: Its Natural History, Geology, and Geochemistry.* Wetland Surveys, Los Alamos, N.M., pp. 132–143.

Brown, S., 1978. *A Comparison of Cypress Ecosystems in the Landscape of Florida.* PhD Dissertation, University of Florida, Gainesville, Fla., 569 pp.

Brown, S., 1981. A comparison of the structure, primary productivity, and transpiration of cypress ecosystems in Florida. *Ecol. Monogr.*, 51: 403–427.

Brown, S. and Lugo, A., 1982. A comparison of structural and functional characteristics of saltwater and freshwater forested wetlands. In: B. Gopal, R.E. Turner, R.G. Wetzel and D.F. Whigham (Editors), *Wetlands: Ecology and Management.* Proceedings of the First International Wetlands Conference. International Scientific Publications, Jaipur, pp. 109–130.

Brown, S., Cowles, S.W. and Odum, H.T., 1984a. Metabolism and transpiration of cypress domes in north-central Florida. In: K.C. Ewel and H.T. Odum (Editors), *Cypress Swamps.* University Presses of Florida, Gainesville, Fla., pp. 145–163.

Brown, S., Flohrschutz, E.W. and Odum, H.T., 1984b. Structure, productivity, and phosphorus cycling of the scrub cypress ecosystem. In: K.C. Ewel and H.T. Odum (Editors), *Cypress Swamps.* University Presses of Florida, Gainesville, Fla., pp. 304–317.

Christensen, N., Burchell, R., Liggett, A. and Simms, E., 1981. The structure and development of pocosin vegetation. In: C.J. Richardson (Editor), *Pocosin Wetlands.* Hutchinson Ross Publishing Company, Stroudsburg, Pa., pp. 43–61.

Cintrón, G., Lugo, A.E. and Martinez, R., 1985. Structural and functional properties of mangrove forests. In: W.G. D'Arcy and M.D. Correa A. (Editors), *The Botany and Natural History of Panama.* Monographs in Systematic Botany 10. Missouri Botanical Garden, St. Louis, Mo., pp. 53–66.

Cohen, A.D., Casagrande, D.J., Andrejko, M.J. and Best, G.R. (Editors), 1984. *The Okefenokee Swamp: Its Natural History, Geology, and Geochemistry.* Wetlands Surveys, Los Alamos, N.M., 709 pp.

Conway, V.M., 1949. The bogs of central Minnesota. *Ecol. Monogr.*, 19: 173–206.

Coultas, C.L. and Duever, M.J., 1984. Soils of cypress swamps. In: K.C. Ewel and H.T. Odum (Editors), *Cypress Swamps.* University Presses of Florida, Gainesville, Fla., pp. 51–59.

Craighead, F.C., 1971. *The Trees of South Florida.* University of Miami Press, Coral Gables, Fla.

Cypert, E., 1961. The effects of fires in the Okefenokee Swamp in 1954 and 1955. *Am. Midl. Nat.*, 66: 485–503.

Dabel, C.V. and Day, Jr., F.P., 1977. Structural comparisons of four plant communities in the Great Dismal Swamp, Virginia. *Bull. Torrey Bot. Club.*, 104: 253–260.

Daniel, C.C., 1981. Hydrology, geology and soils of pocosins: a comparison of natural and altered systems. In: C.J. Richardson (Editor), *Pocosin Wetlands.* Hutchinson Ross Publishing Company, Stroudsberg, Pa., pp. 69–108.

Dansereau, P. and Segadas-Vianna, F., 1952. Ecological study of the peat bogs of eastern North America. *Can. J. Bot.*, 30: 490–520.

Davis, J.H., 1943. The natural features of southern Florida, especially the vegetation, and the Everglades. *Fla. Geol. Surv. Geol. Bull*, 25.

Day, Jr., F.P., 1979. Litter accumulation in four plant communities in the Dismal Swamp, Virginia. *Am. Midl. Nat.*, 102: 281–289.

Day, Jr., F.P., 1982. Litter decomposition rates in the seasonally flooded Great Dismal Swamp. *Ecology*, 63: 670–678.

Day, Jr., F.P. and Dabel, C.V., 1978. Phytomass budgets for the Dismal Swamp ecosystem. *Va. J. Sci.*, 29: 220–224.

Deghi, G.S., 1977. *Effect of Sewage Effluent Application on Phosphorus Cycling in Cypress Domes.* Masters Thesis, University of Florida, Gainesville, Fla.

Deghi, G.S., Ewel, K.C. and Mitsch, W.J., 1980. Effects of sewage effluent application on litter fall and litter decomposition in cypress swamps. *J. Appl. Ecol.*, 17: 397–408.

Dierberg, F.E. and Brezonik, P.L., 1984. Water chemistry of a Florida cypress dome. In: K.C. Ewel and H.T. Odum (Editors), *Cypress Swamps.* University Presses of Florida, Gainesville, Fla., pp. 34–50.

Dierberg, F.E. and Ewel, K.C., 1984. The effects of wastewater on decomposition and organic matter accumulation in cypress domes. In: K.C. Ewel and H.T. Odum (Editors), *Cypress Swamps.* University Presses of Florida, Gainesville, Fla., pp. 164–170.

Duncan, D.P., 1954. A study of some of the factors affecting the natural regeneration of tamarack (*Larix laricina*) in Minnesota. *Ecology*, 35: 498–521.

Ewel, K.C. and Mitsch, W.J., 1978. The effects of fire on species composition in cypress dome ecosystems. *Fla. Sci.*, 41: 25–32.

Ewel, K.C. and Odum, H.T. (Editors), 1984. *Cypress Swamps.* University Presses of Florida, Gainesville, Fla., 472 pp.

Gomez, M.M. and Day, Jr., F.P., 1982. Litter nutrient content and production in the Great Dismal Swamp. *Am. J. Bot.*, 69: 1314–1321.

Gore, A.J.P. (Editor), 1983. *Mires: Swamp, Bog, Fen, and Moor. Part A.* Ecosystems of the World, Vol. 4A. Elsevier Scientific Publishing Co., Amsterdam, 440 pp.

Heilman, P.E., 1966. Change in distribution and availability of nitrogen with forest succession on north slopes in interior Alaska. *Ecology*, 47: 825–831.

Heilman, P.E., 1968. Relative availability of phosphorus and cations to forest succession and bog formation in interior Alaska. *Ecology*, 49: 331–336.

Heimburg, K., 1984. Hydrology of north-central Florida cypress domes. In: K.C. Ewel and H.T. Odum (Editors), *Cypress Swamps.* University Presses of Florida, Gainesville, Fla., pp. 72–82.

Heinselman, M.L., 1970. Landscape evolution, peatland types, and the environment in the Lake Agassiz Peatlands Natural Area, Minnesota. *Ecol. Monogr.*, 40: 235–261.

Hemond, H.F., 1980. Biogeochemistry of Thoreau's Bog, Concord, Massachusetts. *Ecol. Monogr.*, 50: 507–526.

Jaworski, E. and Raphael, C.N., 1979. Historical changes in natural diversity of fresh water wetlands, glaciated region of northern United States. In: P.E. Greeson, J.R. Clark and J.E. Clark (Editors), *Wetland Functions and Values: the State of Our Understanding.* Proceedings of the

National Symposium on Wetlands. American Water Resources Association, Minneapolis, Minn., pp. 545–557.

Jiménez, J.A. and Lugo, A.E., 1985. *Avicennia germinans* In: *Silvics of Forest Trees of the American Tropics.* SO-ITF-SM-4, USDA Forest Service, Southern Forest Experiment Station, New Orleans, La., 6 pp.

Johnson, R.L., 1979. Timber harvests from wetlands. In: P.E. Greeson, J.R. Clark and J.E. Clark (Editors), *Wetland Functions and Values: the State of Our Understanding.* Proceedings of the National Symposium on Wetlands, American Water Resources Association, Minneapolis, Minn., pp. 598–605.

Kramer, P.J., Riley, W.S. and Bannister, T.T., 1952. Gas exchange of cypress knees. *Ecology*, 33: 117–121.

Kurz, H., 1933. Cypress domes. *Annu. Rep. Fla. State Geol. Surv.*, 23–24: 54–56.

Kurz, H. and Demaree, D., 1934. Cypress buttresses and knees in relation to water and air. *Ecology*, 15: 36–41.

Kurz, H. and Wagner, K.A., 1953. Factors in cypress dome development. *Ecology*, 34: 157–164.

Larsen, J.A., 1982. *Ecology of the Northern Lowland Bogs and Conifer Forests.* Academic Press, New York, N.Y., 307 pp.

Lewis, F.J., Dowding, E.S. and Moss, E.H., 1928. The vegetation of Alberta. *J. Ecol.*, 16: 19–70.

Lugo, A.E., 1981. The inland mangroves of Inagua. *J. Nat. Hist.*, 15: 845–852.

Lugo, A.E. and Snedaker, S.C., 1975. Properties of a mangrove forest in southern Florida. In: G.E. Walsh, S.C. Snedaker and H.J. Teas (Editors), *Proc. of the International Symposium on Biology and Management of Mangroves.* Institute of Food and Agriculture Sciences, University of Florida, Gainesville, Fla., pp. 170–212.

Lugo, A.E., Cintrón, G. and Goenaga, C., 1981. Mangrove ecosystems under stress. In: G.W. Barrett and R. Rosenberg (Editors), *Stress Effects on Natural Ecosystems.* J. Wiley and Sons, Chichester, pp. 129–153.

Lugo, A.E., Evink, G., Brinson, M., Broce, A. and Snedaker, S.C., 1975. Diurnal rates of photosynthesis, respiration and transpiration in mangrove forests of south Florida. In: F.B. Golley and E. Medina (Editors), *Tropical Ecological Systems.* Springer-Verlag, New York, N.Y., pp. 335–350.

Lugo, A.E., Twilley, R.R. and Patterson-Zucca, C., 1980. *The role of black mangrove forests in the productivity of coastal ecosystems in South Florida.* Final Report to US Environmental Protection Agency, Corvallis Environmental Research Laboratory, Corvallis, Oreg., Contract No. R806079010. Center for Wetlands, University of Florida, Gainesville, Fla.

McCaffrey, C.A. and Hamilton, D.B., 1984. Vegetation mapping of the Okefenokee Swamp ecosystem. In: A.D. Cohen, D.J. Casagrande, M.J. Andrejko and G.R. Best (Editors), *The Okefenokee Swamp: Its Natural History, Geology, and Geochemistry.* Wetland Surveys, Los Alamos, N.M., pp. 201–211.

McKinley, C.E. and Day, Jr., F.P., 1979. Herbaceous production in cut-burned, uncut-burned and control areas of a *Chamaecyparis thyoides* (L.) BSP (Cupressaceae) stand in the Great Dismal Swamp. *Bull. Torrey Bot. Club*, 106: 20–28.

Marois, K.C. and Ewel, K.C., 1983. Natural and management-related variation in cypress domes. *For. Sci.*, 29: 627–640.

Mattoon, W.R., 1915. *The Southern Cypress.* Bull. 272, US Department of Agriculture, Washington, D.C.

Monk, C.D., 1966. An ecological study of hardwood swamps in north-central Florida. *Ecology*, 47: 649–653.

Monk, C.D. and Brown, T.W., 1965. Ecological consideration of cypress heads in northcentral Florida. *Am. Midl. Nat.*, 74: 126–140.

Novitzki, R.P., 1979. Hydrologic characteristics of Wisconsin's wetlands and their influence on floods, stream flow, and sediment. In: P.E. Greeson, J.R. Clark and J.E. Clark (Editors), *Wetland Functions and Values: the State of Our Understanding.* Proceedings of the National Symposium on Wetlands, American Water Resources Association, Minneapolis, Minn., pp. 377–388.

Odum, H.T., 1984. Summary: cypress swamps and their regional role. In: K.C. Ewel and H.T. Odum (Editors), *Cypress Swamps.* University Presses of Florida, Gainesville, Fla., pp. 416–444.

Parker, G.R. and Schneider, G., 1974. Structure and edaphic factors of an alder swamp in northern Michigan. *Can. J. For. Res.*, 4: 499–508.

Parker, G.R. and Schneider, G., 1975. Biomass and productivity of an alder swamp in northern Michigan. *Can. J. For. Res.*, 5: 403–409.

Pool, D.J., Lugo, A.E. and Snedaker, S.C., 1975. Litter production in mangrove forests of southern Florida and Puerto Rico. In: G.E. Walsh, S.C. Snedaker and H.J. Teas (Editors), *Proc. of the International Symposium on Biology and Management of Mangroves.* Institute of Food and Agriculture Sciences, University of Florida, Gainesville, Fla., pp. 213–237.

Pool, D.J., Snedaker, S.C. and Lugo, A.E., 1977. Structure of mangrove forests in Florida, Puerto Rico, Mexico, and Costa Rica. *Biotropica*, 9: 195–212.

Reader, R.J. and Stewart, J.M., 1972. The relationship between net primary production and accumulation for a peatland in southeastern Manitoba. *Ecology*, 53: 1024–1037.

Reiners, W.A., 1972. Structure and energetics of three Minnesota forests. *Ecol. Monogr.*, 42: 71–94.

Reiners, W.A. and Reiners, N.M., 1970. Energy and nutrient dynamics of forest floors in three Minnesota forests. *J. Ecol.*, 58: 497–519.

Richardson, C.J. (Editor), 1981. *Pocosin Wetlands.* Hutchinson Ross Publishing Company, Stroudsburg, Pa., 364 pp.

Richardson, C.J., 1983. Pocosins: vanishing wastelands or valuable wetlands? *Bioscience*, 33: 626–633.

Richardson, C.J., Evans, R. and Carr, D., 1981. Pocosins: an ecosystem in transition. In C.J. Richardson (Editor), *Pocosin Wetlands.* Hutchinson Ross Publishing Company, Stroudsberg, Pa., pp. 3–19.

Schlesinger, W.H., 1978. Community structure, dynamics and nutrient cycling in the Okefenokee cypress swamp-forest. *Ecol. Monogr.*, 48: 43–65.

Schlesinger, W.H. and Chabot, B.F., 1977. The use of water and minerals by evergreen and deciduous shrubs in Okefenokee Swamp. *Bot. Gaz.*, 138: 490–497.

Scholander, P.F., VanDam, L. and Scholander, S.I., 1955. Gas exchange in the roots of mangroves. *Am. J. Bot.*, 42: 92–98.

Schwintzer, C.R., 1977. Vegetation changes and water levels in a small Michigan bog. In: C.B. DeWitt and E. Soloway (Editors), *Wetlands: Ecology, Values, and Impacts*. Proceedings of the Waubesa Conference on Wetlands. Institute for Environmental Studies, University of Wisconsin-Madison, Madison, Wisc., pp. 326–336.

Schwintzer, C.R., 1978. Nutrient and water levels in a small Michigan bog with high tree mortality. *Am. Midl. Nat.*, 100: 441–451.

Schwintzer, C.R. and Williams, G., 1974. Vegetation changes in a small Michigan bog from 1917 to 1972. *Am. Midl. Nat.*, 92: 447–459.

Small, E., 1972. Photosynthetic rates in relation to nitrogen recycling as adaptation to nutrient deficiency in peat bog plants. *Can. J. Bot.*, 50: 2227–2233.

Spangler, D.P., 1984. Geologic variability among six cypress domes in north-central Florida. In: K.C. Ewel and H.T Odum (Editors), *Cypress Swamps*. University Presses of Florida, Gainesville. Fla., pp. 60–66.

Straub, P. and Post, D.M., 1978. Rates of growth and nutrient concentration of trees in cypress domes. In: K.C. Ewel and H.T. Odum (Editors) *Cypress Wetlands for Water Management, Recycling and Conservation*. Report PB-282-159, National Technical Information Service, Springfield, Va., pp. 271–318.

Tilton, D.L., 1977. Seasonal growth and foliar nutrients of *Larix laricina* in three wetland ecosystems. *Can. J. Bot.*, 55: 1291–1298.

Triska, F.J. and Cromack, Jr., K., 1980. The role of woody debris in forest and streams. In: R.H. Waring (Editor), *Forests: Fresh Perspectives from Ecosystem Analysis*. Oregon State University Press, Corvallis, Oreg., pp. 171–190.

Twilley, R.R., 1982. *Litter Dynamics and Organic Carbon Exchange in Black Mangrove* (Avicennia germinans) *Basin Forests in a Southwest Florida Estuary*. PhD Dissertation, University of Florida, Gainesville, Fla., 260 pp.

Twilley, R.R., Lugo, A.E. and Patterson-Zucca, C., 1986. Litter production and turnover in basin mangrove forests in southwest Florida. *Ecology*, 67(3): 670–683.

Vernon, R.O., 1947. Cypress domes. *Science*, 105: 97–99.

Verry, E.S., 1975. Streamflow chemistry and nutrient yields from upland-peatland watersheds in Minnesota. *Ecology*, 56: 1149–1157.

Vitousek, P.M., 1982. Nutrient cycling and nutrient use efficiency. *Am. Nat.*, 119: 553–572.

Watt, R.F. and Heinselman, M.L., 1965. Foliar nitrogen and phosphorus level related to site quality in a northern Minnesota spruce bog. *Ecology*, 46: 357–361.

Yates, R.F. and Day, Jr., F.P., 1983. Decay rates and nutrient dynamics in confined and unconfined leaf litter in the Great Dismal Swamp. *Am. Midl. Nat.*, 110: 37–45.

Zoltai, S.C. and Pollett, F.C., 1983. Wetlands in Canada: their classification, distribution and use. In: A.J.P. Gore (Editors), *Mires: Swamp, Bog, Fen and Moor. Part B. Ecosystems of the World 4B*. Elsevier, Amsterdam, pp. 245–268.

Chapter 8

THE GREAT DISMAL SWAMP: AN ILLUSTRATED CASE STUDY

VIRGINIA CARTER

The Great Dismal Swamp is an 84 000-ha forested wetland located on the Virginia–North Carolina border in the southern Atlantic Coastal Plain of the United States (Fig. 8.1). The organic soils of the swamp range in depth from 4 m in ancient drainage channels to less than 0.3 m along the outer edges. Lake Drummond, approximately 4 km in diameter, is almost centrally located within the swamp, which slopes gently to east and south. The flora of the Great Dismal Swamp includes individual species and plant assemblages otherwise scattered widely to the north and south along the Coastal Plain (Fig. 8.2). The canopy and subcanopy consist of mixed evergreen and deciduous, broad-leaved and needle-leaved trees, and there are a variety of evergreen and deciduous shrubs and vines (Figs. 8.3–8.8)[1]. Anthropogenic disturbance of the natural vegetation has resulted in a wide diversity of wildlife habitats.

Remotely-sensed data (aerial photographs and satellite images) provide an invaluable tool for studying the complexity of the swamp ecosystem and the diversity of vegetation communities (Carter et al., 1977; Garrett and Carter, 1977). Satellite images provide a regional overview of the swamp and its setting; aerial photographs are important for mapping and for identifying features of interest in the more remote areas. High- and low-altitude seasonal color infrared photographs have been used to locate a variety of wetland habitats and plant assemblages, and to map the vegetation of the swamp at a scale of 1:100 000 (Figs. 8.9, 8.10) (Gammon and Carter, 1979).

Present research in the Great Dismal Swamp includes studies of the dynamics of the wetland-to-upland transition zone (Figs. 8.11–8.13), wetland hydrology (Lichtler and Walker, 1974), litter production and nutrient studies in individual communities (Day, 1982), water quality and phytoplankton populations in the lake and ditches (Marshall, 1979), vegetation trends and regeneration strategies (Levy and Walker, 1979), organic soil development (Whitehead, 1972) and wildlife-habitat requirements (Fig. 8.14). Remotely-sensed data are among the many tools required for these intensive ground studies; they provide an overview and level of detail needed to put small areas into perspective with the total ecosystem represented by Great Dismal Swamp.

Landsat digital data also provide a basis for identifying temporal changes in the Great Dismal Swamp vegetation communities, and for map revision and update (Gammon et al., 1979). Such repetitive data have been used to (1) study flooding dynamics during the dormant winter period when the deciduous trees are leafless, and (2) to monitor seasonal and long-term changes within the swamp as a result of natural successional trends or anthropogenic influences. Black and white aerial photographs from 1937, 1951, and 1972 have provided information on succession in vegetation communities (Fig. 8.15).

---

[1]The scientific equivalents of the vernacular names used in this chapter are as follows: ash: *Fraxinus* spp.; Atlantic white cedar: *Chamaecyparis thyoides*; bay, red: *Persea borbonia*; bay, sweet: *Magnolia virginiana*; beech: *Fagus grandifolia*; cypress: *Taxodium distichum*; greenbrier: *Smilax laurifolia*; hollies: *Ilex* spp.; inkberry: *Ilex glabra*; maple: *Acer rubrum*; oak: *Quercus* spp.; pine, loblolly: *Pinus taeda*; pine, pond: *Pinus serotina*; sweetgum: *Liquidambar styraciflua*; tupelo (watertupelo): *Nyssa aquatica*; tupelo, black: *Nyssa sylvatica* var. *biflora*; yellow poplar: *Liriodendron tulipifera*; cane: *Arundinaria gigantea*.

THE GREAT DISMAL SWAMP

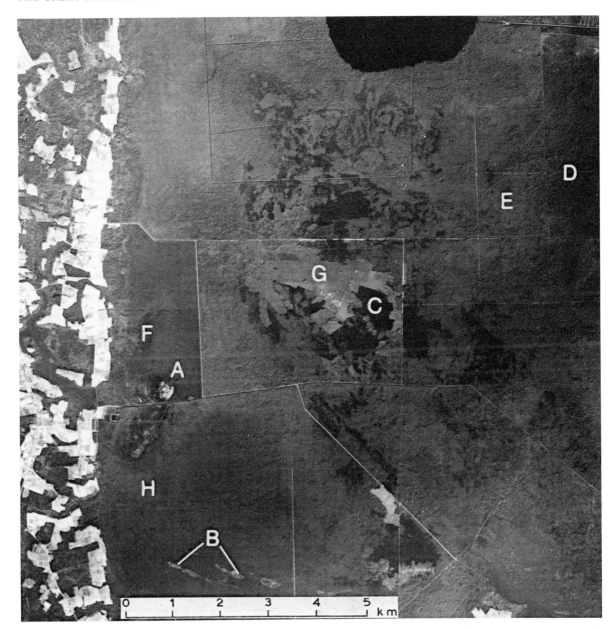

Fig. 8.2. Black and white copy of a high-altitude color infra-red photograph (February 1975) showing the southern part of the Great Dismal Swamp including Lake Drummond. Lightest tones are agricultural fields and roads. Vegetation units identified are (A) marsh, (B) mesic island community, (C) Atlantic white cedar (*Chamaecyparis thyoides*, (D) pine, (E) evergreen shrubs, (F) cypress (*Taxodium*), (G) clearcut, (H) maple–gum (*Acer–Nyssa*).

Fig. 8.1. This Soil Conservation Service Landsat MSS 7 mosaic of coastal Virginia and North Carolina shows the Great Dismal Swamp surrounded by highly productive agricultural lands and small patches of upland and wetland forest. North of the swamp is the urban–industrial complex which fringes on the mouth of the Chesapeake Bay. To the south are Albemarle Sound and the Outer Banks of North Carolina.

Fig. 8.3. *Juncus repens* growing along the margin of Lake Drummond. This black-water lake contains almost no vegetation except for a few cypress trees and some herbaceous aquatic plants near the mouth of Washington Ditch.

Fig. 8.4. Roadsides and ditches throughout the swamp provide a variety of wildlife habitats. These man-made features have modified the hydrology of the swamp, damming up the water or moving it rapidly into Lake Drummond, the Dismal Swamp Canal, or the Pasquotank River.

THE GREAT DISMAL SWAMP

Fig. 8.5. The cypress–gum community contains surface water until midsummer.

Fig. 8.7. Atlantic white cedar has been extensively clear-cut within the swamp. However, this species will benefit from new forest management practices to improve the vitality of the remaining stands.

Fig. 8.6. Upland vegetation grows on mesic "islands" in the southern part of the swamp.

Fig. 8.8. The swamp contains many hectares of southeastern evergreen shrub wetland (*Ilex*) or "pocosin".

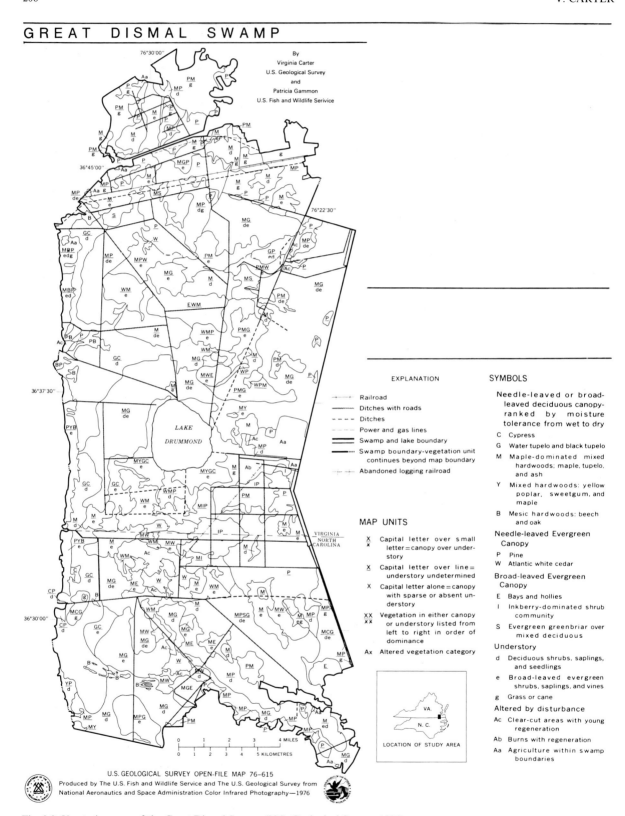

Fig. 8.9. Vegetation map of the Great Dismal Swamp (U.S. Geological Survey, 1976).

THE GREAT DISMAL SWAMP

Fig. 8.10. Information from National Aeronautics and Space Administration color infrared photographs was transferred to an overlay on a black and white orthophotoquadrangle mosaic to make a vegetation map.

Fig. 8.12. Line-intercept transects are used to collect vegetation data on the transition zone. The change in vegetation from wetland to upland can be correlated with changes in the soil and with fluctuations in ground and surface water.

Fig. 8.11. Soil–oxygen relationships in the root zone are studied along the transition zone.

By use of a digital data base developed for the Great Dismal Swamp and other similar areas, Landsat digital images can be combined with ground- and surface-water levels (Figs. 8.13, 8.16), peat depth, surface elevations (Fig. 8.17), and many other types of information. A digital data base of this type aids in the long-term management and protection of the swamp.

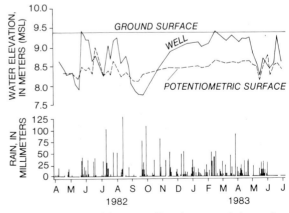

Fig. 8.13. Water-level data are collected at nests of observation wells on the transition zone. These data from the western end of the transect show seasonal fluctuations in water-table and potentiometric surface.

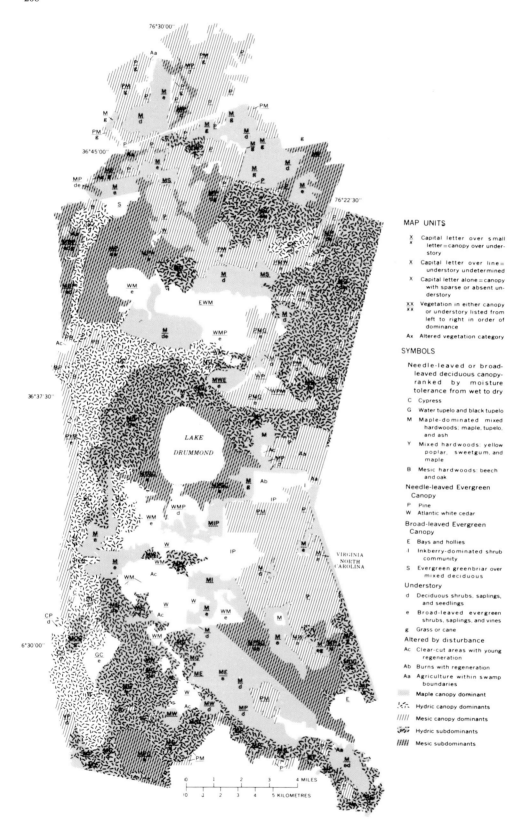

THE GREAT DISMAL SWAMP

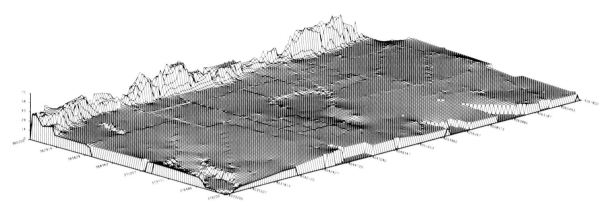

Fig. 8.15. These maps made from archival photographs show the changing vegetative communities south of Lake Drummond.

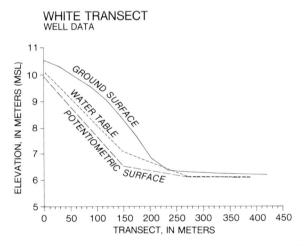

Fig. 8.16. The configuration of the water table and potentiometric surface along the western transition zone changes seasonally. This example is from August 1982.

Fig. 8.14. The vegetation map can be used to look at the location and wildlife management potential of various vegetation communities. The shading shows the maple, hydric and mesic classes.

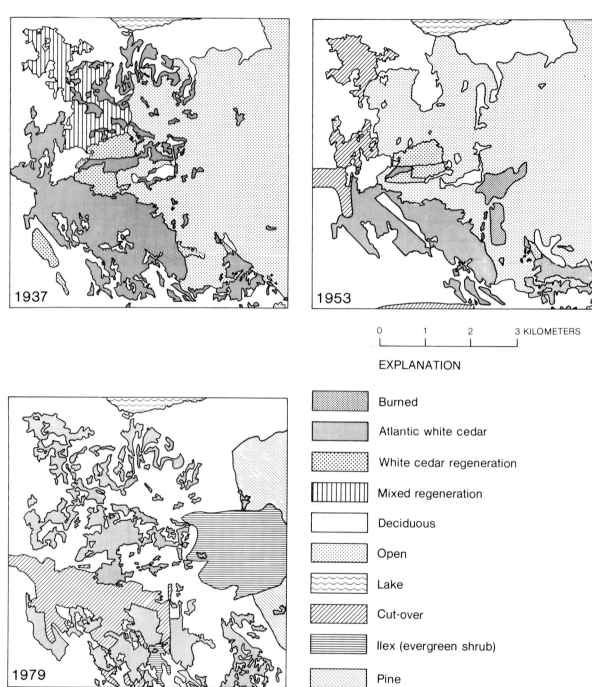

Fig. 8.17. Topographic contour model of the Great Dismal Swamp. Lake Drummond is centrally located, and the transition zone can be seen along the back edge of the map. Vertical exaggeration × 200 (from Caruso and Paschal, 1980).

REFERENCES

Carter, V., Garrett, M.K., Shima, L. and Gammon, P., 1977. The Great Dismal Swamp, management of a hydrologic resource with the aid of remote sensing. *Water Resour. Bull.*, 13: 1–12.

Caruso, V.M. and Paschal, J.E., Jr., 1980. Computer modeling of the Great Dismal Swamp. In: *Proceeding of the*

Conference Coastal Zone '80. Hollywood, Florida. American Society of Civil Engineers, New York, pp. 2308–2321.

Day, F.P., Jr., 1982. Litter decomposition rates in the seasonally flooded Great Dismal Swamp. *Ecology*, 63: 670–678.

Gammon, P.T. and Carter, V., 1979. Vegetation mapping with seasonal color infrared photographs. *Photogramm. Eng. Remote Sensing*, 45: 87–97.

Gammon, P.T., Carter, V. and Rohde, W.G., 1979. Landsat digital classification of the vegetation of the Great Dismal Swamp with an evaluation of classification accuracy. In: M. Deutsch, D.R. Wiesnet and A. Rango (Editors), *Satellite Hydrology, Proceedings of the Fifth Annual William T. Pecora Memorial Symposium on Remote Sensing*. Sioux Falls, S.D., pp. 463–473.

Garrett, M.K. and Carter, V., 1977. Contribution of remote sensing to habitat evaluation and management in a highly altered ecosystem. In: *Transactions of the 42nd North American Wildlife and Natural Resources Conference*. Atlanta, Ga., pp. 56–65.

Levy, G.F. and Walker, S.W., 1979. Forest dynamics in the Dismal Swamp of Virginia. In: P.W. Kirk, Jr. (Editor), *The Great Dismal Swamp*. University Press of Virginia, Charlottesville, Va., pp. 101–126.

Lichtler, W.F. and Walker, P.N., 1974. Hydrology of the Dismal Swamp, Virginia–North Carolina. *U.S. Geol. Surv. Open-File Rep.* 74–39. U.S. Govt., Printing Office, Washington, D.C., 50 pp.

Marshall, H.G., 1979. Lake Drummond: with a discussion regarding its plankton composition. In: P.W. Kirk, Jr. (Editor), *The Great Dismal Swamp*. University Press of Virginia, Charlottesville, Va., pp. 169–182.

U.S. Geological Survey, 1976. Great Dismal Swamp. *U.S. Geol. Surv. Open-File Map* 76–615. U.S. Govt. Printing Office, Washington, D.C.

Whitehead, D.R., 1972. Developmental and environmental history of the Dismal Swamp. *Ecol. Monogr.*, 42: 301–315.

Chapter 9

ECOLOGY AND MANAGEMENT OF SWAMP FORESTS IN THE GUIANAS AND CARIBBEAN REGION

PETER R. BACON

INTRODUCTION

Wet forested ecosystems are a regular feature of the low-lying coastal areas commonly referred to as swamps in the West Indies, Central America and northern South America. Apart from the brackish-water mangrove swamps which are discussed elsewhere in this series (Chapman, 1977), two broad facies of freshwater forests have been recognized in the region (Fig. 9.1) by a number of authors (Beard, 1946; Fanshawe, 1952; Lindeman, 1953; Asprey, 1959; Vann, 1959). These are (a) swamp forest, found on soils which are permanently inundated or have perennially high water table, and (b) marsh forest in areas subjected to periodic inundation.

There is less agreement, however, in the descriptions of the plant communities contained in these categories. As a result the literature contains a multiplicity of sub-divisions, giving the impression that these wetland communities differ significantly from one country to the next (Table 9.1).

Although it is generally agreed that in the process of *Verlandung*[1] the freshwater forest hydroseres occupy an intermediate stage between the brackish-water communities and the terrestrial rain forest, the successional relationships between the numerous sub-units are less well documented. In most areas the swamp forests and marsh forests are associated with herbaceous swamp or marsh communities, introducing further complexity into the successions. The study of swamp forest ecology must, therefore, take into account the associated marsh forest and herbaceous vegetation, and the structure and development of swamp environments.

This paper attempts to resolve some of the problems with the terms "swamp" and "marsh" and some confusion regarding succession. It is necessary first to review the swamp plant communities in different parts of the region, their relation to swamp topography and the development of swamp conditions before returning to questions of terminology.

Because of the inaccessibility of Neotropical wetlands and the large number of plant species involved, there is still a paucity of good autecological data, which hinders the accurate description of the vegetation and its history. The vegetation has received more attention than the swamp animals, however, as only a provisional faunal checklist (Aitken et al., 1973) could be found for the freshwater wetlands of one country in the Caribbean and Guianas region. The neighbouring mangrove faunas are better known (Rutzler, 1969; Bacon, 1970), many species being of proven economic value, and voluminous literature exists on terrestrial faunas in the region. The faunal relationships between the mangal, freshwater swamp and terrestrial habitats appear not to have been investigated.

Little attention has been paid to the productivity of swamp forests or to sylvicultural and economic aspects of management, even though these ecosystems occupy a significant area of Neotropical wetlands.

THE VEGETATION

Trinidad

In his account of the vegetation of Trinidad, Beard (1946) provided one of the earliest detailed

---
[1] "In this process of *Verlandung*, or formation of new land from water, vegetation plays an essential part." (Richards, 1979.)

TABLE 9.1

Plant communities described for Neotropical swamps

Trinidad (Beard, 1946)	Guyana (Fanshawe, 1952)	Surinam (Lindeman, 1953)	Guiana coastal plain (Vann, 1959)
1. *Swamp communities* a. Mangrove woodland b. Swamp forest formation 1. *Pterocarpus officinalis* consociation c. Palm swamp 1. *Roystonea oleracea* consociation 2. *Mauritia setigera* consociation d. Herbaceous swamp 2. *Marsh communities* a. Marsh forest 1. *Manicaria – Jessenia – Euterpe* association b. Palm marsh 1. *Mauritia – Chrysobalanus* association c. Savanna	1. Mangrove forest 2. Swamp forest a. *Pterocarpus – Macrolobium* assemblage *Virola surinamensis* community 3. Swamp woodland a. *Bombax – Pterocarpus* community 4. Marsh forest a. Palm marsh forest 1. *Symphonia – Tabebuia – Euterpe* association *Manicaria saccifera* faciation *Symphonia globulifera* community *Pentaclethra macroloba* community *Vatairea guianensis* community *Pterocarpus – Maximiliana* assemblage b. Marsh forest 1. *Iryanthera – Tabebuia* assemblage *I. macrophylla* facies *Inga – Gustavia* assemblage c. Palm marsh woodland 1. *Clusia – Tabebuia* assemblage *C. fockeana* facies d. Palm marsh 1. *Mauritia flexuosa* consociation *Chrysobalanus icaco* community 5. "Swamp rain forest" — *Mora* forest.	1. Mangroves 2. Herbaceous swamps 3. Swamp forest a. Mixed swamp wood b. *Erythrina glauca* swamp wood c. *Machaerium lunatum* scrub (d. Palm swamp) 4. Marsh forest a. *Triplaris surinamensis – Bonafousia tetrastachya* type b. *Symphonia globulifera* type c. *Hura crepitans* forest d. *Mauritia – Chrysobalanus* association	1. Mangrove and strand 2. Herbaceous swamp 3. Swamp forest a. Mixed swamp wood b. Coral tree swamp wood *Erythrina glauca*-dominant c. Brantimakka scrub *Machaerium lunatum*-dominant 4. Marsh forest a. Tidal marsh forest Mangrove-passing to *Bombax aquaticum* and *Pterocarpus officinalis* b. Floodplain forest *Symphonia globulifera*-dominant c. Ridge marginal forest *Triplaris surinamensis* and *Bonafousia tetrastachya* - dominants d. *Hura crepitans* forest

Caribbean area (Asprey, 1959)
1. Swamp forest formations
   a. Brackish swamp forest
      Mangrove association
   b. Freshwater swamp forest
      Pterocarpus association
   c. Swamp woodland ⎫ (Guyana)
   d. Swamp thicket ⎭
2. Seasonal swamp (marsh) formations
   a. Seasonal swamp woodland
      Clusia – Tabebuia association (Guyana)
      Amanoa association (Dominica)
      Mauritia – Chrysobalanus association (Trinidad)
   b. Seasonal swamp thicket
   c. Savanna

Panama (Porter, 1973)
Swamp formations
   a. Salt water swamp forest ⎫
   b. Salt water riparian forest ⎬ Mangroves
   c. Brackish swamp forest ⎭
      Mora — local canopy dominant
   d. Brackish riparian forest
      Tabebuia — local canopy dominant
   e. Freshwater marsh
   f. Freshwater swamp forest
      Pachira, Pterocarpus officinalis, Tabebuia, local canopy dominants
   g. Freshwater riparian forest
      Prioria copaifera, Pterocarpus officinalis, Tabebuia, local canopy dominants

Guadeloupe (Portecop and Crisan, 1978)
1. Mangrove
2. Herbaceous marsh (Marais à herbe coupante)
3. Swamp forest (Forêt marécageuse)
   a. Transitional swamp forest (Faciès de transition)
      Pterocarpus–Pavonia facies
   b. Swamp forest (Faciès normal)
      Pterocarpus – Inga facies
   c. Dry swamp forest (Faciès sec)
      Pterocarpus – Tabebuia facies
4. *Acrostichum* fern sub-association

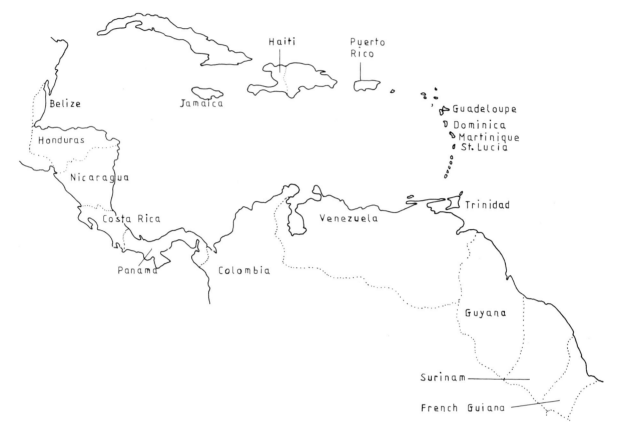

Fig. 9.1. The Guianas and Caribbean region. Countries named contain *Pterocarpus* forests.

descriptions of regional wetland vegetation. He suggested that it was useful for ecological purposes to differentiate between areas of "swamp" and "marsh". In the freshwater swamp habitats in areas of low relief, a series of plant formations was recognized according to increasing depth of inundation, as follows:

(a) the greatest depths were found in herbaceous swamps. These ranged from tall swamp with *Cyperus giganteus*, *Gynerium sagittatum* and *Montrichardia arborescens* in land perpetually inundated to depths of 1.5 m to swamp savanna of *Leersia hexandra* forming a floating grass mat in very deep water.

(b) Palm swamp was found on land perpetually inundated with 30 to 100 cm of water. This included a consociation of *Roystonea oleracea*, which Beard considered similar to *Roystonea* palm swamps in the Greater Antilles, and a consociation of *Mauritia setigera*, similar to the morichales (see p. 219) of the Orinoco Delta and the *M. flexuosa* swamps which Martyn (1934) described behind mangroves in Guiana.

(c) Swamp forest of a single consociation of *Pterocarpus officinalis* (Fig. 9.2) was found in areas with water depths from a few centimetres to one metre.

The palm swamp and swamp forest formations were reported from the Nariva, Los Blanquizales and North Oropuche swamps, although, apart from *Roystonea* being on the seaward side and *Mauritia* on the landward side of the first two areas, data on distribution and extent were not given. Furthermore, even though inundation appeared as Beard's chief ecological factor, it was imprecisely used. No records of water depth at any season were given, and it was not clear whether a high soil water-table or standing water over the soil was the criterion of swamp conditions. The designation of swamp forests as areas "more or less permanently inundated" appears insufficient to separate this formation from Beard's marsh forest

Fig. 9.2. *Pterocarpus officinalis* buttress roots, near Cayenne.

found in areas "more or less waterlogged for the greater part of the year".

The marsh forest, one of Beard's three marsh communities, was reported to be confined to Pleistocene alluvial terraces in northern Trinidad, rather than occurring in any low-lying wetland environment. It was composed of the single *Manicaria–Jessenia–Euterpe* association, although other ecotones were found between this association and the evergreen seasonal forest in better drained areas. Fifty percent of the trees were palms, forming the lower storey between 3 and 9 m. Among the upper-storey trees, *Symphonia globulifera* and *Virola surinamensis* were fairly common, and Beard referred to these as swamp species, although they were not listed in his swamp communities.

Similarly, a number of species with special adaptations for growing in waterlogged and inundated places were classified under climatic as well as edaphic formations. In the evergreen seasonal forest, a *Carapa–Eschweilera* association occupying lowlands up to 300 m, Beard stated that *Crudia glaberrima* and *Pterocarpus rohrii* frequented swampy hollows, and a society of *S. globulifera* and *V. surinamensis* (Fig. 9.3) was found in swampy places along streams and also in hollows. *Manicaria saccifera* occupied the understorey in this society, being densest in waterlogged areas. On the western side of the east coast Nariva Swamp, two societies of the *Carapa–Eschweilera* association were listed, one with pure stands of *C. glaberrima* in waterlogged areas and a widely distributed *Carapa guianensis – Bactris major* society inundated to depths of 60 cm for at least 5 months of the year. Although containing the swamp species *Pterocarpus officinalis*, *Virola surinamensis* and *Inga* sp., and although classified previously by Marshall (1934) as freshwater swamp forest, Beard considered this simply as evergreen seasonal forest growing under extreme moisture conditions. A similar community containing *P. officinalis* and *Bactris minor* was reported in the tidal zone in Honduras by Standley (1931), who assigned it to a swamp forest type.

Beard's (1946) second marsh community, that of palm marsh, similarly lacked precise habitat definition. It was reported as a *Mauritia setigera – Chrysobalanus icaco* association fringing savannas where drainage was too adverse for forest. Although more waterlogged than marsh forest, it was considered a stage between this formation and savanna, rather than between marsh forest and the swamp communities. No relationship was sug-

Fig. 9.3. Mixed swamp wood, O'Meara Swamp, Trinidad.

gested between palm swamp and palm marsh, probably because the latter did not occur in an area which Beard would have called a swamp. In addition, in palm marsh *Mauritia* did not develop pneumatophores as it did in palm swamp and there was a greater number of associated species more typical of the marsh forest, such as *Chrysobalanus icaco*.

Beard considered that the swamp formations constituted an hydrosere developing towards the evergreen seasonal forest climatic climax as the slope of the land increased and inundation decreased. He did not discuss why the plant communities of swampy places in lowland forest should be distinguished from those with similar habit and habitat conditions in coastal swamps.

No successional relationship was suggested between swamps and marshes, however — that is, the swamp formations did not develop to rain forests through marsh formations, although it is obvious that marshy conditions will be present at the intermediate stages. Beard considered his marsh forest and palm marsh to be edaphic postclimaxes, formed as soil development led to deteriorating drainage conditions causing retrogression from climatic climax to edaphic marsh. Although it is theoretically possible for the same retrogressive processes to produce long-term inundation, Beard did not suggest that swampy conditions might also arise in this way.

The Guianas

A slightly different approach to the study of wetlands vegetation was used in Guiana by Fanshawe (1952), who included the marsh forest formations in the hydrosere. A number of vegetation zones were recognized in different parts of the coastal lowlands, representing successions starting from different points (Table 9.2).

Pterocarpus officinalis was dominant in the swamp forest, although associated frequently with *Macrolobium bifolium*, and *Virola surinamensis* was locally abundant. The forest was largely leguminous, although *Euterpe* and other palms shared the 18 to 24 m canopy, and there was a tendency for trees to form small, monospecific stands on the hummocky ground. Along the lower courses of rivers, as soon as the mangroves thinned out, a community of *Bombax aquaticum* and *P. officinalis*

TABLE 9.2

Successions in wetland plant communities in Guiana (after Fanshawe, 1952)

---

1. *Away from open water*
Water→herbaceous swamp→swamp forest→swamp woodland→*Mora* forest

2. *Higher up rivers*
Swamp woodland→swamp forest→*Mora* forest

3. *In delta areas of northwestern district*
Herb swamp→swamp woodland→palm marsh forest→rain forest

4. *Riverbank to uplands*
Palm marsh forest→marsh forest→palm marsh woodland→palm marsh→savanna

---

took over, and Fanshawe called this riparian fringe swamp woodland. Standley (1931) had recognized this swamp forest type in Honduras also.

The marsh communities were developed in areas of impeded drainage, and on alluvial silt or pegasse (partially decayed organic material) where a clay pan was present at the surface or to 3 m deep. The palm marsh forest, in which 60% to 70% of the trees were palms, was considered the equivalent of Beard's (1946) marsh forest, although *P. officinalis* was present. There was a decreasing percentage of palms and a more open aspect in the other marsh formations, and it was suggested that the palm marsh was probably of edaphic – biotic origin as there was evidence of fire.

Fanshawe's (1952) classification of the swamp and marsh forests is of interest in that he emphasized the diversity of habitats in which these communities develop.

In Surinam, Lindeman (1953) gave permanent inundation as the criterion for swamp forest, although noting that these communities could be superficially dry in excessively dry years. A mixed swamp forest, with *Andira inermis*, *Annona glabra*, *Chrysobalanus icaco*, *Ilex guianensis*, *Tabebuia insignis*, and *Triplaris surinamensis* with a subgrowth of *Acrostichum*, *Blechnum*, *Heliconia*, and *Nephrolepis* occurred directly behind the coastal mangroves or the levee forest. Its density decreased away from the levees as herbaceous sub-growth became dominant and tree species of the marsh forest penetrated for some distance from the levees into the mixed swamp wood.

Pterocarpus officinalis was also in scattered, small stands, but not as a characteristic species of the mixed swamp woods (Lindeman, 1953). However, *Erythrina glauca* was the absolute dominant in extensive tracts of swamp forest, especially in areas more brackish than those supporting mixed swamp wood. A few lianes and herbs and occasional *Blechnum indicum*, *Desmonchus horridus* and *Dryopteris gongylodes* were found in this *Erythrina glauca* swamp wood, and invariably the *Erythrina* groves had a narrow fringe of *Montrichardia arborescens*.

Thirdly, a *Machaerium lunatum* scrub occurred in brackish swamps and on river banks in the tidal zone. *Machaerium lunatum* was found in pure stands or associated with *Annona glabra* and *Ficus* sp. This type of swamp forest was common but rarely extensive.

Lindeman's marsh forests occupied intermediate stages between the swamp communities and terrestrial forest. A number of species which inhabited swamp forests played an important part floristically, as did many rain-forest trees which could tolerate periodic flooding. The *Triplaris surinamensis – Bonafousia tetrastachya* type was found on the margins of ridges and depressions in the younger parts of the coastal plain. The *Symphonia globulifera* type, although containing several species more adapted to inundation than those of the previous type, was found in more mature regions. Where creeks were present it resembled swamp forest, with *Pterocarpus*, *Symphonia*, *Tabebuia* and *Virola* well developed.

A *Mauritia – Chrysobalanus* association occurred in Surinam along the borders of freshwater swamps, and Lindeman equated this with Beard's palm marsh, supporting Fanshawe's (1952) conclusion that it was not merely a retrogressive product of rain-forest soil deterioration. A wet sub-association in which *Bonafousia tetrastachya*, *Genipa americana*, *Montrichardia arborescens*, *Pterocarpus officinalis*, *Triplaris surinamensis* and *Virola surinamensis* were present was differentiated by Lindeman, who considered this equivalent to Beard's palm swamp, but included it in his marsh forest types.

Lindeman (1953) also distinguished a marsh association dominated by *Hura crepitans*, but this was of very limited extent.

Vann (1959) followed Lindeman (1953) closely in describing the swamp communities of the three Guianas, recognizing mixed swamp wood, *Erythrina glauca* swamp wood and *Machaerium lunatum* scrub. However, he classified the marsh forests according to habitat types (Table 9.1) in a similar way to Fanshawe (1952), on the grounds that the coastal wetlands topography and tidal fluctuations had a major effect on vegetation community development.

Venezuela

Müller (1959) related the distribution of vegetation types in the Orinoco Delta to topography, soil organic content, and distance from water channels. He recognized four communities in waterlogged situations behind the mangroves and strand vegetation (Fig. 9.4).

Mixed swamp forest formed extensive stands in the lower delta, often directly behind the mangrove belt. This was a tall, mixed forest with a rather restricted number of dominant trees including *Bombax aquaticum*, *Pterocarpus officinalis*, *Symphonia globulifera* and the palms *Euterpe* sp. and *Manicaria saccifera*. Tidal influence was restricted and seasonal variation in water level was probably small. *S. globulifera* was especially abundant in the extensive forests of the easternmost delta, although occurring regularly in the swamp and marsh areas elsewhere, and in marginal forests along lower river courses. This association might be compared with Lindeman's *S. globulifera* marsh forest, a formation not present in Trinidad according to Beard (1946), who said that *Symphonia* was locally abundant in swamps but scattered frequently in terrestrial forests.

On levees in the lower delta, although more typical of the upper delta, Müller reported an *Erythrina glauca* swamp wood forming an open, mixed forest with heavy undergrowth. Associated with the dominant *Erythrina* were some *Euterpe* and *Mauritia* palms.

Palm swamp was very common in swampy and marshy situations. In the central and western Orinoco Delta it occurred as open, bush-like vegetation with emergent clusters of *Euterpe* sp., *Manicaria saccifera* and *Mauritia flexuosa* over a thick undergrowth; Müller called this mixed palm swamp. Large expanses of the lower and central delta were occupied by the morichales, dense

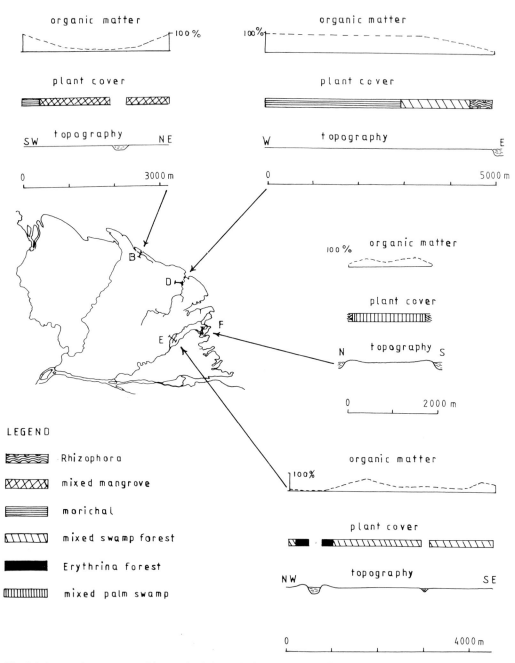

Fig. 9.4. Swamp forest communities on the Orinoco Delta, Venezuela (after Müller, 1959).

stands of *M. flexuosa* with few associated species, in more permanently waterlogged pegasse areas.

Although the swamp forest types found on the delta were closely comparable with those described for the Guianas, the precise distribution and successional relationships between the types are not readily deduced from Müller's data.

The Antilles

Woody swamp vegetation is reported from the Greater and Lesser Antilles, although detailed information is available only for Guadeloupe and Puerto Rico (see Ch. 10, this volume).

Eggers (1882) reported *Pterocarpus draco*

(= *P. officinalis*) in Puerto Rico in stagnant fresh waters bordering mangroves between Luquillo and Bergen on the north and in some areas of the southeast. It was associated with the palm *Oreodoxa regia* and herbaceous vegetation. Britton and Wilson (1922–1930) added that *Pterocarpus* was abundant and of large size in the marsh forests north of Playa de Humacao, which Gleason and Cook (1926) described in more detail. *Pterocarpus officinalis* was dominant, its place in the upper storey being shared by a few scattered *Roystonea borinquena*, with small, relatively uncommon associates including *Machaerium lunatum*, a few *Clusia rosea* on the *Pterocarpus*, and many ferns and other epiphytes on the buttresses. According to Holdridge (1940), the *Pterocarpus* swamp forest was a successional stage which replaced communities of *Conocarpus erectus* or of *M. lunatum* and *Hibiscus* sp. Chapman (1976) reported *Hibiscus tiliaceus*, *M. lunatum* and *Malache scabra* in brackish areas behind mangroves, a landward fringe of *P. officinalis* with *Annona glabra*, *Bucida buceras*, and *M. lunatum*, and an intervening reed swamp of *Mariscus jamaicensis* and *Typha domingensis*. The herbs thus occupy the backswamp regions, while swamp forest was adjacent to terrestrial forest, suggesting that this was an accreting coastline. More recent descriptions of *Pterocarpus* swamps in Puerto Rico are given by Alvarez-Lopez in Ch. 10 of this volume.

Barker and Dardeau (1930) listed *Pterocarpus officinalis* in the flora of Haiti. Britton and Wilson (1922–1930) thought that this species was present in Jamaica, although *Pterocarpus* forests were not reported by Asprey and Robbins (1953). These authors classified woody swamp communities as palm – sedge marsh and marsh forest. Their statement that the former had a cover of *Cyperus giganteus* and scattered *Roystonea princeps* was corrected by Proctor (1964), who gave *Cladium jamaicensis* as the dominant herbaceous species and placed the palm strictly in the marsh forest association. A further correction entailed the removal of *Haematoxylon*, *Piscidia* and *Spondias* from the list of marsh forest species. Proctor (1964) described marsh forest in the Black River Morass area of Jamaica as a jungle-type forest dominated by *Hibiscus elatus*, *Roystonia princeps*, *Symphonia globulifera* and *Typha latifolia*, with an understorey of *Enallagma latifolia*, *Ficus maxima*, *Grias cauliflora*, *Homalium racemosum*, *Tabebuia riparia*, the shrubs *Eugenia* and *Pavonia*, and the palm *Calyptronema occidentalis*. He noted that *Hibiscus elatus*, *Homalium racemosum*, *S. globulifera* and *T. folia* were typically species of upland forest, they are found in areas of high moisture co This marsh forest in Jamaica has a composition comparable to that of simila tions in Trinidad and the Guianas, where the same marsh species are found in w upland areas also.

Proctor's (1964) other woody associ riparian forest, mainly found in the up of the Morass and dominated by C ta and *Lonchocarpus sericeus* with ass lira *inermis*, *Annona glabra*, *Eugenia* and *Nectandra antillana*. The introduc toxylon campechianum had become do beside the rivers, reaching a large size.

Adams (1972) reported *Pterocarp cinalis* in Jamaica, this species being rare local in swamps and from sea level to 120 m. *Symphonia globulifera* was also present locally in the east of the island, in wet submontane forest and marsh forest up to nearly 1200 m.

Small stands of swamp forest occur in the Lesser Antilles and have been described by Beard (1949). In Dominica, pure groves of *Pterocarpus* occur on the northern part of the island, with trees up to 18 m high growing closely with little undergrowth, except for *Acrostichum*. James (1979) found these forests developed on alluvial flats, lagoons and river sides. They consisted of 90% *P. officinalis*, with optimal development on sites not constantly flooded, and stunted growth in *Acrostichum* areas. Palms were absent, but associated plants included *Annona glabra*, *Chimarrhis cymosa*, *Pavonia scabra*, *Sapium carabeum*, *Simarouba amara*, *Sloanea dentata*, *Sterculia caribaea* and *Tabebuia pallida*, while closer to the shore *Coccoloba uvifera*, *Hippomane mancinella* and *Terminalia catappa* were common. At some sites *Calophyllum calaba*, *Carapa guianensis* and *Symphonia globulifera* occurred also. In Dominica, Richards (1979) reported that the *Amanoa* association on podzolic soils contained *Amanoa caribaea* and *Oxythece pallida* with stilt roots, whereas they were buttressed in the lower montane rain forest. A subtype of rain forest on flat, poorly drained soils consisted of almost pure stands of *S. globulifera* and *Tovomita plumieri*. The development of dense stilt roots in these upland situations was a response to waterlogging, as it is in lowland swamps.

In St. Lucia, Beard (1949) found a few riverine swamps on the windward coast, and also some restricted areas in Martinique, whereas *Pterocarpus* dominated large areas of Guadeloupe. Feldmann and Lami (1936) had described the Guadeloupe swamps, in which they considered *Pterocarpus* to be the climax of the mangrove succession following *Conocarpus erectus*. They found the submerged parts of *Pterocarpus* colonized by dense mats of *Rhizoclonium hookeri*.

In a more detailed account, Stehlé (1937) described from Martinique an analogue of the Guadeloupe *Pterocarpus* swamp forest. This was an *Annona glabra* – *Dalbergia ecastophyllum* association dominated by *Annona* and growing in coastal lagoons with stagnant, fresh, acid water. *Dalbergia* apparently took the place of the *Chrysobalanus* present in other areas, while the vines *Brachypteris ovata*, *Cissus sicyoides*, *Hippocratea volubilis* and *Paullinia pinnata* were a conspicuous component of the vegetation. Stehlé (1946) later recorded *Pterocarpus officinalis* in Martinique also, in a restricted area near Galion and Trinité.

Recent work in Guadeloupe has added greatly to the information on Caribbean swamp ecology. Montaignac (1978) reported that the *Pterocarpus* marsh forest contained 55 species from 27 families, compared to the maritime mangroves with four arborescent species, two herbs and one epiphyte. This greater complexity led to greater variety, so that there was a less obvious zonation or regularity in the species distribution. Despite Montaignac's figures, the island swamp forests are less diverse than those of Trinidad and the tropical American mainland; 64 species from 31 families at Nariva (Bacon et al., 1979) and 87 species from 39 families in coastal Surinam (Lindeman, 1953), respectively. In the islands, swamp forests develop in narrow, more or less stable maritime environments rather than wide river flood plains or rapidly accreting coastal areas.

Portecop and Crisan (1978) divided the *Pterocarpus* forest of Guadeloupe into three sub-zones on the degree of inundation and salinity (Fig. 9.5; the degraded zone shown is a result of human impact). They separated a Transition Facies, in more saline environments behind mangroves, with a sub-association of *P. officinalis* – *Pavonia scabra*, from the Normal Facies of *P. officinalis* – *Inga laurina*, which appeared also in riparian vegetation inland. A "Facies sec", containing numbers of dead trees, developed on raised soils which were less marshy than the previous facies in areas inundated to less than 2 m seasonally. This was a facies of *P. officinalis* – *Tabebuia pallida*, but is probably formed by encroachment of the bordering *Cladietum*. Associated species in all three sub-associations included *Calophyllum calaba*, *Eugenia ligistrina*, *Montrichardia arborescens*, *Myrcia splendens* and *Symphonia globulifera*. *Annona palustris* occurred in the two wetter zones and *Chrysobalanus icaco* was absent from the *Pterocarpus* – *Inga* facies although common in the inland zone.

Portecop and Crisan (1978) classified the *Pterocarpus* forests as marsh forest and suggested that the zonation described represented the succession from brackish mangrove areas to the herbaceous marsh inland. These forests are much disturbed by human activity, so that the distribution pattern shown in Fig. 9.5 is almost certainly not due entirely to natural successional processes.

SWAMP STRUCTURE AND DEVELOPMENT

Topography and plant distribution

The description of the physiography of the Orinoco Delta deposits by Van Andel (1967) showed that a variety of habitats were available for colonization by wetland plants (Fig. 9.6). The upper delta was formed by normal fluviatile processes while the outer delta was a featureless swamp modified by mudflat accretion and chenier complexes. Between these, the lower delta had a well-developed system of distributary channels, levees, crevasses and washover fans. The levees were narrow and consisted mainly of fine sand and sandy loam. These graded away from the channels to silty loam, then clay, peaty clay, and finally peat in the intervening back-swamps. The lowest parts of the back-swamps were permanently under water, while during flood season all but the levee crests were submerged. The outer levee margins dried out to form ripened[1] clays in the low-water season.

---

[1] "Ripening" refers to the chemical, physical and biological changes that occur in soil after air penetrates previously waterlogged, reduced material.

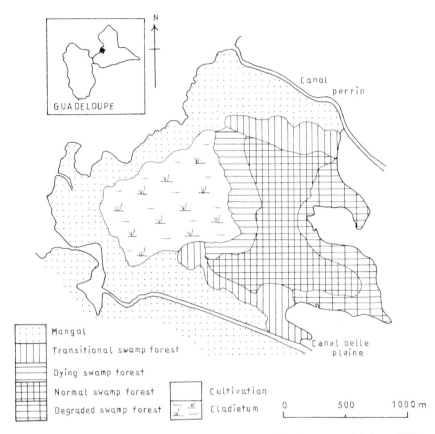

Fig. 9.5. Zonation of *Pterocarpus* forest in Guadeloupe (after Portecop and Crisan, 1978).

Although no further study appears to have been published on the vegetation of the Orinoco Delta, plant community distribution must be related to these physiographic features, as Müller (1959) had hinted, and a complex zonation pattern may be expected as different sequences of habitat conditions exist. The major sequences are: (a) from outer delta to lower delta to upper delta, (b) from lower delta channel to back-swamp, and (c) from river levee to back-swamp. Superimposed on this zonation pattern are variations in salinity, soil types and the degree of inundation.

Furthermore, Van Andel (1967) noted that delta progradation became possible c. 8000 B.P., when postglacial rise in sea level decreased markedly, and, because present sea level was reached c. 3000 years ago, delta front advance had been not less than 2 km each century. A knowledge of the recent geological history of the delta is of importance in understanding the development of the habitat conditions which determine plant distribution.

Recent studies in Trinidad have supplemented Beard's (1946) account of the island's wetlands. The vegetation of Nariva Swamp does not show a regular zonation from seaward to landward, nor from the mangal through swamp forests to terrestrial rain forest, but a mosaic of scattered plant communities (Fig. 9.7).

In northwestern Nariva, the mangal, which occupies a brackish lagoonal situation behind the modern sand beach barrier, is succeeded by an extensive area in which two swamp forest types are recognizable. A widespread mixed swamp wood with *Annona glabra*, *Calophyllum lucidum*, *Crudia glaberrima*, *Pterocarpus officinalis*, *Virola surinamensis* and *Symphonia globulifera* is found directly inland from the mangroves. *Pterocarpus* is nowhere common, but occurs occasionally in small pure stands as do the other tree species. *Blechnum indicum* is occasional in the sparse ground vegetation, and the palms *Bactris major*, *Desmonchus horridus*, *Euterpe oleracea* and *Maximiliana elegans*

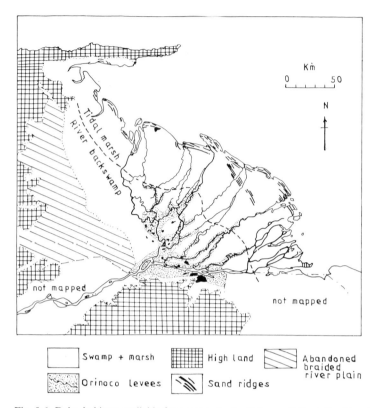

Fig. 9.6. Delta habitats available for swamp plants, Orinoco Delta (after Van Andel, 1967).

occur at intervals but do not form extensive or pure stands.

The mixed swamp wood forms a discontinuous belt of trees broken at intervals by stands of *Montrichardia arborescens*. This aroid becomes dominant over large areas to westward, although small, isolated stands of swamp forest are still conspicuous. The majority of these, and much of the western margin of the large forest area, are dominated by *Erythrina fusca* (= *E. glauca*). As *Erythrina* is the absolute dominant, and the plant form and habitat conditions are virtually identical with those reported for Surinam by Lindeman (1953), this Trinidad community is assigned to *Erythrina glauca* swamp wood. Standley (1931) reported extensive *Erythrina glauca* swamp woods in Honduras and Panama and suggested that, as this species is indigenous to all tropical America, this community may be expected in all coastal areas. Collections in the herbarium of the Office de la Recherche Scientifique et Technique Outre-Mer (ORSTOM) in Cayenne indicate, however, that this species is uncommon in French Guiana.

Some further taxonomic work appears to be required, and the distribution of the various *Erythrina* species in the American tropics should be clarified.

Machaerium lunatum is present in the mixed swamp wood and the drier parts of the mangal in Nariva Swamp, but does not form a distinct community as it does in the Guianas. Bartlett and Barghoorn (1973) reported this species as forming extensive stands in swamps in Panama, however. Small patches of mixed swamp wood are present throughout the swamp, and include pure stands of *Pterocarpus officinalis* only where standing water is present. The *Erythrina glauca* swamp wood is not found in the southern half of the swamp.

Palm-dominated vegetation is found scattered throughout the southern half of the Nariva Swamp. *Roystonea* occurs on both seaward and landward sides and *Mauritia* largely in the central and southern areas (Fig. 9.7). A stand of the *Carapa guianensis* – *Bactris major* society of Marshall (1934) and Beard (1946) occupies a depression between two areas of evergreen sea-

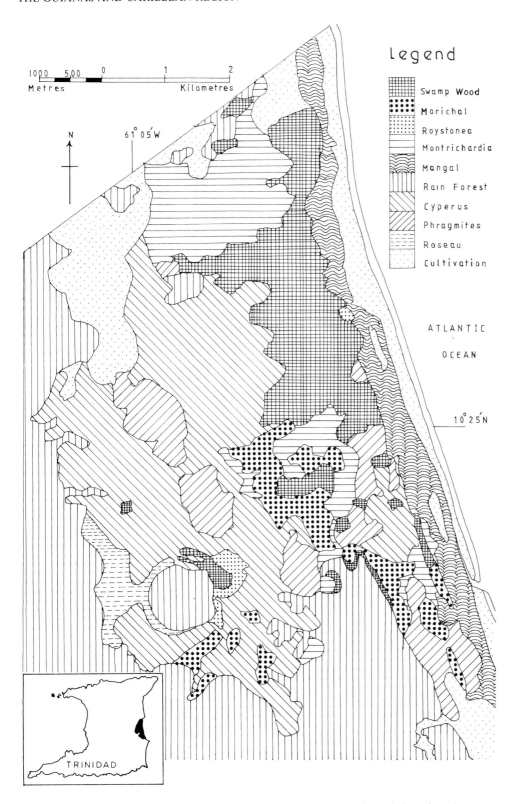

Fig. 9.7. Vegetation map of the Nariva Swamp, Trinidad (based on a map by E.K. Ramcharan).

sonal rainforest on the southwestern margin of the swamp basin.

All these woody vegetation types are associated with herbaceous vegetation, which occupies more than half the swamp. Four communities are conspicuous, three of large, rooted herbs, dominated respectively by *Montrichardia arborescens*, *Cyperus giganteus* and *Phragmites australis*, and a fourth covering open water amongst the others. Here *Leersia hexandra* and *Paspalum repens* grow over floating mats of *Eichhornia* sp., *Pistia* sp. and *Salvinia* spp.

The major parts of the *Montrichardia* and *Phragmites* stands do not dry out in the dry season, although considerable areas in the western swamp and most of the *Cyperus* area become dry, with the aerial parts of the plants dying back. These rooted herbaceous communities are difficult to classify in terms of depth of inundation, although the *Montrichardia* and *Phragmites* stands might be considered as herbaceous swamp and the *Cyperus* area as herbaceous marsh. The *Cyperus* areas of the western swamp have been affected by fire and sporadic cultivation to a great extent, however. The distribution of these plant communities is determined by the physiography of Nariva Swamp which, rather than being a featureless basin, is a complex of low ridges, hills, depressions, and channels formed through Holocene and recent depositional and erosional processes.

Speth (1961) stated that the sea level around Trinidad has remained unchanged for the past 6000 years, although a minor regression and subsequent advance occurred some 1500 years ago. This was supported by Koldewijn (1958), who noted that a fall of approximately 6 m occurred *c.* 3000 B.C., but added that there had been a recent rise of about 50 cm since 850 A.D. Bartlett and Barghoorn (1973) have discussed regional post-Pleistocene sea-level fluctuations and considered that the evidence for changes during the last 6000 years is unsatisfactory, although slightly higher-than-present levels probably occurred between 2000 and 1000 B.P. Nonetheless, old beach sand ridges and hills, such as Sand Hill and Bush Bush (Fig. 9.8), mark the position of former shorelines in the Nariva Swamp, indicating marine regression during the last few thousand years. Although not yet accurately dated, it appears that the shoreline retreated in a number of stages to the present position (Fig. 9.8) with a wide beach barrier broken only by tidal lagoonal inlets and blocking drainage from the swamp basin.

A series of north – south ridges with depressions between was thus produced, and these determine the present pattern of vegetation distribution (Fig. 9.7 and 9.8). A low hill was produced by a mud volcano at Bois Neuf in the southwest, and what appear to be low levee systems of old distributary channels have produced elevated areas in the northeast. These, like the sand ridges, have been colonized by woody communities, with herbaceous swamp and marshes occupying the lower ground surrounding them.

Previous history and present topography can be used to explain plant community patterns in Surinam (Fig. 9.9 and 9.10), as these appear to be major determining factors in the development of swamp vegetational diversity.

Palynological evidence suggests that freshwater swamps replace mangals when the sea recedes or sea-level falls (Van der Hammen, 1963). Coastal accretion may be up to 100 m annually at constant sea level on the Surinam coast (Brinkmann and Pons, 1968), so paludification and *Verlandung* on this dynamic coast are rapid processes. During the last part of the Holocene, sea level was approximately the same as today, except for minor transgressions and regressions (Van der Hammen, 1969). Sand-ridge formation took place at this time, followed by lagoon and swamp formation. Delaney (1969) reported that the older shell beach ridges were formed between 3500 and 3000 years B.P. and the younger ones *c.* 2000 to 1000 years B.P. These produced a series of ridges and depressions in the coastal lowlands supporting tree swamp formations on the elevated portions and herbaceous vegetation between (Figs. 9.9, 9.11).

Where rivers cross these swamps, further complexity is introduced by fluviatile processes such as levee formation. Estuarine portions of rivers, such as the Nickerie (Fig. 9.10), are colonized by mangroves. When the levees are produced, swamp forest displaces the mangal and develops into marsh forest as the levees mature. The levee deposits show soil ripening and mottling as leaching, oxidation and desalination progress. Brinkmann and Pons (1968) thought that the type of vegetation present influenced soil formation on the levees, but it is more likely that the elevation

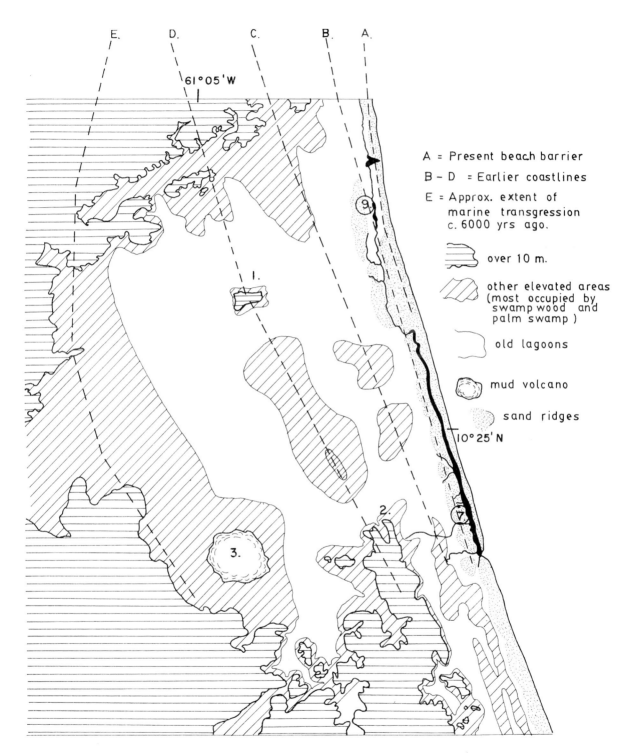

Fig. 9.8. Elevated areas of the Nariva Swamp, Trinidad. *1* = Sand Hill; *2* = Bush Bush Island; *3* = Bois Neuf; *7* and *9* = water sample stations from Table 9.3.

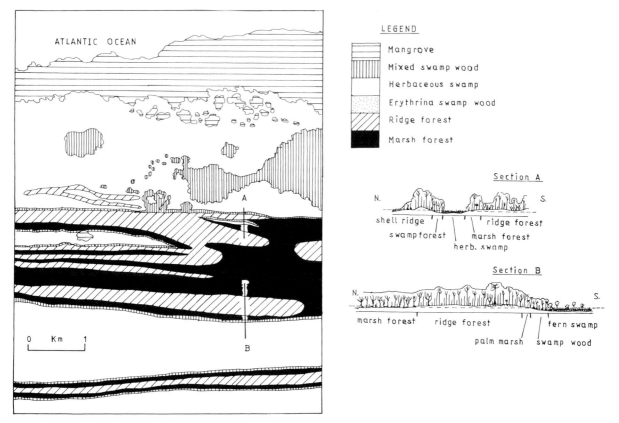

Fig. 9.9. Vegetation map of the Coronie District, Surinam (after Lindeman, 1953).

and degree of soil ripening determine the nature of the colonizing vegetation. The younger levees along the present Nickerie River support mixed swamp wood, but south of Wageningen (Fig. 9.10) the higher levees and the abandoned levees of Moleson age (2500–1300 B.P.), which may be up to 1 m above the surrounding swamp clays, support Lindeman's (1953) marsh forest type.

In the coastal lowlands of the Guianas, the Orinoco Delta, and Trinidad, *Verlandung* produces a variety of marine and fluvial habitats which are responsible for the complex, mosaic nature of the wetland vegetation patterns. The relationships between the plant communities there can be deduced from the sequence of changes which take place as the swamps develop.

The physico-chemical environment

Data on the nature of flood water and soils in regional swamp environments are difficult to assess as they are mostly imprecise. Even those from the major study by Brinkmann and Pons (1968) of the Holocene sediments of the Guiana coastal plain do not distinguish between soils in terms of major swamp habitats or plant-community types.

According to Lindeman (1953), *Erythrina glauca* swamp wood can grow in saline areas, and the position of *Pterocarpus* forests immediately behind the mangal (Fanshawe, 1952; Müller, 1959) implies that *Pterocarpus* is capable of growing in brackish situations (but see Ch. 10). De Wit (1960) stated that mixed swamp wood soils required desalination before they were suitable for rice in Surinam. Some aspects of the woody swamp environment have been analysed in Guadeloupe by Febvay and Kermarrec (1978a), who found salinity in the *Pterocarpus* zone less than 15‰, although varying with seasonal freshwater input.

Spot-sampling is of little value because water is rarely stagnant in swamp forests throughout the year and flushing is normal with seasonal flooding. Analysis of water flowing in channels through mixed swamp wood in Trinidad (Bacon et al.,

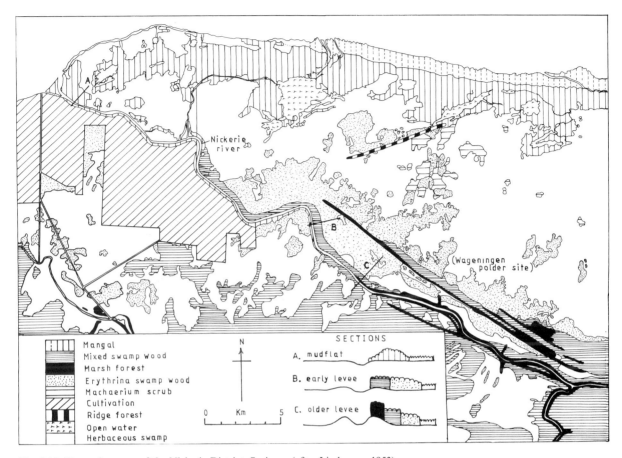

Fig. 9.10. Vegetation map of the Nickerie District, Surinam (after Lindeman, 1953).

1979) showed that, although freshwater conditions predominated, brackish water was present in the dry season. Acidity increased at this time, suggesting penetration of seawater into the swamp as it dried out. Slight seasonal and habitat differences in pH levels were reported by Aitken et al. (1968) in Bush Bush, Trinidad, with swamp forest ranging from 6.4 (wet season) to 6.0 (dry season) compared with 6.2/6.2 in the mangrove zone and 6.4/6.05 in open herbaceous swamp regions.

Compared with Trinidadian mangrove swamps (Bacon, 1970), levels of total iron, phosphates and silicates in mixed swamp wood were higher. Dissolved oxygen was lower, and frequently reached levels which Carter and Beadle (1930) found lethal to South American swamp animals. Dissolved oxygen and nutrient levels showed an increase in June and July, which was probably correlated with flooding of the swamp at the start of the wet season. No information was available on the effects of water chemistry on woody swamp plants, however.

No significant differences in chemistry were found between water from mixed swamp wood channels (Table 9.3) and that flowing from other

Fig. 9.11. Two old beach ridges with *Mauritia* separated by a depression with herbaceous vegetation, French Guiana.

TABLE 9.3

Chemical content (mg l^{-1}) of water in channels through swamp forest, Nariva Swamp, Trinidad, 1978

Month	pH	Specific conductivity µS cm^{-1}	Methyl orange (CaCO$_3$)	Methyl orange (HCO$_3$)	Total hardness	Calcium (Ca)	Total iron (Fe)	Soluble iron (filtered)	Magnesium (Mg)	Chlorides (Cl)	Phosphate (PO$_4$)	Silicates (SiO$_2$)	Dissolved oxygen (O$_2$)
(a) *Station 7* (mixed swamp wood and morichal)													
Jan	6.43	1010	42	51	790	20.0	3.30	0.590	179.8	470	0.025	6.4	1.05
Feb	6.94	—	105	128	3600	200.0	1.41	0.307	753.0	10 800	0.063	5.0	1.20
Mar	8.05	—	123	150	6250	320.0	0.25	0.007	1324.3	11 500	0.040	0.5	5.10
Apr	8.02	46 000	127	154	7300	440.0	0.34	0.020	1507.0	20 600	0.004	1.3	6.21
May	6.31	2600	37	45	1107	91.0	2.10	0.630	213.8	3480	0.008	7.7	0.85
Jun	6.41	320	44	53	66	16.0	5.04	2.400	6.4	48	0.280	9.0	0
July	6.30	230	34	41	32	11.2	3.11	1.410	1.0	35	0.099	5.4	3.69
Aug	6.73	178	43	52	30	5.6	1.82	1.312	3.9	38	0.039	6.9	0.60
Sept	6.52	1125	39	47	52	8.0	—	—	7.8	100	0.015	—	0.30
Oct	7.51	140	33	40	38	12.0	1.66	1.100	1.9	34	0.030	—	0.78
Nov	6.03	240	60	73	40	9.6	0.42	0.350	3.9	72	0.010	—	2.10
Dec	6.30	540	39	47	66	8.8	1.00	0.705	10.9	56	0.068	—	0.48
Mean	6.79	5238	60	73	1614	95.2	1.86	0.803	336	3936	0.056	5.3	1.86
(b) *Station 9* (Mixed swamp wood and mangrove)													
Jan	7.56	1800	88	106	336	39.2	1.26	0.530	57.8	500	0.028	11.5	3.25
Feb	6.74	—	65	79	60	16.0	2.52	0.500	4.8	48	0.040	7.1	1.00
Mar	6.87	—	55	67	69	16.0	2.12	0.047	7.0	60	0.004	6.5	0.85
Apr	6.89	420	68	82	73	18.0	2.03	0.230	6.8	77	0.020	7.7	2.08
May	5.64	1400	30	36	445	110.0	5.88	0.610	41.3	1040	0.018	9.0	0.28
Jun	6.27	390	77	93	84	24.0	6.59	3.140	5.8	51	0.190	13.5	0
July	6.50	355	55	67	70	20.8	6.13	4.450	4.4	64	0.172	18.0	3.49
Aug	6.90	280	192	234	72	17.6	2.67	1.930	6.8	58	0.050	11.5	0.20
Sept	6.57	270	56	69	57	16.8	—	—	3.6	49	0.010	—	0.45
Oct	7.25	230	57	69	76	18.4	4.69	2.760	7.3	51	0.030	—	1.12
Nov	6.33	285	64	78	74	21.6	1.11	0.490	4.9	66	0.010	—	0.40
Dec	6.64	410	70	85	86	16.0	1.72	1.064	11.2	57	0.033	—	0.38
Mean	6.6	448.88	71	87	106	26.8	3.55	1.582	9.5	147	0.052	10.5	0.93

types of vegetation. This was expected as there was little separation hydrologically between different areas of the Nariva Swamp, except in restricted areas during the season of lowest water. Higher chloride levels at station 7 were probably related to the proximity of a small tidal creek through the barrier beach. All channels draining the Nariva Swamp carried black water, rich in humic acids but with little sediment, as Van Andel (1967) reported for back-swamp streams in the Orinoco Delta. Koldewijn (1958) pointed out that negligible amounts of mineral matter were brought to the littoral region of east Trinidad by rivers such as those draining Nariva, the major part of the suspended material derived from the western catchment being deposited in the swamp basin.

Soil ripening leading to colonization by progressively more terrestrial plant communities is a complex process. Augustinus and Slager (1971) recognized two swamp soils following initial colonization by *Avicennia*; those that were always wet, and those that dried out periodically. They found that soil compaction due to ripening depended on the depth and intensity of desiccation, the length of the desiccation period each year, and the number of years in which alternate drying and wetting took place. In general, this was a function of the fluctuations in groundwater and surface water levels, these being determined physiographically either by the absolute height of the swamp above sea level or by the degree of enclosure of ridges and barriers. Understanding this ripening process requires long-term records of inundation and drying which are not available in the region.

Further data on regional swamp soils are summarized in Table 9.4. Although giving a general idea of soil differences under the main plant communities, it shows some degree of overlap and indicates that certain swamp types, such as marsh forest, can grow on a range of soils.

The very detailed account by Léveque (1962) of lowland soils in French Guiana, although not classifying vegetation types at the sample sites, does list the dominant plant species. Communities with *Symphonia globulifera* and *Virola surinamensis* which were inundated in the dry season grew on pegasse on the banks of the River Kaw, on sulphurous soil on a thin pegasse in what appears to be a back-swamp area near the River Cour-

TABLE 9.4

Soil conditions in forested swamp habitats

1. *Levee forest*
    Mottling (De Wit, 1960)
    Ripened clays (Van Andel, 1967)
    Silty clay (Brinkman and Pons, 1968)
    Ripening, mottling, humus thin
      or absent (Lindeman, 1953)
3. *Mixed swamp wood and Pterocarpus forest*
    Grayish clay covered with 30–50 cm
      soft humic clay (Müller, 1959)
    Thick pegasse, high N-content (De Wit, 1960)
    Pegasse (Brinkmann and Pons, 1968)
    Low total sulphur and organic
      sulphur (Chabrol, 1978)
    Organic matter to 30 cm; pH 6.5;
      water content 350%; oxidation to 1 m;
      augmented nitrogen content (Turenne, 1978)
6. *Marsh forest*
    Alluvial silt or pegasse (Fanshawe, 1952)
    Structureless peat; black-brown layer rich
      in humus; impermeable sub-soil (Lindeman, 1953)
    River clay (Léveque, 1962)
    Gley with 4% organic material (under prolonged inundation);
      pseudogley with high organic content (short inundation)
      (Blancaneaux, 1976)

2. *Erythrina swamp wood*
    Saline (De Wit, 1960)
    Stiff clay (Müller, 1959)
    Surface humus thin
      (Lindeman, 1953)
4. *Machaerium lunatum scrub*
    Soft organic mud
      (Lindeman, 1953)
5. *Morichal*
    Soft peaty soil (Müller, 1959)

rouaie, and on the clay banks of the Inery Creek. An area dominated by *Pterocarpus* at the confluence of the River Orapu and the Boulanger Creek had sandy mud on buried peat.

In the coastal area southeast of Cayenne (Fig. 9.12), the distribution of swamp communities shows some relationship to soil type, particularly thickness of the pegasse layer (Table 9.5).

Müller (1959), Lindeman (1953) and Brinkmann and Pons (1968) all found peat layers or thick pegasse below herbaceous swamp communities.

More information on the requirements of the individual swamp plants is needed before the influence of soil type on plant community composition and distribution can be determined, although the influence of geological history and topography on soil maturation is apparent. De Wit's (1960) suggestion that the high nitrogen content of mixed swamp wood soils may be due to enrichment by legumes may help explain why this family plays such an important role in the ecology of wetlands in this region.

Successions

The development of swamp conditions

The successional development of swamp vegetation has been implied above in relation to zonation patterns and changes in soil type, but needs further clarification.

It seems logical that there should be development from open water through increasing elevation to the formation of dry land, whether on accreting areas of sea coast or in lagoons or freshwater swamps. Richards (1979) pointed out that the early stages of tropical hydroseres are dominated by herbaceous vegetation, and later taken over by woody vegetation to give a succession from open water to herbaceous swamp to herbaceous marsh to swamp forest to mesophytic forest. In Neotropical situations the first colonizers

TABLE 9.5

Vegetation and soils in coastal swamps near Cayenne (after F. Colmet-Daage and E. Sieffermann, Office de la Recherche, Scientifique et Technique Outre-Mer, Cayenne, Guyane, 1953)

Dominant vegetation	Soil type	Soil characteristics	Land use
Avicennia	1	Soft, saline mud	Unusable
Avicennia – Rhizophora	2	Consolidated mud, saline throughout profile	Difficult to use
Avicennia – Rhizophora – Euterpe	3	Grey consolidated clay, slightly saline at surface, saline below 40–50 cm	Little present value
	4	Grey or grey-blue clay, consolidated — firm, very slightly or non saline at surface, saline below 1 m	—
Euterpe – Symphonia – Pterocarpus – Virola – Carapa	4a	with very little pegasse, only 5–10 cm	Good soil — cacao, citrus, bananas
Euterpe – Virola – Chrysobalanus	4b	with 30–50 cm pegasse	Excellent soil — cacao, citrus, bananas, drainage difficult
Chrysobalanus	4c	with 50–100 cm pegasse	Good — moderate soil, drainage difficult
Chrysobalanus – Cyperaceae	4d	with more than 1.2 m pegasse	Poor — drainage very difficult
Euterpe – Carapa – Symphonia	5	Grey clay on river borders, partially oxidised, with yellow staining. Non-saline to 1.5 m depth	Good soil — citrus, cacao. Drainage easy, land clearance costly

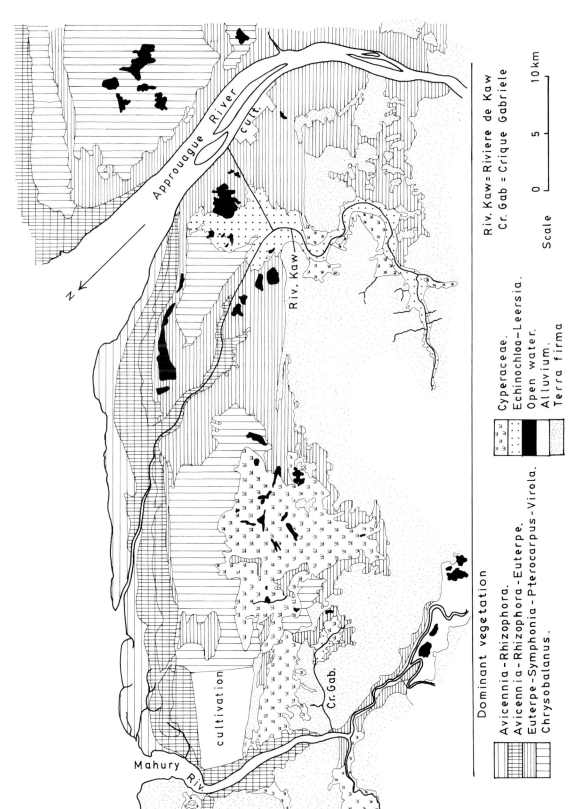

Fig. 9.12. Vegetation map of the area between the Mahury and Approuague Rivers, French Guiana (after F. Colmet-Daage and G. Sieffermann, Office de la Recherche Scientifique et Technique Outre-Mer, Cayenne, French Guiana, 1953).

of open water, or mudflats, on the coast are generally mangroves, so the herbaceous stages may be omitted and the mangroves pass via freshwater *Pterocarpus* swamps to mesophytic forest, as Holdridge (1940) found in Puerto Rico (see also Ch. 10).

Van der Hammen's (1974) palynological studies in tropical South America showed a sequence from *Rhizophora* through *Avicennia* and palm swamp forest to grassy savanna on regressive or stable shorelines as the wetlands dried out.

Lindeman's (1953) vegetation map of the Nickerie District (Fig. 9.10) and the Nariva Swamp map (Fig. 9.7) do not show such a sequence of vegetation types from sea to land, however; and Gleason and Cook (1926) pointed out that, when damming of a coastal area by sand bars forms a swamp, there may be increased flooding rather than gradual drying out of the alluvial land. Furthermore, Blancaneaux (1972) found that, where sand bars had blocked water movement in the Savanne Sarcelle in northwestern French Guiana, there had been accumulation of peat capped by herbs in the deeper inland marsh.

J.P. Lescuré (pers. commun., 1979) had pointed out an area south of Cayenne where extension of the mangrove coastline had been followed by increased backing-up of river water under tidal action, causing increased flooding. Savanna areas near the Crique Gabrielle (Fig. 9.12), cultivated by the Jesuits in the seventeenth century, were now deeply flooded and covered by tall herbaceous swamp with *Montrichardia*. The surrounding marsh forests were changing to swamp forests as the frequency and depth of inundation increased.

In this same region of French Guiana (Fig. 9.12), the mangal on rapidly accreting coastal mudflats is followed inland by freshwater swamp forest. This passes into herbaceous vegetation before meeting rain forest on the more elevated land. Increase in both water depth and pegasse thickness occurs from the mangal to the freshwater back-swamps (Table 9.5). An *Avicennia – Rhizophora – Euterpe* association near the coast appears to be replaced by a swamp forest type in which the characteristic "pinot" palm, *Euterpe oleracea*, is associated with *Carapa guianensis*, *Pterocarpus officinalis*, *Symphonia globulifera* and *Virola surinamensis*. Although *Euterpe* is conspicuous and characteristic of this association, *Pterocarpus* and *Symphonia* are sufficiently common to make the community comparable with mixed swamp wood in Surinam and Guiana. With increase in soil pegasse there is replacement inland by communities dominated by *Chrysobalanus icaco*, *Euterpe* and *Virola*, then by *Chrysobalanus* alone, which merges into a community of *Chrysobalanus* with Cyperaceae, mainly *Scleria* sp. Where the inundation period is extensive, *Echinochloa polystachia*, *Leersia hexandra* and *Panicum purpurescens* are dominant, with patches of *Genipa americana*, *Montrichardia arborescens* and *Pterocarpus*. These areas of herbaceous swamp pass into narrow zones of swamp forest with *Euterpe*, *Mauritia* or *Pterocarpus*, which fringe the hinterland forests.

On more sandy levee areas beside large rivers, such as the Mahury and Approuague (Fig. 9.12), a *Euterpe – Carapa – Symphonia* forest occurs. The smaller Rivière de Kaw flows through an extensive area of herbaceous swamp dominated by *Echinochloa* and *Leersia* which developed in a river channel dominated by the effects of backed-up fresh waters. At the wet limit of the swamp forest, in the back-swamps and along the margins of channels like the Crique Gabrielle, a zone of *Montrichardia* is common. Where levees are developing along the lower course of the Mahury River, stands of dying, epiphyte-covered *Avicennia* occur behind a narrow gallery forest with vines and patches of *Acrostichum aureum*.

The largest areas of swamp in Colombia occur between the distributaries of the lower Magdalena River. On the southwest of the Cienaga Grande, mixed swamp woods of *Bombax*, *Pterocarpus*, *Symphonia* and *Virola* were observed, interspersed with morichales and wide areas of herbaceous vegetation. Beach barrier and levee systems have led to back-swamp flooding, in which open water and herbaceous swamp have extended at the expense of mangroves and swamp forests. Kaufmann and Hevert (1973) found extreme flooding of the Cienaga Grande to be a cyclic phenomenon, with a period of 6 to 7 years, which caused mortality among swamp organisms such as the commercial oyster species.

In the Nickerie district (Fig. 9.10) the presence of a beach barrier had caused flooding behind the mangroves, so it appears that mangrove and possibly swamp forests have been replaced by herbaceous vegetation which is able to grow in deeper water. Such back-swamp production has occurred also in Nariva, Trinidad (Fig. 9.5), so that the most deeply and most permanently inundated areas occur behind the mangrove and swamp forest zones, and are occupied by *Cyperus*, *Montrichardia* and *Phragmites*. A similar situation exists in the North Oropuche Swamp in Trinidad, where herbaceous vegetation occupies the landward swamp basin.

The situation in fluviatile regions is similar. As levees become higher, flooding increases behind them, and increasingly deeper water becomes trapped for longer periods. The swamp vegetation on mudflats is gradually replaced by herbs, as indicated in Section C and the area south of this in Fig. 9.10. A levee forest develops from a swamp type through marshy conditions to terrestrial forest as the levees rise and their soil dries out. This fluviatile situation shows similarities to the várzea – igapó complex of the Amazon basin (Sioli, 1975; Junk, 1983).

Müller (1959) described how levee formation led to the production of inundated back-swamps between the Orinoco distributaries (Fig. 9.4). In these areas, when dry-season river levels fell, the levees were exposed but back-swamps remained waterlogged for a longer portion of the year. The morichales occupied seasonally inundated regions, while herbaceous swamp of *Montrichardia* or floating grasses and stunted trees and palms replaced these in areas of permanent inundation and marked fluctuations in depth, as in the central delta region.

The succession in a flooded back-swamp can only proceed when accumulation of peat raises the soil level. The herbaceous or palm swamp communities on this pegasse will pass through marsh communities to savanna or mesophytic forest as elevation and drying-out increase.

A hydrosere succession may thus move from the early colonization stage to deeper water conditions before entering the phase of elevation and drying out (Fig. 9.13), as the hydrologic sequence may not proceed only in one direction as Richards' (1979) simple scheme suggests. The point of successional initiation is important for the subsequent course of the vegetational succession.

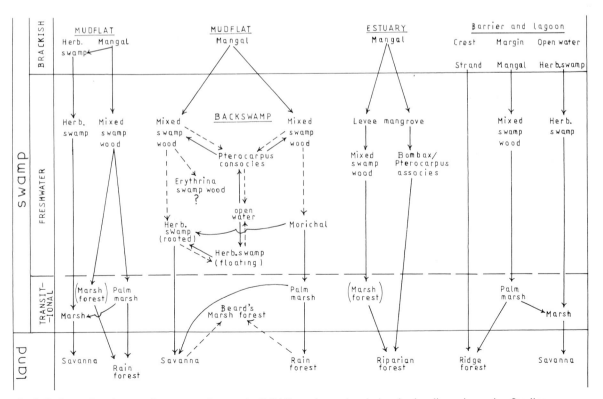

Fig. 9.13. Successions in coastal swamp environments. Solid lines = increasing drying; broken lines = increasing flooding.

Seedling establishment

A critical factor in the successions is seedling establishment by swamp plants. The development of the brackish *Rhizophora* swamp forest in inundated situations is possible because of specialized propagules. None of the freshwater swamp forest trees appears to have comparable adaptations.

Although *Pterocarpus officinalis* is characteristic of inundated situations, it is incapable of setting seed under these conditions. The winged fruits (Fig. 9.14) can germinate when afloat but rooting does not occur in water depths greater than 3 to 4 cm (see also Ch. 10, p. 261). Guppy (1917) found that the water-dispersed seeds of the common swamp trees *Andira inermis*, *Annona palustris*, *Carapa guianensis*, *Manicaria saccifera* and *Symphonia globulifera* would germinate when afloat, but some do not last long once germinated. For instance, *Symphonia* seeds have fleeting vitality, while *Andira* seeds sink readily. Whether they will root in deep water is unknown, but it is more likely that these species normally establish seedlings during periods of seasonal drought, rather than under permanent inundation. Establishment is possible on elevated areas, such as mudflats, sand bars, and levees, in the low-water season, or by stranding along the margins of channels, these being the habitats normally occupied by swamp forest trees. In Nariva Swamp, germinating seeds of *Euterpe*, *Pterocarpus* and *Symphonia* were found stranded among *Montrichardia* stems in the stands surrounding the trees, but rooting was not observed. Lescuré (1980) noted that litter was trapped and transformed to soil between the buttresses of *Pterocarpus* in Guadeloupe, thus raising the local soil level. This process may allow some seed germination if raised above water level. Hummocky ground is a feature of mixed swamp wood soils (Fanshawe, 1952) and these small increases in elevation must be important in diaspore stranding, as is the density and form of old root networks. Alvarez-Lopez (Ch. 10, this volume) has prepared a model to explain differences in seedling establishment by *P. officinalis* under flooded and dry soil conditions.

Information on fruiting and seed production in swamp trees in the Caribbean and Guiana region is minimal. Beard (1946) found that there was a preponderance of dry-season fruiting in seasonal evergreen forest trees in Trinidad, many swamp species occurring in these forests. Janzen (1975) noted that *Ceiba*, *Pterocarpus*, *Spondias* and *Tabebuia* all flowered when trees were leafless in the dry season, and produced mature fruits in less than 1 month. He suggested that this was a strategy for avoiding seed predation — although, as is noted below, wetland forest areas are most accessible to animals in the low-water season. Flowering in *Tabebuia serratifolia* is promoted by water stress (Hallé et al., 1978), and this would be a significant factor at the start of the dry season for marsh forest trees, although of less importance in inundated swamp species.

The report by James (1979) that in Dominica *Pterocarpus* bears fruit from April to November, and that parrots feed on the fruit in February, should be treated with caution as it suggests year-long reproduction. Fruit fall from *P. officinalis* was recorded in the Salybia and Nariva swamps, Trinidad during May and June 1979, but fruit were found in a relatively undecayed condition floating or on the ground during most of the year. In Puerto Rico, *P. officinalis* was observed to have two fruiting periods: from March to June, and from late June to September (Ch. 10).

Although more information is required, fruiting in swamp trees during the dry season appears to be essential if seedling establishment is to occur in low-water conditions. Adams (1972) reported *P. officinalis* in Jamaica to flower from July to August and to fruit from July to September.

A contrary view was put forward by Gottsberger (1978), who found that water dispersed fruit

Fig. 9.14. *Pterocarpus* fruits.

ripened in the high-water season in Amazon várzea when ichthyochory was most likely. He proposed that seeds of swamp plants with edible fruits, such as *Virola surinamensis*, were dispersed by fish, which were major agents in producing monospecific stands of vegetation in inundated regions. While ichthyochory may account for seed dispersal, it does not explain how the dispersed seeds are able to germinate in inundated situations.

The *Pterocarpus officinalis* consociation

Pterocarpus is a conspicuous element of swamps throughout the region (Fig. 9.1), although not confined to it. It is reported from many plant communities in a habitat range from deeply inundated pegasse (Léveque, 1962) to well drained, upland clay soils with thin litter (Knight, 1975). This apparent broad niche occupancy may in part be due to the misidentification of *P. rohrii* Vahl. or *P. santalanoides* L'Herit. whose distribution in the region is poorly documented, although Knight (1975) found that many of these tree species in Panama had considerable tolerance for a range of soil conditions. It may be that, as with *Nyssa sylvatica* in the southern U.S.A. (Keeley, 1979), morphological and physiological adaptations produce ecotypes capable of survival in swampy, marshy and lowland forest habitats.

It appears that *P. officinalis* Jacq. is one of the better adapted of the swamp trees, as a distinct consociation is recognisable only in deep-water habitats. In the southern Nariva Swamp, water in excess of 1 m was recorded in one *Pterocarpus* stand throughout the year, with greater flooding occurring in the wet season.

Recruitment and population maintenance by *P. officinalis* in habitats at the extremes of inundation cannot be brought about by seedling establishment unless the environmental conditions change drastically. It appears, therefore, as a terminal community in the swamp succession by fringing, but unable to further colonize open water, or by competing with herbaceous species able to reproduce vegetatively in deep water (Fig. 9.13). The consistent association of *Pterocarpus* and fringing *Montrichardia* in Trinidad and the Guianas is noteworthy.

Lindeman (1953) noted that swamp forest may become superficially dry in excessively dry years, and he considered that the establishment of trees in a swamp required a series of conditions fulfilled but once in several years and then only locally. Superimposed on annual dry-season changes are long-term fluctuations in rainfall regime such that the degree of flooding in swamps will be highly variable. Wijmstra (1967) reported such phenomena in the pollen diagram of the lower Magdalena Valley, Colombia, with forests extending along rivers and caños during these periods.

Such sporadic occurrence of suitable germination conditions, if very infrequent, should produce tree stands of uniform age. Not all the trees were old or of the same size in Trinidadian *Pterocarpus* stands. Individuals of from 19 to 149 cm (dbh) were measured in one area, and it is suggested that the permanence of such stands is maintained by vegetative reproduction.

Pterocarpus officinalis shows strong resprouting from the roots (Figs. 9.15 and 9.16), occasionally producing several new stems. Lescuré (1980) found with *Pterocarpus* forests in Guadeloupe that most of the trees showed "reiteration" or resprouting on reaching 5 to 10 cm diameter, at which size the old trunk rotted, thus influencing the individual and population architecture. He thought hurricane damage a possible cause of the resprouting, this being an important ecological factor in the Antilles but not in French Guiana, where he believed resprouting did not occur. He considered, however, that the relative stability of the Guadeloupe swamps compared to the rapidity of sedimentation along the Guianese rivers was possibly of greater

Fig. 9.15. Mature *Pterocarpus* showing regrowth, Salybia, Trinidad.

Fig. 9.16. Regrowth in *Pterocarpus*, drawn from life at Salybia, Trinidad, June 1979.

importance in influencing the architectural pattern of *Pterocarpus* forests.

In Dominica, where James (1979) noted that forked trees were common and that *Pterocarpus* had a high coppicing rate, the architecture of the swamp forests may result from cutting timber. Logging has been more prevalent in Guadeloupe than in the swamp forests of French Guiana and may explain differences between these two areas. Strong resprouting from the roots was, however, observed in an inundated *P. officinalis* stand near Sinnamary, French Guiana, in July 1979.

The production of new stems, sometimes several, from an established trunk was observed also in *Euterpe oleracea* in French Guiana and in *Erythrina glauca* in Nariva Swamp, Trinidad. *Erythrina* has the ability to maintain small stands in inundated areas surrounded by *Montrichardia*, resprouting occurring readily from fallen stems (Fig. 9.17), while seedling germination is minimal. In Surinam, Lindeman (1953) pointed out that *Erythrina* inhibits germination of other species in the stands and appears to grow in deeper water than the surrounding *Montrichardia* as it prevents peat accumulation. Although these processes are poorly understood, the production of secondary stems, which he also reported, is important for the continued existence of *Erythrina* in inundated environments. Like *Pterocarpus*, *Erythrina* forms a distinct, virtually monospecific, forest type only in deeply inundated situations, although it rarely occurs elsewhere. Janzen's (1974) suggestion that monospecific stands of swamp trees, such as those of *Parkinsonia aculeata*, *Prioria copaifera* and *Pterocarpus officinalis* in Costa Rica, resulted from reduced predation through resins, toxic woods and protection of their seeds by water, while undoubtedly important, does not explain how stands are

Fig. 9.17. Regrowth in *Erythrina glauca*, drawn from life in Nariva Swamp, Trinidad, June 1979.

maintained in inundated situations where unfavourable germination conditions prohibit recruitment to the population.

Vegetative reproduction in *Pterocarpus* and *Erythrina* allows these tree species to occupy extremes of the swamp habitat otherwise occupied only by herbaceous plants showing similar adaptations. The response to changing inundation conditions thus determines the successions and the resulting distribution of vegetation communities in freshwater swamps (Fig. 9.13).

The terms "swamp" and "marsh"

Although Beard (1946) found it useful to retain the terms "swamp" and "marsh" for description of existing ecological conditions, he thought that some swamp forest communities should be considered as facies of the evergreen seasonal forest growing under extreme edaphic conditions, thus rendering separate terminology unnecessary. Later (Beard, 1955), he revised his classification to place marsh into "seasonal swamp", so his original marsh forest became "seasonal-swamp forest with palms". He retained the distinction between "swamp" and "seasonal swamp", however, and noted that the *Amanoa* consociation of Dominica, originally described as a swamp phase of Montane Thicket, should be considered as "seasonal-swamp woodland".

As establishment and reproduction of woody swamp vegetation largely requires periodic low water conditions, the separation of swamp forest from marsh forest on grounds of permanent or seasonal inundation is questionable.

Mixed swamp wood and palm swamp would have been established initially under waterlogged or marshy conditions, and so could have been classified at that time as marsh forests. Later, following changes in the hydrologic regime or back-swamp flooding, inundation would transform these into swamp forests with different environmental and floristic characters. Those species with special adaptations for survival in more or less permanent water cover, such as the pneumatophores and respiratory prop-roots which give the characteristic physiognomy to swamp forest (Figs. 9.18 and 9.19), would dominate the vegetation increasingly.

The suggestion is made here that inundation conditions are likely to change from seasonal to

Fig. 9.18. Roots of *Pterocarpus officinalis*, Salybia, Trinidad.

Fig. 9.19. Knee roots of *Symphonia globulifera*, O'Meara Swamp, Trinidad.

more permanent during the lifetime of particular swamp forest stands. Further studies on inundation depth and duration in relation to floristic composition would probably show a gradual reduction in the number of tree species with increasing flooding and vice versa, thus making a division between marshy-ground and swampy-ground communities indistinct.

One can, thus, return to Burtt Davy's (1938) scheme and classify all periodically inundated Neotropical forests as swamp forest, whether they occur in a swamp basin or in localised, waterlogged, upland situations — a scheme supported by Fosberg (1958). This approach removes many of the difficulties in the swamp/marsh distinction introduced and used by Beard (1946), Fanshawe (1952) and Lindeman (1953). The marsh forest formations of Fanshawe (1952), Lindeman (1953) and Vann (1959) are best considered as ecotones between swamp forest and the terrestrial communities. They could be omitted from the suggested succession diagram (Fig. 9.13). Beard's (1946) marsh forest is a result of a type of back-swamp flooding and should be considered a transitional facies, the direction of development depending on whether local drying or flooding is taking place.

In the simplified classification suggested (Table 9.6), a number of vegetational units can be grouped together without disguising the variation within and between regional swamp forest habitats. The *Machaerium lunatum* scrub is largely considered as a brackish community and is omitted, while the relationships of the *Erythrina glauca* swamp wood type to other swamp forests is unclear.

A further objection to the use of the term "marsh" for a forest type is the restriction of this term in temperate regions to herbaceous vegetation. The regional *Cyperus* and *Phragmites* communities are floristically and edaphically marshes in this sense, and communities with these herbs and scattered *Mauritia* have been classified here as palm marsh (Fig. 9.13).

Herbaceous communities are considered only superficially in this paper, but those dominated by *Montrichardia*, and some *Cyperus*, *Phragmites*, and *Typha* communities, occur commonly in more permanent deep water. Provisional investigation of *Montrichardia* stands in Trinidad suggests that, compared with *Pterocarpus* and other swamp trees, vegetative reproduction is more important for establishment and maintenance of this species. It can, thus, colonize open water, as do floating species like *Eichhornia* and *Pistia*, and contrasts markedly with the more typical marsh plants. The designation herbaceous swamp is retained here (Fig. 9.13) until more detailed ecological data are available.

HUMAN INFLUENCE ON SWAMP VEGETATION

Throughout the region, human influence on swamp vegetation has been significant. There is evidence of Amerind occupation of sand ridges in the beach areas bordering Nariva Swamp since c. 810 B.C., with cultivation of the drier areas and hunting in the freshwater swamp (Newson,

TABLE 9.6

Classification of freshwater swamp forests

A. Mixed swamp wood
    1. *Annona glabra – Andira inermis – Triplaris surinamensis* type
    2. *Symphonia globulifera – Virola surinamensis* association
    3. *Symphonia globulifera* type
    4. *Bombax aquaticum – Pterocarpus officinalis* association (Riparian Swamp Woodland)
    5. *Triplaris surinamensis – Bonafousia tetrastachya* type
    6. *Annona glabra – Dalbergia ecastophyllum* association (Martinique)
    7. *Amanoa caribaea – Oxythece pallide* association (Dominica)
    8. *Pterocarpus – Pavonia scabra* facies (Guadeloupe)
       Pterocarpus – Inga laurina facies (Guadeloupe)
       Pterocarpus – Tabebuia pallida facies (Guadeloupe)
    9. *Crudia glaberrima* society (Trinidad)
    10. *Pterocarpus officinalis* consociation

B. *Erythrina glauca* swamp wood

C. Palm swamp
    1. Morichal – *Mauritia flexuosa* consociation
       M. setigera consociation (Trinidad)
    2. *Mauritia – Chrysobalanus icaco* association
    3. Mixed palm swamp — *Euterpe – Manicaria* type
    4. *Pterocarpus – Bactris minor* society (Honduras)
    5. *Carapa guianensis – Bactris major* society (Trinidad)
    6. *Roystonea oleracea* consociation (Trinidad)

1976). Amerind settlement in the Guianas is much older, and some of the small mounds they constructed for village settlements in inundated coastal swamps may now be occupied by isolated forest stands. The Indians cultivated the ridges, and Blancaneaux (1976) found depauperate secondary scrub on these near the Maroni River, in western French Guiana, which reduced seedling availability to neighbouring areas of swamp.

Fire has tended to encourage herbaceous plants at the expense of trees. Fanshawe (1952) considered his palm marsh to be fire-affected; Lindeman (1953) reported that a fire in Guiana in 1837 converted an area of *Mora* and *Mauritia* to open grass and reed swamp which had persisted until his examination in 1933. Frequent dry-season fires, associated with temporary cultivation in the southern Nariva Swamp, have cleared large areas of *Roystonea* and have been responsible for the retreat of a morichal known from photographs to have been more extensive 10 years previously (Fig. 9.20). Once herbaceous growth occupies such an area, regeneration of trees is unlikely to occur.

Great care must be taken when interpreting swamp vegetation in the region, particularly with respect to degree of inundation and forest composition and architecture, because drainage and damming for agriculture and irrigation, introduction of weed species, and fire effects are superimposed on natural processes responsible for swamp development.

Fig. 9.20. Fire-affected morichale, Nariva Swamp, Trinidad.

FAUNA

The small amount of information published on freshwater wetland faunas in the Caribbean and Guiana region has not permitted detailed analysis.

Investigations of the macrofauna of the 24,000 ha Nariva Swamp, Trinidad, have produced a faunal checklist of over 600 species (Aitken et al., 1973; Bacon et al., 1979). These can be divided into two broad categories: (a) the aquatic animals; and (b) those found on raised ground, or on vegetation, or seasonally during dry periods, which are referred provisionally to the non-aquatic category.

The majority of the aquatic macrofaunal species (Table 9.7) were found in other freshwater environments throughout Trinidad, particularly in the floodplains of local rivers. It appears that the swamp fauna cannot be readily distinguished from that of other freshwater environments.

The mussels *Anodontites leotaudi* and *Mycetopoda pittieri* have been collected only from Nariva Swamp, but *A. leotaudi* is probably synonymous with *A. irisans* from rivers in northern South America, and *M. pittieri* has been described from inland rivers in Venezuela (Marshall, 1926). The three species of conch (*Pomacea*), and most of the fish, frogs and turtles are found also in the Guianas and Venezuela (Pain, 1950; Kenny, 1969; Lowe-McConnell, 1975; Singh, 1978; Fretey and Renault-Lescuré, 1978), suggesting that a degree of uniformity should be expected in the faunal composition of Neotropical swamps, although the complex palaeogeographical history of the area (UNESCO, 1978) exerts a profound influence on species distribution.

Within Nariva Swamp there was no obvious distinction between the fauna in inundated areas of woody and herbaceous vegetation. The availability of water was of greater importance than the nature of the plant cover; therefore, the fauna below swamp-forest formations was in no way characteristic (Bacon et al., 1979).

Additional aquatic elements entered the Nariva Swamp from the mangal and neighbouring marine environments. Among these were fish, such as *Centropomus* spp., *Megalops atlanticus*, *Mugil* spp., and *Sphaeroides testudineus*, which followed brackish-water penetration of the swamp forest and herbaceous areas during the late dry season. In

TABLE 9.7

Aquatic macrofauna from freshwater habitats in Nariva Swamp, Trinidad

Group	Family	Species
Crustacea		
Decapoda	Trichodactyliidae	*Trichodactylus dentatus*
	Gecarcinidae	*Cardisoma guanhumi*
Insecta		
Hemiptera	Gerridae	*Gerris aduncus*
	Veliidae	*Rhagovelia* sp.
	Belostomatidae	*Belostoma malkini*
		Lethocerus maximus
	Nepidae	*Ranatra* sp.
	Corixidae	*Tenagobia signata*
Mollusca		
Gastropoda	Ampullariidae	*Pomacea urceus*
		P. glauca
		P. cornuarietis
	Thiaridae	*Melanoides tuberculatus*
	Succineidae	*Omalonyx felinus*
Pelecypoda	Mytilidae	*Anodontites leotaudi*
		Mycetopoda pittieri
	Dreissenidae	*Mytilopsis domingensis*
Vertebrata		
Pisces	Megalopidae	*Megalops atlanticus*
	Characinidae	*Aphyocarax axelrodi*
		Astyanax bimaculatus
		Corynopoma riseii
		Hemigrammus unilineatus
		Hoplias malabaricus
		Pristella sp.
		Roeboides dayi
	Gymnotidae	*Gymnotus carapo*
	Callichthyidae	*Corydoras aeneus*
		Hoplosternum littorale
		H. thoracatum
	Loricariidae	*Hypostomus robinii*
	Pimelodidae	*Rhamdia sebae*
	Cyprinodontidae	*Rivulus hartii*
	Poeciliidae	*Poecilia picta*
	Mugilidae	*Mugil* spp.
	Serranidae	*Mycteroperca* sp.
	Cichlidae	*Aequidens pulcher*
		Cichlosoma bimaculatum
	Polycentridae	*Polycentrus schomburgkii*
	Synbranchidae	*Synbranchus marmoratus*
Anura	Pipidae	*Pipa pipa*
	Bufonidae	*Bufo marinus*
		Eleutherodactylus urichi
		Eupemphix pustulosus trinitatis
	Leptodactylidae	*Leptodactylus bolivianus*
		L. petersi
		L. sibilatrix
	Hylidae	*Hyla maxima*
		H. minuta
		H. misera

(*continued*)

TABLE 9.7 (*continued*)

Group	Family	Species
	Hylidae	*H. punctata*
		H. rubra
		Phrynohyas zonata
		Phyllomedusa trinitatis
		Sphaenorhynchus eurhostus
	Microhylidae	*Elachistocleis ovalis*
		E. surinamensis
	Pseudidae	*Pseudis paradoxa caribensis*
Reptilia		
Chelonia	Kinosternidae	*Kinosternon scorpiodes*
	Chelydidae	*Chelys fimbriata*
		Phrynops gibbus
	Pelomedusae	*Podocnemis* sp.
Crocodilia	Crocodilidae	*Caiman sclerops*
Mammalia		
	Mustelidae	*Lutra enudris*
	Trichecidae	*Trichecus manatus*

addition, larval Hemirhamphidae, Megalopidae, Pleuronectidae and Sciaenidae were collected inside the swamp in slightly brackish and fresh water, as were penaeid larvae, suggesting that the swamp might function as nursery habitat.

The major problem facing the aquatic fauna is periodic desiccation of the marsh, and sometimes the swamp, zones. During the dry season many species retreat to the remaining inundated areas or to neighbouring streams. Aestivation followed by rapid reproduction at the start of the wet season and dispersal through the swamp as water levels rose appears to be a major adaptation of the common swamp species.

Pomacea glauca laid eggs in the early dry season on stems of *Montrichardia* and other swamp plants, and the pneumatophores of *Symphonia* and buttresses of *Pterocarpus*. These became exposed as the water level fell, with larval development continuing inside protective cases. The young snails were released with the flooding at the start of the wet season and dispersed rapidly as flooding spread. The related species, *P. urceus*, buried itself in the sediments as the dry season progressed, and the larvae developed in egg cases in the mantle cavity. Release of young occurred as water level increased during the wet season. E. Jaikaransingh (pers. commun., 1978) found that rapid growth rate produced individuals capable of spawning in just over 12 months. He also found that, in the Nariva Swamp, *P. urceus* was most abundant in stands of *Montrichardia*, on which the conch feeds, where there were more permanently wet conditions. *Pomacea urceus* congregated also in swamp forests dominated by *Pterocarpus* or *Symphonia* in Nariva and at O'Meara in central Trinidad.

Singh (1978) found that the feeding activity of *Hoplosternum littorale* (one of the most abundant and widespread fish in the Nariva Swamp) decreased during the dry season. It congregated in permanently inundated areas, but in other drying water bodies its capacity for surface breathing aided survival, as has been found in many New World swamp fish (Carter and Beadle, 1930; Lowe-McConnell, 1975). Maturation of the gonads in *H. littorale* was stimulated by increased water temperature during the dry season when water was more stagnant, while nest-building behaviour and release of gametes were probably stimulated by sudden temperature reduction following inundation of the swamp by flood water. Nests were constructed after the first rains when inundation depth reached from 25 to 30 cm. The young fish grew rapidly and were mature enough to spawn by the following wet season. Young and adult *Hoplosternum* spread rapidly throughout the swamp as flooding progressed, and adult feeding intensity was highest during the inundation season.

The synchronization of breeding, dispersal, and other activities to the water regime is advantageous under seasonal swamp conditions (Lowe-McConnell, 1975). Junk (1978) showed that, in the floating weed fauna of várzea lakes in Amazonia, animals with short life cycles and high reproductive rates were at an advantage following dry-season low water.

The permanently inundated parts of Nariva occupied by *Montrichardia* or other herbaceous swamp and swamp forest areas were important as faunal reservoirs from which the remaining marshy areas could be recolonized during the wet season.

Of the more than 400 non-aquatic animals reported from Nariva Swamp, none was restricted to swamp habitats in Trinidad, although many were most abundant in swamps. Whereas the aquatic species lived in permanently flooded areas and spread throughout the swamp during the wet season from June to December, the non-aquatic fauna was restricted to drier ground on raised areas during the rains, but was able to disperse in the dry season between January and May. The populations were supplemented by seasonal transhumance from the surrounding forested high land to the south and west.

Thirteen snakes, 17 lizards and 27 land mammals were recorded on Sand Hill, Bois Neuf and Bush Bush and the small sand islands near the eastern mangrove area (Fig. 9.8). Some species, such as *Eunectes murinus*, *Helicops angulatus* and other snakes, are able to move freely through water, while *Agouti paca*, *Dasyprocta agouti*, *Mazama americana* and *Tayassu tajacu* will cross shallow water. However, all are essentially terrestrial forest animals whose presence in the swamp is made possible by islands, ridges and seasonal drought. Similarly, in Surinam, although *Odocoileus virginiana* is found commonly in coastal swamps, this "swamp deer" lives largely on the sand ridges or in drier marshy forests rather than inundated areas (H. Reichart, pers. commun., 1979). The Nariva Swamp vertebrates are mostly species common also in coastal northern South America.

Arboreal species are less affected by the water regime. The two primates, 16 bats and 170 birds recorded from Nariva were not confined to the raised-ground areas, although frequently they roost there. Neville (1972) found that, although *Alouatta seniculus* preferred virgin rain forest with tall trees, this species frequently used the *Pterocarpus* swamp forest for feeding, even coming to the ground and crossing shallow water. None of the birds was restricted to the Nariva Swamp, although *Anhima cornuta* may have been so before its extinction in the early part of this century. Most species were terrestrial forest birds, while 74 were common also in mangals in Trinidad (Ffrench, 1966, 1977) as in Surinam (Haverschmidt, 1965). There appears to be a regular interchange between the bird populations in different swamp areas in Trinidad and between the mangrove and freshwater swamp birds. Vermeer et al. (1974) reported *Rostrhamus sociabilis* feeding on *Pomacea* spp. in freshwater swamps in Surinam and returning to roost in the neighbouring mangal.

As with the Trinidadian west coast mangals (Bacon, 1970), the Nariva freshwater swamp forests provide night-roosting sanctuaries for many island forest and pasture birds.

Nottebohm and Nottebohm (1969) stated that *Amazona amazonica* nested commonly in decaying palms in Nariva Swamp. They found a degree of genetic difference between the nesting populations in the morichales, possibly related to the occurrence of islands of these trees being isolated from one another by large areas of non-woody vegetation.

Biting insects formed a conspicuous part of the Nariva Swamp fauna, and included 84 species of mosquito. These, together with tabanids and ceratopogonids, were present throughout the year and most abundant in the forested areas where standing water or damp soil were present.

Aitken et al. (1968) reported that the buttress habitat of *Pterocarpus* was favoured by certain *Culex* species. Of the 92 culicines recorded in freshwater swamps in Trinidad only 24 were common. These showed two periods of activity after the dry season, one after the initial rains in June and the other following the heavy rains of the Petit Careme, the latter being the period of greatest activity. *Aedeomyia*, *Anopheles* spp., *Culex amazonensis* and *Uranotaenia* were all found to be swamp and swamp forest breeders, as were 10 phlebotomines, 10 species of *Culicoides*, and 18 horseflies.

The fauna of swamps in other countries shows many similarities in species composition and ecology. In Guadeloupe, Pointier (1978) found that, whereas the mollusc fauna of the mangals was poor, several species occurred in the zone of *Pterocarpus officinalis*. These included *Biomphalaria glabrata*, *Drepanotrema cimex*, *D. kermatoides*, *Physa marmorata*, and *Pomacea glauca*. Their numbers increased landwards through the *Colocasia* and grass prairie zones, while *B. glabrata* had dense populations at the edge of the swamp forest but disappeared rapidly under the *Pterocarpus* trees. They appeared to be concentrated there by retreat of flood waters from the landward zones in the dry season.

Inundation of the *Pterocarpus* zone provoked a migration of rats into neighbouring cultivated regions (Delattre, 1978). *Rattus rattus* was common and was dispersed in the swamp forests before wet-season flooding occurred. Rising water levels would be expected to have an adverse effect on burrowing animals, particularly rodents.

Neither the mangrove nor *Pterocarpus* zones in Guadeloupe were major breeding sites for mosquitoes (Rioux, 1978) because of the high level of inundation. *Deinocerites magnus*, *Culex inflictus*, *Aedes taeniorhynchus*, and rarely *A. tortilis* were found breeding in crab holes, particularly of *Cardisoma guanhumi*, in swamp forests and riverine forests, however. Kermarrec (1978) found certain nematodes grouped in the swamp soils according to plant communities, *Meloidogyne incognita* being characteristic of the *pterocarpetum*. However, their numbers were reduced by increasing salinity and period of inundation.

In the swamp forests of Dominica, James (1979) reported *Amazona arausiaca* feeding on *Pterocarpus* fruits during its nesting season. Common marsh birds, such as *Ardea herodias*, *Butorides virescens* and ducks, also frequented these habitats. The crabs *Cardisoma guanhumi*, *Goniopsis cruentata*, and *Guinotia dentata* and the common lizards *Anolis bimaculatus oculatus* were listed also; all of these are species commonly reported from other parts of the island (Underwood, 1962; Chace and Hobbs, 1969). The presence of the crab species suggests that these habitats may be brackish.

The freshwater swamp forest faunas are thus intermediate in character between the brackish mangrove and terrestrial/freshwater habitats, and many show seasonal transitory trophic habits and distribution patterns.

MANAGEMENT

Exploitation is minimal and management virtually non-existent for swamp forests in Neotropical regions. These wetland situations are generally inaccessible for much of the year, are inhospitable due to myriads of biting insects, and contain few commercially important resources. A recent review of tropical forests (UNESCO, 1978) suggests that coastal swamp forests should be separated from exploitable forest land, and that one management strategy should be their assignment to the "protective forest" category.

The light brown wood of *Pterocarpus officinalis* is weak, non-durable and has a specific gravity of only 0.6 according to Britton and Wilson (1922–1930). James (1979) quoted a specific gravity of 0.3 for Dominican *Pterocarpus*, and noted that the wood stains easily on drying and is susceptible to termite damage. In a forestry evaluation of the North Oropuche Swamp, Trinidad, Kitson and Bell (1971) listed this species as Class 4 timber of limited value, while in Guadeloupe it is used only for charcoal, boxes, and plywood manufacture (Renard, 1976; Montaignac, 1978). In Puerto Rico, *P. officinalis* is used for fishing floats and for firewood, particularly when burning coral to make lime (James, 1979). Trees are cut for fuel when between 50 and 120 cm (dbh) and have high coppicing ability (Gleason and Cook, 1926); but Richards (1979) pointed out that much timber is wasted whenever this tree is harvested, because it must be cut above the buttresses. Standley (1928) reported that *Pterocarpus* sap was once exported as "sangre de draco" from Guatemala, Nicaragua, Colombia, and Puerto Rico for use as a haematic and astringent.

Several associated swamp-forest trees are of greater commercial value than *P. officinalis*. Vann (1959) lists the following swamp species in the three Guianas: *Triplaris surinamensis*, in swamp wood, prized for its resistance to decay; *Eschweilera longipes*, in swamp wood particularly in Guiana, with very hard wood due to the high silica content; *Bonafousia tetrastachya*, with leaves valued for medicinal properties; *Ceiba pentandra*, a valuable timber tree of marsh forest regions; and *Mora*, with very hard wood, occupying some swamp wood and some marshy forest areas in the western coastal lowlands.

Symphonia globulifera and *Virola surinamensis* are among the most valuable timber trees harvested in coastal French Guiana. Here also the terminal bud region of the palm *Euterpe oleracea* is collected as a culinary delicacy. Exploitable timber species found in Trinidadian mixed swamp woods include *Andira inermis*, *Calophyllum calaba*, *Carapa guianensis*, *Ceiba pentandra*, *Eschweilera subglandulosa*, *Hura crepitans*, *Hymenaea courbaril*, *Symphonia globulifera* and *Virola surinamensis*.

The value of these timber resources is diminished by difficulties of extraction and by their widely scattered distribution. In a Surinamese forest inventory (UNESCO, 1978) the average volume of commercial species was only 20 m^3 ha^{-1}, compared with 50 m^3 ha^{-1} for dry-land forest. No other records were available to show the amount of timber extracted or its market value in any of the region's swamp forests.

J.P. Lescuré (pers. commun., 1979) suggested that the economic value of swamps could be enhanced by modifying the ecosystem so that *Pterocarpus* and other low-value species were replaced by *Eperua jenmani*, *Symphonia globulifera* and *Virola surinamensis*, and that *Euterpe oleracea* could be cultivated for food. The natural coppicing of *Euterpe* would be an advantage under such a plantation forestry system.

This idea is supported by Montaignac (1978) for Guadeloupe, where scheduled sylvicultural practices for swamp forests will include plantations of *Pterocarpus* in the zone presently dominated by this species, and cultivation of *Calophyllum calaba* and *Symphonia globulifera*, to the detriment of *Pterocarpus*, in zones rich in these more valuable species. On an experimental basis, some plots of *Pterocarpus* swamp forest will be planted with introduced *Carapa guianensis* and *Virola surinamensis* from closely similar habitats in South American swamps. In the Black River Morass, Jamaica, the introduced species *Haematoxylon campechianum* has become dominant beside rivers and is the main source of timber in this swamp (Proctor, 1964).

Cultivation of commercial tree species in the swamp forests will depend on successful seed germination. As mentioned earlier, the current environment may not be suitable for seedling establishment because of the dynamic nature of geogenesis and pedogenesis in the swamp habitats.

Greater use of the regenerative capacity of many of these swamp trees might be an important strategy for plantation establishment until more is known of seedling ecology.

The non-forestry values of swamp forests are equally difficult to assess. Many animals are exploited, but all are species which range through herbaceous and tree swamp areas or are essentially part of the surrounding forest fauna. The Nariva Swamp is important for bird and mammal hunting, but catch records are not available. Sport and food hunting for birds and mammals such as the swamp deer, *Odocoileus virginiana*, and marsh deer, *Blastocerus dichotomus*, occurs throughout the Guiana coastal plain, swamp faunas making a significant contribution to the economy of many Amerind villages. Fretey and Renault-Lescuré (1978) showed that the Galibi in western French Guiana, though favouring marine turtles when in season, hunted traditionally in the swamps isolated by coastal sand bars to supplement their diet. The skins of *Eunectes*, the meat and skins of *Caiman sclerops*, *Melanosuchus niger*, and other crocodilians and fish are among the animal products commonly exploited in freshwater swamps near Cayenne.

In Nicaragua (Nietschmann, 1972) the hunting and fishing range of the Tasbapauni included palm swamp and herb marsh, from which 15% of the village meat yield was obtained. During flood periods the availability of game and fish changed in location and density. The greatest amount of hunting occurred in the wet-season months, as flood-trapped animals could be killed easily along river banks and on low hills in the swamps. Some hunting for *Odocoileus virginiana* was carried on all year in the swamps, gallery forests and marshes.

Fishing is an important occupation in Nariva Swamp, Trinidad, particularly during the dry season. The value of the annual harvest of *Hoplosternum littorale* was estimated by Singh (1978) to be approximately US$120,000, although this is now considered to be a very conservative estimate. *Cichlosoma bimaculatum* and *Hoplias malabaricus* are used also, although of less importance, and *Pomacea urceus* is harvested for food locally. In Guadeloupe, Therezien (1978) caught *Centropomus undecimalis* and other edible fish in a canal bordered by *Pterocarpus* and *Acrostichum*.

The existence of *Pterocarpus* forests in Guadeloupe provides a reservoir for rodent pests and for the snail hosts of *Schistosoma (Bilharzia)*, while the forest cycle of yellow fever in Trinidad involves the Nariva Swamp howler monkey population. Over 20 arboviruses affecting man have been isolated from insects in the latter area (Jonkers et al., 1968a).

Natural transmission cycles involved a period of multiplication and viremia in forest-floor rodents, and a period of multiplication in mosquitoes, especially *Culex portesi*, which was responsible for subsequent transmission. When severe reduction in rodent numbers occurred in 1964, infection rates were reduced as no alternative host was available (Jonkers et al., 1968b).

Such negative values of swamp forests require assessment in the management strategy. The value of swamps in their natural or managed state requires more accurate documentation. Sastre (1978) and James (1979) consider that *Pterocarpus* swamps are an endangered resource worthy of conservation concern, although no reasons are advanced to support this view. Areas of freshwater swamp are already set aside as conservation sites in Bush Bush, Trinidad; the Parc Naturelle de Guadeloupe; the Black River Morass, Jamaica; at Cabrite and Indian River, Dominica; and in the Guianas. The area of swamp forests or *Pterocarpus* forests in these parks is not known, and management is mostly in the hands of forestry officials.

Little attention has been paid to the role of swamp forests in freshwater wetland productivity. Waters flowing from stands of *Pterocarpus* and morichales in Nariva Swamp contained high nutrient levels (Table 9.3), although their origin was unknown.

Litter production under *Pterocarpus* in Guadeloupe ranged from 10.5 to 14.2 t ha^{-1} yr^{-1} (dry weight of leaves and seeds) compared with 7 t ha^{-1} yr^{-1} for rain forest in that island (Febvay and Kermarrec, 1978b). As no further data were available, it is apparent that the study of productivity in Neotropical swamp forests is in its infancy (see, however, Ch. 10).

Back-swamp habitats and sand bar/lagoonal situations act as traps for nutrients and particulate matter rather than being important sources of output. Direct access to channels and inshore regions is often more restricted than in neighbouring mangals, which probably contribute more to the productivity of neighbouring maritime regions.

Valuable agricultural land has been created by swamp reclamation in northern South America and the Caribbean. In Trinidad's Nariva area only herbaceous marsh and palm swamp areas have been reclaimed, and these have been cultivated successfully for rice and vegetables. In Surinam and French Guiana most of the rice polders have been constructed on land originally covered by herbaceous vegetation, so the response of swamp forest areas to reclamation is less well documented. In the Wageningen Rice Project area in western Surinam (Fig. 9.10), De Wit (1960) reported that forested wetlands were more than twice as expensive to reclaim as herb swamp, because the trees had to be removed. Removal of the timber remains was a lengthy process, and *Pterocarpus* forest soils frequently had thicker pegasse which interfered with rice growth through its high nitrogen content and tendency to acidity. De Wit recommended that, after ditching the polders, swamp forest soils be allowed to dry out first by leaving the natural vegetation in situ. However, the large rodent populations present in the remaining forest stands caused havoc to rice crops at Wageningen. Whereas the early rice yields were best on herb-swamp soils, later differences were less marked.

Turenne (1968) considered the lowland freshwater swamp soils to be of high chemical fertility in French Guiana, although their clay texture needed amelioration. Intensive water-level management was stated to be essential during empoldering if these soils were to be used successfully in agriculture.

Table 9.5 indicates that soils under swamp forest with *Euterpe*, *Pterocarpus*, *Symphonia* and *Virola* in French Guiana will support tree crops, although there are problems with drainage, and land clearance is costly.

Recent studies on rice growing in the Wageningen and Nickerie districts of Surinam indicate that there is value in leaving stands of swamp forest and marsh vegetation associated with the polders (H.N. Van Dyke, pers. commun., 1979). Modern water-management methods include recycling of water used in polder irrigation. If this water is stored between flood stages by pumping into neighbouring swamp areas, the natural vegetation aids in the removal of pesticide and fertilizer residues. The management of such reservoir/purification areas under plantation forestry might significantly improve the economic value of swamp forest areas preserved from reclamation.

ACKNOWLEDGEMENTS

The help of the following persons and institutions is acknowledged: J.P. Lescuré, Office de la Recherche Scientifique et Technique Outre-Mer (ORSTOM), French Guiana; H. Reichart, STINASU, Surinam; H.N. Van Dyke, Stichting Machinale Landbouw, Surinam; T.B. Singh, University of Guyana; E. Jaikaransingh, University of the West Indies, Trinidad; E.K. Ramcharan, Institute of Marine Affairs, Trinidad and Tobago. The Ministry of Agriculture, Jamaica, kindly made available a report by Proctor (1964) on the Black River Morass and the study of Nariva Swamp was supported partly by funds from the Ministry of Agriculture, Trinidad and Tobago.

REFERENCES

Adams, C.D., 1972. *Flowering Plants of Jamaica*. University of the West Indies, Trinidad, 848 pp.

Aitken, T.H.G., Worth, C.B. and Tikasingh, E.S., 1968. Arbovirus studies in Bush Bush Forest, Trinidad, W.I., September 1957–December 1962. III Entomologic studies. *Am. J. Trop. Med. Hyg.*, 17: 253–268.

Aitken, T.H.G., Worth, C.B., Downs, W.G. and Tikasingh, E.S., 1973. Bush Bush Forest and Nariva Swamp. *J. Trinidad Field Nat. Club*, pp. 1–6; 13–33.

Asprey, G.F., 1959. Vegetation in the Caribbean area. *Caribb. Quart.*, 5(4): 245–263.

Asprey, G.F. and Robbins, R.G., 1953. The vegetation of Jamaica. *Ecol. Monogr.*, 23: 359–412.

Augustinus, P.G.E.F. and Slager, S., 1971. Soil formation in swamp soils of the coastal fringe of Surinam. *Geoderma*, 6: 203–211.

Bacon, P.R., 1970. *The Ecology of Caroni Swamp, Trinidad*. Special Publication, Central Statistical Office, Trinidad, 68 pp.

Bacon, P.R., Kenny, J.S., Alkins, M.A., Mootoosingh, S.N., Ramcharan, E.K. and Seeberan, G.S., 1979. *Studies on the Biological Resources of Nariva Swamp, Trinidad*. Occasional Papers No. 4. Zoology Dep., Univ. of the West Indies, Trinidad, 455 pp.

Barker, H.D. and Dardeau, W.S., 1930. *Flore d'Haiti*. Service Technique du Department de L'Agriculture et de l'Enseignement Professionell, Port-au-Prince, Haiti, 456 pp.

Bartlett, A.S. and Barghoorn, E.S., 1973. Phytogeographic history of the Isthmus of Panama during the past 12,000 years. In: A. Graham (Editor), *Vegetation and Vegetational History of Northern Latin America*, Elsevier, Amsterdam, pp. 203–293.

Beard, J.S., 1946. *The Natural Vegetation of Trinidad*. Clarendon Press, Oxford, 152 pp.

Beard, J.S., 1949. Natural Vegetation of the Windward and Leeward Islands. *Oxford For. Mem.*, 21: 1–192.

Beard, J.S., 1955. The classification of tropical American vegetation types. *Ecology*, 36: 89–100.

Blancaneaux, P., 1972. *Notes Pedo-geomorphologiques sur la Savanne Sarcelle au lieu de Project Sodalg, Nord-Ouest de la Guyane Française*. Circular P. 131, Office de la Recherche Scientifique et Technique Outre-Mer, Cayenne, French Guiana, mimeographed 9 pp.

Blancaneaux, P., 1976. *Caracteristiques Pedo-agronomiques des Terrasses Fluviatiles de la Forestière*. Circular P. 146, Office de la Recherche Scientifique et Technique Outre-Mer, Cayenne, French Guiana, mimeographed 37 pp.

Brinkmann, R. and Pons, T.L., 1968. *A Pedo-geomorphological Classification and Map of the Holocene Sediments in the Coastal Plain of the Three Guianas*. Soil Survey Paper 4, Netherlands Soil Survey Institute, Wageningen, 41 pp.

Britton, N.L. and Wilson, P., 1922–1930. *Botany of Porto Rico and the Virgin Islands. Scientific Survey of Porto Rico*. New York Academy of Sciences, New York, N.Y., 5 and 6, 626 and 663 pp.

Burtt Davy, J., 1938. *The Classification of Tropical Woody Vegetation Types*. Institute Paper 13, Imperial Forestry Institute, Oxford, 85 pp.

Carter, G.S. and Beadle, L.C., 1930. The fauna of the Paraguayan Chaco in relation to its environment. 1. Physico-chemical nature of the environment. *Linnaean Soc. J. Zool.*, 37(251): 209–258.

Chabrol, L., 1978. Approches bactériologiques en mangrove: les germes lies au soufre. *Bull. Liaison Groupe Trav., Mangroves Zones Côtières, Guadeloupe*, 4: 83–84.

Chace, F.A. and Hobbs, H.H., 1969. The freshwater and terrestrial decapod crustaceans of the West Indies with special reference to Dominica. *Bull. Smithsonian Inst., Washington*, 292: 258 pp.

Chapman, V.J., 1976. *Mangrove Vegetation*. J. Cramer, 447 pp.

Chapman, V.J., 1977. *Wet Coastal Ecosystems. Ecosystems of the World, 1*. Elsevier, Amsterdam, 428 pp.

Delaney, P.J.V., 1969. Late recent geological history of coastal Surinam. *Verh. K. Ned. Geol. Mijnbouwkd. Genoot.*, 27: 65 (abstract).

Delattre, P., 1978. Étude des populations des rats en arrière-mangrove au lieu dit "Devarieux". *Bull. Liaison Groupe Trav., Mangroves Zones Côtières, Guadeloupe*, 4: 95–96.

De Wit, T.P.M., 1960. *The Wageningen Rice Project in Surinam; a Study of Mechanised Rice Farming Project in the Wet Tropics*. Mouton, The Hague, 293 pp.

Eggers, H.F.A., 1882. Die poyales des oestlichen Porto Rico. *Bot. Centralschr.*, 11: 331–332.

Fanshawe, D.B., 1952. *The vegetation of British Guiana (A preliminary review)*. Institute Paper 29, Imperial Forestry Institute, Oxford, 134 pp.

Febvay, G. and Kermarrec, A., 1978a. Quelques paramètres physicochimiques de la fôret littorale (mangrove et forest palustre) et leur évolution. *Bull. Liaison Groupe Trav., Mangroves Zones Côtière, Guadeloupe*, 4: 70–73.

Febvay, G. and Kermarrec, A., 1978b. Formation de la litière en fôret littorale et en forêt de montagne humide. *Bull. Liaison Groupe Trav., Mangroves Zones Côtière, Guadeloupe*, 4: 79–80.

Feldmann, J. and Lami, R., 1936. Sur la végétation de la mangrove à La Guadeloupe. *C. R. Acad. Sci., Paris*, 203: 883–885.

Ffrench, R.P., 1966. The utilisation of mangroves by birds in Trinidad. *Ibis*, 108: 423–424.

Ffrench, R.P., 1977. Birds of the Caroni Swamp and marshes. *J. Trinidad Field Nat. Club*: 42–44.

Fosberg, F.R., 1958. On the possibility of a rational general classification of humid tropics vegetation. In: *Proceedings of the Symposium on Humid Tropics Vegetation*. Council for Sciences of Indonesia and UNESCO Science Cooperation Office for South East Asia, Tjiawi, Indonesia, pp. 34–59.

Fretey, J. and Renault-Lescuré, O., 1978. Presence de la tortue dans la vie des Indiens Galibi de Guyane Française. *J. Agric. Trop. Bot. Appl.*, 25(1): 3–23.

Gleason, H.A. and Cook, M.T., 1926. *Plant Ecology of Porto Rico. Scientific Survey of Porto Rico and the Virgin Islands*. New York Academy of Sciences, New York, N.Y., 7, vols. 1–3.

Gottsberger, G., 1978. Seed dispersal by fish in the inundated regions of Humaita, Amazonia. *Biotropica*, 10: 170–183.

Guppy, H.B., 1917. *Plants, Seeds, and Currents in the West Indies and Azores*. Williams and Norgate, London, 531 pp.

Hallé, F., Oldeman, R.A.A. and Tomlinson, P.B., 1978. *Tropical Trees and Forests; An Architectural Analysis*. Springer-Verlag, New York, N.Y., 441 pp.

Haverschmidt, F., 1965. The utilization of mangroves by South American birds. *Ibis*, 107: 540–542.

Holdridge, L.R., 1940. Some notes on the mangrove swamps of Puerto Rico. *Caribb. For.*, 1(4): 19–29.

James, A., 1979. *The Occurrence of Freshwater Swamp Forest and Maritime Mangrove Species in Dominica*. Forestry and Wildlife Division, Ministry of Agriculture, Dominica, mimeographed 24 pp.

Janzen, D.H., 1974. Tropical blackwater rivers, animals and mast fruiting by the Dipterocarpaceae. *Biotropica*, 6: 69–103.

Janzen, D.H., 1975. *Ecology of Plants in the Tropics*. Studies in Biology, 58. Edward Arnold, London, 66 pp.

Jonkers, A.H., Spence, L., Downs, W.G., Aitken, T.H.G. and Tikasingh, E.S., 1968a. Arbovirus studies in Bush Bush Forest, Trinidad, W.I. September 1957 – December 1962. V. Virus isolations. *Am. J. Trop. Med. Hyg.*, 17: 276–284.

Jonkers, A.H., Spence, L., Downs, W.G., Aitken, T.H.G. and Worth, C.B., 1968b. Arbovirus studies in Bush Bush Forest, Trinidad, W.I. September 1957 – December 1962. VI. Rodent-associated viruses: isolations and further studies. *Am. J. Trop. Med. Hyg.*, 17: 285–298.

Junk, W.J., 1978. Faunal ecological studies in inundated areas and the definition of habitats and ecological niches. *Anim. Res. Dev.*, 4: 47–54.

Junk, W.J., 1983. Ecology of swamps on the Middle Amazon. In: H.J.P. Gore (Editor), *Mires: Swamp, Bog, Fen and Moor, Part B. Ecosystems of the World 4B*. Elsevier, Amsterdam, pp. 269–294.

Kaufmann, R. and Hevert, F., 1973. El régimen fluviométrico del Río Magdalena y su importancia para La Cienaga Grande de Santa Marta. *Mitt. Inst. Colombo-Alemán Invest. Cient. Punta de Betin*, 7: 121–137.

Keeley, J.E., 1979. Population differentiation along a flood frequency gradient: physiological adaptations to flooding in *Nyssa sylvatica*. *Ecol. Monogr.*, 49(1): 89–108.

Kenny, J.S., 1969. The Amphibia of Trinidad. *Stud. Fauna Curaçao other Caribbean Islands*, 29: 77 pp.

Kermarrec, A., 1978. Analyse synécologique des nematofaunes télluriques en fôrets littorales inondées de Guadeloupe. *Bull. Liaison Groupe Trav., Mangroves Zones Côtière, Guadeloupe*, 4: 90–91.

Kitson, A. and Bell, T.I.W., 1971. *Manzanilla Windbelt Reserve, Mini Working Plan; 1971–1980*. Forestry Division, Trinidad, mimeographed, 10 pp.

Knight, D.H., 1975. A phytosociological analysis of species-rich tropical forest on Barro Colorado Island, Panama. *Ecol. Monogr.*, 45(3): 259–284.

Koldewijn, B.W., 1958. Sediments of the Paria-Trinidad shelf. *Rep. Orinoco Shelf Exp.*, III, 109 pp.

Lescuré, J.P., 1980. Aperçu architectural de la mangrove Guadeloupéene. *Acta Ecol. Gen.*, 1(3): 249–265.

Léveque, A., 1962. *Memoire Explicatif de la Carte des Sols de Terres Basses de Guyane Français*. Office de la Recherche Scientifique et Technique Outre-Mer, Cahier, 88 pp.

Lindeman, J.C., 1953. The vegetation of the coastal region of Surinam. *Vegetation Surinam*, 1(1): 135 pp.

Lowe-McConnell, R.H., 1975. *Fish Communities in Tropical Freshwaters*. Longman, London, 337 pp.

Marshall, R.C., 1934. Physiography and vegetation of Trinidad and Tobago. *Oxford For. Mem.*, 17: 1–56.

Marshall, W.B., 1926. New land and fresh-water mollusks from Central and South America. *Proc. U.S. Natl. Mus.*, 69(2638): 1–11.

Martyn, E.B., 1934. A note on the foreshore vegetation in the neighbourhood of Georgetown, British Guiana, with special reference to *Spartina brasiliensis*. *J. Ecol.*, 22: 292–298.

Montaignac, P., 1978. Principaux resultats des recherches. *Bull. Liaison Groupe Trav., Mangroves Zones Côtière, Guadeloupe*, 4: 37–39.

Müller, J., 1959. Palynology of recent Orinoco Delta and shelf sediments. Reports of the Orinoco Shelf Expedition. *Micropalaeontology*, 5(1): 1–32.

Neville, M.K., 1972. The population structure of Red Howler Monkeys (*Alouatta seniculus*) in Trinidad and Venezuela. *Folia Primatol.*, 17(1–2): 56–86.

Newson, L.A., 1976. *Aboriginal and Spanish Colonial Trinidad*. Academic Press, New York, N.Y., 344 pp.

Nietschmann, B., 1972. Hunting and fishing focus among the Miskito Indians, eastern Nicaragua. *Hum. Ecol.*, 1(1): 41–67.

Nottebohm, F. and Nottebohm, M., 1969. The parrots of Bush Bush. *Anim. Kingdom*, 72: 18–23.

Pain, T., 1950. Pomacea (Ampullariidae) of British Guiana. *Proc. Malacol. Soc. London*, 28: 63–74.

Pointier, J.P., 1978. La schistosomose intestinale dans la

mangrove de Guadeloupe. *Bull. Liaison Groupe Trav., Mangroves Zones Côtière, Guadeloupe*, 4: 24–26.

Portecop, J. and Crisan, P.A., 1978. Cartographie phyto-écologique de la mangrove et des zones annexes. *Bull. Liaison Groupe Trav., Mangroves Zones Côtière, Guadeloupe*, 4: 45–50.

Porter, D.M., 1973. The vegetation of Panama: a review. In: A. Graham (Editor), *Vegetation and Vegetational History in Northern Latin America*. Elsevier, Amsterdam, pp. 167–201.

Proctor, G.R., 1964. *The Vegetation of the Black River Morass*. Appendix H. Report to Ministry of Agriculture, Jamaica, 15 pp.

Renard, Y., 1976. *La Mangrove*. Bulletin 2, Parc Naturelle de Guadeloupe, 32 pp.

Richards, P.W., 1979. *The Tropical Rain Forest*. Cambridge University Press, Cambridge, 450 pp.

Rioux, J.A., 1978. Les culicides des mangroves Guadeloupéenes relations gites larvaires-phytocénoses. *Bull. Liaison Groupe Trav., Mangroves Zones Côtière, Guadeloupe*, 4: 98–100.

Rutzler, K., 1969. The Mangrove community, aspects of its structure, faunistics and ecology. *Lagunas Costeras, Un Simposio, UNAM–UNESCO, México*, pp. 515–536.

Sastre, C., 1978. *Plantes Rares ou Menacées de Martinique, II. Le Mangle-Médaille*. Groupe de Travail, Mangroves et Zones Côtière, Guadeloupe, mimeographed, 5 pp.

Singh, T.B., 1978. *The Biology of the Cascadura, Hoplosternum Littorale (Hancock 1828) with Reference to its Reproductive Biology and Population Dynamics*. Ph.D. Thesis, University of the West Indies, Trinidad, 298 pp.

Sioli, H., 1975. Amazon tributaries and drainage basins. In: A.D. Hasler (Editor), *Coupling of Land and Water Systems, Ecological Studies, 10*. Springer-Verlag, New York, N.Y., pp. 199–213.

Speth, J., 1961. Trinidad's changing shoreline. *Texaco Quart.*, 4: 12–15.

Standley, P.C., 1928. Flora of the Panama Canal Zone. *Contrib. U.S. Natl. Herbarium*, 27: 6 pp.

Standley, P.C., 1931. Flora of the Lancetilla Valley, Honduras. *Field Museum of Natural History, Publication* 283, Series Botany 10.

Stehlé, H., 1937. Esquisse des associations végétales de la Martinique. *Bull. Agric., Martinique*, 6: 194–264.

Stehlé, H., 1946. Taxonomic and ecological notes on Leguminosae; Papilionaceae of the French Antilles. *Bull. Mus. Hist. Nat., Paris*, 18: 98–117.

Therezien, Y., 1978. Études hydrobiologiques en mangroves. *Bull. Liaison Groupe Trav., Mangroves Zones Côtière, Guadeloupe*, 4: 91–92.

Turenne, J.F., 1968. Soils of French Guiana. *Proc. Caribbean Food Crops Soc.*, 6: 141–147.

Turenne, J.F., 1978. Recherches entreprises par l'ORSTOM (1977–78) en mangrove. *Bull. Liaison Groupe Trav., Mangroves Zones Côtière, Guadeloupe*, 4: 40–43.

Underwood, G., 1962. Reptiles of the Eastern Caribbean. *Caribbean Affairs, New Ser.*, 1: 1–192, mimeographed.

UNESCO, 1978. *Tropical Forest Ecosystems, a State of Knowledge Report, Natural Resources Research 14*. UNESCO/UNEP/FAO, Rome, 683 pp.

Van Andel, J.H., 1967. The Orinoco Delta. *J. Sediment. Petrol.*, 37(2): 297–310.

Van der Hammen, T., 1963. A palynological study on the Quaternary of British Guiana. *Leidse Geol. Meded.*, 29: 125–180.

Van der Hammen, T., 1969. Introduction and short outline of the history of the "younger" areas of the Guianas. *Verh. K. Ned. Geol. Mijnbouwkd. Genoot.*, 27: 9–12.

Van der Hammen, T., 1974. The Pleistocene changes in vegetation and climate in tropical South America. *J. Biogeogr.*, 1: 3–26.

Vann, J.H., 1959. *The Physical Geography of the Lower Coastal Plain of the Guiana Coast*. Tech. Rep. 1, Office of Naval Research, Washington, D.C., mimeographed, 91 pp.

Vermeer, K., Risebrough, R.W., Spaans, A.L. and Reynolds, L.M., 1974. Pesticide effects on fishes and birds in rice fields of Surinam, South America. *Environ. Pollut.*, 7: 217–236.

Wijmstra, T.A., 1967. A pollen diagram from the Upper Holocene of the lower Magdalena Valley. *Leidse Geol. Meded.*, 39: 261–267.

[Submitted November 1979; revision submitted January 1981; accepted November 1983.]

Chapter 10

ECOLOGY OF *PTEROCARPUS OFFICINALIS* FORESTED WETLANDS IN PUERTO RICO

MIGDALIA ALVAREZ-LOPEZ

INTRODUCTION

The distribution of *Pterocarpus officinalis* forests throughout Central and South America has been well documented by Bacon (Ch. 9, this volume). For Puerto Rico the existence of *Pterocarpus* forests has been known since 1882, although it was not until very recently that their specific localities and areas have been determined. Woodbury (1979) reported a total of fifteen forest locations in Puerto Rico, most of them being concentrated in the northeastern region (Fig. 10.1). Cintrón (1980) estimated their coverage to be approximately 238 ha. *Pterocarpus* trees or stands were reported from Utuado and Cabo Rojo (near Boquerón) (Bates, 1929), but have apparently disappeared since that time. In the past, the mountainous region was covered by *Pterocarpus* trees or stands in swamps as reported by Bates (1929), Gleason and Cook (1926) and Beard (1949), but at present only those in the Luquillo Mountains remain.

Most of the literature on this forest type is qualitative and speculative because of the few existing data. As a result, important questions remain unanswered. The objective of this study was to address several of these questions, specifically:

What is their species composition?
How are species and individuals distributed in the forests?
How do stands vary in structure throughout the island?
What constitutes a climax habitat for this ecosystem?
How are *Pterocarpus* forests related to mangrove and terrestrial communities, and what types of ecotones appear between them?

What are the differences in structure and function of *Pterocarpus* forests based on a physiographic classification?
What are the environmental factors regulating the structure of *Pterocarpus* forests?

DESCRIPTION OF STUDY AREAS

Study sites are located between latitude 17°15′N and 18°28′N and longitude 65°45′W to 67°10′W (Fig. 10.1 and Table 10.1). These five areas cover 167 ha or 70% of the *Pterocarpus* forest area of Puerto Rico. The Luquillo and Dorado forests were surrounded by terrestrial forests, while those of Mayagüez and Patillas were associated with coastal mangrove formations. All but the Luquillo forest are coastal, and located in the subtropical moist forest life zone. Luquillo falls in the subtropical wet forest life zone.

In general, basin forests are flooded for large periods of time and to greater depths than riverine forests (Fig. 10.2). Basins reflect changes in rainfall in their hydroperiods. In the Patillas basin, a relatively slow rise in water level occurred between February (28 cm below ground surface) and October (5 cm above ground) with a peak value in the September rainy season (Fig. 10.2A). In the natural depressed basin of Dorado, severe flooding conditions were observed during 6–8 months in 1981 (Fig. 10.2B).

In the four coastal, pure *Pterocarpus* stands in Puerto Rico interstitial salinities were below 5‰ (Fig. 10.3). Febvay and Kermarrec (1978a) found salinities in the range of 2 to 10‰ in *Pterocarpus* stands of Guadaloupe. As expected, the Luquillo mountains had the lowest value

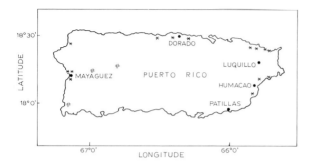

Fig. 10.1. Map of Puerto Rico showing past and present distribution of *Pterocarpus officinalis* forested wetlands and location of study sites (●). # = *Pterocarpus* forests reported in the past and not found in a 1983–84 field survey; * = actual *Pterocarpus* forest distribution.

(0.03‰) because that area is not influenced by saline intrusions.

In an ecotone between a mangrove forest and a *Pterocarpus* swamp in Patillas, the following zones were observed: (a) *Rhizophora* (soil salinity = 25.91‰), (b) *Laguncularia* (soil salinity = 19.57‰), (c) mixed *Laguncularia* and *Pterocarpus* (soil salinity = 12.49‰), and (d) pure *Pterocarpus* stand where salinity dropped abruptly to 0.74‰. Aerial photography and field experience revealed that isolated *Laguncularia* trees are found inside the pure *Pterocarpus* zones, but the reverse situation does not occur. It was evident that the presence of *Pterocarpus* indicated low soil interstitial salinity irrespective of the close proximity of each community type to each other. Apparently, *Pterocarpus* is not able to survive when the salinity in the subsoil increases abruptly. An unexpected saline intrusion caused by pumping freshwater out of a *Pterocarpus* forest in Palmas del Mar (Humacao, Puerto Rico), caused a progressive die-out of trees. This forest recovered only when the pumping stopped (Woodbury, 1979).

FLORISTIC COMPOSITION

Although *Pterocarpus* tends to grow in monospecific stands and is gregarious, a wide diversity of life forms and strata were found in the five study areas.

In coastal *Pterocarpus* forests classified as riverine and basin types, some arborescent and palm species were widespread, and they were a

TABLE 10.1

Summary of general characteristics of *Pterocarpus officinalis* forests in Puerto Rico

Site	Total area[1] (ha)	Mean annual temperature[2] (°C)	Total annual precipitation[2] (mm)	Forest type	Soil type
Luquillo[3]	3.7	18.3	4169	Montane riverine	Organic soil over clay down to 0.4 m over a rock base
Mayagüez	6.3	25.2	2021	Coastal riverine	Clay and organic materials from decaying swamp vegetation
Humacao	150.0	25.2	1789	Coastal riverine	Alluvial origin with moderate amounts of clay and organic matter
Patillas	4.6 ha of *Pterocarpus* in 44.0 ha of mangrove	26.2	730	Coastal basin	Very organic soil composed of organic debris and mangrove peat
Dorado	2.4	25.4	1688	Coastal basin	Mixed type of organic sediments, sand, clay, shells and corals

[1]Cintrón (1980).
[2]Briscoe (1966).
[3]At 400 m elevation.

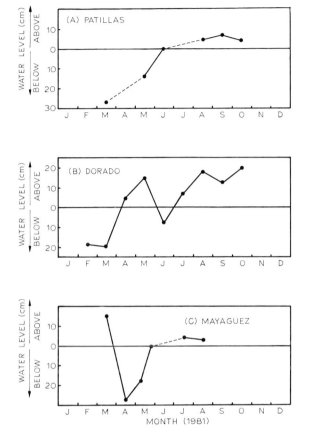

Fig. 10.2. Variations in water levels in three *Pterocarpus* forests in Puerto Rico. (A) and (B) are basin forests and (C) is a riverine forest. The horizontal line represents the soil surface.

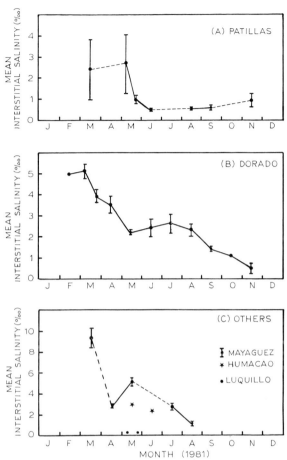

Fig. 10.3. Mean interstitial salinity in five *Pterocarpus* forests in Puerto Rico. Bars are ±1 standard error of the mean.

common component inside the freshwater swamps. These species are *Andira inermis*, *Annona glabra*, *Bucida buceras*, *Calophyllum calaba*, *Clusia rosea*, *Ficus citrifolia*, *F. sintenisii* and *Roystonea borinquena* (Table 10.2). *Clusia rosea* and the two *Ficus* species begin as epiphytes at the top or at the base of the multiple stems that form in *Pterocarpus* trees. Other less common species in this type of environment are *Coccoloba venosa* and *Randia aculeata*. In most salt-influenced forests, close to the coast, mangrove species, *Laguncularia racemosa*, *Conocarpus erectus* and *Rhizophora mangle*, were found, apparently in response to differences in brackish or saltwater input in the subsoil. Clusters or individual seedlings of these species were found scattered on the forest floor. Other arborescent and tree-like shrubs reported inside *Pterocarpus* forests in Puerto Rico are *Hibiscus tiliaceus*, *Machaerium lunatum* and *Pavonia scabra* (mallow shrub) (Stehlé, 1945; Woodbury, 1979).

In the montane riverine forest of Luquillo, species such as *Casearia arborea*, *Cordia borinquensis*, *Inga laurina*, *Manilkara bidentata*, the arboreal fern *Nephelea portoricensis* and *Prestoea montana* among others were found. However, in all of the study sites, *Pterocarpus officinalis* was the most important species (Importance Value[1] varied from 43.6 to 100%, Table 10.2).

The number of forest strata in the *Pterocarpus* forests varied from two in the montane riverine forest (Fig. 10.4a) to one in the coastal systems (riverine and basin) (Fig. 10.4b–10.4e). The single

---

[1] "Importance Value" is the mean of relative frequency, relative density and relative basal area.

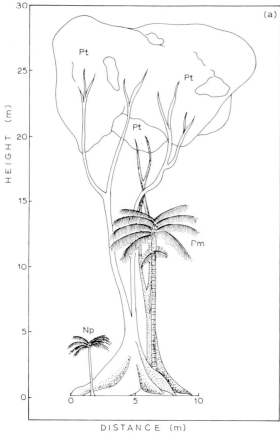

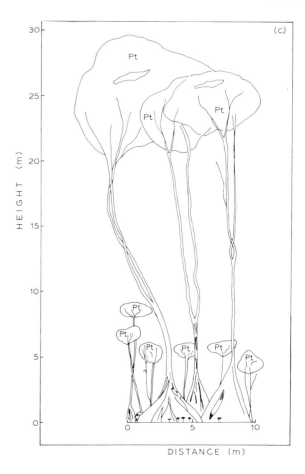

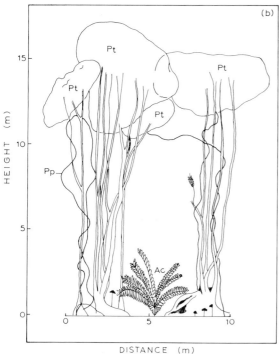

and irregular canopy in the coastal forest ranged from 15 to 20 m in height, with some emergent trees reaching 22.5 m. Two well-defined strata characterized the montane forests at Luquillo: an upper canopy 25–30 m tall with maximum tree heights of 32 m, and a second stratum at 15 m. In the understory, a healthy sapling growth from 3 to 7 m tall occurred in areas where stagnant conditions did not prevail for long periods.

Epiphytes and woody vines were also abundant. The presence of either or both depended on the degree of canopy closure. Epiphytes were conspicuous in relatively young forests where the canopy was open, the forest well illuminated, and the trees were single-stemmed. At least one epiphytic species was carried by most of the trees (58–100%) in an area of 0.1 ha. Atmospheric humidity is an important factor that influences the establishment of these plants. In those forests where the multiple-

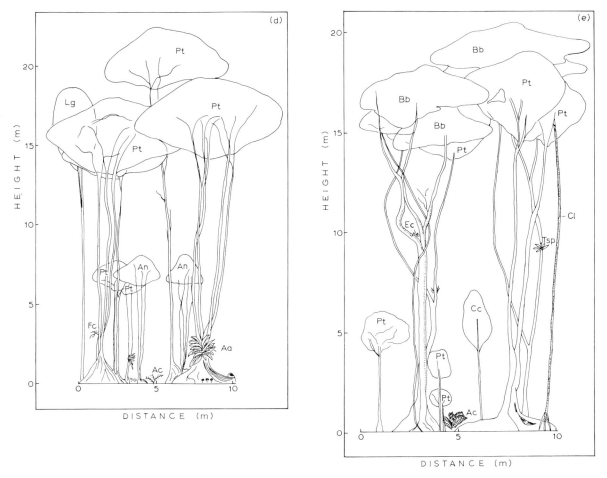

Fig. 10.4. Idealized vegetation profiles for five *Pterocarpus* forests in Puerto Rico. Aa = *Anthurium acaule*, Ac = *Acrostichum* sp., An = *Annona glabra*, Bb = *Bucida buceras*, Cc = *Calophyllum calaba*, Cl = *Clusia rosea*, Ec = *Epidendron ciliare*, Fc = *Ficus citrifolia*, Lg = *Laguncularia racemosa*, M = *Marcgravia* sp., Np = *Nephelea portoricensis*, Pm = *Prestoea montana*, Pt = *Pterocarpus officinalis*, T = *Tillandsia* sp. (a) Luquillo, (b) Mayaguez, (c) Humacao, (d) Patillas, and (e) Dorado.

stem growth habit predominates in the *Pterocarpus* trees, humus epiphytes like *Anthurium acaule* (Araceae) dominate, as well as the stranglers *Ficus* sp. and *Clusia rosea*, and also *Pterocarpus* seedlings. Although the coppicing phenomenon may provide decaying detritus between trunks and substrate, humus epiphytes were not always conspicuous. Instead, sometimes large woody vines covered as much as 40% of the trees in the forests, and there were few or no epiphytes. About twelve species of vines are reported from these forests (Woodbury, 1979).

In almost all the study sites the most abundant species in the understory were the ferns *Acrostichum aureum* and *A. danaefolium* which reach 3–4 m in height. *Pterocarpus officinalis* seedlings either covered the soil in clusters close to the trunk, or might form an extended carpet on the forest floor. The particular distribution and growth of the seedlings varied in each forest and depended on the particular regime of water input and the occurrence of stagnant conditions.

The buttress root system may reach 5–8 m in width and 5 m high. Live roots are easily identified by their clear yellow color, while dead ones are clear to dark grey.

Pterocarpus is often present, but of low importance, within *Raphia*, *Mora* and mixed swamp forests in, for example, Costa Rica (Holdridge, 1967; Holdridge et al., 1971; see also Ch. 11).

TABLE 10.2

Importance values of tree species in *Pterocarpus officinalis* forests of Puerto Rico. Values are for trees with a diameter at breast height ⩾ 2.5 cm

Site	Taxon	Relative value (%)			Importance value(%)
		basal area	density	frequency	
Luquillo (montane riverine)	*Pterocarpus officinalis*	72.4	36.1	22.2	43.6
	Prestoea montana	6.2	21.3	20.0	15.9
	Nephelea portoricensis	1.8	17.6	18.0	12.5
	Cordia borinquensis	9.1	3.7	4.4	5.7
	Casearia arborea	1.3	5.6	6.8	4.6
	Inga laurina	3.0	1.8	4.4	3.0
	Manilkara bidentata	0.1	2.9	4.4	2.5
	Andira inermis	1.1	1.8	4.4	2.4
	Henriettea fascicularis	0.2	1.8	4.4	2.1
	Citrus sinensis	2.4	1.8	2.2	2.1
	Vitex divaricata	0.7	1.8	2.2	1.6
	Ocotea spathulata	0.8	0.9	2.2	1.3
	Unknown species	0.9	2.9	4.4	2.7
Mayagüez (coastal riverine)	*Pterocarpus officinalis*	99.6	96.8	83.3	93.2
	Coccoloba venosa	0.4	3.2	16.7	6.8
Humacao (coastal riverine)	*Pterocarpus officinalis*	100.0	100.0	100.0	100.0
Patillas (coastal basin)	*Pterocarpus officinalis*	85.5	79.1	38.5	67.7
	Annona glabra	3.9	12.1	19.3	11.8
	Laguncularia racemosa	10.0	2.1	11.5	7.9
	Ficus sintenisii	0.2	3.1	11.5	4.9
	Ficus citrifolia	0.4	2.6	11.5	4.8
	Calophyllum calaba	0.1	1.0	7.7	2.9
Dorado (coastal basin)	*Pterocarpus officinalis*	47.6	73.5	40.0	53.7
	Bucida buceras	43.3	13.9	16.0	24.4
	Calophyllum calaba	5.3	10.2	28.0	14.5
	Roystonea borinquena	2.8	0.6	4.0	2.5
	Ficus citrifolia	0.7	0.6	4.0	1.8
	Andira inermis	0.2	0.6	4.0	1.6
	Randia aculeata	0.1	0.6	4.0	1.5

STRUCTURAL CHARACTERISTICS OF *PTEROCARPUS* FORESTS

The total basal area for all species (diameter ⩾ 2.5 cm measured above buttress level in all cases) ranged between 27.7 to 55.0 m² ha⁻¹ (Table 10.3). Stand height varied from 16.9 to as much as 32.0 m in very well-developed freshwater swamps. Stem density was 950–1910 trees ha⁻¹. In all forests, the number of tree species varied from 1 to 13 in 0.1 ha.

Structural complexity of the whole forest (defined as all trees with diameter ⩾ 2.5 cm) decreased in order as shown in Table 10.3: montane riverine (Luquillo); coastal basins (Patillas and Dorado); and the coastal riverine forests (Humacao and Mayagüez). For the mature trees (defined as all trees with diameter ⩾ 10 cm) the order was almost identical. On the coast the most complex systems were the basin forests rather than the riverine types, as is typical for saltwater forested wetlands (Brown and Lugo, 1982).

Among previous descriptions of *Pterocarpus* forests of the Caribbean (Stehlé, 1945; Woodbury, 1979), the Luquillo forest appears to have the most unusual composition and number of species (Tables 10.2 and 10.3). This forest is more closely related to the surrounding subtropical wet forest than to the typical coastal *Pterocarpus* forests. In

TABLE 10.3

Structural characteristics of *Pterocarpus officinalis* forests in Puerto Rico (dbh = diameter at breast height)

Forest site	Basal area (m² ha⁻¹)		Density (ha⁻¹)		Number of species in 0.1 ha		Stand height (m)	Complexity index[2]	
	⩾2.5 cm dbh	⩾10 cm dbh	⩾2.5 cm dbh	⩾10 cm dbh	⩾2.5 cm dbh	⩾10 cm dbh		⩾2.5 cm dbh	⩾10 cm dbh
Luquillo	55.0	50.8	1080	650	13	10	32.0	247	106
Mayagüez	27.7	25.0	950	660	2	1	16.9	9	3
Humacao	43.0	38.6	1770	540	1	1	27.2	21	6
Patillas	54.8	48.6	1910	1140	6	3	18.7	117	31
Dorado[1]	44.6	42.0	1680	860	7	6	19.0	100	45

[1] Mean values from two transects (0.1 ha) were used. Data from Alvarez et al. (1983).
[2] After Holdridge (1967); see Table 5.2.

coastal floodplain forests, the maximum number of species was seven and the tendency towards monospecificity was confirmed. In all coastal forests described herein *P. officinalis* occurred as the dominant species.

GROUND BIOMASS

Root biomass

Total root biomass ranged from 2.6 to 12.4 kg m^{-2} (Table 10.4). Significant differences existed between forests ($P<0.001$). A Student–Newman–Keul (SNK) test (Zar, 1974) showed that root biomass for the five forests can be classified into at least two groups in respect of population mean: low (Patillas and Dorado) and high (Luquillo and Mayagüez). The data available for Humacao were insufficient to conclude whether it belongs to the high biomass group or if it forms, by itself, an intermediate isolated group. Root biomass was higher in the montane and coastal riverine forests, and lower in the coastal basin forests. Riverine forests have more root biomass, probably because of problems associated with seasonal salinity input and/or strong pulsed hydroperiods. Hydroperiods exert the greatest impact on the root system. This is indicated by the lower root biomass in the basin forests with longer hydroperiod.

A significant decrease in root biomass with depth to 0.5 m was found in four of the five forest sites (Fig. 10.5). Only in Dorado was the sampled

TABLE 10.4

Total root biomass, litter standing crop and total soil organic matter in five *Pterocarpus officinalis* forests in Puerto Rico (the figures in parentheses are one standard error)

Forest	Root biomass[1] (kg m^{-2})	Litter standing crop[2] (kg m^{-2})	Soil organic matter[3] (kg m^{-2})
Luquillo	9.4 (1.7)	0.40 (0.04)	16.9 (1.6)
Mayagüez	12.4 (0.9)	0.94 (0.15)	23.8 (2.4)
Humacao	7.4 (0.7)	0.28 (0.03)	26.2 (1.6)
Patillas	4.5 (2.0)	0.98 (0.25)	56.5 (1.0)
Dorado	2.6 (0.4)	0.82 (0.18)	22.7 (3.0)

[1]Sampling consisted of five replicates (25 samples), each pit was 0.25 m^2 (0.5m × 0.5m) in area, and consecutive layers 10 cm thick were collected to a depth of 50 cm.
[2]Ten loose litter samples of 0.25 m^2 were collected in each forest.
[3]Mean of five samples, each being a set of seven subsamples to a depth of 50 cm.

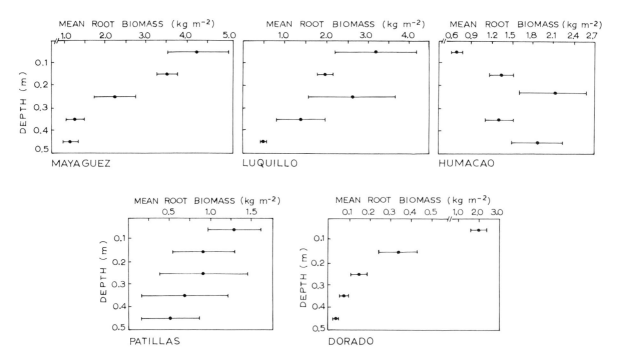

Fig. 10.5. Root biomass variation with depth in five *Pterocarpus* forests in Puerto Rico. Each bar shows the mean ±1 standard error of the mean.

depth deep enough to sample the entire root profile. It is clear, however, that the *Pterocarpus* root system is not shallow as has been proposed (Woodbury, 1979; Cintrón, 1980), but may penetrate to a depth of 0.80–1 m.

Conditions for deep root penetration, as illustrated by the Patillas site, appear to be slowly rising water levels and an organic soil stratum, which extends to a depth of over 2.60 m. This is reflected by the presence of larger and less dense roots (with conspicuous nodules present) throughout the soil profile, and low total root biomass as compared with other forests.

A shallow root system was found in the Dorado coastal basin, and here the soil profile revealed a thin organic layer in the upper stratum (7 cm) and a deeper mixed horizon of sandy clay and silt to approximately 2 m. The highest mean root biomass was obtained in the top 10 cm of soil (2.0 kg m^{-2}) decreasing abruptly to 0.03 kg m^{-2} between 40 and 50 cm (Fig. 10.5). Here, uprooting was observed to be the cause of some tree mortality. This was not observed in the other forests. Shallow roots may represent a response to the anoxic conditions of waterlogged soils, and assist in trapping the nutrients released by decomposition of the plant material near the soil surface. Shallow roots, in any case, occur in waterlogged and nutrient-limited basins.

In general, the distribution pattern of roots with depth, and the total root biomass in *Pterocarpus* forests, are influenced by seasonality of flooding, depth of the organic stratum in the soil, and the input of underground fresh or brackish water. The latter is, in turn, governed by the general topography of the forest floor.

Soil organic matter

In the five forests studied, soil organic matter per unit area increased in the following order: Luquillo, Dorado, Mayagüez, Humacao, and Patillas (Table 10.4). A factorial analysis of variance showed significant differences among soil organic matter values found in the five forests ($F = 62.35$, $n_1 = 4$, $n_2 = 18$; $P < 0.001$). The SNK test separated samples into two groups according to population means, one composed of Luquillo, Mayagüez, Humacao and Dorado, and the other with Patillas alone.

In general, soil organic matter values found in *Pterocarpus* forests are very high compared with the great majority of terrestrial communities.

Litter production

Litter production in the three *Pterocarpus* freshwater swamps in Puerto Rico ranged from 8.7 to 14.1 t ha^{-1} yr^{-1} (Table 10.5), similar in magnitude to riverine salt-water wetlands (8.8–14.5 t ha^{-1} yr^{-1}: Brown and Lugo, 1982). Riverine forests tend to show higher litter production rates (14.1 t ha^{-1} yr^{-1}) than basins (11.9 and 8.7 t ha^{-1} yr^{-1}). A similar trend was found for freshwater forested wetlands (including the scrub type) in the southern United States, but the values were lower (2.5–8.2 t ha^{-1} yr^{-1}: Brown and Lugo, 1982). Litter production in *Pterocarpus* swamps in Guadeloupe was 10.5–14.2 t ha^{-1} yr^{-1} (Febvay and Kermarrec, 1978b).

Litter production of a riverine *Pterocarpus* freshwater swamp is approximately two to four times higher than that of freshwater forested wetlands from North Carolina to South Florida (U.S.A.). Basin *Pterocarpus* forests in Puerto Rico produce from 1.2 to 1.4 times the litter of their basin counterparts in the temperate zone.

PHENOLOGY

According to Bacon (Ch. 9, this volume), information on fruiting and seed production of swamp trees in the Caribbean region and the Guianas is minimal. Little information is available for *Pterocarpus officinalis*, and sometimes it is contradictory in nature.

From field observations and evidence collected in litter baskets during 1981, small differences in the flowering and fruiting periods among five *Pterocarpus* forests were found. Meteorological data collected in the localities suggest that the phenological variation corresponded to differences in rainfall regime or seasonality on different coasts of the island.

In general, *Pterocarpus officinalis* in Puerto Rico appears to have two flowering and fruiting periods during the year instead of year-long production as suggested by James (1979). The first flowering period extends from February to late May, and the

TABLE 10.5

Litter production in three *Pterocarpus officinalis* forested wetlands in Puerto Rico[1]

Type	Date (1981)	No. of days	Litter production			
			$g\,m^{-2}\,day^{-1}$ (SE)	n	$t\,ha^{-1}$	$t\,ha^{-1}\,yr^{-1}$
Coastal riverine	April 7	15	6.13 (0.94)	10		
(Mayagüez)	July 11	43	1.55 (0.19)	5		
	August 19	39	2.84 (0.44)	5		
			Mean 3.50 (1.36)		3.7 (97 days)	14.1
Coastal basin	April 2	15	2.46 (0.25)	10		
(Patillas)	May 5	33	3.14 (0.37)	6		
	May 24	19	1.68 (0.23)	8		
	June 20	27	2.57 (0.34)	8		
	August 23	64	3.37 (0.52)	8		
	September 21	29	4.57 (0.80)	8		
			Mean 2.96 (0.40)		6.1 (187 days)	11.9
Coastal basin	March 25	15	2.86	1		
(Dorado)	April 15	21	7.33	1		
	May 10	25	2.46 (0.31)	5		
	May 24	14	2.18 (0.41)	5		
	June 22	29	1.83 (0.29)	5		
			Mean 2.15[2] (0.18)		2.5[2] (104 days)	8.7

[1]This is a partial data set and it does not cover the dry season.
[2]Based on May and June data only.

first fruiting period from March to June. The second flowering peak occurs from mid-June to mid-August, with a fruiting period extended from mid-July to late September. Apparently, the initiation of both cycles is related to the end of the dry season and beginning of the wet one, and also to the driest month of the wet season. As noted by Reich and Borchert (1984), variations in soil water may affect the relative timing of the tree's endogenous growth periodicity and thus cause it to deviate from prevailing climatic periodicity.

What is apparently essential is the production of flowers during the end of the dry (or relatively dry) season. The fruiting period then occurs during the beginning of the wet season. At this time, water levels are generally at ground level, providing adequate moisture for seedling germination and establishment. Flower opening in *Pterocarpus* may be triggered by leaf fall and the subsequent rehydration of previously water-stressed trees, as may be inferred from evidence presented by Janzen (1975). The small variance in yearly temperatures suggests that this factor is not nearly as important as rainfall in triggering the reproductive phase in *Pterocarpus* in Puerto Rico and the Caribbean.

SEEDLING DENSITY, GERMINATION AND REGENERATION

The factors involved in seedling establishment in flooded environments are not clear. From my observations, I believe that seedling density, establishment and population maintenance are affected by the water regime and differences in soil microtopography, as well as other environmental conditions. Adequate soil moisture, low salinity, absence of flood and moderately shaded environments are conditions conducive to successful seed germination and seedling establishment in *Pterocarpus*.

The *Pterocarpus* fruit is a small ($\simeq 5$ cm in diameter), lightweight pod ($\simeq 1.60$ g), dispersed by air and water currents. Its buoyancy makes it possible for the seed to travel (Beard, 1949) via rivers, streams and the open sea. In Puerto Rico, *Pterocarpus* seedling densities in the five forests ranges from 4 to 44 m^{-2} based on measured areas of 9 to 10 m^2 in each forest (Table 10.6). Mean seedling density was 71.9 m^{-2} in Costa Rica (Janzen, 1978) and 124 m^{-2} in Dominica (James, 1979).

Highest mean density of seedlings was related to well-flushed and aerated soils, perched topography and continuous freshwater input (Luquillo, Humacao and Patillas). Sloping topography and slowly rising water levels provide adequate soil conditions for germination and growth of seedlings, as long as stagnant or anaerobic conditions do not occur throughout the year. In coastal systems, the effect of soil salinities at greater depths is excluded because of the relative soil elevation.

Low mean density of seedlings was related to natural depressed topography, lack of strong water inputs, and human disturbance (Mayaguez and Dorado). Extreme flooding conditions promote anoxic soil conditions that preclude successful seedling establishment at most times. Seedling germination was only possible in the more elevated terrain that surrounds the oldest, largest *Pterocarpus* trees and in more elevated areas at the margins. In some *Pterocarpus* forests in Puerto Rico, artificially built walkways and drainage from sugar cane fields interfered with natural water circulation and allowed flooding to persist, so that the water stagnated, and thus prevented seedling establishment.

PROPOSED *PTEROCARPUS OFFICINALIS* SEED IMPLANTATION MECHANISM

I have developed a model for seed implantation from field observations and data analysis of previous works. The general principles are applicable with few modifications to other freshwater swamp species that form monospecific stands or coexist in swamp forests. The model is dichotomous, establishing differences in seed implantation under flooded and dry soil conditions (Fig. 10.6).

Flooded conditions

Bacon (Ch. 9, this volume) reports that *Pterocarpus* is incapable of establishing seedlings under inundated conditions; the winged fruit can germinate when afloat, but rooting does not occur if water is deeper than 3–4 cm. This is in agreement with my observations in the *Pterocarpus* forests in Puerto Rico.

The growth of the embryo depends on internal and external factors such as water, oxygen, and temperature (Levit, 1972). During the early stages of germination respiration may be entirely anaerobic, but as soon as the seed coat is ruptured the seed switches to aerobic respiration. If the soil is severely waterlogged when fruit fall occurs, the amount of oxygen available to the seed may be inadequate for aerobic respiration to take place, and the seed will fail to germinate. If flooding persists, only those seeds transported to more elevated areas may survive. The longevity and seed

TABLE 10.6

Mean height and abundance of seedlings in *Pterocarpus officinalis* forests (the figures in parentheses are one standard error)

Site	Date	Sample size (m^2)	Mean height (cm)	Mean density (m^{-2})
Puerto Rico				
Luquillo	April 1981	10.0	24.7 (1.2)	44 (26.5)
Mayagüez	August 1981	10.0	36.8 (1.6)	9 (4.5)
Humacao	June 1981	10.0	41.4 (1.3)	17 (3.8)
Patillas	April 1981	9.0	44.1 (1.7)	28 (5.0)
Dorado	February 1981	10.0	39.1 (1.6)	4 (0.8)
Costa Rica	March 1979	11.7	28.35 (0.7)[1]	72
Dominica	1979	3.0		124[2]

[1]Calculated from Janzen (1978).
[2]Calculated from James (1979).

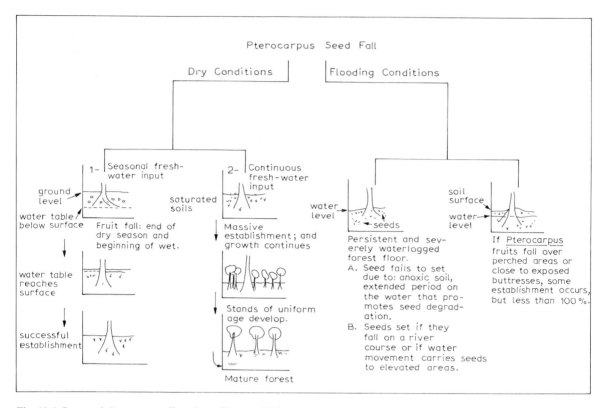

Fig. 10.6. Proposed *Pterocarpus officinalis* seedling establishment mechanisms.

viability after fruit fall in flooded or dry conditions is unknown.

If fruit fall occurs at or close to the time of decreasing water tables (that is late September), some more elevated areas are exposed and implantation can occur. These conditions apply mostly to basins and back-swamps of riverine systems where stagnant conditions appear to prevail most of the year.

Dry soil conditions

A massive successful implantation may occur if the fruits fall at the start of the seasonal freshwater input, when the water table rises slowly to reach the surface and the forest floor does not become suddenly inundated. In some forests a continuous underground freshwater input maintains adequate soil moisture without flooding even during the dry season. This factor, in combination with relatively elevated soil, may provide the ideal conditions for a massive and successful establishment, leading to continuous *Pterocarpus* seedling growth and stands of apparently uniform age. In Puerto Rico,

these conditions were observed in the *Pterocarpus* forests in Dorado, Patillas, Mayagüez and the best developed and extensive ones in Humacao. These phenomena were also observed in the *Pterocarpus* forests of Trinidad (Bacon, Ch. 9). Lindeman (1953) considered that the establishment of trees in swamps required fulfillment of a complex series of conditions, which occurs only once in several years and then only locally.

The stands of juvenile, uniform-aged *Pterocarpus* trees in the Puerto Rican swamps suggest that massive and successful establishment only occurred as a result of optimal conditions such as adequate soil moisture, low soil and water salinity, absence of flood, shaded environment and total lack of disturbance in successive years.

ECOTONAL ZONES AND *PTEROCARPUS* CLIMAX FORESTS

Ecotonal zones from *Pterocarpus* stands to neighboring mangrove communities may extend for 100 m or more (for example Humacao and Mayagüez) or a few meters only (Patillas, Fig. 10.7). Changes from the *Pterocarpus* community to a more terrestrial type, as in Dorado and Luquillo, take place in a few meters only.

The transition zones between *Pterocarpus* forests and coastal mangrove formations had relatively low complexity indices (Table 10.7). They were characterized by a low number of species (3 to 5 in 0.1 ha), relatively low basal areas (28.7 to 33.6 m^2 ha^{-1}), but a high tree density (1160–3230 ha^{-1}). Those forests surrounded by subtropical moist or wet forests exhibited, in general, a high complexity index, due principally to a large number of species (19–20) and high basal area (48.8 and 59.6 m^2 ha^{-1}).

The *Pterocarpus* ecotone zones were characterized by either abrupt or blending transitions, associated with either abrupt or blending salinity regimes, between the *Pterocarpus* and mangrove

Fig. 10.7. An example of a schematic profile of a basin *Pterocarpus* forest (data from Patillas). Ag = *Annona glabra*, An = *Anthurium acaule*, Fc = *Ficus citrifolia*, Lg = *Laguncularia racemosa*, Pt = *Pterocarpus officinalis* and Rh = *Rhizophora mangle*. The right-hand side of the diagram is towards the sea and the left-hand side is towards land.

TABLE 10.7

Complexity index (C.I.) calculated for five ecotonal areas in *Pterocarpus officinalis* forests in Puerto Rico (all trees ⩾ 2.5 cm diameter at breast height)

Forest site	Basal area (m^2 ha^{-1})	Density (ha^{-1})	Number of species in 0.1 ha	Mean stand height (m)[1]	C.I.
Luquillo	59.6	1460	20	20	348
Mayagüez	33.6	1160	5	15	29
Humacao[2]	30.7	3230	4	14	56
Patillas	28.7	2150	3	16	30
Dorado	47.8	1660	19	15	226

[1]Estimated values based on observations.
[2]Mean of two evaluations by point centered quarter plot and prism methods.

communities. Where abrupt transitions occur, ecotones were less complex than the *Pterocarpus* stands themselves. Blending ecotones exhibited higher species richness, tree densities, and basal areas than *Pterocarpus* forests. This occurs because, apart from the conspicuous presence of species of both communities, a number of other adaptable species tend to colonize transitional areas in blending ecotones (Daubenmire, 1968). In general, the type of ecotone that occurs depends on soil drainage and changes in the soil relief.

DISCUSSION

Differences in hydroperiod, coupled with the presence or absence of salinity appeared to influence forest structure. In basins, hydroperiods are long (still water), while in riverine forests they are shorter (moving water). In general, the complexity of *Pterocarpus* forests increases from the basin forest type to the montane riverine types. However, this trend reverses when coastal basin forests are compared with coastal riverine forests. The complexity index decreases because soil salinity becomes an important factor stressing the coastal riverine forests — thus reducing diversity and, therefore, complexity.

The montane riverine *Pterocarpus* forest had the tallest canopy height, greatest basal area and greatest number of species. Among coastal forests, basins support the highest species richness, tree density and basal area. The coastal riverine forests of Mayaguez and Humacao had low complexity indices, largely because they were virtually monospecific.

The observed differences in complexity and structure between the coastal basin forests and the coastal riverine types are related mainly to differences in soil salinities and flooding, and less to nutrient-rich upland runoff. Higher salinities were found in the coastal riverine forests (mean of 3.36‰) than in the coastal basins (mean of 2.34‰). The weak tidal forces coupled with evapotranspiration cause higher salinities in basin mangrove forests. Because tides do not normally flood *Pterocarpus* basins, but do flood riverine sites (they are closer to the sea), inverse salinity conditions develop. As a result, a contrasting structural pattern emerges in these two forest types. Some *Pterocarpus* basin forests may also have small, but continuous, freshwater inputs (as in Patillas forest) where freshwater underground springs keep salinity low.

In general, wetlands are characterized more by their large storage and export of organic matter in the soil than by the large aboveground biomass. In *Pterocarpus* forests, the riverine types store less organic matter in the soil, but more in root biomass. This may be due to problems associated with seasonal salinity input and/or strong pulsed hydroperiods in riverine forests. It is possible that higher root biomass compensates for the potentially lower efficiency of individual roots.

The storage of soil organic matter is higher in *Pterocarpus* forests than in surrounding terrestrial forests. Because of their geographical location on floodplains, *Pterocarpus* forested wetlands act as catchments, retaining large quantities of organic matter of their own production and that exported from upland systems.

My study suggests that *Pterocarpus* forests are climax systems developing under the following set of conditions: tropical location; flooding, with periods of drought to allow for seedling establishment; and low soil salinities of 0 to 12‰.

These conditions may occur from the coast to the mountains, in depressions and along coastal rivers away from the salt-water wedge (dominated by mangroves). Conditions along river courses with gentle slopes, and where pulsed hydroperiods occur, may be appropriate for successful colonization by *Pterocarpus*.

ACKNOWLEDGEMENTS

This study was completed thanks to the support provided by the Institute of Tropical Forestry and the Center for Energy and Environmental Research of Puerto Rico. In the different phases of the study A.E. Lugo, G. Cintrón, J.M. Lopez, R. Woodbury, P. Acevedo and S. Brown gave logistical support, critical comments and technical guidance.

REFERENCES

Alvarez, M., Santiago, J. and Quevedo, V., 1983. Estructura de 3 bosques de *Pterocarpus officinalis* en Puerto Rico. In:

A.E. Lugo (Editor), *Los Bosques de Puerto Rico*. Servicio Forestal de los E.E.U.U. y Departamento de Recursos Naturales, San Juan, P.R., pp. 283–308.

Bates, C.Z., 1929. Efectos del huracán del 13 de septiembre de 1928 a diferentes árboles. *Rev. Agric. Puerto Rico*, 23: 113–117.

Beard, J.S., 1949. *The Natural Vegetation of Windward and Leeward Islands*. Oxford Forestry Memoirs 21. Clarendon Press, Oxford, 192 pp.

Briscoe, C.B., 1966. *Weather in the Luquillo Mountains of Puerto Rico*. U.S. Department of Agriculture Forest Service Research Paper, Institute of Tropical Forestry-3. Institute of Tropical Forestry, U.S.D.A. Forest Service, Rio Piedras, P.R., 40 pp.

Brown, S. and Lugo, A.E., 1982. A comparison of structural and functional characteristics of saltwater and freshwater forested wetlands. In: B. Gopal, R.E. Turner, R.G. Wetzel and D.F. Whigham (Editors), *Wetlands: Ecology and Management. Proceedings of the First International Wetlands Conference*. National Institute of Ecology, Jaipur, pp. 109–130.

Cintrón, B., 1980. El bosque de *Pterocarpus*. *Conferencia en Primer Simposio sobre Anegados en Puerto Rico*. Departamento de Recursos Naturales, San Juan, P.R.

Daubenmire, R.F., 1968. Soil moisture in relation to vegetation distribution in the mountains of northern Idaho. *Ecology*, 49: 431–438.

Febvay, G. and Kermarrec, A., 1978a. Quelques paramètres physicochimiques de la fôret littorale (mangrove et fôret palustre) et leur évolution. *Bull. Liaison Groupe Trav., Mangroves Zones Côtières, Guadaloupe*, 4: 70–73.

Febvay, G. and Kermarrec, A., 1978b. Formation de la litière en fôret littorale et en fôret de montagne humide. *Bull. Liason Groupe Trav., Mangroves Zones Côtière, Guadeloupe*, 4: 79–80.

Gleason, H.A. and Cook, M.T., 1926. *Plant Ecology of Puerto Rico. Sci. Surv. Porto Rico Virgin Islands*, 7(2), 173 pp.

Holdridge, L.R., 1967. *Life Zone Ecology*. Tropical Science Center, San Jose, C.R., 206 pp.

Holdridge, L.R., Grenke, W.C., Hatheway, W.H., Liang, T. and Tosi, Jr., J.A., 1971. *Forest Environments in Tropical Life Zones: A Pilot Study*. Pergamon Press, Oxford, 747 pp.

James, A., 1979. *The Occurrence of Freshwater Swamps and Maritime Mangrove Species in Dominica*. Forestry and Wildlife Division, Ministry of Agriculture, Dominica (mimeo), 24 pp.

Janzen, D., 1975. *Ecology of Plants in the Tropics*. Edward Arnold, London, 66 pp.

Janzen, D., 1978. Description of a *Pterocarpus officinalis* (Leguminosae) monoculture in Concorvado National Park, Costa Rica. *Brenesia*, 14: 305–309.

Levit, J., 1972. *Responses of Plants to Environmental Stresses*. Academic Press, New York, N.Y., 697 pp.

Lindeman, J.C., 1953. The vegetation of the coastal region of Suriname. *Vegetation Suriname*, 1(1), 135 pp.

Reich, P.B. and Borchert, R., 1984. Water stress and tree phenology in a tropical dry forest in the lowlands of Costa Rica. *J. Ecol.*, 72(1): 61–74.

Stehlé, H., 1945. Forest types of the Caribbean Islands. *Caribb. For.*, 6: 273–414.

Woodbury, R.O., 1979. *Notes from Pterocarpus Forest — A Freshwater Swamp in Puerto Rico*. University of Puerto Rico, Rio Piedras, P.R., (mimeo) 2 pp.

Zar, J.H., 1974. *Biostatistical Analysis*. Prentice-Hall, Englewood Cliffs, N.J., 620 pp.

Chapter 11

PALM SWAMPS

RONALD L. MYERS

INTRODUCTION

Palms are the dominant plant life form in many tropical wetlands, and dominance by palms frequently increases with increasing hydroperiod. In some areas, a season of flooding is followed by a dry season. In such cases flooding and fire act in concert to produce palm-dominated forests. The first part of this chapter describes three major categories of palm-dominated swamps: monospecific swamps, species-rich swamps, and seasonal swamps. Palm swamps are largely unstudied, descriptions in the literature are sparse, and what is presented here more than likely only scratches the surface concerning palm swamp types, their areal extent, and species involved. The second part of the chapter includes detailed descriptions and data on the structure of palm swamps in Costa Rica, Central America, dominated by *Raphia taedigera* and *Manicaria saccifera*.

PALM SWAMP TYPES

Monospecific palm swamps

Zonation in many tropical freshwater wetlands from herbaceous marsh to mixed dicotyledonous swamp forest includes an intermediate stage that is dominated by one or several species of palm (Beard, 1944, 1959; Taylor, 1959; Webb, 1968) (Fig. 11.1). The predominant, but by no means exclusive, palm group involved is the Lepidocaryoidae, which includes species which seem particularly adapted to soils subjected to continuous flooding (Moore, 1973). The lepidocaryoid palm genera *Mauritia*, *Raphia*, and *Metroxylon* form extensive nearly monospecific forests on soils which are waterlogged most of the year. *Mauritia* is a South American genus; *Raphia* is a predominantly African genus (it has one species in tropical America); and *Metroxylon* occurs in the Far East and South Pacific Islands (Fig. 11.2). The existence of monospecific forests in the tropics runs counter to the general notion that tropical ecosystems have very high species diversity.

Mauritia swamps

There may be as many as 21 species of *Mauritia*, all of which are restricted to South America. Moore (1973) suggested, however, that this number may be greatly exaggerated. Mallaux (1975) mentioned palm swamps in Peru dominated by *Mauritia vinifera*, but containing some *M. flexuosa*. Wallace (1853) and Braun (1968) reported *Mauritia flexuosa* in the Orinoco River basin in Venezuela. Elsewhere in Venezuela, where *Mauritia* occupies wet depressions and gallery forests in the seasonally wet savannas, it is listed as *Mauritia minor* by Pittier (1938) and Ewel and Madriz (1968). In similar savannas in neighboring Colombia, Verdoorn (1945) and Espinal and Montenegro (1963) mentioned *Mauritia vinifera* as occupying wet depressions, while Blydenstein (1967) stated that the swamps along the rivers are dominated by *Mauritia minor*. Takeuchi (1960) described *Mauritia flexuosa* as forming the canopy of the gallery forests along small rivers in the savannas of the Territory of Rio Branco in northern Brazil, and Richards (1952) mentioned *Mauritia flexuosa* as forming pure stands in the delta of the Amazon.

In eastern Peru, where forested wetlands make up 23.1% of the forested land, *Mauritia* swamps reach their greatest size. More than a million

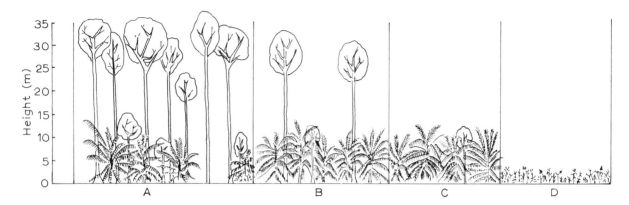

Fig. 11.1. Fluctuating swamp sequence in northeastern Papua with the intermediate stages (A and B) made up almost entirely of a dense layer of *Metroxylon sagu*. A = Seasonal swamp forest; B = open seasonal swamp forest; C = Seasonal swamp woodland; D = herbaceous swamp (from Taylor, 1959, reprinted by permission.)

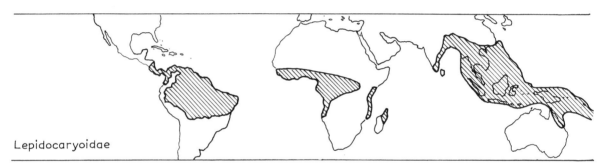

Fig. 11.2. Distribution of the lepidocaryoid group of palms, which includes the genera *Raphia*, *Mauritia* and *Metroxylon* forming palm swamps (Moore, 1973).

hectares, or 6.2% of the wetlands, are classed as *Mauritia* swamps (Mallaux, 1975). These "aguajales" support densities of *Mauritia* palms of 450–500 ha^{-1}, plus scattered individuals of *Ficus* sp., *Symphonia globulifera* and *Virola* sp. The understory is composed of the palms *Euterpe precatoria*, *Socratea* sp., and *Iriartea* sp. Mallaux (1975) recorded 182 *Euterpe* individuals per hectare. In the Rio Ucayali region of Peru, Guerra (1958) distinguished the permanently flooded swamp forests dominated by palms from intermittently flooded forests which parallel the river. The palm swamps were composed of *Mauritia* sp., *Scheelea cephalotes*, and *Scheelea basslerina*. The riverine forest had an open overstory of hardwoods and a dense understory of unidentified palms.

Mauritia flexuosa is said to be the most plentiful palm in South America. Due to its usefulness it is known at the "tree of life" to some indigenous human populations. Wilbert (1976) mentioned sago extraction from the pith by the Warao in Venezuela. Alfred Russell Wallace (1853) noted that at the mouth of the Orinoco River the local people depended almost entirely on the *Mauritia* swamps. They "build their houses elevated on its trunks, and live principally upon its fruit, sap, and fish from the water around them" (p. 7). *Mauritia* is becoming increasingly important economically. The National Research Council (1975) listed it as a potentially important unexploited crop and the Peruvians are contemplating management of their "aguajales" (Salazar, 1967; Mallaux, 1975). Oil can be extracted from the fruit and seed, and the fruit, which is made into a drink, is sold in local markets. The *Euterpe precatoria* understory is harvested for palm heart.

Raphia swamps

Raphia and *Elaeis* (the oil palms) are the only two palm genera that have members spanning the

Atlantic, and their ranges on both continents are practically identical (Moore, 1973). These facts have prompted speculation that both of the American species, *Raphia taedigera* and *Elaeis oleifera*, may have been introduced into the Western Hemisphere (Allen, 1965; Anderson and Mori, 1967). *Elaeis oleifera* does not form large pure stands, but it is commonly found in small aggregations in wetlands throughout northern South America and Central America. *Raphia taedigera*, on the other hand, forms large nearly monospecific swamp forests in the Caribbean lowlands of Colombia, Panama, Costa Rica, and the southern half of Nicaragua. It is also found in the Golfo Dulce region along the Pacific coast of Costa Rica and in the Amazon River delta.

In Colombia, *Raphia* is called "pangano". Espinal and Montenegro (1963) reported large "panganales" in the humid forests of the Golfo de Uraba region and along the lower courses of the Rio Atrato. Golley et al. (1976) also described "semi-aquatic communities" containing *Raphia* thickets along the Rio Atrato. In Costa Rica, where *Raphia* is called "yolillo", 606 km^2 or 1.2% of the forested land area is classified as "yolillales" (Sylvander, 1978).

Golley et al. (1976) working in Colombia, and Nuhn and Perez (1967), working in Costa Rica, mentioned that *Raphia* grows in close association with herbaceous aquatic and semi-aquatic communities. In the tidally influenced delta regions of the Amazon and Para Rivers in Brazil, Richards (1952) described a zonation and successional scheme involving *Raphia*. On new islands of alluvium and on the slip-off slopes of river meanders, the initial invader is either the large aroid *Montrichardia arborescens*, or the leguminous shrub *Drepanocarpus lunatus*, both of which form pure stands. These communities are replaced by *Rhizophora racemosa*, or in some cases *Avicennia tomentosa*, *A. nitida*, or *Laguncularia racemosa*, depending on the amount of brackish-water influence. The mangroves are eventually replaced by palm forest. This forest is either dominated by *Euterpe oleracea* and *Mauritia flexuosa*, which may later develop into typical "várzea" swamp, or by a fringe community of *Raphia taedigera*. Behind the *Raphia* is either a pure stand of *Mauritia flexuosa* or *Mauritia* mixed with other palms such as *Euterpe oleracea*, *Manicaria saccifera*, and *Maximiliana regia*. Common dicotyledonous trees are *Carapa guianensis* and *Virola surinamensis*.

Anderson and Mori (1967) suggested a successional process, in which the *Raphia*-dominated swamps are the pioneer community on sites where drainage has become impeded. During the early stages of development the palm canopy is closed, but as the palms mature the canopy opens. The improved light conditions on the forest floor permit hardwood establishment.

Little information is available on the *Raphia* swamps in Africa. The most extensive palm swamps are located in the humid lowlands of West Africa (Thompson and Hamilton, 1983), but they also occur in the wetlands surrounding Lake Victoria. The literature suggests that *Pandanus* swamps are more prevalent than the palm swamps on the west coast, but the palms *Raphia* spp., *Calamus* sp., and *Phoenix reclinata* are common elements of the *Pandanus* swamps (Richards, 1952). Richards (1952) described an aquatic-to-wetland successional scheme in southwestern Nigeria which starts with a free-floating aquatic community, followed by rooted aquatic plants, a graminoid or papyrus herbaceous marsh which succeeds to *Pandanus* swamp, and then palm swamp dominated by *Raphia*. In West Africa there are stands of *Raphia hookeri*. Behind the papyrus swamps along Lake Victoria, there is an abundance of *Phoenix reclinata* and *Raphia monbuttorum* mixed with tall hardwood trees (Eggeling, 1935). Secondary succession in swamp forests of Ghana produce forests that are dominated by *Raphia vinifera* about 7 years after clearing (Ahn, 1958). Langdale-Brown et al. (1964) mentioned palm swamps of *Phoenix reclinata*, *Elaeis guineensis*, and *Raphia monbuttorum* in Uganda.

Metroxylon swamps

The most extensive palm swamps in the Old World tropics are the *Metroxylon sagu* or sago palm swamps of New Guinea. Like *Raphia*, sago palm forms clumps whose individual stems are monocarpic. Moore (1973) listed eight species of *Metroxylon*, all of which are restricted to Southeast Asia, Oceania, and Australia. In growth form, habitat, and community structure and function, the *Metroxylon sagu* swamps of New Guinea seem analogous to the *Raphia taedigera* swamps of the

Neotropics. The New Guinea swamps have been described or mentioned by Taylor (1959, 1964a, 1964b), Paijmans (1967, 1969, 1971, and Ch. 14, this volume), Robbins (1968), and Heyligers (1972). *Metroxylon sagu* is very important to the marsh-dwelling human populations throughout the South Pacific islands (Barrau, 1959; Townsend, 1974) because, among other uses, it provides the staple carbohydrate source for much of the rural population in Oceania.

Metroxylon is an important component in most of the freshwater swamps along the wide alluvial coastal plain in Papua New Guinea (Taylor, 1959). In the brackish-water swamps it is replaced by *Nypa fruticans*, a nypoid palm. Like the *Raphia taedigera* swamps of the Neotropics, the swamps in Papua New Guinea dominated by *Metroxylon sagu* are very extensive, and occur in both rain-fed basins and on broad floodplains. The basin swamps, on peaty soils, have a sparse overstory of *Campnosperma auriculata* and a dense lower layer of *Metroxylon sagu*. In some cases the overstory is more diverse, with *Campnosperma auriculata*, *Garcinia* spp., *Neuburgia* sp. and *Syzygium* sp. The lower layer may include *Pandanus* spp. and *Schuirmansia henningsii*, in addition to sago palms (Taylor, 1964a).

The *Metroxylon*-dominated swamp forests associated with river systems are on mineral soil. Flooding is frequent and drainage is rapid, but the soils usually remain waterlogged (Taylor, 1959). These forests may consist of a single dense layer of sago palms, 6 to 12 m tall, in which hardwood trees are extremely rare, or they may be two-storied, with the lower layer consisting of dense *Metroxylon* and the overstory composed of sparse to moderately dense hardwoods. In many cases the hardwood *Bischofia javanica* constitutes 50% of the individuals in the canopy. The ground layer consists of spreading palm fronds that reach the ground, fallen fronds, and thin pneumatophores. Ground vegetation is very sparse (Taylor, 1964a).

Sago-palm swamps are frequently associated with herbaceous marshes dominated by *Phragmites karka* 3 to 4 m tall. Taylor (1959) maintained that the palm swamp was actually wetter than the herbaceous marsh. In many places the palms form round clumps surrounded by marsh vegetation. Paijmans (1967) described the boundaries between palm swamp and marsh as sharp, although toward the periphery of the palm swamp the palms become shorter and may be segregated into groves. Robbins (1968) reported a similar patchy pattern in which low circular groves of *Metroxylon*, up to 100 m in diameter, are scattered throughout a dense stand of *Phragmites karka*. Where better drained, *Metroxylon*-dominated swamps gradually grade into mixed hardwood swamp forests.

In the tidal zone near the New Guinea coast there is a zonation of *Metroxylon* swamps, *Pandanus* swamps, and *Nypa* swamps that appears to be associated with salinity (Robbins, 1968). *Nypa* swamps are similar in appearance to the *Metroxylon* swamp, but the canopy of the former is lower, rarely exceeding 6 m (Taylor, 1964b). They cover extensive tidal flats, especially on the more seasonal south coast of the island of New Guinea, and form fringe wetlands at the mouths of the larger rivers and along tidal creeks with mangroves behind them. In other situations the mangroves border the creeks and the *Nypa* community lies behind them.

The *Pandanus* swamps are usually monospecific, but stands mixed with *Metroxylon* also occur. After destructive floods, initial invasion on river banks and bars is by a low scrub of *Pandanus* and *Metroxylon*. The community is eventually overtopped by a forest of *Casuarina cunninghamiana*. In these broad tidally-influenced swamps, *Metroxylon* becomes increasingly more important inland (Paijmans, 1969). On floodplains, where growth is best, the fronds of non-reproductive stems are up to 10 m long and individuals with flowering stalks are over 20 m tall. In basins where the water is stagnant and nutrient-poor, or where there is a brackish-water influence, the palms may be dense but the individuals are seldom over 3 m tall, and reproductive individuals are rare.

Species-rich palm swamps

In broad floodplains and basins where flooding is not prolonged enough to produce monospecific palm forests, more species-rich forests occur. Although such forests contain many hardwood species, one or two species of palm are frequently abundant. These may be *Mauritia*, *Raphia*, *Metroxylon*, or other genera.

In the Amazon Basin a number of palm species are abundant in "várzea" swamps (Verdoorn,

1945; Takeuchi, 1961; Lovejoy, 1975; Irmler 1977). In the coastal regions of Guyana, in northeastern Venezuela, in Trinidad and Tobago, and in the Caribbean lowlands of Central America, there are several types of palm swamps in which the palms form a dense stratum under an overstory of dicotyledonous trees. One of these is the *Manicaria saccifera* swamp type (Fig. 11.3). Others include *Euterpe edulis*, *Jessenia oligocarpa*, *Maximiliana eligans*, and *Roystonea oleraceae* swamps (Fanshawe, 1954).

Not all tropical swamp forests have a prevalence of palms. In the riverine forests near the confluence of the Rio Negro and the Rio Branco palms are rare in "whitewater" swamps and absent in the "blackwater" swamps (Williams et al., 1972). Although Takeuchi (1961) reported that palms are absent in the "igapo" (either blackwater basin wetlands or riverine wetlands along large blackwater rivers that drain nutrient-poor areas) along the main river of the Amazon and the lower Rio Negro and its tributaries, *Leopoldinia pulchra* and several species of *Bactris* have been mentioned as characteristic palms of blackwater swamp forests (Junk, 1983). The paucity of palms in at least some basin wetlands of the Amazon is reminiscent of the East Indian peat swamps where palms are poorly represented and tend to occur only on the periphery (Whitmore, 1975).

Manicaria swamps

In Venezuela, *Manicaria* is called "temiche" and forms large "temichales". In Guyana it is known as "truli" and the swamps "truli bush"; in Brazil it is called "ubussu". In the remote northeastern corner of Costa Rica it is locally known as "palma real", and except for a brief mention of the tree by Nuhn and Perez (1967), the large Costa Rican *Manicaria* swamps have, until recently, been undescribed. On topographic and vegetation maps they are mislabeled *Raphia* swamps or "yolillales", but in this part of Costa Rica they are far more extensive than the *Raphia* type (Myers, 1981).

Manicaria swamps are estuarine, but the influence of brackish water is uncertain. Richards (1952) described *Manicaria* swamps as forming in brackish estuarine areas behind the mangroves, and suggested that they are analogous to the *Nypa fruticans* palm swamps of Southeast Asia and the *Pandanus* swamps on the west coast of Africa. In Guyana he noted that where the water is brackish there is a narrow fringe of *Rhizophora mangle* followed by a brackish *Manicaria* swamp. Budowski (1966) listed *Manicaria* swamps as a freshwater type that is frequently inundated. Lundell (1945) reported that *Manicaria* dominates the swamp areas of the Temesh River estuary in Belize. Wilbert (1976), in describing the use of *Manicaria* sago by the Warao of the Orinoco Delta of Venezuela, stated that the swamps are found in the Intermediate Delta zone behind the mangrove-covered Lower Delta. He noted that there is some tidal influence, but that flooding is largely due to rainfall. The annual flooding of the Orinoco River has little effect on the palm swamp, but during the dry season salt water penetrates the Intermediate Delta, turning the river brackish, and causing a drinking water shortage for the local human populations. During this time *Manicaria* milk and *Mauritia* wine serve as alternative sources of liquid. Fanshawe (1954) described the soils in the Guyana

Fig. 11.3. Distribution of *Manicaria* (Moore, 1973).

swamps as peat from 1 to 2 m deep overlying alluvial clay. The soil becomes inundated for short periods at the height of the rainy season, and at high tide during the times of flood. During dry periods the peat may dry out enough to burn.

In Trinidad, the overstory trees in *Manicaria* swamps are *Calophyllum lucidum*, *Carapa guianensis*, *Clusia* sp., *Ficus* sp., *Parinari coriaceus* and *Symphonia globulifera* (Beard, 1946). In Panama, the palms are overtopped by *Carapa guianensis* (Holdridge and Budowski, 1956). In Guyana, Fanshawe (1954) described a *Manicaria saccifera* faciation of a *Symphonia–Tabebuia–Euterpe* palm association. The two-storied association is named for the dominant overstory trees, *Symphonia globulifera*, *Tabebuia insigna* var. *monophylla*, and the palm *Euterpe edulis*. The middle stratum is almost exclusively *Manicaria saccifera*. The ground layer is sparse except for *Manicaria* seedlings. In places the overstory is so dense that the *Manicaria* understory may be lacking. The emergent trees average 60 cm in diameter and 15 m in height. *Manicaria* averages 20 cm in diameter and 8 to 10 m in height. In these Guyana swamps, two palm species, *M. saccifera* and *E. edulis*, occupy 72% of the canopy and comprise 58% of the individuals (Fanshawe, 1954). Forty-five hardwood tree species and five palm species were recorded. In places where the coastal wetlands extend from one river system to the next, the *Manicaria* swamps, which form along the rivers, are replaced by a swamp forest dominated by the myristicaceous tree *Iryanthera paraensis* away from the rivers. This inland swamp is described as an *Iryanthera–Tabebuia* community (Fanshawe, 1954). Palms are a negligible component, consisting of five species comprising 5% of the individuals.

Seasonal palm swamps

Seasonally flooded, fire-maintained palm swamps are common in depressions and along river courses in the llanos of Colombia and Venezuela, in northeastern Brazil, and along the Río Paraguay. Smaller fire-maintained palm stands are a common feature in poorly drained areas throughout the seasonally dry forests of tropical and subtropical America. A number of species are involved. Species of *Mauritia* are common in the more humid savannas of northern South America. In depressions and along river courses in drier areas of Colombia there are pure stands of *Scheelea magdalena* (Espinal and Montenegro, 1963). *Scheelea rostrata* forms similar stands in the dry forests of Costa Rica (Holdridge and Poveda, 1975). Forests dominated by *Sabal palmetto* are maintained by a combination of fire and flooding in subtropical Florida (Brown, 1976; Myers, 1977).

In the states of Maranhao and Pinaui in northeastern Brazil the most abundant palm vegetation is composed of nearly pure stands of *Orbignya martiana* (babassu). These generally do not flood, and the species seems to have been favored by human activities (Anderson, 1983). Within the same region there are seasonal swamp forests of the wax palm, *Copernicia cerifera*, and, on even wetter sites, *Mauritia flexuosa* and *Euterpe oleracea*. The *Copernicia* forests are favored by the flood–drought–fire regime. The palm grows best on the deep alluvial soils near rivers and coastal areas where it forms sparse pure stands (Johnson, 1972). As the amount of flooding decreases, an understory of brush develops. On sites above the flood level it is replaced by *Orbignya martiana*. The *Copernicia* palms are heavily exploited for wax, construction materials, thatch, and food.

Along the Río Paraguay in the state of Matto Grosso, Brazil, *Copernicia australis* grows in pure stands of low density (Johnson, 1972). Gurvich (1945) reported stands of *Butia capitata* in the lowlands of the Department of Rocha Cruben in Uruguay. To the east of the Río Paraguay there are savannas of *Acrocomia totai* and *Euterpe edulis* in the drier areas; these species are replaced by *Arecastrum remanzoffranum*, *Bactris anizitzii*, *B. bidentata*, *B. inundata* and *Geonoma schottiana* on wetter sites (Rojas and Carabia, 1945; Michalowski, 1958). According to Michalowski (1958), the species of *Bactris* form dense gallery forests which are frequently flooded. He also mentioned other palm wetlands with *Scheelea parviflora* as the major constituent. The existence of fire in these swamps is not mentioned. To the west of the Río Paraguay there are savannas with *Copernicia australis* forming dense stands along the river. Further west it is replaced by *Bactris anizitzii* and *Trithrinax biflabellata* (Rojas and Carabia, 1945).

COSTA RICAN PALM SWAMPS

The region

Costa Rica, a country of 49 132 km² and vertical relief ranging from sea level to 3800 m, contains 12 ecological life zones (sensu Holdridge, 1967). Five of these life zones contain significant areas of wetlands (Tosi, 1972). In 1977, nearly 21 000 km², or 42% of the total land area, was under some form of forest cover. Of this, 1390 km² or 7% was classed as forested wetlands. When herbaceous marshes are included, the total wetland area is 2352 km² or 5% of the national territory (Sylvander, 1978). Most swamps (783 km² or 42%) are in the Tropical Wet Forest life zone (Tosi, 1972), and most of this life zone lies in the Caribbean lowlands of the northeastern quarter of the country.

A large part of the Caribbean lowlands of Costa Rica (Fig. 11.4) is remote and poorly known ecologically. Ecological life zones have been mapped by Tosi (1969). The narrow coastal plain

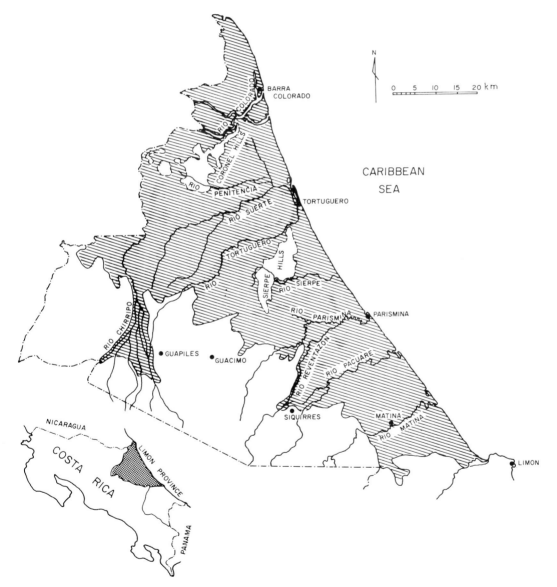

Fig. 11.4. Major cultural and physiographic features of the Caribbean lowlands of Costa Rica. Shaded area indicates terrain subject to inundation.

south of Puerto Limón is Tropical Moist Forest and going northward there is a sequence of transitional zones of increasing humidity, ultimately reaching Tropical Wet Forest. Data accumulated by the Costa Rican National Meteorological Service (1978) show Limón with a 38-yr average rainfall of 3587 mm, but there is considerable variation from year to year. The maximum recorded is 6078 mm and the minimum is 2287 mm. At the northern end of the region, 20-yr records at Barra del Colorado show a mean of 5787 mm. The highest recorded was 8831 mm in 1957. Inland from the coast there is a gradual decrease in precipitation until orographic effects come into play on the eastern slope of the Cordillera Central.

In describing Costa Rican palm swamps, I will present data from surveys, observations, and experiments conducted in the swamp forests of the Tropical Wet Forest life zone near Tortuguero (Lat. 10°37′N, Long. 83°22′W). Forest structure studies and water-level monitoring were undertaken to (a) determine the species composition and stand structure of several swamp forest types (b) compare swamp forest composition and structure with forests on zonal soils, (c) examine some of the edaphic and hydric features of each community, and (d) accumulate evidence that may help ascertain the successional relationships among the swamp-forest types. The methods used for the observations are described in the Appendix (pp. 284–285).

Tortuguero is a small coastal village (approximately 25 dwellings) located near the juncture of the Río Tortuguero, the Laguna Penitencia, and the Laguna Tortuguero. The residents derive their livelihood from fishing, hunting, wood cutting, and subsistence agriculture. Farms are located along the coast, on the banks of lagoons, and on the higher river levees. To the south of the village lies the Tortuguero National Park covering 27 000 ha, established in 1975 to protect the marine turtle nesting beach and examples of the forest types in the region.

The climate in the Tortuguero area is aseasonal, but heavy rainfall and extensive flooding routinely occur in July–August and November–December. Drier months are less predictable, but January through April and September usually experience the least rainfall. The average annual rainfall (over 2 years) was 5061 mm; days without rain amounted to 127 per annum, the longest period without rain being 13 days.

Tidal influence on the Caribbean coast is minimal. The mean tidal range at Tortuguero is about 0.25 m, and it only rarely exceeds 0.5 m (Nordlie and Kelso, 1975). The slight tidal fluctuation and the aseasonal heavy rainfall and runoff interact to reduce salinities in the estuary and to preclude the existence of mangrove vegetation. During a dry period Nordlie and Kelso (1975) measured high-tide salinities in the Tortuguero estuary of 1.8‰ at the surface and 26.4‰ at a depth of 12 m. During rainy periods the surface salinities dropped to about 0.2‰.

Study sites

Most of the data were collected between September 1977 and April 1979 at locations shown in Fig. 11.5. Local names of rivers and channels are used when they differ from those on the 1:50 000 topographic maps produced by the Costa Rican National Geographic Institute. Field studies were concentrated in the following four locations:

(a) Rain-fed *Raphia* swamp. This is drained by Caño Servulo, a blackwater channel. The *Raphia* swamp, 1920 ha in area, is located at the eastern base of the Lomas de Sierpe, a group of volcanic hills. The soils are organic and usually remain permanently flooded.

(b) River-fed *Raphia* swamp No. 1. This is a *Raphia* swamp of 1330 ha bisected by the Río Penitencia, which originates in the piedmont region. The river water is sediment-laden and river levees are well developed. Vegetation on the levees is dominated by *Raphia* with some *Pentaclethra macroloba* and scattered groves of *Prioria copaifera*. Soils are alluvial clays low in organic matter. The swamp usually remains permanently flooded or waterlogged.

(c) River-fed *Raphia* swamp No. 2. This is located on Caño Mora, a small stream, clear to moderately sediment-laden, originating in the upper flood plain. The vegetation includes a small herbaceous marsh surrounded by a small (~30 ha) *Raphia* swamp which grades into mixed swamp forest. *Pentaclethra* is the most common dicotyledonous tree species.

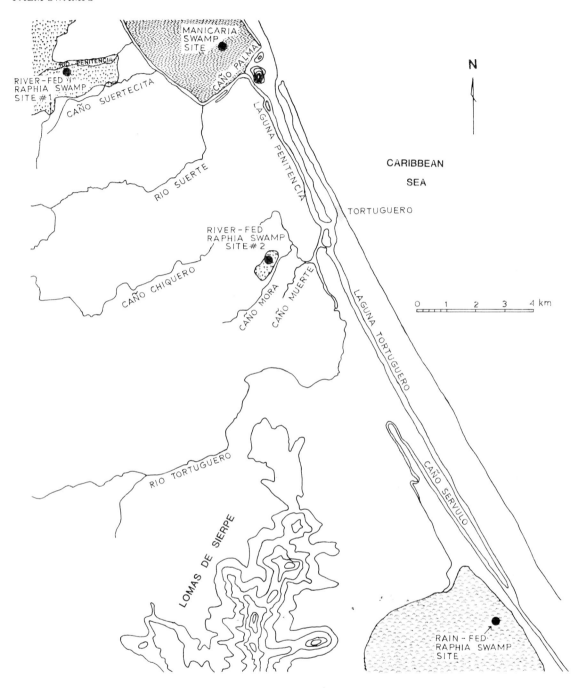

Fig. 11.5. Study site locations near Tortuguero, Costa Rica.

(d) Rain-fed *Manicaria* swamp. This is a wetland of 8215 ha bounded by Caño Penitencia, the Cerros de Coronel, Laguna Samay, and the Caribbean coastal fringe, and is drained by Caño Palma, a blackwater channel. The predominant vegetation type is *Manicaria* palm swamp, with *Raphia* groves occurring in depressions and mixed swamp forests on rises. The hydromorphic soils are variable, being derived from deposition during delta formation and coastal regosols. Organic matter accumulation occurs in poorly drained areas.

Water levels

Water levels in the rain-fed *Raphia* swamp were not monitored, but standing water was always present. The degree of water fluctuation is not known.

Water-level changes in the two river-fed *Raphia* swamps are shown in Fig. 11.6a and b. The Río

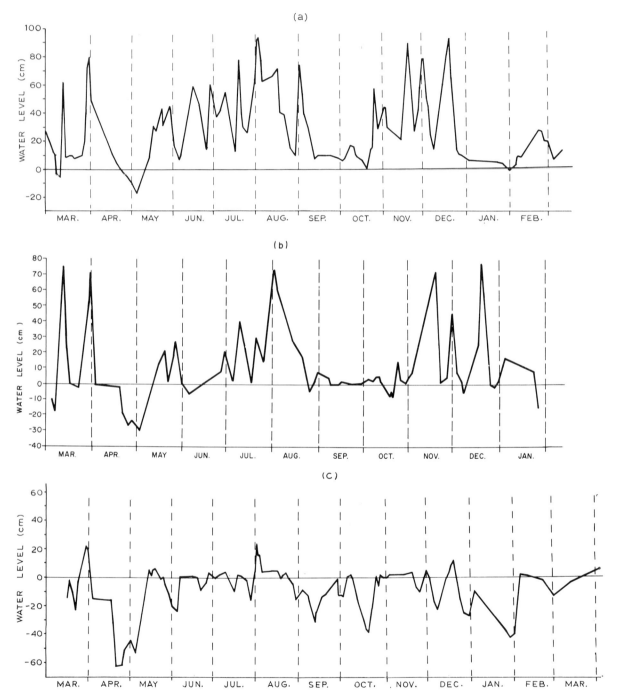

Fig. 11.6. Water levels from March 1978 through February 1979 in (a) river-fed *Raphia* swamp No. 1 on Río Penitencia, (b) river-fed *Raphia* swamp No. 2 on Caño Mora, and (c) rain-fed *Manicaria* swamp drained by Caño Palma.

Penitencia swamp (No. 1) was flooded 92% of the time, and a high water mark of 93 cm was reached in August 1978. The lowest level recorded was −17 cm in May 1978. The Caño Mora swamp (No. 2) was flooded 66% of the time. The high was 76 cm, recorded in December 1978, and the low was −27 cm in April 1978. The adjacent herbaceous marsh was flooded 52% of the time. It had a high water level of 76 cm in August 1978 and a low of −25 cm in April 1978. The marsh was slightly lower than the swamp, but because of its proximity to the channel it drained faster. Occasionally it would be flooded by water backed up by high tides that would not reach into the swamp.

In the rain-fed *Manicaria* swamp the ground surface was hummocky. Mounds were interspersed with depressions and channels. Many of the depressions remained permanently flooded, while some of the mounds were rarely if ever submerged. In the vicinity of the transects, the swamp was flooded 33% of the year (Fig. 11.6c). It was flooded on 16 separate occasions, with a maximum period of inundation of 18 days. The highest water level, 24 cm, occurred in August 1978 and the low, −62 cm, occurred in April 1978. There were many days when the water was at or within 1–2 cm of the soil surface, in effect making the soil saturated or flooded about 53% of the year.

Forest structure

The river-fed *Raphia* swamp No. 1 and the rain-fed *Raphia* swamp had similar species densities, stand basal areas, stand heights, and relative abundances of *Raphia* (Tables 11.1, 11.2 and 11.3); nevertheless, there was only one hardwood tree species (*Grias fendleri*) in common between the two.

Fig. 11.7 shows community similarity among the four swamps. When the dominant species *Raphia* is included, the three *Raphia* swamps are very similar. However, when *Raphia* is excluded the communities that remain are very dissimilar. The river-fed *Raphia* swamp No. 2 and the rain-fed *Manicaria* swamp had similar densities of tree species, but the other structural features were dissimilar. The large difference in basal area between the *Raphia* swamps as a whole and the other two forest types reflects the fact that the diameter of *Raphia* clumps was measured as opposed to single stems.

In general, the number of species increased with improved drainage. The difference in structure was more apparent when palms and hardwoods were considered separately. There was a decrease in the ratio of palm stems to hardwood stems from the most flooded sites through the unflooded site (Fig. 11.8). The *Manicaria* swamp had the greatest number of palm stems (680 ha^{-1}) but it also had a large number of hardwood stems (230 ha^{-1}). In

TABLE 11.1

Summary of data from transects of 0.1 ha in five forests in Costa Rica

Forest type	Location	Sample area (m^2)	Number of species in sample area	Mean species density in 0.1 ha	Stand density (stems ha^{-1})	Stand basal area (m^2 ha^{-1})	Stand height[1] (m)
Rain-fed *Raphia* swamp	Caño Servulo	3000	9	5.0	640	388	14.5
River-fed *Raphia* swamp No. 1	Río Penitencia	3000	9	5.3	637	344	15.3
River-fed *Raphia* swamp No. 2	Caño Mora	3000	11	7.7	527	224	15.1 (24)
Rain-fed *Manicaria* swamp	Caño Palma	3000	13	7.0	910	94	10.9 (22)
Slope forest	Cerro Tortuguero	1000	27	27.0	520	31	39.0

[1]The numbers in parentheses are the mean height of the dicotyledonous trees.

TABLE 11.2

Composition of the rain-fed *Raphia* swamp at Caño Servulo

Species	Relative frequency[1]	Relative abundance[2]	Relative dominance[3]	Importance value[4]
Raphia taedigera	67.27	89.06	99.42	255.75
Grias fendleri	18.18	5.73	0.36	24.27
Cassipoura guianensis	3.64	2.08	0.04	5.76
Campnosperma panamensis	1.82	0.52	0.12	2.46
Guatteria amphifolia	1.82	0.52	0.01	2.36
Posoquiera latifolia	1.82	0.52	0.01	2.35
Rauvolfia sp.	1.82	0.52	0.01	2.35
Ardisia sp.	1.82	0.52	0.01	2.35
Ixora nicaraguensis	1.82	0.52	0.01	2.35
Total	100.01	99.99	100.00	300.00

[1] In 50 m^2.
[2] Based on density.
[3] Based on basal area.
[4] Sum of preceding columns.

TABLE 11.3

Composition of the river-fed *Raphia* swamp No. 1 at Río Penitencia

Species	Relative frequency[1]	Relative abundance[1]	Relative dominance[1]	Importance value[1]
Raphia taedigera	64.87	84.35	98.00	247.22
Pentaclethra macroloba	8.81	4.19	0.92	13.92
Grias fendleri	8.81	3.61	0.27	12.69
Crudia acuminata	7.01	3.14	0.10	10.25
Prioria copaifera	3.53	1.05	0.06	4.64
Leuhea seemannii	1.74	1.57	0.61	3.92
Homalium racemosum	1.74	1.05	0.02	2.81
Cecropia obtusifolia	1.74	0.52	0.01	2.27
Spondias mombin	1.74	0.52	0.01	2.27
Total	99.99	100.00	100.00	299.99

[1] See Table 11.2.

the *Raphia* swamps, hardwood basal area was considerably less than in the other forests (Fig. 11.9). There was little difference in hardwood basal area between the *Manicaria* swamp and the slope forest.

Descriptions and characteristics of each of the swamp types are presented below:

Rain-fed Raphia swamp. The data for the nine species recorded in the Caño Servulo swamp (Table 11.2) show the marked predominance of the *Raphia* palm. Total density of stems (≥ 10 cm dbh) was 640 ha^{-1}. This included 173 *Raphia* clumps with a total of 570 stems. *Grias fendleri*, the next most abundant species, had a stem density of 37 ha^{-1}.

The non-palm tree species were widely scattered, small, spindly, and rarely taller than the *Raphia* palms. The exception was the scattered emergent, *Campnosperma panamensis*. Most of the hardwoods were small in diameter. Only two individuals on the transects, one of *Grias fendleri*, the other *Campnosperma panamensis*, exceeded 40 cm dbh. Basal area of hardwoods was quite low (2.27 m^2 ha^{-1}).

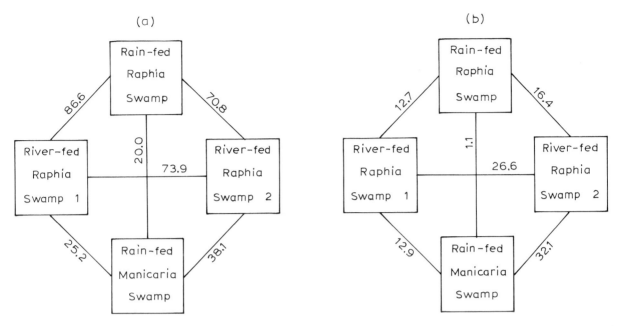

Fig. 11.7. Community similarity (ranging from 0 to 100) calculated using importance values of species (a) similarity calculated using all species ⩾10 cm dbh and (b) similarity when presence of *Raphia* is ignored.

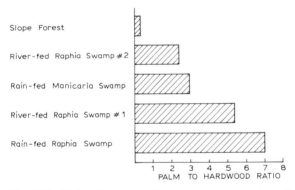

Fig. 11.8. Ratio of number of palm stems to number of hardwood stems (⩾10 cm dbh) in five forest types.

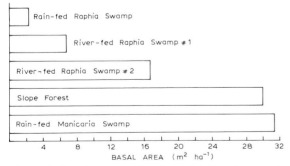

Fig. 11.9. Basal area ($m^2\ ha^{-1}$) of hardwood trees (⩾10 cm dbh) in five forest types.

The forest appeared open and it was fairly easy to walk through. A dense mat of roots, pneumatophores, and an herbaceous ground cover dominated by two sedges, *Becquerelia cymosa* and *Calyptrocarya glomerulata*, provided relatively firm footing. The open canopy may be due to shorter fronds or fewer fronds per stem than in the other *Raphia* swamps which seemed more closed, but the density of the palm clumps and the mean basal area per clump were slightly greater in this swamp than in the others.

A common low shrub in the understory was *Psychotria chugrensis*, and small trees < 10 cm dbh included *Alchornea costarricense*, *Eschweilera* sp., *Mabuetia guatemalensis*, *Posoquiera grandiflora*, and small specimens of those listed in Table 11.2. Other understory species encountered were the ginger *Renealnia aromatica*, the herb *Palicourea fastigiata*, the fern *Adiantum latifolium*, and the shrub *Tabernaemontana chrysocarpa*. Small *Raphia* palms were sparse; only one small palm was recorded on the three 0.1 ha transects, and seeds and seedlings were rarely encountered. Notably absent was the tree *Pentaclethra macroloba*. No individuals of *Manicaria* were observed on the ground, but from the air it appeared quite abundant along the western boundary of the swamp near the base of the Lomas de Sierpe.

River-fed Raphia swamp No. 1. *Raphia* was as important in this swamp as in the previous one (compare Table 11.3 and Table 11.2). The density of *Raphia* clumps was 163 ha^{-1}, with a total of 537 stems. The next most important species was *Pentaclethra macroloba*, with a density of 27 ha^{-1}. Most hardwood trees were widely scattered. Their size ranged from small understory trees to very large emergents. Large emergent specimens of *Pterocarpus officinalis* were observed in the swamp, but they did not fall within the transects.

Unlike the rain-fed *Raphia* swamp, the canopy was closed. Understory vegetation was sparse, and the swamp floor was covered with pneumatophores and fallen fronds. A dense root mat was lacking except around the base of each palm clump. The understory differed from that of the rain-fed swamp and included *Calathea lagunae*, *C. lutea*, *Cyclanthus bipartitus* and *Spathiphyllum friedrichsthalli*. The small palm *Calyptrogyne glauca* was rare (13 individuals ha^{-1}). Newly germinated *Raphia* seedlings clustered around the base of the parent clumps were very common. The density of juvenile *Raphia* was 155 ha^{-1}. Density of woody stems between 2 and 10 cm dbh was 365 ha^{-1}.

River-fed Raphia swamp No. 2. As in the river-fed *Raphia* No. 1, *Pentaclethra* and *Grias* were the most important hardwoods, but there were 25% fewer *Raphia* clumps and twice as many hardwood stems (Table 11.4). Basal area of the hardwood species was 16.72 m^2 ha^{-1} compared to 6.86 m^2 ha^{-1} in the Río Penitencia swamp.

In addition to the species found in the understory of the river-fed *Raphia* swamp No. 1, there was also an abundance of the herbs *Jussiaea latifolia*, *Orthorcade laza*, *Pharus latifolius* and *Urospatha tonduzzii*. Unlike the previous swamp, the understory was relatively dense. Individuals of the palm *Calyptrogyne glauca* were very common (643 ha^{-1}); juvenile *Raphia* were less common (85 ha^{-1}); and juvenile *Manicaria* were also present (85 ha^{-1}). Mature *Manicaria* palms tended to increase in abundance further into the swamp.

The marsh adjacent to this swamp consisted of an impenetrable mass of herbaceous vegetation 1 to 2 m tall. Common species were *Calathea lagunae*, *C. lutea*, *Cyperus giganteus*, *Lasiacis procerruna*, *Montrichardia arborescens*, *Panicum* sp., *Scleria* sp., and *Thalia geniculata*, intertwined with the vines *Ipomoea* sp. and *Solanum lancefolia*.

Rain-fed Manicaria swamp. The structure of this swamp forest is considerably different from that of the *Raphia* swamps (Table 11.5). Density of stems (910 ha^{-1}) was at least 30% greater than in the other swamps. The density of *Manicaria*, which forms a nearly pure second stratum below an open canopy of large hardwoods, was 663 ha^{-1}. *Raphia*

TABLE 11.4

Composition of the river-fed *Raphia* swamp No. 2 at Caño Mora

Species	Relative frequency[1]	Relative abundance[1]	Relative dominance[1]	Importance value[1]
Raphia taedigera	40.54	62.03	92.52	195.09
Pentaclethra macroloba	20.27	15.19	3.54	39.00
Grias fendleri	10.81	5.70	0.71	17.22
Manicaria saccifera	6.76	6.32	0.91	13.99
Pithecolobium latifolium	6.76	3.17	1.34	11.27
Protium pittieri	5.41	3.17	0.12	8.70
Pachira aquatica	2.70	1.27	0.10	4.07
Sickingia maxonii	2.70	1.27	0.03	4.00
Ficus sp.	1.35	0.63	0.66	2.64
Stemmadenia sp.	1.35	0.63	0.03	2.01
Astrocaryum alatum	1.35	0.63	0.01	1.99
Total	100.00	100.01	99.97	299.98

[1]See Table 11.2.

TABLE 11.5

Composition of the rain-fed *Manicaria* swamp at Caño Palma

Species	Relative frequency[1]	Relative abundance[1]	Relative dominance[1]	Importance value[1]
Manicaria saccifera	46.90	72.88	23.77	143.55
Raphia taedigera	4.31	7.69	42.79	55.79
Pentaclethra macroloba	13.27	6.23	6.11	25.61
Pterocarpus officinalis	4.42	1.83	15.90	22.15
Carapa guianensis	6.19	2.56	4.32	13.07
Amanoa potamophila	7.08	2.93	1.41	11.07
Pithocolobium latifolium	6.19	1.10	0.75	8.04
Sickingia maxonii	3.54	1.10	2.31	6.95
Grias fendleri	2.65	1.10	0.59	4.34
Ficus sp.	1.77	0.74	0.11	2.62
Symphonia globulifera	0.88	0.74	0.95	2.57
Calophyllum brasiliense	0.88	0.74	0.67	2.29
Prioria copaifera	0.88	0.36	0.32	1.56
Total	99.96	100.00	100.00	299.96

[1]See Table 11.2.

was the second most abundant species, and tended to be aggregated in depressions. Total palm density was 733 ha^{-1}. *Pentaclethra macroloba* was the most abundant woody non-palm species, with a density of 57 ha^{-1}. Although density was low, the basal area of the hardwoods was high (31.34 m^2 ha^{-1}); 37% of the hardwood stems were greater than 30 cm dbh.

Ground vegetation was almost totally lacking except for numerous *Manicaria* seedlings and juveniles. The density of juvenile *Manicaria* was 850 ha^{-1}. *Raphia* seedlings and juveniles were not encountered. Where *Raphia* clumps were aggregated there was a ground cover very similar to that found in the rain-fed *Raphia* swamp drained by Caño Servulo. The sedges *Becquerelia cymosa* and *Calyptrocarya glomerulata* predominated.

Slope forest. In the slope forest on Cerro Tortuguero, 27 species were recorded on the 10 × 100 mm transect (Table 11.1). *Pentaclethra macroloba* was the most abundant canopy tree (relative abundance 7.7%). In a similar forest 20 km to the north, Holdridge et al. (1971) recorded approximately 12% relative abundance of *Pentaclethra* on the Cerros de Coronel. On three transects in two different forests they recorded twice as many species as found on the single transect on Cerro Tortuguero, but the number of species in 0.1 ha was approximately the same.

The second most abundant tree species on Cerro Tortuguero was *Prioria copaifera*, a common floodplain species. Holdridge et al. (1971) did not encounter *Prioria* on the Cerros de Coronel, and maintained that it was successful only in special alluvial situations. They described a forest on an old sand bar in the Río Colorado where *Prioria* comprised 85% of the stems and 95% of the basal area. On the other hand, Nuhn and Pérez (1967) reported *Prioria* on Cerro Tortuguero and Pittier (1908) reported that it was found in slope forests at elevations of 200 m.

Other species encountered on Cerro Tortuguero included *Apeiba aspera*, *Brosimum costaricanum*, *Guarea* sp., *Luehea seemannii*, *Virola sebifera*, and the palm, *Socratea durissima*. Understory palms included *Asterogyne martiana*, *Astrocaryum alatum* and *Chryosophila albida*. Eleven palms ⩾ 10 cm dbh were recorded on the transect. The shrub layer was dominated by dwarf palms.

DISCUSSION

Although I classed two of the *Raphia* swamps as river-fed, they have distinct water-delivery features. Río Penitencia, which flows through swamp No. 1, originates in the piedmont region above the Llanuras de Tortuguero, and is sediment-laden year-round. The sediment loads are increased by erosion from farming and cattle ranching along its upper reaches. In contrast, Caño Mora flowing through swamp No. 2 originates in the upper floodplain and has very little human activity along its banks; much of the water entering this *Raphia* swamp probably flows out of an extensive mixed swamp forest that extends to the north and west. Sediment inputs into the swamp are far lower than in the Río Penitencia swamp No. 1, and organic accumulation in the soil is greater. During dry periods outflow is hindered by tidal damming. Water budgets would probably show more sheet-flow input to the Caño Mora swamp No. 2 than to the Río Penitencia swamp No. 1. The greater abundance of hardwoods in the Caño Mora swamp No. 2 is probably a reflection of its shorter hydroperiod and smaller size.

The *Raphia* swamp drained by Caño Servulo and the *Manicaria* swamp drained by Caño Palma are rain-fed swamps. There is little or no input of mineral sediment and the soils are organic. Although they appear as basins, they are probably not topographically lower than the surrounding terrain. From the air the Caño Servulo *Raphia* swamp looks like a depression because of the lower stature of the vegetation compared to that of the surrounding mixed swamp forest, but it may be topographically higher. The combination of high rainfall, long distance from main channels, and minuscule gradient may account for slow drainage. The same may be the case in the *Manicaria* swamp. Again, hydroperiod probably accounts for the structural differences.

The Caño Servulo *Raphia* swamp is large (1920 ha); but it is not known how many similar rain-fed swamps dominated by *Raphia* exist in the Atlantic lowlands of Costa Rica. They are isolated and access is difficult. They are usually surrounded by mixed swamp forest that must first be traversed on foot. The rivers make river-fed swamps far more accessible. Another rain-fed *Raphia* swamp was observed from the air at the southeast base of the Cerros de Coronel. It appears that the presence of the hills acts to isolate the swamps from sediment-laden rivers flowing out of the Cordillera and piedmont. Small rain-fed *Raphia* swamps also occur scattered in depressions in the extensive *Manicaria* palm swamp between Tortuguero and Barra del Colorado. The *Manicaria* swamp appears to have formed on part of the Río Colorado delta which became isolated from river floodwaters, and it may have originated from a rain-fed *Raphia* swamp.

Two factors appear to influence the composition and forest structure of these swamps: (1) the nature of the water source, and (2) the hydroperiod. The species in the two rain-fed swamps were quite different from those in the two river-fed swamps. The differences between the two swamps in each category may be related to hydroperiod and/or successional status. The hydroperiod similarities between the Caño Mora *Raphia* swamp (No. 2) and its adjacent marsh suggest that these two communities are successional stages and do not merely represent zonation.

The slope forest represents the climatic potential of Tropical Wet Forest in the Caribbean lowlands of Costa Rica. It is species-rich and structurally complex. The complexity index[1] developed by Holdridge (1967) was used to compare forest structural complexity. The index is the product of number of stems $\geqslant 10$ cm dbh, the number of species, basal area (in m^2), mean canopy height (m) (all measured on 0.1 ha), and the constant 10^{-3}. Table 11.6 compares the complexity indices of (a) three upland forests on the Cerros de Coronel measured by Holdridge et al. (1971), (b) a tropical wet forest on the Pacific coast of Costa Rica, also measured by Holdridge et al. (1971) and (c) my data from Cerro Tortuguero. The aseasonal environment in the Tortuguero area is atypical of the Tropical Wet Forest life zone, and may be the cause of the low complexity indices of slope forests there compared to that of the Pacific Coast forest.

Swamp forests are species-poor and structurally simple, and this is reflected in their complexity index. Table 11.7 compares data on swamp forests for several locations in Costa Rica and the *Manicaria* swamp data (excluding the *Raphia* palms).

The indices of the three *Raphia* swamps are shown in Table 11.8. It appears that complexity is associated with hydroperiod and is independent of water source. However, the higher complexity index of the river-fed *Raphia* swamp No. 2 is due largely to the greater mean canopy height of the dicotyledonous trees and not the palms (see Table

---

[1] See footnote, Table 6.2, p. 152.

TABLE 11.6

Comparison of structural data of several zonal Tropical Wet Forest sites in Costa Rica. Values are means based on one to three samples of 0.1 ha.

Site	Number of stems (ha^{-1})	Number of species in 0.1 ha^{-1}	Basal area (m^2 ha^{-1})	Mean canopy height (m)	Complexity index
Slope forest I[1] (Cerros de Coronel)	436	22.6	32.5	40	128
Ridge forest[1] (Cerros de Coronel)	520	27.5	28.4	40	162
Slope forest (Cerro Tortuguero)	520	27.0	30.6	39	167
Slope forest II[1] (Cerros de Coronel)	513	26.3	57.8	43	355
Slope forest[1] (Osa Península, Pacific side)	690	38.0	53.3	36	503

[1]From Holdridge et al. (1971).

TABLE 11.7

Comparison of structural data of several swamp forest types in Costa Rica

Site	Number of stems (ha^{-1})	Number of species in 0.1 ha	Basal area (m^2 ha^{-1})	Mean canopy height (m)	Complexity index
Rhizophora mangrove swamp[1] (Osa Península)	360	1.0	12.5	34	2
Mora swamp[1] (Osa Península)	235	4.5	35.0	26	10
Prioria swamp[1] (Río Colorado)	290	3.6	54.9	47	27
Pterocarpus — mangrove swamp[2] (Limón)	—	—	96.4	16	85
Palm swamp[3] (Río Colorado)	715	10.0	47.1	40	135
Manicaria swamp[4] (Caño Mora)	910	7.0	93.7	28	167

[1]From Holdridge et al. (1971).
[2]From Pool et al. (1977).
[3]Holdridge et al. (1971) misidentified this as an *Astrocaryum* palm swamp, but the dominant species is *Manicaria*.
[4]*Raphia* clumps (averaging 20 ha^{-1}) were excluded from the calculation because the growth form of the palm does not permit direct comparison of basal area.

11.1). In swamps with equivalent hydroperiods, decreased water flow and nutrients influence species composition. *Raphia* is able to dominate both rain-fed and river-fed wetlands which have appropriate hydroperiods. In doing so it demonstrates a wide ecological amplitude and may be an early successional invader of marsh vegetation.

Manicaria appears to be an understory species which relies on the development of a shade-providing canopy. Though it occurs in both, it is more abundant in rain-fed swamps than in river-fed swamps.

The similarity in structure of the *Metroxylon sagu* swamps found along the north coast of Papua

TABLE 11.8

Comparison of structural data of several *Raphia* swamps. Each value is a mean of three 0.1 ha samples

Forest type	Number of stems (ha^{-1})	Number of species in 0.1 ha	Basal area (m^2 ha^{-1})	Stand height (m)	Complexity index	% of time flooded
Rain-fed *Raphia* swamp (Caño Servulo)	640	5.0	388	15	186	100
River-fed *Raphia* swamp No. 1 (Río Penitencia)	637	5.3	344	15	174	92
River-fed *Raphia* swamp No. 2 (Caño Mora)	527	7.7	224	24	218	66

New Guinea and the *Raphia taedigera* swamps of Costa Rica is striking. The life histories and growth forms of the dominant palms are nearly identical. Both occur in rain-fed and river-fed systems. In both cases they are frequently associated with herbaceous marsh vegetation, and the boundary between the marsh and palm swamp is often abrupt even though there is little or no difference in hydroperiod between the two. Rain-fed *Metroxylon* swamps have a sparse overstory of *Campnosperma auriculata*. The emergent in similar *Raphia* swamps is *Campnosperma panamensis*. In river-fed *Metroxylon* swamps the abundance of *Bischofia javanica* is reminiscent of *Pentaclethra macroloba* in Costa Rican swamps. Both of these hardwoods are also common on upland sites surrounding the swamps. The structural similarities between *Metroxylon* swamps and *Raphia* swamps points to their being ecologically equivalent systems determined by their physical context. Further study may show their dynamics and succession to be nearly identical.

ACKNOWLEDGEMENTS

I thank the Organization of American States, The Conservation and Research Foundation, the Center for Tropical Agriculture at the University of Florida, and Sigma Xi for financial support, and the Caribbean Conservation Corporation, the Costa Rican National Park Service, the Tropical Science Center, and the Center for Wetlands, University of Florida for logistical support. Special thanks to Drs. John Ewel and Ariel Lugo for their continued encouragement and advice.

APPENDIX: METHODS USED FOR OBSERVATIONS AT TORTUGUERO

Rainfall

From August 1977 until April 1978, I recorded rainfall at the Green Turtle Research Station located north of the village of Tortuguero. From May 1978 until June 1980 I used data collected by the Costa Rican Meteorological Service's newly installed pluviometer, just south of the village, at the Tortuguero National Park headquarters. I suspect that the recorded values at this station were somewhat less than the actual rainfall due to monitoring problems.

The total rainfall for 1978 (5977 mm) was considerably more than for 1979 (4145 mm). The mean number of days per year without rainfall, calculated from the 35-month recording period, was 129 days. The number of days per year without rain was practically the same for both 1978 (128 days) and 1979 (126 days) in spite of the 1832 mm difference in total rainfall. The longest period of consecutive days without rainfall was 13 days, in January 1979.

Water level

Water level fluctuations and hydroperiod were determined for the river-fed *Raphia* swamp No. 1, river-fed *Raphia* swamp No. 2 and its adjacent marsh, and the rain-fed *Manicaria* swamp by monitoring staff gauges. The distance to the rain-fed *Raphia* swamp precluded the possibility of measuring water levels there. An attempt was made to read the gauges at least twice a week. The recording period covered 13 months, from March 1978 through March 1979. The gauges were monitored an average of 8.1 times each month. Staff gauge placement differed from site to site, but a location representing a mean water level was chosen in each case.

Vegetation

Detailed inventories of trees were made in each swamp using three 10 × 100 m transects. Transects were placed at 100 m, 200 m and 300 m along, and perpendicular to, a line transect through each swamp. Each transect was divided into 20 cells, 5 × 10 m, and the diameters of all trees ⩾ 10 cm dbh were recorded by cell. Canopy height was determined by measuring the heights of ten arbitrarily selected canopy trees in each transect using a Haga altimeter. In cases where more than one stratum was evident, mean heights of each were determined. For comparison a single 10 × 100 m transect was measured in a well-drained slope forest on Cerro Tortuguero, one of the isolated hills.

In the case of *Raphia* the diameters of entire clumps were determined, and the number of live and dead stems in each clump counted. The numerous sprouts around the bases of the clumps were not counted, but the sprouts and the empty spaces where old dead stems had decayed and disappeared were included in the total bulk of the clump. This method of determining the diameter of *Raphia* did not lead to direct comparison with single-stemmed trees.

Density, basal area, relative frequency, relative abundance, relative dominance, and importance values of each species were calculated for each swamp. A similarity index (Coefficient of Community using importance values) was calculated between each swamp type. Common understory and ground-cover species were collected and identified. In the *Manicaria* swamp, juvenile *Manicaria* (i.e. without a developed stem), were counted in each transect. In the *Raphia* swamps on Río Penitencia and Caño Mora, *Raphia* juveniles and the understory palm *Calyptrogyne glauca* were also counted along with woody stems < 10 cm dbh. Common species in the marsh on Caño Mora adjacent to the river-fed *Raphia* swamp No. 2 were recorded but no attempt was made to quantify their abundance. Species were identified at the herbarium of the National Museum of San José.

REFERENCES

Ahn, P., 1958. Regrowth and swamp vegetation in the western forest areas in Ghana. *J. West Afr. Sci. Assoc.*, 4: 163–173.

Allen, P.H., 1965. *Raphia* in the western world. *Principes*, 9: 66–70.

Anderson, A.B., 1983. *The Biology of Orbignya martiana (Palmae), a Tropical Dry Forest Dominant in Brazil*. Ph.D. Dissertation, University of Florida, Gainesville, Fla., 194 pp.

Anderson, R. and Mori, S., 1967. A preliminary investigation of *Raphia* palm swamps, Puerto Viejo, Costa Rica. *Turrialba*, 17: 221–224.

Barrau, J., 1959. The sago palm and other food plants of marsh dwellers in the South Pacific Islands. *Econ. Bot.*, 13: 151–163.

Beard, J.S., 1944. Climax vegetation in tropical America. *Ecology*, 25: 127–158.

Beard, J.S., 1946. The *Mora* forests of Trinidad, British West Indies. *J. Ecol.*, 33: 173–192.

Beard, J.S., 1959. Climax vegetation in tropical America. *Ecology*, 36: 89–100.

Blydenstein, J., 1967. Tropical savanna vegetation of the llanos de Colombia. *Ecology*, 48: 1–15.

Braun, A., 1968. Cultivated palms of Venezuela. *Principes*, 12: 39–103.

Brown, K.E., 1976. Ecological studies of the cabbage palm, Sabal palmetto. *Principes*, 20: 3–10.

Budowski, G., 1966. Los bosques de los trópicos húmedos de América. *Turrialba*, 16: 278–284.

Costa Rican National Meteorological Service, 1978. *Anu. Meteorol.*, San José, C.R., 242 pp.

Eggeling, W.J. 1935. The vegetation of Namanve Swamp, Uganda. *J. Ecol.*, 23: 422–435.

Espinal, L. and Montenegro, E., 1963. *Formaciones Vegetales de Colombia: Memoria Explicativa Sobre el Mapa Ecológico*. Instituto Geográfico "Augustín Colazzi, Bogotá, 201 pp.

Ewel, J.J. and Madriz, A., 1968. *Zonas de Vida de Venezuela: Memoria Explicativa Sobre el Mapa Ecológico*. Ministerio de Agricultura y Cría, Caracas, 265 pp.

Fanshawe, D.B., 1954. Forest types of British Guiana. *Caribb. For.*, 15: 73–111.

Golley, F.B., Ewel, J.J. and Child, G.I., 1976. Vegetation biomass of five ecosystems in northwestern Colombia. *Trop. Ecol.*, 17: 16–22.

Guerra, S.W., 1958. *Estudio Preliminar Sobre las Asociaciones Forestales del Bosque Nacional de Iparia*. Tesis para Ingeniero Agrónomo, Escuela Nacional de Agricultura, Lima, 104 pp.

Gurvich, B.R., 1945. La vegetacíon del Uruguay. In: F. Verdoorn (Editor), *Plants and Plant Science in Latin America*. Chronica Botanica Company, Waltham, Mass. pp. 142–143.

Heyligers, P.C., 1972. Vegetation and ecology of the Aitape–Ambunti Area. *CSIRO Land Res. Ser.*, 30: 73–99.

Holdridge, L.R., 1967. *Life Zone Ecology.* Tropical Science Center, San Jose, C.R., 260 pp.

Holdridge, L.R. and Budowski, G., 1956. Report of an ecological survey of the Republic of Panama. *Caribb. For.*, 17: 92–111.

Holdridge, L.R. and Poveda, A.L.J., 1975. *Árboles de Costa Rica, Volume I.* Centro Científico Tropical, San José, C.R., 546 pp.

Holdridge, L.R., Grenke, W.C., Hatheway, W.H., Liang, T. and Tosi, J.A., Jr., 1971. *Forest Environments in Tropical Life Zones: A Pilot Study.* Pergamon Press, Oxford, 746 pp.

Irmler, U., 1977. Inundation forest types in the vicinity of Manaus. *Biogeographica*, 8: 17–29.

Johnson, D., 1972. The carnauba wax palm (*Copernicia prunifera*). II. Geography. *Principes*, 16: 42–48.

Junk, W.J., 1983. Ecology of swamps on the middle Amazon. In: A.J.P. Gore (Editor), *Mires: Swamp, Bog, Fen and Moor.* Ecosystems of the World 4B. Elsevier, Amsterdam.

Langdale-Brown, I., Osmaston, H.A. and Wilson, J.G., 1964. *The Vegetation of Uganda and its Bearing on Land Use.* Government of Uganda, Entebbe.

Lovejoy, T.E., 1975. Bird diversity and abundance in Amazon forest communities. In: *The Living Bird.* Cornell Laboratory of Ornithology, Ithaca, N.Y., pp. 127–191.

Lundell, C.L., 1945. The vegetation and natural resources of British Honduras. In: F. Verdoorn (Editor), *Plants and Plant Science in Latin American.* Chronica Botanica Company, Waltham, Mass., pp. 270–273.

Mallaux, O.J., 1975. *Mapa Forestal del Perú, Memoria Explicativa.* Universidad Nacional Agraria La Molina, Lima, 161 pp.

Michalowski, M., 1958. The ecology of Paraguayan palms. *Principes*, 2: 52–58.

Moore, H.E., 1973. The major groups of palms and their distribution. *Gentes Herbarium*, 11: 27–141.

Myers, R.L., 1977. *A Preliminary Study of the Sabal Palmetto Forest on Little Corkscrew Island, Florida.* Interim Report, National Audubon Society Ecosystem Research Unit, Naples, Fla., 47 pp.

Myers, R.L., 1981. *The Ecology of Low Diversity Palm Swamps Near Tortuguero, Costa Rica.* Ph.D. Dissertation, University of Florida, Gainesville, Fla., 300 pp.

National Research Council, 1975. *Underexploited Tropical Plants with Promising Economic Value.* National Academy of Sciences, Washington, D.C., 188 pp.

Nordlie, F.G. and Kelso, D.P., 1975. Trophic relationships in a tropical estuary. *Rev. Biol. Trop.*, 23: 77–99.

Nuhn, H. and Perez, R.S., 1967. *Estudio Geográfico Regional de la Zona Atlántica Norte de Costa Rica.* ITCO, San José, Costa Rica, 360 pp.

Paijmans, K., 1967. Vegetation of the Sufia–Pongani area, Papua-New Guinea. *CSIRO Land Res. Ser.*, 17: 142–167.

Paijmans, K., 1969. Vegetation and ecology of the Kernma–Vailala area, territory of Papua and New Guinea. *CSIRO Land Res. Ser.*, 23.

Paijmans, K., 1971. Vegetation, forest resources, and ecology of the Morehead–Kiunga area, Territory of Papua and New Guinea. *CSIRO Land Res. Ser.*, 29.

Pittier, H., 1908. *Ensayo Sobre Plantas Usuales de Costa Rica.* Editorial Universitaria, San José, C.R., 264 pp.

Pittier, H., 1938. *Clasificación de Los Bosques de Venezuela.* Cartelo de Silvicultura, Caracas, 20 pp.

Pool, D.J., Snedaker, S.C. and Lugo, A.E., 1977. Structure of mangrove forests in Florida, Puerto Rico, Mexico, and Costa Rica. *Biotropica*, 9: 195–212.

Richards, P.W., 1952. *The Tropical Rain Forest.* Cambridge University Press, London, 450 pp.

Robbins, R.G., 1968. Vegetation of the Wewak–Lower Sepik area, Territory of Papua and New Guinea. *CSIRO Land Res. Ser.*, 22: 109–124.

Rojas, T. and Carabia, J.P., 1945. Breve reseña de la vegetación paraguayana. In: F. Verdoorn (Editor), *Plants and Plant Science in Latin America.* Chronica Botanica Company, Waltham, Mass., pp. 121–125.

Salazar, A., 1967. El aguaje (*Mauritia vinifera*) recurso forestal potencial. *Rev. For. Perú*, 1: 65–68.

Sylvander, R.P., 1978. *Los Bosques del País y Su Distribución por Provincia.* PNUD/FAO-COS/72/013, Documento de Trabajo No. 5, San José, C.R.

Takeuchi, M., 1960. The structure of the Amazonian vegetation. *J. Fac. Sci. Univ. Tokyo*, 7: 523–533.

Takeuchi, M., 1961. The structure of the Amazonian vegetation. II. Tropical rain forest. *J. Fac. Sci. Univ. Tokyo*, 8: 1–25.

Taylor, B.W., 1959. The classification of lowland swamp communities in northeastern Papua. *Ecology*, 40: 703–711.

Taylor, B.W., 1964a. Vegetation of the Buna–Kokoda area, Territory of Papua and New Guinea. *CSIRO Land Res. Ser.*, 10: 89–98.

Taylor, B.W., 1964b. Vegetation of the Wanigela–Cape Vogel area, Territory of Papua and New Guinea. *CSIRO Land Res. Ser.*, 12: 69–83.

Thompson, K. and Hamilton, A.C., 1983. Peatlands and swamps of the African continent. In: A.J.P. Gore (Editor), *Mires: Swamp, Bog, Fen and Moor, Part B.* Ecosystems of the World 4B. Elsevier, Amsterdam, pp. 331–373.

Tosi, J.A., Jr., 1969. *República de Costa Rica, Mapa Ecológico.* Tropical Science Center, San José, C.R.

Tosi, J.A., Jr., 1972. *Los Recursos Forestales de Costa Rica.* Tropical Science Center, San José, C.R., 16 pp.

Townsend, P.K., 1974. Sago production in a New Guinea economy. *Hum. Ecol.*, 2: 217–236.

Verdoorn, F. (Editor), 1945. *Plant and Plant Science in Latin America.* Chronica Botanica Company, Waltham, Mass., 381 pp.

Wallace, A.R., 1853. *Palm Trees of the Amazon and Their Uses.* John Van Voorse, London, 129 pp.

Webb, J.L., 1968. Environmental relationships of the structural types of Australian rainforest vegetation. *Ecology*, 49: 296–311.

Whitmore, T.C., 1975. *Tropical Rain Forests of the Far East.* Clarendon Press, Oxford, 282 pp.

Wilbert, J., 1976. *Manicaria saccifera* and its cultural significance among the Warao Indians of Venezuela. *Bot. Mus. Leafl., Harv. Univ.*, 24: 275–335.

Williams, W.A., Loomis, R.S. and Alvim, P. de T., 1972. Environments of evergreen rain forests on the lower Rio Negro, Brazil. *Trop. Ecol.*, 13: 65–78.

Chapter 12

FORESTED WETLANDS OF MEXICO

ANTONIO LOT and ALEJANDRO NOVELO

INTRODUCTION

Most forested wetlands in Mexico are concentrated in the coastal plain of southern Veracruz, Tabasco, and Campeche (Fig. 12.1). Mangroves and riverine forests extend over large areas, but other wetland types cover small areas. Herbaceous wetlands are important further north in the Tampico region towards the state of Tamaulipas, in the Pacific coastal plain of Nayarit, and in northern Michoacan to the center of Jalisco (Rzedowski, 1978). Riparian forests extend from sea level to about 2800 m elevation, with species of *Populus* predominating in arid and semi-arid regions of northern Mexico and species of *Alnus* in more temperate zones with cooler climates. Trees of the genus *Astianthus* grow along lowland river courses with long dry seasons, while in moist and wet regions the genera *Bambusa*, *Ficus*, *Inga* and *Pachira* are predominant.

The emphasis in this chapter is on the tropical lowland region of southern Veracruz, Tabasco, and Campeche, the region of Mexico where forested wetlands have been most studied. Elevations in these wetland regions do not exceed 40 m, topographic relief is slight, and well developed fluvial systems such as those of the Grijalva–Usumacinta river deltas are typical. Annual rainfall ranges from 2000 to 5000 mm, and mean annual temperature between 23 and 26°C. The nomenclature used to classify vegetation is original; however, we have attempted to maintain traditional nomenclature, particularly that of Miranda and Hernández (1963), as far as possible. Vegetation profiles (Fig. 12.2) and soil characteristics (Table 12.1) have been summarized for the types of forested wetlands described below.

HIGH FORESTED WETLANDS

Evergreen and deciduous riparian woodlands

These woodlands (*bosques perennifolios y deciduos riparios*) frequently consist of pure stands of *Salix chilensis* (Fig. 12.3), which grows in periodically inundated riverine sites in the lowlands of Veracruz, Tabasco, and parts of Campeche. The communities are usually monospecific, and restricted to the river banks. Trees may exceed 20 m in height and cover extensive areas of alluvial soils derived from calcareous material. These soils have low concentrations of soil organic matter and slow drainage.

Salix chilensis is usually deciduous in temperate or high-elevation climates, but evergreen in the lowland tropical zones. In subhumid and drier environments, *Salix* and other genera such as *Fraxinus* and *Populus* become the dominant elements of the deciduous forests in river floodplains (Miranda and Hernández, 1963). Rzedowski (1978) listed *Platanus*, *Salix* and *Taxodium* as important genera in the evergreen gallery forests. He considered these genera to be tolerant of a wide range of conditions including flooding. Other authors have added other species as components of the semi-evergreen and deciduous riparian forest (for example, *Liquidambar macrophylla*, *Platanus lindeniana*, and *Taxodium mucronatum*).

High to medium riparian forest (*Selva alta–mediana riparia*)

Associations of *Pachira* and *Ficus* are common along river or lagoon fringes without marine influence. These communities have greater species

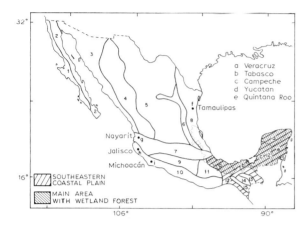

Fig. 12.1. Physiographic provinces (1–15) of Mexico (after Rzedowski, 1978) showing main areas with wetland forest (a–e).

diversity than the riparian woodlands. Vegetation attains 30 m in height in deep alluvial soils of igneous or sedimentary origin. Tree height decreases with proximity to the ocean. Vegetation composition and height vary with water motion and the length of the hydroperiod (Fig. 12.4). A list of the dominant species in many of the rivers and streams of the coastal plain of the Gulf of Mexico includes the following (Rovirosa, 1894; Miranda and Hernández, 1963; West, 1966; Chavelas, 1967, 1968a, b; Sousa, 1968; Menéndez, 1976): *Combretum laxum*, *Ficus insipida*, *Inga vera* subsp. *spuria*, *Lonchocarpus cruentus*, *L. guatemalensis* var. *mexicanus*, *L. hondurensis*, *L. pentaphyllus*, *L. unifoliolatus*, *Machaerium falciforme*, *M. lunatum*, *Muellera frutescens*, *Pachira aquatica*, *Pithecellobium belizense*, *P. calostachys*, *P. recordii* and *Vatairea lundellii*.

Species dominance may change depending upon the type of habitat. Seasonally flooded lagoon fringes (Fig. 12.5) or river fringes support tall evergreen forests. Regardless of habitat, the following are among the commonly observed tree species that tolerate short-term flooding in these environments (Gómez-Pompa, 1965; Sarukhán, 1968a, b; Puig, 1972; Orozco and Lot-Helgueras, 1976): *Andira galeottiana*, *Calophyllum brasiliense*, *Miconia argentea*, *Tabebuia rosea*, *Terminalia amazonia*, *Vochysia guatemalensis* and *Xylopia frutescens*.

The *Bravaisia integerrima* association is found mostly in southern Tabasco and northern Chiapas as an intermediate-sized evergreen wetland forest. This association covers extensive swampy areas characterized by flat terrain with shallow calcareous soils, high in organic matter and available phosphorus (Table 12.1). This species may form pure stands, with tree heights reaching 25 m. In portions of the forest where flooding is less frequent, certain flood-tolerant species from the tall evergreen rainforest can be found. Some of these species include: *Andira inermis*, *Calophyllum*

TABLE 12.1

Physico-chemical analyses of soil samples at 30 cm depth

Soil types (according FAO)	Wetland forest association (example)	Locality	Soil texture sand/clay/silt (%)	Organic carbon (%)	Total nitrogen (%)	C/N (by weight)	P available (mg kg^{-1})	pH [1:5]
Eutric Histosol (fs)	*Rhizophora*	La Mancha, Veracruz	70/26/4	11.2	0.89	12.58	0.4	8.3
	Metopium	Seybaplaya, Campeche	54/42/4	14.9	1.170	12.82	1.2	7.8
Fluvisol	*Bravaisia*	San Cayetano, Tabasco	51/37/12	4.3	0.162	26.54	4.1	7.4
	Bucida	Balancán, Tabasco	65/8/27	3.1	0.123	25.20	3.1	7.4
	Calophyllum	Las Choapas, Veracruz	54/32/14	3.6	0.142	25.35	4.0	4.6
	Acoelorrhaphe Pachira	Frontera, Tabasco	55/15/30	2.5	0.060	41.66	2.5	4.6
	Inga Andira	Río Tonalá, Veracruz	74/12/14	6.1	0.263	23.19	7.1	5.2
Gleysol	*Haematoxylum*	Teapa, Tabasco	26/53/21	2.5	0.159	15.72	1.1	5.4
Organic Eutric Histosol	*Annona*	La Mancha, Veracruz	66/26/8	22.3	1.561	14.28	7.4	4.2

FORESTED WETLANDS OF MEXICO

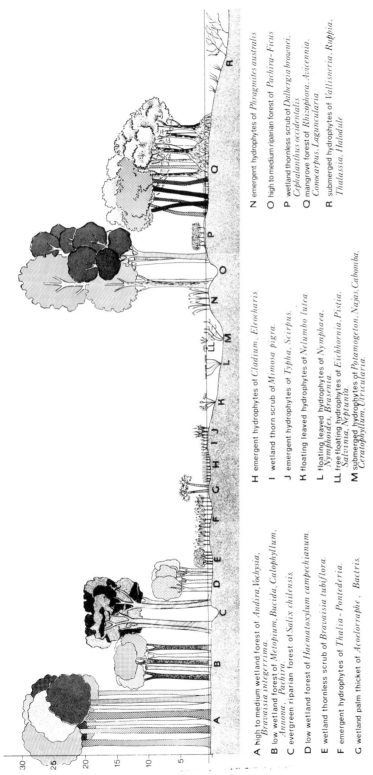

A high to medium wetland forest of *Andira*, *Vachysia*, *Bravaisia integerrima*.
B low wetland forest of *Metopium*, *Bucida*, *Calophyllum*, *Annona*, *Pachira*.
C evergreen riparian forest of *Salix chilensis*.
D low wetland forest of *Haematoxylum campechianum*.
E wetland thornless scrub of *Bravaisia tubiflora*.
F emergent hydrophytes of *Thalia - Pontederia*.
G wetland palm thicket of *Acoelorraphe*, *Bactris*.
H emergent hydrophytes of *Cladium*, *Eleocharis*.
I wetland thorn scrub of *Mimosa pigra*.
J emergent hydrophytes of *Typha*, *Scirpus*.
K floating leaved hydrophytes of *Nelumbo lutea*.
L floating leaved hydrophytes of *Nymphaea*, *Nymphoides*, *Brasenia*.
LL free floating hydrophytes of *Eichhornia*, *Pistia*, *Salvinia*, *Neptunia*.
M submerged hydrophytes of *Potamogeton*, *Najas*, *Cabomba*, *Ceratophyllum*, *Utricularia*.
N emergent hydrophytes of *Phragmites australis*
O high to medium riparian forest of *Pachira - Ficus*
P wetland thornless scrub of *Dalbergia brownei*, *Cephalanthus occidentalis*
Q mangrove forest of *Rhizophora*, *Avicennia*, *Conocarpus*, *Laguncularia*
R submerged hydrophytes of *Vallisneria*, *Ruppia*, *Thalassia*, *Halodule*

Fig. 12.2. Typical profiles of the main plant associations in the forested wetlands of southeastern Mexico.

Fig. 12.3. Riparian forest dominated by *Salix chilensis*, a common association along the coastal-plain rivers of southeastern Mexico.

brasiliense, *Ceiba pentandra*, *Diospyros digyna*, *Ficus panamensis*, *Licania platypus*, *Lonchocarpus cruentus*, *Mortoniodendron guatemalense*, *Tabebuia rosea* and *Vatairea lundellii*. This forested wetland is known locally as "canacoital" and has been described by many authors (Miranda, 1952; Miranda and Hernández, 1963; Gómez-Pompa, 1965; Sarukhán, 1968; Pérez and Sarukhán, 1970; Puig, 1972; Rzedowski, 1978).

LOW FORESTED WETLANDS

This community type (*selva baja inundable*) comprises some of the most widely distributed flooded ecosystems of southeastern Mexico. *Annona glabra* is the dominant species, and may form extensive pure stands of trees 3 to 8 m tall. These associations cover permanently flooded sites from the center of Veracruz to western Tabasco and in isolated patches in Campeche and south of Quintana Roo. Some associations of this type may support high species richness. Species richness decreases with increasing length of hydroperiod. Trees grow in shallow alluvial soils (fluvisols) with a high percentage of organic matter and a loamy texture (Table 12.1).

Chrysobalanus icaco is a frequent codominant with *A. glabra* in floodplains of southern Veracruz (Orozco and Lot-Helgueras, 1976). Other woody species in these types of wetlands have been reported by West (1966), Orozco and Lot-Helgueras (1976), and Novelo (1978) and include: *Acoelorrhaphe wrightii*, *Calophyllum brasiliense*, *Calyptranthes millspaughii*, *C. perlaevigata*, *Dalbergia brownei*, *Diospyros digyna*, *Ficus cotinifolia*, *F. padifolia*, *Lonchocarpus pentaphyllus* and *Randia mitis*.

An association of *Haematoxylum campechianum* (known locally as "tintal") covers large areas of southeastern Mexico. These areas flood during the rainy season, but are dry during the dry season.

Fig. 12.4. A *Pachira aquatica* association near the mouth of the Sontecomapan lagoon in southern Veracruz, Mexico.

Trees grow on shallow soils derived from calcareous materials, with a slightly acid to alkaline reaction, and a high percentage of organic matter. These forests are under intense human impact and are being converted to pastures. Most of the conversions are occurring between Tabasco and Quintana Roo. The area covered by "tintales" amounts to a third of the central and southern portion of Campeche and the central portion of Quintana Roo (Miranda, 1958). Among other trees associated with *H. campechianum* are: *Bucida buceras*, *Cameraria latifolia*, *Coccoloba reflexiflora*, *Crescentia cujete*, *Curatella americana*, *Eugenia lundellii*, *Hampea trilobata*, *Hyperbaena winzerlingii* and *Metopium brownei*.

Some of these species may become dominant in other types of forested wetlands. Lundell (1934) considered these wetlands as wooded swamps, and observed that they are flooded between 6 and 9 months of the year. Locally they may be called "bajos" or "acalches". Species lists of these types of wetlands including epiphytes, climbers, herbaceous species, and other plant components were given by Bravo (1955), Vázquez (1963), Chiang (1967–8), Sarukhán (1968a, b), Telléz and Sousa (1982), and Rico-Gray (1982).

Metopium brownei forms variants of the low wetland forest known locally as "chechenal". The species has wide tolerance to flooding and salinity, and may grow at the ecotone with mangroves in northern Yucatán where trees attain 12 m. It also grows in tall to medium-sized forested wetlands, where it attains heights of 40 m on better drained soils. In the state of Campeche where the high and medium forested wetlands occur, this species is associated with *Manilkara zapota*. *Bucida buceras* behaves similarly and its associations are locally known as "puktal". Associations of *B. buceras* with *Metopium brownei* and *H. campechianum* grow isolated as low forested wetlands between the Mamantel and Candelaria rivers in Campeche (Vázquez, 1963), in southern Champotón, Cam-

Fig. 12.5. Flooded fringe of Catemaco lagoon, Veracruz. Tall evergreen trees are the dominant vegetation.

peche (Rzedowski, 1978), and around the Ascensión and Espíritu Santo Bays in Quintana Roo.

In mangrove ecotones with freshwater seeps, *Pachira aquatica* forms small but stable associations locally known as "apompal". Most trees are about 8 to 10 m in height, but isolated trees may reach 15 m. These have been considered transitional associations (West, 1966; Chavelas, 1967, 1968b; León and Gómez-Pompa, 1970; Vázquez-Yanes, 1971).

Wetland palm thickets (*Palmar inundable*)

Here we differentiate between palm associations that are tolerant of flooding and those that, although growing at the ecotone of floodplains, cannot withstand more than 6 months of flooding during any year.

Among flood-tolerant palm associations are those of *Acoelorrhaphe wrightii*, also known as "tasistal" throughout southeastern Mexico (Fig. 12.6). Trees reach 8 m in height, but are frequently only 2 to 5 m tall. They grow in strips forming fringes around other wetlands, including the landward sectors of mangroves, and in the interior of flooded savannas. Soil may be saline with high percentages of organic matter. Water depths in this association may reach 1.5 m, but plants also tolerate dry soils and even periodic burning. As a result, *tasistales* may replace other less tolerant plant communities and expand their territorial coverage. Other woody species associated with this palm community are *Annona glabra*, *Bactris balanoidea*, *Dalbergia glabra* and *Haematoxylum campechianum* (West, 1966; Orozco and Lot-Helgueras, 1976; Telléz and Sousa, 1982).

Roystonea dunlapiana forms taller (18–30 m) palm associations from Quintana Roo to Veracruz, while *R. regia* is restricted to northern Yucatán (H. Quero, pers. commun., 1983). *Roystonea* also grows in tall and medium-sized evergeen forests in areas with poorer drainage (Miranda and

Fig. 12.6. Flooded palm forest dominated by a fringing *Acoelorrhaphe wrightii* association in a flooded savanna matrix in Tabasco, Mexico.

Hernández, 1963; Sarukhán, 1968a, b; Rzedowski, 1978).

Bactris balanoidea and *B. trichophylla* form low and spiny forested wetlands in shallow-water areas below 200 m elevation throughout the Yucatán peninsula. They form small narrow fringes (H. Quero, pers. commun., 1983). The palms may be associated with herbaceous vegetation or be found in the interior of evergreen forests (Miranda, 1958; West, 1966). Other palms known to tolerate a certain degree of flooding are *Sabal mauritiiformis*, *S. yapa*, *Scheelea liebmannii* and *Orbignya cohune* (Lundell, 1934; Miranda, 1958; Miranda and Hernández, 1963; Gómez-Pompa, 1965, 1973; Sarukhán, 1968a, b; Rzedowski, 1978; Telléz and Sousa, 1982).

Mangrove forests

Mangroves grow on low-energy coast-lines, in protected bays, coastal lagoons, and river deltas. They are absent from rocky areas, zones with active dunes, and areas under strong wave or tidal activity (Fig. 12.7).

They form fringes that may range from a few meters to kilometers in width. Tree height ranges from a few meters up to 30 m in the coastal zones of southeastern Mexico. In these humid coastal areas mangrove communities are more complex in terms of species and life forms that those at higher latitudes. With increasing latitude mangrove ecosystem complexity decreases (Lot-Helgueras et al., 1975).

Areas of exceptional mangrove development are those in the deltas of the Usumacinta river towards the Laguna de Términos; in the Atasta and Pom lagoons in Campeche (Fig. 12.8); in the Sontecomapan lagoon in southern Veracruz (Menéndez, 1976); and in the Agua Brava lagoon in Nayarit (Rollet, 1974) and the Tecapàn lagoon in Sinaloa on the Pacific side (Pool et al., 1977). *Rhizophora mangle* has been found in the riparian community

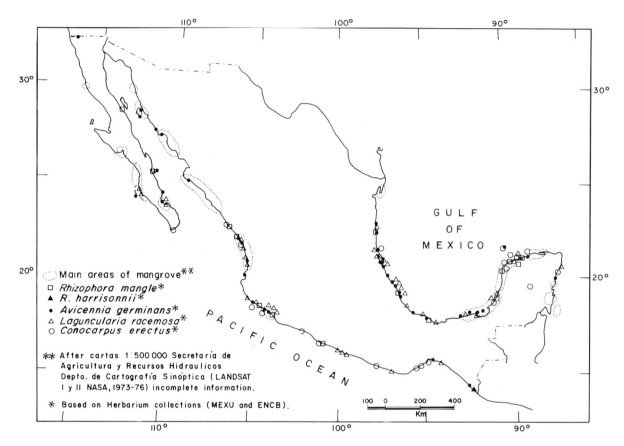

Fig. 12.7. Generalized distribution of mangroves and the dominant mangrove species along the coastlines of Mexico.

of the Río San Pedro in Tabasco, 200 km from the ocean (Lundell, 1942), and 100 km from the ocean in Río Hondo at Quintana Roo (Miranda, 1958).

Conocarpus erectus is a common mangrove species in areas of low frequency of inundation, but it may also grow in other associations far from the ocean. Examples are its dominance in the littoral fringes of interior lagoons in western Chichancanab and Muyil in Quintana Roo (Miranda, 1958; Telléz and Sousa, 1982).

Five mangrove tree species grow in Mexico: *R. mangle*, *Avicennia germinans*, *Laguncularia racemosa*, and *C. erectus* (on both Pacific and Atlantic coasts), and *R. harrisonnii* on the Pacific coast of Chiapas (Rico-Gray, 1981). Species associated with mangroves may be halophytes or hydrophytes adapted to short freshwater hydroperiods through short life cycles. *Ficus involuta*, *Hibiscus tiliaceus* and *Thrinax parviflora* are usually associated with landward ecotones or occur towards the riparian ecotones. Other associated species tolerate short periods of inundation only.

Low soil salinity, intermediate texture, superficial drainage, and a high percentage of organic matter are edaphic factors associated with the tallest mangroves. Calcareous soils with high sulfide content support short mangroves (1.5–2.5 m), as is seen near the coastal lagoons of Bacalar in Quintana Roo.

In the northeastern coastal plain of the Yucatán peninsula of the state of Campeche, there is a wetland 1300 km^2 in area known as "Los Petenes". This wetland is long and narrow (10–15 km wide), and supports most of the forested wetland types described in this chapter (Rico-Gray, 1982). Mangrove forests grow as fringe and dwarf forests, and also grow around circular islands. They act as ecotones with other wetland associations.

Fig. 12.8. Fringe mangrove forest with tall (25 m) *Rhizophora mangle* trees in the Atasta lagoon in Campeche, Mexico.

SHRUB WETLANDS

These formations are shorter than 4 m; trees are generally branched from the base. Most of the scrubby species also grow in the higher forested wetlands already described, but do so as understory or ecotonal species. We found few stable communities in this grouping. Apparently, human or natural disturbances of the original association leave these species as the main residual plant community. These communities are included in this chapter because they appear to be increasing in area.

Thorn-shrub wetland (*Matorral espinoso inundable*)

The most common thorn-shrub formation is dominated by *Mimosa pigra*, which is common in flooded pastures, in the fringes of artificial water bodies, or in fringes of water bodies modified by humans. Because of cattle activity, this species, locally known as "zarzal", is extending its range. The wetland is usually monospecific and only herbaceous plants may be associated with the dominant *zarzal* (Fig. 12.9).

Thornless-shrub wetland (*Matorral inerme inundable*)

Bravaisia tubiflora ("julubal") is a thornless shrub, attaining heights of 2 to 3 m with few branches. This species forms pure stands along ecotones with low forests, mangroves, and sedge-dominated flooded savannas with few woody species. The species may also fringe channels and mangroves growing far from the ocean. In savannas dominated by *Crescentia cujete* near Laguna de Términos in Campeche, islands of *Bravaisia tubiflora* and *C. erectus* have been observed (Vázquez, 1963).

Another shrubby plant community which is

Fig. 12.9. Flooded *Mimosa pigra* scrub, an indicator of human disturbance.

barely 3 m tall is the "mucal" or "mucaleria" which is dominated by *Dalbergia brownei*. This species forms pure stands of irregular size (ranging from a few meters to kilometers in width) in flooded zones behind the mangroves, particularly in those regions where the mangroves have been converted to pastures. West (1966) considered this community as transitional between freshwater wetlands and mangroves. *Cephalanthus occidentalis* also forms dense pure stands, about 2 m in height, in chronically flooded, cleared areas inside low-statured forested wetland ecosystems in southern Veracruz and Tabasco.

DISCUSSION

In several soil samples made in various wetland associations in southern Mexico, four soil types based on geomorphology, physicochemical characteristics of sediments and hydroperiods were encountered (Table 12.1). However, it is important to note that these communities are not always restricted to the soil types given in Table 12.1. In the Eutric Histosol (halomorphic) category we group those communities that develop on heavy soils with pH > 7, and high soluble and exchangeable sodium content, and with low concentrations of organic matter and macro-nutrients. Forested wetland communities in this category are generally poor in species and of low stature. Species-rich communities develop on fluvisols which are characterized by intermediate clay content, a high percentage of organic matter in the top 5 cm, and long hydroperiods. Gley soils (hydromorphic) support flooded savannas and lowland "encinares" susceptible to flooding. Wetland formations dominated by *Haematoxylum campechianum* are examples. These soils have a high clay content in the top horizons but low concentrations of nutrients and organic matter. On organic alluvial soils (Eutric Histosol) mostly herbaceous wetlands with few woody plants are found. These wetlands contain large quantities of partially decomposed organic matter in the sediments but are chronically flooded, occasionally to depths of up to 2 m.

ACKNOWLEDGEMENTS

We thank Agustín Quiróz Flores of the Instituto de Biología of the Universidad Nacional Autónoma de México (UNAM) for his contribution to the manuscript in relation to edaphic conditions in forested wetlands. Elvia Esparza from the Instituto de Biología did the art work. This work was sponsored by Project PCECNAL-790236 under the National Ecology Program of the CONACYT. The volume editors translated the article from a Spanish manuscript.

REFERENCES

Bravo, H., 1955. Algunas observaciones acerca de la vegetación de la región de Escarcega, Campeche y zonas cercanas. *An. Inst. Biol. Univ. Nac. Autón. Méx.*, 26: 283–301.

Chavelas, P.J., 1967–68a. Estudio florístico-sinecológico del Campo Experimental Forestal "El Tormento", Escárcega, Campeche. In: *Comisión de Estudios sobre la Ecología de Dioscóreas*, V Informe Inst. Nac. Invest. For., México, D.F., México, pp. 130–221.

Chavelas, P.J., 1967–68b. La vegetación de San Lorenzo Tenochtitlán, estado de Veracruz. In: *Comisión de Estudios sobre la Ecología de Dioscóreas*, V Informe. Inst. Nac. Invest. For., México, D.F., México, pp. 64–87.

Chiang, C.F., 1967–68. La vegetación de Villahermosa, estado de Tabasco. In: *Comisión de Estudios sobre la Ecología de Dioscóreas*, V Informe. Inst. Nac. Invest. For., México, D.F., México, pp. 88–105.

Gómez-Pompa, A., 1965. La vegetación de México. *Bol. Soc. Bot. Méx.*, 29: 76–120.

Gómez-Pompa, A., 1973. Ecology of the vegetation of Veracruz. In: A. Graham (Editor), *Vegetation and Vegetational History of Northern Latin America*. Elsevier, Amsterdam, pp. 73–148.

León, C.J.M. and Gómez-Pompa, A., 1970. *La vegetación del sureste de Veracruz*. Instituto Nacional de Investigaciones Forestales. Publ. Espec. 5, México, D.F., México, pp. 13–48.

Lot-Helgueras, A., Vázquez-Yanes, C. and Menéndez, L.F., 1975. Physiognomic and floristic changes near the northern limit of mangroves in the Gulf Coast of Mexico. In: G.E. Walsh, S.C. Snedaker and H. Teas (Editors), *Proceedings of the International Symposium on the Biology and Management of Mangroves*. Institute of Food and Agricultural Sciences, University of Florida, Gainesville, Fla., pp. 52–61.

Lundell, C.L., 1934. *Preliminary Sketch of the Phytogeography of the Yucatan Peninsula*. Carnegie Institution of Washington, Publ. No. 436: 255–321.

Lundell, C.L., 1942. Flora of eastern Tabasco and adjacent Mexican areas. *Contrib. Univ. Mich. Herbarium*, 8: 1–74.

Menéndez, L.F., 1976. *Los Manglares de la Laguna de Sontecomapan, Los Tuxtlas, Veracruz, Estudio Florístico-ecológico*. Tesis, Facultad de Ciencias, Universidad Nacional Autónoma de México, México, D.F., México, 115 pp.

Miranda, F., 1952. *La Vegetación de Chiapas*. Ediciones del Gobierno del Estado, Tuxtla Gutiérrez, Vol. 1, 334 pp.

Miranda, F., 1958. Estudios acerca de la vegetación. In: E. Beltrán (Editor), *Los Recursos Naturales del Sureste y su Aprovechamiento*, 2. Instituto Mexicano de Recursos Naturales Renovables A.C., México, D.F., México, pp. 215–271.

Miranda, F. and Hernández X.E., 1963. Los tipos de vegetación de México y su clasificación. *Bol. Soc. Bot. Méx.*, 28: 29–179.

Novelo, R.A., 1978. La vegetación de la Estación Biológica El Morro de la Mancha, Veracruz. *Biotica*, 3(1): 9–23.

Orozco, A. and Lot-Helgueras, A., 1976. La vegetación de las zonas inundables del sureste de Veracruz. *Biotica*, 1(1): 1–44.

Pérez, J.L.A. and Sarukhán, K.J., 1970. *La vegetación de la región de Pichucalco, Chiapas*. Inst. Nac. Invest. For., Publ. Espec. 5, México, D.F., México, pp. 49–123.

Pool, D.J., Snedaker, S.C. and Lugo, A.E., 1977. Structure of mangrove forest in Florida, Puerto Rico, Mexico and Costa Rica. *Biotropica*, 9: 152–212.

Puig, H., 1972. La sabana de Huimanguillo, Tabasco, México. In: *Memorias del Ier Congreso Latinoamericano de Botánica*. México, D.F., México, pp. 389–411.

Rico-Gray, V., 1981. *Rhizophora harrisonnii* (Rhizophoraceae), un nuevo registro de las costas de México. *Bol. Soc. Bot. Méx.*, 41: 163–165.

Rico-Gray, V., 1982. Estudio de la vegetación de la zona costera inundable del noreste del estado de Campeche, México: Los Petenes. *Biotica*, 7(1): 171–190.

Rollet, B., 1974. Introduction à l'étude des mangroves du Mexique. *Bois Fôr. Trop.*, 156: 3–26; 157: 53–74.

Rovirosa, J.N., 1894. Bosquejo de la flora tabasqueña. *Naturaleza*, 2: 438–441.

Rzedowski, J., 1978. *Vegetación de México*. Limusa, México, 432 pp.

Sarukhán, K.J., 1968a. *Análisis sinecológico de las selvas de Terminalia amazonia en la Planicie Costera del Golfo de México*. Tesis, Colegio de Postgraduados, Escuela Nacional de Agricultura, Chapingo, México, 300 pp.

Sarukhán, K.J., 1968b. Los tipos de vegetación arbórea de la zona cálido-húmeda de México. In: T.D. Pennington and J. Sarukhan (Editors), *Manual para la identificación de campo de los principales árboles tropicales de México*. Instituto Nacional de Investigaciones Forestales and Food and Agricultural Organization, México, D.F., México, pp. 3–46.

Sousa, S.M., 1968. Ecología de las leguminosas de los Tuxtlas, Veracruz. *An. Inst. Biol. Univ. Nac. Autón. Méx., Ser. Bot.*, 39: 121–160.

Telléz, O. and Sousa, S.M., 1982. *Imágenes de la flora Quintanarroense*. Cent. Invest. Quintana Roo, A.C., México., 224 pp.

Vázquez, S.J., 1963. *Clasificación de las masas forestales de Campeche*. Inst. Nac. Invest. For., Bol. Téc. 10, México, D.F., México, 30 pp.

Vázquez-Yanes, C., 1971. La vegetación de la Laguna de Mandinga, Veracruz. *An. Inst. Biol. Univ. Nac. Autón. Méx., Ser. Bot.*, 42: 49–94.

West, R.C., 1966. The natural vegetation of the Tabascan lowland, Mexico. *Rev. Geogr.* (Rio de Janeiro), 64: 107–122.

Chapter 13

OLIGOTROPHIC FORESTED WETLANDS IN BORNEO

EBERHARD F. BRUENIG

INTRODUCTION

Borneo is within the Malesian region (Fig. 13.1) unique in that this third largest island in the world has a long history of uninterrupted evolution under conditions which combined excellent suitability for plant growth with a high degree of geological and geomorphic heterogeneity. The warm and humid, only weakly seasonal, but periodically variable and episodically extreme climate adds further to a scenario which is unusually favourable for the development and maintenance of an extremely high species richness and diversity of the vegetation at three levels: within the stand, between stands, and between geographical landscape and regional units.

Unique within this island are the forested oligotrophic rain-flooded wetlands which occupy areas which may be small and scattered, or very extensive, continuous tracts of land, and occur on a great variety of sites and substrates. Among the first naturalists to mention these peculiar forests was the famous Italian botanist Beccari, who visited Sarawak in northwestern Borneo in 1865–1868. During his extensive travels he discovered various types of oligotrophic forested wetlands. He commented on the tea-coloured, acid "black water" which he ascribed to "the quantity of dead leaves and humus accumulated in the forests through which they flow" and compared this limpid, humic water to the humic-acid coloured water in South American rivers (Beccari, 1904).

In the appendix to his book, Beccari described the forests from which these waters issue in Borneo as being of two major types: one is the "great primeval forest in lowlands constantly watered, immensely rich in species and of enormous extent in the neighbourhood of Kuching, of the delta of the Rejang...". The other is the "so-called Mattang" which occurs as stunted forests on white sands, issues black water and has a "certain analogy with the 'campos' of Brazil, which might also be considered ancient islands which have been surrounded with alluvial land of recent formation".

Oligotrophic peaty swamp forests in Kalimantan and Sumatra were described by Polak (1933) and Endert (1920). Polak gives a detailed account of a raised peat bog. Endert describes an oligotrophic forest in wet hollows between sandy terraces on which Winkler (1914) found a forest which reminded him in some ways of heath and which he termed "Heidewald", translated to "heath forest" by Richards (1952).

During the 1950s, extensive, quantitative and comprehensive ecological studies began with the work of Ashton (1964) in the dipterocarp forests of Brunei and Sarawak, of Anderson (1961a) in the oligotrophic peat-swamp forests of Sarawak, and of Bruenig (1966) in the "Heidewald" of Sarawak and Brunei. Bruenig distinguished two major subdivisions: the "kerangas" forests on mineral soils, corresponding with Winkler's *Heidewald*, and forests on wet, peaty soil for which he adopted the native name "kerapah". The results were reported in a number of publications, some of which are listed in the references. The *kerangas* forests have also been discussed in another volume of this series (Specht and Womersley, 1979). In neighbouring Malaysia, ecological studies by Wyatt-Smith (1959) concurrent with those in Borneo resulted in a comprehensive description and practical classification of the main forest types. Research

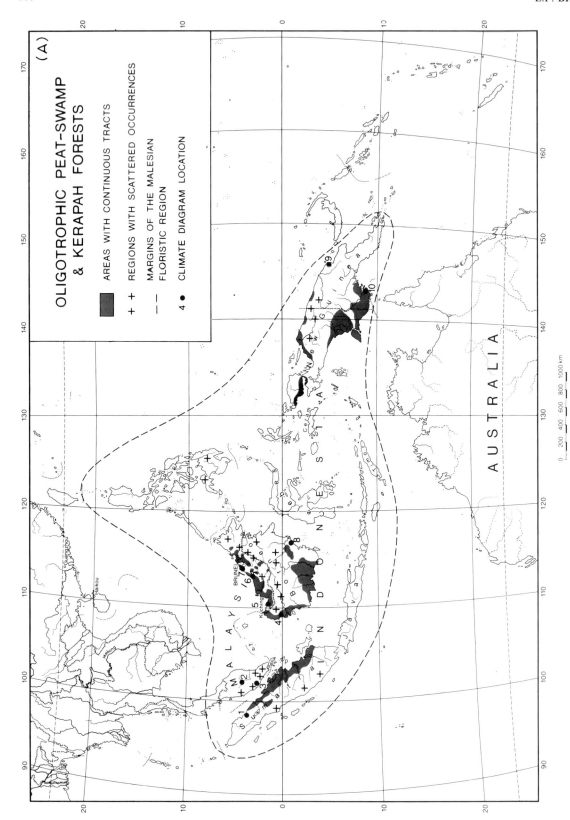

Fig. 13.1. A. Oligotrophic forested wetlands in the Malesian floristic region.

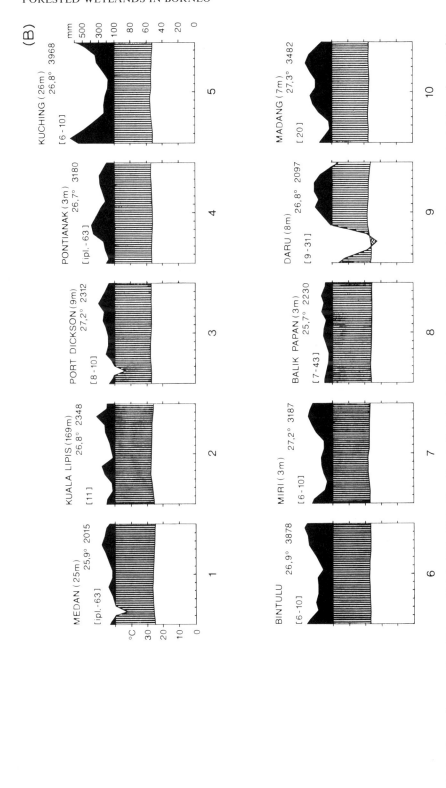

Fig. 13.1. B. Climate diagrams (from Walter and Lieth, 1960–1967) of locations indicated in A. Altitude, mean annual temperature and mean annual rainfall are shown to the right of the place name.

by others in Kalimantan, Sumatra, Mindanao and the other eastern islands of the Malesian region has recently been succinctly reviewed and summarized by Whitmore (1975) in a regional monograph on the tropical rain forests of the Far East.

According to the present state of our knowledge, two major formations of oligotrophic forested wetlands exist in the lowlands and one comparable formation in the montane zone in Borneo. All three formations are characterized by extreme, but interrupted, waterlogging and low levels of mineral nutrient availability. The drainage of all formations produces tea-coloured, very acid water. The two oligotrophic wetland formations in the lowlands are the coastal or deltaic peat-swamp forests previously discussed by Anderson (1983) in another volume of this series and the kerapah forests described by Bruenig (1966, 1968, 1974).

The oligotrophic forested wetlands cover approximately 2×10^5 km^2 in the Malesian region. Centres of occurrence are Borneo, Sumatra and parts of New Guinea (Fig. 13.1), the latter also possessing large tracts of other types of swamp formations with sago palm and herbaceous vegetation.

Recently, a typical montane kerangas forest with black water streams and small natural padang (see Bruenig, 1961, 1965) was found at the northern margin of the dipterocarp rainforests in Bawang Ling Nature Conservation in Hainan. In hollows, a trend toward kerapah was noticeable (Bruenig field notes, 1988).

SITE CONDITIONS

The climate

The island of Borneo has an equatorial, oceanic climate which is characterized by small seasonal fluctuation of temperature but somewhat more marked fluctuations of humidity, rainfall, and wind velocities and occasional episodic climatic extremes.

The interplay of varying trade-wind strength, width and often erratic movement of the intertropical convergence zone, intensity of land–sea atmospheric interchange, and the effects of physiographic barriers causes a marked variability and a more or less distinct seasonality and regionality of this apparently uniform humid equatorial climate on Borneo. While tropical cyclones do not hit the island, their trails may affect the weather conditions. However, violent thunderstorms occur, which are associated with heavy convective rainfall, extremely high-speed descending cold airmasses, so-called micro-bursts which inflict devastation on the forest canopy, and lightning. These storms may be particularly frequent during the passage of the tropical trough region in some years when the converging trade winds are unusually strong and produce long lines of heavy cumulo-nimbus formations (Koteswaram, 1974).

According to Thornthwaite's classification, all of the Bornean lowlands have a perhumid climate with "adequate rainfall" at all seasons. Bruenig (1968, 1969a), however, showed that this uniformly "adequate" rainfall is an artefact caused by the calculation of monthly means. Rather than uniformly wet, this equatorial, "perhumid" climate has frequent periods of dry, bright weather which may occur at all seasons and episodically over long periods of several months. During such spells, solar radiation is high, humidity low, atmospheric saturation vapour pressure deficits extreme, especially about noon, and, with extremely low soil water potentials, drought conditions may develop — especially in soils with low water storage capacity for plant-available water. The severity of such prolonged droughts and their ecological effects are illustrated by the forest fires in kerangas and lime-stone forests, e.g. Bako National Park and Mulu National Park, reported for the 19th century, and the deep-burning peat-swamp forest fires in Anduki Forest Reserve, Brunei, in 1958 and in Kalimantan in 1983. In 1865, Beccari (1904) commented that "abnormalities in the prevailing course of the monsoon are not rare in Sarawak" after reporting on a 10-days spell of dry, bright weather at the end of December in Kuching, which is the month of peak rainfall in that area.

The humidity and rainfall distributions become more uniform with increasing altitude and average values are higher especially in the montane zone above 1000 m, while temperatures are lower by about 0.6 to 0.5°C per 100 m.

The soils

Development of peat

The hot equatorial lowland climate favours decomposition of litter, and consequently is generally not conducive to the accumulation of humic matter and to peat formation. The possibility of Podzol (Spodosol) and related soils occurring in the equatorial lowlands and developing a thick raw humus horizon has not been readily accepted, and has not become common knowledge until fairly recently. The reports on the occurrence of oligotrophic peat swamps in the tropical lowlands were more readily recognized on the grounds of the irrefutable evidence provided by the pioneering work of Polak (1933). However, it remained obscure until recently whether or not the development of these peat-swamp soils, apparently permanently wet, possessed any analogies to the dynamics of raw humus formation on the Spodosols on more or less waterlogged but periodically dry sites.

Initiation of peat formation in tropical lowlands may be due to a number of interacting factors. Of major importance in this respect appear to be: permanent or intermittent waterlogging; oxygen deficiency during waterlogging; chemical soil conditions which are unfavourable and in some cases directly toxic to micro-organisms, such as high content of sulphur compounds in mangrove clay, free aluminium in upland clay, high acidity and high contents of polyphenols and fulvo-acids in litter and humus, and allelopathic secondary compounds; and extreme dryness on certain sites during prolonged rainless periods, which may accentuate the effects of allelopathic substances.

The "deltaic peats" accumulate in seaward-expanding deltas subsequent to the final stages of the mangrove series on humic, acid clay under waterlogged conditions, locally interspersed with pockets or low terraces of sand. As the beach progresses further seaward, and the peat depth increases, phasic plant communities may develop which are characterized by the gregarious occurrence of tree species, such as *Shorea albida*, and low decomposition rates of the litter which possibly accelerate the growth of peat.

Peat accumulation may eventually reach a considerable thickness (Fig. 13.2, section A). Depths of 20 m and more have been recorded. The rather small size and dense forest cover of the hilly catchment areas in relation to the extensive coastal plains and the high capacity of the river systems keep the deltaic, frequently dome-shaped raised peat swamps out of reach of river floods except for narrow strips along the river banks. More than a few hundred metres away from the river, the peat swamps are exclusively rain-flooded.

Oligotrophic peaty soils also may develop on sandy groundwater podzolic gley soils in the low intervals between successive recent raised beaches. They also occur locally at the base of older marine or riverine terraces. In these locations, peaty soils often grade into mesotrophic muck soils of adjacent recent alluvium. Sometimes, at the base of slopes of siliceous sedimentary parent material, they may form an intermediate zone between alluvial soils of adjacent valley bottoms and sandy tropaquod. In both cases these soils are intermediate between oligotrophic peat and eutrophic muck soils (Fig. 13.2, section B, right-hand side).

Peat may also develop on the nearly flat tops of extensive Pleistocene terraces of redeposited and therefore extremely impoverished sediments (Fig. 13.2, section B, centre). Other sites of peat formation are plateaux of older, often Tertiary, siliceous sedimentary parent materials. The podzolic soils which develop under these conditions may become seasonally or almost permanently waterlogged, because of impeded internal subsoil drainage and/or topographic drainage barriers. A periodically high water table may be perched on an impervious humus-iron pan, on a textural clayey B-horizon or on hard, siliceous, dense, flat-bedded bedrock. The tendency to waterlogging, initial mineral deficiency of the parent material, and the biochemistry of the vegetation interact to initiate the formation of raw humus and eventually of peat. The peaty wetlands of this type are called *kerapah*, to be distinguished from the humid, deltaic peat-swamps and high-altitude peat soils, and from the dry Spodosols and related soils which may accumulate fibrous peaty material on the surface and which on non-calcareous substrates are known as *kerangas* (Bruenig, 1968, 1974).

At higher altitude, climatic conditions permit peat development even on very steep slopes and dissected summits with excessive drainage on a variety of parent materials, including relatively fertile igneous rocks such as grano-diorites and basalts. These mountain soils in the cloud belt are

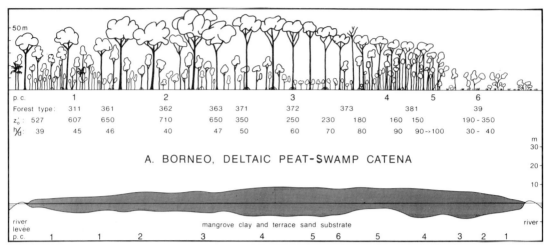

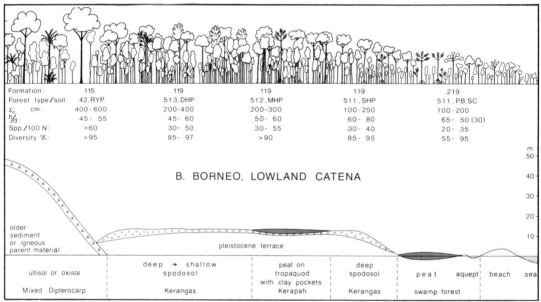

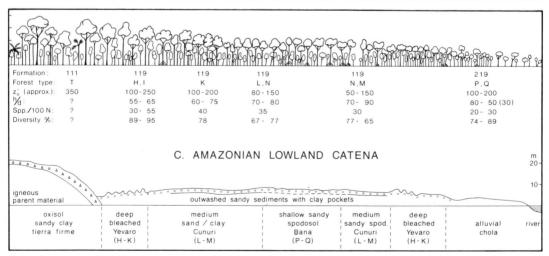

wet, peaty, gleyed podzols with a thin iron pan, and are equivalent to the Placaquods of temperate regions (Bruenig, 1974; Burnham, 1975). On topographically and lithographically similar sites in the lowlands, fibrous, red-brown raw humus may accumulate. For example, on sedimentary quartzitic rocks, soils develop which are extremely acid and are alternately waterlogged and extremely dry. This condition favours the accumulation of undecomposed organic matter which, on siliceous sandstone and equally on limestone summits with free drainage, develops to a considerable thickness. Examples are the rocky headlands formed of thick-bedded sandstone in Bako National Park, and the limestone hills south of Kuching and in Mulu National Park, Sarawak (Bruenig, 1961, 1965, 1966). The landscape relationships of these soil groups are illustrated in Fig. 13.2.

Types of soils

Polak (1933) distinguished two main types of peat swamp in Southeast Asia: an ombrogenous moor due to waterlogging in the coastal plains with more than 2000 mm annual rainfall, and a topogenous moor due to impeded drainage as a result of topographic conditions, such as lake shores, upland basins and hollows. The vegetation of the latter often is herbaceous or more rarely mossy (*Sphagnum* peat). Polak considered the ombrogenous moor analogous to the temperate raised bog (*Hochmoor*) and the topogenous moors as eutrophic and equivalent to temperate fen.

Richards (1952) distinguished two basically different swamp soils which correspond to two distinct types of hydroseres leading to different types of edaphic climax forest. The "normal" swamp is non-peaty and supplied with relatively eutrophic water. The other type is inundated with very oligotrophic drainage water which initiates peat formation, finally leading to an ombrogenous raised bog forest of bi-convex, lens-shaped masses of peat with depths exceeding 7 m. The accumulation of peat is "clearly due partly to waterlogged, and therefore anaerobic conditions". These slow the rate of breakdown of plant debris below the rates of production of fine and coarse litter, and organic matter consequently accumulates. But some additional edaphic factor would seem necessary to sustain peat formation because not all waterlogged soils accumulate acid humus. Richards suggested that inundation by oligotrophic water is this additional condition.

Jacobs (1974) distinguished "true swamp forests" and "peat forests". The former were considered to be more or less permanently flooded by fresh river water, and included *Gonystylus bancanus* forests in Borneo and Sumatra and *Metroxylon* swamps in New Guinea. The peat forests were considered specialized variants, which occur "in such places where peat could be formed (largely absent in true swamp forest)". These peats are often lens-shaped, exclusively rain-flooded and "very poor in nutrients as these are washed out by the rain". In reality, in Borneo, the *G. bancanus*

Fig. 13.2. Sections through mature natural forest stands along steep ecological gradients in Borneo (A and B) and Amazonia (C) constructed from research data and airphoto interpretation. *Section A.* Transect through a deltaic peat swamp with complete series of phasic communities p.c. 1 to 6. The corresponding peat (histosol) dome is shown cross-hatched. The numbers of the phasic communities p.c. 1 to 6 are given below the two profiles. *Section B.* Transect from mixed dipterocarp forest (115, 42) on sandy-loamy Ultisol through a sequence of types of kerangas (formation 119, forest types 513–511) on Spodosol (podzol) to finally stunted, closed to open kerapah on Tropaquod-Spodosol or Histosol, or kerangas on dry peat woodland (219, 511). The terrace flank and peat-swamp forest p.c. 1 following toward the beach is not shown in the forest transect. *Section C.* Transect through the natural forest study area of the International Amazon Rainforest Ecosystem MAB-Pilot Project at San Carlos de Río Negro. To the left "tierra firme" (Mixed Tropical Climatic Climax) forest (111; T), followed by various association groups of tall forest: Yevaro (119; H, I), Cunuri (119; K, L, N and M), closed bana forest (219, P) and finally open (219, Q) bana woodland. The vegetation and soil sequences correspond ecologically and structurally closely to those of section B, except for the lack of formation of peat sensu strictu and the lower stature of the Amazonian vegetation. *Explanations:* Formation: major classificatory unit according to Bruenig (1972). Forest type: subunits according to Bruenig (1969b, 1979), corresponding to phasic communities of Anderson (1961a) in section A (311 = p.c. 1, 361–3 = p.c. 2, 371–3 = p.c. 2, 381 = p.c. 4, 39 = p.c. 6). Soil in section B: red yellow podzolic = RYP = Ultisol; deep, medium and shallow humus podzol = DHP, MHP, SHP = Spodosol. Tropaquod; PB = kerapah peat bog, SC = skeletal soil. z_0: estimator of the aerodynamic roughness parameter of canopy (Bruenig, 1970). h/d: mean height/diameter ratio of top-canopy trees in the A and B strata (indicates vigour and stability). Spp./100 N: species number in 100 individual trees = species richness. Diversity %: value of the McIntosh within-stand diversity index in percent, of the maximum possible for the given number of individuals (see footnote, p. 316). *Yevaro:* vernacular name for *Eperua purpurea*. *Cunuri:* vernacular name for *Micrandra sprucei*.

forests are peat-swamp forests on medium deep peat which are rarely flooded by river water. Anderson (1961a, 1963, 1983) described a series of phasic communities of peat-swamp forest (see below pp. 309–316), designated as p.c. 1, ..., p.c. 6; this *G. bancanus* community is p.c. 1 in the series, termed mixed peat-swamp forest.

It is well known that rain introduces substantial amounts of nutrients into the peat-swamp ecosystem. The amounts may exceed those imported by any flood water from blackwater rivers. The nutrients imported in rain are absorbed so effectively by the ecosystem that the outflowing tea-coloured "black" water is almost free of any electrolytes. Rain is the major vector of supply to these open systems of mineral nutrients, except possibly nitrogen. An example of the effectiveness of this mechanism in an Amazonian forest ecosystem, which is intermediate between kerangas and kerapah, has been described by Herrera and Jordan (1980) and Jordan et al. (1980).

Anderson (1961a, 1964a) delineated the soils of his deltaic peat-swamp forests by the standard definition of peat, and commented on the diversity of their geomorphic features, especially of the surface and substrate topography. He did not define these peat-swamp soils in relation to Bruenig's (1968, 1974) kerapah peat soils. Anderson suspected variation of drainage to be the major condition which differentiates peat-swamp soils. This variation may be related to the gradient of the seral "phasic peat-swamp forest communities". It is most noticeable and ecologically effective in peat swamps in which the dome-shaped structure of the peat is well developed. Anderson thought, on the evidence of chemical analyses of peat sample, that the peat soils were uniformly oligotrophic with little chemical variability. The loss on ignition is usually very high, mostly above 90% and often approaching 99%. The only noticeable trend he was able to establish by chemical analysis was a decline of phosphorus content from the margin to the centre of the peat swamp.

The existence of physiologically and ecologically important chemical differences between the early phase of peat development (p.c. 1) and the later phases (e.g. p.c. 3) is indicated by an example from a peat swamp in Lassa Forest Reserve, Rejang delta, Sarawak (Fig. 13.3). Peat samples were collected from a depth of 40 cm below the peat surface in p.c. 1 (*Gonystylus bancanus* mixed peat-swamp association) and p.c. 3 (*Shorea albida* peat-swamp consociation). Ten samples were taken in each phasic community.

The results of the chemical analysis (Salfeld, 1974) showed not only significant differences in total nitrogen content and correspondingly in C/N ratios (Fig. 13.4), but also in the chemical nature of the nitrogen compounds (Table 13.1). The peat in p.c. 1 is more heterogeneous than the peat from p.c. 3. The two peats differ in some important features. The ratio of HCl-hydrolysable nitrogen to total nitrogen is somewhat higher in p.c. 1, particularly in a group of seven adjacent samples (Table 13.1 and Fig. 13.4, group 1(1)). These samples also have a distinctly lower content of total nitrogen, a higher carbon content and a consequently higher C/N ratio (Fig. 13.4). The

TABLE 13.1

Chemical nature of nitrogen in two batches and three groups of ten composite samples each of forest peat collected in p.c. 1 and p.c. 3, Lassa F.R., Sarawak (Salfeld, 1974). p.c. 1 (1) = subgroup of seven composite samples, p.c. 1 (2) = subgroup of the remaining three composite samples (compare Fig. 13.4 and 13.5) (Nitrogen in percent of total dry matter can be read from Fig. 13.4)

Phasic community (p.c.) and group	Number of samples	Hydrolysable N (% of total N)	Ammonium and amide N[1]	Hexosamine N[1]	α-amino N[1]	C/N (mean)
1 (1+2)	10	54.3	22.7	13.2	47.9	42
1 (1)	7	60.2	22.6	13.4	49.7	46
1 (2)	3	51.7	22.7	13.2	47.1	30
3	10	48.0	27.8	20.6	18.9	38

[1] % of hydrolysable N.

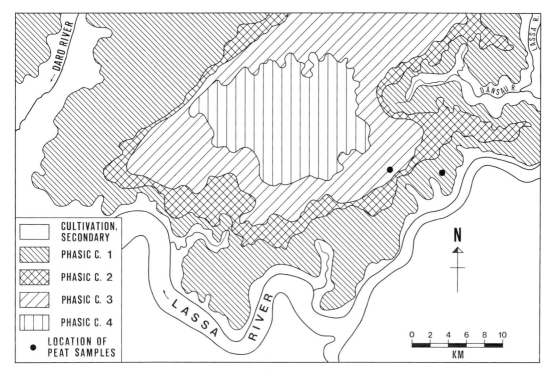

Fig. 13.3. Typical pattern of the concentric zonation of phasic communities in peat-swamp forest in relation to the river drainage system, Lassa Forest Reserve, Rejang Delta, Sarawak. Phasic community 1, p.c. 1: *Gonystylus–Dactylocladus–Neoscortechinia* association, forest types 31–34; p.c. 2: *Shorea albida–Gonystylus–Stemonurus* association, forest types 361–363; p.c. 3: *Shorea albida* consociation, forest types 371–373; p.c. 4: *Shorea albida–Litsea–Parastemon* association, forest types 38; p.c. 5 and 6: the peat development has not reached these phases yet; but given suitable conditions of drainage, geomorphological development and tectonic movement p.c. 6 will develop in time. The location of the two batches of peat samples (Table 13.1 and Figs. 13.4 and 13.5) is indicated by circles.

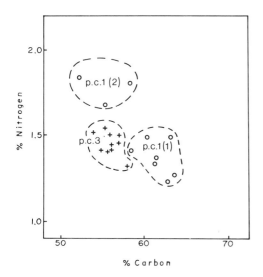

Fig. 13.4. Carbon and nitrogen concentrations of peat samples from phasic community 1 (*Gonystylus–Dactylocladus–Neoscortechinia* association) (circles) and from p.c. 3 (*Shorea albida* consociation) (crosses) in Lassa Forest Reserve, Rejang Delta, Sarawak (from Salfeld, 1974). For location of samples see Fig. 13.3.

composition of the hydrolysable nitrogen is almost identical within sample p.c. 1, but it is different in p.c. 3, the latter having a considerably higher content and proportion of ammonium and amide and hexosamine nitrogen and a much lower content and proportion of α-amino nitrogen.

The three groups of peat samples are also distinguished by their humic acid fractions, which show distinctly different extinction rates at different wave lengths. As shown in Fig. 13.5 the ratios of the extinction rates at 400 and 500 nm, and those at 600 and 700 nm, separate the three groups rather clearly, and the resulting graph closely reflects that for carbon and nitrogen content (Fig. 13.4). These tentative results cannot yet be fully interpreted in ecophysiological terms, and more research in the field and the laboratory is required. However, I conclude that the peats in the two phasic communities differ in chemical properties. The growth of *Gonystylus bancanus* in response to peat chemistry and the floristic

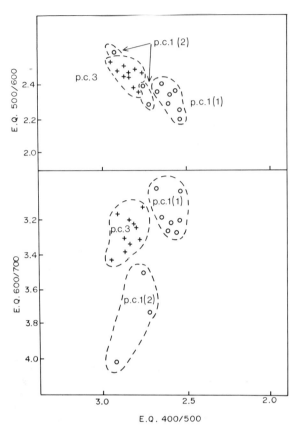

Fig. 13.5. Ratios of spectral extinction at different wave-lengths by the humic acid fractions of peat samples from p.c. 1, subgroups (1) and (2) (circles) and from p.c. 3 (crosses) in Lassa Forest Reserve, Rejang Delta, Sarawak (from Salfeld, 1974).

differences between the forests in p.c. 1 and 3 will be described further below.

The concentrations of the major minerals (Ca, Na, Mg, K, Zn, Al, Fe, Mn, P) in the same 20 samples from p.c. 1 and 3 were also determined. Significant differences among the means were found for aluminium, iron, phosphorus, potassium and zinc, the values in p.c. 3 being consistently lower (Waughman, 1974). In concluding, I hypothesize that there are differences in the concentrations and the biochemical nature of nutrients and organic compounds among peats in different phasic communities, which are related to the floristic vegetation structure.

The oligotrophic wetland soils of the kerapah formation in Sarawak and Brunei have been described in some detail by Bruenig (1965, 1968, 1974) on the basis of extensive ecological field sampling and of reports by Dames (1956, 1962), but without analytical data. This information suggests that no clear edaphic distinction is yet possible between permanently waterlogged kerapah bog soils and Anderson's peat-swamp soils. Both groups overlap in Munsell colour notations and other field characteristics such as thickness of peat and its consistency. Drier types of peat in kerangas have a generally browner hue than the wet more reddish-brown peat-swamp and kerapah peat soils and are more fibrous. Peat-swamp and wet kerapah soils both have a predominance of lateral drainage of water over an impervious horizon or substrate, which is strongly coloured brown by unsaturated humic acids, and has a pH of around 3.

Generally, kerapah peat of comparable seral stages is shallower than peat of the deltaic peat-swamp soils. Both forest formations and their soil groups may develop from groundwater humus podzol (aquod) or humus (iron) podzol (humod and orthod in the Spodosol order) or from heavy clay soils. In all cases peat accumulation may reach considerable thickness. Consequently it is impossible to separate the two formations simply by edaphic characteristics. A distinction is only possible by the whole vegetation–soil complex and its history, resulting in broad transitional zones of overlap between kerangas, kerapah, the late phasic peat-swamp communities, and the corresponding montane forest types which seem to be intermediate between peat-swamp and kerapah forests (Bruenig, 1966, 1974, 1986).

The peat-swamp phases originate predominantly from the latest stage of mangrove clay development under geomorphic conditions which favour waterlogging and rain-flooding, while preventing continued inflow of eutrophic river sediments. Pockets of sandy Spodosols will usually be engulfed and eventually drowned by the peat-swamp formation. The development of kerapah peat originates on dry-land kerangas soils under the influence of strong lateral inflow of oligotrophic humic water, or by deteriorating drainage and in-situ rain-flooding, or both. The deterioration may be caused by a growing B_b horizon or by a build-up of barriers to lateral drainage, such as a raising kerapah peat bog, or a combination of the two. Examples and a montane summit peat for comparison are described in detail by Bruenig (1974) and arranged in a catena concept as regular sequences of phases distinguished by increasing podzolization, accumulation of acid organic matter and deteriorating drainage, and distinguished by substrate, geomorphic conditions and species–soil interaction. Fig. 13.2 (section B) illustrates these conditions in a simplified, schematic manner.

VEGETATION

Structural variation

There is a notable and similar change of forest structure in the peat-swamp and kerapah forests along the gradients of phasic development and of decreasing site quality. This change is very consistent and repeated throughout Borneo in similar catenary sequences of soil and site conditions. This suggests that there may be a causal interdependence between vegetation structure and site and soil conditions, and that the structural variation expresses an adaptation to the physical and chemical site conditions. These changes of geometric or architectural vegetation structure are illustrated schematically as catenary sequences in Fig. 13.2. Section A shows this sequence in the peat-swamp phasic communities, section B in the upland mixed dipterocarp forests – kerangas/kerapah catena, and section C, for comparison, in the catena of oligotrophic forests in the Amazon basin at San Carlos de Río Negro in Venezuela.

In spite of the general consistency of the structural change, there is considerable diversity in detail. This is partly due to differences in floristic composition, which may be the result of the particular, restricted range of some species and the tendency to vicarious occurrence of some of the dominant gregarious species. It may also be due to the changes which govern the colonization of new sites, regeneration of gaps and the development of communities. Finally, sites and soils are neither uniform nor identical sensu strictu, but will always differ in some aspects.

Bornean peat-swamp forests and woodlands

The catenary vegetation sequence on the oligotrophic lowland peat swamps has been subdivided by Anderson (1961a, 1963, 1983) into six phasic communities, the term designating a forest type in a certain phase of primary seral development. The term is particularly appropriate because pollen analyses (Anderson and Müller, 1975; see also Anderson, 1983) have shown that the horizontal sequence of the vegetation types is usually repeated in the vertical sequence of the phasic vegetation types from the bottom to the surface of the peat. This correspondence is particularly well illustrated in well-developed raised peat domes. The six phasic communities have been differentiated by floristic composition and structure of the vegetation. Two phasic communities (p.c. 5 and 6) occur only in a few areas where the development of the peat dome has reached a certain maturity, and possibly the final stage of a near-steady state. Examples are the highly raised bogs in the Baram river basin in Sarawak (Fig. 13.6).

There is a general trend of change along this successional gradient of such parameters as canopy height, size of crowns and integrated groups of crowns, and of the basal area stocking of the most important timber species and species groups. This trend is consistent throughout the region, and so regular that it can be used for classifying the forests into strata in forest inventories (Bruenig, 1969b). The spatial sequence of the phasic communities, and their subdivisions, are illustrated in Fig. 13.2 (cross-section with peat and forest profiles), Fig. 13.3 (arrangement of phasic communities in concentric zones corresponding to the sequence in time along the vertical peat profile) and Fig. 13.7 (change of structural parameters from p.c. 1 to p.c. 4).

Phasic community 1 (formation 113.11 and forest type 311 in Fig. 13.2).

Gonystylus bancanus–Dactylocladus stenostachys–Neoscortechinia kingii association, mixed peat-swamp forest.

This is the initial phase of the primary peat-swamp forest, and occurs as a more or less extensive, often narrow zone on the perimeter of peat swamps or in some areas of a geomorphologically younger development throughout the expanse of a coastal peat swamp. It is the structurally most complex and species-rich phase of the peat-swamp forests. The top canopy is aerodynamically rough (see Figs. 13.6 and 13.8) and on average 40 to 45 m high. Principal species of the emergent A and main-canopy B strata are *Gonystylus bancanus* (often more in B than in A), *Copaifera palustris*,

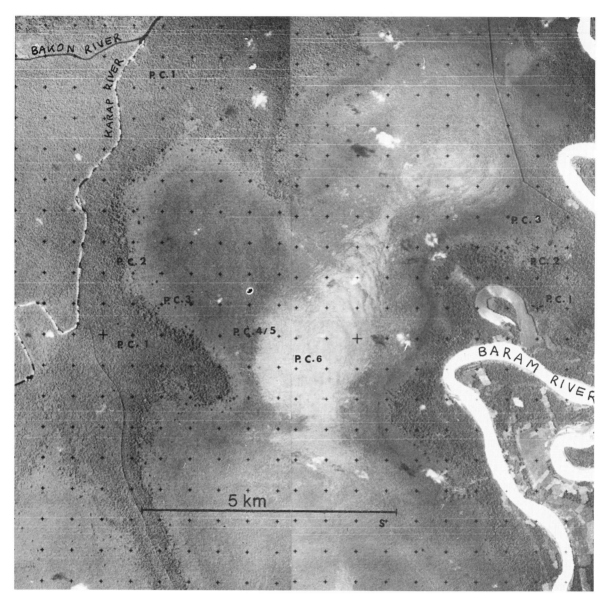

Fig. 13.6. Small-scale aerial photograph of a peat swamp showing the whole sequence of phasic communities between the Karap and Baram rivers in northern Sarawak, East Malaysia. Mixed peat-swamp forest, p.c. 1, at the perimeter; along the river courses successive belts of *Shorea albida* forest types p.c. 2, 3 and 4, in some places followed by a zone of darker forest of type p.c. 5, then by more extensive areas of open p.c. 6, in which surface water reflecting sunlight can be noticed (Courtesy of Sarawak Government).

Dactylocladus stenostachys and *Shorea* spp. (4 species, but not *S. albida*). The dipterocarp *Dryobalanops rappa* may locally become gregarious. The distribution pattern of species and total stocking is generally markedly patchy. The first two species tend to local gregariousness or alternatively rare occurrence. The well integrated C stratum resembles structurally and floristically, some types of mixed dipterocarp and kerangas forests on well-drained sandy soils (Fig. 13.2). Species of the families Clusiaceae, Euphorbiaceae, Myrtaceae and Sapotaceae are particularly prominent. The D stratum is rich in herbs, sedges and ferns (Fig. 13.9). The humic, oxygen-deficient soil water, with usually a high water table, causes characteristic adaptions of the roots (Fig. 13.10).

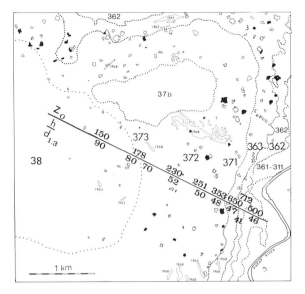

Fig. 13.7. Vegetation map of the peat-swamp forest shown in Figs. 13.11 and 13.12. The phasic communities and forest types are: p.c. 1, 311: $z_0 = 500$, $h/d = 46$; p.c. 2, 361–363: $z_0 = 712$–650, $h/d = 41$–47; p.c. 3, 371–372: $z_0 = 353$–178, $h/d = 48$–70; p.c. 4, 38: $z_0 = 178$–150, $h/d = 80$–90. Lightning gaps in 1963 (open), new gaps in 1968 (inked black) and windthrow areas (broken lines), with year of appearance on the aerial photographs, are shown. The type 37 D indicates the area in 37 destroyed by "ulat bulu" (a hairy caterpillar of a moth) during 1947. $z_0 =$ estimator of the aerodynamic roughness of the canopy, $h/d =$ (height of tree)/(stem diameter at 1.3 m above ground).

A distinct subtype occurs locally as a narrow transitional zone between the perimeter of this phasic community and the final stage of the mangrove forest succession. This subtype is particularly variable, but usually rich in sedges, herbs and stemless, climbing or tree-like palms. Among trees, species of the genera *Alstonia, Campnosperma, Cratoxylum, Lophopetalum* and *Parishia* are particularly common. Similar transitional subtypes occur between the peripheral peat-swamp forests and the base of kerangas terraces and hills, the margins of alluvial swamp forests and the various types of littoral swamps.

Phasic community 2 (formation 113.21, forest type 361–363 in Fig. 13.2).

Shorea albida–Gonystylus bancanus–Stemonurus secundiflorus association, Alan forest.

This phasic community may, on more steeply rising, old, peat domes, occupy a usually rather narrow zone between p.c. 1 and 3 (Figs. 13.2, 13.6, 13.11, 13.12) but may in some areas cover large tracts or zonal belts. The A and B canopy strata are composed of tall, large diameter *S. albida* of the "alan"-type. Their stems are almost invariably hollow, consisting of a pipe of very hard, dark red-brown "alan" timber of "red selangan" type. The boles are heavily buttressed. The large, cauliflower-shaped crowns are often stagheaded. Intermediate trees of *S. albida* are rare. Seedlings are abundant after the infrequent mass fruiting, but rapidly disappear. The species and its forest, therefore, had been considered moribund, a view which has not held true. The moderately dense middle and lower storeys are mostly composed of species of p.c. 1, *Stemonurus secundiflorus* var. *lanceolatus* being a characteristic species of the C stratum. *G. bancanus* occurs in the C/B stratum. The D stratum is less rich in herbs and ferns than in p.c. 1.

The surface of the main canopy is uneven and irregular which is a very conspicuous feature on aerial photographs. Emergent trees and groups of trees protrude to a height of 10 to 20 m above the plane of the intermediate tree canopy. This makes the canopy aerodynamically very rough and increases the turbulent air exchange between canopy and atmosphere. The tree form is sturdy with low ratios of tree height to diameter (h/d), ranging around 45 to 55. This sturdiness improves their stability against wind, which is important because of the high aerodynamic roughness of the canopy. The mesophyllous (50–182 cm^2) leaves of *S. albida* are very coriaceous, their upper side strongly reflectant (as shown by the light grey tone on the aerial photographs in Figs. 13.6, 13.8, 13.11 and 13.12, indicating a high albedo) and the underside pubescent with slightly in-rolled margins.

Phasic community 3 (formation 111.22, forest type 371–373 in Fig. 13.2).

Shorea albida consociation, Alan bunga forest.

The p.c. 3 covers extensive areas especially in Brunei and northern Sarawak. The A and B strata merge to a dense, uniform and aerodynamically moderately rough (type 371) to moderately smooth (type 373) canopy almost exclusively composed of *S. albida*. The trees are of the "alan bunga" type. The tree boles are rarely piped, and possess brittle heart wood. The timber is much lighter in weight and colour than in the piped boles of the "alan"

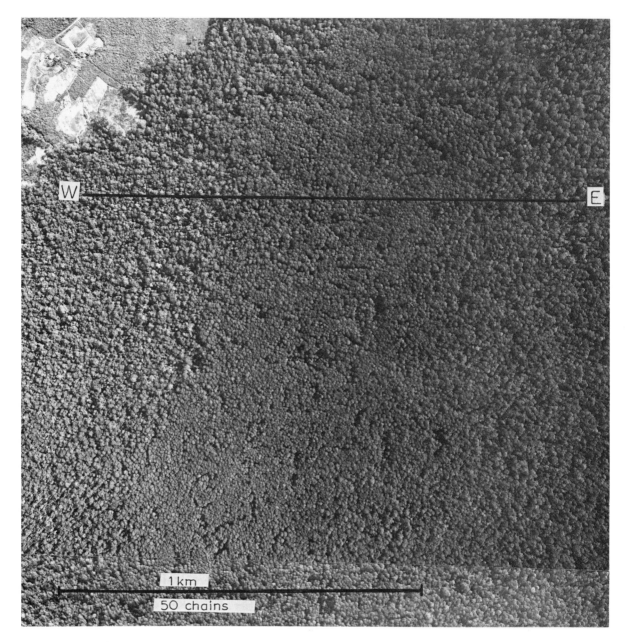

Fig. 13.8. A coastal Pleistocene terrace above a Holocene floodplain in Sungai Dalam Forest Reserve, Sarawak. Mixed peat-swamp forest (p.c. 1) with rough, irregular canopy on either side is followed by small-crowned, dense and uniform-canopied kerangas and kerapah forests on the terrace. The two types cannot be distinguished on the aerial photograph. In the lower right corner, a transition to mixed dipterocarp forest. In the top left corner, currently cultivated areas and secondary forest (Courtesy of Sarawak Government).

type in p.c. 2. The buttresses are much lower and narrower than in p.c. 2. The top canopy height ranges 50–60 m, to as high as 70 m (locally in type 371). The B stratum is very poorly developed or absent. *Gonystylus bancanus* is extremely rare to absent (Bruenig, 1969b). The C stratum is generally low and poorly developed and, frequently dominated by a single species (especially in types 372 and 373), such as *Cephalomappa paludicola, Ganua curtisii* or *Tetractomia parvifolia*. Herbaceous vegetation is practically absent. *Pandanus andersonii* frequently forms dense thickets in the D stratum.

Fig. 13.9. Typical coastal mixed peat-swamp forest, p.c. 1, near Tatau, Sarawak, East Malaysia. A large and an intermediate tree of *Gonystylus bancanus* in the B-stratum without buttresses to the left and a large *Parishia* sp. of the A/B-stratum with high and spreading buttresses. Note the leaves of a large, epiphytic pandan on the left and the sedges and numerous herbs on the ground.

Fig. 13.10. Pneumatophores of a big *Dactylocladus stenostachys* in coastal mixed peat-swamp forest, p.c. 1, near Tatau, Sarawak. Trees adapt themselves in diverse ways to the exacting environment of peat swamp and kerapah.

A notable feature of this p.c. 3 is the abundance of damage from lightning and storms (Anderson, 1964b; Bruenig, 1964, 1973a) and insects (Anderson, 1961b) (see Figs. 13.11 and 13.12). This combination of moderate aerodynamic roughness of the tall canopy, 50 to 70 m high, and a high h/d ratio of the trees make the stand extremely susceptible to wind damage. The height and roughness of the canopy cause an increase of wind speed and turbulence, while the slenderness (high h/d ratio) of the tall trees substantially reduces their resistance to bending and causes them to sway heavily and finally fall or break under storm impact. The reason for the concomitant high incidence of lightning gaps is unknown, but may also be related to architectural features of the canopy. For instance, it is conceivable that the combination of tallness and roughness of the canopy influences the electric field during thunder strikes. The maximum frequency and areal proportion of lightning gaps and windthrow gaps occurs in types 371 and 372, which also have the maximum canopy height in the phasic series.

Further instability is introduced by the high density of trees of one species (*S. albida*) in p.c. 3 and 4 which predisposes these forest types to epidemic spread of sudden insect pest outbreaks (Figs. 13.7, 13.11, 13.12). These catastrophes, destroying *S. albida* over hundreds to many thousands of hectares, however, do not seem to arrest or divert the long-term phasic development from p.c. 3 to finally p.c. 6, but may play a role in regulating the speed of transition from p.c. 3 to 4, and from 4 to either 5 or 6.

The leaf-size spectrum of the main canopy

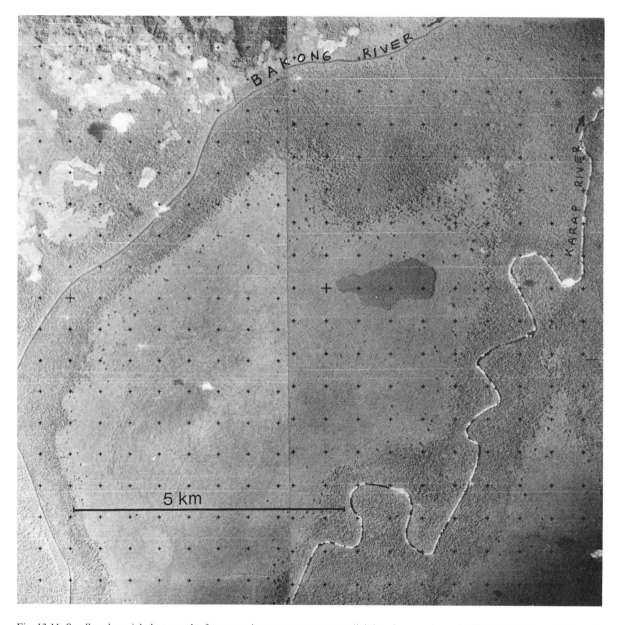

Fig. 13.11. Small-scale aerial photograph of an extensive peat-swamp area adjoining the area shown in Fig. 13.6 to the west. Near the rivers dark-grey, rough mixed peat-swamp forest, p.c. 1, followed by a narrow zone of light grey, rough p.c. 2, a broad belt of dense, light grey, gap-pitted p.c. 3 and finally by an extensive central area of p.c. 4. Note (1) the kidney-shaped dark gap which is shown and explained in Fig. 13.12; (2) the cobweb pattern of vegetation in the "epicenter" below the darkish area and just above "km" in p.c. 4 in the lower left part of the area (Courtesy of Sarawak Government).

changes from predominantly mesophyll in p.c. 2 through mesophyll/notophyll in p.c. 3 (type 371), to predominantly notophyll (20–50 cm^2 leaf surface) in type 373. This trend exemplifies a change toward a greater sclerophyllic-xeromorphic physiognomy which is continued in p.c. 4.

Phasic community 4 (formation 113.23, forest type 381 in Fig. 13.2).

Shorea albida–Litsea–Parastemon association, Padang Alan bunga forest or Padang Medang forest.

This p.c. is transitional between p.c. 3 and 4 as a

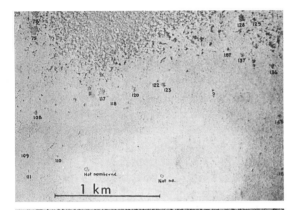

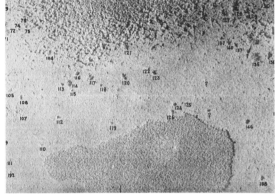

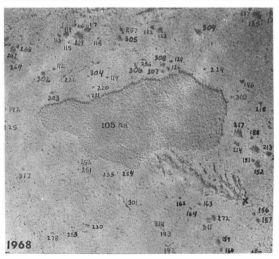

Fig. 13.12. The forest around the kidney-shaped gap area shown in Figs. 13.7 and 13.11. Numbered small gaps are lightning holes. The forest communities are p.c. 1 at the top centre followed by large-crowned, irregular, light grey p.c. 2, then dense, medium crowned p.c. 3 with numerous lightning gaps, then dense, small-crowned p.c. 4 with few, small gaps around the kidney-shaped area which is partly in p.c. 3 and 4. *Top:* condition in 1947, attack by the unidentified hairy caterpillar "ulat bulu" had just started. The *Shorea albida* trees appear very light-coloured probably due to physiological damage and partial defoliation. *Middle:* condition in 1961. The trees in the top canopy have died, the former understorey is exposed. Some new lightning gaps have occurred. *Bottom:* condition in 1968. Little change yet in the "ulat bulu" area, a few light-tones *Shorea albida* trees are recognizable and indicate beginning of recovery by succession. Noticeable increase of lightning gaps in p.c. 3 since 1961, windthrow has originated in p.c. 3 in a small gap marked by a cross and extended to close to the edge of the "ulat bulu" gap.

narrow zone on old peat domes or as an extensive central area where development has not proceeded to later stages, as in the area shown in Fig. 13.11 (in contrast to the area shown in Fig. 13.6). The forest is dense, the trees are relatively small and boles are rarely larger than 40 to 60 cm dbh. The canopy of *S. albida* (Alan bunga) or *Litsea crassifolia* (Medang Padang) is dense, uniform and aerodynamically relatively smooth at 30 to 40 m top height. Few other species are present in the canopy (*Calophyllum obliquinervum, Combretocarpus rotundatus* and *Parastemon spicatum*). The understoreys are also poor in species number and the diversity of mixture is low. The D layer consists of sclerophyllous shrub and sedge species which also occur commonly in kerangas, kerapah, and montane forests. The flora includes species which are especially adapted to poor nutrient availability, such as the insect-consuming *Nepenthes* and symbiotic myrmecophytes.

Phasic community 5 (formation 113.23, forest type 382, not included in Fig. 13.2).

Tristania–Parastemon–Palaquium association, Padang Selunsor forest.

This is a locally occurring, but characteristic and distinct, narrow transition zone between p.c. 3 and p.c. 4 and p.c. 6. *Tristania beccarii* or *T. obovata, Parastemon spicatum, Palaquium cochlearifolium, Combretocarpus rotundatus* and *Dactylocladus stenostachys* are the most abundant species in a species-poor pole forest with low (15–20 m high), smooth and dense canopy. The herbaceous flora is sparse. The D layer is dominated by *Pandanus sigmoideus* and the sedge *Thoracostachyum bancanum*.

Phasic community 6 (formation 213.11, forest type 39).

Combretocarpus–Dactylocladus association, Padang Keruntum woodland.

This association is the final known phase in raised-bog development. The thickness of the peat is very variable but may exceed 20 m. In some areas p.c. 6 covers extensive areas (Fig. 13.6) but its occurrence is rather rare and local. The forest is open and xeromorphic. Trees and shrubs are stunted. *Combretocarpus rotundatus* is the only tree species attaining diameters of 30 cm and heights of 12 m. *Dactylocladus stenostachys, Garcinia cuneifolia* and *Litsea crassifolia* are abundant, but mostly little more than shrubs. Myrmecophytes of the families Melastomataceae and Apocynaceae and *Nepenthes* spp. are particularly numerous. *Pandanus ridleyi* and *Thorachostachyum bancanum* are abundant in the D stratum. Sphagnum moss (*Sphagnum junghuhnianum*) also occurs, which is the only known habitat type in the Bornean lowlands outside kerapah of this habitually high-altitudinal species.

The predominantly notophyll/microphyll leaves are more sclerophyllous, coriaceous, and more vertically oriented than in the p.c. 4 and 5. For example, *C. rotundatus* has noticeably smaller and more upright leaves than in the other peat-swamp communities, resembling more closely its appearance in open padang in Bako National Park and in poor kerangas and kerapah vegetation on the Merurong plateau (Bruenig, 1961, 1965, 1968).

Bornean kerapah forests and woodlands

The oligotrophic forested wetlands of the kerapah forest and woodland formations are much more diverse within and between stands and between geographic regions than the peat-swamp formation. Initial edaphic and physiographic conditions are much more varied, but, even on initially almost identical sites in different localities, vicarious dominant species may cause very different canopy aspects. Emergent A stratum species are particularly notable in this respect. A certain soil/site type and seral stage may be characterized by the gregarious occurrence of such endemic species as *Dryobalanops fusca* in Pueh Forest Reserve, western Sarawak, which may be absent on identical sites elsewhere. Vicarious occurrence of species and local endemism could be expected in kerapah and kerangas because they occur in patches and small tracts, are isolated by the effective biological barriers of other formation types, and are characterized by the accumulation of biochemically very complex compounds in and on the soil.

The species richness of kerapah communities is between 20 and 45 tree species per 100 individuals above 1 cm dbh and markedly and consistently lower than that of kerangas (30–60) or mixed dipterocarp forest (>60). Both species richness (number of species) and diversity[1] decline with progressive waterlogging and peat accumulation along the gradient of kerapah peat bog development (Bruenig, 1974). The sequence of forest types along this gradient is illustrated in Fig. 13.2, section B, and Figs. 13.13–13.17. Generally, and in both kerapah and peat-swamp formations, the relative values of the species diversity index (Fig. 13.18)[1] are high and scatter around the 90% line, which is the maximum value arithmetically possible with equal proportions among the existing number of species. This means that the mixture is fairly balanced and the tendency to single-species dominance moderate if the number of individuals per species is used as a criterion.

The index value is nearly equally high in mature communities with high numbers of species and those which are poor in species. It varies with the regeneration and building phases of stand development. High values are typical for late-building and maturing phases. They occur in stands developed from large gaps, in transitional ecotones where two floristically different forest vegetation types overlap and integrate, and in relatively dynamic stands with high rates of turn-over, mortality and tree fall (Figs. 13.13, 13.19). Low values are found in stands dominated by a single species with predominantly small-gap regeneration or, if large gaps occur, with effective, possibly allelopathic, species interaction preventing the successful invasion by further species. Examples are kerapah consociations such as *Dacrydium* kerapah (Fig. 13.18, solid circle with cross) and *Casuarina* kerapah

---

[1]The diversity index used is that of McIntosh (1967):

$$\frac{N - \sqrt{\sum_{i=1}^{s} n_i^2}}{N - \sqrt{N}}$$

where N is the total number of individuals recorded, s the number of species, and n_i the number of individuals of the ith species. Its value ranges from zero where all individuals are of one species to unity where all are of different species.

Fig. 13.13. Kerapah forest with dominant *Shorea albida* (top-height 35 m) and intermediate *Casuarina nobilis* (25 m), much wind-throw in the A and B layers, irregularly grouped understorey (C layer), peaty sandy clayey gley soil near the edge of an extensive peat bog on the Merurong plateau, Sarawak 730 m altitude (see Fig. 13.20).

Fig. 13.14. Kerapah forest on the Merurong plateau, Sarawak, at 730 m altitude. The soil is 70 cm peat over sand and clay. The tall emergent is *Shorea albida*, associated with intermediate *Diospyros evena* and *Casuarina nobilis*. The undergrowth contains shrubby *Vaccinium* sp. nov., Melastomataceae, pandans, and sedges.

(Figs. 13.20–13.23). Species diversity, but not necessarily species richness, declines if a species becomes gregarious in the top canopy or, more rarely, in the under-canopy layers, provided that this is not linked to changes in soil conditions caused by the vegetation, and provided there are no effective allelopathic interactions which cause the extermination of species from the stand.

The complexities of interrelationships obscure the expression of any correlations between species richness, species diversity, and stability of the forest as an ecosystem, and make it more difficult to generalize on phasic development than for the peat-swamp forests. As a result, species richness and diversity values of single stands (study plots) scatter widely within the classification units of the kerapah formations. In spite of this scatter of species richness and diversity, mean values for forest/soil type units which were calculated from the study plots in Sarawak and Brunei show distinct trends in the species richness/diversity matrix of coordinates (Fig. 13.18). There is a noticeable trend of declining species richness as the soils become shallower, the water regime more fluctuating or the peaty soil more often waterlogged, and the soil generally more oligotrophic. The decline of the diversity value as a percentage of the possible maximum is less marked, except in the case of the *Dacrydium* kerapah in which soil changes with the single-species dominance. The values for the *Casuarina* kerapah are not included in Fig. 13.18, but would be close to the position of the *Dacrydium* kerapah. It is noteworthy in this

Fig. 13.15. Transition from the tall, closed kerapah forest with large emergent *S. albida* (Fig. 13.14) to irregularly open kerapah woodland (Fig. 13.17) on 1–2 m deep, red-brown peat, with high-stilted *Calophyllum rhizophorum*. Canopy height up to 30 m, codominant *Dacrydium beccarii* var. *beccarii*. Merurong plateau, Sarawak, 730 m altitude.

Fig. 13.16. Kerapah forest with dominant *Tristania obovata* (light, eucalypt-type bark) and *Calophyllum rhizophorum*. The stand is more open than that shown in Fig. 13.15 and grades into open kerapah woodland (Fig. 13.17). Sedges (*Thoracostachyum bancanum*) and *Pandanus scandens* are common. The soil is red-brown peat about 2 m deep. Merurong plateau, Sarawak, 730 m altitude.

respect, that both these dominant and, at the same time, differentiating species, possess particularly high contents of condensed tannins in the leaves, and that leaf extracts have a very strong gelatine-precipitating power. This would prevent rapid litter decomposition and increase peat forming tendencies on the site. Possibly it also would accentuate allelopathic interactions generally — particularly if the site is subject to occasional drought, as is the rule in kerapah.

Concomitantly with floristic richness, the geometric complexity and aerodynamic roughness of the canopy also decline. The cause of these changes, generally toward xeromorphy, has been the subject of much speculation and dispute. The major causes seem related to increasing oligotrophy and the unfavourable edaphically controlled water regime, which alternates between water-logged and drought conditions. The nature of change of tree stature, crown geometry, canopy architecture and zonation along ecological gradients is illustrated by the fairly typical catena from mesic red-yellow Ultisols or Oxisols (Acrisols or Ferralsols) to oligotrophic shallow Spodosols and peats shown in Fig. 13.2.

These changes of vegetation structure are analogous to those along the gradient of phasic communities in the deltaic peat-swamp forests (Anderson, 1961a; Bruenig, 1970) and similarly noticeable, if not quite as conspicuous, on aerial photographs (Figs. 13.8, 13.20–13.22). At the same time sclerophylly and general xeromorphic features increase. This trend is visually very noticeable in Borneo as a result of the presence and local dominance of very small-leaved species of such

Fig. 13.17. Open mossy kerapah pole woodland with few *Calophyllum* spp., *Ploiarium alternifolium* and *Tristania obovata* on red-brown peat 2 m deep. The closed stand in the back is on a better drained patch at the edge of a small stream. Merurong plateau, Sarawak, 730 m altitude, centre of the extensive kerapah peat-bog area.

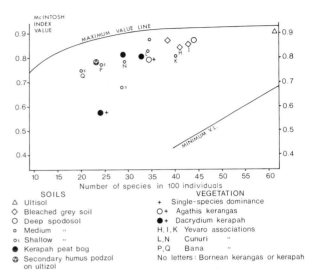

Fig. 13.18. Mean values of the diversity index (McIntosh, 1967; see footnote, p. 316) calculated from 40 stands (ecological study plots). In Sarawak and Brunei, the strata (vegetation/soil units) are Ultisols with mixed dipterocarp forest, bleached grey clayey soils with transitional forest between mixed dipterocarp, mixed peat-swamp and kerangas forest, kerangas on Spodosols of three soil depth classes (deep, medium, shallow), kerapah and secondary kerangas. The Bornean stands are *Agathis* kerangas, *Dacrydium* kerapah and other types, without lettering in the drawing. In the Amazon Rainforest Ecosystem area, means are calculated for the complex *yevaro (Eperua purpurea)* association groups (H, I, K) on deep, well-drained Spodosols; the dense, uniform *cunuri (Micranda sprucei)* association groups (L, N) on medium Spodosol; and the closed (P) and open (Q) *bana* association groups on shallow Spodosol with very fluctuating water regime. Species richness and diversity between the two regions correspond very closely on similar sites unless the forests differ in phasic development (e.g. *Dacrydium* kerapah) or are disturbed by man. The values are calculated for all trees above 2 cm dbh in Borneo and above 1 cm dbh in the Amazon area.

genera as *Casuarina*, *Dacrydium*, *Falcatifolium*, *Podocarpus* and *Tristania*, and because many species develop smaller leaves on less favourable sites.

This xeromorphic diminution of leaf sizes and the tendency to upright or pendulous orientation and bunching of leaves and twigs produces a very characteristic physiognomy. The aerodynamic roughness, diffusivity and optical properties of the tree crowns and canopy as a whole are also affected. This has important ecophysiological implications, which have been discussed by Bruenig (1968, 1970, 1971, 1974, 1976) in terms of adaptation to the physical and chemical site conditions. Particularly important seems to be the ability to photosynthesize and to grow fairly rapidly if environmental conditions are relatively favourable, to survive without damage during periods of water deficiency and strong solar radiation, and to recover quickly if conditions change. Physiological research into these phenomena is only just beginning and is still in its infancy.

The mixed dipterocarp communities on the more mesic sites with Ultisols are predominantly mesophyllous (individual leaf surface area 50–182 cm^2), complex, tall, and species-rich. The crowns in the A stratum are large. The canopy is aerodynamically rough, similar to that in mixed peat-swamp forest community p.c. 1. More unfavourable kerangas and kerapah sites carry stands which are poorer in species, smaller in stature, and more distinctly separated into layers. Emergent trees of

Fig. 13.19. Transition from kerangas to kerapah forest. The soil changes from a groundwater humus podzol to peat 2 m deep, which becomes increasingly waterlogged and periodically heavily rain-flooded towards the centre. Typical is frequent storm-throw and high rates of mortality, irregular pattern of horizontal and vertical forest stand structure, medium species richness and high diversity. The single dominant is emergent *Shorea albida* (two large trees in the photograph), typical codominants are *Dacrydium elatum* and *D. pectinatum*. The site is a flat Holocene terrace with clayey subsoil impeding drainage, 6 m altitude, Pueh F.R., West Sarawak.

particularly adapted species, such as *Shorea albida*, often occur as single trees or small groups, usually widely scattered and isolated. The habit becomes markedly xeromorphic, leaves are strongly coriaceous to pachyphyllous, frequently with a distinctly vertical orientation, and the leaves are often sparse and distinctly clumped. The steeply upright leaves are mostly microphyllous ($2.25-20.25$ cm^2) or leptophyllous (less than 0.25 cm^2) needle- or scale-like. The leptophyllous twigs are typical of trees growing on the most oligotrophic and physiologically xeric sites. Trees on these sites possess higher contents of condensed tannins and pigments (Bruenig, 1966, 1971). The crowns are small, the canopy low and aerodynamically smooth (Figs. 13.21 and 13.22). The kerapah forests (Fig. 13.2, section B, formation 219 and transitions to 119) are very similar in general physiognomy to the Padang forests of p.c. 4 and 5 in the peat-swamp forest formation.

The final phases of kerapah development closely resemble p.c. 6 in structure. They possess the same high degree of xeromorphy — the same leaf size spectra, sclerophyllous leaf anatomy and vertical leaf orientation. A difference to the corresponding peat-swamp community is that the kerapah forests are richer in species (Bruenig, 1968, 1974).

Root systems in kerapah are often superficial, dense and widely spreading. This compact mat of lateral roots often forms perpendicular broom-like roots which penetrate vertically into the peat. This habit is similar to that often found in kerangas on well-drained, nutrient-poor sandy soils. The common stilt-roots may show this behaviour even above ground (Fig. 13.24). Contributing to the formation of such "brooms" may be death of the tip of the original leader root as a result of periodical waterlogging in the ground, and possibly drought or mechanical damage, especially above-ground. Bünning (1947) in a simple experiment obtained some proof of direct uptake of water from rain and stem-flow by such aerial tree roots. This mechanism would be an important advantage, not only to the water supply but also to the nutrient supply in a nutrient-poor environment, by direct scavenging of nutrients from throughfall and stemflow. Stilt roots are generally more common in kerapah than in well-drained kerangas. For example, *Casuarina nobilis* produces very high stilt-roots in kerapah and waterlogged kerangas (Fig. 13.24), but few or none on well-drained sites, where instead it produces narrow, spreading buttresses. Species which easily form high stilt-roots, such as *Calophyllum sclerophyllum*, *Palaquium leiocarpum*, *Ploiarium alternifolium* and *Xylopia corrifolia*, are particularly common in kerapah forests, but stilted and non-stilted trees of the same species and size may stand side by side. The formation of aerial roots in oligotrophic wetlands is obviously controlled by a number of interacting internal and external factors, and is not simply a reaction to oxygen deficiency in the soil or to softness of the ground.

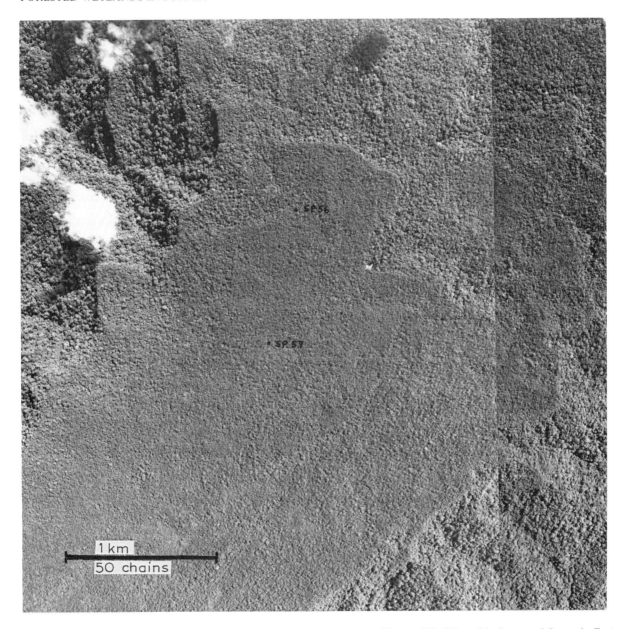

Fig. 13.20. Kerangas and kerapah forest on a sandstone plateau, Merurong Plateau, 700–750 m altitude, central Sarawak, East Malaysia. The plateau sides (lower left and upper left corner) carry rich lowland to submontane mixed dipterocarp forest with rough canopy and large emergents on scarp-slopes and smaller crowned kerangas on dip-slopes. The table-land of the plateau carries kerangas on slopes and kerapah with peat-bog formation on level parts. The latter can be distinguished by its darker tone. The kerapah forest structure along the catenary sequence of forest association types is illustrated in Figs. 13.13 to 13.17.

Relationship between peat-swamp and kerapah forests

An ordination by a similarity-of-species-stocking index of 55 kerangas and kerapah stands in Sarawak and Brunei produced a distinct grouping of stands on kerapah soils on a plane representing drainage (X-axis) and fertility (Y-axis) gradients (Bruenig, 1968, 1974). The lowland kerapah was positioned between kerangas on the drier sites (shallow Spodosols and podzolic gley soils, alternately saturated and periodically waterlogged and rapidly draining) on the left, the phasic communi-

Fig. 13.21. Oblique aerial view of a typical upper lowland to montane landscape in the interior of central Sarawak. Mixed dipterocarp forest covers the steeper slopes, especially scarp-slopes, kerangas forest the gentle dip-slopes and the rims of the plateau-like flats. Kerapah forest covers the centre of the more extended flats with impeded drainage. The summits carry gnarled mossy "elfin" forest. The catena sequence and relationships are described by Bruenig (1974). The structure of the kerapah forests along the catena from rim to centre is illustrated in Figs. 13.13 to 13.17.

Fig. 13.22. View into the aerodynamically smooth canopy of kerapah forest with gregarious leptophyll *Casuarina nobilis* forest and a few *Shorea albida* in the flat centre of the Pleistocene terrace, Sungai Dalam Forest Reserve, Sarawak, shown in Fig. 13.8.

ties 2 to 6 of the peat-swamp formation to the top right-hand corner of the ordination (wetter and poorer in nutrients) and the montane kerapah equivalents to the right (wetter but equally fertile).

Comprehensive calculations of species richness per 100 individuals and species diversity have not been made for peat-swamp forests and suitable field data are not available. An assessment by comparing summarized sample-plot data of Anderson (1964a) with those from kerangas (Bruenig, 1966) gives some indication of the location of the peat-swamp phasic communities in relation to kerapah and kerangas in the matrix of McIntosh diversity index and species number in 100 individuals (Fig. 13.25). The kerangas and kerapah forests appear then floristically intermediate between mixed dipterocarp forest and the later phases of peat-swamp forest (p.c. 2 to 6). The peat-swamp community p.c. 1 is positioned close to kerangas communities on moist grey-white podzolic and bleached gley soils which are located in the transition zone to mixed dipterocarp forest. The floristic relationships, therefore, appear to be in part governed by such site conditions as water regime and soil fertility.

The kerangas and kerapah forests have 948 tree species (428 genera and 71 families), whereas the peat-swamp forests have 242. In the different peat-swamp forest phases, the proportions in common with kerangas and kerapah are 85% in p.c. 1, 75%

Fig. 13.23. Natural regeneration of light-demanding *Casuarina nobilis* in a small gap in kerapah on the terrace in Dalam F.R. (Figs. 13.8 and 13.22). The two trees to the right are *C. nobilis*, *Diospyros evena* on the left. *Dacrydium* spp. also react favourably to increased light in small to large gaps. *C. nobilis* and *Dacrydium* spp. have retained their dominant status in the forest for about 4000 years in spite of the habitually sparse and scattered regeneration — often, indeed, absent (Bruenig, 1966, 1974).

Fig. 13.24. Stilt-roots of *Casuarina nobilis* in a wet kerangas–kerapah transition, forming horizontal broom-like roots and partly upward-growing roots which embrace the trunk of the neighbouring tree and possibly scavenge nutrients from the rain-water flowing down the mossy trunk. Sabal Forest Reserve, Sarawak.

in p.c. 2, 95% in p.c. 4 and 100% in p.c. 5 and 6. Also, a 100% agreement exists between kerapah and Anderson's "aberrant" *Dacrydium–Casuarina* peat-swamp forest near Lawas. This latter forest is in fact a kerapah which has developed on a potential peat-swamp site as a result of an initial invasion and persistent occupation of the site by biochemically particularly well-armed species of *Dacrydium* and *Casuarina* which produce litter with a very high content of polyphenols.

The percentage of species in common between kerapah/kerangas and mixed dipterocarp forest on Ultisols (Ferralsols, Acrisols) lies between 40 and 50%. Few of the species common to peat-swamp forest and kerangas/kerapah forest also occur in mixed dipterocarp forest. There is no species known to be shared by peat-swamp forest and mixed dipterocarp forest which does not also occur in kerangas/kerapah forest. This means that the latter formation is indeed floristically intermediate between the former two, as shown by the ordination.

Anderson (1963) hypothesized that "some of the species such as *Swintonia glauca*, *Stemonurus scorpioides*, *Mangifera havilandi*, and *Quassia borneensis*, which in peat-swamp forest have only been recorded in the Rejang Delta, may be considered more accurately as constituents of dry-land forest, and only in the Rejang Delta have they become adapted to growing and regenerating in peat-swamp". These "dry-land" species are widely spread; they are common in kerangas and also occur in kerapah. It may well be that the

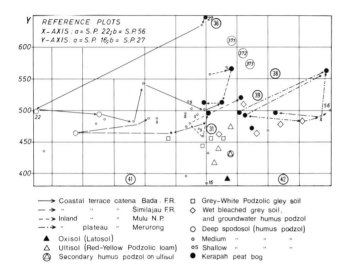

Fig. 13.25. Ordination of 55 kerangas and kerapah forest stands by means of a similarity matrix. The ordination is 3-dimensional, but only two axes are shown: the X-axis relating to a moisture and altitudinal gradient (both increasing to the right) and the Y-axis relating to a nutrient gradient (fertility decreasing upward). The reference stands of maximum dissimilarity are for the X-axis stand 22 = deep Spodosol, and 56 = high-altitude medium Spodosol, for the Y-axis stand 16 = medium Spodosol transition Ultisol near the coast, and 27 = Tropaquod transition Histosol (kerapah with *Shorea albida* dominant). The kerapah stands (solid circles) occupy sites of ample to superfluous water supply and moderate to poor fertility. The location of the peat-swamp forest stands (large, numbered circles) indicates a close relationship between *S. albida*-dominated kerapah (no. 27 and the stand at $X = 540$, $Y = 570$), p.c. 2 (type 36) and p.c. 3 (371–373), and between the high-altitude kerapah and the final peat-swamp phases p.c. 4 (38) and 6 (39). Arrows indicate catenary and successional relationships in kerangas and kerapah. Forest types (41) and (42) are mixed dipterocarp forest on Ultisol.

kerangas/kerapah has been and still is the steppingstone for "dryland forest" species to be selected and to adapt for invasion and occupation of peat swamps. *Shorea albida*, for example, appears to be more vigorous and healthy, produces sounder timber, and regenerates more freely in kerangas forests. It appears as if the species is better adapted to kerangas than to peat-swamp where it is much more common and gregarious, and where, according to the results of pollen analyses, it dominates phasic development over many thousands of years.

Conditions and relationships in Kalimantan seem to be generally very similar if the floristic differences, which are to be expected at this scale of area, are taken into account. Similarly, detailed and comprehensive studies like those for Sarawak and Brunei are not available. Anderson (1963) stated: "Further understanding of the distribution of the peat-swamp flora is only likely to be achieved after greater knowledge has been gained of the Bornean flora as a whole. Later, a comprehensive study of the peat-swamp flora in the Malesian region might yield some interesting and valuable information on the phytogeography of Malesia."

Comparison with oligotrophic forested wetlands outside Borneo

Corner (1978) considered the peat-swamp forests he studied in southern and western Malaysia "to represent Anderson's first phasic community more or less integrated, at least peripherally, with freshwater swamp forest". At Pontian, Johore, in southern Malaysia, the occurrence in peat swamp of ramin (*Gonystylus bancanus*) and other species characteristic of p.c. 1 clearly indicates correspondence. Other southern Malaysian swamp forests, such as those of the Sedili rivers, similarly resemble p.c. 1 physiognomically, even if such species as ramin are missing. In the Sedili peat swamps between 18 and 48% of the species in the various storeys and life-form categories are shared with peat swamps in Sarawak and Brunei, and about 50% of the species at Pontian. This is a fairly large proportion if the tendency to local endemism in the peat-swamp flora and the role of the Batang Lupar, southern Sarawak, as a floristic dividing line (Ashton, 1972) are considered.

In the Hutan Melingtang peat-swamp forest, *Shorea uliginosa* appears to play a role in Malaysia similar to that of *S. albida* in northwestern Borneo.

The density increases from 4.8 large trees ha^{-1} in tall mixed forest at the perimeter to 30 to 51 pole-sized trees ha^{-1} in the centre of the peat swamp. In another area in southern Pahang, *Vatica wallichi* forms a pole forest resembling pole stages of Bornean peat-swamp (Beveridge, 1953).

Bünning (1947) noted dense stands of *Sonneratia acida* at the inland fringe of the mangrove on the northeastern coast of Sumatra which were growing on "already rather peaty, acid soil ... fringing well developed cyperaceous bogs". Further inland and up-river, Bünning found a peat bog several kilometres broad with forest 20 to 30 m tall consisting partly of the same species which he had previously found on an acid, alluvial swamp above the tidal flood zone. Major components noticed were species of the genera *Canarium*, *Eugenia*, *Melastoma*, *Myristica*, *Pandanus* and several species of Euphorbiaceae and Myrtaceae.

In the peat-swamp forest on the Paneh peninsula of Sumatra, the central zone is a low, open woodland dominated by a species of the somewhat confusing genus *Tristania* (Polak, 1933). *Tristania* spp. also typify locally some central areas of peat swamps in Borneo, but are more generally typical of kerapah forests on deep peat between sea level and 1200 m altitude. Examples are the forests in the basins of Pueh Forest Reserve in West Sarawak, of the Binio river basin and on the Merurong plateau in central Sarawak (Bruenig, 1974).

Phasic peat-swamp forest communities occur in the superhumid climate of coastal Colombia. The peat-swamp vegetation includes phases with single-species dominance of *Iryanthera juruensis* and of *Campnosperma panamensis*. Christen (1973) suspects that oxygen availability in the peat is a major factor which controls the vegetation pattern.

Oligotrophic mixed peat-swamp forest occurs in a few small areas in southern Viet Nam on peat up to 3 m deep. The natural forest is less species-rich than in Borneo, and contains *Alstonia spatulata*, *Eugenia* spp, *Melaleuca leucadendron* and several other species of the Malesian peat-swamp flora. Most of the natural forest has disappeared and the vegetation now is mostly secondary forest, often dominated by *Melaleuca*.

There are a few reports which substantiate the existence of kerapah outside Borneo. In Malaysia, the kerangas on a waterlogged soil over sandstone on Gunung Panti is possibly equivalent to kerapah (Whitmore, 1975). Corner (1978) described the forests on the sandstone crest over the granitic base of Gunung Panti as a "mixture of montane, lowland and freshwater swamp forest ... comparable perhaps with some range in west Borneo". This seems to indicate that this forest could be ecologically equivalent, or similar, to Bornean kerapah. Recently, montane forests on wet Spodosols were found in the Bawang Ling Nature Conservation area, Hainan, which resemble Bornean kerangas and transitions to kerapah in hollows. Common species are *Dacrydium pierrei*, *Pinus fenzeliana* and several species of *Lithocarpus* (Bruenig, unpubl. notes, 1988).

The poorly drained and extremely oligotrophic soil and species association groups in the Amazon Rainforest Ecosystem Project[1] area near San Carlos de Río Negro, Venezuela, could be the nearest equivalent to kerapah outside Malesia, except that the characteristic peat formation is lacking (Fig. 13.2). Otherwise, the floristic and geometric structure of the forest and woodland types along the ecological gradient is strikingly similar, except for the lack of stenophylls and the consistently lower tree and stand height at otherwise identical stocking characteristics. Floristic structure (species richness and diversity, Fig. 13.18) and ecologically meaningful features of geometric structure (Fig. 13.2) are very similar, or practically identical, on similar sites and soils in the oligotrophic forests in Borneo and at San Carlos in the Amazon basin. The almost identical floristic structural characteristics are the more remarkable as the tree flora in the Amazon is much poorer in species. Apparently, the floristic forest-stand structure develops towards a state of species richness and diversity which is not determined by the total number of available species in the region, but possibly primarily by other factors which seem to be related to site and soil conditions and the sustained functioning of the forest ecosystems. The kerapah forests, and similarly the kerangas forests, of the Malesian region and their equivalents in the Amazon basin appear to have developed a floristic and geometric-physiognomic structural adaption to the environment which shows many fundamen-

---

[1] Of the Man and the Biosphere (MAB) programme of UNESCO.

tal features in common. But, before sound comparisons can be made, much more ecophysiological research is needed, particularly in the other countries of the Malesian region outside Malaysia.

Ecosystem functions and dynamics

Ecological significance of structure

The structural changes in the vegetation along the gradients of declining site quality for plant life affect the functioning of the system in various ways and in a very complex manner. Many structural features interact with others, and the same result may be brought about in different ways. The following is a very simplified demonstration of the more important features and interactions.

I have earlier described the change of site conditions along the catena mixed dipterocarp forest→kerangas forest→kerapah forest→kerapah woodland (Fig. 13.2); generally, the trend is from relatively favourable conditions (deep soils with good drainage and large storage capacity of plant-available water, and physical conditions generally favourable for deep rooting and microbial life) to relatively unfavourable conditions (extremely fluctuating water availability from excessive to deficient, very acid raw-humus and peat accumulation). The Amazonian catena from "tierra firme" forest to *bana* woodland shows similar features of structural change (Fig. 13.2, bottom).

Common to these examples are a decline of geometric and floristic complexity and of the aerodynamic roughness of the canopy with a simultaneous increase of the albedo of leaves, crowns, and canopy. The result is a reduction of the radiation balance (net intake) and of the intensity of exchange of air by forced and free convection. The simultaneous decrease in leaf size, the steeper orientation and the more pronounced bunching of leaves improve ventilation and couple leaf and air temperatures more closely. This reduces the risk of leaf damage during hot spells with high radiation, low wind speeds, and declining availability of soil water. The advantage of these structural adaptations is the possibility of reducing transpiration rates on the more oligotrophic and xeric sites without excessive risk of damage from high leaf temperatures.

In the Bornean peat-swamp forest community p.c. 2, dominated by *S. albida*, or locally vicariously by *S. pachyphylla*, the aerodynamic roughness increases (values of the estimator z_0 above 600, Fig. 13.2) and leaf size in the top canopy remains in the mesophyll class. This is contrary to expectation and contrasts with the comparable sequences of the kerangas–kerapah gradient. The reason may be that these *Shorea* species are able to compensate for the effects of mesophylly by having coriaceous leaves with very high albedo due to the glossy surface, pubescent under-surface, and high pigment content of the leaves. The protective value against excessive evaporation, and against leaf predators, of a thick cuticle and a leaf surface cover of waxes, scales, and hairs in Bornean forests was already suspected by Beccari (1904).

The hypothesis that the water regime is involved as a causal factor in the gradient of increasing sclerophylly and general xeromorphy of vegetation in Borneo (Bruenig, 1968, 1970) is supported by similar observations of Herrera (1977) and Medina et al. (1978) in the Amazon basin. Herrera measured the soil water table with piezometers during 20 days in October/November 1975 along the ecological gradient in the Amazon Rainforest Ecosystem Study area at San Carlos de Río Negro which is illustrated in Fig. 13.2. Rain fell on 6 days during this period. The fluctuation of the water table was minimal (between 7 and 20 cm depth) in a facies of the *yevaro (Eperua)* association to which the moisture-loving tree species *E. leucantha* and *E. purpurea* are restricted, intermediate (between 40 and 100 cm) in the cunuri association groups dominated by *Micrandra spruceana* and greatest in the *bana* (from 10 cm to below 100 cm — the maximum recording depth). Along this gradient, structural changes have been described by Bruenig et al. (1979) and Medina et al. (1978) similar to those previously described as a characteristic for comparable vegetation–site catenas in Borneo (Anderson, 1961a; Bruenig, 1966, 1970). In all cases the highest degrees of xeromorphy and sclerophylly occur in the centre of the "dome-shaped" Spodosol or peat sites. The higher aerodynamic roughness in the open woodland vegetation types is not a contradiction. The increase of z_0 on the most unfavourable sites is caused by the wider spacing of the trees, which is related to an interaction between the extremely shallow, wide-spreading rooting and the occasional dryness of the site. Herrera concluded that the peculiar scleromorphic features and open structure of the *bana* vegetation are related to both the oligotrophy of the site and the extremely fluctuating soil water regime. This is similar to the hypothetical interpretation by Anderson (1961a) and Bruenig (1966), of results of field studies of the structure–site relationships in the Bornean peat-swamp and kerangas–kerapah forests.

The increase of sclerophyllous anatomy together with the increase in content of secondary biochemical compounds, such as polyphenols, terpenes, aromatic oils, and alkaloids, also could be important for preserving essential mineral nutrients on oligotrophic sites. Bruenig (1966) reported very high contents of polyphenols (non-hydrolysable condensed tannins) in the leaves of tree species on oligotrophic soils in kerangas and kerapah forests. He suggested that the tannins and the toughness of sclerophyllous leaves may increase longevity and also deter herbivores and thereby reduce the rates of leaf consumption. This and masking of proteins would retard litter decomposition, which would reduce leaching and loss of nutrients. The possible existence of a complex of feed-back interactions between nutrient availability, tannin content of the plants, accumulation of undecomposed organic matter, raw humus and peat, soil water regime, activity of soil flora and fauna, and podzolisation and peat formation has been suggested by Bruenig (1966), and is illustrated in Fig. 13.26.

Related to this is the interdependence between xeromorphic features (low aerodynamic roughness, sclerophylly, pigmentation, reflectivity), reduction of transpiration, and the consequent increase of periodical waterlogging and alternating soil leaching when water drains off. This in turn reduces nutrient availability. There is some evidence (Bruenig, 1966) that low nutrient availability is related to higher content of polyphenols and to stronger sclerophylly. Both features act toward podzolization and paludification, which again induces xeromorphic vegetation structures, which in turn reduces transpiration, consumption, and decomposition rates, thus closing the circle of complex interactions (see also Ch. 4, Sect. 8).

Proctor et al. (1983), in discussing field data from a study in the area previously investigated by Bruenig (1966), doubted that drought was the prime cause of the small sclerophyllous leaves in kerangas in Mulu National Park, Sarawak, because the study site grades into kerapah forest on waterlogged soils in which sclerophyllous species also are well represented. They hypothesized that sclerophyllous leaves may reduce transpiration of acid soil solutions with relatively high concentrations of toxic aluminium ions, but that the low nitrogen content observed in litter during the study may also be involved in sclerophylly. This would support the earlier conclusions of Bruenig. More information on the physiological, biochemical, and edaphic features and interacting processes in kerangas, kerapah, and peat-swamp forests, especially in p.c. 3 and 6, are needed to explain the factors controlling sclerophylly and xeromorphy in these Borneo forests.

Dynamic changes

The simultaneous change of aerodynamic roughness and of the stature of trees, indicated by the height and h/d ratio of trees in the A and B canopy layers along the gradient of peat development represented by Anderson's phasic communities (Fig. 13.2 and 13.7), coincides with a peculiar pattern of windthrow and lightning gaps. Both types of gap are more frequent and larger in stands of p.c. 2 and 3 which possess a relatively tall and rough canopy and a high h/d ratio. Consequently, stand dynamics in these phasic communities are more strongly affected by mortality and gaps due to lightning and storms than in p.c. 1 and 4 to 6. In addition, these dense, single-species canopies provide ideal conditions for pest outbreaks of the moth caterpillar "ulat bulu", as yet unidentified, which may destroy the top canopy completely over large areas up to about 100 km^2 (Anderson, 1961b, 1964a, b; Bruenig 1964, 1973a). Bruenig measured the change in number and size of gaps caused by lightning and wind throw in an area of 3815 ha of forests in phases p.c. 2, 3 and 4 on aerial photographs taken in 1947, 1961, 1963 and 1968, and finally in 1983. The general area is shown on the aerial photograph in Fig. 13.11. Part of the area with particularly impressive changes is shown in Fig. 13.12 and in the map in Fig. 13.7.

Lightning gaps in 1968 covered 0.7% of type 362, 3% of type 363 (both p.c. 2), 1.0% of types 371–373 (p.c. 3) and 0.2% of type 38 (p.c. 4). Many old gaps were enlarged and many new gaps formed during the 21 years of observation to 1968. The

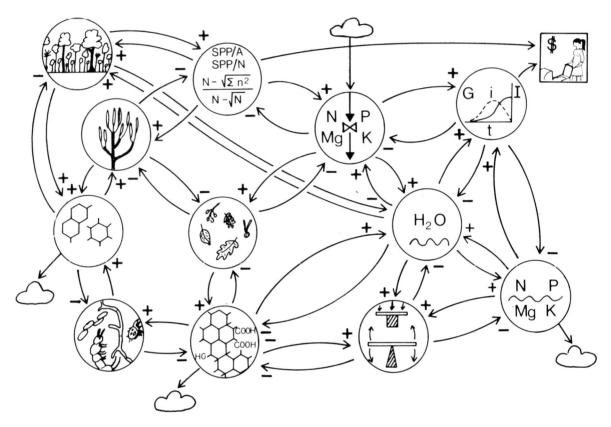

Fig. 13.26. Some of the major features and interactions in a natural forest ecosystem which have been described in the text and which determine structure, functions and reactions of the system and the forest's eventual economic and social values to mankind: (in circles) geometric ecosystem complexity, species richness per area or number of individuals and species diversity, anatomy, orientation and distribution of leaves, production of biochemical aromatic compounds in plants (e.g. polyphenols), litter production, soil fauna and flora, decomposition and humus formation, nutrient filtering capacity of the whole system and efficiency of uptake from inflow (e.g. biological N-fixation and N, P, Mg, K in rainfall), water regime and balance, nutrient regime and balance, static and dynamic stability, growth and increment; (in square) spiritual, material, and monetary benefits to mankind. The terms or elements each aggregate a number of separate structural properties and functions of the stand ecosystem. Variation of levels of one or several terms or elements affects many or all of the other components positively or negatively. The overall dynamics, stability and reactions of the whole oligotrophic tropical forest ecosystem is strongly affected by the efficiency of the microbial activities, especially nutrient fixing and scavenging, but changes in any other section of the system may become critical through positive feed-backs. This interaction diagram is being computerized as a flow diagram for a tropical forest ecosystem in an amended form.

increase in gap area was 100% and that in the number of gaps was 116%. This may indicate a particularly high intensity of storms and lightning during the period of observation. Windthrow gaps covered 0.9% of the total area in 1968, mainly occurring in type 37 between 1963 and 1968. Similarly high windthrow and mortality, often combined with high values of species richness and diversity, have been reported for transitional communities in kerapah forests (Bruenig, 1968, 1974; see also Figs. 13.13, 13.15 and 13.19).

These observed differences in gap frequency and areal extent in the peat-swamp example indicate that it may be dangerous to use these indices as a basis for calculating the average life or residence time of trees and turnover rates of the biomass of forest stands. The data, however, indicate that growth rates and general ecosystem dynamics may be lower in the central parts (p.c. 4) than in the perimeter zone of the peat swamps (p.c. 2 and 3).

The age of the deepest peat deposits which had been sampled in the Baram valley in Sarawak, about 50 km inland of the present coastline, has been calculated by C_{14}-tests to be 4270 ± 70 years (Anderson and Müller, 1975). These peats were from 10 m to more than 20 m thick and must, therefore, have accumulated at considerable speeds of approximately 0.2 to 0.5 m in a hundred years. Comparable information is not available on the rates of accumulation of dead organic matter on the Spodosol soils and in the kerapah peats in Borneo or elsewhere. It is likely that the rates are somewhat lower because observed moor and peat layers are much thinner. The thickness of surface raw humus and peat on kerangas and kerapah sites which are not permanently water-saturated show only a narrow range of variation. This indicates that the production–decomposition–accumulation system may balance, at some level of organic matter accumulation, through the interaction of biochemical litter constituents, physical soil conditions and microbial activity. Topogenic drainage conditions may impose additional constraints through waterlogging and oxygen deficiency. This would shift the balance towards peat accumulation, which may account for the development of raised peat domes in some localities and not in others.

Nutrient conservation

Little direct information exists on the mechanisms of nutrient cycling and conservation in oligotrophic forested wetlands in Malesia or elsewhere. Microbial activity above and below ground level is possibly as important a part of the nutrient cycle in these systems as in kerangas (Bruenig, 1966) and Amazonian oligotrophic forests (Herrera and Jordan, 1980) but there is no field or experimental evidence to substantiate this. Research into the microbial activities in the different soils of natural evergreen humid tropical forests in the Bawang Ling Nature Conservation area, Hainan, China, and in Sarawak have begun during 1988 in the Chinese–German Cooperative Ecological Research Project (CCERP). In both areas, the study includes the comparison between mixed dipterocarp forest and kerangas/kerapah. Also, the nature and role of possible interactions between decomposing and conserving mechanisms in sustaining a dynamic equilibrium of the nutrient store and cycle are not sufficiently known to permit more than the formulation of hypotheses as a basis for further research (Bruenig and Sander, 1983). Also, the actual and potential effects of the increasing global spread of such pollutants as sulphur oxides on the nutrient cycle, microbes, and roots, and finally on the viability of these obviously fragile ecosystems, are yet unknown.

Biomass and productivity

There is no useful information on the gross and net productivities of phytomass, on the rate of production of live and dead phytomass, or on decomposition rates, in forests on oligotrophic wetlands in Borneo or elsewhere. Commercial estimates of volume and increment are of little value as indicators of volumes and growth of growing stock. They cover only a small and variable fraction of the total biomass, and are generally unreliable. There exist few reliable estimates of total basal area and tree volumes. The few available data on peat-swamp forest were summarized by Anderson (1961a).

The only reliable estimate of tree phytomass is for a *Shorea albida* stand in p.c. 3, subtype 371, in the Batang Lupar peat swamps, Sarawak. Complete harvesting of an area of 0.4 ha in 1957 (E.F. Bruenig, unpubl. data) gave the following data. The basal area was $38.2 \, m^2 \, ha^{-1}$, the mean top height 52 m, and the biomass above-ground 1150 t dry matter ha^{-1}. An even higher value was obtained in a *S. albida* stand in kerapah on a permanently wet coastal groundwater podzol (Tropaquod) in the Baram delta, Brunei. The basal area was $67.7 \, m^2 \, ha^{-1}$, canopy height 41 m, the biomass above-ground just over $1400 \, t \, ha^{-1}$. These values appear high. Comparison of the basal areas and the top heights with data from commercial enumerations of *S. albida* forests in central Sarawak, however, show that the two stands are by no means unusual and that higher biomass stockings can be expected locally in stands of p.c. 3 — top height between 60 and 70 m and basal areas above $60 \, m^2 \, ha^{-1}$. But high stockings are by no means an indication of high productivity. In fact, site productivity is probably highest in p.c. 1, where the basal area and stand height are distinctly lower than in p.c. 2 or 3 (Bruenig, 1969b).

In kerapah forests, basal area and stand heights vary as widely as in peat-swamp forests. The lowest basal area reported by Bruenig (1966, 1974) was $16.9 \, m^2 \, ha^{-1}$ and the highest $88.0 \, m^2 \, ha^{-1}$. Stand heights ranged from less than 20 m to above 60 m. No information exists on biomass stocking and productivity of kerapah forests, but estimates (using the formula $0.5 \times \text{height} \times \text{basal area} \times 0.7$)

give a range from less than 100 to a rare maximum of 1500 t ha^{-1} for mature fully stocked stands.

Due to the potentially high concentrations of organic compounds in the black waters draining the peat-swamp and kerapah forests, these systems could be an important link in the assessment of the function of the tropical oligotrophic wet- and dryland forests as a carbon sink in the global carbon balance (Bruenig and Schmidt-Lorenz, 1985). However, no such information is presently available.

Regeneration

Pollen analysis and intensive studies of vegetation structure in kerangas, kerapah (Bruenig, 1974) and peat-swamp (Anderson, 1961a) forests have given sufficient evidence that the actual stand curve of a species — that is, the distribution of individuals in the various size classes from seedling to large tree — is no useful indication of the social status of a species. Species such as *Shorea albida* in the peat swamps, or *Casuarina nobilis, Dacrydium beccarii*, and *S. albida* in the kerapah forests, may only be represented by few individuals in the small size classes and lower canopy layers, but still maintain their social position or even be on the increase. Small gaps and death of a single tree will admit sufficient light in these sclerophyllous forests to permit the few seedlings of light-demanding trees to grow up rapidly (Fig. 13.23). Growth is even more rapid in the larger gaps which are characteristic of the phases, p.c. 2 and 3 and of transitional ecotones in kerapah (Figs. 13.15 and 13.23). Where more light, moisture and nutrients from decaying organic matter become available, on the other hand, large gaps of several hundred hectares caused by the unidentified caterpillar "ulat bulu" may remain without regeneration of *Shorea albida* and with little growth of the surviving understorey for many decades (Forest Department Serawak, Bakong Ulat Bulu Survey, 1983, unpubl. pers. commun., 1985; Figs. 13.11, 13.12).

The regeneration and recruitment of dominant top-storey species in the various phasic communities in peat-swamp forests, and possibly similarly in kerangas forests, apparently are controlled by many factors which are little understood as yet. An example of the complexity of the problem is given by Müllerstael and Bruenig (1976). Plants of *Gonystylus bancanus*, a characteristic species in p.c. 1 and rare in p.c. 3, were grown in the greenhouse on peat which had been collected in p.c. 1 and p.c. 3 in Lassa P.F., Rejang Delta, Sarawak. After 3 years, the carbon dioxide exchange was monitored at different light intensities and temperatures. The plants grown in peat from the community p.c. 3 reacted to high illumination and air temperatures with increased gross photosynthesis, but also with even more increase in respiration.

As a result, the net primary production rate was lower than in plants which grew in their native peat from p.c. 1. Such behaviour in its natural environment would mean reduced competitiveness, and could explain the failure of the species to maintain itself in p.c. 3. There is some reason to suspect that the causal factor is some organic compound which is present in the peat under *Shorea albida* and interferes directly or indirectly with the metabolism of *G. bancanus*, but experimental proof is not yet available. Similar possibly allelopathic interactions may be involved in establishing the dominance of species which have a high content of polyphenols, such as *Casuarina nobilis* and *Dacrydium* spp., in some kerapah forests, in Anderson's "abberrant" peat swamp near Lawas, and in many cases of secondary succession with secondary podzolization on sandy Ultisols and Oxisols after destruction of the original mixed dipterocarp forest.

UTILIZATION AND SILVICULTURE

Since the development of settled agriculture, perimeter peat-swamp forests of the p.c. 1 type have been converted to sustained rainwater-irrigated agriculture, mainly rice, sago, rubber and pineapple. Some problems arose through peat subsidence, but, on the whole, agricultural use of these shallow peats has been successful. More recently, intensive and mechanized commercial timber exploitation, mainly of *Dactylocladus stenostachys* and *Gonystylus bancanus*, has extended beyond the riparian belt where selective logging of timber had been customary for centuries. The result of commercial girth-limit fellings has been a complete change in the forest structure and, usually, a drastic shift of species composition in favour of species which are tolerant to sudden change, such as *Cratoxylon arborescens*, but not

G. bancanus. The latter is a naturally slow-starting species and, in practical silvicultural trials, has reacted poorly to felling and release operations. Also, as we have seen, this species is, at the margin of its range in p.c. 3, very sensitive to sudden increases in illumination and temperature.

Clear-felling of stands for the logging of *Dacrydium* timber has been practised since the 1920s in the "aberrant" *Dacrydium–Casuarina* peat-swamp forest near Lawas. Since the 1950s, throughout Sarawak, clear-felling of *Shorea albida* has spread in p.c. 3. In the former case, regeneration of *Casuarina nobilis* and *Dacrydium beccarii* var. *subelatum* has been ample and responsive. However, the dense, slender pole stands which developed were extremely unstable and prone to wind damage. In contrast, *S. albida* regeneration in p.c. 3 failed from the start. Regeneration is either absent or, if profuse after seed years, responds poorly to the sudden and practically complete removal of the top canopy. The nearly unchanged aspect and the persistent lack of *Shorea albida* regeneration of the "ulat bulu" area between 1947 and 1983 (Fig. 13.12) indicates the problems which are the result of extremely fluctuating seedling population and poor response to sudden opening. This would make it seem feasible to adopt the Tropical Shelterwood System, or at least leave seed trees after logging. But the great slenderness (high h/d ratio) of the trees makes such systems extremely risky. The drastic opening of the canopy by complete utilization in the Malayan Uniform System seems equally risky and uncertain of success in view of the poor chances of obtaining adequate seedling numbers and the poor responsiveness of seedlings. Planting of well-conditioned plants has been reported to be successful in some cases (Lai, 1976) but has failed in others.

Another hazard of utilization and silviculture in the oligotrophic forested wetland ecosystems would seem to be the risk of loss of nutrients by interrupting the cycle. Results of experiments on groundwater podzols (Tropaquods) at San Carlos de Río Negro indicate that the loss of nutrients, even from these infertile, sandy soils, is slight after clear-felling if the slash is not burned. The resprouting stumps, surviving small trees and seedlings, and germinating seeds, and the apparently more or less unaffected micro-flora capture the nutrients released from decomposing slash. In contrast, considerable losses are incurred if the slash and surface litter are burned (Herrera and Jordan, 1980; Jordan et al., 1980). Information on the effects of selective girth-limit logging, clear-felling, and burning on the nutrient cycles in peat-swamp and kerapah forests in Borneo is not available.

The formation of gaps not only increases illumination and through-fall of rain, but it also increases the release of nutrients from the decaying trees. This is probably an important ecological factor because it may accelerate growth of intermediate trees and small seedlings. Immature trees at the margins of the gap may rapidly expand their crowns and close the gap (Bruenig, 1973a). Young regeneration may respond by accelerated height growth. Both responses may proceed simultaneously. Decomposition of dead trees, crown expansion in the top canopy, and height growth in the sub-canopy would then be interdependent processes which, in a stand dominated by a single species, would lead to the formation of a closed, uniform top canopy with few intermediate trees of the top canopy species below. This complex interaction may be one of the reasons for the puzzling fact discussed by Anderson (1961a) that, during regeneration of *Shorea albida* which becomes established beneath a gap in the canopy in p.c. 3 and grows very rapidly, few pole-sized trees can be found, the forest is uniform and even-sized, and appears to be even-aged in spite of the probably very wide and patchily patterned distribution of age-ranges between trees, tree groups and larger stands. Analyses of patterns have not produced any indication that a grouping of size classes exists which could be interpreted as the result of regeneration of the main canopy trees in groups in forests dominated by a single species, such as *S. albida*. This extraordinary uniformity of the architectural structure of the *S. albida* peat-swamp forests in p.c. 2, 3 and 4, therefore, does not even reflect age but may be attributable to the rapid responses to additional light, water and nutrients under a regime of small-to medium-sized gap regeneration.

This poses a serious obstacle to silviculture. A uniformly even-sized, but uneven-aged crop with hardly any intermediates and a strongly fluctuating population of seedlings and saplings of the main crop species which are periodically super-abundant, but alternatingly rare to absent, is difficult to manage by any silvicultural system with natural regeneration, while planting, though possible, is expensive (Lai, 1976).

Kerapah vegetation and sites are unsuitable for productive forestry or agricultural use, except for a few site types of localized occurrence which could be converted to intensive irrigation agriculture. Otherwise, the naturally low productivity and fragile fertility of kerapah sites preclude silvicultural or agricultural uses. That this is reasonable is demonstrated by the secondary vegetation in areas of wet *Dacrydium* and *Ploiarium* padang in Bako National Park (Fig. 13.27), in which trees of all species have remained almost static during more than 20 years of observation. The interruption of the functioning of the natural ecosystem obviously has led to severe degradation and loss of fertility. Throughout Borneo, stagnating low scrub and woodlands occupy similar sites of periodically waterlogged and dry humic sands and shallow peats, after transient agricultural use or repeated accidental burning during unusually severe droughts. Particularly common and extensive are the sapling and pole stands of arrested secondary succession, often dense, which are dominated by *Ploiarium alternifolium*. These stands are an important source of domestic firewood and poles if lightly used, but will deteriorate to unproductive waste-land if over-used. How serious this degradation can be is illustrated by some of the secondary padang types in Bako National Park (Bruenig, 1961). These types are degraded by over-use to such an extent that they have shown no sign of growth and development or any other sign of recuperation, in spite of complete protection between my first survey in 1957 and a re-visit of the areas in 1985.

In conclusion the peat-swamp forests are suitable for silvicultural or agricultural use in the peripheral phase, p.c. 1, where conditions are more favourable for plant growth and for controlling water and nutrient supplies. Major problems are connected with management for the production of certain timber species, such as *Gonystylus bancanus*, while production of unspecified wood fibre is relatively easy. The silvicultural management of the succeeding phasic communities in p.c. 2 to 4 is as yet an ecologically and economically unsolved problem. The following stages of the

Fig. 13.27. *Dacrydium beccarii* var. *subelatum* with a few shrubby *Baeckia frutescens*, *Ploiarium alternifolium*, *Tristania* spp., *Calophyllum* spp. and *Eugenia* spp. on a very humic bleached grey-white sand, mostly very wet, occasionally waterlogged and alternately dry, which overlies sandstone at 60 cm depth. Note the upright, bunched twigs of *Dacrydium* and the relatively steeply oriented leaves of the broadleafed species. Myrmecophytic epiphytes and *Nepenthes* are common (see the ant-harbouring bulbs of *Hydnophytum formicarium* and *Myrmecodia tuberosa* on the leaning trunk in the picture). The vegetation is most probably secondary and has remained unchanged with no growth for the last 25 years. Bako National Park, Sarawak, East Malaysia.

primary peat-swamp succession (p.c. 5 and 6) are too poor and restricted to have any economic potential.

The growing stock of kerapah forests is rarely attractive for commercial use. Exceptions are some coastal areas with dense and easily accessible stands of *Casuarina nobilis*, *Dacrydium beccarii*, *D. fusca*, *Dryobalanops rappa*, *Shorea albida* or *S. pachyphylla*. The tendency to degradation of the fragile soils precludes sustained management as a commercial proposition even in these cases, and kerapah forests should be generally completely protected and maintained as natural reserves of wetland forest (Bruenig, 1974).

REFERENCES

Anderson, J.A.R., 1961a. *The Ecology and Forest Types of the Peatswamp Forests of Sarawak and Brunei in Relation to Their Silviculture*, Ph.D. Thesis, University of Edinburgh, Edinburgh, 117 pp.

Anderson, J.A.R., 1961b. The destruction of *Shorea albida* forest by an identified insect. *Empire For. Rev.*, 40(103): 19–28.

Anderson, J.A.R., 1963. The flora of the peatswamp forests of Sarawak and Brunei, including a catalogue of all recorded species. *Gardens Bull., Singapore*, 20: 131–228.

Anderson, J.A.R., 1964a. The structure and development of the peatswamps of Sarawak and Brunei. *J. Trop. Geogr.*, 18: 7–16.

Anderson, J.A.R., 1964b. Observations on climatic damage in peatswamp forest in Sarawak. *Commonw. For. Rev.*, 43(116): 145–158.

Anderson, J.A.R., 1983. The tropical peatswamps of western Malesia. In: A.J.P. Gore (Editor), *Mires: Swamp, Bog, Fen and Moor*. Part B. Ecosystems of the World 4B. Elsevier, Amsterdam, pp. 181–199.

Anderson, J.A.R., and Müller, J., 1975. Palynological study of a holocene peat and a miocene coal deposit from NW Borneo. *Rev. Palaeobot. Palynol.*, 19: 291–351.

Ashton, P.S., 1964. *Ecological Studies in the Mixed Dipterocarp Forests of Brunei State*, Clarendon Press, Oxford, 75 pp.

Ashton, P.S., 1972. The quarternary geomorphological history of Western Malaysia and lowland forest phytogeography. In: *Transactions 2nd Aberdeen–Hull Symposium on Malesian Ecology*. Univ. of Hull, Dep. Geogr., Misc. Ser., 13: 35–62.

Beccari, O., 1904. *Wanderings in the Great Forests of Borneo*. (Transl. by E.H. Giglioli and revised by F.H.H. Guillemard.) Archibald Constable, London, 11 + 424 pp.

Beveridge, A.E., 1953. The Menchali forest reserve. *Malay. For.*, 16: 87–93.

Bruenig, E.F., 1961. *An Introduction to the Vegetation of Bako National Park*. Report of the Trustees of National Parks, 1959–1960. Government Printer, Kuching, Sarawak.

Bruenig, E.F., 1964. A study of damage attributed to lightning in two areas of *Shorea albida* forest in Sarawak. *Commonw. For. Rev.*, 43(2): 134–144.

Bruenig, E.F., 1965. Guide and introduction to the vegetation of the kerangas forests and the padangs of the Bako National Park. In: *Unesco Symposium on Ecological Research in Humid Tropical Vegetation*. Kuching, 1963. Unesco Regional Office, Tokyo, pp. 289–318.

Bruenig, E.F., 1966. Der Heidewald von Sarawak und Brunei — eine Studie seiner Vegetation und Ökologie. Habilitation Thesis, University of Hamburg, Hamburg, IV + 117 pp.

Bruenig, E.F., 1968. Der Heidewald von Sarawak und Brunei. (The heath forests of Sarawak and Brunei). I. Standort und Vegetation (Site and vegetation). II. Artenbeschreibung und Anhänge (Species descriptions and appendices). *Mitt. Bundesforschungsanst. Forst. Holzwirtsch.*, 68.

Bruenig, E.F., 1969a. On the seasonality of droughts in the lowlands of Sarawak (Borneo). *Erdkunde*, 23(2): 127–133.

Bruenig, E.F., 1969b. Forest classification in Sarawak. *Malay. For.*, 32(2): 143–179.

Bruenig, E.F., 1970. Stand structure, physiognomy and environmental factors in some lowland forests in Sarawak. *Trop. Ecol.*, 11(1): 26–43.

Bruenig, E.F., 1971. On the ecological significance of drought in the equatorial wet evergreen (rain) forest of Sarawak (Borneo). In: *Transactions First Aberdeen–Hull Symposium on Malesian Ecology*. Univ. Hull, Dep. Geogr., Misc. Ser., 11: 66–97.

Bruenig, E.F., 1972. *A Physiognomic–Ecological Classification of Tropical Forests, Woodlands and Scrublands*. Xylostyled Manuscript, Chair of World Forestry, Hamburg, 68 pp.

Bruenig, E.F., 1973a. Some further evidence on the amount of damage attributed to the lightning and wind-throw in *Shorea albida* forest in Sarawak. *Commonw. For. Rev.*, 52(153): 260–265.

Bruenig, E.F., 1973b. Species richness and stand diversity in relation to site and succession in forests in Sarawak and Brunei. *Amazoniana*, 4: 293–320.

Bruenig, E.F., 1974. *Ecological Studies in Kerangas Forests of Sarawak and Brunei*. Borneo Literature Bureau for Sarawak Forest Department, Kuching, 250 pp.

Bruenig, E.F., 1976. Classifying for mapping of kerangas and peat swamp forest examples of primary forest types in Sarawak/Borneo. In: P.S. Ashton, (Editor), *The Classification and Mapping of South-East Asian Ecosystems*. Univ. Hull, Dep. Geogr., Misc. Ser., 17: 57–75.

Bruenig, E.F., 1986. Lowland–montane ecological relationships and interdependencies between natural forest ecosystems. In: *Tropical and Subtropical Ecosystems*, 4: 1–21, Acad. Sin., Guangzhou. (Abstr. in *Intecol Bull.*, 13: 13–17.)

Bruenig, E.F. and Schmidt-Lorenz, R., 1985. Some observations on the humic matter in Kerangas and Caatinga soils with respect to their role as sink and source of carbon in the face of sporadic episodic events. *Mitt. Geol.-Palaeontol. Inst. Univ. Hamburg, SCOPE/UNEP Sonderband*, 58: 107–122.

Bruenig, E.F. and Sander, N., 1983. Ecosystem structure and functioning: Some interactions of relevance to agroforestry. In: P.A. Huxley (Editor), *Proceeding of a Consultative Meeting, Nairobi, Kenya*. ICRAF, Nairobi, pp. 221–248.

Bruenig, E.F., Alder, D. and Smith, J.P., 1979. The International MAB Amazon Rainforest Ecosystem Pilot Project at San Carlos de Rio Negro: Vegetation classification and structure. In: S. Adisoemarto and E.F. Brünig (Editors), *Transactions of the Second International MAB-IUFRO Workshop on Tropical Rainforest Ecosystems Research*. Chair of World Forestry, Hamburg–Reinbeck, Spec. Rep. No. 2, IV + 295 pp.

Bünning, E., 1947. *In den Wäldern Nordsumatras*. F. Duemmlers Verlag, Bonn, 187 pp., 64 plates.

Burnham, C.P., 1975. The forest environment: Soils. In: T.C. Whitmore (Editor), *Tropical Rainforests of the Far East*. Clarendon Press, Oxford, XIII + 282 pp.

Christen, H.V., 1973. *Climatic Classification and Land-use of the Humid Parts of Colombia with Particular Consideration of*

Forestry. Unpublished Manuscript, available at Chair of World Forestry, Hamburg.

Corner, E.J.H., 1978. The freshwater swampforest of South Johore and Singapore. *Garden's Bull. Suppl. No. 1, Singapore*, IX + 266 pp., 40 pl.

Dames, T.W.G., 1956. *Preliminary Classification of Some of the Soils of Sarawak and Brunei*. Misc. Rep. For. Dep., Kuching, Mimeographed, 9 pp.

Dames, T.W.G., 1962. *Soil Research in the Economic Development of Sarawak*. FAO/EPTA Rep. No. 1512, Rome.

Endert, F.H., 1920. De Woudboomflora van Palembang. *Tectone*, 13: 113–160.

Herrera, R., 1977. Soil and terrain conditions in the International Amazon Project at San Carlos de Río Negro, Venezuela, correlation with vegetation types. In: E.F. Bruenig (Editor), *Transactions of the International MAB–I–UFRO Workshop on Tropical Rainforest Ecosystems Research, Hamburg–Reinbek, 12.–17.5.1977*. Chair of World Forestry, Hamburg, Spec. Rep. No. 1, VI + 364 pp.

Herrera, R.A. and Jordan, C.F., 1980. Nitrogen cycle in a tropical rainforest of Amazonia: the case of low mineral nutrient status in the Amazon caatinga. In: Institute of Ecology, Athens (Editors), *Nutrient Dynamics of a Tropical Rain Forest Ecosystem and Changes to the Nutrient Cycle due to Cutting and Burning*. Sixth Year Report, Institute of Ecology, Athens, Ga., pp. 86–114.

Jacobs, M., 1974. Botanical panorama of the Malesian archipelago. In: *UNESCO, Natural resources of humid tropical Asia*. Natural Resources Research XII. UNESCO, Paris, pp. 263–294.

Jordan, C.F. and Centro de Ecología, Instituto Venozolano de Investigaciones Científicas, 1980. Nutrient cycling in an Amazon rain forest. In: Institute of Ecology, Athens (Editor), *Nutrient Dynamics of a Tropical Rain Forest Ecosystem and Changes to the Nutrient Cycle due to Cutting and Burning*. Sixth Year Report, Institute of Ecology, Athens, Ga., pp. 7–85.

Koteswaram, P., 1974. Climate and meteorology of humid tropical Asia. In: *UNESCO, Natural Resources of Humid Tropical Asia*. Natural Resources Research XII. UNESCO, Paris, pp. 27–85.

Lai, K.K., 1976. Performance of planted *Shorea albida*. *For. Dep. Sarawak, Malaysia, For. Res. Rep. S.R.* 14: 13 pp.

McIntosh, R.P., 1967. An index of diversity and the relation of certain concepts of diversity. *Ecology*, 48: 392–403.

Medina, E., Sobrado, M.A. and Herrera, R., 1978. Significance of leaf orientation for temperature in Amazonian sclerophyll vegetation. *J. Radiat. Environ. Biophys.*, 15: 131–140.

Müllerstael, H. and Bruenig, E.F., 1976. Wachstumsverhalten von Ramin in verschiedenen Waldgesellschaften. *Jahresber. Bundesforschungsanst. Forst. Holzwirtsch.*, pp. N 11–13.

Polak, B., 1933. Über Torf und Moor in Niederländisch Indien. *Verh. K. Akad. Wet., Amsterdam*, 30: 1–85.

Proctor, J., Anderson, J.M., Chai, P. and Vallack, H.W., 1983. Ecological studies in four contrasting lowland rain forests in Gunung Mulu National Park, Sarawak. I. Forest environment, structure and floristics. *J. Ecol.*, 71: 237–260.

Richards, P.W., 1952. *The Tropical Rain Forest*. Cambridge University Press, Cambridge, 450 pp.

Salfeld, J.-Chr., 1974. *Interim Report on the Chemical Nature of Peat Samples from Sarawak*. Unpublished Research Report, Brunsvig, 18.6.1974, available at Chair of World Forestry, Hamburg.

Specht, R.L. and Womersley, J.S., 1979. Heathlands and related shrublands of Malesia (with particular reference to Borneo and New Guinea). In: R.L. Specht (Editor), *Heathlands and Related Shrublands. Descriptive Studies*. Ecosystems of the World 9A. Elsevier, Amsterdam, pp. 321–338.

Van Steenis, C.G.G.J., 1950. The delimitation of Malesia and its main plant geographical divisions. *Flora Malesiana, Ser. I, Spermatophyta*, 1: LXX–LXXV.

Walter, H. and Lieth, H., 1960–1967. *Klimadiagramm-Weltatlas*. Fischer, Jena, 245 pp.

Waughman, J., 1974. *Contents of Some Major Elements in Replicate Peat Samples from Two Phasic Communities in the Peat Swamp Forest in Sarawak*. Unpublished Research Data, University of Durham, letter of 9.2.1974.

Whitmore, T.C., 1975. *Tropical Rainforests of the Far East*. With a Chapter on Soils by C.P. Burnham. Clarendon Press, Oxford, 282 pp.

Winkler, H., 1914. Die Pflanzendecke Südost-Borneos (The vegetation cover of Southeast Borneo). *Bot. Jahrb. Syst. Pflanzengesch. Pflanzengeogr.*, 50 (Suppl.): 188–208.

Wyatt-Smith, J., 1959. Peat swamp forest in Malaya. *Malay. For.*, 22(1): 5–32.

Chapter 14

WOODED SWAMPS IN NEW GUINEA

K. PAIJMANS

INTRODUCTION

The island of New Guinea lies wholly within the tropics, between the Equator and just over 10°S latitude. Extensive alluvial plains flank both sides of a central mountain range which runs roughly east–west along its length for almost 2000 km. Several peaks reach over 4000 m, the highest ones occurring in the western half of the island.

The broad weather pattern is determined by two seasonal wind systems, the southeast "trade winds" and the northwest "monsoon". The air of both these systems is laden with moisture picked up during their long passage over warm oceans, and this, together with the topographic lay-out, leads to a relatively high annual precipitation, averaging 2500 to 3500 mm over most of the island. However, when and where rain falls is greatly influenced by the local height and trend of the ranges. Two wetter regions, north and south of the main cordillera, receive between 5000 and 10 000 mm yr^{-1}. On the other hand, some coastal regions backed by ranges running parallel to the prevailing winds, and a number of inland "rain-shadow" valleys, have average rainfall of between 1000 and 2000 mm yr^{-1} only, and strongly marked dry seasons with monthly totals of less than 50 mm. The capital, Port Moresby, lies within the most prominent of the drier coastal regions and has an annual rainfall less than 1000 mm yr^{-1}.

As a result of the generally high rainfall and its markedly uneven distribution over the year for much of the island, the major rivers flood periodically, inundating thousands of square kilometres in their lower reaches for shorter or longer periods each year. Swamps are maintained wherever land gradients and drainage outlets are insufficient to disperse the rain and run-on waters which become impounded at the foot of hills, between higher-lying, more rapidly aggrading riverine tracts, or behind beach ridges. Swamps are also present in very gently shelving estuaries where river water is banked up by the incoming ocean tide through a maze of creeks and channels.

Differences in flooding regime and drainage conditions and, in some regions, seasonal drought stress have given rise to a variety of swamp-vegetation types. Woody freshwater swamp vegetation generally occupies relatively shallow swamps where the water table is at or near the surface at least for part of the year. In deep, permanent, stagnant swamps, and in riverine tracts subject to deep and prolonged seasonal flooding by fast-moving water, wooded swamp gives way to herbaceous and grass swamp. Mangroves, the characteristic vegetation of the tidal coastal zone, range from a low scrub to a tall forest, depending on tidal regime, salinity, climate, soil, and age of the stands.

Most of the information presented here was collected during resource reconnaissance surveys in Papua New Guinea by teams of the CSIRO Division of Land Use Research. As New Guinea's vegetation is insufficiently known and in some areas has yet to be explored, particularly in parts of West Irian, the account is bound to be incomplete.

"New Guinea" is used to mean the whole island of which the independent Papua New Guinea and the Indonesian province West Irian are political divisions. When Papua New Guinea is used it refers to the eastern half that bears that name.

AREA

Wooded freshwater swamps of one kind or another occupy an area estimated at 100 000 km², or some 10% of mainland New Guinea. In Papua New Guinea the largest areas are found along the Fly River and in the lower reaches of the Bamu, Turama, Kikori, Purari, and Lakekamu rivers on the south side, and north of the ranges in the plains of the lower Mambare–Kumusi and middle and lower Sepik rivers. In West Irian the main areas are the deltas of the Pulau (Eilanden) and Digul (Digoel) rivers, the plains around the bay south of Cendrawasik (Vogelkop), and those of the lower Mamberamo River, and the Meervlakte, a huge structural depression between the central range and the Pegunungan van Rees Mountains to the north (Fig. 14.1). Pockets of coniferous swamp forest are found above 1000 m altitude in intermontane basins scattered throughout the ranges.

Mangroves occupy between 1 and 2% of the total land mass of New Guinea. The largest areas are found in the estuaries of the major rivers.

CLASSIFICATION

Woody swamp vegetation has been classified into three categories: forest, woodland and scrub (Table 14.1). Forest has a canopy cover of more than 30%, and woodland less than 30%. Scrub is a usually dense vegetation of trees and shrubs, with or without scattered emergent trees, and less than 10 m high (Specht et al., 1974). The categories have been subdivided into types, which are based on floristics. Because of their special structural features, sago palm and pandan vegetation do not fit readily in any of the three categories. Because they are most akin to swamp woodland, they have been included as types in the woodland category. Swamp woodland and scrub are seen as closely related to swamp forest, but impoverished as a result of an increasingly unfavourable environment. Variants within the types can be related with some confidence to particular environmental stresses (Table 14.1).

SWAMP FOREST

Freshwater swamp forests occur under a wide range of flooding and drainage conditions. Flood-

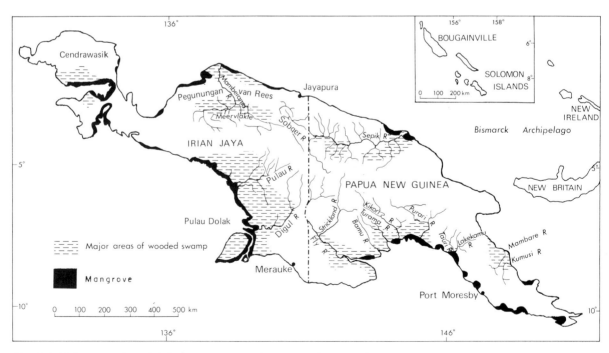

Fig. 14.1. Major areas of wooded freshwater swamp and mangrove in New Guinea.

TABLE 14.1

Classification and habitat of woody swamp vegetation

Category	Type	Habitat	Variant	Likely main stressors
Swamp forest canopy cover >30%	Mixed swamp forest	Widespread lowlands; flooding shallow to deep; water moving or stagnant	Open swamp forest with palms	Prolonged flooding by initially fast-moving water
			Coastal swamp forest	Water table slightly brackish
			Monsoonal swamp forest	Seasonal drought
			Coastal monsoonal swamp forest	Seasonal drought and brackish water table
	Campnosperma swamp forest	Widespread lowlands; flooding shallow to deep; water stagnant		
	Melaleuca swamp forest	Seasonally dry lowland swamps; flooding up to about 1 m; water stagnant	Low thin-stemmed *Melaleuca* swamp forest	Prolonged inundation by stagnant water
	Terminalia brassii swamp forest	Islands east of mainland New Guinea; flooding up to about 2 m (?); water moving		
	Coniferous swamp forest	Mountain valleys; irregular, probably shallow inundation		
	Swamp forest of other species	Very localized; various flooding conditions		
	Mangrove forest	Widespread lowland tidal coasts; daily to infrequent saline or brackish flooding	Low *Avicennia* forest	Periodic high salinity
Swamp woodland canopy cover <30%	Mixed swamp woodland	Widespread lowlands; flooding generally shallow; water generally stagnant		
	Sago swamp woodland	Widespread to about 1200 m altitude; shallow inundation; water moving or stagnant	Stunted sago swamp woodland	Prolonged inundation, or brackish water table, or seasonal drought
	Pandan swamp woodland	Lowlands and mountains; flooding shallow to deep; water moving or stagnant		
Swamp scrub usually dense, <10 m high	Mixed swamp scrub	Seasonally dry, brackish coastal swamps; probably shallow flooding; also mountain streams	Rheophytic scrub	Irregular flooding by fast-moving water
	Mangrove scrub	Monsoonal coasts; infrequent tidal flooding		

ing can be gradual and prolonged, as in backswamps of wide river plains, or sudden, short-lived, and in cases up to 4 m deep, in relatively narrow valleys. Water tables can be almost permanently above the surface, or may periodically drop to 1 m or more below in seasonally dry regions.

Soils are also variable. Gleyed, weakly acid to

neutral silts and clays are the rule in lowland riverine tracts. Away from present river courses, clay soils may be interbedded with organic layers, and in more or less permanent swamps under conditions of prolonged waterlogging and minimal deposition of silt there is usually a top layer of peat, which is over 2 m thick in places (Haantjens, 1964). Peat soils and peaty soils also occur locally under permanently flowing water (Scott, 1967), and are invariably present under coniferous swamp forest above 2000 m. Near the coast, soils tend to be alkaline, and subsoils are often sandy.

Mixed swamp forest is the general and probably most widespread type of swamp forest, but quite a number of swamp-dwelling trees tend to occur in pure stands. The reasons for this tendency can often be related to flooding conditions, soil, or climate, but in other cases they are not clear.

Mixed lowland swamp forest

This forest has a rather irregular open canopy which is 30–35 m high under favourable conditions. Scattered emergents may reach a height of over 40 m. The lower tree layers are generally open, but sago palms or pandans form a dense second stratum in places. The shrub and ground layers are commonly sparse and discontinuous, particularly in forest subject to periodic short-lived flooding (Fig. 14.2). Trees with high and wide buttresses are present, but are not as common as in forest on better drained alluvial plains. Prop roots, peg-like and knee-shaped breathing roots, and sprawling surface roots are a feature of this forest. Thin woody climbers, climbing ferns, and herbaceous climbers often thickly cover tree trunks and reach up into the crowns. Some of the most common climbers are the robust, large-leaved *Stenochlaena palustris* and various species of *Pothos*. Climbing rattan palm is generally rare, but clumps of it are sometimes present in more open forest, together with patches of the tall monocotyledonous *Hanguana malayana* and coarse sedges such as *Thoracostachyum sumatranum* and species of *Mapania* and *Rhynchospora*. Movement through swamp forest is generally easy once the floor is dry, but can be very difficult and slow where undergrowth is dense.

Mixed swamp forest, though floristically not as diverse as tropical rain forest, comprises a large number of tree species including, for example, species of *Campnosperma*, *Neonauclea* and *Syzygium*, *Nauclea coadunata* and *Terminalia canaliculata*. Common in the lower strata are *Barringtonia racemosa*, *Diospyros ferrea*, and species of *Garcinia*, *Horsfieldia* and *Myristica*. It must be noted, however, that most trees occurring in swamp forest also grow on relatively well-drained sites. Even the characteristically swamp-dwelling *Campnosperma brevipetiolata*, for example, is occasionally found on dry land not subject to inundation.

A particular type of mixed swamp forest occurs in parts of estuaries that are flooded regularly by fresh water from a maze of creeks and channels as the water in these is backed up by the incoming ocean tide. Flooding occurs daily near the coast, but becomes less frequent inland. This type differs from ordinary mixed swamp forest in having a lower and more even canopy of small-crowned trees, and by the frequency of the stilt-rooted *Myristica hollrungii* (Fig. 14.3). The type has some features in common with mangrove forest, such as the coastal location, the many stilt-rooted trees, and the even, dark-toned, small-crowned appearance from the air. Because of this it is sometimes referred to as "freshwater mangrove" (Paijmans, 1975, 1976). A structurally similar type of mixed swamp forest, but different in floristic composition, covers extensive areas of flat swampy valley floor on a dissected plateau between the Fly and Strickland rivers. Frequent trees here are *Mangifera* sp., *Vatica papuana* and, in the lower stratum, *Barringtonia acutangula*.

Less favourable environmental conditions are clearly reflected in the structure and composition of the forest. On riverine sites subject to prolonged wet-season inundation by moving water the forest has a very irregular and broken canopy some 20 m high at the most. Tree palms may comprise up to 50% of the canopy (Robbins, 1968), and in the lower storeys *Areca*, *Arenga*, and *Caryota* palms are prominent, together with pandans. Rattan growth is abundant in openings. Isolated pockets of such forest occupy meander scrolls[1] of old river courses and low levees, commonly within grass swamp. Away from the river, towards more

---

[1]Scrolls are groups of parallel curved ridges separated by swampy swales, situated in inner curves of rivers.

Fig. 14.2. Mixed swamp forest in flooded river valley, western Papua New Guinea. The sparse undergrowth consists mainly of coarse sedges. Epiphytes abound.

Fig. 14.3. "Freshwater mangrove" swamp forest in Purari delta about 5 km inland from the coast.

permanent quiet back-plain swamps, the forest becomes gradually lower and smaller-crowned, and commonly passes into woodland as the water becomes deeper and more stagnant. Near the coast where the water table is often slightly brackish, the forest is lower and features rear-mangrove elements, such as *Brownlowia argentata*, *Bruguiera gymnorrhiza*, *B. sexangula*, *Dolichandrone spathacea* and *Heritiera littoralis*. An occasional nipa palm (*Nypa fruticans*) forms part of the undergrowth, and the tall lily *Crinum asiaticum* is often present in the ground layer.

In regions with a marked dry season the water table may fall well below surface level, causing temporary drought stress. Such conditions are reflected in a proportion higher than normal of deciduous trees in the canopy, including *Erythrina* sp., *Planchonia papuana*, *Pterocarpus indicus* and species of *Terminalia* and *Ficus*. During the dry season such trees are leafless for a short period.

P. papuana and *Terminalia canaliculata* provide a sight unusual for tropical swamp forest when their foliage turns red just before leaf fall. *Carallia brachiata* and species of *Acacia* are characteristic lower-stratum trees. In strongly monsoonal regions *Melaleuca* swamp forest (see later) is prominent. In environments that are brackish as well as monsoonal, the number of tree species is greatly reduced. *Melaleuca* commonly dominates the canopy of the low and open forest, *Excoecaria agallocha* often forms a lower storey, and *Hibiscus tiliaceus* may dominate in the shrub layer. A peculiarity of this forest is that *Melaleuca* and *E. agallocha* both grow bunches of adventitious roots at the bases of their trunks (Heyligers, 1965).

It is interesting to note that in seasonally dry swamps the proportion of fire-tolerant trees appears to be higher than average. Examples are *Melaleuca* and *C. brachiata*. *Melaleuca* is protected against fire damage by a thick bark made up of a large number of paper-thin layers. In *C. brachiata* thick cork ribs on the outer bark fulfil the same function.

Campnosperma swamp forest

Two species of swamp *Campnosperma*, *C. brevipetiolata* and *C. coriacea*, are present in this vegetation type. *C. brevipetiolata*, by far the more common of the two, is widespread throughout lowland swamp forest to an altitude of at least 800 m, growing on sandy, clayey and peaty soils. *C. coriacea* appears to have a more limited distribution, and within its range is less frequent than *C. brevipetiolata*, except in western Papua New Guinea where stands consisting of both species in about equal proportions are not uncommon.

Campnosperma forms pure, dense or more or less open stands up to 30 m high in stagnant permanent swamps of river back-plains, basins and inland margins of beach plains. The large-leaved trees develop long straight boles and umbrella-shaped crowns with a frame of horizontal branches. Pure stands are readily recognizable from the air by their smooth, even, light-toned canopy (Fig. 14.4). Many trees are buttressed and stilt-rooted, and have kneed pneumatophores. Scattered sago palms and pandans are almost invariably present in the lower strata. A thick layer of leaves and organic debris covers the forest floor, and the underlying soil is peaty and saturated with water. Regeneration appears to take place through irregularly spread groups of seedlings and saplings.

There is a full range of transitions from pure *Campnosperma* forest to mixed swamp forest with scattered *Campnosperma*, and from dense *Campnosperma* forest to an open forest or woodland with emergent *Campnosperma*. The open forest has a lower storey of sago and pandans, and a ground layer of tall sedges, *Hanguana malayana*, ferns, *Lycopodium*, and scrambling *Nepenthes*. The open forest type is widespread throughout New Guinea and covers extensive areas of floodplain in the Sepik River valley.

Melaleuca swamp forest

Pure stands of swamp *Melaleuca* are characteristic of seasonally dry swamps in the Fly River–Merauke area and the Lakekamu–Tauri River delta, and occur along rivers, creeks and lagoons on sites that are flooded to a depth of about 1 m in the wet season (Fig. 14.5). As the annual rainfall increases above about 2000 mm, the presence of swamp *Melaleuca* becomes restricted to trees scattered in mixed swamp forest, and at rainfall higher than about 2500 mm they tend to disappear. Several species of *Melaleuca* are present in swamp forest, but most pure stands consist of either *M. leucadendron* or *M. cajuputi*. *M. leucadendron* predominates in the Lakekamu–Tauri delta, and *M. cajuputi* appears to be somewhat more frequent in the Fly River area.

Swamp *Melaleuca* species typically grow in stands of closely spaced, slender trees, rarely reaching a diameter at breast height (dbh) of 40 cm. They form a one-layered even canopy with a closure seldom more than 50%, because crowns are small relative to tree size. The white-barked trunks are usually irregularly fire-blackened, and from a distance show a darker band indicating the level of wet-season flooding. The trees often stand on little mounds, and masses of adventitious roots may grow up to flood level from the bases of big trunks. An impression of monotony caused by the scarcity of trees other than *Melaleuca* is sometimes heightened by a virtual absence of shrubs and herbs. Climbers are rare generally, but *Stenochlaena palustris* completely encases the trunks in places. Associated trees include *Mitragyna spe-*

Fig. 14.4. *Campnosperma*-dominated swamp forest and sago palm vegetation, lowlands of southern Papua New Guinea.

ciosa, *Nauclea coadunata*, the palm, *Livistona*, and species of *Acacia* and *Neonauclea* and, in the Fly River–Merauke area, *Barringtonia acutangula*, *Dillenia alata*, and *Erythrina fusca*.

The shape and size of the *Melaleuca* dominants, and the kind and frequency of associated trees and ground flora, depend on drainage conditions and length of inundation. *Melaleuca* trees grow straight and up to 30 m tall on sites that are flooded by river water and are dry for several months each year. By contrast, stands of low, thin-stemmed crooked trees are the rule in back-swamps with more or less permanently stagnant water impounded by slow-draining grass swamps. *Barringtonia acutangula*, *Hanguana malayana*, swamp grasses such as *Leersia hexandra* and *Phragmites karka*, and small aquatic herbs are found in stands on very low-lying, more or less permanently

Fig. 14.5. Wet-season view of *Melaleuca* swamp forest, Lakekamu area, southern Papua New Guinea. Blackened tree trunks show in the open, fire-damaged part of the forest.

swampy sites. Prolonged inundation is indicated by the presence of the grass *Pseudoraphis spinescens* and various semi-aquatic herbs. *Acacia*, *Carallia brachiata*, and *Dillenia alata* are most frequent on slightly elevated terrain within and along the edges of the swamp, and on such sites *Imperata cylindrica* may form a sparse ground cover.

The status of *Melaleuca* swamp forest is not clear. It seems likely that much of it is a disclimax replacing mixed swamp forest after repeated fire damage. In many places it is probably a fire-arrested seral stage in the development towards mixed forest, but in some places it may be a true edaphic–climatic climax. Charred tree trunks indicate that fires sweep through most stands, at least in some years. Sharp boundaries, unrelated to topography, between *Melaleuca* swamp forest and grass swamp are common on the Fly River plains, while a gradual reduction in tree height and density in transitions from forest to only moderately fire-prone herbaceous swamp is the rule in the Lakekamu–Tauri area. It is likely that sharp boundaries are fire-induced, while gradual transitions would indicate an overriding influence of flooding conditions.

Terminalia brassii swamp forest

This forest appears to be restricted virtually to swamps with moving water. *T. brassii* has been recorded only from Bougainville, New Britain and New Ireland, but may be present on other islands of the Solomon group and the Bismarck Archipelago. Extensive stands in southern Boungainville mainly occupy coastal, low-lying riverine tracts, swamp margins and drainage intake zones, while smaller stands line narrow swampy valley floors in the adjoining hills and low mountains. The tree occurs on peaty, sandy, and stony alluvia as well as loams derived from volcanic ash. Water levels range from about 1.80 m above surface to just below in mountain valleys.

Terminalia brassii grows into a tall, straight-boled, flange-buttressed tree. Its wide, cauliflower-shaped crown is very distinctive from the air (Fig. 14.6). Towards stagnant swamps its place is taken by *Campnosperma brevipetiolata*, a common associate in *T. brassii* forest. Large stands were seen from the air to be dying (Heyligers, 1967). The cause could not be determined, but may be impeded drainage resulting from volcanic activity and changing stream courses. The status of *T. brassii* forest is not clear; it is perhaps a seral stage in the development towards mixed forest.

Coniferous swamp forest

This type occurs on flat to gently sloping valley floors, and fringes intermontane basins occupied by grass or sedge swamp. Stands are usually no larger than 5 km² and are mainly at altitudes between 2000 and 3000 m. The sites are subject to irregular inundation, and pools of water remain standing on the hummocky forest floor throughout the year.

The forest has a canopy at 14–18 m and is dominated in many places by various species of *Dacrydium* and *Podocarpus*, which may emerge above the canopy (Fig. 14.7) and reach a height of

Fig. 14.6. *Terminalia brassii* forest on valley floor on Bougainville island.

Fig. 14.7. Log causeway leading into coniferous swamp forest on valley floor at about 2000 m above sea level.

30 m. Myrtaceous trees, pandans and, locally, *Nothofagus grandis* and *N. perryi* also form part of the upper storeys. Many of the smaller trees and shrubs belong to the families Ericaceae, Myrsinaceae and Myrtaceae.

Most coniferous swamp forest sites are subject to cold air drainage and occasional severe frosts. Repeated frost damage may have prevented the establishment or re-establishment of broad-leaved trees and resulted in the local predominance of a relatively frost-hardy coniferous element.

Swamp forest of other species

Many other swamp trees tend to form monospecific stands of smaller size. Examples from both ever-wet and seasonally dry regions are mentioned below. The list of such species is likely to grow as New Guinea's swamp vegetation becomes better known.

Two subspecies of *Barringtonia acutangula* form forests of compact bushy trees up to 6 m high on occasionally deeply flooded, but seasonally dry, sites fringing lagoons and rivers in the Fly–Merauke area, Meervlakte, and Sepik River plains (Brass, 1938; Rand and Brass, 1940; Archbold et al., 1942; Van Royen, 1963; Payens, 1967). Associated trees and shrubs are *Leptospermum abnorme* and *Mangifera* sp. A sparse ground layer of sedges and herbs is present mainly in open spots, and *Stenochlaena palustris* and various epiphytes abound in the trees.

Forests of a stilt-rooted *Calophyllum* occur in more or less permanent river back-swamps. The horizontal branches form a flat-topped canopy, which resembles that of *Campnosperma* from the air and can be mistaken for it.

Dillenia papuana, a tall, often buttressed tree, forms pure stands on riverine sites subject to deep seasonal flooding. The tree is readily recognized by its reddish-brown bark peeling off in thin papery scales. Archbold et al. (1942) described such a forest from the Idenburg River floodplain, when it was inundated to a depth of about 2.5 m.

Erythrina fusca swamp forest is present in generally swampy but seasonally dry parts of floodplains in the Fly–Merauke area (Brass, 1938; Rand and Brass, 1940; Van Royen, 1963). Unlike the *Melaleuca* swamp forests in the area, *E. fusca* forest has a broken canopy, beneath which is an open layer of smaller trees, palms, and *Pandanus*. Tall sedges, swamp grasses, and aquatics form the ground layer, and numerous climbing ferns and epiphytes adorn the trees.

Open-forest patches of the fan palm *Livistona brassii* are a typical feature of the vast stretches of grass swamp along the middle Fly and Strickland rivers.

Slender trees of *Mitragyna speciosa* form an open forest on ground covered to a depth of about 4 m by the highest flood (Archbold et al., 1942). The species is a common inhabitant of New Guinea lowland swamps, and may be one of the very few trees confined to permanent swamp.

Nauclea coadunata is a widespread tree scattered in the canopy of mixed swamp forest in both permanent and seasonally dry swamps, but occasionally forms pure stands. Swamp forest of the smaller-leaved *N. tenuiflora* has been described by Archbold et al. (1942). The forest covered banks and islands of a lagoon deeply flooded by the Sobger River, and was said to resemble *Sonneratia alba* mangrove forest.

Other swamp species colonize newly formed low river scrolls and mud islands in pure even-aged stands. A variety of *Sonneratia caseolaris* with long pendulous leaves and drooping branches acts as a pioneer on such sites in brackish estuaries, and reaches upstream to the limit of tidal influence. Further upstream its place is taken by *Timonius timon*, *Althoffia* sp., or *Nauclea coadunata* (Fig. 14.8). As such stands grow older and the forest floor rises through continuous sedimentation, an assortment of other trees grows up underneath and eventually takes over to form mixed rain forest.

Mangrove forest

Young and middle-aged forests of most mangrove species have only one tree layer, forming a dense, small-crowned even canopy. The ground layer is very sparse, and climbers and epiphytes are absent. In these forests circular openings are a common occurrence throughout New Guinea, particularly in stands of *Rhizophora*. They are

Fig. 14.8. Dense stand of *Nauclea coadunata* and *Mitragyna speciosa* colonizing mud scroll of a recently cut-off meander of the Fly River, western Papua New Guinea.

caused by lightning strikes killing the trees over areas 20–30 m across (Paijmans and Rollet, 1977). Mangrove seedlings thickly colonize the ground in such openings and eventually restore the forest.

In old mangrove forests, which are commonly dominated by *Bruguiera gymnorrhiza* and *Rhizophora apiculata*, the more open canopy admits sufficient light for an undergrowth to develop, consisting of mangrove seedlings and saplings, the fern *Acrostichum aureum*, the prickly shrub *Acanthus ilicifolius*, and an occasional nipa palm. Thin woody climbers and epiphytes are common in places, and include species of the two asclepiads *Hoya* and *Dischidia* or button plant, and orchids, ferns, and the ant-house plant *Myrmecodia*. The *Bruguiera* trees have buttresses and knobbly or knee-shaped breathing roots, and *Rhizophora* has high and wide, arched and branched stilt roots, as well as aerial roots that hang down from the crown. Both mangroves reach a diameter (dbh) of 90 cm, and the largest *Rhizophora* trees are over 35 m tall.

Low forest dominated by *Avicennia marina* is characteristic of upper tidal flats in areas experiencing a low and seasonal rainfall. Like *Bruguiera* and *Rhizophora*, *Avicennia* trees reach large dimensions, but they do not grow as tall and straight, and mature *Avicennia* forest commonly consists of widely spaced, short-boled, gnarled, and broad-crowned trees. Where tidal flooding is relatively frequent, the peg-like breathing roots of *Avicennia* form the only ground cover. Where tidal flooding is infrequent, a sparse layer of salt-tolerant grasses, sedges, and succulent herbs is developed. Short and straight-boled trees and shrubs of *Aegiceras corniculatum*, *Ceriops tagal*, and, on sandy sites, species of *Lumnitzera* are usually present and locally grow in pure stands. Strips of dead *Ceriops* are quite common, suggesting that this mangrove is very sensitive to changes in salinity and flooding conditions.

SWAMP WOODLAND

The main type, mixed swamp woodland, commonly forms the transition between herbaceous swamp vegetation in deeper water and open swamp forest or swampy rain forest on higher ground. Two other types of swamp "woodland", sago palm vegetation and pandan vegetation, are not strictly woody and often have dense canopies, but are included here because of their close relation to true swamp woodland.

Mixed swamp woodland

This type has a more open and also usually lower canopy than the swamp forest, and is floristically less diverse. The upper storey is formed of scattered individuals and groves of trees such as *Campnosperma brevipetiolata*, *Mitragyna speciosa*, species of *Syzygium*, and many others. Sago and other palms, and pandans, fill in much of the space below the trees, and *Hanguana malayana*, the fern *Cyclosorus*, *Phragmites karka*, and sedges of the genera *Mapania*, *Rhynchospora* and *Thoracostachyum* form a dense ground cover. This lowest layer is often up to 3 m tall, but higher still where *P. karka* dominates. Tree trunks are usually clothed in fleshy climbers and climbing ferns, which include *Flagellaria indica*, *Nepenthes* sp., and *Stenochlaena palustris* (Fig. 14.9).

Artocarpus incisus (*A. altilis*) and *Octomeles sumatrana* are characteristic emergent and upper-storey trees in swamp woodland on low, frequently flooded river banks. Species prominent in coastal localities and on sites that are seasonally dry for shorter or longer periods are those mentioned under swamp forest occurring under similar conditions.

Sago swamp woodland

Sago palm, *Metroxylon sagu*, is a tall shrub palm covering hundreds of square kilometres of flood-plain swamp, and many small areas in beach-plain swales and swampy mountain valleys. It occurs naturally from almost sea level to an altitude of roughly 1000 m, but has spread beyond its natural range through planting. The palm grows best in shallow swamps where there is a regular inflow of fresh water, such as in inlet zones of swamp margins, and where the water table is below the surface outside the flooding season. On such sites, its fronds are up to 14 m high, and the flowering, starch-producing tree-like stems reach over 25 m. The trunks, which form when the palm is 10–15 years of age, die after flowering. The palm multiplies, commonly after flowering, by growing

Fig. 14.9. Mixed swamp woodland bordering herbaceous swamp vegetation, lowlands of southern Papua New Guinea. *Stenochlaena palustris* densely covers the tree trunks.

suckers around the base of the old stems, thus forming stools of 3–5 palms. Both smooth and spiny varieties are present, but the variety with needle-like spines along the frond midrib is more common than the unarmed type.

All gradations occur from stands of pure sago virtually without trees to woodland with a rather dense layer of trees and an open lower tier of sago (Fig. 14.4). Associated trees are species tolerant of wet conditions, such as *Alstonia scholaris, Bischofia javanica, Campnosperma brevipetiolata, Nauclea coadunata, Planchonia papuana*, and species of *Syzygium* and *Neonauclea*.

The ground layer of a sago palm stand varies with the density of the palms and the water table. In dense, well-developed stands the long fronds, overarching from the base, form a closed canopy, the interior of the stand is gloomy, and there is no undergrowth. The peaty soil is layered with fallen dead fronds, and the palm's numerous pneumatophores form the only live "cover". Open stands have a ground layer of grasses, gingers and ferns where the water table is below ground level at least for part of the year, and an undergrowth of shrub pandans, tall coarse sedges, *Hanguana malayana* or *Phragmites karka* where the water table is permanently at or above the surface.

Sago stands become permanently stunted in transitions to deep herbaceous swamp, in brackish environments, and on sites where the water table temporarily sinks deep enough to cause drought stress. On such sites the palms do not flower, but sucker strongly. In transitions to permanent herbaceous swamp and reed swamp, sago grows in scattered circular groves which are up to about 100 m across.

Pandan swamp woodland

Swamp pandans occupy a habitat similar to that of sago palm, but have a wider range. They form rather open to quite dense pure stands 5–15 m high in shallow, fresh to brackish, and stagnant to frequently flooded swamps of river back-plains, cut-off meanders, break-through splays[1] (Fig. 14.10), swales of river scroll complexes, and low-lying scrolls and river banks. Stands are usually small, but air photos indicate that pandans cover extensive areas in the floodplains of the middle Sepik River. An upper layer of scattered trees may be present. In open stands, herbaceous creepers often cover the trunks, branches, and prickly prop roots, and there may be a ground layer of coarse sedges (Fig. 14.11), *Hanguana malayana*, or Zingiberaceae.

In mountain swamps above 2000 m a much taller *Pandanus* with thick, high prop roots dominates over small areas, in association with stunted swamp forest trees.

SWAMP SCRUB

Swamp scrub has been recorded only in seasonally dry, slightly brackish to saline coastal swamps near Port Moresby, and along streams in the hills and mountains. Other types of swamp scrub are probably present, most likely in the high mountains.

Mixed swamp scrub

Mixed swamp scrub of the coastal lowlands appears to be confined to alkaline soils in seasonal swamps of innermost tidal flats, back-plains, and shallow beach swales in the Port Moresby region. The most frequent species *Hibiscus tiliaceus* and *Pluchea indica* dominate the type in many places. The scrub is usually 2–6 m high, and is densely tangled by *Flagellaria indica*, scrambling shrubs and thorny vines. Low trees such as *Excoecaria agallocha, Melaleuca*, and the palm *Livistona* may emerge above it. The ground cover is formed by scattered tufts of *Acrostichum aureum*, grasses, and sedges (Heyligers, 1965). As the emergent trees increase in numbers the scrub grades into woodland.

A variant of mixed "swamp" scrub, termed rheophytic scrub, is peculiar to sandy, gravelly and rocky banks and beds of mountain streams, sites that are subject to sudden brief flooding by fast-running water. The shrubs and low trees have flat crowns formed by almost horizontal branches spreading in the direction of the stream, and narrow, willow-like leaves. They are flood-resistant, tough, and firmly anchored by a wide root system. Common genera are *Ficus, Nauclea* and *Syzygium*, but many others show the same habit.

---

[1] A splay is a pattern of braided channels which results from a river breaking through its bank and fanning out.

Fig. 14.10. Dense pandan vegetation colonizing break-through splay of Musa River near north coast, Papua New Guinea. The river broke its banks in 1943, and the photo was taken in 1963. Many trees have died, probably as a result of increased swampiness.

Rheophytes, a term introduced by Van Steenis (1952, 1954), also include several ferns, grasses, sedges, and water plants.

Mangrove scrub

In coastal monsoonal low-rainfall areas, mangrove scrub forms the landward transition from low mangrove forest to unvegetated saline flats. The scrub ranges from a height of 10 m on the forest side to 30 cm or less where it borders bare salt-flats. Mangroves associated with the main species *Avicennia marina* are *Ceriops tagal*, *Lumnitzera racemosa*, and, in southwestern Papua New Guinea, *Batis argillicola*.

Fig. 14.11. Pandan vegetation on flat, swampy valley floor, Sogeri Plateau east of Port Moresby. Sedges dominate in the ground layer.

SWAMP SUCCESSIONS AND SEQUENCES

Sedimentation on coasts and lowland river plains, and changes in river courses, are causing continuous changes in land form, which in turn give rise to a succession of vegetation types. Because many of these processes take place in wet environments, swamp vegetation types play a large part in such successions (Table 14.2). Species of *Avicennia* and *Sonneratia* are commonly the first mangrove pioneers. In sheltered spots they are accompanied, or are followed shortly after, by *Rhizophora*, and are shaded out by it at a later stage. Species of *Bruguiera* establish themselves when the environment becomes less saline, and either grow up in pure stands, or form a mixed forest with *Rhizophora*. As tidal influence decreases, they are joined by trees such as *Brownlowia*

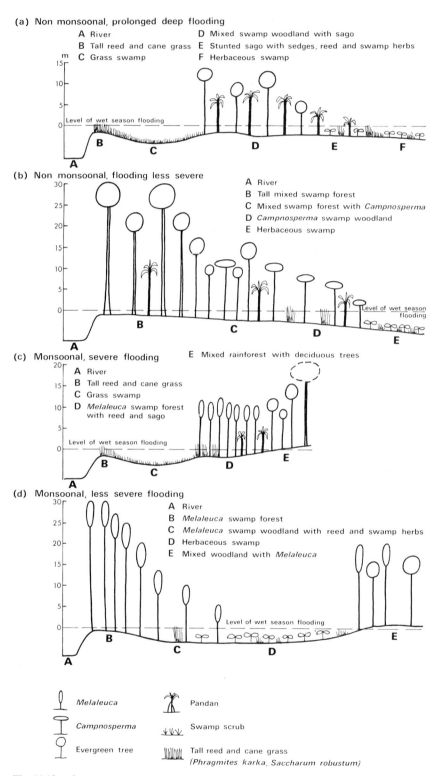

Fig. 14.12.a–d.

WOODED SWAMPS IN NEW GUINEA

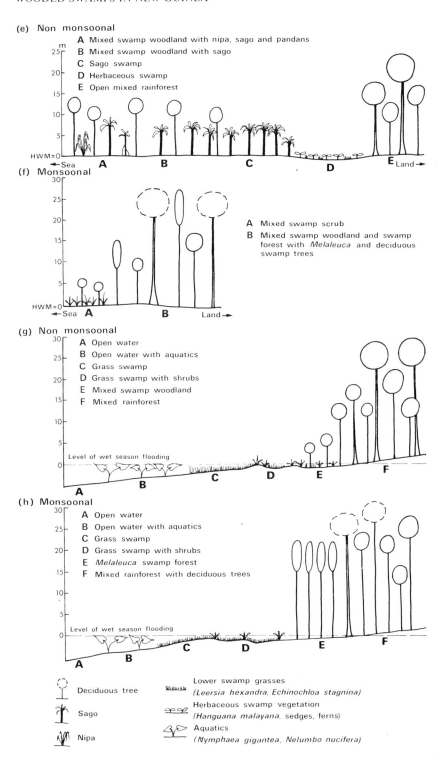

Fig. 14.12. Cross-sections of some common freshwater swamp sequences: (a–d) riverine; (e–f) coastal; (g–h) lagoonal.

TABLE 14.2

Freshwater swamp successions

River scroll succession
Reed/canegrass: *Phragmites karka* (lowest sites) and *Saccharum robustum* (higher sites), or *Timonius timon*, or *Althoffia* sp., or *Nauclea coadunata* → Open mixed swamp forest with palms → Open mixed rain forest with rattan

River bank succession
Reed/canegrass → Reed/canegrass with trees of *Artocarpus incisus* and *Octomeles sumatrana* → Mixed swamp woodland with emergent *A. incisus* and *O. sumatrana* → Mixed rain forest

argentata, *Heritiera littoralis*, *Intsia bijuga*, and species of *Xylocarpus*, which together with palms and pandans form a multi-layered forest below the now overmature emergent *Rhizophora* and *Bruguiera*. Young mangroves are still present in places, but do not appear to develop beyond the seedling stage.

Where land accretion is relatively rapid, as on aggrading shores and lowlying river scrolls, banks, and mud islands, vegetation succession is readily perceived (Table 14.2). Where sedimentation and accumulation of organic matter do not raise the level of the land in relation to the water table, or do so only very slowly, virtually stable edaphic climax communities develop, which are determined mainly by the prevailing flooding regime. Such communities are found in freshwater as well as mangrove swamps; they undergo changes only as a result of natural or man-induced environmental changes. Abrupt changes in a river course are a common natural cause of renewed successional development, while fire is probably the commonest man-related agent. Recurrent dry-season fires lit in grass swamps may encroach upon adjoining mixed woody swamp vegetation and result in its impoverishment and eventual destruction and replacement by *Melaleuca* swamp or grass swamp. Depending on fire frequency, the vegetation may remain permanently degraded, or gradually revert to the original climax community.

Cross-sections of common freshwater swamp sequences are shown in Fig. 14.12 in a somewhat idealized manner.

USES

For many thousands of lowland-dwellers, the starch from the sago palm is the staple diet, or forms a supplementary food when garden produce is in short supply due to drought or otherwise. The starch is produced from trunks felled just before flowering. After removal of the woody outer layer of the trunk the pith is chiselled out, pounded and washed, and the sago flour obtained by sedimentation. In some parts of New Guinea, sago stands are regularly cultivated, and suckers transplanted and tended.

On the island of Pulau Dolak, off the south coast of West Irian, small artificial islands are built in the freshwater swamps by cutting the natural reed growth, plastering it with clay and then building up the island height with alternate layers of drift grass and clay. Low islands are planted with sago palms, and higher islands with coconuts, fruit trees, and food crops (Powell, 1976).

A phenolic oil in the sapwood of the swamp *Campnosperma* species, called tigasso, is locally extracted from the trunks and traded by the indigenes for use as a body oil on ceremonial occasions (Robbins and Pullen, 1965; Powell, 1976).

Large-scale exploitation of swamp forests is commonly hampered by difficulties in access, but stands of *Campnosperma* and *Terminalia brassii* are being logged locally for the manufacture of light construction timber, plywood and packing cases. A number of other trees scattered in forested swamps and ill-drained plains yield excellent timber — for instance *Intsia bijuga*, *Pterocarpus indicus*, and *Vitex cofassus*.

The mangroves of New Guinea have been little used for commercial purposes until recently, but they form a rich source of food and raw materials for the local indigenous population. Fish, eels, the large mud crab *Scylla serrata*, the bivalve mollusc *Polymesoda coaxans*, and sprouting fruits of *Bruguiera* are common and much valued sources of food. Nipa palm fronds are widely used for thatching, and mounds thrown up by the mud lobster, *Thalassina anomala*, are planted with food crops after they have been cleared of their former mangrove and nipa vegetation.

ACKNOWLEDGEMENTS

I am grateful to Dr G.S. Hope, Geography Department of the Australian National University, Canberra, and to colleagues in the Division of Land Use Research, for helpful comments on the draft.

REFERENCES

Archbold, R., Rand, A.L. and Brass, L.J., 1942. Results of the Archbold Expeditions. No. 41. Summary of the 1938–1939 New Guinea Expedition. *Bull. Am. Mus. Nat. Hist.*, 79: 199–283.

Brass, L.J., 1938. Botanical results of the Archbold Expeditions, XI. Notes on the vegetation of the Fly and Wassi Kussa Rivers, British New Guinea. *J. Arnold Aboretum*, 19: 174–190.

Haantjens, H.A., 1964. Soils of the Buna–Kokoda area. In: *Lands of the Buna–Kokoda Area, Territory of Papua and New Guinea*. CSIRO Aust. Land Res. Ser., 10: 69–88.

Heyligers, P.C., 1965. Vegetation of the Port Moresby–Kairuku area. In: *Lands of the Port Moresby–Kairuku Area, Papua–New Guinea*. CSIRO Aust. Land Res. Ser., 14: 146–173.

Heyligers, P.C., 1967. Vegetation of Bougainville and Buka Islands. In: *Lands of Bougainville and Buka Islands, Papua–New Guinea*. CSIRO Aust. Land Res. Ser., 20: 121–145.

Paijmans, K., 1975. *Explanatory Notes to the Vegetation Map of Papua New Guinea*. CSIRO Aust. Land Res. Ser., 35: 1–25.

Paijmans, K., 1976. Vegetation. In: K. Paijmans (Editor), *New Guinea Vegetation*. Aust. Natl. Univ. Press, Canberra, A.C.T., pp. 23–105.

Paijmans, K. and Rollet, B., 1977. The mangroves of Galley Reach, Papua New Guinea. *For. Ecol. Manage.*, 1: 119–140.

Payens, J.P.D.W., 1967. A monograph of the genus *Barringtonia* (Lecithidaceae). *Blumea*, 15: 157–263.

Powell, J.M., 1976. Ethnobotany. In: K. Paijmans (Editor), *New Guinea Vegetation*. Aust. Natl. Univ. Press, Canberra, A.C.T., pp. 106–183.

Rand, A.L. and Brass, L.J., 1940. Results of the Archbold Expeditions. No. 29. Summary of the 1936–1937 New Guinea Expedition. *Bull. Am. Mus. Nat. Hist.*, 77: 341–380.

Robbins, R.G., 1968. Vegetation of the Wewak–Lower Sepik area. In: *Lands of the Wewak–Lower Sepik area, Papua New Guinea*. CSIRO Aust. Land Res. Ser., 22: 109–124.

Robbins, R.G. and Pullen, R., 1965. Vegetation of the Wabag–Tari area. In: *Lands of the Wabag–Tari area, Papua New Guinea*. CSIRO Aust. Land Res. Ser., 15: 100–115.

Scott, R.M., 1967. Soils of Bougainville and Buka Islands. In: *Lands of Bougainville and Buka Islands*. CSIRO Aust. Land Res. Ser., 20: 105–120.

Specht, R.L., Roe, E.M. and Boughton, V.H. (Editors), 1974. Conservation of major plant communities in Australia and Papua New Guinea. *Aust. J. Bot. Suppl. Ser.*, 7: 1–667.

Van Royen, P., 1963. Sertulum Papuanum 7. Notes on the vegetation of South New Guinea. *Nova Guinea, Bot.* 13: 195–241.

Van Steenis, C.G.G.J., 1952. Rheophytes. *Proc. R. Soc. Queensl.* 62: 61–69.

Van Steenis, C.G.G.J., 1954. Vegetatie en flora. In: W.C. Klein (Editor), *Nieuw Guinea. De ontwikkeling op economisch, sociaal en cultureel gebied, in Nederlands en Australisch Nieuw Guinea. Deel II*. Staatsdrukkerij- en Uitgeversbedrijf, The Hague, pp. 218–275.

Chapter 15

ENVIRONMENT AND ECOLOGY OF FORESTED WETLANDS OF THE SUNDARBANS OF BANGLADESH

M. ISMAIL

INTRODUCTION

Bangladesh is located on the south slope of the world's highest mountain and along the northern extremity of the Bay of Bengal (Fig. 15.1). Its abundance of large rivers, rainfall, flora, and fauna make it an ecologically interesting country. The extensive mangrove forest of the Sundarbans with many endangered species, the highly interwoven river system, and large tracts of wetlands with their rich avifauna, are matters of great significance. They were covered to some extent, along with other mangrove forests of the Indian sub-continent, in an earlier volume of this series (Blasco, 1977); but their extent and importance justify more detailed treatment here. The present lack of knowledge concerning the ecosystems of this vast area is primarily due to the great expense involved in research in the Sundarbans.

Tidal woodlands of the Sundarbans occupy an area of about 7800 km^2, the world's most extensive single tract of mangrove. This area is fed by snowmelt as well as heavy monsoonal rainfall. A large amount of sediment is transported by an annual run-off of about 1357 km^3 from a catchment area of more than 1,6 million km^2. The eastern part of the Bay of Bengal has undergone accretion of more than 20 000 km^2 in recent times (Choudhury, 1973). The region can be described as a sink for about 500×10^6 t y^{-1} of silt washed from the extensive Ganges catchment alone. Fertile topsoil eroded in the catchment area, nutrients leached from the soil, and organic matter falling into the river floodplain basins, all may eventually reach the Sundarbans. In addition, nutrients contributed from the Bay of Bengal itself enter the Sundarbans with the tides and mix with upland runoff to form a rich interface of brackish water.

The Sundarbans environment oscillates between salt and fresh water, and between excess water at high spring tides during the monsoon season and drought conditions during the long dry season. Distinct habitats are found adjacent to each other (e.g., riparian, estuarine, shallow marine, coastal shelf, and deep ocean). Terrestrial habitats adjoin aquatic ones, separated by extensive areas with characteristics of both. Superimposed on these distinct habitats are the fluxes of tides and salt wedges, monsoons, cyclones, currents, waves, and winds with their occasional devastating intensity.

The Sundarbans are an unusual mangrove ecosystem. The forests are flushed year-round with upland river water, and salinity remains relatively low. A dominant plant species, *Heritiera fomes* (*Sundri*), a species practically unknown in mangrove forests elsewhere in the world, does not possess an adaptive tolerance to high salinity. This species is dominant in the north eastern part of the forest where salinity is lowest. All Sundarbans species appear to be distributed according to long-term salinity patterns influenced by tidal inundations. Nipa palm (*Nypa fruticans*) is the exception (Ismail, 1973a). This palm occurs predominantly in the most frequently inundated, low-salinity areas along the river banks.

The dependence of the Sundarbans on freshwater runoff is cause for concern because upstream uses curtail run-off to the mangroves. Recently the inadequate inundation of the forest during the dry season, particularly in its central and western parts, has been suggested as the cause of mangrove deterioration (Ismail, 1973b, 1976a, 1977). Historic abundance of river water causes the Sundarbans to

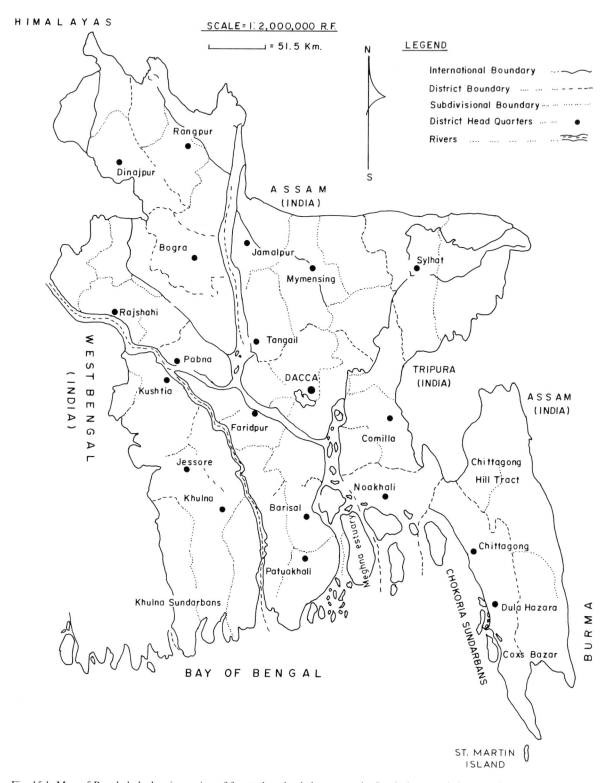

Fig. 15.1. Map of Bangladesh showing region of forested wetlands known as the Sundarbans, and the extensive Meghna estuary.

differ widely from most other tropical mangroves, even from the nearby tidal forest of Chokoria Sundarbans in the Chittagong district, located in the less riverine eastern part of the coast (Ismail, 1975). Chokoria mangroves are more saline and show a closer similarity to most mangroves in Southeast Asia.

The aesthetic value of the forest is well recognized all over the world. The Sundarbans forest is used for recreation and provides ideal habitat for a great variety of fish and game animals. The fauna of the Sundarbans forests is very rich, having some three hundred species of terrestrial vertebrates, as well as unknown numbers of insect, earthworm, and other invertebrate groups (see Appendix). Such richness may be attributed first to the exceptionally varied environment, and second to the extraordinary richness of its soils. In the cyclone-prone monsoon region, the mangrove forests are also significant in mitigating tidal bores and many other environmental hazards. There is no doubt that in the near future these forests will be valued more for such benefits other than as a source of timber, firewood, and many other products.

REGIONAL CLIMATE

There appear to be alterations in the Sundarbans, probably as a result of changes in discharge from the Ganges, which are causing climatic changes of major ecological significance. In this section on climate, I bring together climatic and hydrologic data which argue for a strong interdependence among dry-season flows, newly accreted deltaic deposits and regional rainfall patterns.

Most of the rivers of Bangladesh have their origin in the Himalayas, and conjointly flow through the Meghna estuary to the Bay of Bengal, from where the water is recycled to a large extent in late summer and during the monsoon. The water mass of the estuary is thus significantly related to the humid nature of a vast region. However, these mountains are under heavy pressure of terrace cultivation. Moreover, in addition to the siltation of most rivers of the region, an extensive tract of land has been appearing in the Bay of Bengal adjacent to the Meghna estuary, as evidenced by satellite imagery and ground surveys in recent years (Ismail, 1975).

The long-known rainfall pattern of some areas of Bangladesh, particularly during the dry season, has changed during the last decade (Table 15.1); there has been a progressive reduction of river discharge since the early 1960's (Fig. 15.2). There was heavier dry-season rain in some stations during 1967–78 than had been recorded during the previous decade (Table 15.1). For example, the stations nearer the Bay of Bengal and Meghna estuary (e.g., Morrelgonj) do not show much higher incidence of rain as expected from earlier rainfall patterns in Fig. 15.3. The northern stations of Pabna and Jessore, which are relatively far from the dominating water source of the Bay of Bengal and the large estuary, have comparatively less change in the incidence of dry-season rain than the station in Khulna nearer the Bay of Bengal (Table

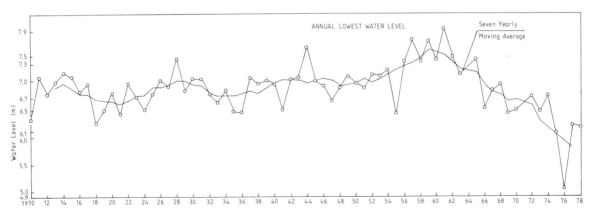

Fig. 15.2. Hydrograph showing the lowest water level of the Ganges near Hardinge Bridge for each year from 1910 to 1978. (After Bangladesh Water Development Board.)

TABLE 15.1

Dry season rainfall (November–April) in six stations located in and around the Sundarbans. Data are 7-year totals for two time intervals (1961–1967 and 1971–1977) and are expressed in mm

Site and year interval	January	February	March	April	November	December	Total
Pabna							
61–67	49	36	133	505	82	24	829
71–77	18	196	68	535	128	123	1068
Faridpur							
61–67	57	93	112	933	84	15	1294
71–77	17	174	243	1604	149	118	2305
Jessore							
61–67	92	117	174	467	89	13	952
71–77	31	126	265	453	211	139	1225
Barisal							
61 67	65	120	230	626	71	99	1211
71–77	95	149	207	853	495	171	1970
Khulna							
61–67	173	37	152	733	71	51	1217
71–77	24	340	239	664	226	135	1628
Morrelgonj							
61–67	71*	72*	198*	621+	15	52	1029
71–77	44	59	199	375+	209	0+	886

*Missing 2 years.
+Missing 1 year.

15.1). Natural incidence of occasional dry-season high river floods in 1974–75, and also 1978, illustrates this physical principle by showing a condition of scanty rainfall in the dry season, particularly in Khulna and Satkhira.

Wind patterns for the dry season (e.g., February) at stations located in Comilla, Khulna, Faridpur and Jessore show marked changes in the direction of wind, when one compares recent years and those of the past decade. The present course of wind is dominantly from southerly, southeasterly and easterly directions, whereas in Barisal (lying close to the maritime front of the Meghna estuary) southerly winds have been substantially reduced in recent years. The silted lands in the southern part of the country and the shrinking Meghna estuary are the new and noteworthy features in the directions from which recent winds are dominantly blowing in all stations except Barisal. That silted lands in the Bay of Bengal and in the Meghna estuary are creating conditions of low pressure (i.e. high temperature) is shown in the extra dry months of February. (The earlier months of the dry season, and March and April, are often transitional, and it is not generally very appropriate to regard them as extremely dry periods.) The frequencies of moisture-laden southerly wind in most stations increased remarkably in the massive diversion of Ganges water at Farakka in 1976 and 1977, and other years when extreme reduction of river water flow is known to have occurred. These southerly winds caused increased dry-season rain in some of the stations noted above. Frequencies of northerly and northwesterly winds in the general dry season in these stations (except Barisal) were lower in the recent decade when siltation and exposure of river beds became prominent (except in the years of high river water flows noted above).

The recent reduction of river water in Bangladesh exposes extensive river beds, and there are frequent massive siltations in the river as very often experienced by the obstruction of water transport in most of the Ganges-fed southern rivers near the coast (Ismail, 1976b). I noted that the formation of differential temperature fronts, primarily induced by the newly developed continental climatic zone in the south near the Meghna estuary (Ismail, 1976b), significantly affects established hydrological features of monsoon winds because of the backward "pull" of the wind to the south instead

THE SUNDARBANS OF BANGLADESH

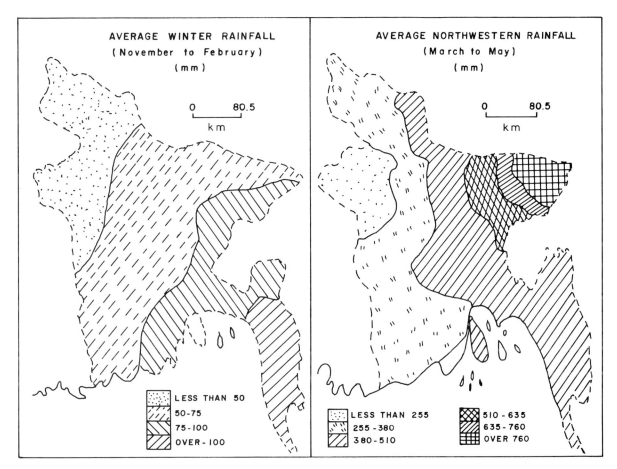

Fig. 15.3. Normal dry-season rain pattern in Bangladesh during the early 1960s. (After Khan and Islam, 1966.)

of its normal direction to the north in the summer. This backward "pull" will happen more prominently as the progressive drying-up of the estuary continues in association with its rapid siltation. A reason for this change is the development of a new continental climatic zone (Ismail, 1976b, 1985).

MANGROVES OF THE SUNDARBANS AND SINGAPORE

Study areas

The mangrove stands described in this chapter were studied beginning in 1965 in Khulna and Chokoria in the Sundarbans. Some comparisons will be made with the mangroves in the Pandan Nature Reserve in Singapore.

The Sundarban forests of Bangladesh are located at the southern extremity of the Gangetic delta bordering the Bay of Bengal (Fig. 15.1). The forests extend about 80 km north of the Bay of Bengal and are bounded on the east by the Baleswar River and on the west extend across the international boundary with India (lat. 20° 30′N and long. 89°E). The total land area of the Khulna Sundarbans is 4071 km². In the Chokoria Sundarbans, the Chittagong Forest Range in the Cox's Bazar Forest Division is situated about 32 km north of Cox's Bazar and 97 km south of the city of Chittagong. The total area of the Chittagong Forest Range is 12 181 ha of which about 80% is frequently inundated by tidal water. The climate of Khulna Sundarbans is warm and humid (Fig. 15.3 and 15.4). Due to the proximity of the Bay of Bengal, air temperature is fairly stable. The average of maximum daily temperature is about 25.8°C in January and 37.2°C in May and June.

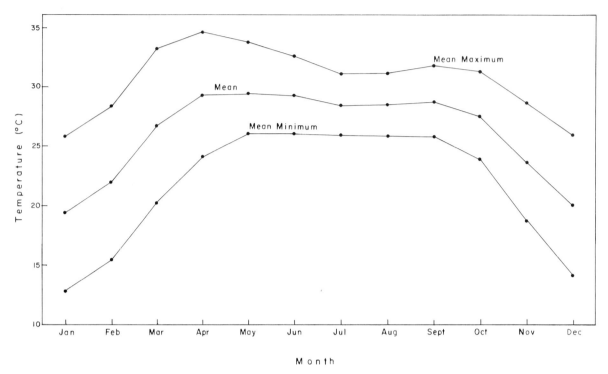

Fig. 15.4. Mean monthly temperature for Khulna (9 m altitude) during the 20-year period ending in 1940. (After Champion et al., 1965.)

The daily range of temperature does not vary significantly throughout the year except for occasional decreases following large storms. The hot weather starts from about the middle of March and ends about the end of September. Cooler weather begins in early November and lasts until the middle of February. Annual rainfall averages between 1650 and 1780 mm, and is irregularly distributed (Fig. 15.3). From July to September rainfall is comparatively abundant. The true monsoon season starts about the middle of June and ends about the middle of September, at which time the prevailing northeasterly winds change to a southwesterly direction causing occasional violent thunderstorms. These storms develop into cyclones which may severely damage ships, buildings, and forests.

Chokoria Sundarbans lie on the coastal region of the Bay of Bengal and, therefore, enjoy what is essentially a moist tropical maritime climate, with a low range of temperature variations and a short cold season. During the winter months when rainfall is low, there are heavy night dews that sometimes persist until 9 or 10 a.m. The prevailing winds are from the south and west between March and May, from the south and east between June and September, and from the north and west between October and February. Cyclones of considerable severity occur periodically, generally in October and again in March and April, but they also develop in other months.

In the open ocean, tidal amplitude is not more than about 0.3 or 0.6 m, but in the shallow seas bordering the continents it may be up to 12 to 15 m. In the rainy season (June to September) the river carries large volumes of runoff. Consequently, during this period, the water level is higher than in the comparatively drier seasons (October to May).

The Pandan Nature Reserve, Singapore (lat. 1°N, long. 105°E) occupies an area of 154 ha extending from Tanjong Penjuru Road up to the eastern bank of the Jurong river of Singapore. The Reserve was initially protected as a Botanical Reserve in 1937 and then as a Nature Reserve in 1939.

The climate of Singapore is warm and humid with little annual variation (mean temperature in the lowlands is about 26.6°C). Temperatures may fall as much as 5.5°C following a storm. There is no regular dry season on the island. The average annual rainfall is nearly 2540 mm, and is uniformly distributed. Atmospheric humidity rises to about 90% soon after sunset and may fall to about 60% in dry daytime weather, but remains at about 85% during rainy periods.

Vegetation

The composition of the mangrove vegetation of the areas investigated in Bangladesh is different from that of Singapore in spite of both countries being in the humid tropical region. Stands in the Sundarbans had higher species richness and higher plant density than those in Singapore (Table 15.2). The abundance of species such as *Heritiera fomes*, *Phoenix paludosa*, *Sonneratia apelata*, and *Dalbergia spinosa* in the mangrove forests of Bangladesh contrasts with their absence from the Pandan Nature Reserve in Singapore (Table 15.3). Results for the Pandan Nature Reserve show a large number of dominants having high density in the interior of the forest, and fewer dominants on the banks of rivers and canals.

Salinity in Bangladesh mangrove areas is low (Tables 15.4 and 15.5) although variable. Almost identical vegetation was observed in habitats having wide differences in salinity. Soil factors likely to influence species distribution, such as carbon content and texture, are more or less identical (Table 15.5). Tracts of land having lower salinity sustain a few freshwater species (e.g., *Cyperus* sp., *Eleocharis* sp., and *Imperata cylindrica*) along with mangroves. Again, certain species such as *Clerodendrom inerme*, a small grass, *Eleocharis* sp. and *Pongamia pinnata* have fair percentages of occurrence and abundance, and occasionally dominate.

Nipa palm (*Nypa fruticans*) is distributed along the banks of rivers with low salinity within the Sundarbans mangrove forests. In other places, *Heritiera fomes* is a tall dominant in the community. The species was noted to occur in the margin separating freshwater zones from higher-salinity mangrove areas in Malaysia and Singapore.

Typical mangrove species are poorly represented in the driest localities of the region. Often soils of the same textural class, such as clay soils, predominate at both Chittagong and Khulna. Sandy soils, however, influence the composition of vegetation in the open drier beaches. Here, typical mangrove species are practically absent. *Casuarina equisetifolia*, *Ipomoea pescapre*, *Pongamia pinnata*, *Tamarix dioica*, and *Vitex trifolia* are generally dominants.

Most soils of Chokoria Sundarbans (Chittagong) were noted to contain sand in variable proportions (Table 15.5). Grasses and *Acrostichum aureum* tend to cover the ground here. In Singapore, however, all mangrove areas are tidally inundated at least once a month, and regular rainfall keeps the soil moist, creating a situation favourable for the growth of tree species including those with viviparous germination (Table 15.A-8).

Human disturbance of mangrove forests in Bangladesh is not serious in general because of unusual difficulty in access. Buffaloes occasionally graze in Chokoria Sundarbans, Chittagong, where some areas are occasionally cleared in the forests for housing. Vegetation in such disturbed areas is identical to that of elevated ground already discussed. The exposure of the ground contributes to the creation of moisture condition similar to those on more elevated ground.

Light and moisture conditions explain the failure of seedling establishment under old *Bruguiera gymnorrhiza* trees in Malaysia (Watson, 1928) and *Sonneratia apetala* in Bangladesh (Choudhury, 1968). In his account of mangrove species, Watson (1928) noted that *B. gymnorrhiza* grows in elevated places that are inundated by spring tides only. Extensive heaps of mud raised by crabs are one of the characteristic features of these elevated places, unlike the condition of most tidal areas of Bangladesh where the land surface is mostly flat (except sometimes near the coast). These heaps of mud, formed generally from soils of lower horizons, are clayey and cone-like in shape. In between the cones, water-logged depressions occur where few mangrove species can grow. These dry elevated soil cones themselves are not a convenient ground for the establishment of viviparously germinated species such as *B. gymnorrhiza*. This species has an elongated extruded radicle which easily slides downwards into the depressions. Under these conditions, small-seeded species of *Acrostichum aureum*, *Derris heterophylla* and *Excoecaria agallocha* occupy the heaps of mud. These species have

TABLE 15.2

Site description and plant species richness and density in three study areas. Each study site was 250 m². Value in parentheses is the abundance of grasses as % of plant density

Site	Description	Plant species (in 250 m²)		Plant density (number ha^{-1})
		total	grasses	
Pandan reserve, Singapore				
1	Southeastern section, 10 tides per month, open	9	0	6400
2	Northern section, 50 tides per month, dense	5	0	9600
3	Forest interior, 21–40 tides per month, open	4	0	15 680
4	Northern bank of Jurong river, 21–40 tides per month, open	9	—	—
5	Inner curve of a creek, 41–60 tides per month, open	4	0	3000
6	Southeastern portion of reserve, 1–20 tides per month, open	7	—	—
Chokoria Sundarbans, Chittagong				
1	Mithachuri near Dula Hazara, 21–40 tides per month, open	10	3 (84)	118 960
2	Edge of Charandwip area along canal, 1–20 tides per month, rather dense	12	1 (52)	9280
3	Southern bank of Mitachuri canal, 21–40 tides per month, open	10	3 (99)	625 920
4	Newly accreted soil on the bank of Sunati Khal, 41–60 tides per month, open	4	2 (95)	2 040 360
5	Soajonya opposite Malumghat, 1–20 tides per month, open	16	3 (28)	15 560
6	Disturbed bank opposite Malumghat, 21–40 tides per month, open	11	1 (48)	12 240
7	From canal bank to interior, opposite Missionary Hospital, Dula Hazara, dense	7	1	—
Khulna Sundarbans				
8	Plot 51/12/1966, Sharankhola range, 1–20 tides per month, dense	7	2 (2.4)	11 800
9	Plot 19/39/1966, Nalianala range, 1–20 tides per month, dense	5	1 (1)	10 800
10	Compartment 39 by canal bank, Nalianala range, 1–20 tides per month, open	23	3 (32)	26 080

(*continued*)

TABLE 15.2 (continued)

Site	Description	Plant species (in 250 m²)		Plant density (number ha^{-1})
		total	grasses	
11	Compartment 39 by bank of Hardura canal, Nalianala range, 1–20 tides per month, rather dense	18	2 (3)	14 400
12	Compartment 12, Northern bank of Ulta Khal canal, Sharankhola range, 1–20 tides per month, open	16	6 (60)	29 160
13	Compartment 39, bank of Aor-Shibsa in Hardura, Nalianala range, 1–20 tides per month, sunny and exposed	17	6 (84)	151 320
14	Compartment 39, degraded stand opposite to Aor-Shibsa near Shibsa river, Nalianala range, 1–20 tides per month, open	7	1 (52)	50 440
15	Plot 17/39B/1966, Chalabogi river bank, Nalianala range, open	8	2	—
16	Ganderkhali, Nalianala range, 21–40 tides per month, open	5	1 (31)	12 360

no extensive, shade-casting, dense crowns, which results in a more or less exposed condition of the habitat. Consequently, conditions are too dry for the establishment of seedlings of *B. gymnorrhiza*, *Rhizophora* sp., and most other viviparous species. Because the soil remains dry for most of the year, stilt-rooted mangrove species are rare.

Sonneratia apetala occurs in sunny habitats close to the water in rivers or canals, and in newly accreted soils in Bangladesh. As in *B. gymnorrhiza*, seedlings of *S. apetala* cannot grow under its own shade. Its fruits are eaten by most herbivorous animals, especially deer.

Community coefficients of similarity (Kulczynski, 1937) of the mangroves of Singapore, Chokoria Sundarbans (Chittagong), and Khulna Sundarbans, indicate that little similarity exists between these communities. The difference between Singapore and Khulna Sundarbans mangroves is more (12.5) than that between Singapore and Chokoria Sundarbans (20.0). The two Bangladesh mangrove forests are floristically not very close as evidenced by the low similarity coefficient (24.0). Thus, the tidal forest of Sundarbans is remarkably different from that of the Malayan Archipelago and represents an unusual mangrove ecosystem.

USES OF THE MANGROVES

The notable economic importance of the Bangladesh mangrove forest is based on its potential to furnish timber, thatching material, pulp, and fuel. Since the forest is under-utilized (only primitive extraction and natural regeneration are practised), there is an attractive and major opportunity to augment the value of the forest, even though artificial regeneration of mangroves is still in the experimental stage. At present, 45 000 people work in the Sundarbans, and most of these are seasonally employed. The inability of agriculture and industries to absorb currently severe unemployment and under-employment emphasizes the opportunity to do so in mangrove-related activities.

The Sundarbans produce the main supply of raw materials to a number of industries (e.g., Forest Industries Development Corporations' industrial

TABLE 15.3

Relative density (%) of species in quadrats measuring 250 m². Site numbers correspond to those in Table 15.2

Species	Singapore Pandan Reserve						Chokoria								Khulna									
	1	2	3	4*	5	6†	1	2	3	4	5	6	7†	8	9	10	11	12	13	14	15†	16
Rhizophora apiculata	14	62	6	100	14	12	20	16	14		6	2	2	24	21	11	7	75	31	0.3	9	66
Excoecaria agallocha	25	25	65	60	14																	
Bruguiera cylindrica	34	11	28	50								15				0.4						
Xylocarpus granatum	8.7																					5
Bruguiera gymnorrhiza	5.6																					
Avicennia officinalis	4.3	0.4		50	60		14	3	17		1		7			0.2	0.6		0.3			
Scyphiphora hydrophyllacea	3.7			30																		
Ceriops tagal	2.5	1.2	2	30																		
Xylocarpus moluccensis	1.2					18																
Sonneratia apetala				10			18	4	16	94		1	17									
Rhizophora mucronata				20	11																	
Aegiceras majus				20								3										
Thespesia populnea						10																
Pongamia pinnata						10										0.2					3	
Pandanus fascicularis						10										45	0.3	1.0				
Randia patula						6																
Guettarda speciosa						2																
Ceriops decandra								2			16	3		48	22	7	7	10	8	6	37	17
Heritiera fomes								1			19	15		26	29	13					20	
Nypa fruticans														2		1	0.3	1.0	49	1	3	
Amoora cucullata														0.3								
Phoenix paludosa											50	1			28	0.4	57					
Xylocarpus mekongensis															11		0.3	0.3		0.2		
Dalbergia spinosa																						
Finlaysonia obovata																						
Tamarix dioica							43	36	18	6	1	48	13				0.3		0.3	90		
Eugenia sp.							8	5			1	9										
Cynometra ramiflora							3		25													
Hibiscus tiliaceus							1				3					8	16	6	2	2		
Sarcolobus globosus								1	3		1					7	3	2	2		9	
Derris heterophylla								17			1	8				3		4	2			
Phyllantus sp.																2	0.3					
Sapium indica																2	0.6					
Barringtonia racemosa																0.4		1.0				
Dolichandrone rheedii																0.4						
Kandelia candel									5							0.2						
Oxystelma esculentum																0.2						
Afzelia retusa																0.2	0.3					

Species				
Unknown 1				4
Unknown 2				1.4
Brownlowia lanceolata	10			0.9
Dalbergia torta		0.3		0.3
Clerendendrom nerifolium	4	0.3		
Flagellaria indica		1		4
Clorodendron inerme				3
Bruguiera sexangula			2	2
Dipterocarpus turbinatus			7	
Maesa ramentacea				

*Values represent % frequency of occurrence in 5 quadrats.
†Values represent % of ground area covered in 5 quadrats.

TABLE 15.4

Chloride content of water samples collected from Khulna Sundarbans and Chokoria Sundarbans (Chittagong). All values are the mean of three replicates, taken at 7.2 cm depth during ebb tide, with standard deviation

Location	Chloride content (%)
Khulna	
Ulta Khal, Compartment 12, Sharankhola Range	0.12 ± 0.26
Bhola River, Sharankhola Range	0.14 ± 0.35
Creek near Sela River, Sharankhola Range	0.52 ± 0.28
Chelabogi Khal, near Hardura area, Nalianala Range	0.46 ± 0.16
Adhachaki Khal, Nalianala Range	0.94 ± 0.08
Chokoria	
At the mouth of Noakhal, joining with the River Ringbong	1.59 ± 0.37
At the mouth of Noakhal, joining the River Ringbong	1.02 ± 0.22
Domkhali Khal	1.01 ± 0.22

TABLE 15.5

Analyses of soils from Khulna Sundarbans and Chokoria Sundarbans (Chittagong). All values are the mean of three replicates taken at a 15.2 cm depth, with standard deviation

Location	Carbon content (%)	Chloride content (%)	Soil color	Texture class
Khulna				
Northern bank of Ulta Khal, Compartment 12, Sharankhola Range	0.09 ± 0.01	0.07 ± 0.04	Grey	Clay loam
Plot 51/1966, Compartment 12 Sharankhola Range	0.34 ± 0.02	0.30 ± 0.09	Grey	Clay
Plot 19/1966, Compartment 39(b), Nalianala Range	0.33 ± 0.02	1.00 ± 0.28	Grey	Clay
Highly elevated place under grasses near Hardura area, Nalianala Range	0.49 ± 0.03	0.38 ± 0.08	Grey	Clay loam
Plot 3/1966, Compartment 39, a degenerated area, Nalianala Range	0.54 ± 0.02	0.24 ± 0.14	Reddish grey	Clay loam
Highly elevated place, full cleared spot along the river bank near Hardura area, Nalianala Range	0.33 ± 0.03	0.43 ± 0.08	Grey	Sandy clay
Eastern bank of Shibsa, Nalianala Range	0.25 ± 0.06	0.75 ± 0.09	Grey	Sandy clay
Chokoria				
Interior of Sunati Khal forest	0.38 ± 0.03	0.85 ± 0.04	Pale yellow	Sandy loam
Soajonya near Sunati Khal	0.71 ± 0.03	1.17 ± 0.17	Whitish	Clay
Site opposite Domkhali	0.30 ± 0.02	1.05 ± 0.09	Whitish	Loam
Near Shibsa, well-exposed elevated spot, extensive degeneration in progress, gentle slope	0.54	0.23		Loam
Kaira, well-exposed, sporadic tree death, gentle slope	0.79	0.31		Clay loam
Nalianala, not exposed, degeneration rare, low flat land	0.90	0.28		Clay loam
Extensively degenerated spot along the Shibsa, well-exposed and elevated	—	0.23		Loam
Hardura, well-exposed, gentle sloping	—	0.39		Clay loam
Near Shibsa, fully exposed elevated land, poor vegetation, gentle slope	—	0.60		Clay

complexes and other small industries established in Khulna, Patuakhali, Jessore, and Bogra). Jessore district alone produces a few thousand tons of molasses from date-palm juice. These industries use firewood for the conversion of juice to molasses. In addition to the 45 000 people working in the Sundarbans, millions of people depend on the Sundarbans-based factories and in the trade of the products of these factories.

About 22 250 m^3 of *gewa* (*Excoecaria agallocha*) wood are extracted from the Sundarbans for newsprint mills, match factories, and packingbox industries. A total of some 20 000 boats are engaged in this operation. For the extraction of timber other than *gewa*, three to four thousand boats each of about 37 t capacity work year-round. Almost all the boat-building industry in Bangladesh is dependent on *Sundri* (*Heritiera fomes*) timbers. About 50% of other constructional timbers, posts, bridge piles, transmission poles, agricultural implements, etc., come from the Sundarbans.

Practically all of southern Khulna's economy is dependent on the Sundarbans. Many of the landless people of Khulna, Patuakhali, Barisal, and part of Faridpur and Jessore districts earn their livelihood by working in the Sundarbans. About 65 000 houses are roofed with *Nypa* leaves from the Sundarbans every year. Once the quality and the yield deteriorate, the pressure on the economy for importing roofing materials for the ever-increasing population of the area will be high, and will require foreign exchange. About 129 000 kg of honey is collected annually from the Sundarbans.

The abundance of nutrients stimulates the growth of phytoplankton (Islam and Aziz, 1975), the base of many food chains. Where river water meets the brackish front, coagulation or flocculation of nutrient-rich particles occurs, and this material acts as a substrate for the proliferation of fungi and bacteria. Such energy-rich food is rapidly consumed by particle-feeders, and this helps to explain why there is such an abundance of shrimps, prawns, lobsters, and crabs, all of which are particle-feeders. A large group of birds (e.g., Anatidae; Rallidae; Charadriidae) forage in the extensive muddy zones, either filtering out rich organic materials or feeding on small animals (e.g. worms and snails) that live in the mud. Many of the other birds prey on fish.

At the other major site of production, the terrestrial ecosystem, trees are the dominant primary producers. These are irrigated with nutrient solution from floodwater flushes and high tides, and they are rooted in a rich alluvial substrate. Leaves appear to be the beginning of most terrestrial food chains in the Sundarbans, and sustain the abundant deer and many other animal populations.

MANAGEMENT

The currently deteriorating Sundarbans need proper investigation to manage this important forest successfully. Moreover, the need for afforestation of newly-formed land is being keenly felt, in order to check environmental degradation in this humid monsoon region. The task is not an easy one. Some of the problems facing successful mangrove management are: public education; availability of transportation on water and land; short working days; problems with tides, cyclones, and other natural phenomena; lack of trained staff and labor; animal grazing; political problems with squatters; and lack of silvicultural information on mangrove species.

Vegetative propagation by transplanting branches of *Excoecaria agallocha* (*gewa*) and *Tamarix dioica* (*banjau*) in moist soil appears to be a satisfactory and quick method for their regeneration, as observed in the vicinity of Chokoria Sundarban, Chittagong. More transplantation experiments with diverse species may prove useful in this regard.

ACKNOWLEDGEMENTS

This research was conducted in the Botanic Gardens, Singapore and in the Department of Botany, University of Dacca, Bangladesh, when the author was associated with the Bangladesh Water Development Board (1976 to 1978). I thank all these institutions, the herbarium of Singapore, and the government of Singpaore for the use of facilities. The writer expresses his gratitude to Mr. H.M. Burkill, Dr. Chew Wee-Lek, Dr. C.X. Furtado, and Mr. Juraimi of the Botanic Gardens, Singapore for their kind cooperation. I also thank the Government of Singapore for the award of a Colombo Plan Fellowship in 1965. The generous assistance received from the officials of the Government Forest Department, Khulna Newsprint Mill and Bangladesh Water Development Board is acknowledged. I am also grateful to the staff of Bangladesh Meteorology Department for their cooperation in supplying and verifying the meteorological data.

APPENDIX: BIOTIC COMPOSITION OF THE SUNDARBANS

The following are tentative and preliminary lists of the fauna and flora of the area. The literature on specific groups of organisms varies from thorough (e.g., fish studies of Rahman, 1974, 1975), to scanty (e.g., insects, except mosquitoes) or non-existent (e.g., nematodes and earthworms except some estuarine nematodes: Timm, 1967). Khan (1982) has presented an illustrated account of some of the animals of Bangladesh.

Although still not fully developed, commercial exploitation of tiny segments of this system demonstrate how effective and potentially important the Sundarbans are. Shrimps and prawns, to mention but two components of this remarkable ecosystem, already account for a full half of total seafood exports from Bangladesh. Nearly all (90%) of Bangladesh's famous tiger prawn (*Penaeus indicus*) are taken from Khulna.

Mammals

In 1727, Alexander Hamilton reported that rhinoceros were "common" in Khulna district (O'Malley, 1908). Nearly 130 years later, the main comment on wildlife by a long-term resident of the Sundarbans was a complaint about the abundance of tigers, leopards, and crocodiles. Beveridge (1857) wrote: "A very moderate amount of organization and of expense might suffice to stamp out the plague in noxious animals", although he adds that "tigers and lions are numerous" (even as far from the Sundarbans as) "in Bakarganj, but it is somewhat surprising how few human beings are killed by them".

Since that time the situation has changed drastically. The rhinoceros, categorized as "rare" in 1908 (O'Malley, 1908), is now almost certainly extinct in the Sundarbans. The royal Bengal tiger (*Panthera tigris*) became so depleted that it was one of the first animals to be classified as "endangered" by the International Union for the Conservation of Nature (IUCN) in 1969 (Mountfort, 1970). Similarly, the wild buffalo (*Bubalus bubalis*) was fast disappearing at the turn of the century (O'Malley, 1908) and has since been obliterated. The gaur (*Bos gaurus*) can no longer be found in the Sundarbans, but it is difficult to say whether it has become extinct, since the records are inadequate. The leopard (*Panthera pardus*), alleged to occur "in surrounding districts" by Curtis (1933), may not have been exterminated. Husain (1974) estimated that 8% of the mammal species of Bangladesh have been lost in the last century. Noteworthy mammals including 12 that are at present either endangered or critically depleted are listed in Table 15.A-1 (cf. Mukherjee, 1966; Mountfort and Poore, 1967; Schaller, 1967; Mountfort, 1970). It is apparent that human pressure was largely responsible for the extinction of the wild buffalo and the rhinoceros, and it is almost the sole factor contributing to the extermination of the royal Bengal tiger, and the endangerment of other species (Table 15.A-1).

At present, spotted deer (*Axis axis*), the prime sustenance of the tiger population, are abundant, and number 7800 in Hendrichs' (1975) study area at Katka. The supply of deer fodder, mainly leaves, is an important consideration. Geological and other changes already have resulted in a deterioration of the western portion of the Sundarbans forest, and these changes are now said to be spreading eastward (Strickland, 1940; Bagchi, 1944). The superior quality of the eastern part of the Sundarbans forest probably accounts for the greater numbers of wildlife there. Environmental change resulting in a degradation of the forest, shortage of browse, lack of cover, and reduction of fresh water all lead to depletion of the deer herd and their predators.

Reptiles

Reptiles in and around the Sundarbans include 18 turtles, 17 snakes, 3 crocodilians, and 10 lizards (Table 15.A-2). Lizards are probably

TABLE 15.A-1

Noteworthy mammals of forested wetlands of Bangladesh

Order and family	Scientific name	Bengali name	English name	Status
Artiodactyla				
Suidae	*Sus salvanius**		Pygmy hog	Extinct ?
	S. scrofa	Shukar	European boar	Frequent
Bovidae	*Bos frontalis**		Gayal	Endangered
	*B. gaurus**		Gaur	Endangered
	Bubalis arni	Ban mohish	Wild buffalo	Extinct
Cervidae	*Axis axis*	Chital	Spotted deer	Abundant
	*A. porcinus**		Hog deer	Endangered
	*Cervus eldi ?**			
	Muntiacus muntjak	Kukuri horin	Barking deer	Uncommon
	*Rucervus (Cervus) duvauceli**	Barasingha	Swamp deer	Endangered or extinct
Carnivora				
Canidae	*Canis aureus*	Shial	Asiatic or golden Jackal	Present
	Vulpes bengalensis	Khek-shial	Bengal fox	Present
Felidae	*Felis chaus*		Jungle cat	Four seen in 1971
	Panthera pardus	Chita Bagh	Leopard	(Absent) ?
	*P. tigris**	Tigris Bagh	Royal Bengal tiger	Endangered
	Prionailurus bengalensis	Ban Biral, Khatash, Bagdus	Leopard cat	
	P. viverrinus (*Felis viverrina*)	Mach Biral	Fishing cat	Rare, depleted
Mustelidae	*Amblonyx cinerea*	Udh	Small-clawed otter	Present
	Lutra lutra	Udh	Smooth otter	(Absent) ?
	L. perspicillata	Bodor, Daria	Smooth-coated otter	
Viverridae	*Herpestes auropunctatus*	Bezi	Small mongoose	
	H. edwardsii	Bezi	Mongoose	
	Paradoxurus hermaphroditus		Palm civet	
	*Prionodon pardicolor**		Oriental linsang	Endangered
	Viverra indica	Bagdasha	Small civet	Present
	V. zibetha	Gandha gokul	Civet	
Cetacea				
Delphinidae	*Delphinus delphis*		Dolphin	Present
	Neomeris phocanoides		Little porpoise	Abundant
	Stenella malayana	Sushuk, Shishu	Malay dolphin	Less common
Platanistidae	*Platanista gangetica**	Sushuk, Shishu	Ganges dolphin	Endangered
Chiroptera				
Emballonuridae	*Taphozous longimanus*		Sheath-tailed bat	
Megadermatidae	*Megaderma lyra*	Chanchara	Indian false vampire	
Pteropodidae	*Pteropus giganteus*	Badur	Flying fox	Abundant
Vespertilionidae	*Coelops frithii*		Tailless bat	
	Nycticejus (*luteus*)		Bengal yellow bat	
	Pipistrellus (*coromandra*)		Pipistrelle	
	Scotophilus (*heathi*)		Yellow bat	
Lagomorpha				
Leporidae	*Caprolagus hispidus**		Hispid hare (?)	Extinct ?
	Lepus nigricollis		Indian hare	Uncertain
Perissodactyla				
Rhinocerotidae	*Rhinoceros sundaicus*	Gondar	Lesser rhinoceros	Extinct

(*continued*)

TABLE 15.A-1 (*continued*)

Order and family	Scientific name	Bengali name	English name	Status
Primates				
Cercopithecidae	*Macaca assamensis*	Banor	Assamese macaque	Frequent
	M. mulatta	Banor	Rhesus monkey	Frequent
Hominidae	*Homo sapiens*	Manush	Humans	45 000, seasonal in Sundarbans alone
Rodentia				
Hystricidae	*Hystrix hodgsoni*	Sajaru	Crestless Himalaya porcupine	Present
Muridae	*Bandicota bengalensis*	Chika	Mole rat	Present
	B. indica	Khorgosh	Bandicot rat	Present
	Mus cervicolor		Fawn field mouse	Uncertain
	M. musculus	Bati idur	House mouse	Present
	Rattus rattus	Idur	House rat	Common
	Vandeleuria oleracea		Tree mouse	Uncertain
Sciuridae	*Callosciurus pygerythrus*	Katbiral	Beautiful squirrel	Frequent
	Funambulus pennanti	Katbiral	Five-striped palm squirrel	Present

*Endangered, or extinct as per IUCN.

the most effective reptiles in controlling insect pests.

Terrestrial snakes prey upon small rodents and amphibians, some of which damage embankments, canals, and irrigation ditches. The python is afforded some protection. The marsh or snub-nosed crocodile (*Crocodylus palustris*) is the commonest of the three crocodilians in the Sundarbans. The gavial (*Gavialis gangeticus*) is a vulnerable or depleted species, as is also the case with the estuarine crocodile (*Crocodylus porosus*). Most of the turtles eat fish or mud-inhabiting organisms. Human population pressure in the rivers in and around the Sundarbans have depleted sensitive species of turtles and crocodiles.

Birds

The avian population of the Sundarbans is rich and varied (Husain, 1974). The number of bird species (Table 15.A-3) in this forest is probably well over 160.

Fish and fisheries

Fish abound in the Sundarbans and represent a rich renewable protein resource. Because the extensive Sundarbans ecosystem acts as a slow-release mechanism for stored foods, the sustainable fish harvest potential is likely to be enormous and may be, at present, under-exploited. Fish and prawn exports contribute so enormously to the national economy in Bangladesh that improvement of fisheries merits careful attention.

The characteristic of the Sundarbans fish fauna is its richness in species (Table 15.A-4). Freshwater, brackish, estuarine, oceanic, surface-feeding, bottom-dwelling, carnivorous, phytophagous, lotic, lentic, and even some amphibious species, are represented. The highly prized cyprinid carps (e.g., *Catla catla* and *Labeo calbasu*) are economically the most valuable. These inhabit rivers, low wetlands, and sluggish streams. *Catla catla* also thrives in brackish waters.

Three fish families are diadromous in Bangladesh: Mastacembelidae (eels), Clupeidae (*Hilsa ilisha*, etc.), and Latidae (*bhetki*), of which the clupeids are economically significant. *Hilsa ilisha* are obligate migrants (Swarup, 1959). They ascend large rivers during the summer monsoon to spawn upstream. River currents exert a necessary stimulus to migration and to reproductive activity in this group of fish.

TABLE 15.A-2

Noteworthy reptiles of Bangladesh wetlands (after Shafi and Kuddus, 1976)

Family	Scientific name	Bengali and/or English name
Chelonia		
Cheloniidae	*Chelonia amboinensis*	Sea turtle
	C. emys	Sea turtle
	C. mydas	Green turtle
	Eretmochelys imbricata	Hawksbill turtle
Dermochelidae	*Dermochelys coriacea*	Marine leatherneck turtle
Emydidae	*Emyda granosa*	Matia kaitta
	Geoclemys hamiltoni	Magam kachim
	Hardella thurji	Kali kaitta
	Kachuga smithi	Bhaital kaitta
	Kachuga tectum	Kari kaitta, roof turtle
	Morenia petersi	Holdi kaitta
Testudinidae	*Testudo elongata*	
	Chrysemys pieta (doubtful)	
Trionychidae	*Chitra indica*	Chim kachim
	Lissemys punctata	Sundi kachim
	Pelochelys bibroni	Gata kachim
	Trionyx gangeticus	Khalua kachim, soft-shelled turtle
	T. hurum	Dhum kachim, black mud turtle
	T. nigricans	
Crocodilia		
Crocodilidae	*Crocodylus palustris*	Kumir, marsh crocodile
	C. porosus	Kumir, salt-water crocodile
Gavialidae	*Gavialis gangeticus*	Gharial, gavial
Squamata		
Agamidae	*Calotes versicolor*	Calotis
Boidae	*Python molurus*	Python
Colubridae	*Dryophis mycterizans*	Tree snake
	Fordonia leucobalia	
	Natrix piscator	
	Ophiophagus hannah	
	Ptyas mucosus	Dhaman, dhamni, rat snake
Elapidae	*Naja naja*	Gekh-khar, cobra
Gekkonidae	*Eublepharis hardwickii*	
	Gecko gecko	Takh-khak, tokay gekko
	Hemidactylus flaviviridis	Tiktiki, wall lizard
Hydrophidae	*Enhydris enhydris*	
	E. schistosa	
	Hydrophis nigrocinctus	
	H. obscura	
Typhlopidae	*Typhlops diardi*	Blind snake
Varanidae	*Varanus bengalensis*	Grey Indian monitor
	V. flavescens	Ruddy snub-nosed monitor
	V. salvator	Ocellated water monitor
Viperidae	*Echis carinata*	Carpet viper
	Vipera russelli	Russell's viper

Decapods

Shrimps, crayfish, prawns, lobsters, and crabs are economically and ecologically among the most important groups of animals in the Sundarbans. These decapod crustaceans are particle-feeders, and may be an important and effective component of the estuarine nutrient-trap. Because the Sundar-

TABLE 15.A-3

Noteworthy birds of Bangladesh wetlands (after Ripley, 1961; Rashid, 1967; Mukherjee, 1969; Ali, 1972; Ali and Ripley, 1974; Husain and Sarker, 1974; Gusen, 1976). Asterisk (*) indicates that hunting is permitted under licence

Family	Scientific name	Bengali and/or English name
Accipitriformes		
Accipitridae	(*Aquila pomarina*)	Lesser-spotted eagle
	(*A. rapax*)	Tawny eagle
	Elanus caeruleus	Black-winged kite
	Gyps bengalensis	White-backed vulture
	G. fulvus	Griffon vulture
	Haliaeetus leucogaster	White-bellied sea eagle
	H. leucoryphus	Pallas' sea eagle
	Haliastur indus	Lal chil, Brahmini kite
	(*Icthyophaga icthyaetus*)	Grey-headed fishing eagle
	Ictinaetus malayensis	Black eagle
	Milvus migrans	Chil, black kite
	Pernis ptilorhynchus	Crested honey buzzard
	Spilornis cheela	Crested serpent eagle
Falconidae	(*Falco chicquera*)	Red-headed falcon
Anseriformes		
Anatidae	*Anas* sp.*	Ducks, wigeon, teal
	*A. acuta**	Pintail
	A. poecilorhyncha	Spot-billed duck
	Carina scutulata	White-winged wood duck
	Dendrocygna bicolor	Fulvous tree duck
	*D. javanica**	Lesser whistling duck
	Nettapus coromandelianus	Cotton pygmy goose
	Sarkidiornis melanotos	Knob-billed goose
Ardeiformes		
Ardeidae	*Ardea cinerea*	Gray heron
	A. purpurea	Purple heron
	*Ardeola grayii**	Pond heron
	*Bubulcus ibis**	Cattle egret
	Butorides striatus	Striated heron
	Dupetor flavicollis	Black bittern
	Egretta alba	Great egret
	*E. garzetta**	Little egret
	E. intermedia	Intermediate egret
	Gorsachius melanolophus	Tiger bittern
	Ixobrychus sinensis	Chinese little bittern
	Nycticorax nycticorax	Black-crowned night heron
Ciconiidae	*Anastomus oscitans*	Asian openbill stork
	Ciconia ciconia	White stork
	C. episcopus	White-necked stork
	Ibis leucocephalus	Painted stork
	Leptoptilos dubius	Greater adjutant stork
	L. javanicus	Lesser adjutant stork
	Xenorhynchus asiaticus	Black-necked stork
Threskiornithidae	*Platalea leucorodia**	White spoonbill
	Plegadis falcinellus	Glossy ibis
	Pseudibis papillosa	Black ibis
	Threskiornis melanocephala	Oriental ibis
Apodiformes		
Apodidae	*Apus* sp.	Swift
	Cypsiurus parvus	Palm swift

(*continued*)

TABLE 15.A-3 (*continued*)

Family	Scientific name	Bengali and/or English name
Caprimulgiformes		
Caprimulgidae	*Caprimulgus affinis*	Franklin's nightjar
	C. asiaticus	Indian nightjar
	C. indicus	Jungle nightjar
Hemiprocnidae	*Hemiprocne longipennis*	Crested tree swift
Charadriiformes		
Burhinidae	*Esacus magnirostris*	Beach stone-curlew
Charadriidae	*Charadrius dubius**	Little ringed plover
	C. hiaticula	Ringed plover
	*Pluvialis squatarola**	Grey plover
	Vanellus indicus	Red-wattled lapwing
	V. spinosus	Spurwinged lapwing
Glareolidae	*Glareola lactea*	Little pratincole
	G. pratincola	Collared pratincole
Haematopodidae	*Haematopus ostralegus*	Oystercatcher
Jacanidae	*Metopidius indicus*	Bronze-winged jacana
Laridae	*Larus ridibundus*	Black-headed gull
	Sterna albifrons	Least tern
	S. aurantia	Gang chil, river tern
	S. bengalensis	Lesser crested tern
	S. bergii	Greater crested tern
	S. fuscata	Sooty tern
	S. nilotica	Gull-billed tern
Recurvirostridae	*Himantopus himantopus*	Blackwinged stilt
	Recurvirostra avosetta	Avocet
Rhynchopidae	*Rhynchops albicollis*	Indian skimmer
Rostratulidae	*Rostratula bengalensis*	Painted snipe
Scolopacidae	*Calidris minutus*	Little stint
	Gallinago sp.	Snipe
	*Numenius arquata**	Curlew
	N. phaeopus	Whimbrel
	Tringa sp.	Sandpiper
	*T. hypoleucos**	Common sandpiper
Columbiformes		
Columbidae	*Chaleophaps indica*	Emerald dove
	Columba livia	Rock dove
	Streptopelia chinensis	Spotted dove
	S. decaocto	Collared turtle dove
	S. tranquebarica	Red turtle dove
	Treron bicincta	Orange-breasted green pigeon
	T. phoenicoptera	Yellow-footed green pigeon
	T. pompadora	Pompadour green pigeon
Coraciiformes		
Alcedinidae	*Alcedo atthis*	Common kingfisher
	Ceryle rudis	Lesser pied kingfisher
	Halcyon chloris	White-colored kingfisher
	H. pileata	Black-capped kingfisher
	H. smyrnensis	White-breasted kingfisher
	Pelargopsis amauroptera	Brown-winged kingfisher
	P. capensis	Stork-billed kingfisher
Coraciidae	*Coracias bengalensis*	Indian roller
Meropidae	*Merops orientalis*	Little green bee-eater
	M. philippinus	Blue-tailed bee-eater
Cuculiformes		
Cuculidae	*Cacomanthis merulinus*	Grey-breasted brush cuckoo
	C. sonneratii	Banded bay cuckoo

(*continued*)

TABLE 15.A-3 (*continued*)

Family	Scientific name	Bengali and/or English name
	Centropus sinensis	Common coucal
	C. toulou	Black coucal
	Cuculus micropterus	Short-winged cuckoo
	C. varius	Common hawk cuckoo
	Eudynamis scolopacea	Koel
	Phaenicophacus leschenaultii	Sirkeer cuckoo
Galliformes		
Phasianidae	*Coturnix* sp.	Quail
	C. coturnix	Common quail
	Francolinus francolinus	Assam black partridge
	Gallus gallus	Red jungle fowl
Gruiformes		
Rallidae	*Amaurornis phoenicurus*	White-breasted water-hen
	Fulica atra	Coot
	Gallicrex cincrea	Kora, water cock
	Gallinula chloropus	Gray moorhen
	Porphyrio porphyrio	Purple swamp hen
	Porzana sp.	Crake
	Rallina curizonoides	Branded crake
	Rallus striatus	Blue-breasted banded rail
Passeriformes		
Alaudidae	*Calandrella raytal*	Sand lark
Artamidae	*Artamus fuscus*	Ashy shallow shrike
Campephagidae	*Coracina novaehollandiae*	Large cuckoo shrike
	Hemipus picatus	Pied flycatcher-shrike
	Pericrocotus cinnamomeus	Small minivet
	Tephrodornis pondicerianus	Common wood shrike
Corvidae	*Corvus macrorhynchos*	Jungle crow
	C. splendens	House crow
	Dendrocitta vagabunda	Indian tree pie
Dicaeidae	*Dicaeum agile*	Thick-billed flowerpecker
	D. erythrorhynchos	Tickell's flowerpecker
Dicruridae	*Dicrurus adsimilis*	Fork-tailed drongo
	D. paradiseus	Greater racked-tailed drongo
Estrildidae	*Amandava amandava*	Red avadavat
Irenidae	*Aegithina tiphia*	Common iora
	Chloropsis aurifrons	Gold-fronted leaf bird
Laniidae	*Lanius schach*	Black-headed shrike
Motacillidae	*Anthus novaeseelandiae*	Richard's pipit
Muscicapidae	*Chaetornis striatus*	Bristled grass warbler
	Cisticola exilis	Golden-headed cisticola
	Copsychus erimelas	Magpie robin
	C. saularis	Dyal
	Megalurus palustris	Striated marsh warbler
	Monarcha azurea	Black-naped monarch flycatcher
	Niltava tickelliae	Tickell's niltava
	Orthotomus sutorius	Common tailorbird
	Pachycephala cinerea	Mangrove whistler
	Prinia flaviventris	Yellow-bellied wren-warbler
	P. hodgsoni	Franklin's wren-warbler
	P. subflava	Tawny prinia
	Rhipidura albicollis	White-throated fantail flycatcher

(*continued*)

TABLE 15.A-3 (*continued*)

Family	Scientific name	Bengali and/or English name
	Saxicola caprata	Pied stone chat
	S. ferrea	Gray bush chat
	Terpsiphone paradisi	Asian paradise flycatcher
	Trichastoma abbotti	Abbott's babbler
	Turdoides longirostris	Slender-billed babbler
Nectariniidae	*Nectarinia zeylonica*	Purple rumped sunbird
Oriolidae	*Oriolus xanthornus*	Black-headed oriole
Paridae	*Parus major*	Grey tit
Ploceidae	*Lonchura malacca*	Black-headed munia
	L. punctulata	Spotted munia
	L. striata	White-rumped munia
	L. subundulata	Spotted munia
	Passer domesticus	House sparrow
	Ploceus bengalensis	Black-throated weaver bird
	P. burmanicus	Baya
	P. manyar	Streaked weaver
Pycnonotidae	*Pycnonotus cafer*	Red-vented bulbul
	P. jocosus	Red-whiskered bulbul
Sturnidae	*Acridotheres fuscus*	Jungle myna
	A. ginginianus	Bank myna
	A. tristis	Common myna
	Sturnus contra	Pied starling
Zosteropidae	*Zosterops palpebrosa*	Oriental white-eye
Pelecaniformes		
Anhingidae	*Anhinga rufa*	African darter
Pelecanidae	*Pelecanus philippensis*	Spot-billed pelican
Phalacrocoracidae	*Phalacrocorax carbo*	Great cormorant
	P. fuscicollis	Shag
	P. niger	Pygmy cormorant, pan kawri
Piciformes		
Picidae	*Chrysocolaptes lucidus*	Crimson-backed woodpecker
	Dinopium bengalensis	Lesser golden-backed woodpecker
	Micropterus brachyurus	Rufous woodpecker
	Picoides (*Dendrocopos*) *nanus*	Pygmy woodpecker
Podicipediformes		
Hydrobatidae	*Fregetta tropica*	Black-bellied storm petrel
Phaethontidae	*Phaethon lepturus*	Long-tailed tropicbird
	P. rubricauda	Red-tailed tropicbird
Podicipedidae	*Podiceps cristatus*	Great-crested grebe
	*P. ruficollis**	Little grebe
Psittaciformes		
Psittacidae	*Psittacula eupatria*	Alexandrine parakeet
	P. krameri	Roseringed parakeet
Strigiformes		
Strigidae	*Athene brama*	Spotted little owl
	Bubo bubo	Great eagle-owl
	B. coromandus	Dusky eagle-owl
	Glaucidium radiatum	Jungle owlet
	Ketupa (*Bubo*) *zelonensis*	Brown fish-owl
	Ninox scutulata	Brown hawk-owl
	Otus bakkamoena	Collared scops owl
	O. scops	Scops owl
	Strix leptogrammica	Brown wood-owl
Tytonidae	*Tyto alba*	Barn owl
	T. capensis	Grass owl

TABLE 15.A-4

Noteworthy wetland fish of Bangladesh. Freshwater fish after Bhuiyan, 1964; FAO, 1972; Takagi, 1972; Rahman, 1974, 1975; Beckman, 1975. Brackish-water fish from Ahmed (1966)

Family	Scientific name	Bengali and/or English name
	Freshwater fish	
CHONDRICHTHYES		
Galeoidea		
Carcharhinidae	*Carcharhinus melanopterus*	Shark
	Scoliodon sorrakowah	Doyan, dogfish
Rajiformes		
Dasyatidae	*Himantura uaranak*	Sting
OSTEICHTHYES		
Clupeiformes		
Clupeidae	*Gudusia chapra*	Chapila, khoira, goori, herring
	Hilsa ilisha	Hilsa, ilisha, hilsha
	H. kanagurta	Chandana ilisha, shad
	Ilisha filigera	Choukya, jewelled shad
Engraulidae	*Setipinna phasa*	Phausa, feoah, tel tempari, anchovy
Notopteridae	*Notopterus notopterus*	Kanla, phali, phole, featherback
Cypriniformes		
Cyprinidae	*Catla catla*	Katla, katal, major carp
	Cirrhinus mrigala	Mirga, major carp
	Labeo calbasu	Kal bausch, major carp
	L. rohita	Rohit, rui, major carp, minnow
Mastacembeloidei		
Mastacembelidae	*Macrognathus aculeatum*	Goichi, kota, bauir, spiny eel
	Mastocembelus pancalus	Baim, eel
Perciformes		
Anabantidae	*Anabus testudineus*	Koi, climbing perch
	Trichogaster (Colisa) fasciatus	Khalisha, khaila
Centropomidae (Latidae)	*Lates calcarifer*	Bhetki, sea bass, barramundi
Gobiidae	*Glossogobius giuris*	Baillya, bele, belia, goby
Leiognathidae (Carangidae)	*Equula (Leiognathus) edentula*	Tak chanda, taka mach, pony fish
Mugilidae	*Mugil corsula*	Halla, kalla, freshwater mullet
Nandidae	*Nandus nandus*	Bheda, bhera, boza, mu-perch
Sciaenidae	*Pama (Sciaenoides) pama*	Koi, bhola, pia, jewfish, croaker
Stromateidae	*Stromateus cinereus*	Chanda, pomfret
Siluriformes		
Bagridae	*Mystus aor*	Aeir, aor, catfish
	M. seenghala	Air
	Rita rita	Rita, rida
Clariidae	*Clarias batrachus*	Jagur, magur, catfish
Heteropneustidae	*Heteropneustes fossilis*	Singhi, singh, jiol, stinging catfish
Schilbeidae	*Pangasius pangasius*	Pangas

(*continued*)

TABLE 15.A-4 (*continued*)

Family	Scientific name	Bengali and/or English name
Tetraodontiformes		
Tetraodontidae	*Tetraodon* (*Chelonodon*) *patoca*	Bogo, photka, taptepa, puffer-fish
	Brackish-water fish	
Atheriniformes		
Scomberesocidae	*Belone strongylurus*	Bara-tunchi
	Hemirhamphys cantori	Ekthute, gangturi
Clupeiformes		
Clupeidae	*Chatoessus* (*Anodontostoma*) *chacunda*	
	C. (*Nematalosa*) *nasus*	Barang
	Clupea (*Hilsa*) *toli*	Chandana, norhilsha
	C. (*Sardinella*) *fimbriata*	Chandana, takya
	Coilia (*Thryssa*) *dessumieri*	Carialli
	Engraulis telara	Teoach, teltempri
	E. (*Thryssa*) *hamiltonii*	Phyasa
	E. (*Thryssa*) *mystax*	Phansa
	E. (*Stalephorus*) *indicus*	Kagaja
	Hilsa ilisha	Hilsa, ilish
	Opisthopterus tartoor	Tarture
	Pellona (*Ilisha*) *elongata*	Ramgasha
	P. indica	
	Raconda russelliana	Kura phasa, phyasa
Cypriniformes		
Plotosidae	*Plotosus canius*	Gang-jhagur, gang-magur
Galeoidea		
Carcharinidae	*Carcharias gangeticus*	Hungur, kamat
	C. laticaudus	
Mastacembeloidei		
Mastacembelidae	*Mastocembelus armatus*	Bamosh, bam
	M. unicolor	Baim, turi
	Rhynchobdella (*Macrognathus*) *aculeatum*	Goichi, baim
Perciformes		
Carangidae	*Chorinemus moadetta*	
	Chornemus lysan	Karsia, mattia mach
	Equula edentula	Tanka chanda
	E. (*Trachinotus*) *blochii*	
	Patax teira	
	Psettus argenteus	Chanda
Centropomidae	*Lates calcarifer*	Kekti, bhatki
Chaetodontidae	*Holacanthus anularis*	Dudkamal
Ephippidae	*Drepane punctata*	Rupi-chanda
Geneidae	*Gerres filamentosus*	Jagiri, tak chanda
Gobiidae	*Apocryptes bato*	Chiring, ruta
	A. (*Pseudopocryptes*) *lanceolatus*	
	Boleophthalmus (*Scartelaos*) *viridis*	
	Eleotris (*Bathygobius*) *fusca*	Budha baillya

(*continued*)

TABLE 15.A-4 (*continued*)

Family	Scientific name	Bengali and/or English name
	E. amboinensis	Bele
	Gobiodes anguillaris	Bele
	G. buchanani	Bele
	G. rudicundus	Bele, chenna
	Periophthalmus schlosseri	
	Peropphthalmus koelreuteri	
Lobotidae	*Datnioides quadrifasciatus*	Bedha, patta
	Lobotes surinamensis	Dora chanda, kat koi
Mugilidae	*Mugil (Liza) tade*	Bhanga, tarui
	M. oeur	Ash bhangan
	M. (Rhinomugil) corsula	Bata, arwari
	M. (Valamugil) speigleri	Bhangan
Nandidae	*Nandus nandus*	Bheda, meni, boka
Scatophagidae	*Scatophagus argus*	Bishtara, paira chanda
Sciaenidae	*Otolithus maculatus*	
	O. ruber	
	Sciaena (Johnius) coitor	Bhola, Degripao
	S. (Johnius) sina	
	S. (Macrospinosa) cuja	Dantina, nuna, koi
	S. (Pama) pama	Bhola, poa, poma
	S. (Protonibea) diacanthus	
Serranidae	*Cromileptes altivelis*	
	Serranus lanceolatus	Bhol-karal, bole
	S. sonnerati	Bhetki, sal
Sparidae	*Crenidens indicus*	Kala khuramti
	Chrysophrys datnia	Kharanti datina
	C. (Mylia) berda	Datina
	Pagrus (Argyrops) spinifer	Datina, khuramti
Theraponidae	*Therapon jarbua*	Berguni, jirpai
	T. theraps	Berguni, jirpai
Toxitidae	*Toxotes chatareus*	
Trichiuridae	*Trichiurus (Eupleurogrammus) muticus*	
	T. haumela	Churi, rupa patia
	T. (Lepturacanthus) savala	Rupa patia
Pleuronectiformes		
Cynoglossidae	*Cynoglossus bengalensis*	Kukurjib
	C. brevis	
	C. hamiltonii	Banspata
	C. macrolepidotus	
Soleidae	*Solea ovata*	Kantal patta
	Synaptura (Euryglossa) pan	
Siluriformes		
Ariidae	*Arius (Tachysurus) dussumieri*	
	A. (Tachysurus) gagora	Gagla, gagra
	Ketengus typus	Med, meda
Bagridae	*Mystus gulio*	Gula, nuna tengra
Pangasidae	*Pangasius pangasius*	Pangas, Paungwas
Schilbeidae	*Silonia silondia*	Bancha, basa
Synbranchiformes		
Amphipnoidae	*Amphipnous cuchia*	Bangas, kuchla
Tetraodontiformes		
Tetraodontidae	*Tetraodon lunaris*	Photak
	T. patoca	Potka

bans estuary is so extensive, this group of animals contributes significantly to the animal biomass. Since many of these abundant organisms are highly prized as human food, they are one of the most profitable exports of Bangladesh. In the Sundarbans estuary there are probably more than 28 species of prawns (Table 15.A-5) representing only two genera (*Penaeus* and *Metapenaeus*).

Crabs are a significant component of the Sundarbans ecosystem in terms of energy transfer, nutrient trapping, substrate modification, as a food source, and as a potentially lucrative export. The 26 crab species (Table 15.A-6) are all amphibious to varying degrees; some live in the mud, often making extensive burrows slightly above water level, others slightly below. One species of *Paratelphusa* damages rice-field irrigation ditches. The more numerous marine crabs are mainly the Portunidae. These abundant particle-feeders consume rich slime from the mud surface. *Scylla serrata* already has been commercialized and is exported.

The Grapsidae, Potamonidae, and Ocypodidae are commonly seen on the mud banks at low tides. Many species dwell in mud holes. Their action in

TABLE 15.A-6

Noteworthy crabs of Bangladesh wetlands

Family	Scientific name	Bengali name
Calappidae	*Matuta lunanis*	Lonai kakra
	M. planipes	
Grapsidae	*Metaplax* sp.	
	M. indica	
	Sesarma sp.	
	S. minuta	
Maiidae	*Paramithrox aculeatus*	Macorosa kakra
Ocypodidae	*Ocypoda cratopthalma*	Lalkakra
Portunidae	*Chrybdis cruciara*	Jugi kakra
	Neptunus pelagicus	Santaru kakra
	N. sanguinolentus	Neptune kakra
	Portunus pelagicus	
	P. sanguinolentus	
	Scylla serrata	Shila kakra
	Thalamita crenata	
Potamonidae	*Paratelphusa (Barytelphusa) lamellifrons*	Patafalak kakra
	P. spinigera	Katra kakra
	Potamon (Acanthotelphusa) wood-masoni	Mansoni kakra
	P. (Acanthotelphysa) mortensi	Chinta kakra
Xanthidae	*Eurycarcinus orientalis*	

TABLE 15.A-5

Noteworthy prawns and lobsters of Bangladesh wetlands

Mainly estuarine prawns	Mainly freshwater prawns	Lobsters
Metapenaeus affinis	*Acetes erythraeus*	*Panulirus homarus*
M. brevicornis	*A. indicus*	*P. ornatus*
M. dobsoni	*A. japonicus*	*P. penicallactus*
M. monoceros	*Caridina gracilirostris*	*P. polyphagus*
Parapenaeopsis hardwickii	*Hippolysmata ensirostris*	*P. versicolor*
P. sculptilis	*Macrobrachium lamarrei*	*Puerulus sewelli*
P. stylifera	*M. malcolmsonii*	
Penaeus carinatus	*M. mirabili*	
P. indicus	*M. rosenbergii*	
P. merguiensis	*M. rude*	
P. monodon	*M. scabriculum*	
P. semi-sulcatus	*Palaemon concinnus*	
	P. lamarrei	
	P. stylifera	
	P. tenuipes	
	Solenocera indicus	
Deep sea prawns (below 300 m)		
Heterocarpus gibbosus		
H. woodmasoni		
Metapenaeopsis andamanesis		
Plesionika martis		

excavating large numbers of holes, depositing anaerobic mud on the surface, and increasing the circulation of water exerts a profound influence on the Sundarbans ecosystem. Crabs are reported to be one of the main sources of food for many carnivores of the Sundarbans. There are differences in salinity preference among these crab species (Hendrich, 1975).

Molluscs

Primarily filter-feeders, molluscs (Table 15.A-7) act as finer sieves than decapod particle-feeders. They are food sources for birds, carnivorous mammals, and reptiles. Their larval stages contribute greatly to the zooplankton eaten by fish.

A large number of people collect mollusc shells from the Sundarbans, which they convert to lime by burning. The principal genera of molluscs supporting the lime industry are also included in Table 15.A-7.

The famous pink pearls of Bengal, a valuable export, are grown by oysters of the genus *Crassostrea*. The other economically important molluscs are the shipworms, marine borers, or teredos (Teredinidae). These molluscs can bore neat holes in the wood of pilings, lock gates, sluice gates, and in wooden ships. Because the holes can reach 1–2 cm in diameter and more than 20 cm in depth, these molluscs can cause extensive damage. Living mangrove trees are also known to be bored by the shipworms.

Snails and other molluscs are extraordinarily abundant in the cultivated part of wetland areas.

TABLE 15.A-7

Noteworthy molluscs of Bangladesh wetlands

Taxon	Scientific name	English name	Taxon	Scientific name	English name
BIVALVIA			Turbinellidae	*Turbinella pyrum*	Sacred chank
Cyrenidae	*Gafrarium tuimidum*	Cockle clam	Strombidae	*Lomlis (Pterocara) lambis*	Fingered chank
	Paphia malabarica	False clam		*Strombus canarensis*	Winged shell
	P. marmorata		Common disease-vectoring snails		
	Velorita cyphinoides	Black clam	Ampullariidae	*Ampullaria* sp.	
Donacidae	*Donax cuneatus*	Wedge clam	Bythiniidae	*Bithynia* sp.	
Mytilidae	*Mytilus viridus*	Green mussel	Lymnaeidae	*Lymnaea auricularia*	
Ostreidae	*Crassostrea cucullata*	Rock oyster	Melaniidae	*Melanoidis tuberculata*	
	C. discoides	Disc oyster	Planorbidae	*Indoplanorbis exustus*	
	C. gryphoides	Kosture	Viviparidae	*Viviparus bengalensis*	
	C. madrasensis	Backwater oyster	**CEPHALOPODA**		
Teredinidae	*Bactronophorus thoracites*		Decapoda		
	Bankia (Liliobankia) campanellata		Loliginidae	*Loligo affinis*	Squid
	B. (Neobankia) roonwali			*L. duvaucelii*	
	Nausitora lanceolata			*L. hardwickii*	
	N. sajnakhaliensis			*L. indica*	
	Teredo (Kuphus) mannii		Sepiidae	*Sepia aculeata*	Cuttlefish
Unionidae	*Perissia* sp.	Lime source		*S. rostrata*	
Verenidae	*Cyrena bengalensis*	Jhinook		*S. rouxii*	
	Katelysia opima	Inflated clam		*Sepiella intermis*	
	Meretmix casta	Backwater clam		*Sepiotenthis arctipinnis*	
	M. meretmix	Bay clam	**OCTOPODA**		
GASTROPODA (often used as source of mollusc lime)			Octopodidae	*Octopus favonia*	
Cerithiidae	*Cerithium telescopium*	Jongra		*O. herdmanii*	
	Turbo marmoratus	Turban shell		*O. hongkongensis*	
Olividae	*Olivia gibbosa*	Otere shell		*O. incertus*	
Thaisidae	*Thais bufo*	Purple shell		*O. octopodia*	
Trochidae	*Trochus niloticus*	Top shell		*O. rugosus*	
	Umbonium vestiarium	Little butter shell			

They act as vectors of human and animal diseases, and are dealt with in greater detail below.

Water-borne diseases

High population density (averaging over 595 people km^{-2}), lack of education in sanitation, nutrition, and hygiene (20–25% literacy rate), and the low availability of physicians, contribute to the fragility of the health environment. This situation is compounded by the widespread malnutrition of the population. As important as specific disease, yet frequently overlooked, is the pervasive environmental stress in Bangladesh. The laborious work necessary merely for survival, hampered by abject poverty and malnutrition, means that any increase in either effort or expenditure of energy becomes more damaging than such increases would be in other countries. The life expectancy in Bangladesh of only 42 years as compared to the global figure of 59 years reflects this environmental stress.

More than 60% of all diseases in Bangladesh are water-borne, according to UNICEF (Bangladesh Observer, March 17, 1977). The principal water-borne diseases are malaria, amoebiasis, hemorrhagic and dengue fevers, cholera, fasciolopsiasis, filariasis, giardiasis, hepatitis, and typhoid.

Most of the malaria in and near wetlands of Bangladesh is caused by *Plasmodium vivax*, but *P. falciparum* is also increasing in recent years. Malaria caused by the refractory strains of *P. falciparum* is often not curable by the common malaria drugs.

The five major malaria vectors all require water in which to breed. The principal vector of most non-saline areas, *Anopheles philippinensis*, breeds in fresh water, especially in open swamps, pools with water hyacinth (*Eichhornia crassipes*), and rice-fields. This mosquito proliferates in the dry season in rivers that deteriorate into stagnant ponds.

Anopheles sundaicus breeds almost exclusively in brackish water, and is found mostly in the Sundarbans. Sandoshan (1965) found that this species associates with *Hodo* fern (*Acrostichum aureum*), which is abundant in the Sundarbans. It is one of the few mosquitoes able to tolerate polluted and highly organic water. It also prefers stagnant and exposed sites (Sandoshan, 1965), such as pools, green coconut shells, buckets, and cans. Not much is known concerning the ecology of *Anopheles aconitus*, *A. annularis* and *A. minimus*; these species appear less important as vectors in the wetlands of Bangladesh.

Arthropod-borne viral diseases, such as dengue and hemorrhagic fevers, are water-related insofar as their vectors, mosquitoes of the genera *Aedes* and *Culex*, breed in water (Chow, 1966).

Planorbid snails, such as *Indoplanorbis exustus*, are abundant in the wetlands of the country. Fasciolopsiasis, for which these snails are vectors, is less of a threat than schistosomiasis (which has not yet been reported to occur in Bangladesh), because the former is less severe. Filariasis constitutes a significant public-health problem in Bangladesh (Barry et al., 1971). This disease is water-related to the extent that it is transmitted by the mosquito *Culex* and midges that breed in water.

Animal diseases are also of concern in and near the wetlands of Bangladesh. Environmental stress reduces the value of livestock in the country. The most common ailments of livestock are malnutrition, indigestion, and diarrhea (Islam, 1975).

Livestock diseases transmitted by animal vectors which breed in water are fowl pox (transmitted by the mosquitoes *Aedes* and *Culex*), avian malaria (also transmitted by *Aedes* spp.), onchocerciasis (transmitted by a simulid fly), and schistosomiasis (transmitted by snails). Fowl pox and avian malaria conform to the pattern of human malaria, except that it is not yet known if any animal-disease mosquitoes proliferate in brackish water. Schistosomiasis and another largely identical disease, fascioliasis, conform to the pattern of human fasciolopsiasis, and are transmitted by the different species of snail vectors.

Plants

Of the very large number of plants (Tables 15.A-8 and 15.A-9) in the Sundarbans, only a few are economically important. *Heritiera fomes* (*sundri*), the most important species for the supply of timber and firewood (about 170 000 m^3 of timber and 708 000 m^3 of firewood annually), grows dominantly in elevated zones of low salinity located mostly in the northeastern part of the forest.

Another important plant is *Excoecaria agallocha* (*gewa*), because of its extensive use at the Khulna Newsprint Mills, annually manufacturing 50 000 t

TABLE 15.A-8

Principal woody plants of the Sundarbans (after Prain, 1903; Choudhury, 1968)

Family	Scientific name	Bengali name
Acanthaceae	*Acanthus ilicifolius*	Hargoza
Apocynaceae	*Cerbera odollam*	Dakur
Arecaceae	*Nypa (Nipa) fruticans*	Golpata
	Phoenix paludosa	Hantal
Barringtoniaceae	*Barringtonia racemosa*	Hizol
Bignoniaceae	*Dolichandrone rheedei*	
Caesalpiniaceae	*Caesalpinia bonducella*	Nata karanj
	C. nuga	
Casuarinaceae	*Casuarina equisetifolia*	
Combretaceae	*Lumnitzera littorea*	Kripa
	L. racemosa	Kripa
Ebenaceae	*Diospyros embryopteris*	Gab
Euphorbiaceae	*Excoecaria agallocha*	Gewa
	Sapium indicum	Mel, batul
Fabaceae	*Afzelia retusa*	Bhadal
	Cynometra racemosa	Singra
	Dalbergia spinosa[1]	Chulia kanta
	D. torta	
	Erythrina indica	Madar
	Pongamia pinnata	Karanj
Malvaceae	*Hibiscus tiliaceus*	Bhola
	Thespesia populnea	Paras, pipal
Meliaceae	*Amoora cucullata*	Amur
	Xylocarpus (Carapa) mekongensis	Pasur
	X. granatum	Dhundul
Moraceae	*Ficus retusa*	Jir
Myrsinaceae	*Aegiceras majus*	Khalsi
Myrtaceae	*Eugenia fruticosa*	Ban-jam
Plumbaginaceae	*Aegialitis rotundifolia*	Nunia
Rhizophoraceae	*Bruguiera gymnorhiza*[3]	Kankra
	B. sexangula[3]	Kankra
	Ceriops decandra	Goran
	Kandelia candel[3]	Goria
	Rhizophora apiculata[2,3]	Garjan
	R. mucronata[2,3]	Garjan
Sonneratiaceae	*Sonneratia acida*	Ora
	S. alba	
	S. apetala	Keora
	S. caseolaris	Ora
Sterculiaceae	*Heritiera fomes (H. minor)*	Sundri
	H. littoralis	
Tamaricaceae	*Tamarix dioica*	Urussia
Tiliaceae	*Brownlowia lanceolata*	Sundri-lata
Verbenaceae	*Avicennia alba*	Baen
	A. officinalis	Baen
	Clerodendron inerme	Ban-jai
	C. nerifolium	Koek-tita
	Vitex negundo	Nishinda
	V. trifolia	Pani-nishinda

[1] In Chokoria.
[2] Species with distinct viviparous germination.
[3] Rare in the Sundarbans.

TABLE 15.A-9

Principal herbaceous plants of the Sundarbans (after Prain, 1903)

Family	Scientific name	Bengali name
LILIOPSIDA (MONOCOTYLEDONES)		
Cyperaceae	*Cyperus tegetiformis*	
	Eleocharis sp.	
	Fimbristylis ferruginea	
	Mariscus albescens	
	Scirpus maritimus	
Orchidaceae	*Bulbophyllum xylocarpi*	
	Dendrobium rhizophoretti	
Pandanaceae	*Pandanus fascicularis*	Kea
Poaceae	*Andropogon squarrosus*	
	Imperata sp.	Ulu
	Oryza coarctata	
	Phragmites karka	Nal
	Saccharum cylindricum	Ulu
	Zoysia pungens	
Typhaceae	*Typha angustifolia*	Hogla
	T. elephantina	Hogla
MAGNOLIOPSIDA (DICOTYLEDONES)		
Asclepiadaceae	*Finlaysonia maritima*	
	F. obovata	Dhudi-lata
	Hoya parasitica	
	Sarcolobus carinatus	Kharrulla
	S. globosus	Baoli-lata
Fabaceae	*Derris heterophylla*	
Loranthaceae	*Dendropthoe pentandra*	
Mimosaceae	*Entada pursaetha*	Gila
PTERIDOPHYTA		
Blechnaceae	*Stenochlaena palustre*	
Polypodiaceae	*Drymaria* sp.	
	Polypodium irioides	
Pteridaceae	*Acrostichum aureum*	Hodo

of paper from *gewa* timber. The species is also common (21% of the total exploitable tree species, *sundri* being 73%). Like *sundri*, *gewa* grows better in elevated regions of comparatively low salinity in the northeast, as well as in the moderately saline regions in the central part of the Sundarbans.

Rhizophora apiculata and *R. mucronata* occasionally occur in the low southern area of the Sundarbans. These species are often straggling in habit, occasionally shrub-like, and have no profitably exploitable bole. These species are regarded as of minor importance in Bangladesh in contrast to the important economic position of *Rhizophora* spp. elsewhere in the world.

Non-viviparous plants predominate in general in the Bangladesh mangrove forests (Table 15.A-8).

The peculiar climate, with a pronounced dry season, is probably the determinant factor of the distinctness of the mangrove forests here. The dominant plants (*sundri* and *gewa*) are non-viviparous species which mostly occupy high ground. On relatively lower ground with soft shallow muds *Ceriops* sp. (*goran*) occurs. It possesses only slender and short viviparously germinated seedlings capable of anchoring in shallows of the habitat. In sloping elevated lands which are not frequently inundated, plant survival becomes critical. Such landforms are at times badly affected by run-off and erosion. In addition to the gentle slopes generally occurring in the tidal forest, occasional soil subsidence creates uneven or slanting landforms which are adversely affected by runoff and erosion, creating conditions unfit for most organisms.

REFERENCES

Ahmed, N., 1966. Fish and fisheries of the Sundarbans. *Symposium Humid Tropics Research (Dhaka). Proc.*, UNESCO, Paris, pp. 271–276.

Ali, S., 1972. *The Book of Indian Birds*. Bombay Natural History Society, Bombay, 162 pp.

Ali, S. and Ripley, S.D., 1974. *Handbook of the Birds of India and Pakistan Together with Those of Bangladesh, Nepal, Kikkim, Bhutan and Sri Lanka*. Oxford University Press, New York, N.Y. 334 pp.

Bagchi, K., 1944. *The Ganges Delta*. Calcutta University Press, Calcutta.

Barry, C., Ahmed, A. and Khan, A.Q., 1971. Endemic filariasis in East Pakistan. *Am. J. Trop. Med. Hyg.*, 20: 592–597.

Beckman, W.C., 1975. *Fisheries Development in Bangladesh — Fisheries Coordination Report*. FAO, Rome, 7 pp.

Beveridge, H., 1857. *The District of Bakarganj: Its History and Statistics*. Trubner, London 447 pp.

Bhuiyan, A.L., 1964. *Fishes of Dacca*. Asiatic Society (Bangladesh), Dacca, 148 pp.

Blasco, F. 1977. Outlines of ecology, botany and forestry of the mangals of the Indian subcontinent. In: V.J. Chapman (Editor) *Wet Coastal Ecosystems*. Ecosystems of the World 1. Elsevier, Amsterdam, pp. 241–260.

Champion, H.G., Seth, S.K. and Khattak, G.M., 1965. *Forest Types of Pakistan*. Pakistan Forest Institute, Peshwar.

Choudhury, A.M., 1968. *Working Plan of Sundarban Forest Division for the Period From 1960–61 to 1979–80*. Forest Department, Government of East Pakistan, Dhaka.

Choudhury, M.U., 1973. *Remote Sensing of Natural Resources of Bangladesh*. Forest Directorate, Dhaka, 17 pp.

Chow, C.Y., 1966. Mosquito-borne haemorrhagic fevers of South-East Asia and the Western Pacific. *Bull. World. Health Org.*, 35: 17–30.

Curtis, S. 1933. *Working Plan of the Sundarbans Division*. Bengal Government Press, Calcutta.

FAO, 1972. *Survey for the Development of Fisheries, Bangladesh — The Development of Fishery Statistics*. FAO, Rome, 34 pp.

Gusen, E.J., 1976. *Checklist of World's Birds*. Quadrangle, New York Time Book Co., New York, N.Y., 416 pp.

Hendrich, H., 1975. The status of the tiger *Panthera tigris* (Linn., 1758) in the Sundarbans Mangrove Forest (Bay of Bengal). *Saeugetierkd. Mitt.*, 23(3): 161–199.

Husain, K.Z., 1974. *An Introduction to the Wildlife of Bangladesh*. Ahmed Publ., Dacca, 81 pp.

Husain, K.Z. and Sarker, S.U., 1974. Birds of Dacca (with notes on their present status). *Bangladesh J. Zool.*, 2(2): 153–170.

Islam, A.K.M.N. and Aziz, A., 1975. Study of marine phytoplankton from the north-eastern Bay of Bengal, Bangladesh. *Bangladesh J. Bot.*, 4(1): 1–32.

Islam, K.S., 1975. Shistosomiasis in domestic ruminants in Bangladesh. *Trop. Anim. Health Prod.*, 7(4): 244.

Ismail, M., 1973a. An ecologist remark on Sundarban vegetation. *Bangladesh Observer*, 4 and 11 March, Dhaka.

Ismail, M., 1973b. Mangrove ecosystem in Bangladesh. *Newsl. Int. Assoc. Ecol. (INTECOL)*, 3: 5.

Ismail, M., 1975. Environment of the newly accreted land in the Bay of Bengal. Dhaka Univ., Dhaka, (Stencil), 8 pp. Reprinted in *Bangladesh Observer*, 12 Sept., 1975 as "Ecology of new lands in the Bay of Bengal".

Ismail, M., 1976a. Forest industries and Sundarban ecology. *International Seminar on Problems of Transfer of Laboratory Research to Industrial Application*. BCSIR Lab., Dhaka.

Ismail, M., 1976b. Some ecological features of Bangladesh climatology and Water Resource. *4th Commonwealth Human Ecology Conference (CHEC)*, December 13–18, Dhaka.

Ismail, M., 1977. Progressive desertification in Bangladesh. *Proc. 1st. Natl. Conf. For.*, Dhaka, Febr. 11–15, pp. 442–453.

Ismail, M., 1985. Biomass stability in environmentally critical Bangladesh coastal wetlands. In: *Proc. SAARC Seminar on Biomass Production*, April 12–15. Science and Technology Division, Ministry of Education, Govt. of People's Republic of Bangladesh, Dhaka, pp. 158–165.

Khan, F.K. and Islam M.A., 1966. Water balance of East Pakistan. *Oriental Geogr.*, 10(1): 1–10.

Khan, M.A.R., 1982. *Wildlife on Bangladesh*. Dhaka University, Dhaka, 173 pp.

Kulczynski, S., 1937. Zespoly roslin w Pieninach. Die Pflanzenassoziationen der Pieninen. *Pol. Acad. Sci. Leerers, Cl. Sci. Math. Nat. Bull. Int. Ser. B* (Suppl. 11).

Mountfort, G., 1969. *The Vanishing Jungle*. Collins, London, 286 pp.

Mountfort, G., 1970. The Bengal tiger goes into the "Red Book". In: F. Vellmar (Editor), *World Wildlife Year Book, 1969*. World Wildlife Fund, Morges, pp. 295–299.

Mountfort, G. and Poore, M.E.D., 1967. *The Conservation of Wildlife in Pakistan*. Govt. of Pakistan, Rawalpindi.

Mukherjee, A., 1969. *Extinct and Vanishing Birds and Mammals of India*. Indian Museum, Calcutta, 37 pp.

O'Malley, L.S.S., 1908. *Bengal District Gazetter: Khulna*. Bengal Secretarial Book Dept., Calcutta, 209 pp.

Prain, D., 1903. *Bengal Plants*. 2 vols. West, Newman and Co., London, 1319 pp.

Rahman, A.K., 1974. A checklist of the freshwater fishes of Bangladesh. 1. Cypriniformis (Ostariophysi). *Bangledesh J. Sci. Ind. Res.*, 9(3/4): 198–206.

Rahman, A.K., 1975. A checklist of the freshwater bony fishes of Bangladesh. *Chandpur, Freshwater Res. Stn.*, 1: 1–8.

Rashid, H., 1967. *Systematic List of the Birds of East Pakistan*. Asiatic Soc. Publ. No. 20, Dhaka, 144 pp.

Ripley, S.D., 1961. *A Synopsis of the Birds of India and Pakistan*. Natural History Society, Bombay, 702 pp.

Sandoshan, A.A., 1965. *Malarioloy*. Univ. of Malay Press, Singapore, 349 pp.

Schaller, G.B., 1967. *The Deer and the Tiger — A Study of Wildlife in India*. Univ. of Chicago Press, Chicago, Ill. 331 pp.

Shafi, M. and Kuddus, M.A., 1976. *On the Turtles of Bangladesh*. Bangladesh Academy Biggan Patrica, Dhaka, September.

Strickland, C., 1940. *Deltaic Formation with Special Reference to the Hydrographic Processes of the Ganges and Brahmaputra*. Longmans, Green and Co., Calcutta, 157 pp.

Swarup, K., 1959. Seasonal variation in the ovary of *Hilsa ilisha* (Ham) found in Allahabad. *Proc. Natl. Acad. Sci., India Ser. B*, 29(3).

Takagi, Z., 1972. *Survey for the Development of Fisheries, Bangladesh*. Tech. Rep. 2, FAO/UNDP, FI: SF/PAK 22, Rome, 34 pp.

Timm, R.W., 1967. Some estuarine nematodes from the Sundarbans. *Proc. Pak. Acad. Sci.*, 4: 1–13.

Watson, J.G., 1928. *Malayan Forest Records*. Fraser and Neave, Singapore, 6, 275 pp.

Chapter 16

FORESTED WETLANDS IN AUSTRALIA

R.L. SPECHT

INTRODUCTION

Australia is a continental mass, 7.7 million km^2 in area, which lies south of the equator between latitudes 11° and 44°. The climate varies greatly. Monsoonal summer rainfall predominates in the north; winter rainfall is characteristic of the mediterranean climate of the south. On the eastern side of the continent, rainfall has no pronounced seasonality. However, a marked rainfall gradient is found from the high rainfall areas of the southern, eastern and northern coastal regions into the dry interior. Some two-thirds of the continent may be classed as semi-arid to arid.

The humid coastal areas are drained by short, perennial streams which flow towards the coasts discharging almost all of the annual estimated runoff of 344×10^9 m^3 of water into the oceans. Much of the stream flow over the rest of the Australian continent is intermittent or non-existent because of low and unreliable rainfall, high evaporation losses, and flat topography. The largest inland river system is that of the River Murray and its tributaries (Fig. 16.1) which drain about one-seventh of the continent with an average annual rainfall of 430 mm; the annual flow of 15×10^9 m^3 is largely derived from the higher rainfall section of this large catchment area (Dunk and Hutchinson, 1970).

Along these drainage systems form swamps, floodplains, river banks and even seasonally-dry streambeds; these are the habitats of a series of forested wetlands which extend from the most humid areas of Australia into the arid interior. The forested wetlands can be classed as true forests only in the wettest localities. A series of communities which may be termed closed-forest, open-forest, woodland, and open-woodland have been observed from humid to arid regions (Specht, 1970).

Throughout Australia, prolonged periods of waterlogging inhibit the establishment of seedlings of tall shrubs and trees (Specht, 1981b); the treeless wetlands form wet-heathlands, reed swamps, herbaceous swamps, bogs and mire (Specht, 1979; Campbell, 1983).

Forested wetlands, in the broadest sense, are thus found over the whole of Australia, wherever seasonal flooding occurs. As almost every ecological survey made in Australia is likely to include some mention of forested wetlands, only some major ecological reports dealing with significant areas of forested wetlands are listed.

The following studies contain valuable ecological information:

Australia (general): Boomsma (1950), Specht (1970), Specht et al. (1974).

Australia (tropical): C.S.I.R.O. (1952–1977), Specht (1958), Specht et al. (1977).

New South Wales: Beadle (1948), Moore (1953), Jacobs (1955).

Queensland (subtropical): Coaldrake (1961), Queensland Department of Primary Industries, Division of Land Utilization (1974, 1978).

South Australia: Crocker (1944), Specht and Perry (1948), Specht (1951, 1972b), Specht et al. (1961).

Victoria: Incoll (1946), Connor (1966), Dexter (1978).

The taxonomic revision (Blake, 1968) of *Melaleuca leucadendron* and its allies, which are common in forested wetlands of northern and eastern Australia, and the book on *Forest Trees of Australia* (Hall et al., 1970) include summaries of

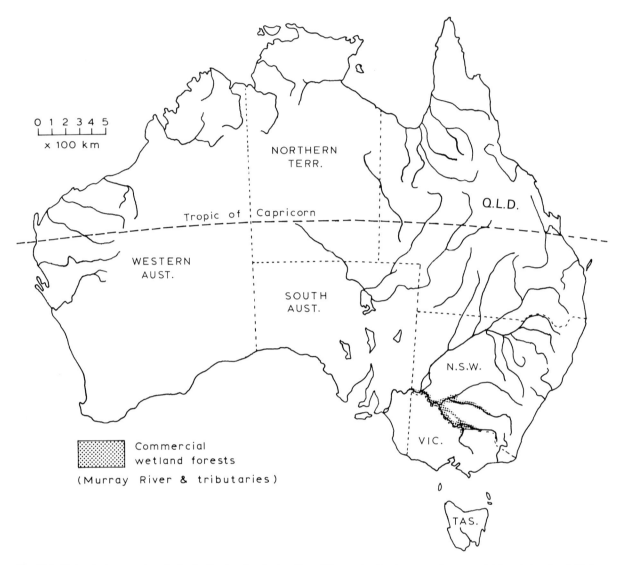

Fig. 16.1. Major drainage systems in Australia, showing the location of the most extensive area of commercial forested wetland. (Lakes of the arid interior are saltpans.)

the ecological relations of forested wetlands throughout Australia.

Until a comprehensive study of Australian wetlands has been made, it is impossible to estimate the proportion of the continent covered by these ecosystems. Wetland areas are so small that it was impossible to map them for the vegetation map of Australia at a scale of 1 : 6 000 000 produced by Carnahan (1976). Possibly the largest area of forested wetland in Australia is on the floodplain of the Murray River (Fig. 16.1) in the region bounded by Tocumwal,
Echuca and Deniliquin, where over 60 000 ha of river red gum (*Eucalyptus camaldulensis*) forest is found (Dexter, 1978).

On a continent as arid as Australia, water becomes limiting for man, his beasts, his crops, and his industries. Because of this demand for water, wetlands in all areas of the continent, from humid to arid regions, have been destroyed or are at risk. Conservation of the diverse plants and animals dependent on wetlands is of vital concern to all Australian biologists. Over the last 200 years, timber, for firewood, fencing and housing, has been extracted from most forested wetlands, and many weedy species have invaded the overgrazed ground stratum.

Very few areas of forested wetlands are large enough to form commercial forests; the major exception is the area of 60 000 ha of river red gum forest, mentioned above (Fig. 16.1). Even this area is endangered, because intensive river regulation has seriously reduced the extent and frequency of winter and spring flooding, and has caused unseasonal summer flooding in low-lying areas (Dexter, 1978).

In both arid inland and coastal regions of Australia, wetlands may become brackish and often markedly saline. The mangrove and salt-marsh vegetation types characteristic of saline, coastal wetlands have been discussed by Saenger et al. (1977) elsewhere in this Series. As the saline wetlands of arid, inland Australia are treeless, they will not be considered here.

In this paper, pertinent ecological notes and biological data for the main tree species of Australian forested wetlands (two species of *Casuarina*, nine species of *Eucalyptus* and ten species of *Melaleuca*) are presented in the first section of the paper. In the second section, generalizations are developed concerning the major ecological principles controlling the distribution, structure, and regeneration of forested wetlands in Australia.

MAJOR TREE SPECIES

Casuarina cunninghamiana (river oak) is a tree of moderate size, commonly attaining 20 to 27 m in height and 45 to 75 cm in trunk diameter. It is found scattered throughout eastern Australia from the south of New South Wales to the north of Queensland, and along larger rivers in the high-rainfall area of the Northern Territory (Fig. 16.2). River oak occurs as pure stands in narrow belts along water courses, especially in the zone between normal water level and maximum flood level. It also extends onto flats adjacent to rivers, where its roots have access to ground moisture additional to that derived from normal rainfall (Hall et al., 1970, p. 274).

Casuarina glauca (swamp she-oak), a tree 10 to 15 m tall, may form pure stands on low-lying, often brackish, wetland near the coast of eastern

Fig. 16.2. Distribution of *Casuarina cunninghamiana*.

Australia south of the Tropic of Capricorn (Fig. 16.3). In most areas, the species is closely associated with the brackish types of mangrove vegetation which have been described by Saenger et al. (1977).

Eucalyptus alba (white gum) varies from a small stunted tree sometimes less than 6 m in height to a tree over 20 m tall and 60 cm in trunk diameter. The species extends from the tropical regions of northern Australia (Fig. 16.4) to southern Papua New Guinea and Timor. It attains its best development towards the northern coast of the Northern Territory, where it is found on flats and water courses seasonally inundated with fresh

Fig. 16.3. Distribution of *Casuarina glauca*.

Fig. 16.4. Distribution of *Eucalyptus alba*.

water during the monsoonal wet season. In this area, it may be found in pure stands or intermixed with *E. polycarpa* (long-fruited bloodwood), *E. papuana* (ghost gum) and *Melaleuca* spp. (paperbarks) (Specht, 1958; Hall et al., 1970, p. 104). A small form, which grows on tablelands, hillsides and ridge tops of drier inland sites, appears to be a distinct subspecies of this variable species.

Eucalyptus camaldulensis (river red gum) is a medium to moderately tall tree, commonly up to 24 to 37 m in height, whilst trunk diameters range from 90 to 210 cm. This is the most widely distributed species of the large genus *Eucalyptus* (Fig. 16.5). It occurs along or near almost all the seasonal water courses in the arid to semi-arid areas. It is common along most streams and rivers in southeastern Australia, mainly on the inland side of the Great Dividing Range. Although mainly a tree inhabiting both river banks and floodplains, in more humid areas the species may extend away from drainage lines onto relatively high hill slopes (Specht and Perry, 1948; Hall et al., 1970, p. 100).

Eucalyptus largiflorens (black box) is a stunted or small tree, usually 10 to 20 m tall and 30 to 60 cm in trunk diameter. The tree is common in the vicinity of water courses on the western plains of New South Wales (Fig. 16.6). It is found on broad river flats, depressions in treeless plains and silted lake beds — low-lying sites subject to periodic but irregular flooding, or just above the edges of floodplains. The most common soil is a grey clay loam, occasionally dark grey self-mulching clays, and less commonly, fine red-brown sands (with heavy clay near the surface). Gilgais[1] (depressions associated with circular mounds or puffs) may develop in the surface of some of these heavy-textured soils (Hall et al., 1970, p. 228). On flat,

---

[1] The soil pattern called gilgai (an Australian aboriginal word) has been recorded from numerous parts of Australia (Stace et al., 1968). The soil surface shows small-scale undulations, the alternate hummocks and hollows of which show some degree of regularity. Considerable differences in magnitude and form of the undulations occur. The lower portion of the soil is termed the depression, crab-hole, or melon-hole; while the hummock or raised part has been termed the mound or puff (from the puffy surface soil of some gilgais).

Fig. 16.5. Distribution of *Eucalyptus camaldulensis*.

Fig. 16.6. Distribution of *Eucalyptus largiflorens*.

poorly drained areas with heavy soils, *E. largiflorens* occurs in almost pure stands. In other areas, it grows a little above the river red gum (*E. camaldulensis*), which may line both permanent and intermittent streams. Towards the arid centre of Australia, black box is found intermixed with coolibah (*E. microtheca*).

Eucalyptus microtheca (coolibah) ranges in stature from a tree (15 to 21 m in height with a bole which may become 1 m in diameter) to a small scraggly tree. The species is widespread in the arid and semi-arid zones of Australia (Fig. 16.7), where it is found on seasonally inundated country around the edges of swampy ground and lagoons or in open belts along watercourses. It is commonly found on heavy alluvial and alkaline clays, but may also be found on a wide range of other soil types (Hall et al., 1970, p. 220).

Eucalyptus ochrophloia (napunyah) is a tree 10 to 17 m high with persistent box-like bark on the trunk as far as the lower limbs. The species replaces black box (*E. largiflorens*) between the Paroo River and Cuttaburra Creek in southwestern Queensland (Fig. 16.8), where it is found on infrequently flooded areas of channels and floodplains (Hall et al., 1970, p. 228; Blake and Roff, 1972, p. 215).

Eucalyptus ovata (swamp gum) is usually a small to moderate-sized tree, 10 to 20 m tall with a bole 60 to 100 cm in diameter; but stunted trees do occur. The species is widely distributed in Tasmania and in coastal areas of southeastern Australia (Fig. 16.9). The species has the ability to grow in

Fig. 16.8. Distribution of *Eucalyptus ochrophloia*.

sites which are inundated for several months of the year, but it is also able to grow on well-drained slopes and in frost hollows (Hall et al., 1970, p. 106).

Eucalyptus robusta (swamp mahogany) is a tree usually attaining only moderate heights of 25 to 30 m, with trunks over 1 m in diameter. The species is found on poorly drained soils in a very narrow coastal belt from north of Fraser Island, Queensland to south of Bega in southern New South Wales (Fig. 16.10). The species often occurs in pure stands. In other areas, it may be intermixed with other eucalypt species such as *E. tereticornis* (forest red gum), *Casuarina glauca* (swamp she-

Fig. 16.7. Distribution of *Eucalyptus microtheca*.

Fig. 16.9. Distribution of *Eucalyptus ovata*.

Fig. 16.10. Distribution of *Eucalyptus robusta*.

oak) or *Melaleuca quinquenervia* (paperbark). Pure forests of paperbark replace *E. robusta* in sites where the water-table is even higher.

Eucalyptus rudis (Western Australian flooded gum) is a small tree 9 to 15 m tall with a trunk diameter of 50 to 100 cm. The species has a limited distribution along the west coast of southwestern Australia (Fig. 16.11). In this area, the species tends to replace the river red gum (*E. camaldulensis*), occupying similar habitats on river flats, along stream banks, and even in the beds of streams which flow only occasionally. In the more humid parts of its distribution, flooded gum may extend away from seasonally flooded areas onto low hilly country — an ecological compensation noted in the closely related *E. camaldulensis* in areas of similar rainfall (Hall et al., 1970, p. 102).

Eucalyptus tereticornis (forest red gum) is a moderately large to very large tree, 30 to 45 m in height, with a trunk 1 to 2 m in diameter. The species extends along the eastern coast of Australia from Warragul and Bairnsdale in Victoria to north of Cairns in Queensland (Fig. 16.12). Within the cooler and drier areas, the species is found on alluvial flats subject to flooding, but not to regular seasonal inundation. In regions of higher rainfall, it grows on the lower slopes of hillsides and extends onto higher slopes of hillsides and plateaux in the wet tropics of northern Queensland. Moderately rich alluvial soils are typical of the edaphic requirements of the species. The soils are moist but seasonally waterlogged, as may be characteristic of the closely related inland species (*E. camaldulensis*). Dense forest communities rarely occur; the species typically forms open forests or grows as scattered trees on alluvial flats and stream banks (Hall et al., 1970, p. 92).

Melaleuca argentea, a tree 5 to 25 m tall, is a characteristic rheophyte of sandy or gravelly stream banks and beds in northern Australia (Fig. 16.13), sometimes occurring as a belt of trees behind the fringing *M. leucadendron* forest (Blake, 1968, pp. 47–50).

Melaleuca bracteata is one of the most widely-spread species of the genus in Australia (Fig. 16.14). It is found as a tree, up to 15 m tall, along streams in eastern Queensland. In contrast,

Fig. 16.11. Distribution of *Eucalyptus rudis*.

Fig. 16.12. Distribution of *Eucalyptus tereticornis*.

Fig. 16.13. Distribution of *Melaleuca argentea*.

Fig. 16.15. Distribution of *Melaleuca cajuputi*.

it forms a low shrub, about 2 m tall, in depressions on heavy black soil and on rocky places of inland Australia (Blake, 1968, pp. 65–66).

Melaleuca cajuputi, a tree up to 25 m tall, sometimes forms pure forest communities on waterlogged soils in northern Australia (Fig. 16.15) (Blake, 1968, pp. 22–27).

Melaleuca dealbata is found as a tall tree up to 24 m in height in waterlogged soils, rarely far from the coast, from Maryborough, Queensland northwards into southern Papua New Guinea (Fig. 16.16) (Blake, 1968, pp. 41–43).

Melaleuca leucadendron is widely spread in northern and northeastern Australia (Fig. 16.17),

Fig. 16.16. Distribution of *Melaleuca dealbata*.

Fig. 16.14. Distribution of *Melaleuca bracteata*.

Fig. 16.17. Distribution of *Melaleuca leucadendron*.

southern Papua New Guinea and thence northwest to Amboina. In Australia it is most commonly found as a rheophyte on sandy or gravelly river banks, often forming a nearly pure fringing forest up to 25 m high (Blake, 1968, pp. 17–22).

Melaleuca preissiana is a paperbark tree, 8 to 10 m tall, restricted to southwestern Australia (Fig. 16.18). The species commonly forms pure stands on sandy sites subject to seasonal flooding.

Melaleuca quinquenervia, a tree up to 25 m tall, is widely spread in eastern Australia near the coast from Sydney northwards (Fig. 16.19), in New Caledonia and southern Papua New Guinea. It commonly occurs in pure communities on waterlogged soil, but may extend to hillsides, where groundwater is close to the surface (Blake, 1968, pp. 28–35). In some areas, the species appears to tolerate mildly saline conditions.

Melaleuca rhaphiophylla is a paperbark tree, 3 to 7 m tall, forming pure stands on sandy sites subject to seasonal flooding in southwestern Australia (Fig. 16.20).

Melaleuca stenostachya, a tree 4 to 25 m high, is common on seasonally waterlogged soil in northern Queensland (Fig. 16.21) (Blake, 1968, pp. 50–52).

Melaleuca viridiflora, a tree up to 18 m tall growing on seasonally swampy ground forms pure

Fig. 16.18. Distribution of *Melaleuca preissiana*.

Fig. 16.20. Distribution of *Melaleuca rhaphiophylla*.

Fig. 16.19. Distribution of *Melaleuca quinquenervia*.

Fig. 16.21. Distribution of *Melaleuca stenostachya*.

Fig. 16.22. Distribution of *Melaleuca viridiflora*.

forest communities (Blake, 1968, pp. 36–41). The species is also found as a small crooked tree, 6 to 8 m tall, on poorly drained slopes and flats over most of its distribution (Fig. 16.22).

STRUCTURE OF FORESTED WETLANDS

Wetlands with a stratum of tree species are widespread across the continent of Australia from the tropics to the cool temperate region, from the humid to the arid zone. Considerable variation in structure is thus observed.

In recent years, classification of Australian plant communities (Specht, 1970, 1972b, 1981c; Specht et al., 1974) has been based on two major structural attributes: (1) characteristic life form of the tallest stratum (tree, shrub, hummock grass, graminoid layer), subdivided into height categories: (2) foliage projective cover of the tallest stratum (the proportion of the landscape vertically covered by photosynthetic tissue).

Using this classification, the forested wetlands of Australia can be subdivided as shown in Table 16.1. Further definition of the various categories of forest or woodland may be made by including the name of the dominant genus and/or species with the name of the structural formation, for example "*Eucalyptus camaldulensis* open-forest". Examples of forested wetland formations are illustrated in Figs. 16.23 to 16.26. Usually, shrubs are rare or absent in these forested wetlands. A ground layer of grasses, sedges and herbaceous plants is always prominent, although, in the more arid areas, this stratum may be seasonal or ephemeral. On infertile soils, the graminoid layer is replaced by low sclerophyllous (heathland) plants; in the tropics and subtropics, this sclerophyllous stratum may be predominantly members of the node-sedge family Restionaceae.

TABLE 16.1

Structural formations of Australian forested wetlands

Foliage projective cover of tallest stratum	Life form/height of tallest stratum	
	tree 10–30 m	tree < 10 m
100–70%	closed-forest	low closed-forest
70–30%	open-forest	low open-forest
30–10%	woodland	low woodland
<10%	open-woodland	low open-woodland

With increasing periods of waterlogging, possibly more than 5 months (Boomsma, 1950), tree and shrub species fail to survive in the seedling phase (see also M.P. Bolton in Specht, 1981b). It would appear that no Australian freshwater forest species can survive long periods of inundation. The treeless understorey of the forested wetland persists as a graminoid wetland or as a wet-heathland.

VARIATION IN STRUCTURE WITHIN A CLIMATIC REGION

Semi-arid zone

Variation in the structure of forested wetlands dominated by *Eucalyptus camaldulensis* has been studied on the floodplains of the Murray River, near Mathoura (35°49′S, 144°54′E) in New South Wales (Boomsma, 1950). Structure appears to be related to the depth of the water-table and seasonal flooding. In this semi-arid area with an annual rainfall of 400 mm and a mean annual Moisture Index (the ratio of monthly actual evapotranspiration to pan evaporation) of 0.3, two distinct structural formations have developed:

A *closed-forest* (foliage projective cover over 70%), with trees up to 40 m tall, forms where the

Fig. 16.23. Closed-forest formation dominated by *Eucalyptus camaldulensis*, Moira State Forest, New South Wales. Photo, Forestry Commission of New South Wales.

depth of the water-table is less than 5 m below ground surface.

A *woodland* (foliage projective cover about 30%), with trees 25 to 30 m tall, forms where the depth of the water-table exceeds 5 m.

A closed-forest develops on sites which receive regular winter floods, irrespective of the depth of the water-table. If flooding occurs during summer (the season of active growth) and is prolonged by artificial manipulation of the river levels, regeneration is prevented and death of mature trees may result. Some trees develop *aerial roots* on the section of their trunk which is submerged by floodwater for several weeks.

A similar structural relationship has been reported on the St. Helena floodplain in the Millewa State Forest, New South Wales (35°50′S, 145°10′E), as shown in Fig. 16.27 (developed from Jacobs, 1955).

Sub-humid zone

In the sub-humid climate of the Mount Crawford Forest Reserve (34°42′S, 138°58′E) in South Australia, with annual rainfall of 720 mm and mean annual Moisture Index of 0.5, *E. camaldulensis* forms a woodland, with less than 25 mature trees ha^{-1}, from valley bottom to ridge top (Boomsma, 1950; see also Specht and Perry (1948) for the distribution of the species in the Mount Lofty Ranges).

A proportion of the annual rainfall is lost by runoff from the slopes, and no permanent water-table exists within reach of the root system. Under these conditions, the woodland formation is widespread. It is only along drainage channels and soaks, which receive additional water as runoff from other areas, that *E. camaldulensis* may grow densely enough to form small patches of closed-forest.

Fig. 16.24. Woodland formation dominated by *Eucalyptus camaldulensis*, in lagoons on the coastal plains of the Gulf of Carpentaria near the Queensland – Northern Territory border. Photo, C.S.I.R.O. Division of Land Use Research.

Humid zone

Along the tropical–subtropical coastline of Queensland, where the mean annual Moisture Index is over 0.8, two distinct structural communities, usually with a sharp line of demarcation, are observed in the forested wetlands dominated by *Melaleuca quinquenervia* and other *Melaleuca* spp. In this climatic region, forested wetlands may be either closed-forest (Fig. 16.25) or open-forest/woodland (Fig. 16.26) in structure. In the driest sites within the forested wetland, the trees of the open-forest/woodland become stunted (Fig. 16.26). No studies have been made on the hydrology of these forested wetlands, but waterlogging patterns discussed above for the sub-humid and semi-arid zones are probable.

ENVIRONMENTAL FACTORS: SOIL FERTILITY

Forested wetlands, because of their position in the landscape, invariably receive periodic additions of alluvial particles. The soils are continually being replenished by nutrients eroded and leached from higher elevations. Thus, it is not surprising that the wetland soils are some of the more fertile in the area. However, as much of the Australian landscape is low (by international standards) in soil nutrients, the nutrient concentrations of the wetland soils may be little above that of the soils of the adjacent dryland landscape.

Table 16.2 compares the total phosphorus and nitrogen concentrations of three surface soils from forested wetlands dominated by *Eucalyptus camaldulensis* with the concentrations recorded in soils from adjacent dryland savannah woodlands; differences are not significant.

Fig. 16.25. Closed-forest formation dominated by *Melaleuca quinquenervia*, on mainland near Bribie Island, southeastern Queensland. Photo, M. Olsen, Botany Branch, Qld Dep. Primary Industries.

In contrast, soils from sclerophyll-forest ecosystems are very low in nutrients, with levels of soil phosphorus 10–15%, and soil nitrogen 30–50%, of the values recorded for the savannah land system. Wetland soils derived from the sclerophyll-forest landscape are inevitably far lower in nutrients than soils derived from the savannah landscape. The wetland areas of Kalangadoo sand in southeastern South Australia (Table 16.2), contain low concentrations of phosphorus, but still support forested wetlands dominated by *Eucalyptus camaldulensis*.

ENVIRONMENTAL FACTORS: TEMPERATURE RESPONSE

Field and laboratory observations of the seasonal growth responses of Australian plant communities are fragmentary. Seasonal records on dryland forest communities have been summarized by Specht and Brouwer (1975) and Specht (1981a). A few observations on wetland forest species indicate that essentially the same seasonal sequences will be observed as in dryland forests.

The major tree species recorded in Australian wetlands are listed in Table 16.3 in three broad categories: (1) tropical/subtropical species, which produce new foliage shoots at mean daily temperatures between 20 and 30°C; (2) subtropical/warm temperate species, which produce new foliage shoots between 15 and 25°C; (3) warm temperate/cool temperate species, which produce new foliage shoots between 10 and 20°C. It must be stressed that this tentative classification is not based on sound observation, but represents the likely eco-

TABLE 16.2

Soil phosphorus and nitrogen (%) of four surface soils supporting the *Eucalyptus camaldulensis* association, compared with surface soils of savannah and sclerophyll land systems (after Boomsma, 1950; Specht et al., 1961)

Soils	Number of soil samples	Total phosphorus (%)	Total nitrogen (%)
E. camaldulensis association			
Savannah land system			
Belalie loam (northern Mount Lofty Ranges, S.A.)	1	0.029	0.088
Tulla clay (Riverina, N.S.W.)	1	0.020	0.113
Niemur clay (Riverina, N.S.W.)	1	0.067	0.063
Sclerophyll land system			
Kalangadoo sand (Southeast of S.A.)	1	0.008	0.102
Savannah land system			
Mount Lofty Ranges, S.A.	24	0.026 ± 0.004[1]	0.127 ± 0.012[1]
Sclerophyll land system (podzolic soils)			
Mount Lofty Ranges, S.A.	18	0.007 ± 0.001[1]	0.074 ± 0.010[1]
Southeastern South Australia	4	0.006	0.033

[1] Standard error of mean.

Fig. 16.26. Low woodland formation dominated by *Melaleuca viridiflora*, on broad sandy outwash plains on the southwestern side of Cape York Peninsula, Queensland. Photo, C.S.I.R.O. Division of Land Use Research.

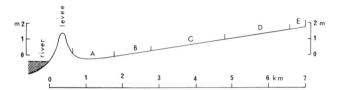

Fig. 16.27. Diagrammatic cross-section of the wetlands dominated by *Eucalyptus camaldulensis* on Millewa State Forest, New South Wales (after Jacobs, 1955). A, treeless swamp. B, poor-quality *E. camaldulensis* open-forest; covered with flood-water too frequently and for too long. C, high-quality *E. camaldulensis* closed-forest; ground covered by flood-water for several weeks in most years. D, fairly open-forest of intermediate-quality *E. camaldulensis*; ground covered by water only during peak floods (see Table 16.4). E, dense groups of saplings of low quality, germinating at the dryland/wetland margin after an extremely high flood in 1931. (Most of the trees in Communities B to D appear to have germinated after floods during the 1870–80 decade.)

physiological responses of the 21 tree species to the seasonal march of temperature.

ENVIRONMENTAL FACTORS: WATER BALANCE

Four major points emerge from the survey of Australian forested wetlands presented above:

(1) Forested wetlands are found all over the continent from the humid coastal belt to the arid interior. In all localities, the foliage projective cover (F.P.C.) of the tree stratum of the major part of the forested wetland is essentially the same as that observed in adjacent dryland forests/woodlands (Figs. 16.24 and 16.26).

(2) It is only in particular habitats, where the annual water balance is supplemented by surface flooding or subsoil seepage to an optimum level, that closed-canopied communities (F.P.C. 70–100%) may develop, usually with sharp discontinuity from the open communities (Figs. 16.23 and 16.25).

(3) Prolonged waterlogging, producing anaerobic conditions, leads to death of tree and shrub seedlings (see M.P. Bolton in Specht, 1981c). Death of adult trees often results if the area is artificially flooded. A treeless wetland results.

(4) A wetland tree species, normally confined to floodplains and watercourses, may extend over dryland hillsides towards the wetter limit of the distribution of the species. Competition with more vigorous dryland tree species may reduce the chances of a wetland tree species from occupying a dryland landscape at the humid end of its ecological range.

The water-balance studies described by Specht (1972a, 1981a) explain all the above observations.

Over the majority of the Australian continent, both the dryland and wetland forests/woodlands experience some period of water deficit.

In these habitats, the monthly water balance for any particular plant community may be expressed as follows:

$$\text{Moisture Index} = E_a/E_o = k(P - R + S_{ext}) \quad (1)$$

where E_a = actual evapotranspiration, E_o = pan evaporation, P = precipitation, R = run-off − run-on (or seepage), and S_{ext} = extractable soil moisture (at the beginning of the month). All terms are expressed in cm of water per month. The value of the constant k (termed the evaporative coefficient) ranges from 0.02 for an open-woodland to 0.10 for a closed-forest; it may be determined experimentally in the field, or by computer optimization which balances water conservation and water usage throughout the year (Specht, 1972a).

Foliage Projective Cover (F.P.C. %) of the climax community appears to be related to the constant k by the following equation

$$\text{F.P.C.} = 896.5k - 6.4 \quad (2)$$

Using these basic equations, the annual Moisture Indices and Foliage Projective Covers of the tree stratum of the dryland forests/woodlands across Australia have been calculated (Figs. 16.28 and 16.29). The ecological ranges, with respect to Moisture Index, of 21 Australian wetland tree species are shown in Table 16.3.

In wetland environments, extra water enters the ecosystem by run-on and seepage. A linear relationship exists between the mean annual value of the monthly Moisture Index (E_a/E_o) and the percentage of annual precipitation (less losses by run-off or supplemented by run-on or seepage)

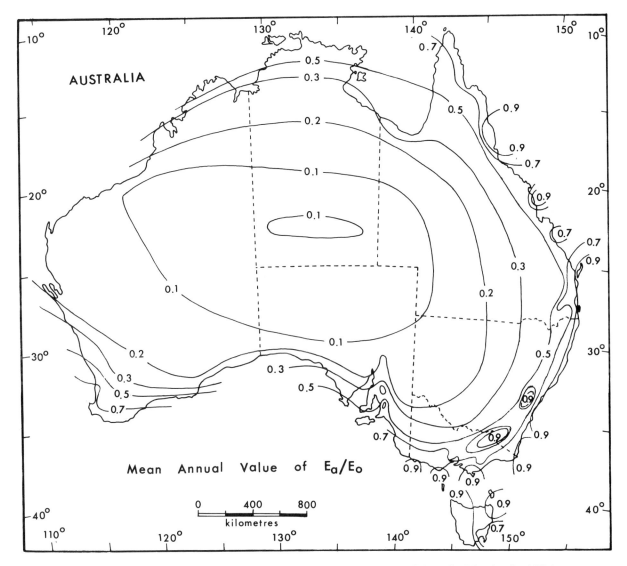

Fig. 16.28. Mean annual values of the Moisture Index (see text) across the continent of Australia (after Specht, 1972a).

which is retained in the soil within the aerated rooting zone of the plant community (Fig. 16.30a). This linear relationship is maintained until, in the wettest micro-habitats, the Moisture Index reaches the optimal value of about 1.0 for every month of the year. When this point is reached, and not before, the magnitude of the evaporative coefficient (k) suddenly switches from a constant to a maximum value of about 0.10 (Fig. 16.30b). A closed community is then produced in a climatic region which would typically support an open-structured plant community (Specht, 1981a). This transition from open to closed communities is clearly demonstrated in many Australian forested wetlands (Figs. 16.23 to 16.26).

It is only in small areas of Australia that dryland communities achieve an annual water balance which would ensure the development of closed communities (Fig. 16.28). Wetland tree species which become dryland species at the humid end of their distribution develop the structure (life form and foliage projective cover) typical of the dryland landscape. In drier areas these species, now confined to floodplains and watercourses, require continual additions of water for their survival; the foliage projective cover of the community remains

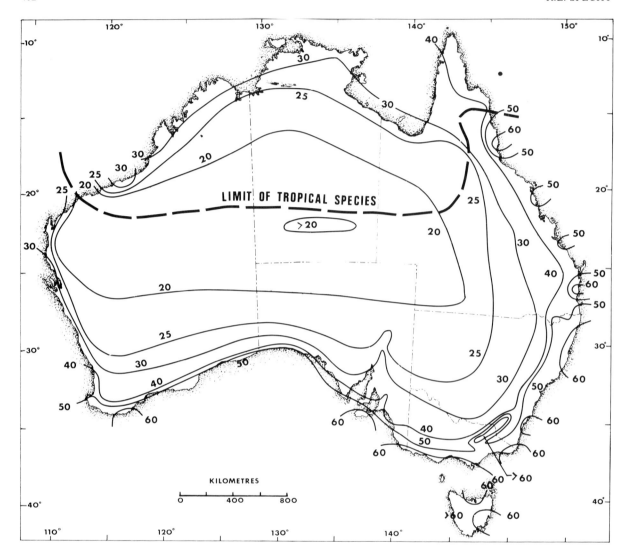

FOLIAGE PROJECTIVE COVER (%) OF CLIMAX VEGETATION

Fig. 16.29. Estimated values of Foliage Projective Cover (%) of the dominant stratum of climax dryland formations across the continent of Australia.

essentially the same as the adjacent dryland community (Figs. 16.29 and 16.30b), but the stature of the trees decreases progressively (Specht, 1972a, 1981a; Fig. 16.30a).

FIRE

It would appear that fire, ignited by lightning, has been an integral environmental factor in most Australian dryland and wetland ecosystems. The incidence of fire has increased considerably since man arrived in Australia. However, all the tree species listed above as common in Australian forested wetlands survive fire, resprouting from epicormic buds buried in trunks and branches. Many myrtaceous species have the added capacity to sprout buds from lignotubers (at the junction of stem and root) should the tops be destroyed by fire. It is only when the tree species is regenerating as a seedling that fire may kill the young plant (see below).

TABLE 16.3

Tentative ecological classification of tree species of Australian forested wetlands, based on seasonal shoot growth and annual Moisture Index (discussed in the text)

Tree species	Annual Moisture Index (dryland landscape)
Tropical/subtropical species (shoot growth 20–30°C)	
Eucalyptus alba	>0.5
E. microtheca	<0.5
Melaleuca argentea	>0.3
M. bracteata	>0.3 (some 0.2 to 0.3)
M. cajuputi	>0.5
M. dealbata	>0.5
M. leucadendron	>0.3
M. stenostachya	>0.4
M. viridiflora	>0.3
Subtropical/warm temperate species (shoot growth 15–25°C)	
Casuarina cunninghamiana	>0.4
C. glauca	>0.8
Eucalyptus camaldulensis	<0.4
E. largiflorens	0.1 to 0.3
E. ochrophloia	0.1 to 0.2
E. robusta	>0.8
E. rudis	>0.4
E. tereticornis	>0.4
Melaleuca preissiana	>0.4
M. quinquenervia	>0.8
M. rhaphiophylla	>0.4
Warm temperate/cool temperate species (shoot growth 10–20°C)	
E. ovata	>0.7

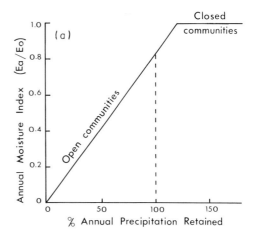

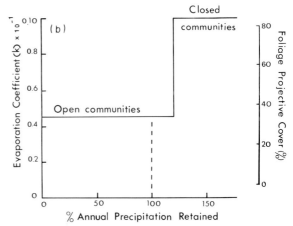

Fig. 16.30. Relationships between the percentage of annual precipitation which is retained in the soil within the aerated rooting zone of the plant community, after loss by run-off or gain by run-on and seepage, and other variables: A. Mean annual values of Moisture Index. B. Evaporative coefficient (k) and Foliage Projective Cover (after Specht, 1981a).

REGENERATION OF TREE SPECIES

Little is known of the regeneration of the many tree species in Australian forested wetlands. Some information has been gathered in the commercial *Eucalyptus camaldulensis* forest in southeastern Australia (Boomsma, 1950; Jacobs, 1955, pp. 219–236; Grose and Zimmer, 1958; Dexter, 1967, 1970).

Treloar (1959) has presented some information on the factors affecting survival of seedlings of *E. largiflorens*. Over a million seeds are produced per tree every second year. In southern Australia, the seeds mature towards the end of the dry Mediterranean-type summer and are then released from the capsules. In storage, seeds may remain viable for 15 years, but, in the field, many fail either to germinate or to survive beyond the seedling stage. Jacobs (1955) estimated that only one seedling in 150 million germinating seeds is necessary to perpetuate the forest.

Soil moisture, suitable temperature and light are necessary to ensure germination. In the field, ideal combinations of these factors tend to be discontinuous, so that, if germination occurs, it takes place in patches, often located where mature trees have been removed.

Germination of *E. camaldulensis* is closely correlated with the incidence of floods (Table 16.4), the best germination following a late spring

TABLE 16.4

Frequency of flood peaks in the Murray River at Mildura, Victoria 1867–1957 (Treloar, 1959)

Month	Frequency of flood peak
June	0
July	1
August	4
September	10
October	15
November	10
December	6
January	0
Total number of floods	46

(October–November) flood. Millions of seeds germinate in the forest every year. If the floodwaters persist for an extended period, germination is concentrated along the edges of the floodwaters — a phenomenon recorded in a number of Australian forested wetlands dominated by *E. camaldulensis*, *E. largiflorens*, *E. microtheca*, *Melaleuca quinquenervia*, etc.

Climatic conditions of late spring, when mean daily temperatures approach 20°C, appear "ideal" for germination. Laboratory tests, using constant conditions, have shown that 50% of river red gum seeds will germinate in about 10 days at that temperature. Maximum germination (almost 100%) has been recorded in 3–4 days with a constant temperature of 35°C (Grose and Zimmer, 1958), but, as mean daily temperatures during the summer months rarely exceed 25°C, optimal conditions for germination seldom occur.

Increased germination of *E. camaldulensis* occurs if the ground level is raised by the settling of silt and detritus in small depressions. The mounds of soil around a veteran tree, which has been destroyed by felling, lightning or fire, is an ideal site for regeneration. The temporary lowering (by pumping) of the water-table on a large nonforested freshwater wetland on North Stradbroke Island, southeastern Queensland, enabled the paperbark tree (*Melaleuca quinquenervia*) formerly confined to the edge, to germinate across the grass – sedge wetland.

Loss of seed by harvester ants is low in the red gum forests compared with the dryland forests, probably due to regular flooding which destroys the ant colonies. Most seedlings fail to survive; the few that develop are, today, usually destroyed by ground fires or by grazing animals (sheep, cattle, horses, rabbits).

The widespread survival of seedlings in the Millewa State Forest following the floods of the 1870–1880 period has puzzled ecologists. Jacobs (1955) speculated that this period may have coincided with both (1) a reduction in the number of aborigines in the area, with fewer hunting-fires which may have destroyed red gum seedlings, and (2) the absence of domestic grazing animals which dominated the locality a decade or two later.

COMMUNITY DYNAMICS

Examples of forested wetlands, seasonally inundated with fresh water, are found all over Australia from humid to arid areas. The structure of the tree stratum in these communities ranges from that of an open-forest, through woodland, to an open-woodland. At least 21 tree species — two species of *Casuarina*, nine *Eucalyptus* spp., and ten *Melaleuca* spp. — are prominent in forested wetland communities. Each tree species has a distinctive ecological range across the continent, often forming pure stands; there is little overlap with adjacent wetland species. Many wetland trees have very limited areas of distribution. Outside such areas, and in apparently very similar habitats, they are replaced by other, often closely related species.

Towards the humid limit of its distributional range, a wetland tree species may extend over dryland habitats. As seasonal water deficits are usual in most Australian ecosystems, an open-structured community results. This open-structured community extends down watercourses and floodplains, seasonally flooded, but still subjected to some seasonal water deficit.

Certain regions of the floodplain or drainage channels may receive a continuous supply of extra water from either surface run-on and storage or subsoil seepage; a portion of the root system remains well aerated for much of the summer growing season. In these sites, the wetland trees form a closed community, often sharply delineated from the adjacent open communities.

In extreme parts of the wetland landscape, the soils become waterlogged (and anaerobic) for long periods, producing a stunted, open-structured forested wetland. Prolonged waterlogging reduces the chances of survival of tree seedlings and may even lead to the death of mature trees — treeless grassland or heathland results depending on the fertility of the soil. If prolonged waterlogging is not too frequent, trees and shrubs tend to re-establish themselves.

Soil nutrient concentrations of forested wetlands are usually high, especially if the alluvial soil is derived from the erosion of relatively rich dryland soils. Wetland soils derived from a landscape low in plant nutrients are also low in nutrients; the resultant forested wetland is usually somewhat stunted, though with essentially the same foliage cover as the adjacent dryland tree stratum.

The structural changes in forested wetlands, discussed above, may be explained by reference to the water-balance model outlined on pp. 400–402. Variation in soil fertility may affect the height of the tree stratum, but not its foliage projective cover. The distribution of individual tree species appears to be defined by a combination of responses to the seasonal march of temperature and to the annual water balance.

Although great numbers of seeds are produced regularly by most wetland tree species, and germination has been recorded annually (at least in red gum forests), almost no seedlings survive. A gap in the mature community is needed for survival; but even the established seedling may be subjected to prolonged waterlogging, grazing or fire, in future years. Exceptionally high floods may float seeds into a zone where competition from mature trees is minimal and subsequent flooding, which may kill established seedlings, is rare. A distinct band of regenerating saplings is developed some distance above the normal level of the floodwaters; without regular flooding, the trees often become stunted in growth.

REFERENCES

Beadle, N.C.W., 1948. *The Vegetation and Pastures of Western New South Wales*. Govt. Printer, Sydney, N.S.W., 281 pp.

Blake, S.T., 1968. A revision of *Melaleuca leucadendron* and its allies (Myrtaceae). *Contrib. Qld. Herbarium*, 1: 114 pp.

Blake, S.T. and Roff, C., 1972. *The Honey Flora of Queensland*. Qld. Dep. Prim. Ind. and Govt. Printer, Brisbane, Qld., 234 pp.

Boomsma, C.D., 1950. The red gum (*E. camaldulensis* (Dehn)) association of Australia. *Aust. For.* 14: 97–110.

Campbell, E.O., 1983. Mires of Australasia. In: A.J.P. Gore (Editor), *Ecosystems of the World.* Vol. 4B. *Mires: Swamp, Bog, Fen and Moor. Regional Studies.* Elsevier, Amsterdam, pp. 153–180.

Carnahan, J.A., 1976. Natural vegetation. In: *Atlas of Australian Resources.* Second Series. Dept. National Resources, Common. Aust., Canberra, A.C.T., 1 map, 26 pp.

Coaldrake, J.E., 1961. Ecosystems of the coastal lowlands ("Wallum") of southern Queensland. *Bull. CSIRO, Melbourne*, No. 283, 148 pp.

Connor, D.J., 1966. Vegetation studies in north-west Victoria. II. The Horsham area. *Proc. R. Soc. Vic.*, 79: 637–647.

Crocker, R.L., 1944. Soil and vegetation relationships in the Lower South East of South Australia. *Trans. R. Soc. South Aust.*, 68: 144–172.

C.S.I.R.O., 1952–1977. *Land Research Series* No. 1–39. CSIRO, Canberra, A.C.T.

Dexter, B.D., 1967. Flooding and regeneration of river red gum *Eucalyptus camaldulensis* Dehn. *Bull. Vic. For. Comm.*, No. 20, 35 pp.

Dexter, B.D., 1970. *Regeneration of River Red Gum* Eucalyptus camaldulensis *Dehn.* M.Sc.For. Thesis, University of Melbourne, Melbourne, Vic., 159 pp.

Dexter, B.D., 1978. Silviculture of the river red gum forests of the central Murray flood plain. *Proc. R. Soc. Vic.*, 90: 175–191.

Dunk, P. and Hutchinson, S.J., 1970. Water and irrigation. In: G.W. Leeper (Editor), *The Australian Environment*. C.S.I.R.O. Australia – Melbourne University Press, Melbourne, Vic., 4th ed., pp. 32–43.

Grose, R.J. and Zimmer, W.J., 1958. Some laboratory germination responses of the seeds of river red gum, *Eucalyptus camaldulensis*, Dehn., syn. *Eucalyptus rostrata* Schlecht. *Aust. J. Bot.*, 6: 129–153.

Hall, N., Johnston, R.D. and Chippendale, G.M., 1970. *Forest Trees of Australia*. Aust. Govt. Publ. Service, Canberra, A.C.T., 334 pp.

Incoll, F.S., 1946. The red gum (*E. camaldulensis* syn. *E. rostrata*) forests of Victoria. *Aust. For.*, 10: 47–50.

Jacobs, M.R., 1955. *Growth Habits of the Eucalypts*. Forestry and Timber Bureau, Govt. Printer, Canberra, A.C.T., 262 pp.

Moore, C.W.E., 1953. The vegetation of the south-eastern Riverina, New South Wales. I. The climax communities. *Aust. J. Bot.*, 1: 485–547.

Queensland Department of Primary Industries, Division of Land Utilization, 1974 and 1978. *Western Arid Region Land Use Study*. Part I and IV. Qld Dept. Prim. Ind. Tech. Bull. No. 12, 131 pp.; No. 23, 92 pp. (plus Appendices and Maps).

Saenger, P., Specht, M.M., Specht, R.L. and Chapman, V.J., 1977. Mangal and coastal salt-marsh communities in Australasia. In: V.J. Chapman (Editor), *Wet Coastal Ecosystems. Ecosystems of the World.* Vol. 1. Elsevier, Amsterdam, pp. 293–345.

Specht, R.L., 1951. A reconnaissance survey of the soils and vegetation of the Hundreds of Tatiara, Wirrega and

Stirling of County Buckingham. *Trans. R. Soc. South Aust.*, 74: 79–107.

Specht, R.L., 1958. The climate, geology, soils and plant ecology of the northern portion of Arnhem Land. In: R.L. Specht and C.P. Mountford (Editors), *Records of the American–Australian Scientific Expedition to Arnhem Land. 3. Botany and Plant Ecology*. Melbourne University Press, Melbourne, Vic., pp. 333–414.

Specht, R.L., 1970. Vegetation. In: G.W. Leeper (Editor), *The Australian Environment*. C.S.I.R.O. Australia – Melbourne University Press, Melbourne, Vic., 4th ed., pp. 44–67.

Specht, R.L., 1972a. Water use by perennial, evergreen plant communities in Australia and Papua New Guinea. *Aust. J. Bot.*, 20: 273–299.

Specht, R.L., 1972b. *The Vegetation of South Australia*. Gov. Printer, Adelaide, S.A., 2nd ed., 328 pp.

Specht, R.L., 1979. The sclerophyllous (heath) vegetation of Australia: The eastern and central States. In: R.L. Specht (Editor), *Ecosystems of the World*. Vol. 9A. *Heathlands and Related Shrublands. Descriptive Studies*. Elsevier, Amsterdam, pp. 125–210.

Specht, R.L., 1981a. Ecophysiological principles determining the biogeography of major vegetation formations in Australia. In: A. Keast (Editor), *Ecological Biogeography of Australia*. Junk, Den Haag, pp. 229–332.

Specht, R.L., 1981b. The water relations of heathlands: Seasonal waterlogging. In: R.L. Specht (Editor), *Ecosystems of the World*. Vol. 9B. *Heathlands and Related Shrublands, Analytical Studies*. Elsevier, Amsterdam, pp. 99–106.

Specht, R.L., 1981c. Structural attributes — foliage projective cover and standing biomass. In: A.N. Gillison and D.J. Anderson (Editors), *Vegetation Classification in Australia*. C.S.I.R.O., Aust. – Aust. Nat. Univ. Press, Canberra, A.C.T. pp. 10–21.

Specht, R.L. and Brouwer, Y.M., 1975. Seasonal shoot growth of *Eucalyptus* spp. in the Brisbane area of Queensland (with notes on shoot growth and litter fall in other areas of Australia). *Aust. J. Bot.*, 23: 459–474.

Specht, R.L. and Perry, R.A., 1948. The plant ecology of part of the Mount Lofty Ranges. (1) *Trans. R. Soc. South Aust.*, 72: 91–132.

Specht, R.L., Brownell, P.F. and Hewitt, P.N., 1961. The plant ecology of the Mount Lofty Ranges, South Australia. 2. The distribution of *Eucalyptus elaeophora*. *Trans. R. Soc. South Aust.*, 85: 155–176.

Specht, R.L., Roe, E.M. and Boughton, V.H. (Editors), 1974. Conservation of major plant communities in Australia and Papua New Guinea. *Aust. J. Bot.* (Suppl. No. 7) 667 pp.

Specht, R.L., Salt, R.B. and Reynolds, S.T., 1977. Vegetation in the vicinity of Weipa, North Queensland. *Proc. R. Soc. Qld.*, 88: 17–38.

Stace, H.C.T., Hubble, G.D., Brewer, R., Northcote, K.H., Sleeman, J.R., Mulcahy, M.J. and Hallsworth, E.G., 1968. *A Handbook of Australian Soils*. Rellim Tech. Publ., Glenside, S.A., 420 pp.

Treloar, G.K., 1959. Some factors affecting seedling survival of *Eucalyptus largiflorens* F. Muell. *Aust. For.*, 23: 46–48.

Chapter 17

FORESTED WETLANDS IN WESTERN EUROPE

J. WIEGERS

GENERAL CHARACTERISTICS OF THE AREA SURVEYED

Geography

The part of Europe extending from 10°W to 13°E long. and from 45°N to 57°N lat. comprises the countries Ireland, United Kingdom, Denmark, The Netherlands, Belgium, Luxembourg, and West Germany, as well as the western part of East Germany and the northern half of France (Fig. 17.1).

Flats and gently sloping plains, dissected by numerous rivers and their tributaries and at an altitude of less than 100 m above sea level, occupy some 30% of this area. Most forested wetlands are confined to the geologically defined river basins. At altitudes around sea level they reach greater extent, just as they do within the range of inland bogs.

Some 20 000 km^2 of Western Europe are situated below sea level, mainly in the western and northern parts of The Netherlands. These low-lying grounds originated from erosion by seawater and subsequent successful reclamation or from peat-digging in places where the peat layers were mainly below the regional groundwater table. In the coastal area of The Netherlands a formerly vast area of bogs had even been covered by marine sediments following transgression of the sea after a relative rise of the sea level. The low lying areas are nowadays commonly protected against renewed penetration of seawater by embankments. This process of confining rivers within man-chosen boundaries started early in the Middle Ages.

The basins of the rivers Thames, Rhine, Meuse, Scheldt, Ems, Weser, Elbe, Seine, and Loire, together with numerous smaller ones, contribute the major part of these lowlands. All the rivers named discharge into the North Sea and the Atlantic Ocean.

Climate

Macroclimatic conditions in the area surveyed may be considered fairly homogeneous (Fig. 17.2, Table 17.1). The climate is temperate, with relatively short, mild winters and warm summers (Walter et al., 1975). Western Europe lies within the oceanic region of this Atlantic climate zone, which means that severe winter frosts are almost absent and humidity is rather high. In the British Isles (except in mountain regions) mean daily temperatures in the coldest winter month do not fall below 0°C, making them differ slightly from the adjacent continent.

Neighbouring zones show somewhat different climatic types (Figs. 17.3 and 17.4). The most westerly parts of Ireland and Scotland and the southwestern part of Norway show extreme oceanic features. Mean precipitation is always over 1000 mm yr^{-1} and mean daily temperatures in winter only occasionally fall below 0°C. The climate diagram for Cork (CO; Fig. 17.2) is an example of these extreme oceanic conditions.

To the northeast and east, oceanic influences diminish and winter temperatures fall rapidly. This zone is characterized by mean daily temperatures in the coldest winter month below -2.5°C, prolonged frost periods and lowest minimum temperatures less than -20°C. The climate diagrams for Kiel and Stuttgart (KL and ST; Fig. 17.2) show transitions to this continental type.

In the southeast and south, the Alps form a sharp geological and climatic boundary of the area surveyed. The southern half of France, bordered by the escarpment of the Pyrenées, shows a more submediterranean type of climate. Mean annual temperatures in this zone are over 10°C and in summer a relatively dry period of 1–3 months is common. The climate diagram for Rennes (RE; Fig. 17.2) shows the influence of this warmer climatic type.

Because mean annual precipitation in Western Europe is higher than 400 mm yr^{-1} and mean

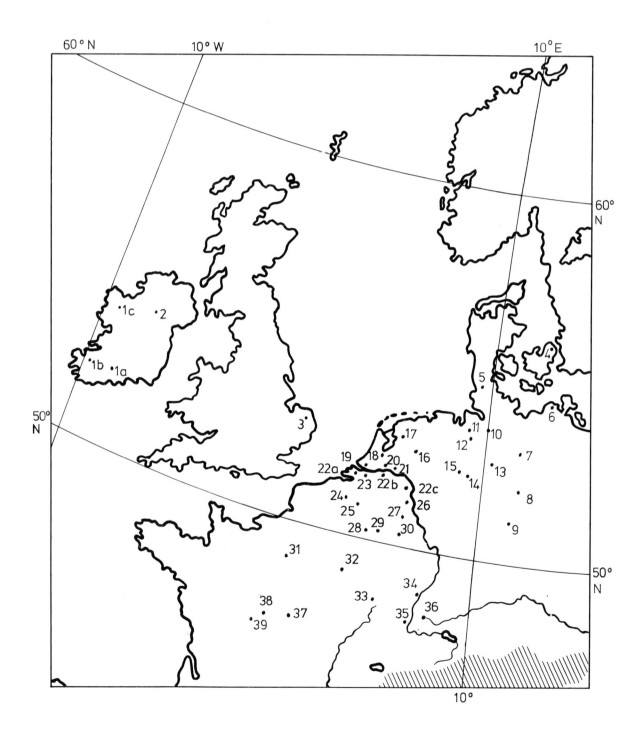

TABLE 17.1

Climatic data for the stations shown in Fig. 17.2 (after Walter et al., 1975)

Code	Station	T^1 (°C)	P^2 (mm)	Daily minimum[3] T (°C)	Lowest T^4 (°C)
AB	Aberdeen (U.K.)	8.4	836	1.7	−11.1
CA	Cambridge (U.K.)	9.8	551	0.5	−17.2
CO	Cork (Ireland)	10.3	1047	2.7	−9.4
DU	Dublin (Ireland)	9.4	753	1.7	−12.3
DIJ	Dijon (France)	10.7	696	−1.6	−20.4
HE	Helgoland (W. Germany)	8.5	596	0.0	−15.6
KL	Kiel (W. Germany)	7.6	717	−1.8	−20.0
OS	Ostende (Belgium)	9.7	777	0.6	−18.9
RE	Rennes (France)	11.2	678	1.7	−18.4
ST	Stuttgart (W. Germany)	10.0	673	−3.5	−25.0

[1] Mean annual temperature.
[2] Mean annual precipitation.
[3] Mean daily minimum temperature of the coldest month.
[4] Lowest minimum temperature measured.

annual evaporation does not exceed precipitation, the overall climate is characterized by a precipitation surplus. This condition, in combination with relatively low summer temperatures, favours the development of peatlands. Extensive areas have been covered with ombrogenous mires (dependent upon the influx of minerals from rain water for the nutrition of the vegetation they support: Du Rietz, 1954; Gore, 1983) since the last Ice Age. On the continent, these bogs already have, for the greater part, been excavated for fuel. In Ireland, this practice is still in full swing and most of the remaining bogs meet their end as fuel for power plants (Westhoff, 1981). The remains of nearly all

Fig. 17.1. Map of Western Europe, showing localities from which data have been used: *1a.* Macroom (Cork, Ireland), *1b.* Killarney (Kerry, Ireland), *1c.* Pontoon (Mayo, Ireland) (Braun-Blanquet and Tüxen, 1952); *2.* Mullingar (Westmeath, Ireland) (O'Connell, 1981); *3.* The Norfolk Broads (U.K.) (Lambert, 1951, 1965; Wheeler, 1978); *4.* Gammelmose (Frederiksborg, Denmark) (Hansen et al., 1978), Maglemose (Frederiksborg, Denmark) (Petersen, 1980); *5.* Gaarde (Schleswig-Holstein, W. Germany) (Schlottmann, 1966); *6.* Darss, peninsula (Rostock, E. Germany) (Fukarek, 1961); *7.* Drömling (Magdeburg, E. Germany) (Buchwald, 1951a); *8.* Harz (Niedersachsen, W. Germany) (Dierschke, 1969a); *9.* Thüringer Wald (Suhl, E. Germany) (Schlüter, 1969); *10.* Lüneburger Heide (Niedersachsen, W. Germany) (Dierschke, 1969b); *11.* Holtum (Niedersachsen, W. Germany) (Dierschke, 1979a); *12.* Syke (Niedersachsen, W. Germany) (Buchwald, 1951b); *13.* Vesbeck (Niedersachsen, W. Germany) (Buchwald, 1953); *14.* Höxter (Nordrhein-Westfalen, W. Germany) (Lohmeyer, 1953); *15.* Sennestadt (Nordrhein-Westfalen, W. Germany) (Tüxen and Dierschke, 1968); *16.* Achterhoek (Gelderland, The Netherlands) (Meijer Drees, 1936); *17.* Linde, river (Friesland, The Netherlands) (Smittenberg and De Roos, 1974); Weerribben (Overijssel, The Netherlands) (Wiegers, 1985); Wieden (Overijssel, The Netherlands) (Wiegers and De Vries, 1982); *18.* Naardermeer (Noord-Holland, The Netherlands) (Van Zinderen-Bakker, 1942); Botshol (Utrecht, The Netherlands) (Westhoff, 1949); Loosdrechtse Plassen (Utrecht, The Netherlands) (De Vries, 1969); Kortenhoef (Utrecht, The Netherlands) (Van Dijk, 1955; Smittenberg, 1976); *19.* Oude Maas, river (Zuid-Holland, The Netherlands) (Van Wirdum-Daan and Van Wirdum, 1977); *20.* Wijk bij Duurstede (Utrecht, The Netherlands) (Pons, 1976, 1977a, b); *21.* Middachten (Gelderland, The Netherlands) (Jeswiet et al., 1933); *22a.* Poortugaal (Zuid-Holland, The Netherlands), *22b.* Tuil (Gelderland, The Netherlands), *22c.* Wanssum (Limburg, The Netherlands) (Kop, 1961); *23.* Biesbosch (Noord-Brabant, The Netherlands) (Zonneveld, 1960); *24.* Dendermonde (Oost-Vlaanderen, Belgium) (Claessens, 1935); *25.* Chapelle-Lez-Herlaimont (Hainaut, Belgium) (Duvigneaud et al. 1971; Kestemont, 1974); *26.* Hilden (Nordrhein-Westfalen, W. Germany) (Hild, 1960); *27.* Aachen (Nordrhein-Westfalen, W. Germany) (Schwickerath, 1933); *28.* Tremblois (Ardenne, France) (Jouanne, 1926); *29.* Wavreille (Namur, Belgium) (Duvigneaud, 1968); *30.* Hautes Fagnes (Liège, Belgium) (Schwickerath, 1944); *31.* Perche (Orne, France) (Lemée, 1937); *32.* Tonnerre (Yonne, France) (Chouard, 1927); *33.* Apance, river (Haute Marne, France) (Chouard, 1932); *34.* Strasbourg (Bas-Rhin, France) (Carbiener, 1970); *35.* Lanterne, river (Haute Saône, France) (Malquit, 1929); *36.* Schwarzwald (Baden-Württemberg, W. Germany) (Bartsch and Bartsch, 1940; Willmanns, 1977); *37.* Preuilly (Indre-et-Loire, France) (Gaume, 1924); *38.* Brigueil (Charente, France) (Chouard, 1924); *39.* Brie (Charente, France) (Gaume, 1925).

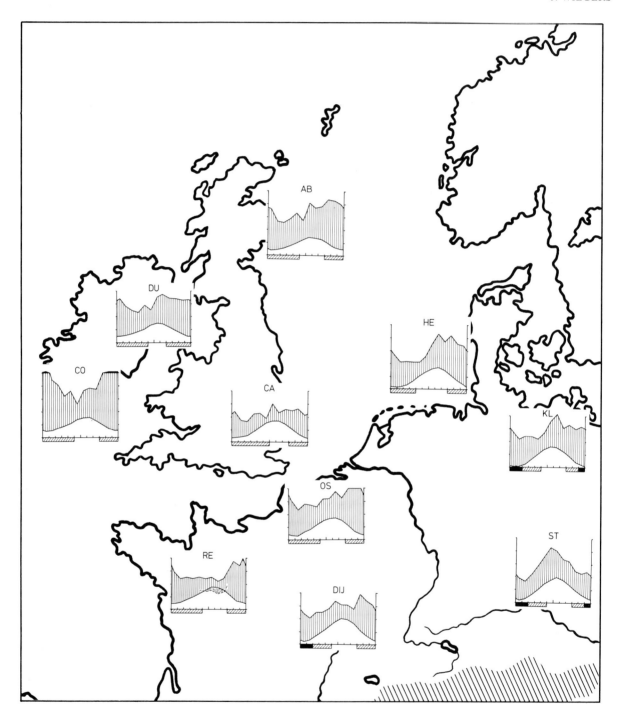

Fig. 17.2. Climatic conditions in Western Europe. Climate diagrams after Walter et al. (1975). For explanation of the abbreviated names of the stations and more detailed climatic data, see Table 17.1. The upper line in the graphs represents the mean monthly precipitation. Above 100 mm the scale is reduced by a factor of 1:10 and the vertical hatching, which denotes a relatively humid season, is replaced by a black field. One scale interval = 20 mm. The lower line is the curve of mean monthly temperatures (one scale interval = 10°C). A horizontally hatched field indicates a relatively dry season. The bars at the bottom of the graph indicate periods with mean daily minimum temperatures under 0°C (black) and months with absolute minimum temperatures under 0°C (angle hatched).

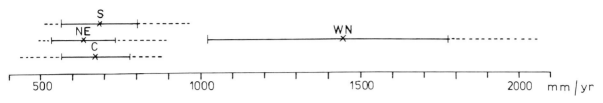

Mean annual precipitation

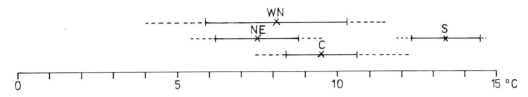

Mean annual temperature

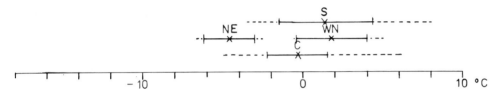

Mean daily minimum temperature of the coldest month

Fig. 17.3. Climatic types in Western Europe and adjacent zones (after data from Walter et al., 1975). C: central study area (47 stations); WN: west and northwest (13 stations); NE: northeast and east (18 stations); S: south (14 stations); x: mean value; ⊢⊣: standard deviation; - -: extreme values.

of these bogs suffer from alteration of their hydrology by the peat cutting and by drainage for agricultural purposes in their surroundings. This drainage results in lower groundwater tables in the bog, and these originally treeless areas (Osvald, 1925; Aletsee, 1967) become invaded by trees, thereby turning into wetland forests.

This transition from treeless to wooded areas, however, may also take place even when hydrological conditions of the bog do not seem to have been disturbed seriously. These vegetation changes, occurring over the course of decades, have been evaluated for two bogs in Denmark, north of Copenhagen (Hansen et al., 1978; Petersen, 1980). The growing dominance of trees in these ombrotrophic bogs may be due to insufficient precipitation for regeneration of the ombrotrophic vegetation (Hansen et al., 1978). This statement suggests that climatic changes are taking place in the area studied. Such changes may be traced with climatograms after Walter (1973). Another cause of changes may be found in the increase in nutrient input by atmospheric deposition (Zwerver et al., 1984), which is attracting considerable attention from woodland ecologists.

Distribution of forested wetlands

In a broad sense, forested wetlands can be found from near the coast to the lower alpine region. Two main vegetation series are discerned. Adjacent to streams and rivers are woodlands that are regularly or periodically inundated; groundwater flow is easily detectable. The latter feature is shared by spring-forests and woodlands on shallow soils situated on slopes with superficial and permanent drainage. The availability of nutrients to these communities is generally good, minerals being supplied by a constant groundwater flow or inundation.

In the upper tracts of the rivers the substrate is stony and poor in organic content, whereas in their middle and lower courses sedimentation by flooding often exceeds erosion and soils

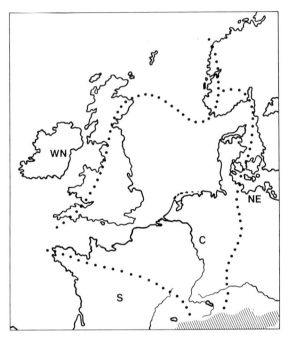

Fig. 17.4. Delimitation of the areas with slightly differing climatic types, as presented in Fig. 17.3. C: central study area; WN: western and northwestern zone; NE: eastern and northeastern zone; S: southern zone.

vary from sandy clay to heavy clay, according to the strength of the current in which sedimentation took place. These soils too are generally poor in organic material. As sediments accumulate, vegetation becomes less susceptible to frequent flooding and these woodlands tend to evolve to a drier type. The soil becomes temporarily aerated and a humus-rich layer builds up. Down-cutting of the stream channel produces comparable effects.

The second main type, peatland forests, may also be found from the lower alpine region down to the coastal area, although woodland vegetation in brackish tidal areas does not occur in Western Europe. In places where drainage is poor and organic matter accumulates, peat begins to form. Peatland forests develop as soon as a thin layer of organic material is deposited. Groundwater flow in these forest types is generally low or virtually absent. Decomposition of litter tends to slow down in the anaerobic environment and nutrient supply often is low. The groundwater level is generally close to the surface throughout the year.

Nearly all forested wetlands in Western Europe show traces of human influence. To protect crops and animals, extensive channelization has occurred in the lowland rivers since the early Middle Ages. In places where regular flooding was prevented by the building of dykes and dams, peatland forests could develop. However, since man settled in permanent dwellings, large areas of wetland forest have been felled and their sites turned into meadows or arable land. Other natural or semi-natural forests were turned into plantations, mainly of *Alnus*[1], *Fraxinus*, *Populus*, and *Salix*. For this purpose the hydrology of the sites was altered in order to meet optimal requirements of the planted species.

Peatlands were drained and the peat mined for fuel, or afforested with conifers, thereby completely destroying the characteristic lagg zone (the border of the mire, which is influenced by both seepage water from the mire and mineral-rich groundwater from outside the mire proper) of the domed ombrotrophic bogs. These lagg zones, unlike the center of the mire, carried special types of wetland forest which are now very rare in Western Europe.

In a few places man created new sites for forested wetlands, especially for swamp forests. In the western parts of The Netherlands and the Broads (eastern England), peat was mined from below groundwater level, and succession in many places led to the establishment of relatively undisturbed swamp woodlands. A general theory for the development of these woodlands has been given by Van Wirdum (1979).

Extensive studies by Godwin (1975) and Overbeck (1975) exist on the history of peatland forests. The relative importance of the area covered by wetland forests in the various vegetation units discerned for Western Europe by Ozenda (1979) is indicated in Table 17.2. A broad survey of all European forest types was recently published by Mayer (1984).

---

[1]For names of higher plants I follow the Flora Europaea (Tutin et al., 1964–1980); for those of Musci, Smith (1978); for those of Hepaticae, Margadant and During (1982); and for those of syntaxonomical units, Rothmaler (1976).

TABLE 17.2

Relative importance of wetland forests in vegetation units in Western Europe. Names and numbers of vegetation units following Ozenda (1979)

Covering extensive areas:
  Polder vegetation (2)
  Vegetation of wide valleys (3)
  Atlantic peat bogs (4)

Widely distributed:
  Acidophilous oakwoods and oak – beechwoods with birch (13)
  Ashwoods and mixed ashwoods with oaks (15)
  Atlantic beechwoods (16)

Mostly confined to stream valleys:
  Acidophilous oakwoods with *Quercus petraea* and *Quercus robur* (14)
  Submontane acidophilous beechwoods (22)
  Submontane neutrophilous beechwoods (23)

Only directly bordering streams:
  Acidophilous beech–fir formations of the greater Hercynian ranges (24)
  Subatlantic oakwoods with *Quercus pubescens* (26)
  Central European oakwoods with hornbeam and beech–oakwoods (27)

Commonly absent:
  Atlantic montane and subalpine vegetation with *Calluna* heaths (17)
  Subcontinental pinewoods and oakwoods (28)
  Calcicolous vegetation complex of the Pre-Alps and Jura (30)

GENERAL ECOLOGICAL CHARACTERISTICS

Riverine forests

Woodlands lying within the confines of river basins or fed by springs or drainage water share a common feature of standing in open connexion with a constantly renewed rich source of nutrients. Flooding of lands adjacent to the river (in the lower courses generally only the river forelands) not only brings in dissolved nutrients and coarse organic material but also mineral-rich sediments. When inundation is not too frequent, the groundwater table falls rapidly after flooding, and oxygen can penetrate the soil for rapid decomposition of the organic material present. These alluvial soils are generally alkaline, and under favourable moisture conditions most bacteria and actinomycetes can reach optimal development (Steubing, 1970). In a forest dominated by *Quercus robur*, *Alnus glutinosa*, *Fraxinus excelsior*, and *Ulmus minor* on a hydromorphic soil with groundwater levels changing from -5 cm in March to -100 cm in September, Lemée (1967) found an increase in nitrogen mineralization [in p.p.m. ($NH_4N + NO_3 - N$) in the organic matter per day] from 4 in April to 17 in September, and a decrease to 6 in November. Nitrogen content of the soil [in p.p.m. ($NH_4 - N + NO_3 - N$) in the organic matter], decreased from 287 in January to 67 in May, followed by a sharp rise to 212 in July, a second fall to 141 in August, and a rise to 203 in October. Aerobic nitrogen mineralization comes to a halt when the groundwater level rises in autumn, and the supply of nutrients is replenished by leaching from the surface layers in the period when consumption of nutrients by the vegetation is minimal.

Duvigneaud and Denayer-De Smet (1970a) and Denayer-De Smet (1970) analysed the mineral content of a wide range of species from the herb layer of riverine forests in Belgium. In nutrient-rich *Alnus* woodlands many species occurred which accumulated potassium to a higher degree (2.5–7.5% dry weight) than in nutrient-poor *Alnus* woodlands (1.0–3.5% dry weight). The content of phosphorus in species from the nutrient-rich woodlands was also higher (0.2–0.5% dry weight) than in species from the nutrient-poor woodlands (0.1–0.3% dry weight). Denayer-De Smet (1970) mentioned a relatively high nitrogen content (3–6% dry weight) and a low potassium content (1.5–2.5% dry weight) in species from alluvial soils. Such differences in mineral contents of plants correspond with differences in mineral content and aeration of the soils studied (Duvigneaud and Denayer-De Smet, 1970b). In soils from alluvial forests, contents of exchangeable calcium and magnesium and total nitrogen were higher (Ca 200 µeq g^{-1}, Mg 53 µeq g^{-1}, N 0.5%) and the exchangeable potassium content was lower (1 µeq g^{-1}) than in peaty soils (Ca 58, Mg 14, N 0.4, K 2). The more favourable soil conditions in the alluvial woodland (pH 6.8 and good aeration) stimulate decomposition, which results in a higher availability of nitrogen, as compared to the peaty, waterlogged soils with a lower pH (5.0). In the wetland forests studied by Duvigneaud and Denayer-De Smet (1970b), estimation of the production of biomass of the herb layer was difficult as some species (e.g. *Ranunculus ficaria*) reached their maximum development early in the growing season, when species such as *Filipendula ulmaria* were just beginning to develop their first leaves. Taking account of these seasonal differences, Duvigneaud and Denayer-De Smet (1970b) reported a decreasing biomass corresponding to a decrease in soil moisture.

The ecology and floristic composition of forests flooded by spring water is strongly controlled by the mineral composition of the spring water. In general these forests show resemblances to riverine forests as far as their species composition is concerned, but they differ from them in having nearly constant high levels of groundwater. The vegetation of the spring-fed forest often contrasts sharply with that of the area immediately outside the reach of the spring water. In summer, soil temperatures remain relatively low and in winter they are relatively high, due to the nearly constant temperature of the upwelling spring water. Because the area covered by spring-fed forests is relatively small and the number of types is great, they will not be treated extensively in this survey. A treatise on the floristic composition and ecology of spring vegetation, including soil and water analysis, has been published by Maas (1959).

Wetland forests in Western Europe, either in their virgin state or altered by light cutting, generally have been neglected by forestry research. Data on basal area, stem density and biomass are commonly lacking. Research in these ecosystems has mainly concentrated on floristic composition, but recently the gathering of ecological data is becoming a more widespread practice. Data on production and decomposition in temperate forest ecosystems, like those presented by Bocock and Gilbert (1957), Ovington (1962), Bray and Gorham (1964), Duvigneaud (1968), Reichle (1970), Kestemont (1974), and Matthews (1976) rarely refer to wetland species, and in cases that such species are involved, the data commonly pertain to plantations. Such managed forests on wetland sites can hardly be compared with natural woodland communities, because the sites have been altered to a great extent.

For a rough estimate of the primary production in these ecosystems one has to rely on Lieth (1965). He estimated the production of organic matter during a single growing season, but without complete adjustment for losses to consumers, decomposers, and substrate, to be in the order of 400–600 g C m^{-2} yr^{-1} in Western Europe. The definition of "growing season", however, is not sharp, and different authors may use non-comparable data, measuring different parameters (Schnelle, 1955; Lieth, 1970).

Peatland forests

An overall characteristic of the substrate in these forest types is the presence of an organic layer. Its thickness may range from a few centimetres to several metres, either over a mineral subsoil or as a floating mat over underlying water. Groundwater movement is generally slow and groundwater levels can be within a few decimetres from the surface throughout the year (Dinter, 1982). In floating woodlands this relative constancy of the groundwater level may be brought about by a vertical movement of the whole forest floor, following changes in the level of adjacent open water that stands in open connection with the water under the floating peat raft (Wiegers, 1988). The spongy structure of recently formed *Sphagnum*-peat also contributes much to this feature. Some of these woodland types are characterized by permanent pools on the peat surface. In these pools, hydrophytes (e.g., *Hottonia palustris* and *Lemna minor*) may add to the diversity of the life form spectra of these communities.

As Western Europe is situated in an area with net precipitation surplus, the overall downward-directed movement of water carries minerals, leached from the superficial layers, down into the permanently anaerobic zone. There, they are out of reach of species not adapted to waterlogging. Species that can extend their roots down into this zone and that are adapted to an anaerobic root environment may utilize this source of nutrients, for instance, *Eriophorum angustifolium* and *Molinia caerulea* (Armstrong, 1964; Hook and Crawford, 1978).

The supply of nutrients in peat soils is ample where recent peat layers were formed under the influence of mineral-rich groundwater, e.g., in sedge peat. *Sphagnum* peat is generally poor in nutrients and the vegetation it supports has largely to rely on the nutrient supply from rainwater and leachates from the tree layer. To a certain extent this holds also for forest ecosystems on permanently wet sandy soils with a low content of humus, such as the *Molinia*-rich variant of the mixed oak–birchwoods occurring in Belgium, The Netherlands, northern Germany, and Denmark.

Nitrogen is generally low in availability in peaty wetlands. Some species, such as *Alnus glutinosa* and *Myrica gale*, can overcome this handicap with the help of nitrogen-fixing microorganisms in their root nodules (McVean, 1955, 1956a, b; Sprent et al., 1978). In addition, the water economy of *Alnus glutinosa* is not reduced by anaerobic conditions (Braun, 1974) because water consumption in oxygen-free soil is more strongly depressed than growth, in this case expressed as increase in cm^2 leaf area. The water economy of *Alnus glutinosa*, expressed as cm^2 increase in basal area per 100 l water consumption, amounted to 2.6–3.8 under aerated soil conditions and to 3.3–5.9 under anaerobic soil conditions. These figures are markedly higher than those for *Salix alba* (cultivar "Liempde"), a species from riverine wetlands. Under aerobic conditions the water economy of this plant, in cm^2 increase in basal area per 100 l water consumption, was 1.2–1.9, whereas under anaerobic soil conditions these figures were slightly depressed to 1.0–1.8.

Whether the net production of organic material in peatland forests is great enough to form recognizable peat layers is strongly dependent on the hydrological conditions of the site. In the western part of The Netherlands, and the northern part of Germany, peat produced under woodland conditions often forms a conspicuous part of the peat body. Pals et al. (1980) discerned, on the basis of macrofossil remains, three zones. The lowermost one represented a local vegetation consisting of *Salix*-scrub, followed by two zones in which *Betula pubescens* apparently was the dominant tree species. A considerable admixture of inorganic material in the peat (25% ignition residue) of the *Salix* zone pointed to regular flooding of the site. This clayey woodland peat is widespread in the lower course of the river Rhine and its distributaries, both those still existing and their former courses (De Bakker and Edelman-Vlam, 1976).

The strong decomposition of all macrofossils in the upper *Betula* layer points to drier local conditions (Pals et al., 1980). The depth of this layer (6 cm), which covers a time span of ca. 100 years, suggests at least a considerably lower rate of peat formation than in the two earlier phases. The *Salix* zone and the first *Betula* zone each have a depth of ca. 20 cm and cover a time span of ca. 150 years each. Comparable data on the rate of peat formation in riverine *Alnus* woodland peat were presented by Van der Woude (1981). His data on the rate of peat formation in the Subboreal period (ca. 0.9–1.1 mm yr^{-1}) are considerably higher than those for ombrogenous *Sphagnum* peat during the same period [0.12–0.84 mm yr^{-1}; data compiled by Overbeck (1975, table 38 and fig. 237)]. Walker (1970) mentioned differences between accumulation of fen peat and bog peat which are in the same order of magnitude (0.70–1.0 mm yr^{-1} for fen peat and 0.12–0.96 mm yr^{-1} for bog peat).

From a core originating from an inland bog, Grosse-Brauckmann (1974) separated three components of a strongly decomposed layer, 25 cm deep, of woodland peat (*Pinus* stumps, *Alnus* roots, and particles <1 mm). These three components and an unpartitioned sample yielded rather different ages on ^{14}C-analysis. The *Pinus* stumps showed an age of 8965±85 years B.P., the *Alnus* roots 7245±85 years B.P., the smallest fraction 8265±110 years B.P., and the whole sample 7700±300 years B.P. From these data Grosse-Brauckmann concluded that the net production of organic material did not occur. He even did not exclude the possibility that under drier local circumstances there might be decomposition of peat layers already deposited. Ellenberg (1978), too, stated that only under special geological conditions (e.g., periods of transgression in coastal areas) considerable peat formation under woodland conditions is possible (see also Ch. 4, Section 15).

In stands of *Betula pubescens*, production of organic matter is low (7–10 t ha^{-1} yr^{-1}) in comparison with other wetland forests (De Sloover et al., 1974; Kestemont, 1974), and decomposition of leaf litter is fast. Bocock and Gilbert (1957), using *Betula*-leaves in a decomposition experiment, found a loss in oven-dry weight on peat soils over a period of 6 months of only ca. 25%, but according to Ovington and Madgwick (1959a, b) litter decomposition amounted to 100% with no accumulation of peat.

Grosse-Brauckmann (1976) stated that the species composition of peat containing fragments of wood was similar to that of samples without such fragments. The wood fragments present must have originated from communities growing on the peat but not producing additional peat layers. Miles (1981) reported increasing nitrogen mineralization under stands of *Betula pendula* of increasing age, ranging from 25 mg dm^{-3} week^{-1} under an 18-year-old stand to 40 mg dm^{-3} week^{-1} under a 90 year-old stand. Both Williams et al. (1978) and Miles (1981) reported no measurable changes in the amounts of nitrogen, phosphorus, and potassium in afforested peat, but contents of exchangeable calcium were lower with increasing age of *Betula* stands (Miles, 1981). Both calcium and magnesium were lower in peat beneath the tree crop than in unplanted areas (Williams et al., 1978). Both Williams and Miles reported enhanced decomposition promoted by aerobic conditions beneath the tree crop.

Special types of peatland forest are those growing on floating root mats. They have their most extensive development in The Netherlands and the Broads (England). Their history and vegetation succession was elaborately treated by Jennings and Lambert (1951), Lambert (1951), and Lambert and Jennings (1951). Analysis of recent succession in these swamp forests by means of macrofossils in the peat was also presented by Smittenberg and De Roos (1974). An extensive study of succession in these fen woodlands, in which vegetation changes in permanent plots covering 50 years of undisturbed woodland development (1931–1981) are treated, has been published by Wiegers (1985). He reported a deflection of the succession, which originally led to increasingly ombrotrophic conditions, towards a greater availability of nutrients, indicated by the re-appearance of *Alnus glutinosa* in communities formerly dominated by *Betula*. Although no detailed data are available, it seems likely that this change in the line of succession can be attributed to an increased nutrient input from atmospheric deposition.

COMPOSITION, STRUCTURE AND ECOLOGY OF DIFFERENT TYPES OF WETLAND FORESTS

Introduction

The separation into riverine forests and peatland forests (both taken in a broad sense), as introduced above, will be maintained in this section. I shall follow at first the wooded banks of rivers from the subalpine region to their estuaries where these ecosystems are influenced by tidal movements. Secondly, those wetlands not directly in connexion with river systems will be dealt with.

Woodland vegetation in the greater part of the surveyed area was studied by Ellenberg (1978) and Mayer (1984). Several regional surveys of vegetation composition, covering whole countries, are also available. The works of Tansley (1939) give valuable information on the vegetation of the British Isles, but, as they lack phytosociological tables, it remains difficult to compare the societies described with phytosociological units from the continent. Wheeler (1980) published more detailed descriptions of peatland forest ecosystems, following the classification made by Klötzli (1970). Braun-Blanquet and Tüxen (1952) presented data on vegetation types in Ireland. Doing (1962) proposed a system for woodland vegetation in The Netherlands and phytosociological units were also treated by Westhoff and Den Held (1969) and Den Held (1979). Vegetation types for Belgium were given by Lebrun et al. (1949). Noirfalise and Sougnez (1961) described riverine forests in Belgium. Scamoni (1960) and Passarge and Hofmann (1968) described in detail forest ecosystems in East Germany. Oberdorfer (1957) and Tüxen (1937), respectively, delineated plant communities in the southern and northern parts of West Germany.

Riverine forests

Alnus thickets

In the subalpine region, *Alnus viridis* thickets border small streams and grow on shallow soils with permanently percolating groundwater. Willmanns (1977) described this community (*Alnetum viridis*) from the Schwarzwald (West Germany) at an altitude of ca. 600 m above sea-level, mainly on slopes with a northern aspect. The most frequent accompanying species there were *Chaerophyllum hirsutum*, *Filipendula ulmaria*, and *Stellaria nemorum*. Other frequent species are listed in Table 17.3. Hemicryptophytes account for nearly 60% of the total species number, geophytes and therophytes are rare, and hydrophytes are absent (Fig. 17.5a). *Alnus viridis* commonly is the only woody species. The bushes are short ($\leqslant 6$ m) and there is a well-

TABLE 17.3

The most frequent species in *Alnus viridis* scrub[1] and *Alnus incana* wood[2]. Only species with a frequency >40% in the tables are mentioned (V: 80–100%, IV: 60–80%, III: 40–60%). Mosses and liverworts are marked with (m)

Alnus viridis scrub		*Alnus incana* wood	
V	*Chaerophyllum hirsutum*	III	
V	*Filipendula ulmaria*	III	
III	*Deschampsia cespitosa*	V	
IV	*Impatiens noli-tangere*	III	
IV	*Primula elatior*	III	
IV	*Urtica dioica*	III	
III	*Stachys sylvatica*	—	
V	*Alnus viridis*	V	*Aegopodium podagraria*
	Stellaria nemorum		*Alnus incana*
IV	*Alchemilla vulgaris*	IV	*Brachypodium sylvaticum*
	Crepis paludosa		*Cirsium oleraceum*
	Galium mollugo		*Lonicera xylosteum*
	Geranium sylvaticum		*Prunus padus*
	Geum rivale		*Rubus fruticosus*
	Holcus mollis	III	*Aconitum napellus*
	Knautia dipsacifolia		*Angelica sylvestris*
	Luzula sylvatica		*Carduus personata*
	Polygonum bistorta		*Festuca gigantea*
	Rubus idaeus		*Fraxinus excelsior*
	Senecio fuchsii		*Geum urbanum*
III	*Athyrium filix-femina*		*Glechoma hederacea*
	Ranunculus aconitifolius		*Humulus lupulus*
			Phalaris arundinacea
			Picea abies
			Plagiomnium undulatum (m)
			Sambucus nigra

[1] 8 relevés; after tables from Willmanns (1977).
[2] 59 relevés; after tables from Bartsch and Bartsch (1940); Oberdorfer (1957); Dierschke (1969a).

developed layer of tall herbs, including *Crepis paludosa*, *Geranium sylvaticum*, *Geum rivale*, *Knautia dipsacifolia*, *Senecio fuchsii*, and *Urtica dioica*.

These *Alnus* thickets form narrow bands along small streams. They grow on stony soils containing little organic material. The absence of a closed tree-layer is necessary for their continued existence. Neither seedlings nor full-grown plants of *Alnus viridis* are able to maintain themselves under a closed canopy. *Alnus viridis* has root nodules which contain nitrogen-fixing symbionts. These may fix up to 250 kg ha^{-1} yr^{-1} nitrogen (Rehder, 1970), thereby creating a continuous rich source of nitrogen. The development of a species-rich herb layer may be stimulated by this surplus of nitrogen, as both the substrate and the percolating groundwater are poor in nutrients.

Alnus woodland

In the montane zone alongside small streams, *Alnus incana* woods are found. *Fraxinus excelsior* and *Picea abies* frequently occur in this community (*Alnetum incanae*). The trees can reach a height of 20 m and the canopy is often dense, suppressing much of the undergrowth. Lianas such as *Humulus lupulus* and *Lonicera xylosteum* form a layer of climbers, while the shrub layer is mostly composed of *Prunus padus*, *Rubus fruticosus* s.l., and *Sambucus nigra*. The commonest species are listed in Table 17.3. Hemicryptophytes account for 50% of the total species composition and the number of geophytes is relatively high (Fig. 17.5b). These woodland types are characterized by coarse, mineral soils with a low content of humus (Oberdorfer, 1957), and a permanently high

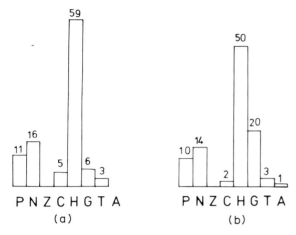

Fig. 17.5. Life-form spectra of *Alnus viridis* thicket (a) and *Alnus incana* wood (b) according to tables mentioned in Table 17.3. P: phanerophytes, N: nanophanerophytes, Z: woody chamaephytes, C: herbaceous chamaephytes, H: hemicryptophytes, G: geophytes, T: therophytes, A: hydrophytes. The height of the bars represents the percentages of the total species number.

groundwater level (Bartsch and Bartsch, 1940). In spring they are deeply inundated after melting of the snow (Dierschke, 1969a). Aeration of the upper soil layers is adequate for rapid decomposition of litter. *Alnus incana* woods are mostly confined to narrow bands on the steeper slopes which are not suitable for regular cultivation. On gentler slopes these communities have mostly been converted into meadows.

At lower altitudes, *Alnus glutinosa* gradually takes its place in the vegetation and the importance of *Alnus incana* decreases. *Fraxinus excelsior* remains the second important tree species. In the herb layer, subalpine species (*Aconitum napellus*, *Carduus personata*) disappear and *Aegopodium podagraria*, *Filipendula ulmaria*, *Geum urbanum*, *Impatiens noli-tangere*, *Lamiastrum galeobdolon*, *Stachys sylvatica*, *Stellaria nemorum*, and *Urtica dioica* become the most frequent species, being present in more than 60% of the relevés recorded, (Allorge, 1922a; Durin, 1951; Lohmeyer, 1957; Oberdorfer, 1957; Tüxen, 1957; Dierschke, 1969a).

Salix thickets

The middle and lower courses of the rivers generally have a fairly wide bed; their width varies according to the quantities of water that have to pass through it, although their profile and maximum width are strongly influenced by human activities, such as those mentioned above. Under natural circumstances, forested wetlands can occupy the whole width of the river bed outside the main current, where establishment of tree seedlings is prevented, and the zone where development of phanerophytes is hampered by too frequent flooding or by the breaking up and drifting of ice in winter and early spring.

In a direction perpendicular to the main line of the river current, frequency of flooding and velocity of the water diminish, giving rise to a gradient in the grain size of the sediment. Erosion may take place locally, sometimes enhanced by heavy shipping movements (Van Wirdum, 1977). Concurrent with the gradient in sedimentation and flooding, several vegetation zones in the land adjacent to the river may be found.

In the most frequently flooded zone *Salix* thickets are found. *Salix alba*, *S.* × *dasyclados*, *S. fragilis*, *S. triandra*, and *S. viminalis* form a mixed scrub with comparatively scanty undergrowth. The most frequent herbs are *Callitriche stagnalis*, *Caltha palustris*, *Cardamine amara*, and *Myosotis scorpioides*. Other species are listed in Table 17.4. The number of shrub species in these communities (*Salicetum albae*) is significantly greater than the number of tree species and more than 10% of the total species number are hydrophytes (Fig. 17.6a).

This type of wetland forest occurs in the slightly brackish tidal zone in the lower estuaries, and also outside the tidal region where the water is permanently fresh. Kop (1961) found no floristic

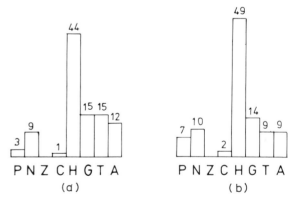

Fig. 17.6. Life-form spectra of frequently flooded *Salix* scrub (a) and more elevated *Salix* scrub (b) according to tables mentioned in Table 17.4. See legend of Fig. 17.5.

TABLE 17.4

The most frequent species in the frequently flooded *Salix* scrub[1] and in the more elevated *Salix* scrub[2]. (See notes in Table 17.3)

Salix scrub, frequently flooded			*Salix* scrub, more elevated
IV	*Salix alba*		IV
IV	*Cardamine amara*		III
III	*Solanum dulcamara*		IV
III	*Urtica dioica*		IV
III	*Valeriana officinalis*		IV
III	*Anthriscus sylvestris*		III
III	*Rumex obtusifolius*		III
III	*Salix viminalis*		III
IV	*Callitriche stagnalis*	V	*Symphytum officinale*
	Caltha palustris	IV	*Calystegia sepium*
	Myosotis scorpioides		*Poa trivialis*
III	*Lythrum salicaria*	III	*Angelica sylvestris*
	Polygonum hydropiper		*Galium aparine*
			Ranunculus repens
			Rumex conglomeratus

[1] 67 relevés; after tables from Oberdorfer (1957), Zonneveld (1960), Kop (1961), and Van Wirdum-Daan and Van Wirdum (1977).
[2] 120 relevés; after tables from Schwickerath, 1933, 1944; Meijer Drees, 1936; Tüxen, 1937; Oberdorfer, 1957; Zonneveld, 1960; Kop, 1961; Van Wirdum-Daan and van Wirdum, 1977.

differences between communities from inland stations and brackish sites near the coast. Trautmann and Lohmeyer (1960) described a similar vegetation from stations more than 100 km from the coast, along the Ems, but there they were present only in narrow bands.

This community reaches optimal development around and just below mean high water levels in the tidal area (Zonneveld, 1960; Van Wirdum-Daan and Van Wirdum, 1977), and occurs on heavy soils that are almost permanently waterlogged. The weakness and instability of the soil and the dense growth of the bushes make these communities difficult to penetrate. Flooding is of almost daily occurrence and the amount of nutrients available is high. Zonneveld (1960) gave data on soils for the freshwater tidal area of De Biesbosch (province of Noord-Brabant, The Netherlands) as follows: organic matter content 2.5–3.5%, C/N ratio 16–20, SO_4^{2-} 25–40 µeq g^{-1}, $CaCO_3$ content 7.5–8%, P_2O_5 – total 19 ppm, pH 7.4–7.6.

Heavy flooding and drifting of ice in winter may cause severe damage to the shrubs. But, as they possess ample means for vegetative reproduction, the canopy is closed again in short time. Loose branches easily form roots when they become lodged in the mud and add to the density of the vegetation.

The most extensive area where this type of vegetation was once found, was in the freshwater tidal marshes of De Biesbosch already mentioned. Since the closure of the dam in the Haringvliet in 1970, the tidal regime controlling the development of the vegetation was removed and the vegetation of these marshes had to adapt to this new situation. No comprehensive study of this area has been published, however, since the survey by Zonneveld (1960). Zonneveld (1961b) expected a development into *Alnus*-dominated forests in the more elevated parts, and lower depressions were likely to be filled in with peaty sediments. Peat layers with a depth up to 2 m are present in these tidal marshes (Zonneveld, 1960; Van Wirdum-Daan and Van Wirdum, 1977), but natural woodland vegetation on these sites is too rare to draw any conclusions about its structure and composition.

These highly productive *Salix* thickets have often been converted into willow coppice. The bushes are cut every 3 to 4 years. This cycle leads to an alternation of shrub vegetation with tall-herb vegetation. As the growth rate of the young shoots is high (at least 1.5 m yr^{-1} for *Salix alba*) the tall-herb stage is of lesser importance in the cycle. Twigs and branches are cut and are used for making baskets, brooms, and fences. On the lowest sites *Salix* × *dasyclados* and *S. purpurea* are the most successful species, and on higher grounds *S. alba* and *S. fragilis* are commonly planted (Zonneveld, 1961a). To enhance physical and chemical weathering of the soil and to improve drainage, a network of dykes has been dug. The vegetation tends to concentrate on the dyke ridges, thereby slackening the speed of the flooding water and enhancing the rate of sedimentation.

Natural communities also undergo this process of sedimentation and tend to develop into a second type of *Salix* thicket. This community can be found in places where flooding is less frequent. Between successive inundations (often only in winter) the upper layers of the sediment become aerated and soil formation is initiated. The substrate still remains basic in character and poor in organic material. Analyses by Zonneveld (1960) showed the following: organic matter 1.5–7.0%, C/N ratio 13, SO_4^{2-} 20–25 μeq g^{-1}, $CaCO_3$ 5.5–7.5%, pH 7.3–7.4. In the uppermost few centimetres alternating layers of inorganic sediment (deposited in winter and spring) and thin sheets of organic material (leaf litter) were observed (Meijer Drees, 1936; Schwickerath, 1954).

The *Salix* species mentioned for the low-lying *Salix* communities remain dominant in the tree layer, but *Populus nigra* may become locally co-dominant. As the development of a herb layer is not blocked by frequent flooding, many species can enter the community. *Calystegia sepium*, *Poa trivialis*, *Solanum dulcamara*, *Symphytum officinale*, *Urtica dioica*, and *Valeriana officinalis* are the most frequent ones. Other common species are listed in Table 17.4. In this community (*Salici–Populetum*) the percentage of hemicryptophytes is higher and those of therophytes and hydrophytes are lower than in the foregoing community (Fig. 17.6b).

When regular flooding is stopped by embankment, a development towards communities dominated by *Alnus glutinosa* takes place (Schwickerath, 1933; Kop, 1961). This type of wetland forest was preferred by foresters for conversion into coppice-wood. In 1937, Tüxen already stated that only small fragments remained. Schwickerath (1944) and Kop (1961) also found it difficult to give good descriptions of these woodlands because of their rarity.

Both in the low lying and in the more elevated *Salix* communities the number of epiphytes may be great. On the stems of the trees a gradient in frequency of flooding exists. This flooding causes the lower part of the trunks to be coated with a layer of clay. This gradient gives rise to a number of vertically zoned bryophyte communities on the trunks and lower branches. Brand and During (1972) noted over 60 species of mosses and liverworts, and distinguished six zones. In frequently flooded thickets, the lowermost zone (0–40 cm) was characterized by the dominance of *Fissidens bryoides*. The second zone, from 20–50 cm, was dominated by *Amblystegium riparium*. This zone was followed by either a *Brachythecium rutabulum*-dominated zone or (on exposed northern sides) a *Cirriphyllum crassinervium*-dominated zone, both occurring from ca. 50–100 cm. The next, fifth zone, was dominated by *Leskea polycarpa*, and was present from 80 to 130 cm. The sixth zone was called the "isolated patches" zone and was not characterized by the dominance of any species. Both in this and the lowermost zone, total cover of the bryophytes was generally less than 25%. In the other zones total cover nearly always was 100%.

In the more elevated *Salix* communities, the lowermost zone was dominated by *Brachythecium rutabulum* (0–70 cm), directly followed by a zone dominated by *Leskea polycarpa* (50–100 cm). Here an "isolated patches"-zone was also present.

In planted and managed willow coppice the gradually decomposing organic material in the broad stools and crowns of pollarded trees offers a foothold to species that normally do not occur as epiphytes. *Cardamine pratensis*, *Geum urbanum*, *Polypodium vulgare*, and several grasses are the commonest.

Other forest types

Leaving the frequently flooded zone, a wealth of vegetation types is found, belonging to the Fraxinetalia, that are not regularly flooded, but are still under the influence of frequently occurring high groundwater levels. Factors governing their composition are mainly the frequency of flooding, the depth of the alluvial layers, and the carbonate content of the flood water. Groundwater levels vary from >1 m in depth in July to <20 cm in depth from October to April (Trautmann and Lohmeyer, 1960; Schlottmann, 1966). The soil contains a higher percentage of organic material (25%; Zonneveld, 1960) and has a lower pH (5.0–6.0; Schlottmann, 1966; Duvigneaud and Denayer-De Smet, 1970b) than under the *Salix* communities. In the drier types, groundwater levels frequently fall below −50 cm and the soil is well aerated for the major part of the growing season. Decomposition of organic material is fast and humus-rich layers build up slowly. The soil, as a rule, is alkaline.

Important tree species in these communities are *Acer pseudoplatanus*, *Fraxinus excelsior*, *Populus alba*, *P. canescens*, *P. tremula*, *Prunus avium*, *Quercus robur*, *Tilia platyphyllos*, *Ulmus laevis*, and *U. minor*. A full-grown mature stand may have two tree layers, a layer of lianas and shrubs, a tall herb layer, and a field layer, giving an intricate structural network (Krausch, 1960; Trautmann and Lohmeyer, 1960; Carbiener, 1970). Carbiener (1970) noted 40 different tree species in such a community and concluded that this most strongly structured woodland type approaches the complexity of temperate primeval forests from other parts of the world.

Species composition and characteristics of these woods, which are only marginal to wetland forests, were also described by Gaume (1924), Jouanne (1926), Chouard (1927), Malquit (1929), Schwickerath (1933), Lemée (1937, 1939a), Tüxen (1937), Bartsch and Bartsch (1940), Braun-Blanquet and Tüxen (1952), Lohmeyer (1953), Oberdorfer (1957), Fukarek (1961), Kop (1961), Schlottmann (1966), Passarge and Hofmann (1968), Tüxen and Dierschke (1968), Westhoff and Den Held (1969), Klein (1975), Dierschke (1979b), Dinter (1982) and Mayer (1984). This list of authors is far from complete but their works present data on these communities well spread over the area surveyed.

Growing conditions for many tree species in these drier ecosystems are optimum, therefore only relatively few stands are left in their natural state due to the demand for firewood and timber. Many sites have been planted with *Alnus glutinosa*, *Fraxinus excelsior*, or *Populus* species, and hybrids or cultivars of the latter. Pons (1976, 1977a, b) studied processes in the field layer of *Fraxinus excelsior* coppice-wood. He compared growth in the woodland plant *Geum urbanum* with that of a plant of felled areas, *Cirsium palustre*. Before coppicing, *Geum* showed a high rate of growth and photosynthesis in the spring light phase and an adaptation to shade conditions in summer. The first year after coppicing, *Geum* showed stunted growth and partly disappeared. *Cirsium palustre*, on the other hand, germinated after coppicing and demonstrated a high assimilation rate and rapid vegetative growth, which led to abundant blooming in the second year of the coppice cycle. The influence of light intensity on vegetative growth in both species could give no sufficient explanation for the observed phenomena in the field, as seedlings of *Cirsium* showed shade-tolerant characteristics and seedlings of *Geum* grew well in full daylight under laboratory conditions. In the generative phase *Geum* adapted rapidly after the plants had been transferred from a high to a low light intensity. Dry weight increase in *Cirsium* ceased after such a transfer and the plants failed to complete their life-cycle.

Because the rotation period of felling is longer in the planted stands on the drier sites than in the willow coppices, these plantations acquire a structure and species composition that becomes comparable to those of more natural woods.

Peatland forests

When drainage of flood waters or rainwater is prevented by geological, topographic, or soil conditions, decomposition rates of litter are reduced, and an organic layer is formed. Under nutrient-poor conditions this may ultimately lead to the establishment of *Sphagnum* bog. In places where run-off water leaves the bog, this water carries nutrients from the center of the bog to its edges, and, although this water will generally still be poor in nutrients, the vegetation relying on it as a source of nutrients is not strongly oligotrophic. A constant flow of low quantities of nutrients may support a vegetation which, without this constant flow, would need a substrate considerably richer in nutrients.

Such vegetation types, however, often form the end of a line of development in a mineral-rich environment. Ample availability of nutrient-rich water and frequent inundation may lead to the establishment of a vegetation dominated by tall grasses like *Phragmites australis* or tall sedges like *Carex riparia*, *C. paniculata*, or *C. elata* (the latter two both tussock-forming species). In base-rich water, *Cladium mariscus* may be the dominant species. While the base of tussocks may be almost permanently inundated, their tops may reach several decimetres above the mean water level. This elevated part of the community offers a foothold to many species that do not thrive in a permanently aquatic habitat. Scamoni (1960) and Fukarek (1961) described such mixed communities which are often treated as a mosaic of aquatic vegetation and swamp-carr. *Alnus glutinosa* is the most important tree species, growing on top of the tussocks. When the trees become larger, their situation becomes very unstable and they have to support one another in order not to topple over. A frequent non-aquatic species that occurs in places where the soil between the tussocks is not permanently inundated, is *Calamagrostis canescens*. *Carex elata*, *C. paniculata*, and *C. pseudocyperus* are the main tussock-building species. On top of the tussocks, *Galium palustre*, *Lysimachia vulgaris*, and *Thelypteris palustris* are very common. Between the tussocks the water level seldom completely falls below the soil surface (Fukarek, 1961) and the aquatic species *Hottonia palustris*, *Lemna minor*, *L. trisulca*, and *Utricularia vulgaris* are often present. The peat between the tussocks is soft and may be > 0.5 m in depth. The soil–water reaction is only slightly acid.

Invasion of such communities by *Carex riparia* brings about the formation of a firm root-mat and a relatively rapid growth of the peat layer. This leads to a strong reduction of the area with open water and the disappearance of the larger aquatic species. The vegetation now resembles *Alnus glutinosa* swamp-carr developed from stands dominated previously by non-tussocky species. However, the tussocks present initially may remain visible for more than 10 years.

This type of community is widespread over Europe. *Alnus glutinosa* is the dominant tree but *Salix atrocinerea* (especially in the British Isles), *S. aurita*, and *S. cinerea* may be frequent and reminiscent of the development of this community from shrub stages in which these species once were dominant.

In the western part of the area surveyed of Europe, *Myrica gale* communities may mark the first stages of the development to wetland forests on the edges of ombrotrophic bogs. Going eastward, *Salix cinerea* becomes more important (Hild, 1960). Surveys of these initial stages (belonging to the Calamagrostio–Salicetalia cinereae) are given, for instance, by Allorge (1921, 1922c), Meijer Drees (1936), Tüxen (1937), Lambert (1951), Braun-Blanquet and Tüxen (1952), Hild (1960), Kop (1961), Westhoff and Den Held (1969), Wheeler (1980), O'Connell (1981), and Mayer (1984).

A great variety of regional community-types can be distinguished. In the British Isles they find their greatest development in the Norfolk Broads (Lambert, 1965; Wheeler, 1978). The frequent occurrence of the species *Bryonia cretica* subsp. *dioica*, *Cladium mariscus*, *Osmunda regalis*, and *Rhamnus catharticus* (Wheeler, 1978, 1980) makes it clear that these communities receive a supply of rather basic water that has a relatively high content of calcium carbonate.

A swamp-forest community, characterized by the presence of many species with an Atlantic distribution (very common in the British Isles and becoming rarer on the continent in easterly direction) is found not only in the British Isles, but also in the western parts of France, Belgium, and The Netherlands. *Athyrium filix-femina*, *Carex laevigata*, and *Juncus effusus* have a high presence in the herb layer of this woodland type. Other frequent species are listed in Table 17.5. These communities (belonging to the *Carici laevigatae–Alnetum*) show (Fig. 17.7a) have a low percentage of hydrophytes and a relatively high percentage of herbaceous chamaephytes (Figs. 17.7 and 17.8), compared with the *Alnus glutinosa* spring-woodland (Fig. 17.7b) and *Alnus glutinosa* woodland on nutrient-poor soils (Fig. 17.8a). Allorge (1922b) gave the composition of such woodlands as follows: 20.5% phanerophytes, 6.5% chamaephytes, 57% hemicryptophytes, 12% geophytes, and 4% therophytes. These data correspond very well with those from Fig. 17.7a.

Chouard (1924) mentioned the occurrence of a high percentage of shade plants in these communities. The woodlands are characterized by neutral or

TABLE 17.5

The most frequent species in nutrient-rich *Alnus glutinosa* woodland on peat[1] and in *Alnus glutinosa* spring-woodland[2]. (See notes in Table 17.3)

Nutrient-rich *Alnus glutinosa* woodland			*Alnus glutinosa* spring-woodland
V	*Alnus glutinosa*		V
III	*Galium palustre*		V
IV	*Athyrium filix-femina*		III
IV	*Juncus effusus*		III
III	*Cirsium palustre*		III
IV	*Carex laevigata*	IV	*Deschampsia cespitosa*
III	*Betula pubescens*		*Ranunculus repens*
	Blechnum spicant		*Urtica dioica*
	Calamagrostis canescens	III	*Caltha palustris*
	Frangula alnus		*Dryopteris carthusiana*
	Osmunda regalis		*Filipendula ulmaria*
	Sphagnum palustre (m)		*Lysimachia vulgaris*
			Mnium hornum (m)
			Poa trivialis
			Solanum dulcamara

[1] 67 relevés; after tables from Chouard (1924, 1932), Jouanne (1926), Claessens (1935), Schwickerath (1944), and Bodeux (1955).
[2] 25 relevés; after tables from Jeswiet et al., 1933; Meijer Drees, 1936; Lemée, 1939b; Schwickerath, 1944; Scamoni, 1960; Dierschke, 1969b.

only slightly acid conditions in the soil (pH 5.8–7.7; Lemée, 1939c) and a fairly good supply of nutrients. The peat layer varies in depth from a few decimetres only to almost 1.5 m, and may consist mainly of woodland peat formed by the *Alnus* woodland itself (Lohmeyer, 1960).

A community type characterized in the herb-layer by *Carex elongata*, *Peucedanum palustre*, and *Thelypteris palustris* can be found in the greater part of the area surveyed. The most frequent associating species is *Calamagrostis canescens*. Other frequent species from this nutrient-poor *Alnus glutinosa* fen woodland are listed in Table 17.6. Hydrophytes make up more than 10% of the total species number of this woodland type (*Carici elongatae–Alnetum*; Fig. 17.8).

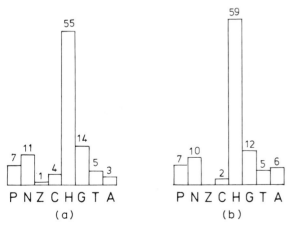

Fig. 17.7. Life-form spectra of nutrient-rich *Alnus glutinosa* woodland on peat (a) and *Alnus glutinosa* spring-woodland (b) according to tables mentioned in Table 17.5. See legend of Fig. 17.5.

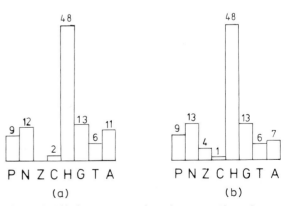

Fig. 17.8. Life-form spectra of nutrient-poor *Alnus glutinosa* woodland on peat (a) and *Alnus glutinosa* – *Betula pubescens* woodland (b), according to tables mentioned in Table 17.6. See legend of Fig. 17.5.

TABLE 17.6

The most frequent species in nutrient-poor *Alnus glutinosa* woodland on peat[1], and in *Alnus glutinosa–Betula pubescens* woodland on peat[2]. (See notes in Table 17.3)

Nutrient-poor *Alnus glutinosa* woodland			*Alnus glutinosa–Betula pubescens* woodland
V	*Alnus glutinosa*		IV
III	*Dryopteris carthusiana*		V
IV	*Calamagrostis canescens*		III
III	*Lysimachia vulgaris*		IV
III	*Mnium hornum* (m)		IV
III	*Rubus fruticosus* s.l.		III
III	*Cirsium palustre*	IV	*Betula pubescens*
	Galium palustre		*Frangula alnus*
	Iris pseudacorus		*Sorbus aucuparia*
	Lycopus europaeus		
	Lythrum salicaria		
	Phragmites australis		
	Salix cinerea		
	Solanum dulcamara		
	Thelypteris palustris		
	Urtica dioica		

[1]278 relevés; after tables from Chouard (1924), Gaume (1925), Meijer Drees (1936), Van Zinderen-Bakker (1942), Westhoff (1949), Buchwald (1951a, b), Bodeux (1955), Van Dijk (1955), Fukarek (1961), Kop (1961), Dierschke (1969b, 1979a), Smittenberg and De Roos (1974), Smittenberg (1976), Wiegers and De Vries (1982), and Wiegers (1982).
[2]130 relevés; after tables from Lemée (1937), Van Zinderen-Bakker (1942), Buchwald (1951a, b, 1953), Bodeux (1955), Dierschke (1969b), Smittenberg and De Roos (1974), Smittenberg (1976), and Wiegers (1984).

The peat layer under these communities may vary from < 50 cm to > 1.5 m in depth. The substrate is poor in nutrients and the pH ranges from 4.0 to 6.5 (Lemée, 1939c). Particularly in The Netherlands, these communities have developed in old peat cuttings on a floating root mat formed by tall helophytes. *Carex riparia*, *Cladium mariscus*, *Phragmites australis*, *Typha angustifolia*, or *T. latifolia* often initiate the developments with dense monospecific stands that gradually change to communities richer in species. On a thin layer of recently formed peat a *Salix–Frangula* community is the precursor of a swamp-carr dominated by *Alnus glutinosa*. When a fairly thick peat layer forms as a result of a protracted use of the initial reed swamp for reed-cutting and subsequent haymaking, the establishment of a *Betula–Frangula* scrub (sometimes with *Myrica gale*) succeeded by a fen-carr dominated by *Betula pubescens* is more likely (Wiegers, 1982, 1985).

A recent survey of woodland communities on floating root mats ("kragge") in The Netherlands indicates that they are related to the *Carici elongatae–Alnetum*, but form a separate association, characterized by a high frequency of *Sphagnum* species, *Dryopteris carthusiana*, *Lysimachia thyrsiflora*, and *Thelypteris palustris*. These woodlands are rich in mosses and liverworts (unpublished data, J. Wiegers).

A fenwood dominated by *Betula pubescens* is reported from all parts of the area surveyed. It differs from the *Betula* woodland on ombrotrophic peat by being similar to the *Alnus glutinosa* community previously mentioned both in floristic composition and in mineral content of the peat layer. The most frequent species are *Alnus glutinosa*, *Betula pubescens*, *Calamagrostis canescens*, *Dryopteris carthusiana*, *Frangula alnus*, *Lysimachia vulgaris*, *Mnium hornum*, *Rubus fruticosus* s.l., and *Sorbus aucuparia* (Table 17.6). This type (*Alno–Betuletum*) is characterized by a higher percentage of woody chamaephytes than the foregoing woodland types (Fig. 17.8b).

Sphagnum-species may cover a considerable

area, both in the *Alnus* woodland and in the *Alnus–Betula* communities. The peat surface is uneven with uprooted tree bases, fallen stems, and rotting stumps, often completely covered with mosses and liverworts. The highest points of the moss cover may reach more than three decimetres over low-lying, wet spots which are often filled with water throughout the year. The peat layer is rather unstable and large trees may become partially or completely uprooted by heavy winds, thereby creating new pools. The trees may even, by their own weight, be pushed down into the peat, so that a circular pool around the tree-base is formed. *Betula* has a smaller chance to survive under such conditions than *Alnus* as it is intolerant of prolonged flooding.

In the moss carpet, *Sphagnum squarrosum* is found in the wettest places. Higher sites are occupied by *S. fimbriatum* and *S. palustre* (both building low hummocks) and, on an already established *Sphagnum*-peat layer, *S. capillifolium* and *S. recurvum* may become the dominant species in almost level carpets. In the end the *Sphagnum* layer may become completely closed, hiding all the micro-relief of the underlying peat. Seedlings of trees are rare in such a closed *Sphagnum* carpet, although germination often takes place. The young plants become quickly overtopped by the mosses and within a few weeks after germination all seedlings have disappeared, being buried in the moss layer (Wiegers, 1985). It is possible that such a community could develop into a treeless bog, since with the increase of the depth of the peat layer the storage of rainwater is improved and the community tends to rely more and more on ombrotrophic conditions.

The opinion that a *Betula–Sphagnum* wood is a successional stage of an *Alnus–Sphagnum* wood has been defended by several authors (e.g., De Vries, 1969). This opinion is, however, commonly based on the comparison of different communities and not on observations from the same site over a sufficiently long period. Repeated mapping of permanent plots during a period of 10 years (unpublished data from J.H. Smittenberg, 1970–1980) or descriptions of relevés with a time lag of 10 to 15 years (Wiegers, 1982, 1985) cannot show clearly that succession actually takes place. Development of both types on different substrates from different communities fed by different types of groundwater, as argued by Westhoff (1949), seems more likely for these communities on floating peat-rafts. Development in these swamp forests over a period of 10 years was depicted by Wiegers (1982) using structure-diagrams (Fig. 17.9) following Dansereau et al. (1966). This type of diagram emphasizes the differences that originate from development in the vegetation by leaving out all the rarely occurring but rather constantly present structure-types that might blur the picture.

The *Betula* fenwoods in the United Kingdom differ from those on the continent by having *Dryopteris cristata* as a frequent species. *Salix atrocinerea*, a common species in fenwoods in the British Isles, has a more southerly distribution on the continent, being common only in fenwoods in the middle and southern parts of France.

Betula fenwoods on floating peat-rafts in The Netherlands differ from those described for Germany and France by the high frequency of *Salix aurita* and *S. cinerea* and their hybrids. Fenwoods with *Betula pubescens* as the dominant tree species may have a closed groundlayer of *Sphagnum* species and occur on firm peat layers with a pH ranging from 3.5 to 5.0 (Buchwald, 1951a; Fukarek, 1961; Van den Berg and De Smidt, 1985). The substrate is poor in nutrients and most species in the herb layer are not able to reach underlying, more nutrient-rich, but anaerobic layers [either mineral soil or free water under the floating peat-raft ("kragge")]. Groundwater levels are, throughout the year, within 30 cm from the surface (Schwickerath, 1944). Movement of water (predominantly rain water fallen on the vegetation) in the peat layer is generally only downward, thereby carrying soluble products, originating from decomposition of leaf litter, down into the anaerobic zone and building up in course of time a body of oligotrophic water. *Betula pubescens* itself has its main root mass in the upper, temporarily aerated zone, and always has a mycorrhizal symbiont. Prolonged inundation lasting for nearly the whole growing season leads to the death of this species. *Alnus* and *Salix* species can survive even when the major part of the root system is inundated throughout the year.

A wetland forest community resembling more closely the type dominated by *Alnus glutinosa*, but occurring on shallow peat, and characterized by a

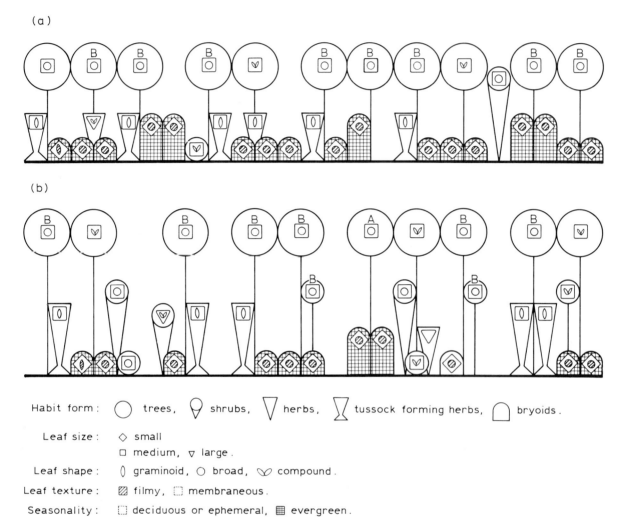

Fig. 17.9. Structure-diagrams of *Alnus glutinosa – Betula pubescens* woodland in the Weerribben (The Netherlands) (from Wiegers, 1982). (a) structure-diagram of a community in 1970, and (b) structure-diagram of the same site in 1980. Symbols according to Dansereau et al. (1966) and Wiegers (1982). In the tree symbols B stands for *Betula* and A for *Alnus*.

horizontal or upwelling groundwater flux, is the *Alnus glutinosa* springwood. It differs from communities with stagnant groundwater in having *Cardamine amara*, *Chrysosplenium alternifolium*, and *C. oppositifolium* as characteristic species. *Athyrium filix–femina*, *Caltha palustris*, *Deschampsia cespitosa*, *Filipendula ulmaria*, *Juncus effusus*, *Poa trivialis*, and *Ranunculus repens* frequently occur. Other common species are listed in Table 17.5. The life-form spectrum of this type (Fig. 17.7b) closely resembles that from the nutrient-rich *Alnus glutinosa* woods.

In their natural state these swamp- and fen-woods are commonly not under forest management. Dead trees remain standing as long as their trunk holds out and there is no room for them to fall lengthwise. The moss and liverwort flora on both dead and living boles is well developed. On living trunks, *Hypnum cupressiforme*, *Lophocolea heterophylla*, and *Mnium hornum* are very common. *Aulacomnium androgynum* and *Tetraphis pellucida* are frequent on rotting stumps. In a more advanced stage of rotting the wood may become completely hidden under dense cushions of liverworts.

Forest management in these woods requires extensive drainage, a practice which leads to mineralization of the upper peat layers. After drainage, species such as *Fraxinus excelsior* and *Populus tremula* can successfully be planted. In the herb layer, *Allium ursinum*, *Mercurialis perennis*, and *Stachys sylvatica* become characteristic species. These communities can, as has been argued before, no longer be considered to be wetland forests.

The shrub stages that precede these woodland types on more or less oligotrophic peat are not only the *Salix–Frangula* shrubland already mentioned, but they may also include a community dominated by *Myrica gale*. This type occurs on peat that is poor in nutrients and has low pH. These communities are inundated in winter, whereas in summer small pools remain behind between the tussocks of *Molinia caerulea*, the dominant species in the herb layer. *Myrica* and *Molinia* may be accompanied by two different species groups. Where nutrients are now readily available, *Cirsium palustre*, *Dryopteris carthusiana*, *Galium palustre*, *Hydrocotyle vulgaris*, *Lysimachia vulgaris*, *Peucedanum palustre*, *Potentilla palustris*, *Salix cinerea*, and *Sphagnum recurvum* are frequent. When nutrient availability is low, *Betula pubescens*, *Calluna vulgaris*, *Erica tetralix*, *Frangula alnus*, *Narthecium ossifragum*, *Sphagnum fimbriatum*, *S. palustre*, and *S. papillosum* are the commonest species (Hild, 1960; Dierschke, 1969b, 1979a).

The *Myrica* shrub community in its nutrient-poor form is often found fringing fenwoods dominated by *Betula pubescens*. Such woods not only occur in isolated and relatively oligotrophic sites in nutrient-rich peatlands but also in the enriched zones of ombrotrophic bogs (lagg-zones and edges of gullies), or at places in these bogs where the hydrology has been disturbed (e.g., by the digging of a drainage system). In these communities, woody chamaephytes form more than 10% of the total species number (Fig. 17.10a). Under young or scattered stands of *Betula*, a closed bryophyte layer may be present (mostly *Sphagnum* spp.), but old and dense stands may be almost completely devoid of such a moss carpet. The layer of raw humus and freshly fallen leaves, twigs, and catkins may be several centimetres thick. Although the raw humus layer generally does not become completely dry in summer, water capacity of the upper litter layer is very low. In this layer the seeds of *Betula pubescens* have insufficient

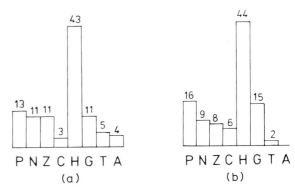

Fig. 17.10. Life-form spectra of *Betula pubescens* fenwoodland (a), according to tables from Hild (1960) and Dierschke (1969b, 1979a), and of *Quercus – Betula* woodland (b), according to tables from Meijer Drees (1936), Tüxen (1937), Schwickerath (1944), Buchwald (1951b), Scamoni (1960), and Fukarek (1961). See legend of Fig. 17.5.

water supply for successful germination and establishment (Wiegers, 1985). Moderate irrigation of an undisturbed site greatly enhanced germination while the numbers of seedlings in unwatered controls remained at zero. Lack of available water may thus be the main reason why such *Betula* stands are nearly always even-aged. In a single year the availability of water permitted the seedlings to establish and extend their roots far enough downward to reach a zone with sufficient water supply. After a young shrub layer had established, the superficial roots of the plants then consumed all the available water in the litter layer, rendering the site even more inhospitable for future seedlings.

The most frequent associating species in these *Betula* communities are *Deschampsia flexuosa*, *Dicranum scoparium*, *Dryopteris carthusiana*, *Frangula alnus*, *Molinia caerulea*, *Pleurozium schreberi*, *Quercus robur*, *Sorbus aucuparia*, *Trientalis europaea*, and *Vaccinium myrtillus* (Tüxen, 1937; Schwickerath, 1944; Buchwald, 1951a, b; Oberdorfer, 1957; Fukarek, 1961; Dierschke, 1969b, 1979a). These *Betula* communities have become commoner during this century because large areas of ombrotrophic bogs have been reclaimed, and the small remnant areas that are left to nature conservation suffer from a seriously disturbed hydrology. After lowering of the groundwater table, *Quercus robur* appears in these communities (Buchwald, 1951b) and development towards a mixed *Quercus–Betula* woodland type seems possible (Dierschke, 1979a).

In montane and continental regions, *Picea abies* or *Pinus sylvestris* are the dominant trees in these forest types (Schwickerath, 1953; Schlüter, 1969; Kaule, 1974). In the shrub layer *Ledum palustre* and *Vaccinium uliginosum* are characteristic species. Groundwater levels are generally within 20 cm from the surface throughout the year (Schlüter, 1969). *Eriophorum vaginatum* and ericacious shrubs form a more or less closed layer over a *Sphagnum* carpet. Trees are often stunted and do not form a closed tree layer. These bog communities with scattered trees gradually pass into treeless bog vegetation.

In the subalpine zone, where shrub communities are the climax of wooded vegetation, *Pinus mugo* is the dominant shrub (Bartsch and Bartsch, 1940; Kaule, 1974). Micropatterns in the dwarf shrub and herb layer originate from the relief in the *Sphagnum* layer, caused by the growth form of the species. *Sphagnum magellanicum* and *S. capillifolium* form low carpets, but *S. fuscum* and *S. papillosum* build hummocks several decimetres in height, just like *Leucobryum glaucum*. The resulting differences in hydrological conditions give rise to an uneven distribution of the species in the herb and dwarf shrub layer.

Because there is only palynological evidence that forest communities can be succeeded by treeless ombrotrophic bogs, the conclusion seems justified that under recent climatic conditions the *Betula*, *Pinus*, and *Picea* woodlands on ombrotrophic peat represent a climatically determined climax in vegetation development.

Forests on wet sandy soils

A last type of wetland forest which is neither a peatland forest nor a riverine forest, is the *Quercus robur–Betula* woodland on wet, sandy soils. Both *Betula pendula* and *B. pubescens* are common in these communities. The upper layers of the mineral soil are acid (pH 4–5; Fukarek, 1961) and their organic content is low. In the upper 15 cm, an admixture of organic material is recognisable (Meijer Drees, 1936; Schwickerath, 1944; Buchwald, 1951b), but there is no sharp transition to the underlying layers. The groundwater table is generally within 50 cm of the surface (Meijer Drees, 1936), but may show appreciable fluctuations. Groundwater flow is practically absent.

Calluna vulgaris, *Carex pilulifera*, *Deschampsia flexuosa*, *Dicranum scoparium*, *Fagus sylvatica*, *Frangula alnus*, *Holcus mollis*, *Leucobryum glaucum*, *Luzula pilosa*, *Maianthemum bifolium*, *Melampyrum pratense*, *Molinia caerulea*, *Pinus sylvestris*, *Pleurozium schreberi*, *Polytrichum formosum*, *Pseudoscleropodium purum*, *Pteridium aquilinum*, *Rubus fruticosus* s.l., *Sorbus aucuparia*, and *Vaccinium myrtillus* are the most frequent associating species (Meijer Drees, 1936; Tüxen, 1937; Schwickerath, 1944; Buchwald, 1951b; Scamoni, 1960; Fukarek, 1961). This forest type is found on wet places in mixed *Quercus* woods, but may also establish outside existing forests (Duvigneaud et al., 1971). A natural development of the shrub layer is often prevented by forestry practices and frequently the sites have been "improved" for better yield. Hydrophytes are completely absent from these woodland communities, whereas the percentage of tree species is high (Fig. 17.10).

THE ANIMAL COMPONENT

The heavy impact of man on nature, all over lowland Europe, has left wetland forests generally only as small units. Most forests, even when forming part of a larger reserve, are too small to house viable populations of larger grazers or carnivores. Smaller animals in these ecosystems have to adapt their way of life to frequent floodings, not only caused by heavy rain on poorly drained soils, but, especially in riverine forests, as a yearly phenomenon which urges most of the soil-inhabiting fauna either to move to safer areas or to be prepared to survive a long period under water.

This section only aims at mentioning a few examples of species belonging to diverse systematic groups, and does not claim to be a full treatment of the fauna characteristic of or occurring in wetland forest ecosystems.

Mammals

The largest indigenous herbivore is the roe deer (*Capreolus capreolus*). In summer deer graze mainly in open vegetation adjacent to woodlands. In winter, shrubs and bark of saplings may constitute a substantial part of their diet. By nibbling all the branch tops from shrubs and peeling the bark from the stems of the saplings, they may exert a distinct influence on the rate of woodland succession. In some cases they may even prevent the establishment of closed woodland for prolonged periods (De Waard, 1979).

In order to influence forest succession and to open up areas too much overgrown by woodland, the Dutch State Forestry Service started in 1972 a grazing experiment with a northern cattle breed, the Swedish *fjäll-ko* in a reserve in the northwestern part of Overijssel (The Netherlands). Initially five pregnant cows were allowed to graze an area of 22 ha, partly consisting of woodland. Already after a few years it became clear that they were able to remove dense undergrowth and to survive in the swampy fen woodland (Hanskamp, 1974). In spring they browse often on the bark of *Salix* spp., which generally form the first successional stage in woodland development there.

In olden times one of the characteristic animals of wetland forests was the beaver (*Castor fiber*). Severe hunting led to almost complete extinction of this species in Western Europe. In The Netherlands the last beaver was shot in 1825 (Van Wijngaarden, 1966). Along the river Elbe (West Germany) a single colony survived. Recently the species was reintroduced in northeastern France and in north- and southeastern Germany (Smit and Van Wijngaarden, 1981). Reintroduction in Belgium or The Netherlands is impeded both by poor water quality and lack of a suitable area of sufficient size to house a viable population (Van Wijngaarden, 1983).

The common otter (*Lutra lutra*) is, like the beaver, an endangered species (Smit and Van Wijngaarden, 1981). It occurs all over Europe in wetland areas, where it uses the woodlands mainly for shelter. Even in suitable areas, however, population numbers have considerably decreased. They seem to suffer from poor water quality.

Both the raccoon dog (*Nyctereutes procyonoides*) and the North American raccoon (*Procyon lotor*) have been introduced into Western Europe, the former from Asia, the latter from America (Lyneborg, 1972). Both are able to survive in Western Europe, and prefer wetland habitats (Bannikov, 1964; Kampmann, 1975).

Introduced rodents are the muskrat (*Ondatra zibethica*) and the nutria (*Myocastor coypus*). The former comes from North America and is intensively hunted, as its habit of digging extensive burrows in embankments of waterways may seriously threaten human safety (Doude van Troostwijk, 1976). It does not occur in the British Isles. The nutria originates from South America (Klapperstück, 1954) and seems to be only of local occurrence. Its numbers are strongly affected by severe winters.

Indigenous rodents that prefer wetland habitats are the European water vole (*Arvicola terrestris*), which is absent from Ireland, the tundra vole (*Microtus oeconomus*) in The Netherlands, Belgium and eastern Germany, and the bank vole (*Clethrionomys glareolus*) (Creveld, 1972).

Insectivorous species found in wetland forests are, among others, the European water shrew (*Neomys fodiens*), the common pipistrelle bat (*Pipistrellus pipistrellus*) and the serotine bat (*Vespertilio serotinus*).

Birds

Wetland forests, being generally left to develop naturally, often have three or more vegetation strata, each of which is preferred by different bird species. According to general surveys of the European avifauna some 40 species are, at least regionally, more or less common breeding birds of wetland forests (Peterson et al., 1965; Makatsch, 1974). In this survey the emphasis will be on regional differences in the species composition of the avifauna in the study area. More detailed data on the ecology of the species can be found in the literature cited.

In the western part of the continent and on the British Isles the black kite (*Milvus migrans*) is generally absent. In winter it may be an occasional visitor. As a breeding bird the hooded crow (*Corvus corone cornix*) is found not only in the eastern part of the study area but also in the northwestern part of the British Isles. A third species which is almost absent from the lowlands is the dipper (*Cinclus cinclus*). It is mainly found in the lower mountain ranges where it forages in shallow streams.

Some species have their main distribution in Europe in the southern or southwestern parts, but have outposts in The Netherlands. The purple heron (*Ardea purpurea*) is rather widespread in southern Europe. Its colonies in The Netherlands are mainly found in the western and northwestern parts of the country. The species prefers low woodland and scrub (Den Held and Den Held, 1976). The smaller night heron (*Nycticorax nycticorax*) is more or less restricted to river valleys. A few small colonies are known in The Netherlands (Van den Bergh, 1983). Unlike these two species, which breed and feed in wetland forest areas or in close vicinity to them, the spoonbill (*Platalea leucorodia*) and the continental cormorant (*Phalacrocorax carbo sinensis*) use forested wetlands only for nesting. Their foraging grounds (fresh and slightly brackish waters) may be more than 10 km away from the breeding colonies. The spoonbill is present in The Netherlands with slightly over 200 breeding pairs. Its nearest colonies elsewhere in the continent are found in southern Spain and southeastern Europe (Poorter, 1983). The continental form of the cormorant does not, like its Atlantic relative, occupy rocky cliff coasts. It has its main distribution in southeastern Europe, but there are over 5000 breeding pairs in The Netherlands, which constitute more than 50% of the total European population (Coomans de Ruiter, 1966; Timmerman, 1983).

Some species which are fairly widespread on the continent do not occur in the British Isles. Amongst these are the golden oriole (*Oriolus oriolus*), the goshawk (*Accipiter gentilis*), the icterine warbler (*Hippolais icterina*), the little bittern (*Ixobrychus minutus*), the short-toed tree creeper (*Certhia brachydactyla*), and the white-spotted bluethroat (*Cyanosylvia svecica cyanecula*).

Even more species are absent from Ireland. Among the birds of prey the buzzard (*Buteo buteo*), the hobby (*Falco subbuteo*), and the tawny owl (*Strix aluco*) reach the western edge of their distribution in the U.K. The same holds for the carrion crow (*Corvus corone corone*), the great and the lesser spotted woodpeckers (*Dendrocopos major* and *D. minor*), the lesser whitethroat (*Sylvia curruca*), the marsh and the reed warblers (*Acrocephalus palustris* and *A. scirpaceus*), the nightingale (*Luscinia megarhynchos*), the redstart (*Phoenicurus phoenicurus*), the tree pipit (*Anthus trivialis*), and the willow tit (*Parus montanus*).

Among the species that are found all over the surveyed area, the grey heron (*Ardea cinerea*) and the rook (*Corvus frugilegus*) nest in colonies. These species, together with the magpie (*Pica pica*), the jay (*Garrulus glandarius*), and the wood pigeon (*Columba palumbus*), nest in the higher tree layers.

The hole-nesting birds, such as the long-eared owl (*Asio otus*), the stock dove (*Columba oenas*), the great, blue, and long-tailed tits (*Parus major*, *P. caeruleus*, and *Aegithalos caudatus*), and the spotted flycatcher (*Muscicapa striata*) are dependent on the availability of suitable nesting places, mainly found in older trees, so that they are generally absent from younger woodland stages.

The lower tree strata are inhabited by such species as the kestrel (*Falco tinnunculus*), the sparrow hawk (*Accipiter nisus*), the song and mistle thrushes and the blackbird (*Turdus philomelos*, *T. viscivorus*, and *T. merula*), the chaffinch (*Fringilla coelebs*), the bullfinch (*Pyrrhula pyrrhula*), the willow warbler and the chiffchaff (*Phylloscopus trochilus* and *P. collybita*), and the blackcap and the golden warbler (*Sylvia atricapilla* and *S. borin*).

In the shrubs and tall herbs one finds the wren (*Troglodytes troglodytes*), the dunnock (*Prunella modularis*), the robin (*Erithacus rubecula*), and the whitethroat (*Sylvia communis*). In the vicinity of open or shrubby reed-marshes they may be accompanied by the reed bunting (*Emberiza schoeniclus*), the sedge warbler (*Acrocephalus schoenobaenus*), and the grasshopper warbler (*Locustella naevia*).

Ground breeders are the woodcock (*Scolopax rusticola*) and the pheasant (*Phasianus colchicus*), which has been introduced from western Asia but has become a common resident bird all over Europe (except the extreme northern and southern parts). At the edges of open water the moorhen (*Gallinula chloropus*), the mallard (*Anas platyrhynchos*), and the teal (*Anas crecca*) may be found nesting. Steep banks along clear streams are the nesting habitat for the kingfisher (*Alcedo atthis*).

The species composition of the avifauna of wetland forest depends strongly on the successional stages of the vegetation. Young and coppiced woodland are only slightly richer in species than old woodland, but have a far higher density of breeding pairs. A second factor influencing the species composition and density is the coverage of the shrub layer (see, for instance, Van Hees, 1978; Wilson, 1978).

Reptiles and amphibians

The climatic conditions in northwestern Europe are not very suitable for the poikolithermous amphibians and reptiles. Of the c. 25 species present in the area studied, at least 15 may be found in wetland forests (Hvass, 1971). Strictly aquatic species will not be mentioned here, as their presence in these ecosystems depends on the presence of a suitable aquatic habitat. Many species mentioned in this section depend — at least for part of their life cycle, mainly reproduction and/or larval stages — on open water, which may vary in size from cart tracks to lakes. A second use of open water for them may be as a place for foraging; and in the third place they may hibernate in the mud on the bottom.

Two newts (*Triturus alpestris* and *T. marmoratus*) and the fire salamander (*Salamandra salamandra*) represent the Caudata in the study area. They are not found in the British Isles. *Salamandra salamandra* is an almost exclusively terrestrial species, with a wide distribution on the continent. *Triturus marmoratus* is restricted to southwestern Europe, whereas *T. alpestris* has its main distribution in the northwestern and central parts of continental Europe.

Among the Anura, *Bufo bufo* has the widest distribution. It is absent only from Ireland. *Alytes obstetricans* prefers mountainous habitats. It is not found in the British Isles or in the northern parts of Belgium, The Netherlands, and Germany. *Bombina variegata* is also absent from these areas, but is not restricted to mountainous habitats. *Pelodytes punctatus* occurs in western and southern France.

The not predominantly aquatic frogs are represented by four species. *Rana temporaria* is found all over the study area. *Hyla arborea* is absent from the British Isles. *Rana arvalis* is restricted to the northern, eastern and central parts of the continent. The fourth species, *Rana dalmatina*, has a predominantly southern distribution. A few scattered populations are found in northern Germany.

The only turtle to be found in the study area is *Emys orbicularis*. Its distribution is discontinuous here, covering the western parts of France and the (north)eastern parts of Germany.

Two lizards also occur in wetland forest areas. Both *Anguis fragilis* and *Lacerta vivipara* have a distribution that covers both the study area and its wide surroundings.

Of the snakes, both *Natrix natrix* and *Vipera berus* are absent from Ireland. *Natrix natrix* prefers generally a wetter and richer environment than *Vipera berus*, which has its main distribution in more nutrient-poor communities.

Other groups of animals

Data about the specific composition of the arthropod, mollusc, and earthworm faunas and the like in forested wetlands are scanty and not easily grouped into a concise overview.

The data of Rabeler (1969), Schiller (1973) and Koth (1974) make it clear that some carabid species might be more common in wetland forests than in neighbouring ecosystems, but this could be influenced more by differences in the structure of the ground layer (both low plants and litter) than by the structure of the taller vegetation.

Mörzer Bruyns (1965) and Schorer (1973) published lists of molluscs found in some vegetation types, but the data of these authors cannot be compared because they worked in widely different ecosystems.

Data on other groups of animals are even more difficult to evaluate. A general rule that can be applied to all invertebrate species is that, the longer the periods of flooding are, the poorer the fauna in wetland forests will be. As a consequence of this it is to be expected that decomposition under such circumstances will be appreciably slower than in woodlands with a well-developed soil fauna. The accumulation of coarse litter may be prevented by strong currents accompanying flooding. These currents transport the litter either higher up the embankment, or downstream where it may be deposited in places with less violent movement of the water. As in such places also fine-grained inorganic material is deposited, this process gives rise to soils with a high organic content.

A second type of woodland with strongly organic soils is the peaty woodlands, often with an extensive ground cover of *Sphagnum* spp. Here it is difficult to tell where the vegetation ends and the "soil" begins. The strongly acid conditions here also prevent the establishment of a rich soil fauna. The conditions in these ecosystems, which are very poor in calcium, are hostile for molluscs. Besides, the groundwater levels, which are often constantly high, lead to a poor soil-inhabiting fauna. Even smaller mammals are generally rare in these fen woodlands.

CONCLUSIONS

Most forested wetlands urgently need protection by nature conservancy organisations in a densely populated area like Western Europe. Their sites often seem promising for agricultural use or forestry now that systems for draining and irrigation can be highly automated, and the need for wood from fast growing species is still rising.

"Normal" development in these woods seems to suffer from the presence of introduced species in some areas. The North American species *Amelanchier laevis*, *Aronia × prunifolia* (Wiegers, 1983, 1984), and *Prunus serotina* were, for several reasons, introduced into Western Europe. These exotics grow vigorously and suppress the shrub and herb layer in many peatland forests, especially in The Netherlands.

The acrocarpous moss *Orthodontium lineare* was introduced into the British Isles in the first decade of this century. It probably came from South Africa, and was inadvertently transported to Europe with the logs on which it was growing (Margadant and During, 1982). It produces dense monospecific carpets and is replacing the rich vegetation on tree bases; it cannot be considered a desirable contribution to the Western European flora.

Animals are rare in these woods. However, such ecosystems poor in species present exquisite research areas where entangling of all the processes in the ecosystem can be studied relatively easy. There are many questions to be solved in the decades to come, in order to make progress in the way of understanding the functioning of whole ecosystems.

ACKNOWLEDGEMENTS

My thanks go to Dr. B.D. Wheeler for putting his tables at my disposal, Drs. J.H. Smittenberg for letting me use his unpublished data, the Misses J. dos Santos and E. IJssel for typing the earlier drafts of this manuscript, and H. Koerts Meijer and G. Oomen for the preparation of the drawings.

Drs. S.M. ten Houte de Lange is thanked for his critical reading of the section on the fauna of forested wetlands.

REFERENCES

Aletsee, L., 1967. Begriffliche und floristische Grundlagen zu einer pflanzengeographischen Analyse der Europäischen Regenwassermoor-Standorte. *Beitr. Biol. Pflanz.*, 43: 117–283.

Allorge, P., 1921. Les associations végétales du Vexin Français. *Rev. Gén. Bot.*, 33: 708–751.

Allorge, P., 1922a. Les associations végétales du Vexin Français. *Rev. Gén. Bot.*, 34: 134–144.

Allorge, P., 1922b. Les associations végétales du Vexin Français. *Rev. Gén. Bot.*, 34: 178–191.

Allorge, P., 1922c. Les associations végétales du Vexin Français. *Rev. Gén. Bot.*, 34: 425–431.

Armstrong, W., 1964. Oxygen diffusion from the roots of some British bog plants. *Nature*, 204: 801–802.

Bannikov, A.G., 1964. Biologie du chien viverrin en URSS. *Mammalia*, 28: 1–39.

Bartsch, J. and Bartsch, M., 1940. Vegetationskunde des Schwarzwaldes. *Pflanzensoziologie* 4. Fischer, Jena, 229 pp.

Bocock, K.L. and Gilbert, O.J.W., 1957. The disappearance of litter under different woodland conditions. *Plant Soil*, 9: 179–185.

Bodeux, A., 1955. Alnetum glutinosae. *Mitt. Floristisch-soziologische Arbeitsgemeinschaft, N.F.*, 5: 114–137.

Brand, M. and During, H., 1972. Verslag van het voorzomerkampje 1969 in de Biesbosch. Epiphyten op knotwilgen in de getijzone. *Kruipnieuws*, 34(1–2): 2–37.

Braun, H.J., 1974. Rhythmus und Grösse von Wachstum, Wasserverbrauch und Produktivität des Wasserverbrauches bei Holzpflanzen. I. *Alnus glutinosa* (L.) Gaertn. und *Salix alba* (L.) "Liempde". *Allg. Forst- Jagdzg.*, 145: 81–86.

Braun-Blanquet, J. and Tüxen, R., 1952. Irische Pflanzengesellschaften. *Veröff. Geobot. Inst. Eidg. Tech. Hochsch. Stift. Rübel Zürich*, 25: 244–415.

Bray, J.R. and Gorham, E., 1964. Litter production in forests in the world. *Adv. Ecol. Res.*, 2: 101–157.

Buchwald, K., 1951a. Bruchwaldgesellschaften im Grossen und Kleinen Moor, Forstamt Danndorf(Drömling). *Angew. Pflanzensoziol.*, 2: 1–46.

Buchwald, K., 1951b. Wald- und Forstgesellschaften der Revierförsterei Diensthoop, Forstamt Syke b. Bremen. *Angew. Pflanzensoziol.*, 1: 1–72.

Buchwald, K., 1953. Erläuterungen zur Naturlandschaftskarte des Naturschutzgebietes "Blankes Flat" bei Vesbeck und seiner näheren Umgebung. *Mitt. Floristisch-soziologische Arbeitsgemeinschaft, N.F.*, 4: 125–136.

Carbiener, R., 1970. Un exemple de type forestier exceptionnel pour l'Europe occidentale: la fôret du lit majeur du Rhin au niveau du Fossé Rhénan (*Fraxino–Ulmetum* Oberd. 53).

Intérêt écologique et biogéographique. Comparaison à d'autres forêts thermophiles. *Vegetatio*, 20: 97–148.
Chouard, P., 1924. Monographies phytosociologiques. I. La région de Brigueil l'Aîne (Confolentais). *Bull. Soc. Bot. Fr.*, 71(9/10): 1130–1158.
Chouard, P., 1927. Monographies phytosociologiques. II. La végétation des environs de Yonnerre (Yonne) et des pays jurassiques au S.-E. du basin de Paris. *Bull. Soc. Bot. Fr.*, 74(1/2): 44–64.
Chouard, P., 1932. Associations végétales des forêts de la Vallée de l'Apance (Haute Marne). *Bull. Soc. Bot. Fr.*, 79(7/8): 617–634.
Claessens, B., 1935. Étude phytosociologique de la région de Termonde. *Bull. Soc. R. Bot. Belg.*, 76(2): 146–169.
Coomans de Ruiter, L., 1966. De Aalscholver, *Phalacrocorax carbo sinensis* (Shaw & Nodder) als broedvogel in Nederland, in vergelijking met andere Westeuropese landen. *Limosa*, 39: 187–212.
Creveld, M.C., 1972. *Het voorkomen van zoogdieren in de uiterwaarden*. Intern Rapport, Rijksinstituut voor Natuurbeheer, Leersum, 23 pp.
Dansereau, P., 1958. A universal system for recording vegetation. *Contrib. Inst. Bot. Univ. Montréal*, 72: 1–58.
Dansereau, P., Buell, P.F. and Dagon, R., 1966. A universal system for recording vegetation. II. A methodological critique and an experiment. *Sarracenia*, 10: 1–64.
De Bakker, H. and Edelman-Vlam, A.W., 1976. *De Nederlandse Bodem in Kleur*. STIBOKA/PUDOC, Wageningen, 148 pp.
Denayer-De Smet, S., 1970. Biomasse, productivité et phytogéochimie de la végétation riveraine d'un ruisseau Ardennais. II. Aperçu phytogéochimique. *Bull. Soc. R. Bot. Belg.*, 103: 383–396.
Den Held, J.J., 1979. Beknopt Overzicht van de Nederlandse Plantengemeenschappen. *Wet. Meded. K. Ned. Natuurhist. Ver.*, 134: 1–86.
Den Held, J.J. and Den Held, A.J. (Editors), 1976. *Het Nieuwkoopse Plassengebied*. Thieme, Zutphen, 314 pp.
De Sloover, J.R., Devillez, F., Dumont, J.M., Lebrun, J. and Van Coppenolle, F., 1974. Biomasse, contenu en eau et productivité d'une boulaie pubescente en Haute-Ardenne. *Bull. Jard. Bot. Natl. Belg.*, 44: 191–218.
De Vries, H.A., 1969. Het moerasgebied ten oosten van de Loosdrechtse Plassen, zomers 1961 en 1962. In: P. Leentvaar (Editor), *De Zuidelijke Vechtplassen*. Stichting Commissie voor de Vecht en het Oostelijk en Westelijk Plassengebied, Weesp, pp. 11–44.
De Waard, J., 1979. *Een onderzoek naar het voorkomen van boomopslag op enkele percelen in het CRM-reservaat "de Weerribben" (NW-Overijssel)*. Intern Rapport Hugo de Vries-Laboratorium No. 72, Amsterdam, 58 pp.
Dierschke, H., 1969a. Pflanzensoziologische Exkursionen im Harz. *Mitt. Floristisch-soziologische Arbeitsgmeinschaft*, *N.F.*, 14: 458–479.
Dierschke, H., 1969b. Natürliche und naturnahe Vegetation in den Tälern der Böhme und Fintau in der Lüneburger Heide. *Mitt. Floristisch-soziologische Arbeitsgemeinschaft*, *N.F.*, 14: 377–397.
Dierschke, H., 1979a. Die Pflanzengesellschaften des Holtumer Moores und seiner Randgebiete (Nordwest-Deutschland). *Mitt. Floristisch-soziologische Arbeitsgemeinschaft, N.F.*, 21: 111–143.
Dierschke, H., 1979b. Laubwald-Gesellschaften im Bereich der unteren Aller und Leine (Nordwest-Deutschland). *Doc. Phytosociol.*, *N.S.*, IV: 235–252.
Dinter, W., 1982. Waldgesellschaften der Niederrheinischen Sandplatten. *Dissertationes Botanicae*, 64. Cramer, Vaduz, Liechtenstein, 111 pp.
Doing, H., 1962. Systematische Ordnung und floristische Zusammensetzung niederländischer Wald- und Gebüschgesellschaften. *Wentia*, 8: 1–85.
Doude van Troostwijk, W.J., 1976. *The Muskrat (Ondatra zibethicus L.) in The Netherlands, Its Ecological Aspects and Their Consequences for Man*. RIN-Verhandeling 7. Rijksinstituut voor Natuurbeheer, Arnhem, 136 pp.
Du Rietz, E., 1954. Die Mineralbodenwasserzeigergrenze als Grundlage einer natürlichen Zweigliederung der Nord-und Mitteleuropäischen Moore. *Vegetatio*, 5–6: 571–585.
Durin, L., 1951. Aperçu général sur la flore du Massif Forestier de Mormal. *Bull. Soc. Bot. Nord. Fr.*, 4(1): 6–14.
Duvigneaud, P., 1968. Recherches sur l'écosystème forêt. La Chênaie-Frênaie à Coudrier du Bois de Wève. Aperçu sur la biomasse, la productivité et la cycle des éléments biogènes. *Bull. Soc. R. Bot. Belg.*, 101: 111–127.
Duvigneaud, P. and Denayer-De Smet, S., 1970a. Phytogéochimie des groupes écosociologiques de Haute-Belgique. I. Essai de classification phytochimique des espèces herbacées. *Oecol. Plant.*, 5(1): 1–32.
Duvigneaud, P. and Denayer-De Smet, S., 1970b. Biomasse, productivité et phytogéochimie de la végétation riveraine d'un ruisseau Ardennais. I. Aperçu sur les sols, la végétation et la biomasse de la strate au sol. *Bull. Soc. R. Bot. Belg.*, 103: 353–382.
Duvigneaud, P., Tanghe, M., Denayer-De Smet, S. and Dubois, F., 1971. Le terril no. 7 de Chapelle-lez-Herlaimont. Site, végétation et principaux biotopes. *Bull. Soc. R. Bot. Belg.*, 104: 301–321.
Ellenberg, H., 1978. *Vegetation Mitteleuropas mit den Alpen*. Eugen Ulmer, Stuttgart, 2nd ed., 982 pp.
Fukarek, F., 1961. Die Vegetation des Darss und ihre Geschichte. *Pflanzensoziologie*, 12. Fischer, Jena, 321 pp.
Gaume, R., 1924. Les associations végétales de la forêt de Preuilly (Indre-et-Loire). *Bull. Soc. Bot. Fr.*, 71(1/2): 158–171.
Gaume, R., 1925. Aperçu sur les groupements végétaux du plateau de Brie. *Bull. Soc. Bot. Fr.*, 72(3/4): 393–416.
Godwin, H., 1975. *The History of the British Flora*. Cambridge University Press, Cambridge, 541 pp.
Gore, A.J.P. (Editor), 1983. *Mires: Swamp, Bog, Fen and Moor*. Ecosystems of the World, 4. Elsevier, Amsterdam, Part A, 440 pp., Part B, 479 pp.
Grosse-Brauckmann, G., 1974. Zum Verlauf der Verlandung bei einem eutrophen Flachsee (nach quartärbotanischen Untersuchungen am Steinhuder Meer). I. Heutige Vegetationszonierung, torfbildende Pflanzengesellschaften der Vergangenheit. *Flora*, 163(3): 179–229.
Grosse-Brauckmann, G., 1976. Ablagerungen der Moore. In: K. Göttlich (Editor), *Moor- und Torfkunde*. Schweizerbart'sche Verlagsbuchhandlung, Stuttgart, pp. 91–133.
Hansen, B., Jensen, J. and Mogensen, G.S., 1978. Botanical

studies in Gammelmose, 1960–1977. Vegetation and ecology. *Bot. Tidsskr.*, 73(1): 1–19.
Hanskamp, B., 1974. *Experimentele begrazing met fjell-koeien in de Weerribben.* Intern Rapport, Rijksinstituut voor Natuurbeheer, Leersum, 32 pp.
Hild, J., 1960. Verschiedene Formen von *Myrica*-Beständen am unteren Niederrhein. *Ber. Dtsch. Bot. Ges.*, 73(2): 41–49.
Hook, D.D. and Crawford, R.M.M., 1978. *Plant Life in Anaerobic Environments.* Ann Arbor Science, Ann Arbor, Mich., 564 pp.
Hvass, H., 1971. *Krybdyr i Farver.* Politikens Forlag, Kobenhavn, 176 pp.
Jennings, J.N. and Lambert, J.M., 1951. Alluvial stratigraphy and vegetational succession of the Bure Valley Broads. I. Surface features and general stratigraphy *J. Ecol.*, 139: 106–119.
Jeswiet, J., De Leeuw, W.C. and Tüxen, R., 1933. Über Waldgesellschaften und Bodenprofile. *Ned. Kruidkundig Arch.*, 43: 293–333.
Jouanne, P., 1926. Quelques associations végétales de l'Ardenne schisteuse. *Bull. Soc. R. Bot. Belg.*, 59: 54–68.
Kampmann, H., 1975. *Der Waschbär: Verbreitung, Ökologie, Lebensweise, Jagd.* Parey, Berlin, 76 pp.
Kaule, G., 1974. Die Übergangs- und Hochmoore Süddeutschlands und der Vogesen. *Dissertationes Botanicae*, 27. Cramer, Lehre, 435 pp.
Kestemont, F., 1974. Production ligneuse de la strate arbustive du terril no. 7 de Chapelle-lez-Herlaimont. *Bull. Soc. R. Bot. Belg.*, 107: 245–257.
Klapperstück, J., 1954. *Der Sumpfbiber (Nutria).* Ziemsen, Wittenberg, 54 pp.
Klein, J., 1975. *An Irish Landscape.* Ph.D. Dissertation, R.U. Utrecht, 268 pp.
Klötzli, F., 1970. Eichen-, Edellaub- und Bruchwälder der Britischen Inseln. *Schweiz. Z. Forstwes.*, 121: 329–366.
Kop, L.G., 1961. Wälder und Waldentwicklung in alten Flussbetten in den Niederlanden. *Wentia*, 5: 86–111.
Koth, W., 1974. Vergesellschaftungen von Carabiden (Coleoptera, Insecta) bodennasser Habitate des Arnsberger Waldes verglichen mit Hilfe der Renkonen-Zahl. *Abh. Landesmus. Naturkd. Münster Westfalen*, 36(3): 3–38.
Krausch, H.D., 1960. *Die Pflanzenwelt des Spreewaldes.* Ziemsen, Wittenberg, 124 pp.
Lambert, J.M., 1951. Alluvial stratigraphy and vegetational succession of the Bure Valley Broads. III. Classification, status and distribution of communities. *J. Ecol.*, 39: 149–170.
Lambert, J.M., 1965. The vegetation of Broadland. In: E.A. Ellis (Editor), *The Broads.* Collins, London, pp. 69–92.
Lambert, J.M. and Jennings, J.N., 1951. Alluvial stratigraphy and vegetational succession of the Bure Valley Broads. II. Detailed vegetational–stratigraphical relationships. *J. Ecol.*, 39: 120–148.
Lebrun, J., Noirfalise, A., Heinemann, P. and Vanden Berghen, C., 1949. Les associations végétales de Belgique. *Bull. Soc. R. Bot. Belg.*, 82(8): 105–207.
Lemée, G., 1937. Recherches écologiques sur la végétation de Perche. *Rev. Gén. Bot.*, 49: 730–751.
Lemée, G., 1939a. Recherches écologiques sur la végétation du Perche. *Rev. Gén. Bot.*, 51: 103–126.
Lemée, G., 1939b. Recherches écologiques sur la végétation du Perche. *Rev. Gén. Bot.*, 51: 380–384.
Lemée, G., 1939c. Recherches écologiques sur la végétation du Perche. *Rev. Gén. Bot.*, 51: 428–448.
Lemée, G., 1967. Investigations sur la minéralisation de l'azote et son évolution annuelle dans des humus forestiers in situ. *Oecol. Plant.*, 2(4): 285–324.
Lieth, H., 1965. Versuch einer kartographischen Darstellung der Produktivität der Pflanzendecke auf der Erde. In: *Geographisches Taschenbuch 1964/1965.* Steiner Verlag, Wiesbaden, pp. 72–80.
Lieth, H., 1970. Phenology in productivity studies. In: D.E. Reichle (Editor), *Analysis of Temperate Forest Ecosystems.* Springer Verlag, Berlin, pp. 29–46.
Lohmeyer, W., 1953. Beitrag zur Kenntnis der Pflanzengesellschaften in der Umgebung von Höxter a.d. Weser. *Mitt. Floristisch-soziologische Arbeitsgemeinschaft*, N.F., 4: 59–76.
Lohmeyer, W., 1957. Der Hainmieren–Schwarzerlenwald (*Stellario–Alnetum glutinosae* (Kästner 1938)). *Mitt. Floristisch-soziologische Arbeitsgemeinschaft*, N.F., 6/7: 247–257.
Lohmeyer, W., 1960. Zur Kenntnis der Erlenwälder in der nordwestlichen Randgebieten der Eifel. *Mitt. Floristisch-soziologischen Arbeitsgemeinschaft*, N.F., 8: 209–221.
Lyneborg, L., 1972. *Wilde Zoogdieren in Europa.* Moussault, Amsterdam, 256 pp.
Maas, F.M., 1959. Bronnen, bronbeken en bronbossen van Nederland, in het bijzonder die van de Veluwezoom. *Meded. Landbouwhogesch. Wageningen*, 59(12): 1–166.
Makatsch, W., 1974. *Thieme's Handboek voor alle Europese Vogels.* Thieme, Zutphen, 536 pp.
Malquit, G., 1929. Les associations végétales de la vallée de la Lanterne. *Arch. Bot.*, II(6): 1–211.
Margadant, W.D. and During, H.J., 1982. *Beknopte Flora van de Nederlandse Blad- en Levermossen.* Thieme, Zutphen, 517 pp.
Matthews, J.D., 1976. The development of forest science. In: T.H. Coaker (Editor), *Applied Biology I.* Academic Press, London, pp. 49–88.
Mayer, H., 1984. *Wälder Europas.* Fischer Verlag, Stuttgart, 691 pp.
McVean, D.N., 1955. Ecology of *Alnus glutinosa* (L.)Gaertn. II. Seed distribution and germination. *J. Ecol.*, 43: 61–71.
McVean, D.N., 1956a. Ecology of *Alnus glutinosa* (L)Gaertn. III. Seedling establishment. *J. Ecol.*, 44: 195–218.
McVean, D.N., 1956b. Ecology of *Alnus glutinosa* (L.)Gaertn. IV. Root system. *J. Ecol.*, 44: 219–225.
Meijer Drees, E., 1936. *De Bosvegetatie van de Achterhoek en enkele aangrenzende Gebieden.* Ph.D. Dissertation, L.H. Wageningen, 171 pp.
Miles, J., 1981. *Effect of Birch on Moorlands.* Institute of Terrestrial Ecology, Cambridge, 18 pp.
Mörzer Bruyns, M.F., 1965. Over de Malacofauna van het Naardermeer. *Basteria*, 29(1/4): 36–43.
Noirfalise, A. and Sougnez, N., 1961. Les forêts riveraines de Belgique. *Bull. Jard. Bot. État Bruxelles*, 31(2): 199–287.
Oberdorfer, E., 1957. Süddeutsche Pflanzengesellschaften. *Pflanzensoziologie*, 10. Fischer, Jena, 564 pp.
O'Connell, M., 1981. The phytosociology and ecology of Scragh Bog, Co. Westmeath. *New Phytol.*, 87(1): 139–187.

Osvald, H., 1925. Die Hochmoortypen Europas. *Veröff. Geobot. Inst. Eidg. Tech. Hochsch. Stift. Rübel Zurich*, 3: 707–723.

Overbeck, F., 1975. *Botanisch-geologische Moorkunde*. Wachholtz, Neumünster, 719 pp.

Ovington, J.D., 1962. Quantitative ecology and the woodland ecosystem concept. *Adv. Ecol. Res.*, 1: 103–192.

Ovington, J.D. and Madgwick, H.A.I., 1959a. The growth and composition of natural stands of birch. 1. Dry-matter production. *Plant Soil*, 10(3): 271–283.

Ovington, J.D. and Madgwick, H.A.I., 1959b. The growth and composition of natural stands of birch. 2. The uptake of mineral nutrients. *Plant Soil*, 10(4): 389–400.

Ozenda, P. (Editor), 1979. *Vegetation Map (scale 1:3 000 000) of the Council of Europe Member States*. Nature and Environment Series, 16. Council of Europe, Strasbourg, 99 pp. + 3 maps.

Pals, J.P., Van Geel, B. and Delfos, A., 1980. Palaeoecological studies in the Klokkeweel bog near Hoogkarspel (prov. of Noord-Holland). *Rev. Palaeobot. Palynol.*, 30: 371–418.

Passarge, H. and Hofmann, G., 1968. Pflanzengesellschaften des nordostdeutschen Flachlandes II. *Pflanzensoziologie*, 16. Fischer, Jena, 298 pp.

Petersen, P.M., 1980. Changes in the vascular plant flora and vegetation in a protected Danish mire, Maglemose, 1913–1979. *Bot. Tidsskr.*, 75(1): 77–88.

Peterson, R.T., Mountfort, G. and Hollom, P.A.D., 1965. *A Field Guide to the Birds of Britain and Europe*. Collins, London, 344 pp.

Pons, T.L., 1976. An ecophysiological study in the field layer of ash coppice. I. Field measurements. *Acta Bot. Neerl.*, 25(6): 401–416.

Pons, T.L., 1977a. An ecophysiological study in the field layer of ash coppice. II. Experiments with *Geum urbanum* and *Cirsium palustre* in different light intensities. *Acta Bot. Neerl.*, 26(1): 29–42.

Pons, T.L., 1977b. An ecophysiological study in the field layer of ash coppice. III. Influence of diminishing light during growth on *Geum urbanum* and *Cirsium palustre*. *Acta Bot. Neerl.*, 26(3): 251–263.

Poorter, E.P.R., 1983. Lepelaar (*Platalea leucorodia*). In: Rijksinstituut voor Natuurbeheer (Editor), *Dieren*. Pudoc, Wageningen, pp. 63–65.

Rabeler, W., 1969. Über die Käfer- und Spinnenfauna eines Nordwestdeutschen Birkenbruchs. *Vegetatio*, 18(1/6): 387–392.

Rehder, H., 1970. Zur Ökologie, insbesondere Stickstoffversorgung subalpiner und alpiner Pflanzengesellschaften im Naturschutzgebiet Schachen (Wettersteingebirge). *Dissertationes Botanicae*, 6. Cramer, Lehre, 90 pp.

Reichle, D.E., 1970. *Analysis of Temperate Forest Ecosystems*. Ecological Studies 1. Springer Verlag, Berlin, 304 pp.

Rothmaler, W., 1976. *Exkursionsflora, Bd. 4*. Volk u. Wissen, Berlin, 811 pp.

Scamoni, A., 1960. *Waldgesellschaften und Waldstandorte*. Akademie-Verlag, Berlin, 326 pp.

Schiller, W., 1973. Die Carabidenfauna des Naturschutzgebietes Heiliges Meer, Kr. Trecklenburg. *Nat. Heimat*, 33(4): 111–118.

Schlottmann, C.P., 1966. Die Pflanzengesellschaften des Gaarder Bauernwaldes (Kreis Südtondern). *Mitt. Arbeitsgemeinschaft Floristik Schleswig-Holstein und Hamburg*, 14: 3–129.

Schlüter, H., 1969. Hochmoorgesellschaften im Thüringer Wald. *Mitt. Floristisch-soziologische Arbeitsgemeinschaft, N.F.*, 14: 346–364.

Schnelle, F., 1955. *Pflanzen-Phänologie*. Akad. Verlagsges., Leipzig, 299 pp.

Schorer, G., 1973. Qualitative und quantitative Untersuchung der Landgastropoden des Siebengebirges und des Rodderberges in ausgewählten Biotopen. *Decheniana*, 126(1/2): 69–90.

Schwickerath, M., 1933. Die Vegetation des Landkreises Aachen und ihre Stellung im nördlichen Westdeutschland. *Aachener Beitr. Heimatkd.*, 13: 1–135.

Schwickerath, M., 1944. Das Hohe Venn und seine Randgebiete. *Pflanzensoziologie*, 6. Fischer, Jena, 278 pp.

Schwickerath, M., 1953. Hohes Venn, Zitterwald, Schneifel und Hunsrück, ein vegetations-, boden- und landschaftskundlicher Vergleich der vier westlichen Waldgebirge des Rheinlands und seines Westrandes. *Mitt. Floristisch-soziologische Arbeitsgemeinschaft, N.F.*, 4: 77–87.

Schwickerath, M., 1954. *Die Landschaft und ihre Wandlung auf geobotanischer und geographischer Grundlage entwickelt und erläutert im Bereich des Messtischblattes Stolberg*. Rudolf Georgi, Aachen, 118 pp.

Smit, C.J. and Van Wijngaarden, A., 1981. *Threatened Mammals in Europe*. Akad. Verlagsges., Wiesbaden, 259 pp.

Smith, A.J.E., 1978. *The Moss Flora of Britain and Ireland*. Cambridge University Press, Cambridge, 706 pp.

Smittenberg, J.H., 1976. De vegetatie van het moerasboscomplex "De Suikerpot" te Kortenhoef. In: P.A. Bakker, C.A.J. van der Hoeven-Loos, L.R. Mur and A. Stork (Editors), *De Noordelijke Vechtplassen*. Stichting Commissie voor de Vecht en het Oostelijk en Westelijk Plassengebied, Vlaardingen, pp. 181–195.

Smittenberg, J.H. and De Roos, G.Th., 1974. Moerasbossen in de Lindevallei. Intern Rapport R.I.N., Leersum, 31 pp.

Sprent, J.I., Scott, R. and Perry, K.M., 1978. The nitrogen economy of *Myrica gale* in the field. *J. Ecol.*, 66(2): 657–668.

Steubing, L., 1970. Soil flora: Studies of the number and activity of micro-organisms in woodland soils. In: D.E. Reichle (Editor), *Analysis of Temperate Forest Ecosystems*. Springer Verlag, Berlin, pp. 131–146.

Tansley, A.G., 1939. *The British Islands and their Vegetation*. Cambridge University Press, Cambridge, 930 pp.

Timmerman, A., 1983. Aalscholver (*Phalacrocorax carbo*). In: Rijksinstituut voor Natuurbeheer (Editor), *Dieren*. Pudoc, Wageningen, pp. 44–48.

Trautmann, W. and Lohmeyer, W., 1960. Gehölzgesellschaften in der Fluss-Aue der mittleren Ems. *Mitt. Floristisch-soziologische Arbeitsgemeinschaft, N.F.*, 8: 227–247.

Tutin, T.G., Heywood, V.H., Burges, N.A., Valentine, D.H., Walters, S.M. and Webb, D.A., 1964–1980. *Flora Europaea, Vol. 1–5*. Cambridge University Press, Cambridge.

Tüxen, R., 1937. Die Pflanzengesellschaften Nordwestdeutschlands. *Mitt. Floristisch-soziologische Arbeitsgemeinschaft Niedersachsen*, 3: 1–170.

Tüxen, R., 1957. Der Geissbart–Schwarzerlenwald (*Arunco–Alnetum glutinosae* (Kästner 1938)). *Mitt. Floristisch-soziologische Arbeitsgemeinschaft, N.F.*, 6/7: 258–263.

Tüxen, R. and Dierschke, H., 1968. Das Bullerbachtal in Sennestadt, eine pflanzensoziologische Lehranlage. *Mitt. Floristisch-soziologische Arbeitsgemeinschaft, N.F.*, 13: 227–243.

Van den Bergh, L.M.J., 1983. Kwak (*Nycticorax nycticorax*). In: Rijksinstituut voor Natuurbeheer (Editor), *Dieren*. Pudoc, Wageningen, pp. 53–55.

Van den Berg, W.J. and De Smidt, J., 1985. *De vegetatie van het Oostelijk Vecht plassengebied, 1935–1980*. Stichting Commissie voor de Vecht en het Oostelijk en Westelijk Plassengebied, 155 pp. + 10 app.

Van der Woude, J.D., 1981. *Holocene Palaeoenvironmental Evolution of a Perimarine Fluviatile Area*. Ph.D. Dissertation, V.U., Amsterdam, 136 pp.

Van Dijk, J., 1955. Bosvegetaties en bosvorming in het Kortenhoefse veengebied. In: W. Meijer and R.J. de Wit (Editors), *Kortenhoef*. Stichting Commissie voor de Vecht en het Oostelijk en Westelijk Plassengebied, Amsterdam, pp. 60–66.

Van Hees, A.F.M. (Editor), 1978. *Bosbeheer, vegetatie en avifauna in enkele bosgebieden in Midden-Brabant*. Rapport 159, Rijksinstituut voor Onderzoek in de Bos- en Landschapsbouw "De Dorschkamp", Wageningen, 83 pp.

Van Wijngaarden, A., 1966. De bever, *Castor fiber* L., in Nederland. *Lutra*, 8(3): 33–52.

Van Wijngaarden, A., 1983. Bever (*Castor fiber*). In: Rijksinstituut voor Natuurbeheer (Editor), *Dieren*. Pudoc, Wageningen, pp. 242–243.

Van Wirdum, G., 1977. Natuurtijd en Zomerklokjes: Maak van de Oude Maas geen Nieuwe. In: Werkgroep Oude Maas (Editors), *De Oude Maas als Groene Rivier*. Openb. Lich. Rijnmond, Rotterdam, pp. 51–78.

Van Wirdum, G., 1979. Veen, venen en moerassen. In: Rijksinstituut voor Natuurbeheer (Editor), *Levensgemeenschappen*. Pudoc, Wageningen, pp. 99–139.

Van Wirdum-Daan, C. and Van Wirdum, G., 1977. Vegetatie. In: Werkgroep Oude Maas (Editor), *De Oude Maas als Groene Rivier*. Openb. Lich. Rijnmond, Rotterdam, pp. 79–105.

Van Zinderen Bakker, E.M., 1942. *Het Naardermeer*. De Lange, Amsterdam, 255 pp.

Walker, D., 1970. Direction and rate in some British Postglacial hydroseres. In: D. Walter and R.G. West (Editors), *Studies in the Vegetational History of the British Isles*. Cambridge University Press, Cambridge, pp. 117–140.

Walter, H., 1973. *Die Vegetation der Erde, Bd. I: Die tropischen und subtropischen Zonen*. Fischer, Stuttgart, 3rd ed., 743 pp.

Walter, H., Harnickell, E. and Mueller-Dombois, D., 1975. *Climate-diagram Maps*. Springer-Verlag, Berlin, 36 pp. + 8 maps.

Westhoff, V., 1949. *Landschap, Flora en Vegetatie van de Botshol nabij Abcoude*. Stichting Commissie voor de Vecht en het Oostelijk en Westelijk Plassengebied, Baambrugge, 102 pp.

Westhoff, V., 1981. Ombrotrofe venen in Ierland. *Versl. Afd. Natuurkd. K. Ned. Akad. Wet.*, 90(6): 41–46.

Westhoff, V. and Den Held, A.J., 1969. *Plantengemeenschappen in Nederland*. Thieme, Zutphen, 324 pp.

Wheeler, B.D., 1978. The wetland plant communities of the River Ant Valley, Norfolk. *Trans. Norfolk Norwich Nat. Soc.*, 24(4): 153–187.

Wheeler, B.D., 1980. Plant communities of rich-fen systems in England and Wales. III. Fen meadow, fen grassland and fen woodland communities, and contact communities. *J. Ecol.*, 68: 761–788.

Wiegers, J., 1982. Untersuchungen zum Verhalten von *Betula pubescens* Ehrh. in Mooren in den Niederlanden. I. Prozesse der Vegetationsentwicklung in Niedermoor-Bruchwäldern in Nordwest-Overijssel. In: H. Dierschke (Editor), *Berichte Internationales Symposium Struktur und Dynamik von Wäldern, Rinteln 1981*. Cramer, Lehre, pp. 275–297.

Wiegers, J., 1983. *Aronia* Medik. in The Netherlands. I. Distribution and taxonomy. *Acta Bot. Neerl.*, 32(5/6): 481–488.

Wiegers, J., 1984. *Aronia* Medik. in The Netherlands. II. Ecology of *A.* × *prunifolia* (Marsh.)Rehd. in the Dutch Haf District. *Acta Bot. Neerl.*, 33(3): 307–322.

Wiegers, J., 1985. Succession in Fen Woodlands in the Dutch Haf District, with special reference to *Betula pubescens* Ehrh. *Dissertationes Botanicae*, 86. Cramer, Vaduz, 152 pp.

Wiegers, J., 1988. Some bryophyte communities in a wetland forest: their species composition, spatial relations and their groundwater characteristics. *International Symposium Dependent Plant Communities*, Wageningen, 1984, pp. 135–149.

Wiegers, J. and De Vries, E.J., 1982. *Salix pentandra* L. in Nederland. *Gorteria*, 11(1): 4–14.

Williams, B.L., Cooper, J.M. and Pyatt, D.G., 1978. Effects of afforestation with *Pinus contorta* on nutrient content, acidity, and exchangeable cations in peat. *Forestry*, 51(1): 29–36.

Willmanns, O., 1977. Verbreitung, Soziologie und Geschichte der Grün-Erle (*Alnus viridis* (Chaix)DC.) im Schwarzwald. *Mitt. Floristisch-soziologische Arbeitsgemeinschaft, N.F.*, 19/20: 323–341.

Wilson, J., 1978. The breeding bird community of willow scrub at Leighton Moss, Lancashire. *Bird Study*, 25(4): 239–244.

Zonneveld, I.S., 1960. De Brabantse Biesbosch. *Bodemkundige Studies*, 4 A(1–210), B(1–396), C(maps and tables). Ph.D. Dissertation, Wageningen.

Zonneveld, I.S., 1961a. Een blik in de toekomst. In: C.J. Verhey (Editor), *De Biesbosch, Land van het Levende Water*. Thieme, Zutphen, pp. 223–226.

Zonneveld, I.S., 1961b. De vegetatie. In: C.J. Verhey (Editor), *De Biesbosch, Land van het Levende Water*. Thieme, Zutphen, pp. 51–82.

Zwerver, S., Bovenkerk, M. and Mak, P.J., 1984. Verzuring: nationale en internationale ontwikkelingen. In: E.H. Adema and J. van Ham (Editors), *Zure Regen*. PUDOC, Wageningen, pp. 194–203.

Chapter 18

FORESTED WETLANDS OF POLAND

ANDRZEJ J. SZCZEPAŃSKI†

INTRODUCTION

Because timber from forested wetlands in Poland has little economic worth, very few investigations have been made on these ecosystems. The knowledge of forested wetlands is mainly limited to phytosociological aspects. However, soil descriptions, litter production, and litter decomposition rates are available for a limited number of sites. Fig. 18.1 shows the majority of sites on which phytosociological relevés of wet forest have been undertaken.

More than 27.5% of Poland is forested (Prończuk, 1979) and about 5% of the country is covered with wetlands, but no more than 1.5 to 2.2% can be recognized as forested wetlands (J.B. Faliński, personal communication, 1979). Three main types are recognized:

(a) Willow–shrub wetland: a stage in the development of alder or birch swamps from marshes or fens;

(b) Alder swamps: the main type of forested wetland;

(c) Pine swamps: found on mineral-poor raised bogs.

There are many transitional stages between these wetlands and adjacent vegetation where flooding is of shorter duration.

Riverside willow thickets and woody fringes bordering water bodies and water-courses are also included. Other forest types in moist places, such as the associations alder–ash carr or Alno–Padion, will be discussed only in relation to the above-mentioned groups. The reason for such an approach is that associations in the Alno–Padion, or forests on river alluvia, are usually flooded every year, but for a short period only.

Thereafter, the water level subsides to more than 50 cm below the soil surface for most of the year.

My description starts with the phytosociological aspects of forested wetlands because these have been reported most extensively.

ALDER SWAMPS

The largest stands of wet alder forests, covering many hundreds of hectares, occupy broad, peaty river valleys in northern Poland, such as the lower Odra river valley, the Szczecin lagoon, and the Biebrza river swamps (Jasnowski, 1975).

Wet alder forests prefer places with stagnant water in fen sites with forest–sedge peat or forest–reed peat with considerable amounts of mineral silt. The thickness of the peat layer is from 12 to 100 cm or more, and it can be underlain with gyttja. The peat of wet alder forests is rich in nutrients, in part due to symbiotic nitrogen fixation; thus primary productivity is rather high. This is the reason why many alder swamps have been drained and converted to farmlands or to meadows. Wet alder forests are now restricted to the shores of lakes, some local depressions of the terrain, and lakes that have been filled in and covered with vegetation.

Alder peats of about 5 to 6 m in thickness found at Dobrzany, on the terrace of Wierzbiczańskie Lake, and at many other places, confirm the statement that wet alder phytocoenoses are rather stable, and under suitable conditions, can persist

---

†The author died in June 1983. His wife, Dr. Wanda Szczepańska, assisted in final revisions.

Fig. 18.1. Map of Poland with main forest areas outlined. Dots represent sites on which the phytosociological relevés of wet forest were undertaken. W, Warszawa.

for thousands of years (Marek, 1965). Marek (1965) stated that hummocks of alder trees under suitable conditions may persist for hundreds of years, enriching the substratum with woody material of root and stem origin. The formation of alder peat deposits started about 3500 B.C. No data are available for the preboreal and boreal period. In spite of their apparent stability, alder forests reach maturity at age 60 to 80 years (Obmiński, 1978), earlier than other forests in Central Europe. Sexual maturity of individual trees is reached considerably earlier.

Medwecka-Kornaś (1972) listed a number of associations, which are related to water level, to water movement, and to soil chemistry. The wettest are wet alder woods — Alnion glutinosae — which have water above ground year-round and trees which grow on hummocks. The association *Salici–Franguletum* is a successional stage of the hydrosere, and the soil is covered with stagnant water for most or all of the year. The most common edificator is *Salix cinerea*, but the birch (*Betula pubescens*) and, to a minor extent, willow (*Salix pentandra*) appear among the commonly present shrubs of *Alnus glutinosa*, *Frangula alnus*, *Salix aurita*, and *S. cinerea*.

The herb layer is composed of the helophytes *Alisma plantago-aquatica*, *Caltha palustris*, *Carex elongata*, *Comarum palustre*, *Dryopteris cristata*, *D. thelypteris*, *Lycopus europaeus*, and *Phragmites australis*. On the hummocks some mosses grow well, partially as epiphytes on basal parts of shrubs.

The top layer of soil is composed of peat, originating from reeds and/or sedges. In the hollows dy, dy-gyttja, or gyttja is deposited.

Sobotka (1967), who studied the filling-in of seepage lakes in the Suwałski district in north east Poland, found *Salici–Franguletum* around some lakes. The forest canopy consisted of *Alnus glutinosa* and *Betula pubescens*. *Salix cinerea*, with some admixture of *S. pentandra* and *Frangula alnus*, dominated the shrub layer. In the herb layer, *Calla palustris*, *Carex lasiocarpa*, *C. stricta* and *Menyanthes trifoliata* played an important role. The moss layer was composed of 17 species.

Krzywański (1974), in a study of plant succession in oxbow lakes of the Warta River, found that in the final stage of overgrowing of old river beds, forest and brushwood communities appear to belong to *Salicetum albae–S. fragilis*, *Salicetum pentandro–S. cinereae*, and *Carici elongatae–Alnetum* associations.

Oberdorfer (1965) reported on the succession of fens and marshes in the Biebrza valley. He stated that birch shrubland is the pioneer association which, on the transition bogs, leads to the wet birch forest belonging also to the alliance Alnion glutinosae.

The *Betulo–Salicetum repentis* is the meso-oligotrophic association on transition bogs, and the *Salix aurita–S. pentandra* association belongs to a more meso-eutrophic low-moor sere. Lorenc (1965) remarked that the difference between *Carici elongatae–Alnetum* and *Betulo–Salicetum repentis* is that the latter is the next successional stage in associations of small sedges which are slightly flooded.

The second stage in succession is *Carici elongatae–Alnetum*, an association of wet alder woods. This is a very dense, high forest. Alder trees grow on hummocks 2 to 3 m in diameter. Depressions between hummocks are wet and covered with water for a long time. The edificator is alder (*Alnus glutinosa*), which sometimes is accompanied by birch (*Betula pubescens*), ash (*Fraxinus excelsior*), and, in northern parts of the country, by spruce (*Picea excelsa*). Among the shrubs are *Frangula alnus*, *Sorbus aucuparia* and *Viburnum opulus*, with some willows in the nutrient-rich, wet alder woods. *Humulus lupulus* and *Ribes nigrum* also occur. The herb layer is very similar to that of the *Salici–Franguletum* association.

Matuszkiewicz et al. (1958) analyzed phytosociological surveys of Poland and divided the association *Cariceto elongatae–Alnetum* into: (a) *Cariceto elongatae–Alnetum medioeuropaeum* and (b) *Carici elongatae–Alnetum dryopterdetosum cristatae*. Matuszkiewicz et al. (1958) did not accept Bodeux's (1955) point of view, and instead distinguished four associations within the *Alnetum glutinosae*. Matuszkiewicz et al. (1958) argued that the *Cariceto elongatae–Alnetum* represents a permanent association ending the successional chain on shallow peat. The stability of the association is due to the variable, but periodically high, level of ground water, which is slightly mobile with a tendency to stagnate. Dudek (1968) suggested that this mobile groundwater contains sufficient oxygen for alder roots, and hence provides conditions for optimum growth of alder. The habitat is mesotrophic, weakly acidic (pH in the top layer of soil is 5 to 6) with inundation by an ombrophilous water source. Because of the low soil pH, the *Alnetum* has a tendency to transform into the *Pineto–Vaccinietum uliginosi* association when water levels decline.

Very similar to the wet alder forest (*Carici elongatae–Alnetum*) is an association of wet forests in which the dominant tree species is birch (*Betula pubescens*). Ellenberg (1978) was of the opinion that the change of dominance was connected with decreased alkali content in the ground water. When the calcium content falls below 0.07 mg l^{-1}, the alder is unable to compete with birch, and wet birch forests will develop, although of very low economic value. The ground water level in birch forests appears to be a little lower than in alder forests.

Because the wet birch forest grows from seed,

density and self-thinning may be very high at the beginning. In most cases it is accelerated by a parasitic fungus (*Polyporus betulinus*), which attacks the trees (Dudek, 1968).

According to Czerwiński (1972), *Dryopteri thelypteris–Betuletum pubescentis* appears on transitional peat bogs or sometimes on low peat bogs where there is ground water flow. The forest stands in northeastern Poland consist mainly of pines and *Betula pubescens* mingled with alders and spruce trees. In Czerwiński's opinion, a satisfactory forest economy is difficult and ineffective with these forests. Draining them appears ineffective in increasing their production. Their use should be limited to a yield of low-grade timber, and they should be allowed to regenerate by self-seeding. Czerwiński (1972) has placed this wet birch forest in the Alnion glutinosae alliance.

Two hardwood wet forests can be found in Poland. They are the *Vaccinio uliginosi–Pinetum* and the *Sphagno–Piceetum* (Polakowski, 1962). Woods from the association *Vaccinio uliginosi–Pinetum* occur in very wet places when the pH of the soil is very low. These woods may be found on the peripheries of raised bogs or on seepage depressions in which the level of ground water is close to the soil surface in spring, but some 50 to 60 cm below in summer.

The tree layer is composed of *Pinus* and some birches. In northeastern Poland spruce may be common. The density of trees is low and they grow poorly.

The low shrub layer is composed of *Vaccinum uliginosum* and *Ledum palustre*, among which *Sphagnum* mosses occur. There are many transitional stages between wet birch forests and wet *Pinus* forests. The wet *Pinus* forest occurs on a deeper peat layer (often >1 m thick) than the wet birch forest (Matuszkiewicz, 1963).

In the Białowieża Forest, the wet alder forest is often replaced by *Vaccinio uliginosi–Pinetum*, which is of minor economic value (Matuszkiewicz, 1952).

In the peat bogs of northeastern Poland, Polakowski (1962) distinguished an association *Piceo–Sphagnetum girgensohnii*, which has been called *Sphagno–Piceetum* in other papers (Medwecka-Kornaś, 1972). In the Białowieża Forest, this association was determined as *Sphagno girgensohnii–Piceetum* by Sokołowski (1966). This type of spruce forest occurs in northeastern Poland in depressions filled with peat. The peat layer can be several metres deep and the groundwater at depths of 20 to 80 cm is acid. The tree layer is composed of spruce (*Picea excelsa*) with some admixture of *Pinus* and birch (*Betula pubescens*). Mosses are abundant, and appear to play an important role on the forest floor. *Lycopodium* and *Vaccinium myrtillus* are also common species on the forest floor.

FORESTS OF OTHER WET SITES

Along streams and on the floodplains of rivers, some species of trees grow in fringes close to the running water or occupy part of the floodplain. Although alder and willow can play an important role in the composition of such plant associations, these cannot be considered as belonging to the alliance Alnion glutinosae because such trees or shrubs are not adapted to stagnant water. Instead, they develop in river valleys where flooding occurs for a few weeks. During the remainder of the time, the soils are moist but without water near the soil surface.

Not all types of floodplain forests or carrs in Central Europe are as wet as the wet alder forest discussed earlier. Among the floodplain forests in Poland, Matuszkiewicz and Borowik (1957) have distinguished the following associations:

(a) *Salici–Populetum*: poplar–willow forest, with *Populus alba*, *Populus nigra*, *Salix alba*, *Phalaris arundinacea*, and *Calystegia sepium*, occurring on the floodplains of large rivers.

(b) *Alnetum incanae*: *Alnus incana* forest, limited to the mountain and submountain regions.

(c) *Fraxino–Ulmetum*: *Acer campestre*, *Ulmus campestris*, *Acer pseudoplatanus*, and *Quercus robur* dominate the tree layer. Many herbs are found on the forest floor. It is found in the drier and warmer parts of central and southwestern Poland.

(d) *Circaeo–Alnetum*: an alder wood occurring mainly in northern Poland, but not as wet as a wet alder forest in the alliance Alnion glutinosae. This association has some tendency to undergo paludification. The herbs and grasses *Calamagrostis canescens*, *Cardamine amara*, *Carex elongata*, *C. remota*, *Circaea alpina*, *Crepis paludosa*, *Dryopteris thelypteris*, *Equisetum silvaticum*, *Mnium undulatum*, and *Scutellaria galericulata* are significant.

The first association is from the alliance Salicion albae and the other three are all from the alliance Alno–Padion. Based on the flora and site, all associations of the alliance Alno–Padion have an intermediate position between swamp forests of the alliance Alnion glutinosae and hornbeam woods of the alliance Carpinion.

The wettest associations from the Alno–Padion alliance are the *Circaeo–Alnetum* in the lowlands and the *Caltho–Alnetum* and the *Alnetum incanae* in sub-mountain and mountain regions, but they are less wet than wet alder forests of the *Alnetum glutinosae* association.

Along water-courses in the Carpathian Mountains the *Alnetum incanae* association occurs. It corresponds to the *Salici–Populetum* association in the lowland. In addition to *Alnus incana*, species such as ash (*Fraxinus excelsior*), bird's cherry (*Padus avium*), spruce (*Picea excelsa*), and willow (*Salix fragilis*) may also occur. These forests extend to about 900 m above sea level, but are concentrated between 400 and 600 m (Medwecka-Kornaś, 1972). The soil is rich and consists of alluvial sediments; the subsoil consists of pebbles from river beds.

During succession, the *Alnetum incanae* association is preceded by willow shrubland (Zarzycki, 1956) (Fig. 18.2). Alder forests of the association *Alnetum incanae* are heavily affected by wood harvesting and cattle grazing, especially near villages.

Wet places in the mountain forests, especially in the deciduous forest zone, are often occupied by the association *Caltho–Alnetum* (Zarzycki, 1963).

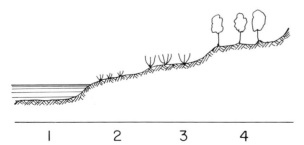

Fig. 18.2. Schematic transect across the river valley. Modified from Zarzycki (1956): (1) river bed with summer water level, (2) *Agrostis stolonifera* stage, (3) *Salix* stage, and (4) *Alnetum incanae*.

Alnus incana dominates this association and *Caltha*, *Carex remota*, *Crepis paludosa*, and *Geum rivale* are found on the forest floor. The moss layer is well developed, and the soil is marshy with peat.

In the lowlands, the small water-courses are very often fringed with alders or willows, which grow either as trees or as shrubs. In spite of their distinctiveness, these fringes can hardly be called forests. On rich alluvial soils in the valleys of streams and rivers, shrublands and floodplain forests once occupied considerable areas. Now they are mostly limited to the area between dikes. Most of the river valleys have been converted into agricultural lands or meadows.

J. Matuszkiewicz (1976) revised the communities of the alluvial forests and thickets in Poland and has distinguished nine associations. These are:

(a) *Myricaria germanica–Salix incana* association, of the alliance Salicion eleagni (Zarzycki, 1956). In this association, shrubs of *Salix purpurea* grow close to the stream on river banks which are elevated 20 to 30 cm above summer water level. Back from the stream, on banks elevated 40 to 100 cm above summer water level, the shrublands are composed of *Salix incana* and *Myricaria germanica*.

(b) *Salicetum triandro–S. viminalis*. This association is composed of shrubs of *Salix triandra*, *S. viminalis*, *S. purpurea*, *S. fragilis*, *S. alba*, and *S. amygdalina*. It occurs on sand and gravel banks within reach of the mean water level of the river. Ice floes often eradicate this association.

(c) *Salici–Populetum*. This association, a lowland alluvial forest on floodplains of large rivers, occurs within the reach of high water and is flooded briefly every year. It is composed of *Populus alba*, *P. nigra*, and some willows such as *Salix alba*, *S. fragilis*, *S. purpurea*, *S. triandra*, and *S. viminalis*. Climbing plants such as *Calystegia sepium* or *Humulus lupulus* also occur.

(d) *Ficario–Ulmetum campestris*. This association grows on flanks of river valleys in places which flood periodically. Horizontal movement of water is important, as it brings nutrients to the fine-grained rich soil.

(e) *Circaeo–Alnetum*. This occurs in moderately wet valleys of small lowland streams, on the bottom of valleys of slowly flowing streams, around springs, and on the edges of *Alnetum glutinosae*. The requirements of this association are wetness and slow water movement in slightly boggy soils. Of all the associations of the alliance Alno–Padion, this one is most similar to the *Alnetum glutinosae*.

(f) *Carici remotae–Fraxinetum*. This association is composed of *Fraxinus excelsior*, *Carpinus betulus*, and *Alnus glutinosa* and occurs in the valleys of fast-flowing streams in the submountain region.

(g) *Alnetum incanae*. This was discussed earlier.

(h) *Caltho–Alnetum*. This was discussed earlier.

(i) *Astrantio–Fraxinetum*, from the alliance Alno–Padion. This association is composed of ash, oak, alder, and sometimes poplar. It grows in the submountain region on rich soils without any signs of paludification.

The area occupied by the forests of the alliance Alno–Padion is becoming more and more limited because the water-courses are diked and soils outside the dikes are converted to agricultural use.

SUCCESSION

On the basis of research on sedimentation and filling-in of waterbodies and on the stratigraphic analyses of peat deposits in mires and wet alder forest, some information about the course of succession is available.

Dubiel (1973) presented a scheme of the succession of oxbow lakes of the Wisła River. *Carici elongatae–Alnetum* is preceded by helophytes, followed by *Circaeo–Alnetum*, and then by *Tilio–Carpinetum*. On the basis of his study of the development of vegetation on the Darss River in the German Democratic Republic, Fukarek (1961) is of the opinion that the *Alnetum* can be a sere in succession but it is more frequently a zonally distributed alder forest.

Fukarek (1961) further divided the *Alnetum* into *Hottonio–Alnetum*, *Carici–Alnetum* and *Urtico–Alnetum*, which are seral stages in succession. In the *Hottonio–Alnetum* community, the water occurs above the ground surface all year. In the *Carici–Alnetum* (Fig. 18.3A) water occurs on the surface only from autumn until spring, and in the *Urtico–Alnetum* (Fig. 18.3B) the water level is, with the exception of a very wet year, below ground surface.

The successional sequence of forests on alluvia may be as follows: initial herb associations on

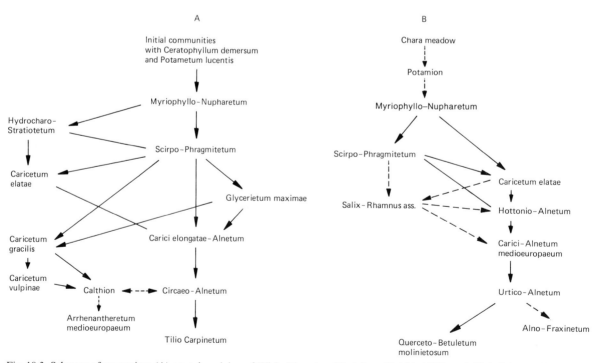

Fig. 18.3. Schemes of succession: (A) on oxbow lakes of Wisła River (modified from Dubiel, 1973); and (B) in lakes on Darss River (after Fukarek, 1961).

alluvia → *Salicetum triandro–S. viminalis* → *Salici–Populetum* → associations from alliance Carpinion → *Pino–Quercetum* (Medwecka-Kornaś 1972).

On the basis of many peat corings, Marek (1965) described some successional sequences which show changes in community development during peat formation:

(a) Potametea → Alnetea;
(b) Phragmitetea → Alnetea;
(c) Potametea → Phragmitetea → Alnetea; Potametea → Phragmitetea → Scheuchzerio–Caricetea fuscae → Alnetea;
(e) Potametea → Scheuchzerio–Caricetea fuscae → Alnetea;
(f) Potametea → Scheuchzerio–Caricetea fuscae → Alnetea → Phragmitetea;
(g) Potametea → Scheuchzerio–Caricetea fuscae → Phragmitetea → Alnetea;
(h) Phragmitetea → Alnetea → Phragmitetea → Alnetea;
(i) Phragmitetea → Alnetea → Phragmitetea → Scheuchzerio–Caricetea fuscae → Sphagnetea → Alnetea;
(j) Alnetea → Phragmitetea → Scheuchzerio–Caricetea fuscae → Sphagnetea → Alnetea;
(k) Potametea → Phragmitetea → Alnetea → Vaccinio–Piceetea;
(l) Potametea → Alnetea → Vaccinio–Piceetea (Oxycocco–Sphagnetea);
(m) Phragmitetea → Alnetea → Vaccinio–Piceetea.

Marek (1965) explained these successions by the changes of water conditions due to flood flush, changes in drainage, siltation, and mineralization of the bog surface.

In northern Poland, Jasnowski (1972) conducted extensive research concerning conditions and human-induced changes in mires. The analysis covered 1162 mires with a total area of 44 664 ha. The natural vegetation in the past was: low bogs on 41 722 ha (93.4%), transitional bogs on 938 ha (2.0%), and raised bogs on 2004 ha (4.6%). These mires are now occupied by the following forest associations: 4776 ha (10.7%) of Alnetalia glutinosae, 545 ha (1.2%) of Betuletalia pubescentia, 1281 ha (2.9%) of Ledo–Pinetalia, and 1285 ha (2.9%) of substituted managed forest. Of the mires analyzed, 54.6% are now covered with a substitute anthropogenic vegetation. In the past, wet alder forests covered an area of 10 757 ha of the study area, compared to 4776 ha today. At present some expansion of wet alder forests is occurring as a secondary association on previously ameliorated mires, as a consequence of the recurrence of paludification (Jasnowski, 1972).

SOILS

The chemical composition of peat and water from wet alder bogs has been reported by Marek (1965). He found that the nitrogen content in peat is reciprocally related to the ash content. The main material of ash consists of minerals insoluble in hydrochloric acid, most probably silica. The calcium content in peat does not correlate with ash content. The pH of water from hollows was between 7.0 and 7.6, generally higher than that of peat water (Table 18.1), and rich in nutrients (expressed in mg l^{-1}) as follows:

$Ca = 23.4 - 174.1$ $Cl = 12 - 42$ $Fe = 0.4 - 20.4$
$K = 0.4 - 8.6$ $Mg = 1.5 - 3.8$ $Mn = 0.07 - 0.60$
$Na = 0.07 - 1.48$ $P = 0.002 - 0.15$ $Si = 6.5 - 205.3$

Wojterski (1960) presented some characteristics of the soil in the association *Circaeo–Alnetum* (Table 18.2), an association similar to the wet alder swamp association *Carici elongatae–Alnetum*. The soil in both profiles is clayey sand. In profile II the bottom layer is heavy clay. Loss on ignition at 550°C from the top layer is low for both profiles. In contrast to the soil in the *Circaeo–Alnetum*

TABLE 18.1

The chemical properties of alder swamp peat (prepared from data in Marek, 1965)

Characteristic	Minimum	Maximum
Bulk density (g cm^{-3})	0.132	0.656
pH of peat water	5.1	7.3
pH of peat from hollows (in KCl solution)	5.4	7.4
pH of peat from hummocks (in KCl solution)	3.5	4.5
Ash content (% dry weight)	6.42	70.24
Organic matter (% dry weight)	29.2	93.6
N (% dry weight)	0.85	3.43
Ca (% dry weight)	0.94	3.72
P (% dry weight)	0.057	0.358
K (% dry weight)	0.017	0.481
Na (% dry weight)	0.007	0.059

TABLE 18.2

Characteristics of the soils in *Circaeo–Alnetum* and in *Vaccinio uliginosi–Pinetum ericetosum tetralicis* (after Wojterski, 1960, 1963)

Depth of horizon (cm)	Soil particle distribution (%)					pH		Loss on ignition (%)
	>1 mm	1–0.1 mm	0.1–0.05 mm	0.05–0.02 mm	<0.02 mm	in water	in KCl	
Circaeo–Alnetum association								
Profile I								
10	3.0	59	11	9	18	7.60	6.75	7.22
40	4.0	59	11	9	17	7.85	7.30	
65	8.0	65	10	5	12	7.90	7.05	
Profile II								
25	1.3	56.8	11.8	10.9	19.2	7.45	6.45	8.74
55	3.3	70.6	16.4	1.5	8.2	8.15	7.25	
90	5.0	24.7	9.0	8.1	53.2	8.15	7.05	
Vaccinio uliginosi–Pinetum ericetosum tetralicis								
	>1 mm	1–0.5 mm	0.5–0.25 mm	0.25–0.1 mm	<0.01 mm			
Variant with *Ledum palustre*								
20	0	0	40	56	4	4.1	3.5	16
40	0	0	38	59	3	4.4	3.7	9
Variant with *Myrica gale*								
5	—	—	—	—	—	3.9	3.1	47
12	0	1	28	71	0	3.6	3.0	63
25	—	—	—	—	—	3.7	3.2	7
40	0	0	33	63	4	4.6	3.9	—
Variant with *Dryopteris austriaca*								
5	—	—	—	—	—	3.6	2.6	88
20	—	—	—	—	—	3.5	2.6	57
35	0	0	11	89	0	4.7	3.5	1

association (Table 18.2), the peat of the *Carici elongatae–Alnetum* association (Table 18.1) is rich in organic substances and often more acidic. The high amount of sand makes the soil in the *Circaeo–Alnetum* association permeable to water, and consequently it drains rapidly after flooding.

Table 18.2 includes some data concerning soil of a wet *Pinus* forest. These associations are named after the occurrence of numerous *Erica tetralix*, a species limited in its Polish distribution to a narrow strip along the Baltic Sea coast. The Atlantic character of this association is also indicated by the occurrence of *Myrica gale*. Patches of *Vaccinio uliginosi–Pinetum ericetosum tetralicis* are located in depressions between coastal dunes on a soil composed of fine sand. The habitat is very oligotrophic. The peat layer, if it occurs, has a measurable amount of sand originating from the surrounding dunes. The peat is very acidic at all depths, with pH increasing slightly with depth from surface to approximately 40 cm. The upper layers of the peat are highly organic in comparison with those of the *Circaeo–Alnetum* association (Table 18.2).

LITTER PRODUCTION

At maturity (60–80 years: Obmiński, 1978), an alder forest of the *Carici elongatae–Alnetum* association produced 412 g m^{-2} yr^{-1} of litterfall, of which 97% was alder leaves (Stachurski and Zimka, 1975a). The difference in total litter production among alder woods is not great. Stachurski and Zimka (1975b) compared the leaf fall of two communities of wet alder forests, *Carici elongatae–Alnetum* and *Circaeo–Alnetum* alder carr. They obtained 337 g m^{-2} yr^{-1} for the first site and 427 g m^{-2} yr^{-1} and 399 g m^{-2} yr^{-1} for the two sites of the *Circaeo–Alnetum* association. Leaf fall for alder trees in the *Circaeo–Alnetum* was slightly lower than in the *Carici elongatae–Alnetum* (281 versus 323 g m^{-2} yr^{-1}, respectively). However, the amount of leaf fall from other tree species was much higher in the *Circaeo–Alnetum* (34.3%) than in the *Carici elongatae–Alnetum* (4.2%). Hornbeam and oak contributed considerably to litterfall in the alder carr (17.8 and 10.0%, respectively). Annual leaf litter production varied little from year to year. Stachurski and Zimka (1975a) gave the following data for three consecutive years: 353.3, 331.1, and 328.8 g m^{-2} yr^{-1}. The variation did not exceed 10% of mean annual leaf litter production.

Biomass production of the herb layer appears to vary little among wet forests (Table 18.3). The wet pine forest of Traczyk and Traczyk (1977) had a lower value of biomass production than the wet pine forest of Moszyńska (1970). Traczyk and Traczyk probably gave information for the true herbs only, while Moszyńska included, in addition to the true herbs, small shrubs such as *Vaccinium uliginosum* which contributed 48% of the production.

TABLE 18.3

Biomass production of the herb layer of different wetland forests

Type of forest	Productivity (g m^{-2} yr^{-1})	Number of species	Production of an "average" individual (g)
Alnion glutinosae[1]	133	21	0.67
Alno–Padion[1]	166	41	0.21
Dicrano–Pinion[1]	36	20	0.13
Circaeo–Alnetum[2]	107.6 ± 14	—	—
Vaccinio uliginosi–Pinetum[3]	103.8	—	—

[1]Traczyk and Traczyk (1977).
[2]Aulak (1970).
[3]Moszyńska (1970).

DECOMPOSITION AND NUTRIENT CYCLING

In a *Carici elongatae–Alnetum* association, saprophages consume as much as 69.9% and the micro-organisms 25.2% of annual leaf fall (Stachurski and Zimka, 1976b). After 1 year, 4.9% of the fallen leaves or 20 g m^{-2} remained undecomposed. The rate of decomposition is shown in Fig. 18.4. Activity of micro-organisms was highest in autumn immediately after leaf fall, whereas the saprophages are most active in the following spring. Alder leaves are rich in organic nitrogen, which constitutes 2.9 to 3.0% of their dry weight (Stachurski and Zimka, 1976a), and consequently provide a good substrate for the micro-organisms.

The dry matter of falling leaves contained 1.74% calcium, 0.29% magnesium and 0.28% potassium (Stachurski and Zimka, 1976b). The annual return of these minerals to the forest floor is 95.6 kg ha^{-1} nitrogen, 71.6 kg ha^{-1} calcium, 12.1 kg ha^{-1} magnesium, and 11.4 kg ha^{-1} potassium.

Zimka and Stachurski (1976a) compared the nutrient economy of a *Pinus* forest (*Vaccinio myrtilli–Pinetum*) with that of a wet alder forest (*Carici elongatae–Alnetum*). The *Pinus* forest is characterized by a high rate of withdrawal of nutrients from the leaves prior to abscission. The cycling of nutrients by litter decomposition is therefore low in comparison to the pool stored in

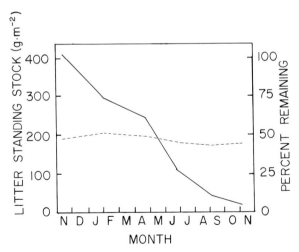

Fig. 18.4. Standing stock of litter mass (left axis; dotted line) and litter decomposition (right axis; solid line) of *Alnus glutinosa* leaves in wet alder forest (after Stachurski and Zimka, 1976a, b).

living tissue. The wet alder forest, however, is characterized by low rate of withdrawal of nutrients from leaves prior to abscission. Thus, the cycling, via the saprobiotic system of the forest floor, is rapid.

REFERENCES

Aulak, W., 1970. Studies on herb layer production in the Circaeo-Alnetum Oberd. 1953 association. *Ekol. Pol.*, 18: 411–427.
Bodeux, A., 1955. Alnetum glutinosae. *Mitt. flor.-soz. Arbeitsgem. Stolzenau/Weser*, 5: 114–137.
Czerwiński, A., 1972. Lasy brzozowe ze związku Alnion-glutinosae w Północno-Wschodniej Polsce. *Rocz. Białostocki*, 11: 101–159.
Dubiel, E., 1973. Zespoły roslinne starorzeczy Wisły w Puszczy Niepołomickiej i jej otoczeniu. *Stud. natur. Zakł. ochr. przyr. PAN, A*, 7: 67–124.
Dudek, Cz., 1968. Lasy i zadrzewienia na terenie Rolniczego Zakładu Badawczego Biebrza. *Zesz. Probl. Postępów Nauk Roln.*, 83: 81–97.
Ellenberg, H., 1978. *Vegetation Mitteleuropas mit den Alpen in ökologischen Sicht*. Verlag Eugen Ulmer, Stuttgart, 2nd ed., 982 pp.
Fukarek, F., 1961. Die Vegetation des Darss und ihre Geschichte. *Pflanzensoziologie*, 12: 1–321.
Jasnowski, M., 1972. Rozmiary i kierunki przekształceń szaty roślinnej torfowisk. *Phytocoenosis Biul. Fitosocjologiczny*, 1, 3: 193–208, W-wa Białowieża.
Jasnowski, M., 1975. Torfowiska i tereny bagienne w Polsce. In: N.J. Kac (Editor), *Bagna kuli ziemskiej*. PWN W-wa, pp. 356–390.
Krzywański, D., 1974. Zbiorowiska roślinne starorzeczy środkowej Warty. *Monogr. Bot.*, 43: 1–80.
Lorenc, K., 1965. Uber das Vorkomen einer Birkengesellschaft auf dem Torfmoor "Czerwone Bagno" (NO Polen). *Mater. Zakł. Fitosoc. Stos. UW.*, 6: 109–112, W-wa Białowieża.
Marek, S., 1965. Biologia i stratygrafia torfowisk olszynowych w Polsce. *Zesz. Probl. Postępów Nauk Roln.*, 57: 5–158.
Matuszkiewicz, J., 1976. Przeglad fitosocjologiczny zbiorowisk leśnych Polski. cz. 3. Lasy i zarośla łęgowe. *Phytocoenosis*, 5, 1: 3–66.
Matuszkiewicz, W., 1952. Zespoły leśne Białowieskiego Parku Narodowego. *Ann. Univ. Mariae Curie — Skłodowska Sect. C Suppl.*, 4: 1–218.
Matuszkiewicz, W., 1963. Zur systematischen Auffasung der oligotropen Bruchwaldgesellschaften im Osten der Pommerschen Seenplatte. *Mitt. Flor. Soc. Arbeitsgem. Stolzenau*, 10: 149–155.
Matuszkiewicz, W. and Borowik, M., 1957. Materiały do fitosocjologicznej systematyki lasów łęgowych w Polsce. *Acta Soc. Bot. Pol.*, 26: 719–756.
Matuszkiewicz, W., Traczyk, H. and Traczyk, T., 1958. Materiały do fitosocjologicznej systematyki zespołów olsowych w Posce. *Acta Soc. Bot. Pol.*, 27: 21–44.
Medwecka-Kornaś, A., 1972. Zespoły leśne i zaroślowe. In: W. Szafer and K. Zarzycki (Editors), *Szata Roślinna Polski*. PWN, Warsaw, pp. 383–440.
Moszyńska, B., 1970. Estimation of the green top production of the herb layer in a bog pine wood *Vaccinio uliginosi–Pinetum*. *Ekol. Pol.*, 18: 779–803.
Oberdorfer, E., 1965. Zur Soziologie der *Betula humilis* und *Betula pubescens*. *Mat. Zakł. Fitosoc. Stos. UW*; 6: 43–52.
Obmiński, Z., 1978. *Ekologia lasu*. PWN W-wa, 481 pp.
Polakowski, B., 1962. Bory świerkowe na torfowiskach (Zespoł *Piceo-Sphagnetum Girgensohni*) w północno-wschodniej Polsce. *Fragm. Florist. Geobot.* Kraków, 8(2): 139–156.
Prończuk, J., 1979. Podział zbiorowisk roślinnych Polski. In: J. Prończuk (Editor), *Świat roślin*. PWN W-wa, pp. 431–473.
Sobotka, D., 1967. Roślinność strefy zarastania bezodpływowych jezior Suwalszczyzny. *Monogr. Botaniczne*, 23, 2: 175–254.
Sokołowski, A.W., 1966. Phytosociological character of spruce woods in Białowieża forest. *Prace Inst. Badaw. Leśń.*, 304: 46–69.
Stachurski, A. and Zimka, J.R., 1975a. Leaf fall and the rate of litter decay in some forest habitats. *Ekol. Pol.*, 23: 103–108.
Stachurski, A. and Zimka, J.R., 1975b. Methods of studying forest ecosystems: Leaf area, leaf production and withdrawal of nutrients from leaves of trees. *Ekol. Pol.*, 23: 637–648.
Stachurski, A. and Zimka, J.R., 1976a. Methods of studying forest ecosystems: Microorganisms and saprophage consumption in the litter. *Ekol. Pol.*, 24, 1: 57–67.
Stachurski, A. and Zimka, J.R., 1976b. Methods of studying forest ecosystems: Nutrient release from the decomposing litter. *Ekol. Pol.*, 24, 2: 253–262.
Traczyk, T. and Traczyk, H., 1977. Structural characteristics of herb layer and its production in more important forest communities of Poland. *Ekol. Pol.*, 25, 3: 359–378.
Wojterski, T., 1960. Lasy liściaste dorzecza Mogilnicy w Zachodniej Wielkopolsce. *Poznan. Tow. Przyj. Nauk Pr. Kom. Biol.*, 23(2): 1–231.
Wojterski, T., 1963. Bory bagienne na Pobrzeżu Zachodnioaszubskim. *Bad. Fizjogr. Pol. Zach.*, 12: 139–191, Poznań.
Zarzycki, K., 1956. Zarastanie żwirowisk Skawicy i Skawy. *Fragm. Florist. Geobot.* Kraków, 2(2).
Zarzycki, K., 1963. Lasy Bieszczadów Zachodnich. *Acta Agrar. Silvestria Ser. Leśna 3*, Kraków.
Zimka, J.R. and Stachurski, A., 1976a. Vegetation as a modifier of carbon and nitrogen transfer to soil in various types of forest ecosystems. *Ekol. Pol.*, 24: 493–514.
Zimka, J.R. and Stachurski, A., 1976b. Regulation of C and N transfer to the soil of forest ecosystems and the rate of litter decomposition. *Bull. Acad. Pol. Sci. Ser. Sci. Biol.*, 24: 127–132.

Chapter 19

SYNTHESIS AND SEARCH FOR PARADIGMS IN WETLAND ECOLOGY

ARIEL E. LUGO, MARK M. BRINSON and SANDRA BROWN

This book contains observations on the forested wetlands in over 50 countries by 16 contributors. Chapters review an extensive literature and discuss many patterns of forested wetland response to environmental factors. The bulk of information on forested wetlands from many parts of the world tends to be floristic (e.g., Ch. 12, 14, 15, 16, and 18), or it focuses on individual species (e.g., Ch. 16) or phenomena (e.g., Ch. 4, Sect. 3, 8, 15, 17 and 18). Descriptions of ecosystem structure and dynamics, hydroperiod, and biogeochemistry are not abundant. Bruenig's work on the physiognomy of peat swamps (Ch. 13) and Myers' study of hydroperiod in Costa Rican palm swamps (Ch. 11) are exceptions. Collection of these kinds of data is critical for advancing the understanding of forested wetlands and for properly managing and assessing their role in the landscape. Experimental approaches on these ecosystems are fewer still. An example is the experiment with sewage enrichment of a basin forested wetland in Florida (Ewel and Odum, 1984).

This chapter reviews paradigms of "forested wetland ecology" and proposes patterns of ecosystem response which could be tested in the future. We refer the reader to Odum (1984) for an overview of what is known about cypress (*Taxodium*) wetlands, including detailed discussions of diurnal, seasonal, and long-term patterns of ecosystem function, regional role of wetlands, wetlands values, and human use. Odum's review covers many paradigms that apply to all types of wetlands and are not repeated here.

Our first proposal is that the principles of forested wetland ecology are the same as those for other wetland types (Brinson et al., 1981; Lugo, 1982). For this reason we use the term "wetlands" in this chapter to mean all wetlands, and we modify the term with "forested" when refering to data sets or ideas unique to them.

THE ROLE OF HYDROLOGY

The available evidence strongly supports the idea that forested wetlands, like all other types of wetlands, develop more structure and are more productive under riverine conditions (Table 19.1). Low values of productivity and structural measures are associated with basin conditions, with the exception of tree density which is high in basin forests. Brown et al. (1979) suggested that this high stem density is a response to poor soil aeration. A higher stem density results in a higher surface area for gas exchange. Alternatively, higher tree density may be a result of lower tree mortality in basin wetlands where hydrologic fluxes are less potentially destructive. Fringe-wetland data are limited to mangrove forests, which are usually intermediate between riverine and basin mangrove forests in terms of structure and functional parameters (Cintrón et al., 1985).

Many investigators have identified relations between hydrological parameters and wetland response (e.g., Mitsch and Ewel, 1979; Connor et al., 1981; Gosselink et al., 1981). Ch. 16 by Specht develops a Moisture Index based on water-balance information to explain wetland physiognomy and regeneration. These relations support the general hypothesis of hydrologic control of wetlands. Even the substrate of wetlands is a function of the hydrologic regime (Ch. 4, Section 15, Fig. 4.10).

The direction and kinetic energy of water flow are critical factors in the regulation of wetland

TABLE 19.1

Summary of structural indices, standing stocks, and rates for forested wetlands[1]

Parameter (units)	Riverine		Fringe		Basin	
	fresh	salt	fresh	salt	fresh	salt
STRUCTURAL INDICES						
Number of species						
Range	1–23	1–4	—	1–3	1–14	1–3
n	30	16	—	32	18	5
Mean	8.3	2.6	—	1.7	5	2
Density (number ha^{-1})						
Range	71–2730	400–4670	—	440–16 760	1440–7820	2468–5130
n	29	16	—	33	15	5
Mean	1076	2131	—	4005	2753	3580
Basal area (m^2 ha^{-1})						
Range	12.0–92.3	11.5–96.4	—	6.0–43.0	9.5–70.8	15.2–23.2
n	32	16	—	33	15	5
Mean	37.8	33.2	—	22.2	39.9	18.5
STANDING STOCKS						
Aboveground biomass (t ha^{-1})						
Leaves						
Range	2.8–12.1	3.6–9.5	—	—	2.3–10.8	—
n	8	3	—	—	12	—
Mean	5.6	5.6	—	—	5.0	—
Total						
Range	79–608	98–279	—	8.1–159	4–345	—
n	17	4	—	8	13	—
Mean	242	170	—	99.8	163.6	—
Belowground biomass (t ha^{-1})						
Range	12–84	154–190	—	14.1–50	7.8–31.0	—
n	9	2	—	2	7	—
Mean	43	172	—	32.1	17.6	—
RATES						
Litter-fall (t ha^{-1} yr^{-1})						
Range	3.15–17.0	11.2–16.0	—	4.3–9.5	2.47–7.57	1.86–9.70
n	16	8	—	11	10	4
Mean	5.72	13.0	—	7.1	5.33	6.08
Wood biomass production (t ha^{-1} yr^{-1})						
Range	1.77–17.88	—	—	—	0.48–5.41	—
n	16	—	—	—	8	—
Mean	6.94	—	—	—	3.25	—
Total aboveground biomass production (t ha^{-1} yr^{-1})						
Range	6.68–21.36	—	—	—	0.72–10.29	—
n	13	1	—	—	10	—
Mean	12.65	10.2	—	—	5.96	—
Litter decomposition (k yr^{-1})						
Range	0.21–4.95	—	—	0.85–8.39	0.25–1.87	1.5–6.0
n	32	—	—	9	7	4
Mean	0.93	—	—	2.5	0.75	3.4
Nutrients in litter-fall (kg ha^{-1} yr^{-1})						
Nitrogen						
Range	60.1–86.7	—	—	15–86	20.8–45.0	53–70
n	4	—	—	11	7	3
Mean	78.6	—	—	46.5	31.0	62.0
Phosphorus						
Range	1.5–8.1	—	—	1.2–9.0	1.2–6.5	—
n	5	—	—	8	8	—
Mean	3.66	—	—	4.5	2.8	—

(*continued*)

SYNTHESIS AND SEARCH FOR PARADIGMS 449

TABLE 19.1 (*continued*)

Parameter (units)	Riverine		Fringe		Basin	
	fresh	salt	fresh	salt	fresh	salt
Potassium						
Range	17.0–30.6	—	—	—	3.3–9.5	—
n	4	—	—	—	5	—
Mean	22.3	—	—	—	5.4	—
Calcium						
Range	29.9–129.2	—	—	65–125	23–91	—
n	4	—	—	7	7	—
Mean	70.8	—	—	96.6	46.3	—
Magnesium						
Range	7.38–37.2	—	—	—	4.7–12.0	—
n	4	—	—	—	7	—
Mean	18.6	—	—	—	7.7	—

[1]Data from Chapter 5 (Tables 5.1, 5.2, 5.3, 5.4, 5.6, 5.7), Chapter 6 (Tables 6.2, 6.3, 6.6, and 6.9) and Chapter 7 (Tables 7.3, 7.4, 7.7, and 7.11).

function. Direction of water flow (unidirectional, vertical, or bidirectional) separates riverine, basin, and fringe wetlands respectively, while the kinetic energy associated with the flow can either stress or subsidize the wetland (Ch. 3 and 6). Water motion is generally a subsidy for most wetlands because of the work it can do in the system (e.g., movement of nutrients, aeration, dispersal of propagules, ventilation of roots, etc.; Ch. 4, Sect. 17). There are few exceptions to this subsidy function, because species distribute themselves around hydrologic conditions to which they are adapted (cf. Fig. 11.8 in Ch. 11 by Myers for examples of how palms adjust to hydroperiod). However, if the hydrologic fluxes are too intense, they become stressful by exporting excessive amounts of materials (Frangi and Lugo, 1985), by causing physical damage to the ecosystem (Chs. 5, 17 and 18 give examples of floating ice damage to floodplain communities), and by reducing the time available to the system to take advantage of the materials in flux (Ch. 4, Sects. 6 and 21). In these cases, the lifetime of the wetland may be limited, the community will function suboptimally, or a less complex system will develop on the site.

While it is obvious that the hydrology of a typical basin wetland is different from that of typical riverine or fringe wetlands, local variations in topography affect water flow, which in turn causes "mixtures" of wetland types. For example, within a large basin wetland (hundreds of hectares) one can identify "water tracks" (Heinselman, 1970), which behave like riverine wetlands. Similarly, protected sectors of riverine floodplains or the landward sectors of fringe wetlands, may behave like basin wetlands. In fact, a given area of wetland might behave as a basin system for part of the year and as a riverine system during another season.

These deviations from the "model" or idealized wetland types are not weaknesses in the concept of wetlands as presented here, but offer instead an excellent opportunity for enhancing understanding of these ecosystems. The three geomorphologic wetland types (fringe, riverine, and basin; Figs. 19.1–19.3) are fundamental functional units which can be identified in the field at scales ranging from individual trees responding to micro-edaphic factors to large regions such as the Great Dismal Swamp in the United States of America (Ch. 8). When the hydrologic regime is identified at the appropriate scale and a wetland type is assigned, the predictable behavior of that wetland allows researchers and managers to stratify research or management actions according to the mix of wetland conditions under consideration.

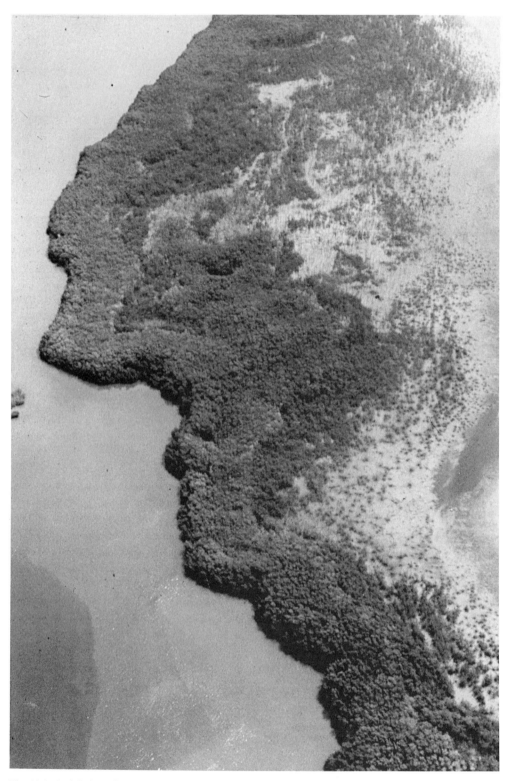

Fig. 19.1. Aerial view of a saltwater fringe forested wetland (mangroves) in a low-energy arid coastline. Tree zonation and ecosystem organization are perpendicular to the incoming tidal energy. Mangrove growth is limited towards the landward portions because of excessive soil salinity (Ch. 6).

Fig. 19.2. Aerial view of a saltwater riverine forested wetland (mangroves). In this type of wetland tree zonation and ecosystem organization are parallel to the river's hydrologic energy (Chapter 5).

THE INFLUENCE OF NUTRIENTS

Hydrology by itself does not explain all the behavior of wetlands. Chapters 9, 12, and 13 use soil chemistry information to explain wetland physiognomy and complexity. Wetlands are part of a larger ecosystem or catchment whose configuration, geomorphology, and size affect the flux of materials (water, sediments, and nutrients) into the wetland. The quality and quantity of surface water and groundwater entering a wetland are a function of the size and landscape diversity of the catchment. The influence of water quality on wetlands has led many to classify wetlands into nutrient-rich and nutrient-poor types (eutrophic vs oligotrophic, or minerotrophic vs ombrotrophic). In fact, models that relate such wetland functions as primary productivity or tree growth to the interaction of hydrological and chemical characteristics of waters yield better correlations than single-factor models (Brown, 1981). Fig. 19.4 summarizes Odum's (1984) interpretation of the synergism between hydrology and nutrient fluxes in Florida cypress wetlands. In this scheme, ecosystem complexity increases with increases in the flows of water and nutrients. Ch. 3 by Kangas elaborates on these concepts.

In spite of the scarcity of information, wetlands are characterized as nutrient-rich or nutrient-poor ecosystems by using water chemistry, peat depth, and physiognomy of vegetation (e.g., presence or absence of xeromorphy or sclerophylly). There is a need to consider the chemical aspects of wetlands in a more holistic context. The chemistry of wetland ecosystems is extremely complex as it involves aerobic as well as anaerobic processes. These processes in turn are closely coupled to atmospheric, hydrologic, and edaphic factors. Furthermore, plants respond to any or all of these chemical environments with a complex array of biotic adaptations that affect the speed and efficiency of nutrient uptake and use. For example,

Fig. 19.3. Aerial view of the Everglades (Florida) basin scrub mangroves with scattered mangrove islands (hammocks). The prevalent direction of hydrologic energy in this type of wetland is vertical (Chs. 3 and 7).

plants conserve nutrients by retranslocation and reuse, affect concentrations in soil through uptake, and return nutrients to the environment through senescence and decay. Because animals play important roles in facilitating the turnover of nutrient elements in wetlands (Irmler and Furch, 1979; Ch. 4, Sect. 4), the task of analyzing nutrient cycling in wetlands is complicated even more. It is thus unrealistic to evaluate the nutrient status of a wetland by simply analyzing the nutrient concentration in its waters, soils, or plant tissues.

The interaction between plants, microbes, and those environmental factors that affect nutrient availability to a wetland is illustrated by the model in Fig. 19.5 (the model is applicable to any forest ecosystem). The leaching potential of rainfall and other environmental stressors contribute to nutrient loss from plant and other ecosystem compartments, while sources, both internal and external to the wetland, supply nutrients. After root uptake, four mechanisms of nutrient recycling operate in a nested fashion and at different speeds and, taken as a group, determine how efficiently the wetland utilizes the available nutrient capital. The four nutrient recycling mechanisms (in decreasing order of speed of cycling) are: (1) liberation of nutrients during respiration, (2) retranslocation, (3) return to the forest floor by leaching, (4) return to the forest floor by litter-fall. The efficiency of each of these cycles is wholly under biotic control, with the exception of leaching which is also controlled by rainfall. In this book we have elaborated on the recycling efficiency of Pathway 4, known as the within-stand nutrient-cycling efficiency index (ratio of litter-fall in kg ha^{-1} yr^{-1} to nutrients in litter-fall in the same units: Vitousek, 1982, 1984), because there are more data available to compute this index. Fig. 19.6 summarizes this information for nitrogen, phosphorus, and calcium for the three wetland types, and for other wetlands which could not be categorized.

Available data on within-stand nutrient-cycling

SYNTHESIS AND SEARCH FOR PARADIGMS 453

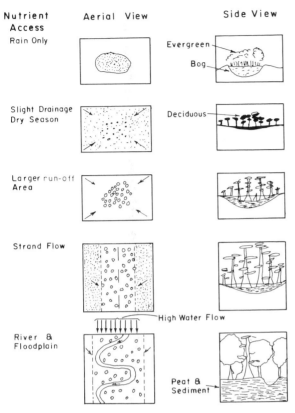

Fig. 19.4. Classification of cypress (*Taxodium*) wetlands of Florida, arranged in order of increasing water and nutrient flows from top to bottom (Ewel and Odum, 1984).

efficiency (Fig. 19.6) can be used to formulate the following generalizations:

(1) The recycling efficiency of calcium is low, suggesting that this element is not limiting to wetlands. However, Igapo riverine wetlands (black-waters) with acid, nutrient-poor waters, are the most efficient in the recycling of calcium. This suggests that in this type of environment calcium may be limiting. The same appears to be the case with black mangrove (*Avicennia germinans*) wetlands.

(2) The pattern of nitrogen recycling efficiency is better defined than that of calcium. Australian mangroves show a higher recycling efficiency than those from Malaysia, basin wetlands have higher recycling efficiency than riverine ones, and systems with *Laguncularia racemosa*, the white mangrove, have higher recycling efficiencies than any other species or wetland type shown. In addition, for a given rate of nutrient return by litter-fall, saltwater wetlands appear to be more efficient than their freshwater counterparts.

(3) Most wetlands exhibit very high efficiencies of phosphorus recycling, suggesting that this element may be the main limiting factor of these ecosystems. However, freshwater wetlands are less efficient than saltwater ones for any given rate of phosphorus return by litter-fall. The pattern for phosphorus recycling efficiency shows less scatter than for calcium or nitrogen and also separates the efficient Australian mangroves from the less efficient ones in Malaysia. Scrub red mangroves (*Rhizophora mangle*) growing on marl in the Florida Everglades (Fig. 19.3), exhibit the highest phosphorus recycling efficiency of all systems studied.

We propose that the efficiency of nutrient cycling in ecosystems can be calculated with a variety of indices derived from Fig. 19.5 (Table 19.2), and that depending upon conditions any or a combination of these indices may be reflecting the "true" efficiency of the ecosystem. For example, Frangi and Lugo (1985) found that an eroding floodplain forest in Puerto Rico was exporting more phosphorus than it received from upstream sources, and was losing large amounts of phosphorus due to leaching from the canopy and soils (i.e., it was inefficient in terms of inputs–outputs and Pathway 3). However, its within-stand efficiency of phosphorus recycling (Pathway 4) and rate of phosphorus retranslocation (Pathway 2) were extremely high. This example illustrates two points: (1) the importance of identifying boundaries and limitations to nutrient-cycling studies; and (2) that high efficiency of biotic recycling may be a response to environments where abiotically-controlled recycling efficiencies are low. A corollary hypothesis is that the biotically-controlled recycling efficiency is low in environments where the abiotically-controlled recycling efficiency is high.

Hydrologic and nutritional factors must be considered when explaining the degree of nutrient recycling in a given type of wetland. As an example, the turnover of litter will be considered. Hydrologic forces will be instrumental in determining the residence time of litter on the forest floor, while nutritional factors affect the rate of microbial degradation. It appears that wetlands with fast water turnover are characterized by high rates of nutrient turnover and low litter accumulation.

TABLE 19.2

Nutrient-cycling efficiency ratios. Letter codes in the "Process" column are keyed to flows in Fig. 19.5. Units have a time and area basis on both numerator and denominator (e.g. g m^{-2} yr^{-1}). Nutrient mass is always in the numerator for consistency but the inverse could be used as well

	Process	Numerator	Denominator	Comments
A.	BIOTIC CONTROL			
a.	Nutrient uptake	Uptake	Transpiration	Chapin, 1980
		Uptake	Root respiration	Difficult to measure root respiration
		Uptake	(Transpiration) × (root respiration)	Would measure total cost of nutrient uptake
b.	Nutrient fixation	Need[1]	Photosynthesis	Chapin, 1980
c.	Liberation by respiration	Mobilization	Leaf respiration	Difficult to measure mobilization in leaves
d.	Retranslocation	Retranslocation	Leaf respiration	
e.	Litter-fall	Nutrient mass	Mass fall	Inverse of the number proposed by Vitousek (1982, 1984)
f.	Decomposition	Uptake plus immobilization	Inputs to soil and litter	
g.	Use of fixed nutrients	Nutrient turnover in vegetation	Organic matter turnover in vegetation	
B.	ABIOTIC CONTROL			
a.	Leaching	Leaching	Rainfall	Parker, 1983
b.	Susceptibility to stressors[2]	Loss	Intensity of forcing function	Many environmental factors could cause loss from any tank in Fig. 19.5
c.	Cycle "tightness"	Sum of inputs	Sum of losses	This ratio is commonly used in the literature

[1]"Need" refers to the nutrient requirement to maintain the normal rate of primary productivity.
[2]This item refers to nutrients lost from the system as a function of an environmental factor acting as a stressor (e.g. an unusual flood, a fire, etc.).

Those with less hydrologic energy may compensate with faster rates of in-situ microbial decomposition, but accumulate more litter (Fig. 19.7). Clearly, a single-factor approach to wetland ecology does not explain all the observations, and multi-factor approaches are necessary for a greater understanding.

THE ENERGY SIGNATURE APPROACH

The energy-signature calculation (Ch. 4, Sect. 9 and Chs. 2 and 3) is a sound approach for understanding to what extent the structure and dynamics of forested wetlands are driven by external environmental factors. The advantage of this approach is that it evaluates all environmental factors converging on the wetland using the same unit of measure (embodied energy). The energy signature weighs the relative importance of each factor in energy units by correcting for the respective energy quality, thus preventing the uncertainty of interpretation that one has when dealing with the many units used to measure environmental factors. Ch. 4, Sect. 9 contains examples of calculations of embodied energy.

Unfortunately, the data available do not permit an extensive review of energy signatures for the many examples of forested wetlands in the world. However, Table 4.7 shows how flooding in a floodplain wetland has twice as much embodied energy as incident light. Table 4.9 illustrates how the energy value of floodplain landform changes depending upon the unit of measure. Using caloric values, landform has little value relative to ecosystem biomass. However, by using embodied energy

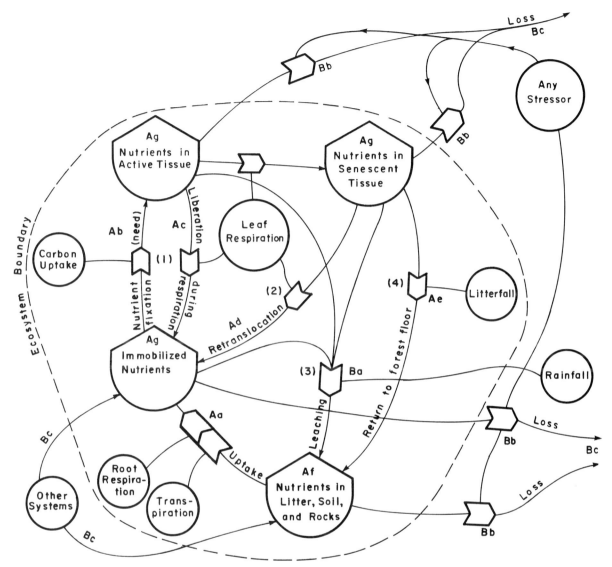

Fig. 19.5. Model of nutrient cycling in forested ecosystems. The ecosystem boundary is the broken line. Tanks represent state variables, circles represent the biotic and abiotic factors believed responsible for the nutrient flux, and the arrow-shaped symbols show interactions of a nutrient flux and its control factor (Ch. 4, Sect. 9 explains the symbols). The ratio of control factor to nutrient flux is an index of recycling efficiency. Number codes relate to pathways mentioned in the text; letter codes are keyed to Table 19.2.

the value of landform is amplified considerably. This latter value is more realistic, as it takes much more energy to replace landform than it does to replace ecosystem biomass. Table 19.3 is another example from Odum (1984) illustrating the embodied energy passing through and being used by the cypress swamp ecosystems illustrated in Fig. 19.4. As energy availability increases, its efficiency of use is lower but more work is done, measured in this case in terms of transpiration (Table 19.3)

or the maintenance of ecosystem complexity (Fig. 19.4).

In spite of the absence of an extensive data base, wetlands can be categorized by a multi-factor approach using the most important variables in the energy signature. We believe that such variables are kinetic energy of water in motion, the chemistry of the ecosystem measured in terms of nutrient availability (exogenous sources) and turnover (internal dynamics), and the hydroperiod (season,

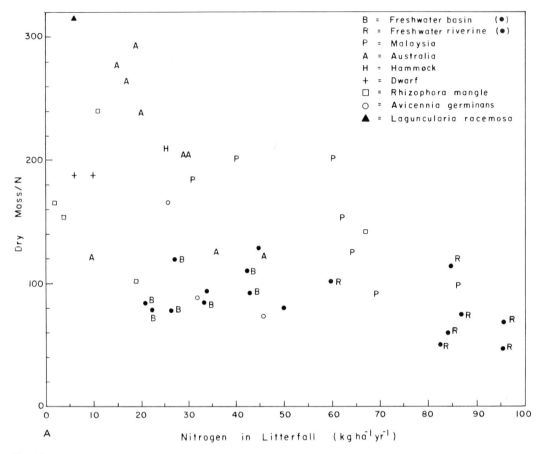

Fig. 19.6. a.

duration, and depth of flooding). These are the "core" factors that define the basic conditions for any type of wetland. Bruenig separates wetland types using fertility and drainage variables in Ch. 13.

The actual type of wetland to be found in a given location depends on the intensity of the core factors, plus that of other subsidy or stress factors that modify wetland behavior (Ch. 4, Sect. 13 and 19). For example, forested wetlands are limited by deep water, long hydroperiods, and high hydrologic energy. They appear to do well in a broad range of nutrient-availability conditions. Subsidy-stress factors may have an overriding effect on wetland systems when they reach certain thresholds of intensity. These factors act singly or synergistically with the core factors just discussed. Examples of these modifying factors are frost (low temperature or short growing season), drought, salinity, fire, hurricanes, and storms. The regulatory effect of frost is illustrated by a reduction in species richness with increasing latitude. Wetlands at lower latitudes can support more tree species than those where frosts occur regularly. Above the frost line, shortened growing season alters community composition. The species richness of trees also appears to decrease from humid environments towards arid ones where droughts are common.

At the ecosystem level, both frost and drought slow metabolism, cause leaf drop, and by effectively shortening the growing season cause reductions in primary productivity and overall ecosystem activity. Salinity also reduces the physiognomic complexity and species richness of wetland ecosystems, as well as altering the physiology of its component species. Complexity of vegetation, measured by the Holdridge index, decreased by 1.25 units for each part per thousand increase in salinity between 35 and 65 parts per thousand. (Brown and Lugo, 1982). The chapter by Alvarez-

SYNTHESIS AND SEARCH FOR PARADIGMS

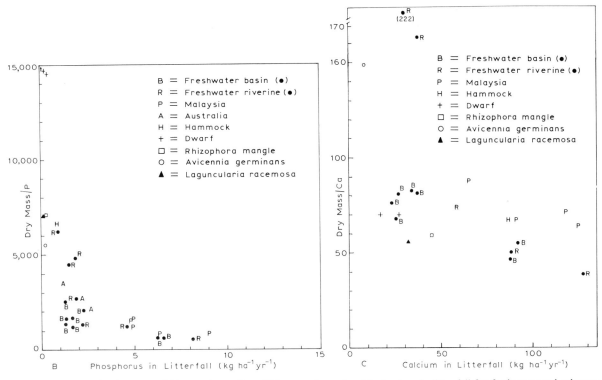

Fig. 19.6. The relation between within-stand recycling efficiency ratio and nutrient return in litter-fall for freshwater and saltwater forested wetlands. Data are mainly for mangrove wetlands, unless stated. Sources are Chapters 5, 7 and 18 for freshwater wetlands and from many sources for mangroves (available from A.E. Lugo).

TABLE 19.3

Embodied energy of swamps with increasing water availability. Wetland types are in the same order as those in Figure 19.4 (This table is from Ewel and Odum, 1984.)

Item	Water processed[1] ($m^3\ m^{-2}\ yr^{-1}$)	Transpiration[2] ($m^3\ m^{-2}\ yr^{-1}$)	Embodied energy passing through[3] ($GJ\ m^{-2}\ yr^{-1}$)	Embodied energy used, global solar[4] ($GJ\ m^{-2}\ yr^{-1}$)
Direct sunlight	—	—	42	36
Rain only, bays	2	0.3[5]	678	100
Dwarf cypress	4	0.34[6]	1364	117
Pond cypress	8	1.12[6]	2720	381
Strand cypress	40	1.5[5]	13 640	510
Floodplain	100	1.89[6]	34 058	640

[1]The amount of water passing through the wetlands.
[2]$1\ m^3\ m^{-2}\ yr^{-1} = 2.74\ mm\ day^{-1}$
[3]Embodied energy is defined in Chapter 3, and is expressed as $GJ\ m^{-2}\ yr^{-1}$ (1 GJ = 239 Mcal) The figures in this column are embodied energy per unit volume multiplied by the water passing. Embodied energy per unit volume = $(1.18 \times 10^{-3}$ cal g actual free energy in rain^{-1}) $(6.9 \times 10^4$ global solar equivalent cal cal^{-1}) = 8.1 global solar cal g water^{-1}.
[4]These figures are embodied energy per unit volume multiplied by the water used in transpiration.
[5]Calculated by interpolation.
[6]Brown (1981).

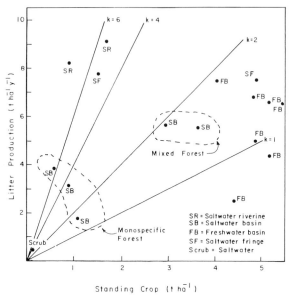

Fig. 19.7. Relation between litter-fall and litter standing crop in forested wetlands. Mangrove data are from Twilley (1982), and the other data are from Chapters 6 and 7. Lines represent the annual rate of litter turnover (k).

Lopez (Ch. 10) illustrates the overriding effect of soil salinity on species composition and type of ecotone in *Pterocarpus* swamps in Puerto Rico. This factor alone appeared to be responsible for a higher complexity index in coastal basin swamps than in coastal riverine ones subject to more seawater inundation. Chapters 9, 10, 11, 14, and 15 review forested wetlands at the ecotone between salt- and fresh-water. A detailed comparison of forested wetlands in salt and fresh water is given by Brown and Lugo (1982) and by Lugo et al. (1988).

Fig. 19.8 illustrates our concept of the distribution of the three wetland types within the dimensions of the core factors just discussed. The display is similar to Hutchinson's (1957) hypervolume niche concept for species populations. Each wetland type not only occupies a particular zone in three-dimensional space, but the configuration also differs. Fringe wetlands are relatively compressed along the hydroperiod and hydrologic-energy coordinates (Fig. 19.8a) because of the strong interdependency of the two, with lateral flushing by tides and seiches (Ch. 6). Basin wetlands segregate in the lower range of hydrologic energy with a broad range of hydroperiods. Riverine wetlands have shorter hydroperiods but a broad spectrum of hydrologic energies that range from swiftly flushed levee forests to stands that receive backwater flushing. Nutrient factors are added in Fig. 19.8b. This factor separates basins from fringe and riverine wetlands. Riverine wetlands are generally in the higher nutrient range because of their geomorphological role in the biosphere. Fringe wetlands range broadly between low and high nutrient richness. More axes need to be added to this model to account for the moderating factors responsible for locating, in the hypervolume, the many variations of wetland types. For example, for bogs of high latitudes, the shortness of the growing season and severe temperature stress reduce the dominance of trees to the point that tree biomass is exceeded by that of the understory (Table 7.4). The basin core factors remain constant, but the modifying influence of temperature is the main cause of a shift from forested to non-forested systems.

ELEMENTAL CYCLES

The task of understanding the complexity of elemental cycling in wetlands is simplified somewhat by clearly identifying ecosystem boundaries and measuring exchanges across those boundaries. Most wetlands are inherently sedimentary systems. Thus, on the average, they accumulate carbon, nitrogen, phosphorus, and other elements. However, these various elements are exchanged to different degrees across the atmosphere–wetland boundary and the wetland–landscape boundary. For carbon, the dominant exchange for all wetlands is across the atmosphere–wetland boundary.

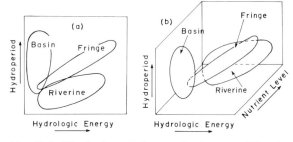

Fig. 19.8. Distribution of three wetland types within the dimensions of hydrologic energy, hydroperiod, and nutrient richness. (a) Projection in two-dimensional space of hydroperiod and hydrologic flow. (b) Nutrient richness represent the third core factor.

Because some wetlands also export significant amounts of organic carbon in run-off (Ch. 4, Sect. 2), they can simultaneously serve as carbon sinks for the atmosphere and carbon sources for downstream ecosystems.

For nitrogen, there can be significant exchanges across both boundaries, and the magnitude may depend on the geomorphic setting. Riverine wetlands exchange water-borne inorganic nitrogen with the landscape, in addition to having significant atmospheric exchanges (Brinson et al., 1983). Basin wetlands are restricted to predominantly wetland–atmospheric exchanges through nitrogen fixation and denitrification pathways, while fringes are probably intermediate between the other two types of wetlands. Because of the difficulty of measuring with confidence the in-situ rates of nitrogen fixation and denitrification, one is far from achieving a holistic perspective on the nitrogen cycle in wetlands. For example, one knows that wetlands have the potential to absorb large amounts of nitrate and to lose this nitrogen source through denitrification. Yet, this potential is seldom realized because of limits in the rate of nitrate supply (Ch. 4, Sect. 18). A more holistic perspective is that basin wetlands accumulate nitrogen through peat-building processes, and that riverine wetlands are dynamic transformers between organic and inorganic forms (Elder, 1985). As with carbon, a wetland can simultaneously serve as a sink for atmospheric and upstream nitrogen (through peat accumulation), and a source of organic nitrogen compounds for downstream ecosystems.

The behavior of phosphorus is simpler at the ecosystem level because it has no atmospheric sink. Atmospheric sources of phosphorus are significant only in strictly ombrotrophic basins. Riverine wetlands rely predominantly on landscape sources, while fringe ones depend on the water body that they fringe, together with any landscape or groundwater sources to which they may be coupled. All types accumulate phosphorus unless they switch to a nonsedimentary regime. In such cases, the lifetime of an eroding wetland is limited.

The significance of sulfur lies not in a potential for nutrient limitation, but in its influence on organic-matter decomposition, its capacity to modify the sediment environment, and its function in detritus food webs. Because sulfur is so much more abundant in saltwater than freshwater wetlands, generalizations about its behavior (Ch. 4, Sect. 18) are best limited at this time to distinguishing between these two types rather than seeking patterns among riverine, fringe, and basin types.

Appreciation of the differing behavior of elements among themselves and among wetland types allows some speculation on the potential for nutrient limitation in primary productivity. Because the source of carbon is strictly atmospheric, and the atmospheric pool is unlimited, arguments for carbon limitation are trivial. For nitrogen and phosphorus, some insight into nutrient limitation is possible at the ecosystem level. Basin wetlands receive both nitrogen and phosphorus mainly from atmospheric sources, by fixation and precipitation in the case of nitrogen and from precipitation only for phosphorus. Strategies for adapting to limitations by these two elements differ. For phosphorus the only strategy is through mechanisms of recycling, because the ecosystem has no control over the abiotic source. This strategy is evident in Fig. 19.6 where the within-stand recycling efficiency of phosphorus is shown to be high in many wetlands.

For nitrogen, the atmospheric source is unlimited (N_2) and the strategy available is nitrogen fixation. Because nitrogen fixation has both an energetic cost and a stimulatory effect on wetland production, there are limits to the amount of nitrogen that can be fixed. Assuming that the ecosystem can compensate for nitrogen deficiency through nitrogen fixation, it is more likely that basins will be limited by phosphorus than by nitrogen. This is supported by the relative within-stand recycling efficiencies of phosphorus and nitrogen in Fig. 19.6, and by the low efficiency of nitrogen recycling by alder wetlands in Poland (the lowest ratio reported, i.e., 43: Ch. 18). Many riverine wetlands have abundant sources of phosphorus through accumulation of inorganic sediments from the landscape (Table 5.8). Like basin wetlands, they also have access to unlimited supplies of atmospheric nitrogen. We postulate that nitrogen and phosphorus limitation in many riverine forested wetlands is not a critical factor. Energy in riverine forests can be allocated toward other adaptive functions. The biomass of many riverine forests (Table 19.1) suggests that plenty of energy is being allocated to the building of structure. In fringe wetlands, nutrient limitations will depend on water chemistry and the hydrology and geomorphology of the fringe relative to the water body (Ch. 6).

REFERENCES

Brinson, M.M., Lugo, A.E. and Brown, S., 1981. Primary productivity, decomposition, and consumer activity in freshwater wetlands. *Annu. Rev. Ecol. Syst.*, 12: 123–161.

Brinson, M.M., Bradshaw, H.D. and Holmes, R.N., 1983. Significance of floodplain sediments in nutrient exchange between a stream and its floodplain. In: T.D. Fontaine and S.M. Bartell (Editors), *Dynamics of Lotic Ecosystems*. Ann Arbor Science, Ann Arbor, Mich., pp. 199–221.

Brown, S., 1981. A comparison of the structure, primary productivity, and transpiration of cypress ecosystems in Florida. *Ecol. Monogr.*, 51: 403–437.

Brown, S. and Lugo, A.E., 1982. A comparison of structural and functional characteristics of saltwater and freshwater forested wetlands. In: B. Gopal, R.E. Turner, R.G. Wetzel and D.F. Whigham (Editors), *Wetlands: Ecology and Management*. International Scientific Publications, Jaipur, pp. 109–130.

Brown, S., Brinson, M.M. and Lugo, A.E., 1979. Structure and function of riparian wetlands. In: R.R. Johnson and J.F. McCormick (Technical Coordinators), *Strategies for Protection and Management of Floodplain Wetlands and Other Riparian Ecosystems*. US Department of Agriculture Forest Service. GTR-WO-12, Washington, D.C., pp. 17–31.

Chapin, F.S. III, 1980. The mineral nutrition of wild plants. *Annu. Rev. Ecol. Syst.*, 11: 233–260.

Cintrón, G., Lugo, A.E. and Martinez, R., 1985. Structural and functional properties of mangrove forests. In: W.G. D'Arcy and M.D. Correa A. (Editors), *The Botany and Natural History of Panama: La Botanica e Historia Natural de Panama*. Monographs in Systematic Botany 10. Missouri Botanical Garden, St. Louis, Mo., pp. 53–66.

Conner, W.H., Gosselink, J.G. and Parrondo, R.T., 1981. Comparison of the vegetation of three Louisiana swamp sites with different flooding regimes. *Am. J. Bot.*, 63: 320–331.

Elder, J.F., 1985. Nitrogen and phosphorus speciation and flux in a large Florida river wetland system. *Water Resour. Res.*, 21: 724–732.

Ewel, K.C. and Odum, H.T. (Editors), 1984. *Cypress Swamps*. University Presses of Florida, Gainesville, Fla., 472 pp.

Frangi, J.L. and Lugo, A.E., 1985. Ecosystem dynamics of a subtropical floodplain forest. *Ecol. Monogr.*, 55: 351–369.

Gosselink, J.G., Bayley, S.E., Conner, W.H. and Turner, R.E., 1981. Ecological factors in the determination of riparian wetland boundaries. In: J.R. Clark and J. Benforado (Editors), *Wetlands of Bottomland Hardwood Forests*. Elsevier Scientific Publ. Co., New York, N.Y., pp. 197–222.

Heinselman, M.L., 1970. Landscape evolution, peatland types, and the environment in the Lake Agassiz Peatlands Natural Area, Minnesota. *Ecol. Monogr.*, 40: 235–261.

Hutchinson, G.E., 1957. Concluding remarks. In: *Cold Spring Harbor Symposia on Quantitative Biology*, 22: 415–427.

Irmler, U. and Furch, K., 1979. Production, energy, and nutrient turnover of the cockroach *Epilampra irmleri* Rocha e Silva & Aguiar in a Central-Amazonian inundation forest. *Amazonia*, 6: 497–520.

Lugo, A.E., 1982. Some aspects of the interactions among nutrient cycles, hydrology, and soils in wetlands. *Water Int.*, 7: 178–184.

Lugo, A.E., Brown, S. and Brinson, M.M., 1988. Forested wetlands in freshwater and saltwater. *Limnol. Oceanogr.*, 33(4), in press.

Mitsch, W.J. and Ewel, K.C., 1979. Comparative biomass and growth of cypress in Florida wetlands. *Am. Midl. Nat.*, 101: 417–429.

Odum, H.T., 1984. Summary: cypress swamps and their regional role. In: K.C. Ewel and H.T. Odum (Editors), *Cypress Swamps*. University Presses of Florida, Gainesville, Fla., pp. 445–468.

Parker, G.G., 1983. Throughfall and stemflow in the forest nutrient cycle. *Adv. Ecol. Res.*, 13: 58–133.

Twilley, R.R., 1982. *Litter dynamics and organic carbon exchange in black mangrove (Avicennia germinans) basin forests in a southwest Florida estuary*. PhD Dissertation, University of Florida, Gainesville, Fla., 260 pp.

Vitousek, P.M., 1982. Nutrient cycling and nutrient use efficiency. *Am. Nat.*, 119: 553–572.

Vitousek, P.M., 1984. Litterfall, nutrient cycling, and nutrient limitation in tropical forests. *Ecology*, 65: 285–298.

AUTHOR INDEX[1]

Aaby, B., 71, *79*
Abernethy, Y., 117, *134*
Abrams, M.D., 96, *134*
Acevedo, P., *264*
Adams, C.D., 221, 236, *248*
Adams, D.E., 94, 102, *134*
Adams, G.D., 14, *169*
Adams, M.S., *168*
Adams, S.N., 3, 5, *10*
Adema, E.H., *436*
Adis, J., 109, 110, 134, *138*
Adisoemarto, S., *333*
Adriano, D.C., 81, 82, *84*
Ahlgren, C.E., 36, *47*
Ahmed, A., *385*
Ahmed, N., 378, *385*
Ahn, P., 3, 10, 269, *285*
Aitken, T.H.G., 213, 229, 241, 244, *248, 249*
Ajtay, G.L., 54, *79*
Alder, D., *333*
Aletsee, L., 411, *432*
Alexander, J., Jr., 23, *49*
Ali, S., 374, *385*
Alkins, M.A., *248*
Allee, W.C., *47*
Allen, J.R.L., 89, 92, *134*
Allen, P.H., 269, *285*
Allison, L.J., *84*
Allorge, P., 418, 422, *432*
Alvarez, L., 106, *137*
Alvarez, M., 257, *264*
Alvarez-López, M., 98, 105, 221, 236, *456*
Alvim, P., *286*
Alvim, P. de T., *14, 141*
Anderson, A.B., 272, *285*
Anderson, B.W., 56, *79*
Anderson, D.J., *406*
Anderson, J.A.R., 4, *10*, 65, *79*, 302, 305, 306, 308, 309, 313, 318, 322, 323, 324, 326, 327, 328, 329, 330, 331, *333*
Anderson, J.M., *334*
Anderson, R., 4, *10*, 269, *285*
Anderson, R.C., 94, 96, 101, 102, *134*
Andrejko, M.J., *11, 47*, 196, *197, 198*
Andrews, J.T., 19, *23*

Andrews, L.S., *138*
Andrews, T.J., *167*
Applequist, M.B., 97, *134*
Archbold, R., 345, 346, *355*
Archer, S.G., 19, *23*
Argent, G., *13*
Armentano, T.V., 53, 54, 55, *80*
Armillas, P., 1, *10*
Armstrong, W., 72, 74, *80*, 414, *432*
Ashendorf, D., *84*
Ashton, P.S., 324, *333*
Askew, J.L., *83*, 137, *139*
Asmussen, L., *83, 138*
Asprey, G.F., 213, 214, 221, *248*
Attiwill, P.M., 162, 164, *167*
Aubin, A., *141*
Aubréville, A., *44*
Aucin, M.J., *49*
Augustinus, P.G.E.F., 231, *248*
Aulak, W., 445, *446*
Austin, G.T., 57, *80*
Austin, T.A., 40, *44*
Ayres, Q.C., 19, *23*
Aziz, A., 369, *385*

Bacon, P.R., 213, 222, 228, 229, 241, 244, *248*, 251, 261, *262*
Bagchi, K., 370, *385*
Bailey, R.G., 115, *134*
Ball, R.C., 35, *51*
Banage, W.B., 58, *80*
Bannikov, A.G., 429, *432*
Bannister, T.T., *198*
Banus, M.D., 144, *166*
Barber, K.R., 39, *44*
Barclay, J.S., *80*
Barghoorn, E.S., 224, 226, *248*
Barker, H.D., 221, *248*
Barnes, W.J., 99, 120, *134, 135*
Barrau, J., 270, *285*
Barrett, G.W., *83*, 168, *198*
Barry, C., 383, *385*
Bartell, S.M., *135, 460*
Bartholomew, J., *14*
Bartlett, A.S., 224, 226, *248*
Bartsch, J., 409, 417, 418, 421, 428, *432*
Bartsch, M., 409, 417, 418, 421, 428, *432*
Bates, C.Z., 251, *265*
Baumann, R.H., 110, *135*
Bay, R.R., 60, 77, *80*, 183, *196*
Bayley, P.B., 133, *135*

Bayley, S., 39, *44*
Bayley, S.E., *460*
Bazilevich, N.I., 162, 163, *168*
Bazzaz, F.A., 39, 40, *46*, 102, 118, *136*
Beadle, L.C., 1, *10*, 56, *80*, 229, 243, *248*
Beadle, N.C.W., 387, *405*
Beals, E.W., 35, *44*
Beard, J.S., 4, 5, 7, 8, 9, *10*, 29, *44*, 165, *166*, 213, 214, 215, 216, 218, 219, 221, 222, 223, 224, 239, 240, *248*, 251, 261, 265, 267, 272, *285*
Beccari, O., 299, 302, 326, *333*
Beckman, W.C., 378, *385*
Bedinger, M.S., 90, 93, *135*
Behnke, A., *83*, 139
Bell, D.T., 44, *47*, 93, 99, 102, 103, 110, 118, *135, 138*
Bell, P.R., 35, *44*
Bell, T.I.W., 245, *249*
Bellamy, D.J., 2, 8, *13*, 22, *23*, 34, *48*, 60, 84, 165, *166*
Bellamy, D.R., 29, *44*
Belt, C.B., Jr., 130, *135*
Beltrán, E., *297*
Bender, M., 31, *48*
Benforado, J., 2, *11*, 80, 94, 117, *135, 460*
Bennett, W.H., *10, 12*
Bernard, J.M., 3, *13*
Bernier, B., 14, *169*
Beschel, R.E., 165, *166*
Best, G.R., *11, 47*, 181, 183, *196, 197, 198*
Beveridge, A.E., 325, *333*
Beveridge, H., 370, *385*
Bhuiyan, A.L., 378, *385*
Binetti, V.P., *49*
Bissell, S.J., *141*
Black, R.A., 37, *44*
Blackburn, W.H., 32, *44*
Blake, S.T., 387, 391, 392, 393, 394, 395, *405*
Blancaneaux, P., 234, 241, *248*
Blasco, F., 357, *385*
Bledsoe, L.J., 40, *44*
Bliss, L.C., 37, *44*, 108, 125, *135*
Blydenstein, J., 267, *285*
Bocock, K.L., 414, 415, *432*
Bodeux, A., 423, *432*, 439, *446*
Bogdan, A.V., 129, *135*
Boelter, D.H., 63, *80*
Boggie, R., 60, 72, 73, *80*

---
[1] Page references to text are in roman type, to bibliographical entries in italics.

Bolin, B., *79*
Bolton, M.P., 395, 400
Bond, T., 29, *44*
Boomsma, C.D., *387*, 395, 396, 399, 403, *405*
Borchert, R., 260, *265*
Bormann, F.H., *141*
Borowik, M., 440, *446*
Bostanoglu, L., 19, *23*
Botkin, D.B., 35, 40, *44*, *45*, 51, *85*
Bottorff, R.L., 57, *80*
Boughton, V.H., *335*, *406*
Bouma, J., *81*
Bovenkerk, M., *436*
Bowden, D.C., *81*
Boyce, S.G., 28, *51*
Boynton, W., *84*
Boyt, R., *14*
Bradshaw, H.D., *80*, *135*, *460*
Brand, M., 420, *432*
Brass, L.J., 345, 346, *355*
Braun, A., 267, *285*
Braun, H.J., 415, *432*
Braun-Blanquet, J., 409, 416, 421, 422, *432*
Bravo, H., 291, *297*
Brawner, C.O., *23*, 51
Bray, J.R., 414, *432*
Brewer, R., *406*
Brezonik, P.L., 175, *197*
Brice, J., 94, *135*
Briese, L.A., *85*
Briggs, S.V., 103, *135*
Brink, V.C., 2, 8, *10*, 60, *80*
Brinkmann, R., 226, 228, 231, 236, *248*
Brinson, M.M., *45*, 56, 59, 68, 70, 75, 76, *80*, *83*, 96, 101, 103, 105, 108, 109, 110, 111, 112, 113, 114, *135*, 143, *167*, *198*, 447, 459, *460*
Brisbin, I.L., Jr., *81*, *82*, *84*
Briscoe, C.B., 252, *265*
Britton, N.L., 221, 245, *248*
Broadfoot, W.M., 61, *80*, 106, *135*
Broce, A., *167*, *198*
Brophy, J.A., *138*
Brouwer, Y.M., 398, *406*
Browder, J.A., 39, *45*, 77, *80*
Brown, B.J., *50*
Brown, B.T., *85*
Brown, C.L., *47*, 72, *82*
Brown, C.M., *84*
Brown, D.E., 6, *10*
Brown, K.E., 272, *285*
Brown, M.S., 140, 162, 164, *168*
Brown, S., 2, 5, 6, 9, *10*, *14*, 26, 32, 41, 42, *45*, 59, 60, 61, 62, 65, 70, 75, 76, 77, 78, *80*, *83*, 95, 97, 103, 104, 105, 108, 109, 110, 112, 114, 124, 131, *135*, *138*, 143, 159, *167*, 171, 174, 176, 177, 182, 183, 184, 185, 186, 187, 188, 190,
192, 193, 195, *196*, *197*, 256, 259, *264*, *265*, 447, 451, 456, 458, *460*
Brown, T.W., 3, *13*, 172, 174, 181, 182, *198*
Broyer, T.C., 108, *136*
Bruenig, E.F., 2, 4, 5, 9, *10*, 62, 80, 127, 299, 302, 303, 305, 306, 308, 309, 312, 313, 316, 318, 319, 320, 321, 322, 323, 325, 326, 327, 328, 329, 330, 331, 332, *333*, *334*, 447, 456
Brugam, R.B., 30, *45*
Bruner, W.E., 120, 121, *135*
Brush, G.A., 114, *135*
Brynildson, O.M., *47*
Bubenzer, G.D., *82*, *138*
Buchholz, K., 94, *135*
Buchwald, K., 409, 424, 425, 427, 428, *432*
Budowski, G., 271, 272, 285, *286*
Buell, H.F., *45*
Buell, M.F., 2, *11*, 29, 30, 36, 37, *45*, *47*, 48, 65, *80*, 95, 119, *135*
Buell, P.F., *433*
Bünning, E., 320, 325, *333*
Bunt, J.S., 160, 162, 164, *167*
Buol, S.W., 78, *81*
Burchell, R., 11, *197*
Burchell, R.B., *80*
Burges, N.A., *435*
Burgess, F.J., *23*
Burgess, R.L., *82*, *138*
Burkill, H.M., 369
Burnett, A.W., 89, *135*
Burnham, C.P., 305, *333*
Burns, G.P., 62, 63, *80*
Burns, L.A., *11*, 38, *45*, 48, *80*, 103, 104, 105, 111, *135*, 159, 160, *167*
Burt, W.H., 35, *45*
Burtt Davy, J., 239, *248*
Bush, J.K., 98, *135*
Butler, T., *23*, *49*

Cain, R.L., 37, *45*
Cairns, D.J., 105, *136*
Cairns, J., Jr., *51*
Cajander, A.K., 33, *45*
Campbell, C.J., 3, *11*, 26, *47*, 94, 101, 122, 123, *135*
Campbell, E.O., 37, *45*, *387*, *405*
Campbell, R.G., 1, 6, *11*
Cantlon, J.E., 108, 125, *135*
Capehart, B.L., 62, *80*
Carabia, J.P., 272, *286*
Carbiener, R., 409, 421, *432*
Carlson, C.A., *85*
Carlson, J.E., *46*, *136*
Carlton, J.M., 32, *45*
Carnahan, J.A., 388, *405*
Carpenter, S.R., 33, *45*
Carr, A.J., Jr., 4, *11*

Carr, D., *198*
Carter, G.S., 229, 243, *248*
Carter, M.R., 1, 2, *11*, 57, 76, *79*, *80*, 152, 159, 160, *167*
Carter, V. *11*, *201*, 210, *211*
Caruso, V.M., *210*
Casagrande, D.J., *11*, *47*, *196*, *197*, *198*
Catenhusen, J., 37, *45*
Caughey, M.C., 62, 63, *80*
Cavanaugh, J.A., *140*
Cavinder, T.R., *11*, *80*, *167*
Cavinder, T.W., *11*
Chabot, B.F., 62, 63, *84*, 191, *198*
Chabrol, L., 231, *248*
Chace, F.A., 245, *248*
Chai, P., *334*
Chamless, L.F., 116, *135*
Champion, H.G., 362, *385*
Chapin, F.S., III *141*, 454, *460*
Chapman, V.J., 1, 2, 8, *11*, 64, *80*, 143, 144, *167*, 213, 221, *248*, *385*, *405*
Chavan, A.R., 4, *11*
Chavelas, P.J., 288, 292, *297*
Chen, H.S., *81*
Cherubini, A., 31, *48*
Chesters, G., 132, *140*
Chiang, C.F., 291, *297*
Chikishev, A.G., *13*
Child, G.I., *82*, *137*, *167*, *285*
Chippendale, G.M., *405*
Chouard, P., 409, 421, 422, 423, 424, *433*
Choudhury, A.M., 384, *385*
Choudhury, M.U., 357, 363, *385*
Chow, C.Y., 383, *385*
Christen, H.V., 325, *333*
Christensen, B., 153, *167*
Christensen, N., 4, 9, *11*, 65, *80*, 174, 176, 178, 181, *197*
Christy, E.J., *140*
Church, M., 91, *135*
Cintrón, B., 251, 252, 259, *265*
Cintrón, G., 5, *11*, *82*, *83*, *139*, 143, 144, 145, 146, 148, 151, 154, 155, 161, *167*, *168*, 183, *197*, *198*, *264*, 447, *460*
Claessens, B., 409, 423, *433*
Clark, J.E., *11*, *13*, *23*, *83*, *84*, *197*, *198*
Clark, J.G.D., 71, *80*
Clark, J.R., 2, *11*, *13*, *23*, *83*, *84*, 94, 117, *135*, *197*, *198*, *460*
Clements, F.E., 29, 30, *45*, *51*
Clements, R.G., *82*, *137*, *167*
Clough, B.F., 2, 4, *11*, *13*, 157, 158, 162, 164, *167*, *168*
Clowes, D.R., *46*
Clymo, R.S., 34, 35, 40, *45*
Coaldrake, J.E., *387*, *405*
Coble, R.W., 92, 93, *135*
Cohen, A.D., 2, 3, *11*, 27, 29, 37, *45*, *47*, 50, 71, *80*, 178, *196*, *197*, *198*
Colby, B.R., *137*

AUTHOR INDEX

Coleman, J.M., *50*
Collins, E.A., 3, *11*
Colmet-Daage, F., 232, 233
Conard, S.G., 6, *11*, 96, 123, 124, *135*
Connell, J.H., 35, *45*
Conner, W.H., 98, 99, 103, 105, 106, 108, 117, *136*, *460*
Connor, D.J., 387, *405*
Conway, V.M., 178, 179, *197*
Cook, M.T., 221, 234, 245, *249, 251*, 265
Cooksey, B., 34, *45*
Cooksey, K.E., 34, *45*
Coomans de Ruiter, L., 430, *433*
Cooper, J.H., *138*
Cooper, J.M., *436*
Cooper, J.R., 113, *136*
Cooper, W.S., 29, *45*
Cope, O.B., 136, *139*
Corner, E.J.H., 324, 325, *334*
Corps of Engineers, 42, *45*
Correa, M.D., *11*, 167, *197*, *460*
Correll, D.L., 79, *84*, 113, *140*
Costa Rican National Meteorological Service, 274, *285*
Coull, B.C., *23*
Coultas, C.L., 177, *197*
Coulter, J.K., 78, *80*
Coulter, S.M., 3, *11*
Coupland, R.T., 126, *136*
Coutts, M.P., 62, 73, *84*
Cowan, I.R., *167*
Cowardin, L.M., 6, *11*
Cowell, D.W., *138, 140*
Cowles, S.W., *80, 197*
Cox, P.W., *140*
Craighead, F.C., 19, 20, *23*, 38, *45*, 176, *197*
Crawford, R.M.M., 2, *12*, 72, 73, *81, 82, 83*, 414, *434*
Creveld, M.C., 429, *433*
Crisan, P.A., 214, 222, 223, 245, *250*
Crisp, D.T., 38, *45*
Crites, R.W., 96, *136*
Crocker, R.L., 387, *405*
Cromack, K., Jr., 188, *199*
Cronin, L.E., *46*
Cropper, W.P., *10*
Crosby, I.B., 20, *23*
C.S.I.R.O., Australia, 387, 397, 399, *405*
Culler, R.C., 133, *136*
Curtis, J.T., 37, *45*
Curtis, S., 370, *385*
Cypert, E., 31, 37, *45*, 176, *197*
Czerwinski, A., 440, *446*

D'Arcy, W.G., 11, 167, *197*, *460*
Dabel, C.V., 3, *11*, 180, 181, 183, 184, *197*
Dachnowski, A., 27, 34, *45*, 62, 65, *81*
Dachnowski-Stokes, A.P., 26, *45*

Dagon, R., *433*
Dahlem, E.A., 133, *136*
Dakshini, K.M.M., 4, *11*, 65, *81*
Dames, T.W.G., 308, *334*
Damman, A.W.H., 119, *139*
Danell, K., 38, *45*
Daniel, C., 175, 176, 197
Daniel, C.C., 77, 78, *81*, 175, 176, *197*
Dansereau, P., 29, 34, *45*, 178, *197*, 425, 426, *433*
Dardeau, W.S., 221, *248*
Darrah, W.C., 27, *45*
Daubenmire, R., 38, *45*, 264, *265*
Davenport, D.C., 26, *46*, 133, *136*
Davie, J.D.S., *84*
Dávila, N., 153
Davis, A., 37, *46*
Davis, C.B., 36, 38, *50, 51*
Davis, J.H., 20, *23*, 65, 71, *81*, 154, 167, 182, *197*
Davis, M.B., 27, *46*
Day, F.P., Jr., 3, *11*, 174, 176, 178, 180, 181, 183, 184, 187, 188, 189, 194, 195, *197, 198, 199, 201, 211*
Day, J.H., Jr., 73, *84*, 152, 161
Day, J.W., Jr., 47, 98, 103, 105, 108, *135, 136*
Dean, J.M., *138*
De Bakker, H., 415, *433*
DeBell, D.S., 83, *137, 139*
Debellevue, E., *14*
Deevey, E.S., 55, *81*
Deevey, E.S., Jr., 30, 34, *46*
Degens, E.T., 79
Deghi, G.S., 188, 189, 190, 194, 195, *197*
DeGraaf, R.M., 85
De la Cruz, A., *80*
Delaney, P.J.V., 226, *248*
De las Salas, J., *11*
Delattre, P., 245, *248*
DeLaune, R.D., 76, *81*, 110, *136*
De Leeuw, W.C., 434
Delfos, A., *435*
Demaree, D., 185, *198*
Denayer-De Smet, S., 413, 414, 421, *433*
Den Held, A.J., 416, 421, 422, 430, *433*, *436*
Den Held, J.J., 416, 430, *433*
Denning, J.L., 58, *81*
De Roos, G.Th., 409, 416, 424, *435*
De Sloover, J.R., 415, *433*
De Smidt, J., 425, *436*
Deutsch, M., *211*
De Vries, E.J., *436*
De Vries, H.A., 409, 424, 425, *433*
De Waard, J., 429, *433*
De Wit, R.J., *436*
Deva, S., 4, *11*
Devillez, F., *433*
DeWit, T.P.M., 228, 231, 232, 247, *248*

DeWitt, C.B., 3, *11*, 39, *46*, 81, 84, *199*
Dexter, B.D., 387, 388, 389, 403, *405*
Dhanaraian, G., *140*
Dick-Peddie, W.A., 3, 6, 8, *11*, 94, 97, 101, 122, 123, *135, 136*
Dickinson, C.H., 34, *46*
Dickson, D.A., *10*
Dickson, J.G., 57, *81*
Dickson, R.E., 108, *136*
Dierberg, F.E., 175, 188, 189, *197*
Dierschke, H., 409, 417, 418, 421, 423, 424, 427, *433, 436*
Dinter, W., 414, 421, *433*
Dirschl, H.J., 126, *136*
Doing, H., 416, *433*
Dolman, J.D., 78, *81*
Dorge, C.L., 83, *139*
Doude van Troostwijk, W.J., 429, *433*
Dowding, E.S., 36, *48, 198*
Downs, W.G., *248, 249*
Drew, M.A., *23*
Drew, M.C., 72, 73, *81, 82*
Drury, W.H., Jr., 19, 23, 26, 28, 33, 36, 44, *46*, 126, *136*
Du Rietz, E., 409, *433*
Dubiel, E., 442, *446*
Dubois, F., *433*
Dudek, Cz., 439, 440, *446*
Duever, L., 14, *46*
Duever, M., 14
Duever, M.J., 27, *46*, 80, 82, 98, 103, 105, 111, *136, 137, 167*, 177, *197*
Dugger, K.R., *11*, 80, *167*
Duisberg, P.C., 85
Dumont, J.M., *433*
Duncan, D.P., 179, *197*
Dunk, P., 387, *405*
Dunn, C.P., 108, *136*
Durin, L., 418, *433*
During, H., 412, 420, *432, 434*
Durno, S.E., 71, *81*
Duvigneaud, P., 79, 409, 413, 414, 421, 428, *433*
Dykeman, W.R., 61, *81*

Ebinger, J.E., 96, *136*
Eckblad, J.W., 71, *81*
Edelman-Vlam, A.W., 415, *433*
Eggeling, W.J., 3, *11*, 65, *81*, 129, *136*, 165, *167*, 269, *285*
Eggers, H.F.A., 220, *248*
Eggler, W.A., 101, *136*
Egler, F.E., 29, 30, 32, 36, 37, *46*
Ehrenfeld, J.G., 96, 99, 101, 103, *136*
Ehrhart, R.L., *140*
Elder, J.F., 105, 113, *136*, 459, *460*
Elkins, J.B., Jr., *135*
Ellenberg, H., 415, 416, *433*, 439, *446*
Ellmore, G.S., 72, *81*
Emerson, F.V., 27, *46*

Encarnación, L., *139*, *168*
Endert, F.H., 299, *334*
Esparza, E., *297*
Espinal, L., *267*, *269*, 272, *285*
Evans, J.O., *140*
Evans, R., *198*
Everitt, B.L., 91, *136*
Evink, G., *167*, *198*
Ewel, J.J., 28, *46*, *48*, *80*, *267*, 284, *285*
Ewel, K.C., 2, 3, *10*, *11*, *14*, 37, *46*, *47*, *48*, *50*, *80*, *82*, *83*, *136*, *138*, *139*, 176, 178, 182, 188, 189, *196*, *197*, *198*, *199*, 447, 453, 457, *460*

Fahey, T.J., 120, *140*
Fail, J., *83*, *138*
Faliński, J.B., 437
Fanshawe, D.B., 4, *11*, 127, *136*, 213, 214, 218, 219, 228, 231, 236, 239, 241, *248*, 271, 272, *285*
Featherly, H.I., 32, *46*, 251, 259, *265*
Febvay, G., 228, 247, *248*, *249*, 251, 259, *265*
Felando, T., *138*
Feldmann, J., 222, *249*
Fenchel, T.M., 75, *81*
Fenton, J.H.C., 71, *81*
Ffrench, R.P., 244, *249*
Firouz, E., 4, *11*, 56, *81*
Fisk, H.N., 89, 91, *136*
Fittkau, E.J., 4, *11*, 58, *81*
Flenley, J.R., *10*, 38, *46*, *80*
Flohrschutz, E.W., *197*
Folster, H., 4, *11*
Fonda, R.W., 43, *46*, 95, 123, *136*
Fontaine, T.D., *135*, *460*
Food and Agriculture Organization (FAO), 54, *81*, 378, *385*
Ford, A.L., 98, 116, *136*
Fore, P.L., *11*, *80*, *167*
Forman, R.T.T., *48*
Forsyth, J.L., 25, *46*
Fortin, J.-A., *141*
Fosberg, F.R., 239, *249*
Fraley, J.J., 131, *136*
Frangi, J.L., 9, *11*, 60, 65, *81*, 98, 99, 103, 105, 109, 110, 111, *136*, 449, 453, *460*
Franklin, M.A., *138*
Franz, E.H., 39, 40, *46*, 102, 118, *136*
Frayer, W.E., 58, *81*
Fredrick, C.M., *46*
Freeman, C.E., 97, 123, *136*
Fretey, J., 241, 246, *249*
Frey, D.G., 28, *46*, 65, *81*, 165, *167*
Friedman, R.M., 39, *46*
Frye, R.J. II, 95, 119, *136*
Fukarek, F., 409, 421, 422, 423, 425, 427, 428, *433*, 442, 446
Fung, I., 53, 55, *83*
Funk, J.L., 131, *136*

Furch, K., 111, *134*, *138*, 452, *460*
Furtado, C.X., 369
Furtado, J.I., 56, 57, 61, *81*, 105, 110, *137*, *139*

Gaddy, L.L., 117, *137*
Gaines, D.A., 57, *81*
Galay, V.I., 132, *137*
Gale, J., 160, *168*
Gammon, P.T., 201, *210*, *211*
Gannon, J.E., 36, *49*
Gardner, L.R., *79*, *81*
Garin, B.E., 14, *85*, *169*
Garrett, M.K., *201*, *210*, *211*
Garrisi, P., 35, *49*
Gates, F.C., 35, *46*, 62, *81*
Gatewood, J.S., 93, *137*
Gaume, R., 409, 421, 424, *433*
Gentry, J.B., 58, *81*, *138*
Gibbons, J.W., *83*, *85*, *167*
Gifford, D.R., *13*
Giglioli, M.E.C., 34, *46*
Gilbert, O.J.W., 414, 415, *432*
Gill, A.M., 31, *50*
Gill, C.J., 94, *137*
Gill, D., 71, 82, *91*, 125, *137*
Gilliam, J.W., 77, 78, *79*, *82*, 113, *136*
Gilliland, M., *23*, *49*
Gillison, A.N., *406*
Gilman, B.A., 101, *137*
Given, P.H., 34, *46*
Gladden, J.B., *139*
Glascock, S., 116, 117, *137*
Glaser, P.H., 60, *84*
Gleason, H.A., 221, 234, 245, *249*, 251, *265*
Gleason, P.J., *45*
Glime, J.M., 34, *46*
Godfrey, M.M., 36, *46*
Godfrey, P.J., 36, *46*, *80*
Godwin, H., 33, 34, 39, *46*, 71, 81, 412, *433*
Goel, P.K., *50*
Goenaga, C., *83*, *167*, *168*, *198*
Golden, M.S., 117, *137*
Golet, F.C., *11*
Golley, F.B., *11*, 28, *46*, *79*, *81*, *82*, 99, 100, 103, 105, 109, 128, *137*, 152, 153, 159, 161, 162, 163, 164, *167*, *168*, *198*, 269, *285*
Golterman, H.L., *45*
Goltz, H.L., *10*
Gomez, M.M., 174, 176, 188, 194, *197*
Gómez-Pompa, A., 288, 290, 292, 293, *297*
Gong, W.K., *140*, *168*
Good, B.J., 97, 102, *137*
Good, R.E., *23*, *82*, *167*, *168*
Goodwin, R.H., 6, *11*
Gopal, B., *10*, *50*, *135*, *197*, *460*
Gore, A.J.H., *50*

Gore, A.J.P., 25, 26, 31, 33, 38, 39, 40, 43, *45*, *46*, *47*, *50*, *137*, *138*, *141*, *168*, 173, 178, *197*, *199*, *286*, *333*, *405*, 409, *433*
Gorham, E., 27, *46*, 414, *432*
Gosselink, J.G., 22, *23*, 60, *82*, *136*, 143, *167*, 447, *460*
Göttlich, K., *433*
Gottsberger, G., 37, *46*, 236, *249*
Gottschalk, M.R., 60, *79*, *84*
Gould, H.R., 71, *84*
Goulder, R., 31, *46*
Goulding, M., 37, *46*, 57, *82*, 133, *137*
Graf, W.L., 32, *47*
Graham, A., 248, *250*, *297*
Grannemann, N.G., 93, *137*
Graul, W.D., *141*
Graybill, F.A., *81*
Green, B.H., 3, *11*, 65, *82*
Green, W., 122, 123, 133
Greeson, P.E., 2, *13*, *23*, *83*, *84*, *197*, *198*
Grenke, W.C., *137*, *265*, *286*
Grime, J.P., 101, 102, *137*
Grittinger, T.F., 19, *23*
Grose, R.J., 403, 404, *405*
Grosse-Brauckmann, G., 415, *433*
Grossenheider, R.P., 35, *45*
Gruson, P.E., 2, *11*, *13*
Guerra, S.W., 268, *285*
Gulick, M., 96, 99, 101, 103, *136*
Gunderson, L.A., *46*
Guppy, H.B., 236, *249*
Gurvich, B.R., 272, *285*
Gusen, E.J., 374, *385*
Gustafson, T.D., *168*

Haantjens, H.A., 338, *355*
Haapala, H., 60, *82*
Haase, E.F., 123, *137*
Haddock, J.D., *139*
Hadley, R.F., 90, *137*
Hagan, R.M., *46*, *136*
Haight, L.T., *82*
Hall, C.A.S., 39, *47*
Hall, N., 387, 389, 390, 391, 392, *405*
Hall, T.F., 3, *11*, *13*, 94, 97, 98, *137*
Hallé, F., 236, *249*
Hallsworth, E.G., *406*
Halpenny, L.C., *137*
Halterman, S.G., *49*
Hamilton, A., 370
Hamilton, A.C., 37, *50*, 129, *141*, 165, *168*, 269, *286*
Hamilton, D.B., 37, *47*, 181, *198*
Hamilton, S.K., 133, *137*
Hanlon, T.M., *139*
Hannan, H.H., 131, *137*
Hansen, B., 409, 411, *433*
Hanskamp, B., 429, *434*
Hanson, R.L., *136*

AUTHOR INDEX

Harcombe, P.A., 94, 99, *139*
Hardin, E.D., 101, 102, 110, *137*
Harms, W.R., 38, *47*, 58, *82*, 108, *137*
Harnickell, E., *436*
Harper, R.M., 3, *12*
Harriss, R.C., 104, *137*
Hart, J.B. Jr., *137*
Hart, M.G.R., 34, *47*
Hartland-Rowe, R., 3, *12*
Hasler, A.D., 31, 35, *47*, 48, *50*, *250*
Hastings, H.M., 40, *47*
Hathaway, E.S., 101, 114, *140*
Hatheway, W.H., *137*, *265*, *286*
Hausler, J.F., *10*
Haverschmidt, F., 244, *249*
Hawk, G.M., 125, *137*
Hay, D., *79*, *82*
Heal, O.W., *45*, *47*
Heald, E., 161, *167*
Heatwole, H., 58, *82*
Heck, K.L., Jr., 32, *48*
Hecky, R.E., 34, *47*
Heede, B.H., 133, *137*
Hegerl, E., *84*
Heikurainen, L., 1, 3, 7, 8, *12*
Heilman, P.E., 29, 35, *47*, 176, 177, 189, 190, 193, *197*
Heimburg, K., 62, *82*, 173, *197*
Heinemann, P., *434*
Heinselman, M.L., 7, *12*, 19, 20, *23*, 29, 33, *47*, *51*, 60, *82*, 120, *137*, 172, 174, 175, 178, 179, 189, *197*, *199*, 449, *460*
Helfgott, T., *46*
Helm, W.T., *47*
Helvey, J.D., *140*
Hem, J.D., *137*
Hemond, H.F., 171, 173, 174, 180, *197*
Hendrich, H., 370, 382, *385*
Hendrickson, O., Jr., *83*, *138*
Hendrix, K.M., *141*
Hepp, G.T., *82*
Hernández, C.J., 106, *137*
Hernández, R.A., 106, *137*
Hernández, X.E., 287, 288, 290, 293, *297*
Herrera, R., 9, *12*, 306, 326, 327, 329, 331, *334*
Herrick, S.M., 27, *47*
Hesse, R., 27, *47*
Hevert, F., 234, *249*
Hewitt, P.N., *406*
Heyligers, P.C., 270, *286*, 341, 344, *355*
Heywood, V.H., *435*
Hibbert, F.A., 71, *82*
Hicks, D.B., *11*, *80*, 159, 160, *167*
Higgins, A.E., *79*
Hild, J., 409, 422, 427, *434*
Hillman, W.S., 31, *47*
Hinckley, T.M., *141*
Hjulström, F., 32, *47*
Hobbie, J.E., *82*

Hobbs, H.H., 245, *248*
Hodges, J.D., 9, *10*, *12*
Hodkinson, I.D., 111, *137*
Hofmann, G., 416, 421, *435*
Hofstetter, R.H., 115, 120, *137*
Holdridge, L.R., 9, *12*, 19, *23*, 98, 100, 128, *137*, 221, 234, *249*, 255, 257, *265*, 272, 273, 281, 282, 283, *286*
Hole, F.D., *81*
Holland, R.F., *11*, *135*
Hollom, P.A.D., *435*
Holmes, R.N., *135*, *460*
Hook, D.D., 2, *12*, *47*, 72, 73, *82*, *83*, *137*, *139*, 414, *434*
Hooper, F.F., 34, *47*
Hope, G.S., *355*
Hopkins, J.M., 56, *82*
Horn, H.S., 28, *47*
Horton, J.S., 26, *47*
Horwitz, E.L., 1, *12*, 57, 58, *79*, *82*
Hosner, J.F., 102, 108, 117, 118, *137*
Hough, A.F., 101, *138*
Houghton, R.A., 55, *82*
Howarth, R.W., 75, *82*, *84*
Howe, V.K., *12*, *83*
Howell, F.G., *138*
Hubbard, J.P., *11*
Hubble, G.D., *406*
Huenneke, L.F., 101, 102, 119, 120, *138*
Hughes, J.H., 1, *11*
Humm, H.J., 38, *49*
Hunt, K.W., 31, *47*
Huntley, B., *46*
Hupp, C.R., 96, 102, 119, *138*
Husain, K.Z., 370, 372, 374, *385*
Hutchinson, G.E., 25, 30, 33, *47*, 458, *460*
Hutchinson, J.J., 77, 78, *82*
Hutchinson, S.L., 387, *405*
Huxley, P.A., *333*
Hvass, H., 431, *434*

Incoll, F.S., 387, *405*
Ingram, H.A.P., 3, *12*, 60, 65, *82*, 165, *167*
Irmler, U., *11*, *81*, 111, 127, *134*, *138*, *271*, *286*, 452, *460*
Irvine, J.R., 123, *138*
Irwin, J.E., *140*
Isaak, D., 33, *47*
Isakov, Y.U., 3, *12*, 56, *82*
Islam, A.K.M.N., 369, *385*
Islam, K.S., 383, *385*
Islam, M.A., 361, *385*
Ismail, M., 357, 359, 360, 361, *385*
Ivarson, K.C., 35, *47*

Jack, W.H., *10*
Jackson, M.B., 73, *82*
Jackson, N.M., Jr., 59, *82*

Jackson, T.A., 34, *47*
Jacobs, M., 305, *334*
Jacobs, M.R., 387, 396, 400, 403, 404, *405*
Jaikaransigh, E., 243, *248*
James, A., 221, 236, 238, 245, 247, *249*, 259, 261, *265*
Janak, J.F., *45*
Jansen, C.R., 120, *138*
Jansson, O., 42, *49*
Janzen, D., 260, 261, *265*
Janzen, D.H., 236, *249*
Jasnowski, M., *14*, 437, 443, *446*
Jasper, S.A., *140*
Jaworski, E., 173, *197*
Jeglum, J.K., *14*, 126, *138*, *169*
Jenik, J., 72, *83*
Jennings, J.N., 416, *434*
Jensen, J., *433*
Jervis, R.A., 43, *47*
Jeswiet, J., 409, 423, *434*
Jiménez, J.A., 59, *82*, 154, *167*, 176, *198*
Johnson, D., 28, *47*, 272, *286*
Johnson, F.L., 41, *47*, 93, 102, 103, 118, *135*, *138*
Johnson, J., 33, *49*
Johnson, K.J., *139*
Johnson, R.L., 172, *198*
Johnson, R.R., 2, *12*, 56, *82*, 85, *139*, *167*, *460*
Johnson, W.C., 71, *82*, 91, 95, 121, 132, *138*
Johnston, C.A., 71, *82*, 112, *138*
Johnston, R.D., *405*
Jones, D.A., 2, *12*
Jones, E., 31, *47*
Jones, H.E., 39, 40, *47*
Jones, M.N., 80, *135*
Jones, R.A., *138*
Jonkers, A.H., 247, *249*
Jordan, C.F., *12*, 306, 329, 331, *334*
Jørgensen, B.B., 75, *81*, *85*
Jouanne, P., 409, 421, 423, *434*
Junk, W.J., *11*, *81*, 127, *138*, 235, 244, *249*, 271, *286*
Juraimi, Mr., 369

Kac, N.J., *446*
Kalliola, R., *140*
Kampmann, H., 429, *434*
Kan, A.Q., *385*
Kane, E.S., *135*
Kaner, N., *13*
Kangas, P., 42, 43, *47*, *51*
Kangas, P.C., 8, 15, 87, 143, 171, 172, 178, 451
Karpiscak, M.M., 132, *141*
Kaufmann, R., 234, *249*
Kaule, G., 428, *434*
Kaynor, E.R., *80*

Keammerer, W.R., *82*, 91, 95, 121, 132, *138*
Keast, A., *406*
Keeley, J.E., 73, *82*, 237, *249*
Kelso, D.P., 274, *286*
Kemp, W.M., 13, 42, 47, 64, *82*, *84*
Kempe, S., *79*
Kenkel, N.C., 126, *138*
Kennard, W.C., *46*
Kennedy, B.J., *46*
Kenny, J.S., 241, *248*, *249*
Kerekes, J., 33, *47*
Kermarr, C.A., 251, 259, *265*
Kermarrec, A., 228, 245, 247, *248*, *249*
Kestemont, F., 409, 414, 415, *434*
Ketner, P., *79*
Khan, F.K., 361, *385*
Khan, M.A.R., 370, *385*
Khanna, P., *11*
Khattak, G.M., *385*
King, C.C., *46*
Kipple, F.P., *136*
Kirk, P.W., Jr., *48*, *211*
Kitchens, W.M., Jr., 112, *138*, *141*
Kitson, A., 245, *249*
Klapperstück, J., 429, *434*
Klein, J., *434*
Klein, W.C., *355*
Klinge, H., *12*, 99, 103, 127, *138*
Klötzli, F., 416, *434*
Knapp, R., 28, *47*
Knight, D.H., 237, *249*
Knight, H.A., 107, *139*
Knight, R.W., *44*
Knights, B., *49*
Koenings, J.P., 34, *47*
Kohlsaat, T.S., *137*
Koldewijn, B.W., 226, 231, *249*
Kolehmainen, S.E., 144, *166*
Kologiski, R.L., 6, *12*
Kondolf, G.M., 123, *138*
Kop, L.G., 409, 418, 419, 420, 421, 422, 424, *434*
Kormanik, P.P., *82*
Kormondy, E.J., 33, *47*
Koteswaram, P., 302, *334*
Koth, W., 431, *434*
Kozlowski, T.T., 2, *12*, 36, *47*, 60, 72, *83*, *169*
Kramer, N., *84*
Kramer, P.J., 185, *198*
Krausch, H.D., 421, *434*
Krenkel, P.A., 131, *138*
Kropp, W., 29, *47*
Krzywanski, D., 439, *446*
Kuddus, M.A., 373, *386*
Kuenen, Ph.H., 27, *51*
Kuenzler, E.J., *79*, *83*
Kulczynski, S., 365, *385*
Kurz,H., 35, 37, *48*, 182, 185, *198*

Kushlan, J.A., 38, *48*

Lagler, K.F., 132, *138*
Lahde, E., 60, *83*
Lai, K.K., 331, 332, *334*
Lai, L., *83*, *139*
Lamb, F.B., 4, *12*
Lambert, J.M., 409, 416, 422, *434*
Lamberti, A., *166*, *167*
Lami, R., 222, *249*
Landers, R.Q., Jr., *44*
Langbein, W.B., 114, *138*
Langdale-Brown, I., 269, *286*
Lanly, J.P., 54, *83*
LaRoe, E.T., *11*
Larsen, J.A., 1, 2, *12*, 178, *198*
Larson, J.S., *85*
Lassoic, J.R., *139*
Lattman, L.H., 89, 90, *138*
Laughnan, P.F., *49*
Laurent, E.A., *137*
Lavine, M.J., 23, *49*
Lawler, D.J., 54, *80*
Lawson, D.L., 111, *139*
Layser, E.F., 6, *13*
Le Barron, R.K., 38, *48*
Leaf, A.L., *137*
Lebrun, J., 416, *433*, *434*
Lee, G.B., *82*, *138*
Lee, G.F., *138*
Lee, L.C., 95, 108, *138*
Lee, S.C., *81*
Lee, T.H., *137*
Leentvaar, P., *433*
Leeper, G.W., *405*, *406*
Lefor, M.W., *46*
Lehman, M., *84*
Leisman, G.A., 71, *83*
Leitman, H.M., 98, *138*
Lemée, G., 409, 413, 421, 423, 424, *434*
Lenk, C., 114, *135*
Leonard, R., *83*, *138*
Leopold, L.B., 42, *51*, 71, *85*, 90, 92, *138*, *141*
León, C.J.M., 292, *297*
Lescuré, J.P., 234, 236, 237, 246, *248*, *249*
Léveque, A., 231, 237, *249*
Levine, E.A., 164, *167*
Levit, J., 261, *265*
Levy, G.F., 201, *211*
Lewis, C.P., 94, *139*
Lewis, F.J., 3, *12*, 36, *48*, 176, 178, *198*
Lewis, W.M., 31, *48*
Lewis, W.M., Jr., 133, *137*
Leyton, L., 72, *83*
Liang, T., *137*, *265*, *286*
Lichtler, W.F., 201, *211*
Lichty, R.W., 71, *84*, 90, *140*
Liegel, L., *80*

Liem, D., 4, *12*
Lieth, H., 301, *334*, 414, *434*
Liew, K.S., *137*
Liggett, A., *11*, *80*, *197*
Likens, G.E., 51, 54, 55, *83*, *85*, *141*
Linacre, E.T., 26, *48*
Lind, E.M., 165, *167*
Lindauer, I.E., 95, 121, *138*
Lindberg, R.D., 74, *83*
Lindeman, J.C., 213, 214, 218, 219, 222, 224, 228, 229, 231, 232, 234, 237, 238, 239, 241, *249*, 262, *265*
Lindeman, R.L., 29, 33, 41, *48*
Lindsey, A.A., 94, 95, 118, 119, *138*, *140*
Litshultz, F., *84*
Little, S., 37, *48*
Livingston, B.E., 34, *48*, 62, *83*
Livingstone, D.A., 30, *48*
Lobeck, A.K., 35, *48*
Lodhi, M.A.K., 113, *138*
Lohmeyer, W., 409, 418, 419, 421, 423, *434*, *435*
Longman, K.A., 72, *83*
Loomis, R.S., *14*, *141*, *286*
López, J.M., *264*
Lorenc, K., 439, *446*
Lorio, P.L., Jr., 9, *10*, *12*, 74, *83*
Lot-Helgueras, A., 287, 288, 290, 292, 293, *297*
Loucks, O.L., *80*
Lovejoy, T.E., 271, *286*
Loveless, A.R., 63, *83*
Loveless, C.M., 20, *23*, 37, *48*
Lowe, C.H., 6, *10*
Lowe-McConnell, R.H., 241, 243, 244, *249*
Lowrance, R., *79*, *83*, *138*
Lowrie, S., *50*
Luckett, L.M., *82*
Lundqvist, J., 19, *23*
Lugo, A.E., 1, 2, 5, 7, *10–12*, 27, 30, 32, 35, 36, 40, 41, 43, *45*, *48–50*, 56, 58, 59, 60, 65–69, 75, 76, *80–83*, *85*, 98, 99, 101, 103–106, 109–111, 114, 127, 131, *135*, *136*, *138–140*, 143–145, 153–161, 163, 165, *167–169*, 176, 185, 187, *196–199*, 256, 259, *264*, *265*, 284, *286*, *297*, 447, 449, 453, 456–458, *460*
Lundell, C.L., 271, *286*, 291, 293, 294, *297*
Lyneborg, L., 429, *434*

Maas, F.M., 414, *434*
MacDonald, B.C., 94, *139*
MacDonald, R.L., *11*, *135*
Mackenzie, F.T., *83*
MacNae, W., 56, *83*
Madgwick, H.A.I., 415, *435*
Madison, F.W., *82*, *138*
Madriz, A., *267*, *285*

AUTHOR INDEX

Maire, A., *141*
Mak, P.J., *436*
Makatsch, W., 429, *434*
Maki, T.E., 1, *13*, 61, *83*
Malecki, R.A., 106, *139*
Mallaux, O.J., *267*, 268, *286*
Malmer, N., 29, *48*
Malquit, G., 409, 421, *434*
Mandelbrot, B.B., 40, *48*
Marek,S., 438, 443, *446*
Margadant, W.D., 412, *432*, *434*
Margalef, R., 39, *48*
Marks, P.L., 94, 99, *139*
Marois, K., 182, *196*, *198*
Marsh, C., 133, *139*
Marshall, H.G., 201, *211*
Marshall, R.C., 216, 224, *249*
Marshall, W.B., 241, *249*
Marshall, W.H., *47*
Martin, C.N., *12*, *83*
Martin, P.E., *46*, *136*
Martin, S.C., *139*
Martínez, R., *11*, 100, *139*, 144, 145, 150–152, 155, 156, 167, *168*, *197*, *460*
Martyn, E.B., 215, *249*
Mattews, E., 53, 55, *83*
Matthews, J.D., 414, *434*
Mattoon, W.R., 3, *12*, 116, *139*, 185, *198*
Matuszkiewicz, J., 439, 440, 441, *446*
May, N., 57, *85*
Mayer, H., 412, 416, 421, 422, *434*
Mayo Meléndez, E., 99, 100, *139*
Mayo, L.R., 94, *140*
McBride, J., 101, 118, 119, *139*
McCaffrey, C.A., 181, *198*
McCann, J.A., *85*
McCarthy, J., 3, *12*, 58, 65, *83*, 129, *139*
McClelland, M.K., 102, *139*
McCless, J.D., 58, *85*
McClure, J.P., 107, *139*
McColl, R.H.S., 79, *83*, 113, *139*
McCormick, I., 28, 39, *48*
McCormick, J.F., 2, *12*, *139*, *167*, *460*
McCoy, E.D., 32, *48*
McDowell, L.L., 71, *83*
McEvoy, T.J., 101, *139*
McFarlan, E., Jr., 89, *136*
McGinnis, J.T., *82*, *137*, *167*
McHargue, L.T., 3, 8, *14*
McHenry, J.R., *82*, *138*
McIntosh, R.P., 28, *48*, 316, 319, *334*
McKee, W.H., Jr., 73, *83*, 108, *137*, *139*
McKinley, C.E., 176, 178, *198*
McLaren, D., *46*
McLellan, J.F., *138*
McManmon, M., 73, *81*, *83*
McMillan, C., 157, *168*
McNaughton, S.J., 34, *48*
McVaugh, R., 32, *48*
McVean, D.N., 415, *434*

Medina, A.L., 123, *139*
Medina, E., *11*, *12*, *81*, 162, *168*, *198*, 326, *334*
Medwecka-Kornás, A., 438, 440, 441, 443, *446*
Meehan, W.R., 6, *12*, 57, *83*
Mehuys, G.R., *84*
Meijer Drees, E., *80*, 409, 419, 420, 422, 424, 427, 428, *434*
Meijer, W., *436*
Meléndez, E.M., 4, *13*
Melillo, J.M., *82*
Menéndez, L.F., 288, 293, *297*
Menges, E.S., 55, *80*, 101, 102, *139*
Merrell, W.D., 65, *85*
Merriam, C.H., 19, *23*
Merritt, R.W., 111, *139*
Meskus, E., *82*
Metzler, K.J., 119, *139*
Michalowski, M., 272, *286*
Miles, J., 416, *434*
Miller, C.A., *135*
Miller, J.P., 92, *138*
Miller, M.A., *23*, *49*
Miller, W.D., 1, *13*, 61, *83*
Millette, J.A., *84*
Minckler, L.S., 102, 117, 118, *137*
Miranda, F., 287, 288, 290, 291, 292, 293, 294, *297*
Misra, R., *50*
Mitsch, W.J., 37, 40, *46*, *48*, 71, *83*, 105, 106, 109, 112, *139*, 176, 178, *197*, 447, *460*
Mogensen, G.S., *433*
Mollitor, A.V., *139*
Monahan, T.J.; *81*
Monasterio, M., 4, *13*
Monefeldt, I., 163
Monk, C.D., 3, 9, *11*, *13*, 172, 174, 176, 177, 181, 182, *198*
Monschein, T.D., 58, *83*
Montague, K.A., 73, *84*
Montaignac, P., 222, 245, 246, *249*
Monteith, J.L., *48*
Montenegro, E., 267, 269, 272, *285*
Monticciolo, R., *47*
Montz, G.N., 31, *48*
Moore, B., *82*
Moore, C.W.E., *387*, *405*
Moore, H.E., *267*, 268, 269, 271, *286*
Moore, P.D., 2, 8, *13*, 34, *48*, 60, 75, *84*
Moorthy, K.K., 4, *13*
Mootoosingh, S.N., 248
Mori, S., 4, *10*, 56, 57, 61, *81*, 269, *285*
Morris, D.R., *168*
Morris, G., *11*, *167*
Morris, L.A., 120, *139*
Mortimer, C.H., 74, *84*
Mörzer Bruyns, M.F., 3, *13*, 431, *434*
Mosquera, R.A., 100, *139*

Moss, E.H., *198*
Moszynska, B., 445, *446*
Mountford, C.P., *406*
Mountfort,G., 370, *385*, *435*
Mueller-Dombois, D., *436*
Mukherjee, A., 370, 374, *385*
Mulcahy, M.J., *406*
Mulholland, P.J., 80, 83, 96, 103, 104, 105, *139*
Mullen, K.P., 106, *137*
Muller, P., *138*
Müller, J., 219, 220, 223, 228, 231, 232, 235, *249*, 309, 328, 333
Müllerstael, H., 330, *334*
Munch, J.C., 74, *84*
Muzika, R.M., 103, 105, *139*
Myers, R.L., *80*, 127, *267*, 271, 272, *286*, 447, 449
Myrick, R.M., *136*

Nadler, C.T., Jr., 132, *139*
Naiman, R.J., 15, *23*
National Oceanographic and Atmospheric Administration, 42, *48*
National Research Council, 26, *48*, 132, *139*, 268, *286*
Neck, J.S., *140*
Nedwell, D.B., 74, *84*
Neess, J.C., *50*
Neetzel, J.R., 38, *48*
Neiring, W.A., 6, *11*
Nessel, J.K., 103, 105, 108, 109, 110, 111, *139*
Neville, M.K., 244, *249*
Newbould, P.J., 72, *84*
Newson, L.A., 240, *249*
Nichols, G.E., 29, *48*
Nickerson, N.H., *81*
Nietschmann, B., 246, *249*
Nisbet, I.C.T., 28, 33, *44*, *46*
Nixon, E.S., 43, *48*, 116, *135*, *140*
Noble, M.G., 32, *48*
Noirfalise, A., 416, *434*
Nordlie, F.G., 274, *286*
Northcote, K.H., *406*
Nottebohm, F., 244, *249*
Nottebohm, M., 244, *249*
Novelo, R.A., 287, 290, *297*
Novitzki, R.P., 6, *13*, 22, *23*, 77, 78, *84*, 173, 174, *198*
Nuhn, H., 269, 271, 281, *286*

O'Connell, M., 409, 422, *434*
O'Malley, L.S.S., 370, *386*
O'Neil, T., 38, *49*
O'Toole, M.A., 35, *49*
Oaks, R.Q., Jr., 35, *48*
Oberdorfer, E., 416, 417, 418, 419, 421, 427, *434*, 439, *446*
Obminski, Z., 438, 444, *446*

Odum, E.P., 15, *23*, 28, 42, *43*, 48, 65, *84*
Odum, H.T., 2, 3, 8, 9, *11*, *13*, 15, 16, 22, 23, 26–28, 32, 33, 39, 40–42, *44*, *46–51*, 57, 63, 64, 67, *80*, *82–84*, *136*, *138*, *139*, 144, 159, *167*, *168*, 178, 185, *196–199*, 447, 451, 453, 455, 457, *460*
Odum, W.E., 39, 51, *84*
Oglesby, P.T., *85*
Ogren, D.E., *50*
Ohmart, R.D., *79*
Oldeman, R.A.A., *249*
Olson, J.S., 40, 43, *46*, 54, *84*
Olson, T.A., *23*
Ong, J.E., 106, *140*, 159, 160, 162, 164, *168*
Orozco, A., 288, 290, 292, *297*
Osmaston, H.A., *286*
Osterkamp, W.R., 102, *138*
Ostlie, K., *81*
Osvald, H., 411, *435*
Ottow, J.C.G., 74, *84*
Outhet, D.N., 91, 92, *140*
Overbeck, F., 71, 412, 415, *435*
Ovington, J.D., 414, 415, *435*
Ozenda, P., 412, 413, *435*

Paerl, H.W., 31, *49*
Paijmans, K., 270, *286*, 335, 338, 347, *355*
Pain, R., 241, *249*
Pallis, M., 31, *49*
Pals, J.P., 71, 415, *435*
Pancholy, S.K., 34, *49*
Paratley, R.D., 120, *140*
Parent, L.E., 78, *84*
Parker, G.G., 454, *460*
Parker, G.R., 178, 179, 180, 183, 184, 186, 187, *198*
Parker, R.E., 5, *13*
Parrish, F.K., 27, *49*
Parrondo, R.T., *136*, *460*
Parsons, S.E., 94, 116, 117, *140*
Paschal, J.E., Jr., *210*
Pase, C.P., 6, *13*
Passarge, H., 416, 421, *435*
Patric, J.H., 90, *140*
Patrick, R., 35, *49*
Patrick, W.H., Jr., 74, *81*, *84*, 112, *136*, *140*
Patten, B.C., *44*, 48, *51*
Patterson-Zucca, C., 2, 5, *12*, 76, *83*, *85*, 144, *167*, *169*, *198*, *199*
Paul, E.A., *46*
Payens, J.P.D.W., 345, *355*
Pearsall, W.H., 7, *13*
Pearson, M.C., 3, *11*, 65, *82*
Pecan, E.V., *85*
Pekelney, R., *47*
Pelezarski, S., *80*
Pendleton, E.C., *141*

Penfound, W.T., 3, 6, *11*, *13*, *14*, 31, *49*, 65, *84*, 94, 97, 98, 101, 114, *137*, *140*
Penman, H.L., 26, *49*
Pennington, T.D., *297*
Peppers, R.A., *49*
Pérez, J.L.A., 290, *297*
Pérez, R.S., 269, 271, 282, *286*
Perkins, D.F., *45*, *47*
Perry, K.M., *435*
Perry, R.A., 387, 390, 396, *406*
Persson, T., 1, *13*
Peterjohn, W.T., 79, *84*, 113, *140*
Petersen, P.M., 409, 411, 429, *435*
Peterson, B.J., 75, *82*, *84*
Peterson, D.L., 95, 104, 105, 109, 110, 111, 112, *135*, *140*
Peterson, N.L., *81*
Petty, R.O., *138*
Pewe, T.L., *23*
Pezeshki, S.R., *81*, *136*
Philipson, J.J., 62, 73, *84*
Phillips, A.J., *49*
Phillips, R.C., 31, *49*
Phillips, T.L., 27, *49*
Phipps, R.L., 39, 40, *49*, 94, 106, *140*
Pigeon, R.F., *13*, *23*, *49*
Pionke, H.B., 132, *140*
Pires, J.M., 9, *13*
Pirkle, W.A., 26, *49*
Pittier, H., *267*, *286*
Pittillo, J.D., 120, *141*
Plantico, R.C., *80*
Pointier, J.P., 244, *249*
Pokkett, F.C., *169*
Polak, B., 299, 303, 305, 325, *334*
Polakowski, B., 440, *446*
Polczynski, A., *14*
Poljakoff-Mayber, A., 160, *168*
Pollett, F.C., *14*, 115, 124, *141*, 178, *199*
Ponce, M., 9, *11*, 65, *81*
Ponnamperuma, F.N., 74, *84*
Pons, J.L., 226, 228, 231, 236, *248*, 409, 421, *435*
Pool, D.J., 6, 11, *13*, 100, 106, 127, *140*, 152, 160, 161, 163, *167*, *168*, 183, 188, *198*, 283, *286*, 293, *297*
Poore, M.E.D., 370, *385*
Poorter, E.P.R., 430, *435*
Portecop, J., 214, 222, 223, 245, *250*
Porter, D.M., 214, *250*
Post, A., 94, *140*
Post, D.M., 190, 191, 192, *199*
Potzger, J.E., 27, 33, *49*
Poveda, L.J., 272, *286*
Powell, J.M., 354, *355*
Prain, D., 384, *386*
Prakash, A., 34, *49*
Prance, G.T., 4, *13*, 127, *140*
Prentki, R.T., 165, *168*
Price, R.J., *137*

Price, W.A., 33, *49*
Proctor, G.R., 221, *248*, *250*
Proctor, J., 327, *334*
Pronczuk, J., 437, *446*
Prouty, W.F., 28, *49*
Puig, H., 288, 290, *297*
Pullen, R., 354, *355*
Pyatt, D.G., *436*
P'yavchenko, N.I., 7, *13*

Qualls, R.G., 113, *140*
Queensland Department of Primary Industries, Division of Land Utilization, 387, 398, *405*
Quero, H., 292, 293
Quevedo, V., *264*
Quinn, J.A., 95, 119, *136*
Quiróz-Flores, A., *297*

Rabeler, W., 431, *435*
Rabinowitz, D., 37, *49*, 144, 156, *168*
Radforth, N.W., 22, *23*, *51*
Ragsdale, H.L., 79, *82*
Rahman, A.K., 370, 378, *386*
Rajpurohit, K.S., *85*, *168*
Ramcharan, E.K., 225, *248*
Rand, A.L., 345, 346, *355*
Randerson, P.T., 39, *49*
Rango, A., *211*
Raphael, C.N., 173, *197*
Räsänen, M.E., 89, *140*
Rashid, H., 374, *386*
Rashid, M.A., 34, *49*
Ratter, J.A., 4, *13*
Reader, R.J., 29, 40, 41, *49*, 174, 178, 179, 183, 184, 186, *198*
Reddy, K.R., 74, *84*
Reese, K.P., *82*
Regan, E.J., Jr., 41, *49*
Rehder, H., 417, *435*
Rehm, A.E., 38, *49*
Reich, P.B., 260, *265*
Reichart, H., 244, *248*
Reichle, D.E., 1, *13*, 414, *434*, *435*
Reid, G.K., 31, *49*
Reily, P.W., *138*
Reiners, N.M., 177, 178, 179, 193, 194
Reiners, W.A., *45*, 174, 177, 178, 179, 183, 184, 186, 188, 189, 193, 194, *198*
Reiss, F., *11*, *81*
Renard, Y., 245, *250*
Renault-Lescuré, O., 241, 246, *249*
Revells, H.L., *11*, *80*, *167*
Revesbech, N.P., *85*
Revilla, C.J., *138*
Reynolds, L.M., *250*
Reynolds, S.T., *406*
Rice, D.L., 75, *84*
Rice, E.L., 3, *13*, 34, *49*, 97, 101, 121, *140*

AUTHOR INDEX

Richards, K., 59, *84*
Richards, P.W., 3, 8, *13*, 31, 35, *49*, *140*, 221, 232, 235, 245, *250*, *267*, 269, 271, *286*, 299, 305, *334*
Richardson, C.J., 2, 5, *11*, *13*, *81–83*, *85*, 173, 174, 176, 178, 181, *197*, *198*
Richley, J.E., *83*
Rico-Gray, V., 291, 294, *297*
Ridd, P., 128, *141*
Riddle, W.F., 44
Rieger, E., *139*
Rieley, J., 165, *166*
Rigg, G.B., 71, *84*
Riley, W.S., *198*
Ringuelet, R.A., 4, *13*
Riopelle, L.A., *136*
Rioux, J.A., 245, *250*
Ripley, S.D., 374, *385*, *386*
Risebrough, R.W., *250*
Robbins, R.G., 221, *248*, 270, *286*, 338, 354, *355*
Roberts, H.H., *136*
Robertson, P.A., 93, 94, 96, 99, 102, 118, *140*
Robinson, J.W., 131, *136*, *137*
Rodin, L.E., 162, *168*
Roe, E.M., *355*, *406*
Roff, C., 391, *405*
Rohde, W.G., *211*
Rojas, T., 272, *286*
Rolfe, G.L., 109, 110, 111, 112, *140*
Rollet, B., 1, *13*, 293, *297*, 347, *355*
Rosenberg, R., *83*, *168*, *198*
Rothmaler, W., 412, *435*
Rousseau, L.Z., 72, *83*
Rovirosa, J.N., 288, *297*
Rowe, J.S., 115, *140*
Ruley, L.A., *83*
Runkle, J.R., 38, *49*
Runnells, D.D., 74, *83*
Russell, R.J., 31, *49*
Rust, W., *83*, *139*
Rutzler, K., 213, *250*
Rykiel, E.J., 27, *49*
Rzedowski, J., 287, 290, 292, 293, *297*
Rzoska, J., *50*

Sabnis, A.D., 4, *11*
Saenger, P., 2, 9, *13*, 53, 54, *84*, 144, 150, 156, *168*, 389, *405*
Sage, B.L., 125, *140*
Salazar, A., 268, *286*
Sale, M.J., *138*
Salfeld, J.-Chr., 306–308, *334*
Salo, J.S., *140*
Salt, R.B., *406*
Sampson, H.C., 38, *49*
Sander, N., 329, *333*
Sandoshan, A.A., 383, *386*
Sands, A., 2, *13*, *135*

Sanger, R., 36, *49*
Santiago, J., *264*
Sarker, S.U., 374, *385*
Sarmiento, G., 4, *13*
Sarukhán, K.J., 288, 290, 291, 293, *297*
Sastre, C., 247, *250*
Savage, T., 32, *49*
Say, E.W., 42, *49*
Scamoni, A., 416, 422, 423, 427, 428, *435*
Scbacher, D.I., 104, *137*
Schacklette, H.T., 26, *49*
Schaeffer-Novelli, Y., 161, *168*
Schaller, G.B., 370, *386*
Scheller, U., 58, *84*
Schenk, E., 19, *23*
Schiller, W., 431, *435*
Schimper, A.F.W., 62, *84*
Schlesinger, W.H., 54, 62, 63, *84*, 174, 176–178, 181, 183, 184, 186, 188, 189, 190–195, *196*, *198*
Schlottmann, C.P., 409, 421, *435*
Schlüter, H., 409, 428, *435*
Schmelz, D.V., 94, 95, 119, *140*
Schmidt, G.W., *11*, *81*
Schmidt, K.P., 47
Schmidt, T.W., 80, *167*
Schmidt-Lorenz, R., 330, *333*
Schneider, G., 178, 179, 180, 183, 184, 186, 187, *198*
Schnelle, F., 414, *435*
Scholander, P.F., 185, *198*
Scholander, S.I., *198*
Scholtens, J.R., 73, *82*
Schorer, G., 431, *435*
Schreuder, H.T., 47, *82*
Schumm, S.A., 71, *84*, 89, 90, *135*, *140*
Schuster, J.L., 44
Schwickerath, M., 409, 419, 420, 421, 423, 425, 427, 428, *435*
Schwintzer, C.R., 36, *49*, 65, *84*, *175*, 183, *196*, *199*
Scoffin, T.P., 32, *49*
Scott, R., *435*
Scott, R.M., 338, *355*
Seamans, T., *139*
Searl, L., *13*
Sedell, J.R., *12*, *83*
Sedlik, B.R., 80
Seeberan, G.S., *248*
Segadas-Vianna, F., 34, 45, 178, *197*
Seischab, F.K., 35, *49*
Sell, M., 48
Sell, M.G., 1, *13*, *84*, 106, *140*, *166*, *168*
Selye, H., 65, *84*
Sen, D.N., *85*, *168*
Sepponen, P., *82*
Serrano, H., 163
Seth, S.K., *385*
Settlemyre, J.L., *81*
Seward, A.C., 28

Sexton, O.J., 58, *82*
Shafi, M., 373, *386*
Shapiro, J., 34, *50*
Sharik, T.L., *139*
Sharitz, R.R., *85*, 101, 102, *138*, *140*
Sharma, K.P., *50*
Sharp, J.M., Jr., 93, *137*
Shaver, G.R., *82*
Sheail, J., 36, *50*
Shelford, V.E., 43, *50*
Sheridan, J.M., 79, *85*
Shevchenko, L.A., 7, *13*
Shima, L., *210*
Shropshire, F.W., 47, *82*
Shugart, H.H., Jr., 39, *50*, *51*, *85*
Shull, C.A., 32, *50*
Shure, D.J., 60, 79, *84*
Sibert, J.R., 15, *23*
Siccama, T.G., *141*
Sieffermann, E., 232, 233
Siegel, D.I., 60, *84*
Sigafoos, R.S., 94, *140*
Silker, T.H., 165, *168*
Simberloff, D., 38, *50*
Simms, E.L., *11*, 80, *197*
Simons, D.B., 131, 140
Simpson, J.M., *82*, *85*
Simpson, R.L., *23*, *82*, *167*, *168*
Sims, R.A., 124, 126, *140*
Singh Aswal, B., 4, *11*
Singh, T.B., 243, 246, *248*, *250*
Sioli, H., 31, *50*, 235, *250*
Sipe, T.W., *141*
Sipp, S.K., 110, *135*
Skaggs, R.W., 77, 78, 79, *82*
Sklar, F.H., *136*
Skoropanov, S.G., 1, 2, *13*, 60, *84*
Slager, S., 231, *248*
Sleeman, J.R., *406*
Small, E., 62, 63, *85*, 190, 191, *199*
Smirnoff, W.A., *141*
Smit, C.J., 429, *435*
Smith, A.J.E., 412, *435*
Smith, C.J., *136*
Smith, D.G., 71, *85*
Smith, D.W., *139*
Smith, H.M., 27, *50*
Smith, J., *135*
Smith, J.E., 27, 46
Smith, J.P., *333*
Smith, M.H., *81*, *85*, *138*
Smith, N.J.H., 133, *140*
Smittenberg, J.H., 409, 416, 424, 425, 432, *435*
Smolyak, L.P., *14*, *85*, *169*
Snasari, T.A., 4, *10*
Snedaker, S.C., 7, *12*, 38, 40, 41, 43, *48*, *50*, 65, *85*, 103, 104, 106, *138*, *140*, 143, 145, 153, 156, 157, 159–162, 164, *166–168*, 185, 187, *198*, *286*, *297*

Sniffen, R.P., *83*
Sobotka, D., 439, *446*
Sobrado, M.A., *334*
Sohm, J.E., *138*
Sokotowski, A.W., 440, *446*
Sollers, S.C., 119, *140*
Soloway, E., 3, *11*, 46, *81*, *84*, *199*
Sørensen, J., 74, *85*
Sougnez, N., 416, *434*
Sousa, A.S., 61, *81*
Sousa, S.M., 288, 291–294, *297*, *298*
Spaans, A.L., *250*
Spangler, D.P., 27, *50*, 176, *199*
Specht, M.M., *405*
Specht, R.L., *80*, 128, 299, *334*, 336, *355*, 387, 390, 395, 396, 398–403, *405*, *406*, 447
Spence, L., *249*
Speth, J., 226, *250*
Spetzman, L.A., 125, *141*
Spielman, R.H., *11*
Sprent, J.I., 415, *435*
Stace, H.C.T., 390, *406*
Stachurski, A., 445, *446*
Staheli, A.C., 29, *50*
Standley, P.C., 216, 218, 245, *250*
Stanford, J.A., *137*, *138*, *140*
Stansell, K.B., *137*
Starrett, W.C., 58, *85*
Staub, J.R., 27, *50*
Stearns, F., 101, 108, *136*, *141*
Stehlé, H., 222, *250*, 253, 256, *265*
Stephens, J.C., 71, 76, 78, *83*, *85*
Sterlin, D.K., *138*
Steubing, L., 413, *435*
Stevens, L.E., 57, *85*
Stevenson, L.H., *138*
Stewart, E.H., *83*
Stewart, J.M., 29, 40, 41, *49*, 174, 178, 179, 183, 184, 186, *198*
Stewart, P.A., 65, *85*
Stoddart, D.R., 32, *50*
Straney, D.O., 58, *85*
Straub, P., 190, 191, 192, *199*
Street, M.W., 58, *85*
Strickland, C., 370, *386*
Stross, R.G., 35, *50*
Sugden, D.E., *137*
Swan, J.M.A., 31, *50*
Swaney, D.P., 43, *50*
Swanson, F.J., *12*, 44, *50*, *83*
Swarup, K., 372, *386*
Swift, B.L., 58, *80*, *85*
Switsur, V.R., 71, *82*
Sylvander, R.P., 273, *286*
Synnott, D.M., 35, *49*

Takagi, Z., 378, *386*
Takeuchi,M., *267*, 271, *286*
Tallis, J.H., 28, 29, 32, 38, *50*

Tanghe,M., *433*
Tansley, A.G., 33, *50*, 416, *435*
Tate, R.L., *80*
Tauber, H., 71, *79*
Taylor, B.W., 4, 7, 8, 9, *13*, *267*, 268, 270, *286*
Taylor, T.N., 27, *51*
Teal, J.M., 34, *50*, 75, *82*
Teas, H.H., *140*, *166*, *167*, *168*, *297*
Teas, H.J., 32, *50*, *198*
Téllez, O., 291, 292, 293, 294, *298*
Tendron, G., 126, *141*
Tenore, K.R., *23*
Teskey, R.O., *141*
Tessier, C., 120, *141*
Therezien, Y., 246, *250*
Thibault, J.-R., 113, *141*
Thimann, K.V., *83*
Thom, B.G., 28, 29, 32, *50*, 154, *168*
Thomas, B.A., 27, *50*
Thompson, J.C., 71, *85*
Thompson, K., 31, 37, *50*, 129, *141*, 165, *168*, *269*, *286*
Thorp, J.H., *83*, *167*
Thorton, I., 34, *46*
Tikasingh, E.S., *248*, *249*
Tilton, D.L., 3, *13*, 174, 178, 179, 189, 190, *199*
Timm, R.W., 370, *386*
Timmerman, A., 430, *435*
Todd, R., *83*
Toliver, J.R., *136*
Tolpa, A., 7, *14*
Tomlinson, P.B., *51*, *249*
Tood, R., *138*
Toole, E.R., 61, *80*
Tosi, J.A., Jr., *137*, *265*, 273, *286*
Townsend, P.K., 270, *286*
Traczyk, H., 445, *446*
Traczyk, T., 445, *446*
Trautmann, W., 419, 421, *435*
Treloar, G.K., 403, 404, *406*
Triska, F.J., 188, *199*
Trivedy, R.K., 31, *50*
Turekian, K.T., *83*
Turenne, J.F., 231, 247, *250*
Turner, J.S., 34, *46*
Turner, R.E., *10*, 22, 23, 55, 57, 60, *82*, *85*, 117, *134*, *135*, 143, *167*, *197*, *460*
Turner, R.M., 131, 132, *136*, *141*
Tusneem, M.E., 112, *140*
Tutin, T.G., 31, *50*, 412, *435*
Tüxen, R., 409, 416, 418–422, 427, 428, *432*, *434–436*
Twilley, R.R., 57, 75, *85*, 152, 160–162, 164, *168*, *169*, 172, 173, 176, 182, 185, 187, 188, 194, 195, *198*, *199*, 458, *460*

U.S. Army Corps of Engineers, 2, *14*
U.S. Geological Survey, 43, *50*, 206, *211*

Ulrich, K., 41, *50*
Ultsch, G.R., 31, *50*
Underwood, G., 245, *250*
UNESCO, 241, 245, 246, *250*
Ungar, I.A., 102, *139*
Ustach, J.R., 31, *49*

Valentine, D.H., *435*
Valiela, I., 34, *50*
Vallack, H.W., *334*
Van Andel, J.H., 222, 223, 224, 231, *250*
Van Asdall, W., *138*
Van Auken, O.W., 98, 116, *135*, *136*
Van Beek, J.L., 130, *141*
Van Cleve, K., 65, *85*
Van Coppenolle, F., *433*
Van den Berg, W.J., 425, *436*
Van den Bergen, C., *434*
Van den Bergh, L.M.J., 430, *436*
Van der Hammen, T., 226, 234, *250*
Van der Valk, A.G., 36, 38, 39, *50*, *51*
Van der Woude, J.D., 71, 415, *436*
Van Dijk, J., 409, 424, *436*
Van Dyke, H.N., 247, *248*
Van Dyne, G.M., *44*
Van Geel, B., *435*
Van Ham, J., *436*
Van Hees, A.F.M., 430, *436*
Van Hildebrand, P., 106, *137*
Van Hylckama, T.E.A., 133, *141*
Van Steenis, C.G.G.J., 5, *14*, *334*, 350, *355*
Van Straaten, L.M.J.U., 27, *51*
Van Wijngaarden, A., 429, *435*, *436*
Van Wirdum, G., 409, 412, 418, 419, *436*
Van Wirdum-Daan, C., 409, 419, *436*
Van Zinderen-Bakker, E.M., 409, 424, *436*
Van Kannon, D., *47*
Van Royen, P., 345, 346, *355*
VanDam, L., *198*
Vann, J., 32, *51*
Vann, J.H., 213, 214, 219, 245, *250*
Vázquez, S.J., 291, 295, *298*
Vázquez,-Yanes, C., 292, *297*, *298*
Vellmar, F., *385*
Verdoorn, F., *267*, 270, 285, *286*
Verghese, S., *137*
Verhey, C.J., *436*
Vermeer, K., 244, *250*
Vernon, R.O., 182, *199*
Verry, E.S., 26, *51*, 175, 183, *199*
Viereck, L.A., 32, *51*, 65, *85*
Viosca, P., 56, 58, *85*
Visser, S.A., 58, *85*, 165, *167*
Vitousek, P.M., 163, 164, *169*, 193, 194, *199*, 452, 454, *460*
Vogl, R.J., 3, 8, *14*, 35, 37, *51*

Volk, B.G., 77, *80*
Vyskot, M., 95, 103, 105, *141*

Wace, N.M., 38, *51*
Wagner, C.A., 27, *51*
Wagner, K.A., 182, *198*
Wainwright, S.J., 160, *169*
Wali, M.K., 121, *141*
Walker, D., 33, 40, *51*, 58, 71, *85*, 415, *436*
Walker, L.R., 125, *141*
Walker, P.N., 201, *211*
Walker, R.E., 29, *51*
Walker, S.W., *201, 211*
Wallace, A.R., 268, *286*
Waller, D.M., 101, 102, *139*
Wallis, J.R., 45
Walmsley, M.E., 34, *51*
Walsh, G.E., 140, *166–168, 198, 297*
Walter, D., *436*
Walter, H., 19, *23*, 301, *334*, 407, 409, 410, 411, *436*
Walters, M.S., 123, *141*
Walters, S.M., *435*
Walton, C.H., 116
Wang, F.C., *23*, 43, 49, *51*
Ward, J.R., *140*
Ward, J.V., *137, 138, 140*
Ware, G.H., 3, *14*
Ware, S., 94, 116, 117, *137, 140*
Waring, R.H., *45, 50, 199*
Warner, R.E., 123, 133, 141
Wasawo, D.P.S., 58, *85*
Washburn, A.L., 26, *51*
Waters, T.F., 35, *51*
Watson, J.G., 143, 144, *169*, 363, *386*
Watt, A.S., 35, *51*
Watt, R.F., 35, *51*, 189, *199*
Watts,, J.A., *84*
Waughman, J., 308, *334*
Weaver, G.T., *140*
Weaver, J.E., 29, 30, *51*
Webb, D.A., *435*
Webb, J.L., *267, 286*
Webb, J.W., *138*
Webb, L.J., 3, 7, 9, *14*
Webber, P.J., 165, *166*
Weber,C.A., 29, *51*
Wee-Lek, C., 369
Weir, W.W., 78, *85*
Welch, B.L., 32, *51*

Welch, P.S., 31, *51*
Welcomme, R.L., 57, *85*, 129, 132, 133, *141*
Wells, B.W., 28, 37, *51*
Wells, T.C.E., 36, *50*
West, D., 28, *51*
West, D.C., 39, *50*, 65, *85*
West, N.E., 123, *138*
West, R., *51*
West, R.C., 288, 290, 292, 293, 296, *298*
West, R.G., *85, 436*
Westhoff, V., 3, *13*, 409, 416, 421, 422, 424, 425, *436*
Wetzel, R.G., *10*, 31, 35, *46*, *51*, 135, *197, 460*
Wharton, C.H., 6, *14*, 20, 21, *23*, 50, 58, *85*, 116, 117, *141*
Wheeler, B.D., 409, 422, 432, *436*
Whigham, D.F., *10*, *23*, 80, 82, *135*, 167, *168, 197, 460*
Whipple, S.A., 97, 102, *137*
White, D.A., 98, 101, 114, *141*
White, J., 96, 101, *134*
White, P.S., 35, *51*
Whitehead, D.R., 3, *14*, 29, 35, 48, *51*, 71, *85, 201, 211*
Whitford, L.A., 28, 37, *51*
Whitmore, T.C., 2, 9, *14*, 27, 38, *51*, 62, *85*, 271, *286*, 302, 325, *333, 334*
Whittaker, R.H., 28, *51*, 54, *85*, 101, 118, *141*
Wickware, G.M., *140*
Wiegers, J., 71, 407, 409, 414, 416, 424–427, 432, *436*
Wiemhoff, J.R., *83, 139*
Wiesnet, D.R., *211*
Wijmstra, T.A., 237, *250*
Wikum, D.A., 121, *141*
Wilbert, J., 268, 271, *286*
Wildi, D., 40, *51*
Wilford, G.E., 29, *51*
Willett, R.L., *140*
Williams, B.L., 416, *436*
Williams, G., 36, 49, *196, 199*
Williams, G.P., 131, 132, *141*
Williams, R.B., 30, *51*
Williams, W.A., 4, *14*, 127, *141*, 271, *286*
Willmanns, O., 409, 416, 417, *436*
Wilson, J., 430, *436*
Wilson, J.G., *286*
Wilson, P., 221, 245, *248*

Wilson, R.E., 32, *51*, 95, *141*
Wilson, R.F., *167*
Winget, C.H., 14, *169*
Winkler, H., 299, *334*
Wistendahl, W.A., 95, 101, 102, 110, 119, *135, 137*
Wojterski, T., 443, 444, *446*
Wolanski, E., 128, *141*
Wolfe, C.B., Jr., 120, *141*
Wollack, A., 30, *47*
Wolman, M.G., 42, *51*, 71, *85*, 90, 131, 132, *138, 141*
Womersley, J.S., 299, *334*
Wong, C.H., *140, 168*
Woodbury, R.O., 251–253, 255, 256, 259, *264, 265*
Woodruff, C.M., Jr., 20, *23*
Woodward, D.K., *82*
Woodwell, G.M., 5, *14*, 82, *84*, 85
Woollett, E.C., 35, *46*
Worth, C.B., *248, 249*
Wright, A.A., 3, *14*, 65, *85*
Wright, A.H., 3, *14*, 65, *85*
Wright, L.D., *50*
Wright, P.B., 3, *12*
Wyatt-Smith, J., 4, 7, *14*, 299, *334*

Yarbro, L.A., 109, 112, 113, *141*
Yates, P., 79, *85*
Yates, R.F., 195, *199*
Yee Mun, L., 57, *85*
Yon, D., 126, *141*
Young, F.N., 58, *85*
Young, H.E., 23, *84, 141*
Yurkevich, I.D., 3, *14*, 60, *85*, 165, *169*

Zach, L.W., 33, *51*
Zar, J.H., 258, *265*
Zarzycki, K., 441, *446*
Zasada, J.C., *141*
Zieman, J.C. Jr., 32, 39, *51*
Zimka, J.R., 445, *446*
Zimmer, W.J., 403, 404, *405*
Zimmerman, M.H., *51*
Zimmerman, R.C., 108, 122, 123, *141*
Zobel, D.B., 125, *137*
ZoBell, C.E., 74, *85*
Zoltai, S.C., 7, *14*, 115, 124, *141*, 165, *169*, 178, *199*
Zonneveld, I.S., 409, 419, 420, 421, *436*
Zwerver, S., 411, *436*

SYSTEMATIC INDEX

Abies amabilis (Dougl.) Forbes, 124
    A. balsamea (L.) Mill. (balsam fir), 126, 178, 179
    A. grandis, 125
Acacia, 129, 341, 342, 343
    A. usambarensis Taub., 129
Acanthaceae, 384
Acanthus ilicifolius L., 347, 384
Accipiter gentilis (goshawk), 430
    A. nisus (sparrow hawk), 430
Accipitridae, 374
Accipitriformes, 374
Acer (maple), 5, 30, 95, 97, 103–105, 120, 180, 183, 188, 201, 203, 209
    A. campestre L., 440
    A. macrophyllum Pursh, 124, 125
    A. negundo L. (box elder), 32, 106, 117, 121, 123, 124
    A. pseudoplatanus L., 421, 440
    A. rubrum L. (red maple), 33, 38, 99, 101, 111, 116, 117, 119, 120, 179–184, 188, 194, 201
    A. saccharinum L., 43, 102, 110, 111, 116–120, 166
    A. saccharum Marsh., 119, 120, 166
Acera, 338
Acetes erythraeus Nobili, 381
    A. indicus H. Milne-Edwards, 381
    A. japonicus Kishinouye, 381
Acoelorrhaphe, 288, 289
    A. wrightii (Griseb. & Wendl.) Beccari (tasistal), 290, 292
Aconitum napellus L., 417, 418
Acridotheres fuscus (Wagler) (jungle myna), 377
    A. ginginianus (Latham) (bank myna), 377
    A. tristis (Linn.) (common myna), 377
Acrocephalus palustris (marsh warbler), 430
    A. schoenobaenus (sedge warbler), 430
    A. scirpaceus (reed warbler), 430
Acrocomia totai Mart., 272
Acrostichum, 218, 221, 246, 255
    A. aureum L., 234, 255, 347, 349, 363, 383, 384
    A. danaefolium Langsd. & Fisch., 255
Adiancum latifolium Domin., 279
Adiantites, 28
Aedeomyia, 244
Aedes, 383
    A. taeniorhynchus Weideman, 245
    A. tortilis, Theobald, 245
Aegialitis, 156
    A. rotundifolia Roxb., 384
Aegiceras, 156
    A. corniculatum (L.), 347
    A. majus Gaertn., 366, 384
Aegithalos caudatus (long-tailed tit), 430
Aegithina tiphia (Linn.) (common iora), 376

Aegopodium podagraria L., 417, 418
Aequidens pulcher, 242
Aesculus glabra Willd., 119
Afzelia retusa Kurz, 366, 384
Agamidae, 373
Agathis, 319
Agouti paca L., 244
Agrostis stolonifera L., 441
Alaudidae, 376
Alcedinidae, 375
Alcedo atthis (Linn.) (common kingfisher), 375, 430
Alchemilla vulgaris auct. s.l., 417
Alchornea costarricense Pax & K. Hoffm., 279
Alisma plantago-aquatica L., 439
Allium ursinum L., 427
Alnetum viridis, 416
Alnus (alder), 5, 30, 38, 71, 95, 125, 179, 183, 184, 186, 187, 287, 412, 413, 415, 417, 419, 424, 425, 437–440, 442, 459
    A. crispa (Ait.) Pursh, 125
    A. glutinosa (L.) Gaertner, 413, 415, 416, 418, 420–426, 439, 442, 445
    A. incana (L.) Moench, 417, 418, 440, 441
    A. oblongifolia, 123
    A. rhombifolia Nutt., 123, 124
    A. rubra, 124, 125
    A. rugosa (DuRoi) Spreng. (speckled alder), 120, 126, 178, 179
    A. serrulata (Ait.) Willd., 120
    A. tenuifolia Nutt., 124
    A. viridis (Chaix) DC., 417, 418
Alouatta seniculus Elliot, 244
Alstonia, 311
    A. scholaris R.Br., 349
    A. spatulata Bl., 325
Alternanthera philoxeroides (Mart.) Griseb (alligatorweed), 31
Althoffia, 346, 354
Alytes obstetricans, 431
Amandava amandava (L.) (red avadavat), 376
Amanoa, 215, 221, 239
    A. caribaea Krug. & Urb., 221, 240
    A. potamophila Croizat, 281
Amaurornis phoenicurus (Pennant) (white-breasted water-hen), 376
Amazona amazonica L. fil, 244
    A. arausiaca Müller, 245
Amblonyx cinerea Illiger (small-clawed otter), 371
Amblystegium riparium (Hedw.) Br. Eur., 420
Amelanchier laevis Wieg., 432
Amoora cucullata Roxb., 366, 384
Amphipnoidae, 380
Amphipnous cuchia Hamilton, 380

Ampullaria, 382
Ampullariidae, 242, 382
Anabantidae, 378
Anabus testudineus (Bloch) (climbing perch), 378
Anacardium excelsum (B. & B.) Skeels, 99
Anas, 374
   A. acuta (Linn.) (pintail), 374
   A. crecca (teal), 430
   A. platyrhynchos (mallard), 430
   A. poecilorhyncha, J.R. Forster (spot-billed duck), 374
Anastomus oscitans (Boddaert) (Asian openbill stork), 374
Anatidae, 374
Andira, 236, 288, 289
   A. galeottiana Standl., 288
   A. inermis (Wright) HBK., 218, 221, 236, 240, 246, 253, 256, 288
Andromeda glaucophylla Link., 178, 180
Andropogon squarrosus L., 384
Angelica sylvestris L., 417, 419
Anguis fragilis, 431
Anhima cornuta L., 244
Anhinga rufa (Daudin) (African darter), 377
Anhingidae, 377
Annona, 222, 288, 289
   A. glabra L., 218, 219, 221–223, 240, 253, 255, 256, 263, 290, 292
   A. palustris L., 222, 236
Anodontites irisans, Marshall, 241
   A. leotaudi Guppy, 241, 242
Anolis bimaculatus oculatus, 245
Anopheles, 244
   A. aconitus Donitz, 383
   A. annularis Walf, 383
   A. minimus Theobald, 383
   A. philippinensis Ludlow, 383
   A. sundaicus (Rodenwald), 383
Anseriformes, 374
Anthriscus sylvestris (L.) Hoffm., 419
Anthurium acaule (Jacq-Schott), 255, 263
Anthus novaeseelandiae Gmelin (Richard's pipit), 376
   A. trivialis (tree pipit), 430
Antocarpus incisus (L.F.) = *A. altilis* (Park.) Fosberg, 347, 354
Anura, 242
Apeiba aspera Aubl., 281
Aphyocarax axelrodi Travassos, 242
Apocryptes bato Hamilton, 379
   A. lanceolatus Bloc., 379
Apocynaceae, 384
Apodidae, 374
Apodiformes, 374
Apus (swift), 374
Aquila pomarina C.L. Brehm. (lesser-spotted eagle), 374
   A. rapax (Temminck) (tawny eagle), 374
Ardea cinerea Linn. (gray heron), 374, 430
   A. herodias L., 245
   A. purpurea Linn. (purple heron), 374, 429
Ardeidae, 374
Ardeiformes, 374
Ardeola grayii (Sykes) (pond heron), 374

Ardisia, 278
Areaceae, 384
Arecastrum remanzoffranum Becc., 272
Arenga, 338
Ariidae, 380
Arius dussumieri Cuvier and Val., 380
   A. gagora Hamilton, 380
Aronia prunifolia (Marsh.) Rehd., 432
Artamidae, 376
Artamus fuscus Vieillot (ashy shallow shrike), 376
Artiodactyla, 371
Artocarpus incisus (L.F.) = *A. altilis* (Park.) Fosberg, 347, 354
Arundinaria gigantea (Walt.) Chapm., 201
   A. gigantea (Walter) Muhlenberg, 180
Arvicola terrestris (water vole), 429
Asclepiadaceae, 384
Asio otus (long-eared owl), 430
Aster, 179
Asterocalamites, 28
Asterogyne martiana Wendl., 281
Astianthus, 287
Astrocaryum, 283
   A. alatum Loomis, 280, 281
Astyanax bimaculatus L., 242
Athene brama (Temminck) (spotted little owl), 377
Atheriniformes, 379
Athyrium filix-femina (L.) Roth, 179, 417, 422, 423, 426
Aulacomnium androgynum (Hedw.) Schwaegr., 426
Avicennia, 37, 66, 144, 148–150, 155, 156, 176, 185, 188, 191, 231–234, 289, 337, 347, 351
   A. alba Bl., 384
   A. germinans (L.) L. (black mangrove), 31, 37, 75, 127, 144, 151, 157–161, 164, 173, 176, 182, 185, 188, 194, 195, 294, 453, 457
   A. marina (Forsk.) Vierh., 164, 347, 350
   A. nitida Jacq., 269
   A. officinalis L., 366, 384
   A. tomentosa Jacq., 269
Axis axix Erxleben (spotted deer), 370, 371
   A. porcinus Zimmermann (hog deer), 371

Bactris, 4, 272, 289
   B. anizitzii Barb. Rodr., 272
   B. balanoidea (Oerst.) Wendl., 292, 293
   B. bidentata Spruce, 272
   B. inundata Mart., 272
   B. major Jacq., 217, 223, 224, 240
   B. minor Jacq., 217, 240
   B. trichophylla Burret, 293
Bactronophorus thoracites Gould, 382
Baeckia frutescens L., 332
Bagridae, 378, 380
Bambusa, 287
Bandicota bengalensis Gray (mole rat), 372
   B. indica Bechstein (bandicot rat), 372
Bankia campanellata Moll. and Roch., 382
   B. roonwali Rajag., 382
Barringtonia acutangula (L.) Gaertn., 338, 342, 345
   B. racemosa Forst., 366, 384
   B. racemosa Payens, 338

SYSTEMATIC INDEX

Barringtoniaceae, 384
Batis argillicola P. Royen, 350
Becquerelia cymosa Kunth, 279, 281
Belone strongylurus V. Hasselt, 379
Belostoma malkini Lauck, 242
Belostomatidae, 242
Betula (birch), 5, 71, 415, 416, 424, 428, 437, 439
    B. lutea Michx. (yellow birch), 179
    B. nana L., 125
    B. nigra L. (river birch), 93, 116, 117, 121
    B. papyrifera Marsh. (paper birch), 126, 178, 179
    B. pendula Roth, 416, 428
    B. pubescens Ehrh., 415, 423–428, 439, 440
    B. pumila L. (bog birch), 178
    B. pumila var. *glandulifera* Regel, 126
Bignoniaceae, 384
Biomphalaria glabrata Say, 245
Bischofia javanica Blume, 270, 284, 349
Bithynia, 382
Bivalvia, 382
Blechnaceae, 384
Blechnum, 218
    B. indicum Burm., 219, 223
    B. spicant (L.) Roth, 423
Bofiodes rudicundus Hamilton, 380
Boidae, 373
Boleophthalmus viridis Hamilton, 379
Bombax, 214, 234
    B. aquaticum (Aubl.) K. Schum., 214, 218, 219, 240
Bombina variegata, 431
Bonafousia tetrastachya (HBK) Mgf., 214, 219, 240, 245
Bos frontalis Lambert, 371
    B. gaurus H. Smith (gaur), 370, 371
Bostrychia, 182
Bovidae, 371
Brachypodium sylvaticum (Hudson) Beauv., 417
Brachypteris ovata (Cav.) Small, 222
Brachythecium rutabulum (Hedw.) Br. Eur., 420
Brasenia, 289
Bravaisia, 288
    B. integerrima (Spreng.) Standl. (julubal), 288, 289
    B. tubiflora Hemsl., 289, 295
Brosimum costaricanum Liebm., 281
Brownlowia argentata Kurtz, 340, 351
    B. lanceolata Benth., 367, 384
Bruguiera, 347, 351, 354, 355
    B. cylindrica (L.), 366
    B. gymnorrhiza Lamk., 340, 347, 363, 365, 366, 384
    B. sexangula (Lour.) Poir., 340, 367, 384
Bryonia cretica L. subsp. *dioica* (Jacq.) Tutin, 422
Bubalis arni L. (wild buffalo), 371
    B. bubalis, 370
Bubo bubo (L.) (great eagle owl), 377
Bubo coromandus (Latham) (dusky eagle owl), 377
Bubulcus ibis (Linn.) (cattle egret), 374
Bucida, 288, 289
    B. buceras L. (puktal), 221, 253, 255, 256, 291
Bufo bufo, 431
    B. marinus, 242
Bufonidae, 242

Bulbophyllum xylocarpi J.J. Smith, 384
Burhinidae, 375
Buteo buteo (buzzard), 430
Butia capitata Becc., 272
Butorides striatus (L.) (striated heron), 374
    B. virescens, 245
Bythiniidae, 382

Cabomba, 289
Cacomanthis merulinus (Scopoli) (gray-breasted brush cuckoo), 375
    C. sonneratii (Latham) (banded bay cuckoo), 375
Caesalpinia nuga Ait., 384
    C. bonducella Flem., 384
Caesalpiniaceae, 384
Caiman sclerops L., 243, 246
Calamagrostis canescens (Weber) Roth, 422–424, 440
Calamites, 28
Calamus, 269
Calandrella raytal (Blyth) (sand lark), 376
Calappidae, 381
Calathea lagunae Woodson, 280
    C. lutea (Aubl.) Mey., 280
Calidris minutus (Leisler) (little stint), 375
Calla palustris L., 439
Callichthyidae, 242
Callitriche stagnalis Scop., 418, 419
Calloscrirus pygerythus Geoffroy (beautiful squirrel), 372
Calluna, 38, 413
    C. vulgaris (L.) Hull, 427, 428
Calophyllum, 288, 289, 319, 332, 345
    C. brasiliense Camb., 281, 288, 290
    C. calaba L., 221, 222, 246, 253, 255, 256
    C. lucidum Benth., 223, 272
    C. obliquinervum Merr., 315
    C. rhizophorum, 318
    C. sclerophyllum Vesq., 320
Calotes versicolor (Daudin), 373
Caltha, 30
    C. palustris L., 418, 419, 423, 426, 439, 441
Calyptranthes millspaughii Urban, 290
    C. perlaevigata Lundell, 290
Calyptrocarya glomerulata (Brongn.) Urban, 279, 281
    C. glauca (Oerst.) Wendl., 280, 285
Calyptronoma occidentalis, 221
Calystegia sepium (L.) R. Br., 419, 420, 440, 441
Cameraria latifolia L., 291
Campephagidae, 376
Campnosperma, 311, 337, 338, 341, 342, 345, 352, 354
    C. auriculata Hook., 270, 284
    C. brevipetiolata Volkens, 338, 341, 344, 347, 349
    C. coriacea = *C. coriaceum* (Jack) Hallier f. ex Steenis, 341
    C. panamensis Standl., 5, 278, 284, 325
Canarium, 325
Canidae, 371
Canis aureus L. (golden jackal), 371
Capreolus capreolus (roe deer), 428
Caprimulgidae, 375
Caprimulgiformes, 375
Caprimulgus affinis Horsfield (Franklin's nightjar), 375

Caprimulgus (continued)
   C. asiaticus Latham (Indian nightjar), 375
   C. indicus Latham (jungle nightjar), 375
Caprolagus hispidus Pearson (hispid hare), 371
Carallia brachiata (Lour.) Merr., 341, 343
Carangidae, 378, 379
Carapa, 217
   C. guianensis Aubl., 217, 221, 224, 234, 236, 240, 246, 269, 272, 281
Carcharhinidae, 378, 379
Carcharhinus melanopterus (Muller and Henle), 379
Carcharias gangeticus (Muller and Henle), 379
   C. laticaudus Muller and Henle, 379
Cardamine amara L., 418, 419, 425, 440
   C. pratensis L., 420
Cardiopteris, 28
Cardisoma guanhumi Latr., 242, 245
Carduus personata (L.) Jacq., 417, 418
Carex, 30, 38, 179
   C. elata, 422
   C. elongata L., 423, 439, 440
   C. laevigata Sm., 422, 423
   C. lasiocarpa Ehrh., 439
   C. paniculata L., 422
   C. pilulifera L., 428
   C. pseudocyperus L., 422
   C. remota L., 440, 441
   C. riparia Curtis, 422, 424
   C. stricta Good, 439
Caridina gracilirostris de Man, 381
Carina scutulata (S. Muller) (white-winged wood duck), 374
Carnivora, 371
Carpinus (hornbeam), 441, 445
   C. betulus L., 442
   C. caroliniana Walt. (American hornbeam), 117, 119, 180
Carya, 30, 98, 105, 116, 117, 120
   C. aquatica (Michx. f.) Nutt., 116, 117
   C. cordiformis (Wang.) K. Koch, 118
Caryota, 338
Casearia arborea (L.C. Rich.) Urban, 253, 256
Cassipourea guianensis Aubl., 278
Castor (beaver), 26, 35, 133
   C. fiber, 429
Casuarina, 317, 319, 323, 331, 389, 404
   C. cunninghamiana Miq. (river oak), 270, 389, 403
   C. equisetifolia L., 363, 384
   C. glauca Sieb. ex Spreng. (swamp sheoak), 389, 391, 403
   C. nobilis (Johns.) Whitm., 317, 322, 323, 330–332
Casuarinaceae, 384
Catla catla (Hamilton), 372, 378
Ceiba, 236
   C. petandra (L.) Gaertn., 290
Celtis, 30, 43, 95, 96, 98, 105, 116, 119, 121
   C. laevigata Willd., 117
   C. occidentalis L., 117, 118
   C. reticulata Torr., 123
Centropomidae (Latidae), 378, 379
Centropomus, 241
   C. undecimalis, 246

Centropus sinensis (Stephens), 376
   C. toulou (P.L.S. Muller), 376
Cephalanthus, 30
   C. occidentalis L., 117, 182, 289, 296
Cephalomappa paludicola Airy-Shaw, 312
Cephalopoda, 382
Ceratocystis ulmi, 118
Ceratophyllum, 30, 289
   C. demensum L., 442
Cerbera odollam Gaertn., 384
Cercis canadensis L., 119
Cercopithecidae, 372
Ceriops, 156, 385
   C. decandra Griff., 366, 384
   C. tagal (Perr.) C.B. Robins, 347, 350, 366
Cerithiidae, 382
Cerithium telescopium L., 382
Certhia brachydactyla (short-toed tree creeper), 430
Cervidae, 371
Cervus eldi McLelland, 371
Ceryle rudis (lesser pied kingfisher), 375
Cetacea, 371
Chaerophyllum hirsutum L., 416
Chaetodontidae, 379
Chaetornis striatus (Jerdon) (bristled grass warbler), 376
Chaleophaps indica (Linn.) (emerald dove), 375
Chamaecyparis thyoides (L.) B.S.P. (Atlantic white cedar), 5, 37, 176, 178, 180, 181, 183, 184, 188, 194, 201, 203, 205
Chamaedaphne calyculata (L.) Moench (leatherleaf), 178, 179, 180
Chara, 30, 442
Characinidae, 242
Charadriidae, 375
Charadriiformes, 375
Charadrius dubius Scopoli (little ringed plover), 375
   C. hiaticula Linn. (ringed plover), 375
Chatoessus chacanda Hamilton, 379
   C. nasus Bloch, 379
Chelonia, 243, 373
Cheloniidae, 373
Chelonia amboinensis Bunther (sea turtle), 373
   C. emys (Schleg and Mull.) (sea turtle), 373
   C. mydas (Linn.) (green turtle), 373
Chelydidae, 243
Chelys fimbriata Schneider, 243
Chimarrhis cymosa Jacq., 221
Chiroptera, 371
Chitra indica (Gray), 373
Chloropsis aurifrons (Temminck) (gold-fronted leaf bird), 376
Chondrichthyes, 378
Chorinemus lysan Forskal, 379
   C. moadetta Cuvier and Val., 379
Chrybdis cruciara (Herbst), 381
Chryosophila albida Bartlett, 281
Chrysemys pieta Plimmer, 373
Chrysobalanus, 214, 215, 219, 222, 232, 233, 234
   C. icaco L., 214, 217, 218, 222, 234, 290
Chrysocolaptes lucidus (Scopoli), 377
Chrysophrys berda (Forskal), 380
   C. datnia Hamilton, 380

SYSTEMATIC INDEX

Chrysosplenium alternifolium L., 426
    C. oppositifolium L., 426
Cichlidae, 242
Cichlosoma bimaculatum L., 242, 246
Ciconia ciconia (Linn.) (white stork), 374
    C. episcopus (Boddaert) (white-necked stork), 374
Ciconiidae, 374
Cinclus cinclus (dipper), 429
Circaea alpina L., 440
Cirrhinus mrigala Hamilton, 378
Cirriphyllum crassinervium (Tayl.) Loeske et Fleisch., 420
Cirsium, 421
    C. oleraceum (L.) Scop., 417
    C. palustre (L.) Scop., 421, 423, 424, 427
Cissus sicyoides L., 222
Cisticola exilis Jerdon, 376
Citrus sinensis Osbeck, 256
Cladietum, 222
Cladium jamaicense Crantz., 19, 38, 221
    C. mariscus (L.) Pohl, 422, 424
Cladophora repens, 182
Clarias batrachus (Linn.) (catfish), 378
Clariidae, 378
Claudium, 289
Clerodendron inerme Gaertn., 363, 384
    C. nerifolium Wall., 367, 384
Clethra alnifolia L. (sweet pepperbrush), 180, 181
Clethrionomys glareolus (bank vole), 429
Clorodendron inerme, 367
Clupea fimbriata Cuvier and Val., 379
    C. toli Cuvier and Val., 379
Clupeidae, 378, 379
Clupeiformes, 378, 379
Clusia, 214, 272
    C. fockeana, 214
    C. rosea Jacq., 221, 253, 255
Coccoloba reflexiflora Standl., 291
    C. uvifera L., 221
    C. venosa L., 253, 256
Coelops frithii Blyth, 371
Coilia dessumieri Cuvier and Val., 379
Colocasia, 245
Colubridae, 373
Columba livia Gmelin (rock dove), 375
    C. oenas (stock dove), 430
    C. palumbus (wood pigeon), 430
Columbidae, 375
Columbiformes, 375
Comarum palustre L., 439
Combretaceae, 384
Combretocarpus, 315
    C. rotundatus (Miq.) Danser, 315, 316
Combretum laxum Jacq., 288
Conocarpus, 289
    C. erectus L. (buttonwood), 144, 157, 158, 182, 221, 222, 253, 294, 295
Copaifera palustris (Sym.) de Wit, 309
Copernicia, 272
    C. australis Becc., 272
    C. cerifera Mart., 272

Copsychus erimelas Oberholser (magpie robin), 376
    C. saularis (L.), 376
Coracias bengalensis (L.), 375
Coraciidae, 375
Coraciiformes, 375
Coracina novaehollandiae (Gmelin) (large cuckoo shrike), 376
Cordaites, 28
Cordia borinquensis Urban, 253, 256
Corixidae, 242
Cornus amomum Mill., 120
    C. stolonifera Michx., 126, 179
Corvidae, 376
Corvus corone cornix (hooded crow), 429
    C. corone corone (carrion crow), 430
    C. frugilegus (rook), 430
    C. macrorhynchos Wagler (jungle crow), 376
    C. splendens Vieillot (house crow), 376
Corydoras aeneus Gill, 242
Corylus cornuta L., 179
Corynopoma riesii Gill, 242
Coturnix, 376
    C. coturnix (Linn.) (common quail), 376
Crassostrea, 21, 382
    C. cucullata (Born.) (rock oyster), 382
    C. discoides, 382
    C. gryphoides (Newton and Smith), 382
    C. madrasensis Preston (backwater oyster), 382
Cratoxylum, 311
    C. arborescens (Vahl.) Bl., 330
Crenidens indicus Day, 380
Crepis paludosa (L.) Moench, 417, 440, 441
Crescentia cujete L., 291, 295
Crinum asiaticum L., 340
Crocodilia, 243, 373
Crocodilidae, 243, 373
Crocodylus palustris Lesson (marsh crocodile), 372, 373
    C. porosus Schn. (salt-water crocodile), 372, 373
Cromileptes altivelis Cuvier and Val., 380
Crudia glaberrima (Steud) Macbride, 217, 223, 240
    C. spicata Willd., 221
Crustacea, 242
Cuculidae, 375, 376
Cuculiformes, 375
Cuculus micropterus Gould (short-winged cuckoo), 376
    C. varius Vahl. (common hawk cuckoo), 376
Culex, 244
    C. amazonensis Lutz, 244
    C. inflictus Theobald, 245
    C. portesi Senevet and Abonnec, 247
Culex, 383
Culicoides, 244
Curatella americana L., 291
Cyanosylvia svecica cyanecula (white-spotted bluethroat), 430
Cyclanthus bipartitus Poit., 280
Cyclosorus, 347
Cynoglossidae, 380
Cynoglossus bengalensis Bleeker, 380
    C. brevis Gunth., 380
    C. hamiltonii Gunth., 380
    C. macrolepidotus Bleeker, 380

Cynometra racemosa Benth., 384
   C. ramiflora L., 366
Cyperaceae, 384
Cyperus, 225, 226, 235, 240, 363
   C. giganteus Vahl., 216, 221, 226, 280
   C. tegetiformis Roxb., 384
Cyprinidae, 378
Cypriniformes, 378, 379
Cypsiurus parvus (Lichtenstein) (palm swift), 374
Cyrena bengalensis Lamarck, 382
Cyrenidae, 382
Cyprinodontidae, 242
Cyrilla racemiflora L., 181

Dacrydium, 316, 317, 319, 323, 330–332, 334
   D. beccarii var. *beccarii*, 318, 330–332
   D. beccarii var. *subelatum*, 332
   D. elatum, 320
   D. fusca, 332
   D. pectinatum de Laub. (Syn. *D. beccarii* var. *subelatum*), 320
   D. pierrei Hickel, 325
Dactylocladus, 307, 315
   D. stenostachys Oliv., 309, 310, 313, 315, 316, 330
Dalbergia, 222
   D. brownei (Jacq.) Urban (mucal), 289, 290, 296
   D. ecastophyllum (L.) Taub., 222, 240
   D. glabra (Miller) Standl., 292
   D. spinosa Roxb., 363, 366, 384
   D. torta Grah., 367, 384
Dasyatidae, 378
Dasyprocta agouti L., 244
Datnioides quadrifasciatus Sevart, 380
Decapoda, 242, 382
Deinocerites magnus Theobald, 245
Delphinidae, 371
Delphinus delphis Linn., 371
Dendrobium rhizophoretti Ridley, 384
Dendrocitta vagabunda Latham, 376
Dendrocopos major (great spotted woodpecker), 430
   D. minor (lesser spotted woodpecker), 430
Dendrocygna bicolor (Vieillot) (fulvous tree duck), 374
Dendrocygna javanica (Horsfield) (lesser whistling duck), 374
Dendropthoe pentandra (L.), 384
Dermochelidae, 373
Dermochelys coriacea Linn. (marine leatherneck turtle), 373
Derris heterophylla (Willd.), 363, 366, 384
Deschampsia caespitosa (L.) Beauv., 111, 417, 423, 426
   D. flexuosa (L.), Trin., 427, 428
Desmonchus horridus Splitg. & Mart., 219, 223
Dialyanthera gordoniaefolia, 5
Dicaeidae, 376
Dicaeum agile Tickell (thick-billed flowerpicker), 376
   D. erythrohynchos Latham (Tickell's flowerpicker), 376
Dicotyledones, 384
Dicranum scoparium Hedw., 427, 428
Dicruridae, 376
Dicrurus adsimilis Bechstein, 376
Dicrurus paradiseus (L.), 376
Dillenia alata (DC.) Martelli, 342, 343

D. papuana Martelli, 345
Dinopium bengalensis (Linn.) (lesser golden-backed woodpecker), 377
Diospyros digyna Jacq., 290
   D. embryopteris L., 384
   D. evena Bakh., 317, 323
   D. ferrea (Willd.) Bakh. f., 338
Dipterocarpus turbinatus, 367
Dischidia, 347
Dolichandrone rheedei See, 366, 384
   D. spathacea (L.f.) K. Schum., 340
Donacidae, 382
Donax cuneatus L. (wedge clam), 382
Dreissenidae, 242
Drepane punctata Gmelin, 379
Drepanocarpus lunatus Mey., 269
Drepanotrema cimex Moricand, 245
   D. kermatoides d'Ord., 245
Dricrurus adsimilis Bechstein, 376
Drosera rotundifolia L., 180
Drymaria, 384
Dryobalanops fusca V. Sl., 316
   D. rappa Becc., 310, 332
Dryophis mycterizans (Daudin) (tree snake), 373
Dryopteris austriaca (Jacq.) Woynar, 444
   D. carthusiana (Villar) H.P. Fuchs, 423, 424, 427
   D. cristata (L.) A. Gray, 425, 439
   D. gongyloides (Schkuhr) Kze., 219
   D. thelypteris (L.) A. Gray, 439, 440
Dupetor flavicollis (Latham) (black bittern), 374

Ebenaceae, 384
Echinochloa, 234
   E. polystachia, 234
   E. stagnina (Retz.) P. Beauv., 353
Echis carinata Schn. (carpet viper), 373
Egretta alba (L.) (great egret), 374
   E. garzetta (L.) (little egret), 374
   E. intermedia (Wagler) (intermediate egret), 374
Eichhornia, 30, 226, 240, 289
   E. crassipes (Mart.) Solms (water hyacinth), 26, 31, 383
Elachistocleis ovalis, 243
   E. surinamensis, 243
Elaeagnus angustifolia L. (Russian olive), 122, 123, 132
Elaeis (oil palm), 268
   E. guineensis Jacq., 269
   E. oleifera (H.B.K.) Cortes, 269
Elanus caeruleus (Latham) (black-winged kite), 374
Elapidae, 373
Eleocharis, 30, 289, 363, 384
Eleotris amboinensis Bleeker, 380
   E. fusca Bloch and Schn., 379
Eleutherodactylus urichi Boelt., 242
Elodea, 30
Emballonuridae, 371
Emberiza schoeniclus (reed bunting), 430
Emyda granosa (Schoepft), 373
Emydidae, 373
Emys orbicularis, 431
Enallagma latifolia Small, 221

Engraulidae, 378
Engraulis hamiltonii Gray and Hardo, 379
   E. indicus Hasselt, 379
   E. mystax Bloch and Schn., 379
   E. telara Hamilton, 379
Enhydrina schistosa Daudin, 373
Enhydris enhydris Schn., 373
Entada pursaetha DC., 384
Eperua, 326
   E. jenmani, 246
   E. purpurea Benth. (yeraro), 305, 319, 326
Ephippidae, 379
Epidendron ciliare L., 255
Equisetum, 125
   E. silvaticum L., 440
Equula blochii Lacépède, 379
   E. edentula Forskall (pony fish), 378, 379
Eretmochelys imbricata (Linn.) (hawksbill turtle), 373
Erica tetralix L., 427, 444
Eriophorum, 180
   E. angustifolium Hockeny, 414
   E. vaginatum, 428
Erithacus rubecula (robin), 430
Erythrina, 219, 220, 224, 228, 229, 238, 239, 340
   E. fusca Lour., 224, 342, 346
   E. glauca Willd., 99, 214, 219, 224, 228, 238, 240
   E. indica L., 384
Esacus magnirostris Vieillot (beach stone-curlew), 375
Eschweilera, 217, 279
   E. longipes, 245
   E. subglandulosa, 246
Estrildidae, 376
Etapteris, 28
Eublepharis hardwickii Gray, 373
Eucalyptus (eucalypt), 165, 389–391, 404
   E. alba Reinw. ex Blume (white gum), 389, 403
   E. camaldulensis Dehnh. (river red gum), 388, 390–392, 395–400, 403, 404
   E. largiflorens F. Muell. (black box), 390, 391, 403
   E. microtheca F. Muell. (coolibah), 391, 403, 404
   E. ochrophloia F. Muell. (napunyah), 391, 403
   E. ovata Labill. (swamp gum), 391, 403
   E. papuana F. Muell. (ghost gum), 390
   E. polycarpa F. Muell. (long-fruited bloodwood), 390
   E. robusta Sm. (swamp mahogany), 391, 392, 403
   E. rudis Endl. (Western Australia flooded gum), 392, 403
   E. tereticornis Sm. (forest red gum), 391, 392, 403
Eudynamis scolopacea L., 376
Eugenia, 105, 110, 221, 325, 332, 366
   E. fadyenii Krug. & Urb., 221
   E. fruticosa L., 384
   E. ligustrina (Sw.) Willd., 222
   E. lundellii Standl., 291
Eunectes, 246
   E. murinus Latr., 244
Euonymus atropurpureus Jacq., 118
Eupemphix pustulosus trinitatis, 242
Eupera simoni Jousseaume, 127
Euphorbiaceae, 384

Eurycarcinus orientalis, 381
Euterpe, 214, 217–219, 232, 234, 236, 240, 247, 268
   E. edulis Mart., 271, 272
   E. oleracea Mart., 223, 234, 238, 246, 269, 272
   E. precatoria, 268
Excoecaria agallocha L., 341, 349, 363, 366, 369, 383, 384

Fabaceae, 384
Fagus grandifolia Ehrh. (American beech), 117–119, 180, 201
   F. sylvatica L., 428
Falcatifolium, 319
Falco chicquera Daudin (red-headed falcon), 374
   F. subbuteo (hobby), 430
   F. tinnunculus (kestrel), 430
Falconidae, 374
Felidae, 371
Felis chaus Guldenstadt (jungle cat), 371
Festuca gigantea (L.) Vill., 417
Ficus, 219, 253, 272, 280, 281, 287, 289, 340, 349
   F. citrifolia P. Miller, 253, 255, 256, 263
   F. cotinifolia HBK., 290
   F. insipida Willd., 288
   F. involuta (Liebm.) Miq., 294
   F. maxima Mill., 221
   F. padifolia HBK., 290
   F. panamensis Standl., 290
   F. retusa L., 384
   F. sintenisii Warb., 253, 256
   F. sycomorus, 133
Filipendula ulmaria (L.) Maxim., 414, 416–418, 423, 426
Fimbristylis ferruginea Vahl., 384
Finlaysonia maritima Backer ex K. Heine, 384
   F. obovata Wall., 366, 384
Fissidens bryoides Hedw., 420
Flagellaria indica L., 347, 349, 367
Fordonia leucobalia Schlegel, 373
Foresteria acuminata (Michx.) Poir., 117
Francolinus francolinus (L.) (Assam black partridge), 376
Frangula alnus Miller, 423, 424, 427, 428, 439
Fraxinus (ash), 30, 98, 105, 106, 117, 119, 120, 163, 201, 287, 412, 442
   F. americana L., 119, 166
   F. caroliniana Mill. (Carolina ash), 117, 180
   F. excelsior L., 413, 417, 418, 421, 427, 439, 441, 442
   F. latifolia Benth., 123, 124
   F. nigra Marsh. (black ash), 111, 120, 126, 166, 178, 179
   F. pennsylvanica Marsh., 73, 116–118, 120, 121
   F. p. var. *subintegerrima* (Vahl) Fern. (green ash), 33
   F. velutina Torr., 123
Fregetta tropica (Gould) (black-bellied storm petrel), 377
Fringilla coelebs (chaffinch), 430
Fulica atra L., 376
Funambulus pennanti Wroughton, 372

Gafrarium tuimidum Roding (cockle clam), 382
Galeoidea, 378, 379
Galium aparine L., 419
   G. mollugo L., 417
   G. palustre L., 422–424, 427
Gallicrex cincrea (Gmelin), 376

Galliformes, 376
Gallinago (snipe), 375
Gallinula chloropus (L.) (gray moorhen), 376, 430
Gallus gallus Robinson & Kloss (red jungle fowl), 376
Ganua curtisii (K. & G.) H.J. Lam, 312
Garcinia, 338
  G. cuneifolia Pierre, 316
Garrulus glandarius (jay), 430
Gastropoda, 242, 382
Gavialidae, 373
Gavialis gangeticus (Gmelin), 372, 373
Gecarcinidae, 242
Gecko gecko (L.) (tokay gekko), 373
Gekkonidae, 373
Geneidae, 379
Genipa americana L., 219, 234
Geoclemys hamiltonii (Gray), 373
Geonoma schottiana Mart., 272
Geranium sylvaticum L., 417
Gerres filamentosus Cuvier, 379
Gerridae, 242
Gerris aduncus, 242
Geum, 421
  G. rivale L., 417, 441
  G. urbanum L., 417, 418, 420, 421
Glareola lactea Temminck, 375
  G. pratincola L., 375
Glaucidium radiantum Tickell, 377
Glechoma hederacea L., 417
Gleditsia triacanthos L., 118, 121
Glossogobius giuris Hamilton & Buch., 378
Glyceria, 179
Gobiidae, 328–380
Gobiodes aneguillaris L., 380
  G. buchanani Day, 380
  G. rudicundus Hamilton, 380
Goniopsis cruentata Latr., 245
Gonystylus, 307
  G. bancanus (Miq.) Kurz, 305–307, 309, 311–313, 324, 330–332
Gordonia, 21
  G. lasianthus (L.) Ellis (loblolly bay), 181
Gorsachius melanolophus (Raffles) (tiger bittern), 374
Grapsidae, 381
Grias, 280
  G. cauliflora L., 221
  G. fendlerii Seem., 277, 278, 280, 281
Gruiformes, 376
Guarea, 281
Guatteria amphifolia Triana & Planch, 278
Gudusia chapra (Hamilton), 378
Guettarda speciosa L., 366
Guinotia dentata Randall, 245
Gustavia, 214
Gymnotidae, 242
Gymnotus carapo Cuvier, 242
Gynerium sagittatum (Aubl.) Beauv., 216
Gyps bengalensis (Gmelin) (white-backed vulture), 374
  G. fulvus (Hablizl.) (griffon vulture), 374

Haematopodidae, 375
Haematopus ostralegus L. (Oystercatcher), 375
Haematoxylon, 221, 288
  H. campechianum L. (tintal), 221, 246, 289, 290–292, 296
Halcyon chloris (Boddaert) (white-colored kingfisher), 375
  H. pileata (Boddaert) (black-capped kingfisher), 375
  H. smyrnensis (L.) (white-breasted kingfisher), 375
Haliaeetus leucogaster (Gmelin) (white-bellied sea eagle), 374
  H. leucoryphys (Pallas) (Pallas' sea eagle), 374
Haliastur indus (Boddaert) (Brahmini kite), 374
Halodule, 289
Hamamelis virginiana L., 120
Hampea trilobata Standl., 291
Hanguana malayana (Jacq.) Merr., 338, 341, 342, 347, 349, 353
Hardella thurji (Gray), 373
Heliconia, 218
Helicops angulatus L., 244
Hemidactylus flaviviridis Castellania and Willey, 373
Hemigrammus unilineatus Gill, 242
Hemiprocne longipennis (Rafinesque) (crested tree swift), 375
Hemiprocnidae, 375
Hemipus picatus Sykes, 376
Hemirhamphidae, 243
Hemirhamphus cantori Bleeker, 379
Henriettea fascicularis (Sw.) Gomez Maza, 256
Heritiera fomes Buch., 357, 363, 366, 369, 383, 384
  H. littoralis Ait., 340, 354
  H. littoralis Dryand., 384
Herpestes auropunctatus Hodgson (small mongoose), 371
Herpestes edwardsii Geoffroy, 371
Heterocarpus gibbosus Bate, 381
  H. woodmasoni Alcock, 381
Heteropneustes fossilis (Bloch) (stinging catfish), 378
Heteropneustidae, 378
Hibiscus elatus Sw., 221
  H. tiliaceus L., 221, 253, 294, 341, 349, 366, 384
Hilsa ilisha Hamilton, 372, 378, 379
  H. kanagurta Bleeker, 378
Himantopus himantopus (L.), 375
Himantura uarnak Forskal, 378
Hippocratea volubilis L., 222
Hippolais icterina (icterine warbler), 430
Hippolysmata ensirostris Kemp, 381
Hippomane mancinella L., 221
Holancanthus anularis Bloch, 379
Holcus mollis L., 417, 428
Homalium racemosum Jacq., 221
Hominidae, 372
Homo sapiens Laveran, 372
Hoplias malabaricus Bloch, 242, 246
Hoplosternum, 243
  H. littorale Hancock, 242, 243, 246
  H. thoracatum Valenciennes, 242
Horsfieldia, 338
Hottonia palustris L., 414, 422
Hoya, 347
  H. parasitica Wall., 384
Humulus lupulus L., 417, 439, 441

SYSTEMATIC INDEX

Hura crepitans L., 214, 219, 246
Hydnophytum formicarium Jack., 332
Hydrobatidae, 377
Hydrocotyle vulgaris L., 427
Hydrophidae, 373
Hydrophis nigrocinctus Daudin, 373
  H. obscura Daudin, 373
Hyla arborea, 431
  H. maxima, 242
  H. minuta, 242
  H. misera, 242
  H. punctata, 243
  H. rubra, 243
Hylidae, 242, 243
Hymenaea courbaril, 246
Hyperbaena winzerlingii Standl., 291
Hypnum cupressiforme Hedw., 426
Hyptostomus robinii Gill, 242
Hystricidae, 372
Hystrix hodgsonii Gray (crestless Himalaya porcupine), 372

Ibis leucocephalus (Pennant) (painted stork), 374
Icthyophaga icthyaetus (Horsfield) (gray-headed fishing eagle), 374
Ictinaetus malayensis (Temminck) (black eagle), 374
  I. cassine L., 181, 182
  I. coriacea (Pursh.) Chapm., 181
  I. decidua Walt., 120
  I. glabra (L.) Gray, 181, 182, 201
  I. guianensis (Aubl.) O.K., 218
  I. opaca Ait. (American holly), 117, 180, 201, 205
Illisha filigera Val. (jewelled shad), 378
Impatiens, 179
  I. noli-tangere L., 417, 418
Imperata, 384
  I. cylindrica (L.) P. Beauv., 343, 363
Indoplanorbis exustus Deshayes, 382, 383
Inga, 214, 217, 222, 287, 288
  I. laurina (Sw.) Willd. [*Inga fagofolia* (L.) Willd.], 253, 256
  I. laurina (Sw.) Willd., 222, 240
  I. vera subsp. *spuria* (Willd.) J. León, 288
Insecta, 242
Intsia bijuga Colebr., 354, 355
Ipomoea, 280
  I. pes capre Sweet., 363
Irenidae, 376
Iriartea, 268
Iris pseudacorus L., 424
Iryanthera, 214, 272
  I. juruensis Warb., 325
  I. macrophylla, 214
  I. paraensis Huber., 272
Itea virginica L., 181, 182
Ixobrychus minutus (little bittern), 430
  I. sinensis (Gmelin) (Chinese little bittern), 374
Ixora nicaraguensis Standl., 278

Jacanidae, 375
Jessenia, 214, 217
  J. oligocarpa Griseb. & H. Wendl. ex Griseb., 271

Juglans hindsii Jeps., 123, 124
  J. major (Torr.) Heller, 122, 123
  J. nigra L., 119, 121
Juncus, 30
  J. effusus L., 422, 423, 426
  J. repens Michx., 204
  J. roemerianus, 21
  J. tracyi Rydb., 111
Juniperus, 43
Jussiaea latifolia Benth., 280

Kachuga smithi (Gray), 373
  K. tectum (Gray) (roof turtle), 373
Kalmia augustifolia L., 180
  K. polifolia Wang, 179
Kandelia candel (L.) Druce, 366, 384
Katelysia opima (Gmelin) (inflated clam), 382
Ketengus typus Bleeker, 380
Ketupa (Bubo) zelonensis (Gmelin) (brown fish-owl), 377
Kinosternidae, 243
Kinosternon scorpiodes L., 243
Knautia dipsacifolia Kreutzer, 417

Labeo calbasu Hamilton, 372, 378
Labeo rohita Hamilton, 378
Lacerta vivipara, 431
Lagomorpha, 371
Laguncularia, 66, 148, 252, 289
Laguncularia racemosa (L.) Gaertn. (white mangrove), 31, 127, 144, 157–159, 161, 182, 253, 255, 256, 263, 269, 294, 453, 457
Lamiastrum galeobdolon (L.) Ehrend. & Polatschek, 418
Laniidae, 376
Lanius schach Linn. (Black-headed shrike), 376
Larix laricina (DuRoi) K. Koch (tamarack), 5, 26, 30, 37, 120, 166, 175, 178, 179, 180, 190
Laridae, 375
Larus rudibundus L. (black-headed gull), 375
Lasiacis procerrima (Hack.) Hitchc., 280
Lates calcarifer Bloch (sea bass, barramundi), 378, 379
Latidae, 378
Ledum groenlandicum (Oeder) (Labrador tea), 178, 179
  L. palustre L., 428, 440, 444
Leersia, 234
  L. hexandra Sw., 216, 226, 234, 342, 353
Leiognathidae (Carangidae), 378
Lemna (duckweed), 26, 30, 31
  L. minor L., 414, 422
  L. trisulca L., 422
Lemnaceae, 26
Lepidocaryoideae, 267, 268
Lepidodendron, 27, 28
Leporidae, 371
Leptodactylidae, 242
Leptodactylus bolivianus Barlenger, 242
  L. petersi Steind., 242
  L. sibilatrix Wied., 242
Leptoptilos dubius (Gmelin) (greater adjutant stork), 374
  L. javanicus (Horsfield (lesser adjutant stork), 374
Leptospermum abnorme F. Muell. ex Benth., 345

Lepus nigrocollis Cuvier (Indian hare), 371
Leskea polycarpa Hedw., 420
Lethocerus maximus De Carlo, 242
Leucobryum glaucum (Hedw.) Ångstr., 428
Leucothoe racemosa (L.) Gray, 181
Libocedrus decurrens, 125
Licania platypus (Hemsl.) Fritsch., 290
Liliopsida (Monocotyledones), 384
Lindera benzoin (L.) Blume, 119
Liquidambar, 95, 97, 98, 105
   L. macrophylla Oersted, 287
   L. styraciflua L. (sweet gum), 73, 116, 117, 121, 180, 182, 201
Liriodendron tulipifera L. (yellow poplar), 43, 73, 102, 117, 119, 120, 180, 201
Lissemys punctata (Bonnaterre), 373
Lithocarpus Blume, 325
Litsea, 307, 314
   L. crassifolia (Bl.) Boerl., 315, 316
Livistona, 342, 349
   L. brassii Burret, 346
Lobotes surinamensis Bloch, 380
Lobotidae, 380
Locustella naevia (grasshopper warbler), 430
Loliginidae, 382
Loligo affinis Koning, 382
   L. duvaucelii D'Orb., 382
   L. hardwickii Gray, 382
   L. indica Pfeffer, 382
Lomlis (Pterocara) lambis L., 382
Lonchocarpus cruentus Lundell, 288, 290
   L. guatemalensis var. *mexicanus* (Pittier) Hermann, 288
   L. hondurensis Benth., 288
   L. pentaphyllus (Poir.) DC., 288, 290
   L. sericeus (Poir) DC., 221
   L. unifoliolatus Benth., 288
Lonchura malacca Linn. (black-headed munia), 377
   L. punctulata (Linn.) (spotted munia), 377
   L. striata (Linn.) (white-rumped munia), 377
   L. subundulata (Godwin-Austen) (spotted munia), 377
Lonicera xylosteum L., 417
Lophocolea heterophylla (Schrad.) Dum., 426
Lophopetalum, 311
Loptodactylus bolivianus Boulenger, 242
Loranthaceae, 384
Loricariidae, 242
Luehea seemannii Triana & Planch, 281
Lumnitzera, 156, 347
   L. littorrea, Voigt, 384
   L. racemosa Willd., 350, 384
Luscinia megarhynchos (nightingale), 430
Lutra enudris F. Cuvier, 243
   L. lutra Linn. (smooth, common otter), 371, 429
   L. perspicillata Geoffroy (smooth-coated otter), 371
Luzula pilosa (L.) Willd., 428
   L. sylvatica (Hudson) Gaudin, 417
Lycopodium, 341, 440
Lycopus europaeus L., 424, 439
Lyginopteris, 28
Lymnaea auricularia Draparnaud, 382
Lymnaeidae, 382

Lyonia lucida (Lam.) K. Koch (fetterbush), 180, 181, 182
Lysimachia thyrsiflora, 424
   L. vulgaris L., 422–424, 427
Lythrum salicaria L., 419, 424

Mabuetia guatemalensis, 279
Macaca assamensis M'Clelland, 372
   M. mulatta Zimmermann (rhesus monkey), 372
Machaerium falciforme Rudd, 288
   M. lunatum (L.f.) Ducke, 214, 219, 221, 224, 240, 253, 288
Macrobrachium lamarrei (H. Milne-Edwards), 381
   M. malcolmsonii H. Milne-Edwards, 381
   M. mirabilii (Kemp), 381
   M. rosenbergii (Deman), 381
   M. rude, 381
   M. scabriculum (Hilgendarf), 381
Macrognathus aculeatum Lacépède (spiny eel), 378
Macrolobium, 214
   M. bifolium (Aubl.) Pers., 218
Maesa ramentacea A. DC., 367
Magnolia, 21
   M. virginiana L. (sweet bay), 180, 181, 182, 201
Magnoliopsida (Dicotyledones), 384
Maianthemum bifolium (L.) F.W. Schmidt, 428
Maiidae, 381
Malache scabra Insh., 221
Malvaceae, 384
Mammalia (mammals), 88, 243, 244, 246
Mangifera, 338, 345
   M. havilandi Ridl., 323
Manicaria (real palm), 214, 217, 240, 271, 272, 275–277, 279–283, 285
   M. saccifera Gaertn., 127, 214, 217, 219, 236, 240, 267, 269, 271, 272, 280, 281
Manilkara bidentata (A. DC.) A. Chev, 253, 256
Mapania, 338, 347
Marcgravia, 255
Mariscus albescens Guad., 384
   M. jamaicense (Crantz) Britton (*Cladium jamaicense* Crantz), 221
Mastacembelidae, 378, 379
Mastacembeloidei, 378, 379
Mastocembelus armatus (Lacépède), 379
   M. pancalus (Hamilton), 378
   M. unicolor Cuvier and Val., 379
Matuta lunaris (Forskal), 381
   M. planipes Fabricius, 381
Mauritia, 214–216, 218, 219, 224, 228, 234, 240, 267–270, 272
   M. flexuosa L. f., 214, 216, 219, 220, 240, 267–269, 272
   M. minor Burret, 267
   M. setigera Griseb. & Wendl., 214, 216, 217, 240
   M. vinifera Mart., 267
Maximiliana, 214
   M. elegans Karst., 223, 271
   M. regia Mart., 269
Mazama americana J.A. Allen, 244
Megaderma lyra Geoffroy (Indian false vampire), 371
Megadermaridae, 371
Megalops atlanticus Cuv. & Val., 241, 242
Megalurus palustris Blyth. (striated marsh warbler), 376

Megalopidae, 242, 243
Melaleuca, 4, 325, 337, 341–343, 349, 352–354, 389, 390, 392, 394, 397, 404
   M. argentea W.F. Fitzg., 392, 393, 403
   M. bracteata F. Muell., 392, 393, 403
   M. cajuputi Powell, 341, 393, 403
   M. dealbata S.T. Blake, 393, 403
   M. leucadendron (L.) L. Mant., 325, 341, 392, 393, 403
   M. preissiana Schauer, 394, 403
   M. quinquenervia (Cav.) S.T. Blake, 392, 394, 397, 398, 403, 404
   M. raphiophylla Schauer, 394, 403
   M. stenostachya S.T. Blake, 394, 403
   M. viridiflora Sol. ex Gaertn., 394, 395, 399, 403
Melampyrum pratense L., 428
Melaniidae, 382
Melanoides tuberculatus Müller, 242
   M. tuberculata J.R. Miller, 382
Melanosuchus niger Spix, 246
Melastoma, 325
Meliaceae, 384
Meloidogyne incognita Kofoid & White, 245
Menyanthes trifoliata L., 439
Mercurialis perennis L., 427
Meretmix casta, 382
   M. meretmix (Linn.) (bay clam), 382
Meropidae, 375
Merops orientalis Latham (little green bee-eater), 375
   M. philippinus Linn. (blue tailed bee-eater), 375
Metapenaeopsis andamensis Wood-Masoni, 381
Metapenaeus, 381
   M. affinis H. Milne-Edwards, 381
   M. brevicornis H. Milne-Edwards, 381
   M. dobsoni Wood-Masoni, 381
   M. monoceros (Fabricius), 381
Metaplax, 381
   M. indica H. Milne-Edwards, 381
Metopidius indicus Latham (bronze-winged jacana), 375
Metopium, 288, 289
   M. brownei (Jacq.) Urban (chechenal), 291
Metroxylon, 267, 270, 284, 305
   M. sagu Rottb. (sago palm), 268–270, 283, 347
Miconia argentea (Sw.) DC., 288
Micrandra spruceana (Baill.) R.E. Schultes Sl., 326
   M. sprucei (cunuri), 305, 319, 326
Microhylidae, 243
Micropterus brachyarus Vieillot (rufous woodpecker), 377
Microtus oeconomus (tundra vole), 429
Milvus migrans (Boddaert) (black kite), 374, 429
Mimosa pigra L., (zarzal), 289, 295, 296
Mimosaceae, 384
Mitragyna speciosa Korth., 341, 346, 347
   M. stipulosa O. Ktze., 3, 65, 129
Mnium hornum Hedw., 423, 424, 426
   M. undulatum, 440
Molinia, 415
   M. caerulea (L.) Moench, 414, 427, 428
Mollusca, 242, 245
Monarcha azurea Boddaert (black-naped monarch flycatcher), 376

Monocotyledones, 382
Montrichardia, 226, 234–238, 240, 243, 244
   M. arborescens Engler, 269, 280
   M. arborescens Schott, 216, 219, 222, 224, 226, 234
Mora, 4, 165, 214, 218, 245, 255, 283
   M. oleifera (Triana) Ducke, 99
Moraceae, 384
Morenia petersi (Anderson), 373
Mortoniodendron guatemalense Standl. & Steyerm., 290
Motacillidae, 376
Muellera frutescens Standl., 288
Mugil, 241, 242
   M. corsula Hamilton (freshwater mullet), 378, 380
   M. ocur, 380
   M. speigleri Bleeker, 380
   M. tade Forskal, 380
Mugilidae, 242, 378, 380
Muntiacus muntiak Zimmermann (barking deer), 371
Muridae, 372
Mus cervicolor Hogdson (fawn field mouse), 372
   M. musculus L. (house mouse), 372
Muscicapa striata (spotted flycatcher), 430
Muscicapidae, 376, 377
Mustelidae, 243, 371
Mycetopoda pittieri Marshall, 241, 242
Mycteroperca, 242
Myocastor coypus (nutria), 429
Myosotis scorpioides L., 418, 419
Myrcia cerifera L., 182
   M. splendens (Sw.) DC., 222
   M. gale L., 444
Myricaria germanica (L.) Desv., 441
Myristica, 4, 325, 338
   M. hollrungii Warb., 338
Myrmecodia, 347
   M. tuberosa Jack, 332
Myrsinaceae, 384
Myrtaceae, 384
Mystus aor Hamilton (eatfish), 378
   M. gulio (Hamilton), 380
   M. seenghala Skyes, 378
Mytilidae, 242, 382
Mytilopsis domingensis, 242
Mytilus viridus L. (green mussel), 382

Naja naja (L.), 373
Najas, 30, 289
Nandidae, 378, 380
Nandus nandus Hamilton, 378, 380
Narthecium ossifragum (L.) Hudson, 427
Natrix natrix, 431
   N. piscator Schn., 373
Nauclea, 349
   N. coadunata Roxb. ex Sm., 338, 342, 346, 349, 354
   N. tenuiflora (Harv.) Merr., 346
Nausitora lanceolata Rajag., 382
   N. sajnakhaliensis Rajag., 382
Nectandra antillana, 221
Nectarinia zeylonica (L.) (purple rumped sunbird), 377
Nectariniidae, 377

Nelumbo lutea, 289
    N. nucifera Gaertn., 353
Neomeris phocanoides Cuvier (little porpoise), 371
Neomys fodiens (water shrew), 429
Neonauclea, 338, 342, 349
Neoscortechinia, 307
    N. kingii, 309
Nepenthes, 315, 316, 332, 341, 347
Nephelea portoricensis (Spreng. ex Kuhn) Tryon, 253, 255, 256
Nephrolepis, 218
Nepidae, 242
Neptunia, 289
Neptunus pelagicus (L.), 381
    N. sanguinolentus (Harust), 381
Nettapus coromandelianus (Gmelin) (cotton pygmy goose), 374
Neuburgia, 270
Neuropteris, 28
Niltava tickelliae Blyth, 376
Ninox scutulata (Raffles) (brown hawk-owl), 377
Nothofagus grandis Steenis, 345
    N. perryi Steenis, 345
Notopteridae, 378
Notopterus notopterus (Pallas), 378
Numenius arquata (L.) (Curlew), 375
    N. phaeopus (L.), 375
Nyctereutes procyonoides (raccoon dog), 429
Nycticejus luteus Blyth (Bengal yellow bat), 371
Nycticorax nycticorax (Linn.) (black-crowned night heron), 374, 430
Nymphaea, 30, 289
    N. gigantea Hook., 353
Nymphoides, 289
Nypa, 270, 369
    N. (*Nipa*) *fruticans* Wurmb. (nipa palm), 270, 271, 340, 347, 353, 357, 363, 366, 384
Nyssa, 5, 21, 96, 97, 98, 101, 103–105, 109, 180, 183, 188, 203
    N. aquatica L. (tupelo gum), 73, 75, 96, 97, 102, 105, 109, 111, 116, 117, 180, 201
    N. sylvatica Marsh, 73, 237
    N. sylvatica var. *biflora* (Walt.) Sarg. (black gum), 98, 105, 111, 116, 117, 121, 180–182, 185, 191, 201

Ocotea spathulata Mez, 256
Octomeles sumatrana Miq., 347, 354
Octopodidae, 382
Octopoda, 382
Octopus favonia Hoyle, 382
    O. herdmanii (Hoyle), 382
    O. hongkongensis Hoyle, 382
    O. incertus Hoyle, 382
    O. octopodia Hoyle, 382
    O. rugosus (Bose), 382
Ocypoda cratopthalma (Pattas), 381
Ocypodidae, 381
Odocoileus virginiana L., 244, 246
Olivia gibbosa (Barn), 382
Olividae, 382
Omalonyx felinus Guppy, 242

Ondatra zibethica (L.) (muskrat), 429
Onoclea sensibilis L., 179
Ophiophagus hannah (Cantor), 373
Opisthopterus tardoor (Cuvier), 379
Orbignya cohune (Mart.) Dahlgren ex Standl., 293
    O. martiana Barb. Rodr. (babassu palm), 272
Orchidaceae, 384
Oreodoxa regia H.B. & K., 221
Oriolidae, 377
Oriolus oriolus (golden oriole), 430
    O. xanthornus (L.) (black-headed oriole), 377
Orthodontium lineare Schwaegr., 432
Orthorcade laza (L. Rich.) Beauv., 280
Orthotomus sutorius Hodgson (common tailorbird), 376
Oryza coarctata Rox., 384
Osmunda cinnamonea L., 179, 182
    O. regalis L., 422, 423
Osteichthyes, 378
Ostreidae, 382
Ostrya virginiana (Mill.) K. Koch (American hophornbeam), 180
Otolithus maculatus Cuvier and Val., 380
    O. ruber (Schn.), 380
Otus bakkamoena Pennant, 377
    O. scops (Linn.) (Scops owl), 377
Oxycoccus quadripetalus (Gilib.), 178
Oxydendrum arboreum (L.) DC. (sourwood), 180
Oxystelma esculentum R. Br., 366
Oxythece pallide (Gaertn. f.) Cronquist, 221, 240

Pachira, 215, 287, 289
    P. aquatica Aubl. (apompal), 280, 288, 291, 292
Pachycephala cinerea (Blyth) (mangrove whistler), 376
Padus avium Mill., 441
Pagrus spinifer, 380
Palaemon concinnus Dana, 381
    P. lamarrei (H. Milne-Edwards), 381
    P. stylifera (H. Milne-Edwards), 381
    P. tenuipes (Anderson), 381
Palaquium cochlearifolium Van Royen, 315
    P. leiocarpum Boerl., 320
Palicourea fastigiata Benth., 279
Pama pama (Hamilton and Buch.) (jewfish, croaker), 378
Pandanaceae, 384
Pandanus, 269–271, 325, 346, 349
    P. andersonii St. John, 312
    P. fascicularis Lamk., 366, 384
    P. ridleyi Solms, 316
    P. scandens St. John, 318
    P. sigmoideus St. John, 315
Pangasidae, 380
Pangasius pangasius (Hamilton), 378, 380
Panicum, 182, 280
    P. hemitomon Shult., 182
    P. purpurescens, 234
Panthera pardus L. (leopard), 370, 371
    P. tigris Linn. (royal Bengal tiger), 370, 371
Panulirus homarus (L.), 381
    P. ornatus, 381
    P. penicallactus, 381

Panulirus (continued)
  P. polyphagus (Herbst), 381
  P. versicolor (Latreille), 381
Paphia malabarica Dil. (false clam), 382
  P. marmorata Reeve, 382
Paradoxurus hermaphroditus Pallas (palm civet), 371
Paramitherox aculeatus H. Milne-Edwards, 381
Parapenaeopsis hardwickii (Meirs), 381
  P. sculptilis (Heller), 381
  P. stylifera H. Milne-Edwards, 381
Parastemon, 307, 314
  P. spicatum Ridl., 315
Paratelphusa, 381
  P. lamellifrons (Alcock), 381
  P. spinigera Wood-Masoni, 381
Paridae, 377
Parinari coriaceus, 272
Parishia, 311
Parkinsonia aculeata, 99, 128, 238
Parus caeruleus (blue tit), 430
  P. major L. (grey, great tit), 377, 430
  P. montanus (willow tit), 430
Paspalum repens Berg., 226
Passer domesticus (L.) (house sparrow), 377
Passeriformes, 376
Patax teira Forskal, 379
Paullinia pinnata L., 222
Pavonia, 215, 221
  P. scabra (B. Vogel) Stehle & Quentin (*Pavonia spicata* Cav.), 253
  P. scabra (B. Vogel) Stehlé, 221, 222, 240
Pelargopsis amauroptera Pearson (brown-winged kingfisher), 375
  P. capensis (L.) (stork-billed kingfisher), 375
Pelecanidae, 377
Pelecaniformes, 377
Pelecanus philippensis Gmelin (spot-billed pelican), 377
Pelecypodae, 242
Pellona elongata Bennet, 379
  P. indica Swains, 379
Pelochelys bibroni (Owen), 373
Pelodytes punctatus, 431
Pelomedusae, 243
Penaeus, 381
  P. carinatus Fabricius, 381
  P. indicus (H. Milne-Edwards), 381
  P. merguiensis (Deman), 381
  P. monodon (Fabricius), 381
  P. semi-sulcatus (Deman), 381
Pentaclethra, 274, 280
  P. macroloba (Willd.) Kuntze, 214, 274, 279–281, 284
Perciformes, 378, 379
Pericrocotus cinnamomeus (L.), 376
Periophthalmus koelreuteri, 380
  P. schlosseri Pall, 380
Perissia, 382
Perissodactyla, 371
Pernis ptilorhynchus Temminck (crested honey buzzard), 374
Persea, 21

P. borbonia (L.) Spreng. (red bay), 180, 181, 182, 201
Peucedanum palustre (L.) Moench, 423, 427
Phaenicophacus leschenaultii Lesson (sirkeer cuckoo), 376
Phaethon lepturus Daudin (long-tailed tropic bird), 377
  P. rubricauda Boddaert (red-tailed tropic bird), 377
Phaetontidae, 377
Phalacrocoracidae, 377
Phalacrocorax carbo (L.) (great cormorant), 377
  P. sinensis, 430
  P. fuscicollis Stephens, 377
  P. niger (Vieillot), 377
Phalaris arundinacea L., 417, 440
Pharus latifolius L., 280
Phasianidae, 376
Phasianus colchicus (pheasant), 430
Phoenicurus phoenicurus (redstart), 430
Phoenix, 3
  P. dactylifera, 165
  P. paludosa Roxb., 363, 366, 384
  P. reclinata Jacq., 269
Phragmites, 30, 226, 235, 240
  P. australis (Cav.) Trin. ex Steud., 226, 289, 422, 424, 439
  P. karka (Retz.) Trin., 270, 342, 347, 349, 352, 354, 384
Phrynohyas zonata, 243
Phrynops gibbus Schweig., 243
Phyllanthus, 366
Phyllomedusa trinitatis Mertens, 243
Phylloscopus collybita (chiffchaff), 430
  P. trochilus (willow warbler), 430
Physa marmorata Guilding, 245
Physocarpus opulifolius (L.) Maxim., 120
Pica pica (magpie), 430
Picea (spruce), 439, 440
  P. abies (L.) Karsten, 417, 428
  P. engelmannii Parry, 95
  P. excelsa (Lam.) LK, 439–441
  P. glauca (Moench) Voss, 120, 125, 126
  P. mariana (Mill.) BSP. (black spruce), 37, 125, 175, 177–180, 183, 189, 190, 193
  P. pungens Engelm., 124
  P. sitchensis (Bong.) Carr., 5, 43, 124
Picidae, 377
Piciformes, 377
Picoides (Dendrocopos) manus (Vigors) (pygmy woodpecker), 377
Pimelodidae, 242
Pinus (pine), 120, 176, 178, 201, 203, 415, 440, 444, 445
  P. contorta Dougl. ex Loud., 73, 111
  P. echinata Mill., 9
  P. elliottii Engelm. (slash pine), 9, 10, 181, 182
  P. fenzeliana Hand.-Mazz., 325
  P. mugo, 428
  P. palustris Mill., 9
  P. ponderosa Lawson, 123
  P. serotina Michx. (pond pine), 178, 181, 201
  P. strobus L. (white pine), 180
  P. sylvestris L., 428
  P. taeda L., 5, 9, 61, 116, 117, 201
Pipa pipa L., 242
Pipidae, 242

Pipistrellus, 371
    P. pipistrellus (pipistrelle bat), 429
Pisces, 242
Piscidia, 221
Pistia, 226, 240, 289
Pithecellobium belizence Standl., 288
    P. calostachys Standl., 288
    P. recordii Standl., 288
    P. latifolium (L.) Benth., 280, 281
Pitys, 28
Plagiomnium undulatum (Hedw.) Kops, 417
Planchonia papuana Knuth, 340, 341, 349
Planera aquatica Walter ex J.F. Gmelin, 117
Planorbidae, 382
Plasmodium falciparum Welch, 383
    P. vivax Grassi and Feletti, 383
Platalea leucorodia L. (white spoonbill), 374, 430
Platanista gantetica Lebeck, 371
Platanistidae, 371
Platanus, 95, 120, 287
    P. lindeniana Mart. & Gal., 287
    P. occidentalis L., 73, 102, 118, 119, 121
    P. racemosa Nutt., 123, 124
    P. wrightii S. Wats, 123
Plegadis falcinellus Linn. (glossy ibis), 374
Plesionika martis H. Milne-Edwards, 381
Pleuronectiformes, 380
Pleurozium schreberi (Brid.) Mitt., 178, 427, 428
Ploceidae, 377
Ploceus bengalensis L. (black-throated weaver bird), 377
    P. burmanicus Ticehurst, 377
    P. manyar (Horsfield) (streaked weaver), 377
Ploiarium, 332
    P. alternifolium (Vahl.) Melch., 319, 320, 332
Plotosidae, 379
Plotosus canius Hamilton, 379
Pluchea indica (L.) Less., 349
Plumbaginaceae, 384
Pluvialis squatarola (L.) (grey plover), 375
Poa trivialis L., 420, 423, 426
Poaceae, 384
Podiceps cristatus (L.) (great-crested grebe), 377
    P. ruficollis salvadori (little grebe), 377
Podicipedidae, 377
Podicipediformes, 377
Podocarpus, 319, 344
Podocnemis, 243
Poecilia picta, 242
Poeciliidae, 242
Polycentridae, 242
Polycentrus schomburgkii, 242
Polygonum, 30
    P. bistorta L., 417
    P. hydropiper L., 419
Polymesoda coaxans, 355
Polypodiaceae, 384
Polypodium irioides Lamk., 384
    P. vulgare L., 420
Polyporus betulinus Fr., 440

Polytrichum, 178
    P. formosum Hedw., 428
Pomacea, 241, 244
    P. cornu arietis L., 242
    P. glauca L., 242, 243, 245
    P. urceus Müller, 242, 243, 246
Pongamia pinnata Merr., 363, 366, 384
Pontederia, 289
Populus (aspen, poplar), 30, 94, 102, 116, 117, 119–121, 287, 412, 421
    P. alba L., 421, 440, 441
    P. angustifolia James, 124
    P. balsamifera L., 94, 124–126, 179
    P. canescens (Aiton) Sm., 421
    P. deltoides Marsh (cottonwood), 32, 95
    P. fremontii S. Wats., 96, 122–124
    P. heterophylla L. (cottonwood), 116
    P. nigra L., 420, 440, 441
    P. sargentii, 121
    P. tremula L., 421, 427
    P. tremuloides Michx., 124
    P. trichocarpa Torr. & Gray, 95, 124, 125
Porphyrio porphyrio (L.) (purple swamp hen), 376
Portunidae, 381
Portunus pelagicus L., 381
    P. sanguinolentus (Hawst), 381
Porzana, 376
Posoquiera grandiflora Standl., 279
    P. latifolia, 278
Potametum lucentis, 442
Potamogeton, 30, 289
Potamon mortensi Wood-Masoni, 381
    P. wood-masoni Ruthbun, 381
Potamonidae, 381
Potentilla palustris, 427
Pothos, 338
Prestoea montana (R. Grah.) Nichols, 105, 253, 255, 256
Primates, 372
Primula elatior (L.) Hill, 417
Prinia flaviventris Delessert (yellow-bellied wren-warbler), 376
    P. hodgsoni (Blyth) (Franklin's wren-warbler), 376
    P. subflava (Hodgson), 376
Prionailurus bengalensis Kerr. (leopard cat), 371
    P. viverrinus Bennet (fishing cat), 371
Prionodon pardicolor Hodgson, 371
Prioria, 4, 281, 283
    P. copaifera Griseb., 58, 98, 99, 128, 215, 238, 274, 281
Pristella, 242
Procyon lotor (North American raccoon), 429
Prosopis juniflora (Sw.) DC., 122
    P. pubescens Benth., 122
Protium pittieri Rose, 280
Prunella modularis (dunnock), 430
Prunus avium (L.) L., 421
    P. mexicana, 43
    P. padus L., 417
    P. serotina Ehrh. (black cherry), 118–120, 180, 432
Psaronius, 28
Psettus argenteus L., 379
Pseudibis papillosa Temminck (black ibis), 374

SYSTEMATIC INDEX

Pseudidae, 243
Pseudoraphis spinescens (R. Br.) Vickery, 343
Pseudoscleropodium purum (Hedw.) Fleisch, 428
Pseudospondias microcarpa Engl., 129
Pseudotsuga menziesii (Mirb.) Franco, 123, 125
Pseudus paradoxa caribensis, 243
Psittacidae, 377
Psittaciformes, 377
Psittacula eupatria (L.) (Alexandrine parakeet), 377
 P. krameri (Scopoli) (rose-ringed parakeet), 377
Psychotria chugrensis Standl., 279
Psygmophyllum, 28
Pteridium aquilinum (L.) Kuhn., 428
Pteridaceae, 384
Pteridophyta, 384
Pterocarpus, 4, 58, 66, 98, 105, 165, 214, 215, 221–223, 228, 231, 232, 234, 236–240, 243–247, 251–253, 255, 258–264, 283, 458
 P. draco, 220
 P. indicus Willd., 340, 355
 P. officinalis Jacq., 214–219, 221–224, 236–240, 245, 251, 253, 255–261, 263, 280, 281
 P. rohii Vahl., 217, 237
 P. santalanoides, 237
Pteropodidae, 371
Pycnonotidae, 377
Pyrrhula pyrrhula (bullfinch), 430

Quassia borneensis Nooteboom, 323
Quercus (oak), 30, 33, 43, 96, 104, 107, 116, 119, 120, 179, 201, 442, 445
 Q. alba L. (white oak), 116–118, 180
 Q. ellipsoidalis E.J. Hill (northern pin oak), 179
 Q. falcata var. *pagodaefolia* Ell., 116, 117
 Q. imbricaria Michaux, 118
 Q. laurifolia Michaux (laurel oak), 116, 117, 180
 Q. lobata Nee, 96, 123, 124
 Q. lyrata Walter, 116, 117
 Q. macrocarpa Michx., 121
 Q. michauxii Nutt. (swamp chestnut oak), 116, 117, 180
 Q. nigra L., 117, 182
 Q. palustris Muenchh., 96
 Q. petraea, 413
 Q. phellos L., 116, 117
 Q. pubescens, 413
 Q. robur L., 95, 103, 105, 421, 427, 428, 440
 Q. schumardii Buckley, 95
 Q. stellata Wang, 43, 96
 Q. velutina Lam., 118
 Q. virginiana Miller, 116, 117

Rachopteris, 28
Rajiformes, 378
Rallidae, 376
Rana arvalis, 431
 R. dalmatina, 431
 R. temporaria, 431
Ranatra, 242
Randia aculeata L., 253, 256
 R. mitis L., 290

 R. patula Miq., 366
Ranunculus aconitifolius L., 417
 R. ficaria L., 414
 R. repens L., 419, 423, 426
Raphia (oil palm), 4, 37, 255, 267–271, 274–277, 279, 280–285
 R. hookeri, 269
 R. monbuttorum Drude, 269
 R. taedigera Mart., 127, 267, 269, 270, 278, 280, 281, 284
 R. vinifera Beauv., 269
Rattus rattus L. (house rat), 245, 372
Rauvolfia, 278
Recurvirostridae, 375
Renealmia aromatica Aubl., 279
Reptilia (reptiles), 88, 243, 253
Restionaceae (node sedges), 395
Rhagovelia, 242
Rhamdia sebae Valenciennes, 242
Rhamnus catharticus L., 422
Rhinocerotidae, 371
Rhizoclonium hookeri, 222
Rhizophora, 37, 66, 100, 127, 144, 146–149, 155, 156, 162–166, 185, 191, 220, 232–234, 236, 252, 253, 283, 288, 289, 346, 347, 351, 354, 365, 384
 R. apiculata Blume, 164, 347, 366, 384
 R. harrisonnii Bleech., 294
 R. lamarckii Montr., 164
 R. mangle L. (red mangrove), 2, 5, 31, 32, 37, 38, 75, 144, 146, 151, 156–164, 182, 185, 195, 263, 271, 293, 294, 453, 457
 R. mucronata Lamk., 163, 366, 384
 R. racemosa Mey, 269
 R. stylosa Griff., 164
Rhizophoraceae, 384
Rhunchospora, 347
Rhus radicans L., 181
Rhynchospora, 338, 347
Ribes nigrum L., 439
Rivulus hartii Boulenger, 242
Robinia pseudo-acacia L., 120, 121
Rodentia, 372
Roeboides dayi, 242
Rosa californica, 124
Rostramus sociabilis Vieillot, 244
Rostratulidae, 375
Roystonea, 216, 224, 225, 241, 292
 R. borinquena O.F. Cook, 221, 253, 256
 R. dunlapiana P.H. Allen, 292
 R. oleracea (Mart.) Cook, 214, 216, 240
 R. oleraceae O.F. Cook, 271
 R. princeps (Becc.) Burret, 221
 R. regia (HBK.) O.F. Cook, 292
Rubus, 124
 R. fruticosus L., 417, 424, 428
 R. idaeus L., 417
Rumex conglomeratus Murray, 419
 R. obtusifolius L., 419
Ruppia, 289

Sabal mauritiiformis (H. Karsten) Griseb. & Wendl., 293
 S. palmetto Rein, 272

Sabal mauritiiformis (continued)

    S. yapa Wright ex Beccari, 293
Saccharum cylindricum Lamk., 384
    S. robustum Brandes & Jeswiet ex Grassl., 352, 354
Salamandra salamandra (fire salamander), 431
Salicornia, 157, 158, 182
Salix (willow), 30, 32, 71, 78, 102, 111, 116, 117, 120, 121, 123–125, 287, 412, 415, 418–421, 425, 440, 441
    S. alaxensis (Anderss.) Coville, 125
    S. alba L., 415, 418–420, 440, 441
    S. amygdalina L., 441
    S. amygdaloides Anderss., 121, 125
    S. arbusculoides Anderss., 125
    S. atrocinerea Brot., 422, 425
    S. aurita L., 422, 425, 439
    S. chilensis Mol., 287, 289
    S. cinerea L., 422, 424, 425, 427, 439
    S. discolor Muhl., 126
    S. exigua Nutt., 121
    S. fragilis L., 418, 420, 441
    S. glauca, 125
    S. gooddingii Ball, 122, 123
    S. incana Schnk., 441
    S. interior Rowlee, 121
    S. lanata L. ssp. *richardsonii* (Hook) A. Skwortz., 125
    S. lasiolepsis, 124
    S. nigra Marshall, 119, 120
    S. pentandra L., 439
    S. planifolia Pursh, 126
    S. pulchra Cham., 125
    S. purpurea L., 420, 441
    S. scouleriana, 124
    S. triandra L., 418, 441
    S. viminalis L., 418, 419, 441
Salix x *dasyclados* Wimmer, 418, 420
Salvinia, 226, 289
Sambucus nigra L., 417
Sapium carabeum Urb., 221
    S. indica Willd., 366
    S. indicum, 384
Sarcolobus globosus Wall., 384, 366
    S. carinitus Griff., 384
Sarkidiornis melanostos (Pennant) (knob-billed goose), 374
Sarracenia purpurea L., 180
Sassafras albidum (Nutt.) Nees, 102
Saxicola caprata (Linn.) (pied stone chat), 377
    S. ferrea Gray (gray bush chat), 377
Scatophagidae, 380
Scatophagus argus Bloch, 380
Scheelea basslerina Burret, 268
    S. cephalotes Karst., 268
    S. liebmannii Beccari, 293
    S. magdalena Dugand, 272
    S. parviflora Barb. Rodr., 272
    S. rostrata Burret, 272
Schilbeidae, 378, 380
Schistosoma (Bilharzia), 247
Schucirmansia henningsii, 270

Sciaena coitor (Hamilton), 380
    S. cuja (Hamilton), 380
    S. diacanthus Lacépède, 380
    S. sina (Cuvier), 380
Sciaenidae, 343, 378, 380
Sciaenoides pama (Hamilton), 380
Scirpus, 30, 78, 289
    S. maritimus L., 384
Sciuridae, 372
Scleria, 234, 280
Scoliodon sorroakowah (Cuvier) (dogfish), 378
Scolopacidae, 375
Scolopax rusticola (woodcock), 430
Scomberesocidae, 379
Scotophilus, 371
Scutellaria galericulata L., 441
Scylla serrata (Forskal), 381
    S. serrata, 355
Scyphiphora hydrophyllacea Gaertn., 366
Selaginella, 101
Senecio fuchsii, 417
Sepia aculeata (F. & d'Orb.), 382
    S. rostrata (F. & d'Orb.), 382
    S. rouxii (F. & d'Orb.), 382
Sepiella intermis (F. & d'Orb.), 382
Sepiidae, 382
Sepiotentis arctipinnis (Gould), 382
Sequoia sempervirens (D. Don) Endl., 124
Serranidae, 242, 380
Serranus lanceolatus Bloch, 380
    S. sonnerati Lacépède, 380
Sesarma, 381
    S. minuta de Man, 381
Sesuvium portulacastrum L., 157–159, 182
Setipinna phasa Hamilton, 378
Shorea, 326
    S. albida Sym., 4, 303, 306, 307, 310, 311, 313–315, 317, 318, 320, 322, 324, 326, 329–332
    S. pachyphylla Ridl. ex Sym., 326, 332
    S. uliginosa Foxw., 324
Sickingia maxonii Standl., 280, 282
Sigillaria, 27, 28
Silonia silondia Hamilton, 380
Siluriformes, 378, 380
Simarouba amara Aubl., 221
Sloanea dentata L., 221
Smilax californica, 124
    S. laurifolia L., 181, 201
Socratea, 268
    S. durissima (Oerst.) Wendl., 281
Solanum dulcamara L., 419, 420, 423, 424
    S. lancefolia Jacq., 280
Solea ovata Richardson, 380
Soleidae, 380
Solenocera indicus Natarey, 381
Solidago, 180
Sonneratia, 351
    S. acida (*S. ovata* Backer), 325
    S. acida Benth., 384
    S. alba Griff., 384

SYSTEMATIC INDEX

Sonneratia (*continued*)

   S. alba J.E. Sm., 346
   S. apetala Ham., 363, 365, 366, 384
   S. caseolaris (L.) Engl., 346, 384
Sonneratiaceae, 384
Sorbus aucuparia L., 424, 427, 428, 439
Sparganium, 30
Sparidae, 380
Spartina, 21
   S. alterniflora, 21
Spathiphyllum friedrichsthalli Schott, 280
Sphaenorhynchus eurhostus, 243
Sphaeroides testudineus L., 241
Sphaeroma terebrans Bate, 38
Sphagnum, 34, 35, 37, 67, 70, 71, 175, 177–181, 190, 193, 305, 414, 415, 421, 424, 428, 431, 440
   S. capillifolium (Ehrh.) Hedw., 425, 428
   S. fimbriatum Wils., 425, 427
   S. fuscum (Schimp.) Klinggr., 428
   S. junghuhnianum Doz. et Molk., 316
   S. magellanicum Brid., 180, 428
   S. palustre L., 423, 425, 427
   S. papillosum Lindb., 427, 428
   S. recurvum P. Beauv., 425, 427
   S. rubellum Wils., 180
   S. squarrosum Crome, 425
Sphenophyllum, 28
Spilornis cheela (Latham) (crested serpent eagle), 374
Spondianthus ugandensis Hutch., 129
Spondias, 221, 236
Squamata, 373
Stachys sylvatica L., 417, 418, 427
Stauropteris, 28
Stellaria nemorum L., 416–418
Stemmadenia, 280
Stemonurus, 307
   S. scorpioides Becc., 323
   S. secundiflorus Bl. var. *lanceolatus* Becc., 311
Stenella malayana Lesson (Malay dolphin), 371
Stenochlaena palustre Bedd., 384
   S. palustris (Burm. f.) Bedd, 338, 341, 345, 347, 348
Sterculia caribaea R. Br., 221
Sterculiaceae, 384
Sterna albifrons Pallas (least tern), 375
   S. aurantia J.E. Grey (river tern), 375
   S. bengalensis Lesson (lesser crested tern), 375
   S. bergii Lichtenstein (greater crested tern), 375
   S. fuscata Linn. (sooty tern), 375
   S. nilotica (Gmelin) (gull-billed tern), 375
Stigmaria, 28
Streptopelia chinensis (Scopoli) (spotted tern), 375
   S. tranquebarica (Hermann) (red turtle dove), 375
Strigidae, 377
Strigiformes, 377
Strix aluco (tawny owl), 430
   S. leptogammica Temminck (brown wood-owl), 377
Stromateidae, 378
Stromateus cinereus Day, 378
Strombidae, 382

Strombus canarensis L., 382
Sturnidae, 377
Sturnus contra Linn. (pied starling), 377
Succineidae, 242
Suidae, 371
Sus salvanius Hodgson (pygmy hog), 371
   S. scrofa L., 371
Swintonia glauca Engl., 323
Sylvia atricapilla (blackcap), 430
   S. borin (golden warbler), 430
   S. communis (whitethroat), 430
   S. curruca (lesser whitethroat), 430
Symphonia, 214, 219, 232–234, 236, 243, 247, 272
   S. globulifera L. f., 214, 217, 219, 221–223, 231, 234, 236, 238, 240, 246, 268, 272, 281
Symphytum officinale L., 419, 420
Synaptura pan (Hamilton), 380
Synbranchidae, 242
Synbranchiformes, 380
Synbranchus marmoratus Bloch, 242
Syzygium, 270, 338, 347, 349

Tabebuia, 214, 215, 219, 236, 272
   T. insignis (Miq.) Sandw., 218
   T. insignis var. *monophylla* Sandw., 272
   T. pallida, 221, 222, 240
   T. riparia, 221
   T. rosea (Bertol.) DC., 288, 290
   T. serratifolia (Vahl.) Nicholson, 236
Tabernaemontana chrysocarpa Blake, 279
Tamaricaceae, 384
Tamarix dioica Roxb., 363, 366, 369, 384
   T. pentandra Pall. (= *T. chinensis*) (salt cedar), 101, 121–123, 131–133
Taphozous longimanus Hardwicke (sheath-tailed bat), 371
Taxodium (cypress), 21, 37, 38, 40, 95–98, 101, 103–105, 107, 109, 174, 176, 178, 180, 181, 183–185, 188, 190, 194, 195, 201, 203–205, 287, 447, 453, 455, 457
   T. ascendens Brongn., 111
   T. distichum (L.) Rich. (bald cypress), 2, 5, 31, 61, 102, 114, 116, 117, 121, 180, 181, 201
   T. distichum var. *nutans* (Ait.) Sweet (pond, scrub or dwarf cypress), 61, 175, 176, 181–188, 191–193
   T. mucronatum Ten., 287
Tayassu tajacu, 244
Tenagobia signata White, 242
Tephrodornis pondicerianus (Gmelin) (common wood shrike), 376
Teredinidae, 382
Teredo mannii Wr., 382
Terminalia, 340
   T. amazonia (Gmel.) Exell, 288
   T. brassii Exell, 337, 343, 344, 354
   T. canaliculata Exell, 338, 340
   T. catappa L., 221
   T. lucida Hoffm., 99
Terpsiphone paradisi (L.) (Asian paradise flycatcher), 377
Testraodon lunaris Bloch and Schn., 380
Testudinidae, 373
Testudo elongata Blyth, 373

Tetractomia parvifolia Ridl., 312
Tetraodon patoca Hamilton (pufferfish), 379, 380
Tetraodontidae, 379, 380
Tetraodontiformes, 379, 380
Tetraphis pellucida Hedw., 426
Thais bufo (Lamarck), 382
Thaisidae, 382
Thalamita crenata (H. Milne-Edwards), 381
Thalassia, 150, 289
Thalassina anomala, 355
Thalia, 289
    T. geniculata L., 280
Thelypteris palustris Schott, 422–424
Therapon jarbua, 380
    T. theraps Cuvier & Val., 380
Theraponidae, 380
Thespesia populnea Corr., 366, 384
Thiaridae, 242
Thithrinax biflabellata Barb. Rodr., 272
Thoracostachyum, 347
    T. bancanum (Miq.) Kurz, 315, 316, 318
    T. sumatranum Kurz, 338
Threskiornis melanocephala (Latham) (oriental ibis), 374
Threskiornithidae, 374
Thrinax parviflora Swartz, 294
Thuja occidentalis L. (northern white cedar), 30, 120, 166, 175, 177–179, 183, 184, 186–189, 193, 194
    T. plicata Donn., 124, 125
Tilia, 95
    T. heterophylla, 43
    T. platyphyllos Scop., 421
Tiliaceae, 384
Tillandsia, 255
    T. fasciculata L., 182
    T. usneoides L. (Spanish moss), 181
Timonius timon (Spreng.) Merr. & Perry, 346, 354
Tovomita plumieri Griseb., 222
Toxicodendron diversilobo, 124
Toxotes chatareus, 380
Toxitidae, 380
Treron bicincta (Jerdon) (orange-breasted green pigeon), 375
    T. phoenicoptera (Latham) (yellow-footed green pigeon), 375
    T. pompadora (Gmelin) (pompadour green pigeon), 375
Trichastoma abbotti (Blyth) (Abbot's babbler), 377
Trichecidae, 243
Trichecus manatus L., 243
Trichiuridae, 380
Trichiurus haumela Froskal, 380
    T. muticus (Gray), 380
    T. savala (Cuvier), 380
Trichodactyliidae, 242
Trichodactylus dentatus Randall, 242
Trichogaster fasciatus (Bloch & Schn.), 378
Trientalis europaea L., 427
Tringa, 375
    T. hypoleucos L. (common sandpiper), 375
Trionychidae, 373
Trionyx gangeticus Cuvier (soft-shelled turtle), 373
    T. hurum Gray (black mud turtle), 373

    T. nigricans Anderson, 373
Triplaris surinamensis Cham., 214, 218, 219, 240, 245
Tristania, 315, 319, 325, 332
    T. beccarii Ridl., 315
    T. obovata R. Br., 315, 318, 319
Trithrinax biflabellata Barb. Rodr., 272
Triturus alpestris, 431
    T. marmoratus, 431
Trochidae, 382
Trochus niloticus L., 382
Troglodytes troglodytes (wren), 430
Tsuga heterophylla, 43, 124, 125
Turbinella pyrum Lamarck, 382
Turbinellidae, 382
Turbo marmoratus L., 382
Turdoides longirostris (Hodgson) (slender-billed babbler), 377
Turdus merula (blackbird), 430
    T. philomelos (song thrush), 430
    T. viscivorus (mistle thrush), 430
Typha (cattail), 30, 34, 240, 289
    T. angustifolia Kurz, 384
    T. angustifolia L., 424
    T. domingensis Pers., 221
    T. elephantina Roxb., 384
    T. latifolia L., 221, 424
Typhaceae, 384
Typhlopidae, 373
Typhlops diardi Schlegel (blind snake), 373
Tyto alba (Scopoli) (barn owl), 377
    T. capensis Smith (grass owl), 377
Tytonidae, 377

Ulmus, 30, 95, 117, 119, 121
    U. alata, 43
    U. americana L. (American elm), 106, 117–121, 126, 166, 179
    U. campestris L., 440
    U. laevis Pallas, 421
    U. minor Miller, 413, 421
    U. rubra Muhl., 118
Ulodendron, 28
Umbonium vestiarium (L.), 382
Unionidae, 382
Uranotaenia, 244
Urospatha tonduzzii Engler, 280
Urtica dioica L., 417–420, 423, 424
Utricularia, 30, 182, 289
    U. vulgaris L., 422

Vaccinium, 179, 181, 182
    V. corymbosum L., 180
    V. macrocarpon Ait., 180
    V. myrtillus L., 427, 428, 440
    V. oxycoccos L., 180
    V. stenanthum Sleum., 317
    V. uliginosum L., 428, 440, 445
Valeriana officinalis L., 419, 420
Vallisneria, 30, 289
Vandeleuria oleracea Bennet (tree mouse), 372

SYSTEMATIC INDEX

Vanellus indicus (Boddaert) red-wattled lapwing), 375
    V. spinosus (L.) (spurwinged lapwing), 375
Varanidae, 373
Varanus bengalensis Daudin (grey Indian monitor), 373
    V. flavescens (Gray) (ruddy snub-nosed monitor), 373
Vatairea guianensis Aubl., 214
    V. lundellii (Standl.) Killip, 288, 290
Vatica papuana Dyer, 338
    V. wallichi (cf. *V. pauciflora* Bl.), 325
Veliidae, 242
Velorita cyphinoides (Gray), 382
Verbenaceae, 384
Verenidae, 382
Vertebrata (vertebrates), 56, 242, 244
Vespertilio serotinus, 429
Vespertilionidae, 371
Viburnum opulus L., 439
Vipera berus, 431
Vipera russelli Shaw (Russell's viper), 373
Viperidae, 373
Virola, 219, 232, 234, 247, 268
    V. sebifera Aubl., 281
    V. surinamensis (Rol.) Warb., 214, 217–219, 223, 231, 234, 237, 246, 269
Vitex cofassus Reinw. ex Bl., 355
    V. divaricata Sw., 256
    V. negundo L., 384
    V. trifolia L., 384
Viverra indica Desmarest, 371

Viverra zibetha L., 371
Viverridae, 371
Viviparus bengalensis (Lamarck), 382
Viviparidae, 382
Vochysia, 289
    V. guatemalensis D. Sm., 288
Vulpes bengalensis Shaw (Bengal fox), 371

Walchia, 28
Washingtonia filifera (Linden ex Andre) H. Wendl., 5, 8
Woodwardia virginica (L.) Small, 182

Xanthidae, 381
Xenorhynchus asiaticus (Latham) (black-nosed stork), 374
Xylocarpus, 354
    X. granatum Koen., 366, 384
    X. mekongensis Pierre, 366, 384
    X. moluccensis (Lamk.) Roem., 366
Xylopia corrifolia Ridl., 320
    X. frutescens Aubl., 288

Yucca, 182

Zenobia pulverulenta (Bartr. ex Willde.), 181
Zizania, 30
Zosteropidae, 377
Zosterops palpebrosa (Temminck), 377
Zoysia pungens Willd., 384

GENERAL INDEX

Aachen (Nordrhein-Westfalen, W. Germany), 409
Aberdeen (Grampian, U.K.), 409
aboriginal, 390
aborigines, 404
aboveground biomass, 102–104, 113, 118, 128, 149, 153
– production, 104, 105
– wood, 109
abscission, 109, 162, 445, 446
acalches, 291
accretion, 71, 90, 130, 221, 222, 226, 232, 234, 357
–, land, 354
–, lateral, 89
–, vertical, 89, 90
accumulation, peat, 69–71
–, sediment, 69–71
Achterhoek (Gelderland, The Netherlands), 409
acid rain, 35, 132
acidic, 439
acidity, 22, 34
–, herbicidal effects of, 34, 37
–, humic, 34
actinomycetes, 413
adaptations, 61–63, 72, 73, 88
Adhachaki Khal (Sundarbans, Bangladesh), 368
adventitious roots, 338, 341
aerated, 412, 420, 425
aeration, 60, 63, 65, 76, 110, 146, 150, 155, 404, 413, 418, 449
–, soil, 93
aerenchymous tissue, 72
aerial photographs, 155
– roots, 338
aerobic, 75, 76, 108
– conditions, 451
– metabolism, 75
aerodynamic roughness, 305, 309, 311, 313, 315, 318, 320, 322, 326, 327
aestivation, 243
Africa (*see also particular countries*), 3, 37, 54, 128, 129, 133, 267, 269, 271, 432
Agathis kerangas, 319
age structure, 130, 133
aggradation, 89, 90, 102
agricultural crop production, 133
– fields, 113
agriculture, 55, 56, 58, 77, 117, 130, 274, 282, 411, 442
– lands, 441
aguajales, 268
Alabama (U.S.A.), 3, 97, 117
Alabama River (U.S.A.), 116
Alan forest, 311

Alaska (U.S.A.) 29, 35, 36, 65, 88, 94, 115, 124, 125, 177, 189, 190, 193
– streams, 108
– tundra, 115, 134
Alaska Range (U.S.A.), 126
albedo, 62, 322, 326
Albemarle Sound (N.C., U.S.A.), 203
Alberta (Canada), 3, 36, 71, 111
Albuquerque (N.M., U.S.A.), 71, 122, 123
alder (*Alnus*), 5, 30, 38, 71, 95, 125, 179, 183, 184, 186, 187, 287, 412, 413, 415, 417, 419, 424, 425, 427–440, 442, 459
–: ash carr, 437
–: carr, 445
– forests, wet, 441
– leaves, 445
–, speckled (*A. rugosa*), 120, 126, 178, 179
– swamps, 437
– trees, 439
Alexandra River (Alta., Canada), 71
algae, 113, 158, 182
algal production, 79
alkali, 439
allelopathic, 113
allelopathy, 303, 316–318
alligator, 38
– hole, 38
alligatorweed (*Alternanthera philoxeroides*), 31
allogenic factors, 143, 160, 165
– forces, 65
– succession, 65
alluvia, 437, 442
alluvial, 127, 397
– deposition, 89, 92
– deposits, 125
– fill, 89
– flats, 392
– forests, 441
– plains, 335, 338
– sediments, 441
– soils, 88, 117, 441
– swamp, 59, 111, 112
– valley, 89, 93
alluviated, 89
alluviation, 91
alluvium, 89, 90, 93, 122, 233, 234, 269
Alnetalia glutinosae, 443
Alnetea, 443
Alnetum, 442
Alnetum glutinosae, 439, 441, 442
– *incanae*, 417, 440–442
– *viridis*, 416

Alnion glutinosae, 438, 440, 441, 445
Alno–Fraxinetum, 442
Alno–Padion, 437, 441, 442, 445
Alps, 407
Altamah River (Ga., U.S.A.), 116
alteration, forest, 130
–, wetland, 129
altitude, 92, 123
Amazon (South America), 4, 9, 37, 57, 58, 108, 109, 127, 133, 269–271
Amazon Basin (South America), 235, 236, 244, 309, 319, 325, 326
Amazonas, 128
Amazonia, 89, 99, 103, 127, 305
–, central, 111
Amboina (Indonesia), 394
Amerind, 240, 241
amino acids, 75
ammonium, 108, 112
amphibians, 88
anaerobic, 65, 68, 73–76, 88, 108, 109, 400, 404, 412, 414, 415, 425
– conditions, 451
– decomposition, 55, 74
– sediments, 112
– soils, 90, 106
– zone, 112
Andes, 89
Anduki Forest Reserve (Brunei), 302
animal(s), 56, 58, 67, 79, 134, 412, 423, 429, 430–432, 452
– biomass, 56, 130, 133
– communities, 56
– life, 56, 60
–, role of, 452
anoxia, 171
anoxic, 60, 62, 74
– sediments, 112
Antarctic, 71
Aor-Shipsa (Sundarbans, Bangladesh), 365
Apalachicola River (Fla., U.S.A.), 98, 105, 116
Apance River (Haute Marne, France), 409
apompal (*Pachira aquatica*), 280, 288, 291, 292
Appalachian Mountains (U.S.A.), 120
Appalachian Uplands (N.Y., U.S.A.), 120
Approuague River (French Guiana), 233, 234
Aqua Brava Lagoon (Mexico), 293
aquaculture, 58, 68
aquatic, 57
– communities, 89
– fauna, 241, 242
– macrophytes, 75, 113
aquatics, 342, 343, 346, 353
aquifers, 21, 143, 172
Aquirre (Puerto Rico), 152
arbovirus(es), 147
arctic, 88, 125, 126
Arctic Zone, 19
Argentina, 4, 58, 71
Argentine Isles, 71
arid, 69, 75
– climate, 88, 94, 102, 112, 120, 123, 133, 134
– environments, 456

arid southwest region (U.S.A.), 128
aridity, 128
Arizona (U.S.A.), 138
Arkansas (U.S.A.), 40, 90, 106
Arkansas River (U.S.A.), 95, 116, 120, 121
Arkhangel'sk (U.S.S.R.), 7
aroid, 269
artesian spring, 16
Ascencion Bay (Mexico), 292
ash (*Fraxinus*), 30, 98, 105, 117, 119, 120, 163, 201, 287, 412, 442
–, black (*F. nigra*), 111, 120, 126, 166, 178, 179
–, Carolina (*F. caroliniana*), 117, 180
–, green (*F. pennsylvanica* var. *subintegerrima*), 33
ash content, 443
Asia (*see also particular countries*), 4, 54, 160, 359, 429
aspen (*Populus tremuloides*), 30, 94, 102, 116, 117, 119, 120, 121, 287, 412, 421
asphyxiation, 60
assimilation number, 22
associations, 439, 441
Astrantio-Fraxinetum association, 442
Atasta Lagoon (Mexico), 293, 295
Atchafalaya River (La., U.S.A.), 130
Atlantic and Gulf Coastal Plain (U.S.A.), 27, 181
Atlantic Coast (U.S.A.), 92, 116
Atlantic Coastal Plain (U.S.A.), 181
Atlantic Ocean, 57, 65, 407, 444
atmospheric conditions, 59
– deposition, 411, 426
– factors, 451
– fronts, 69
– saturation, 8
Australia (*see also particular localities*) 2, 3, 4, 7, 37, 89, 103, 126, 150, 160, 164, 269, 387–395, 400–402, 404, 453, 456, 457
autogenic factors, 162, 165
– forces, 65
– succession, 65
avadavat, red (*Amandava amandava*), 376
avian species, 57
– wildlife, 57
avifauna, 57

babbler, Abbott's (*Trichastoma abbotti*), 377
–, slender–billed (*Turdoides longirostris*), 377
Bacalar Lagoon (Mexico), 294
back-plain swamps, 340, 341, 349
backswamp, 116, 221–223, 231, 234, 235, 239, 240, 247, 340, 342
backwater, 453, 458
bacteria, 413
Bahía de Guanica (Puerto Rico), 152
Bahía de Jobos (Puerto Rico), 152
Bahía Salinas (Puerto Rico), 152
Bahía Sucia (Puerto Rico), 76, 149, 152
Bairnsdale (Vic., Australia), 392
bajos, 291
Bakarganj (Bangladesh), 370
Bako National Park (Sarawak), 302, 305, 316, 332
Bakunda Forest (West Africa), 3
Balancau (Mexico), 288
Baleswar River (Bangladesh), 361

GENERAL INDEX

Baltic Sea Coast, 444
bamboo, 129
Bamu River (Papua New Guinea), 336
bana, 305, 319, 326, 327
Bangladesh (*see also particular localities*), 357–361, 363, 365, 369–374, 378, 381–384
bank, 89, 91
– erosion, 91
–, gravel, 441
–, sand, 441
Baram Delta (Brunei), 329
Baram River (Sarawak), 309, 328
Barisal (Bangladesh), 358, 360, 369
bark, 107
Barra del Colorado (Costa Rica), 274, 282
barramundi (*Lates calcarifer*), 378, 379
barrier island, 19
basal area, 88, 94, 95–97, 99–101, 104, 107, 108, 128, 134, 182, 183, 196, 277–285, 329, 414, 415, 448
– increment, 110
base flow, 77
– level, 89
basin, 26, 33, 124, 127, 226, 231, 239, 251, 264, 270, 271, 282, 305, 345, 447–449, 453, 456, 457–459
– filling, 30
– forested wetlands, 88, 104, 108–110
– wetlands, 147, 150, 158, 171, 177, 182–186, 188, 192, 193, 196
– freshwater, 172, 177, 182–186, 188, 192
– saltwater, 172, 176, 182–188, 196
Batang Lupar (Sarawak), 324, 329
bat(s), 244, 371
–, Bengal yellow (*Nycticejus luteus*), 371
–, pipistrelle (*Pipistrellus pipistrellus*), 429
–, serotine (*Vespertilio serotinus*), 429
–, sheath-tailed (*Taphozous longimanus*), 371
Bawang Ling (China), 302, 329
bay(s), 181
–, loblolly (*Gordonia lasianthus*), 181
–, red (*Persea borbonia*), 180–182, 201
–, sweet (*Magnolia virginiana*), 180–182, 201
bay forest, 181, 182
Bay of Bengal, 357–361, 362
bayheads, 176, 182
beach, 19–21
– barriers, 223, 226, 227, 231, 234
– plains, 341
– ridge(s), 226, 227, 231, 335
– swales, 349
beaver (*Castor*), 26, 35, 133, 429
– dams, 26, 35, 108
– effects on wetland formation, 35
– impoundments, 70
– giant, 35
– pond, 111
Beaverdam Run (Penn., U.S.A.), 89, 90
bee-eater, blue-tailed (*Merops philippinus*), 375
–, little green (*M. orientalis*), 375
beech, American (*Fagus grandifolia*), 117, –119, 180, 201
Bega (N.S.W., Australia), 391
Belgium, 407, 409, 413, 415, 416, 422, 429, 431

Belize, 271
belowground biomass, 104, 153
– production, 104
– wood, 109
Bengal District, 381
berm, 147, 153, 156, 162, 164
Betuletalia pubescentia, 443
Betuletum pubescentis association, 440
Betulo-Salicetum repentis association, 439
Bhola River (Sundarbans, Bangladesh), 368
Bialowieza Forest (Poland), 440
Biebrza River (Poland), 437
Biebrza Valley (Poland), 439
Biesbosch (N-Brabant, The Netherlands), 409, 419
Big Cypress Swamp (U.S.A.), 167
Big Thicket Swamp (Texas, U.S.A.), 19
bilharzia, 247
Binio River Basin (Sarawak), 325
biochemical oxygen demand, 79
biogenic accumulation rate, 69
biogeochemical cycles, 53, 54, 55, 57, 68
biogeochemistry, 447
biogeographic location, 113, 134
biomass, 56, 72, 87, 102, 104, 107–109, 118, 127, 128, 130, 133, 134, 147, 153, 156, 162, 164, 178, 183, 184, 193, 329, 414, 448, 455, 458
–, aboveground, 72, 103, 180, 184, 187, 189, 192, 193, 448
– accumulation, 118
– allocation, 72, 184
–, belowground, 72, 73, 103, 184, 448
– increment, 159
–, leaf, 184
– production, 72, 75, 88, 104–106, 108, 134, 186, 187, 196, 445
–, root, 184, 193
–, total, 448
–, tree, 458
–, understory, 184, 458
biosphere, 458
birch (*Betula*), 5, 71, 415, 416, 424, 428, 437, 439
–, bog (*B. pumila*), 178
–, paper (*B. papyrifera*), 126, 178, 179
–, river (*B. nigra*), 93, 116, 117, 121
–, yellow (*B. lutea*), 179
bird(s), 57, 244–246
– rookeries, 68
Bismarck (N.C., U.S.A.), 71
Bismarck Archipelago, 336, 343
bittern, black (*Dupetor flavicollis*), 374
–, Chinese little (*Ixobrychus sinensis*), 374
–, little (*I. minutus*), 430
–, tiger (*Gorsachius melanolophus*), 374
blackbird (*Turdus merula*), 430
blackcap (*S. atricapilla*), 430
Black Range (N.C., U.S.A.), 97
Black River Morass (Jamaica), 221, 246
blackwater, 9, 231, 299, 301, 302
– rivers, 127, 306
bloodwood, long-fruited (*Eucalyptus polycarpa*), 390
bluethroat, white-spottted (*Cyanosylvia svecica cyanecula*), 430
Boca Barranca (Costa Rica), 100

bog, 19, 21, 30, 33, 34, 37, 40, 63, 70, 71, 126, 172–175, 178, 180, 183, 190, 191, 193, 196, 387, 409, 411, 415, 425, 418, 443, 453, 458
– forest, 30, 38, 41, 56, 58, 62, 70, 175, 178, 183, 184, 186, 325
–, northern, 174, 176
–, ombrogenous, 305
–, herbaceous, 66, 67
–, low, 443
–, peat, 120, 440
–, raised, 70, 437, 440, 443
–, sphagnum, 34, 120
–, transitional, 439
Bogra (Bangladesh), 358, 369
Bois Neuf (Trinidad), 226, 227, 244
boles, 426
Boqueron (Puerto Rico), 251
boreal, 114
– forest, 115, 126
– region, 53, 54, 65
– relict vegetation, 120
boreal zone, 19, 20
Borneo, 89, 127, 299, 302, 305, 309, 316, 324–326, 329, 332
bosque deciduo ripario, 287
bosque perennifolio ripario, 287
Botshol (Utrecht, The Netherlands), 409
bottomland, 117
– forests, 102, 133
– hardwoods, 106, 116, 117
bottoms, 119
Bougainville (Solomon Islands), 336, 343, 344
Boulanger Creek (French Guiana), 232
boulders, 89
boundaries, 458
box, black (*Eucalyptus largiflorens*), 390, 391, 403
box elder (*Acer negundo*), 32
brackish, 269, 270, 271, 389, 412, 418, 419
– swamps, 337, 340, 341, 346
braided channel, 89
– stream, 17, 125, 131
branches, 72
Brazil, 4, 58, 161–163, 183, 267, 269, 271, 272, 299
breaches, 90
breathing roots, 338, 341, 347, 349
Bribie Island (Qld., Australia), 398
Brie (Charente, France), 409
Briguil (Charente, France), 409
British Guiana, 4
British Isles, 407, 416, 422, 430–432
Broads, The (England), 409, 416, 422
bromeliads, 182
Brooks Range (Alaska, U.S.A.), 125
browsing, 133
Brunei, 4, 299, 300, 308, 311, 314, 317, 319, 329
bryophyte, 101, 165, 420, 427
Budongo Forest (Uganda), 129
buds, 69
buffalo, wild (*Bubalis arni*), 371
bullfinch (*Pyrrhula pyrrhula*), 430
bulk density, 443
bunting, reed (*Emberiza schoeniclus*), 430

buoyancy, 76
burial, 94
–, phosphorus, 110
Burma, 358
burning, 292
Bush Bush (Trinidad), 226, 227, 229, 244, 247
buttonwood (*Conocarpus erectus*), 144, 157, 158, 182, 221, 222, 253, 294, 295
buttress, 185, 221, 236, 244, 245, 311–313, 338, 341, 345, 347
– roots, 255
buttressed trees, 127
buttressing, 128, 129
buzzard(s) (*Buteo*), 430
–, crested honey (*Pernis ptilorhynchus*), 374

C/N ratio, 75, 161, 162, 306, 419, 420
Caballo Blanco (La Parguera, Puerto Rico), 147, 152
Cabo Rojo (Puerto Rico), 251
Cabrite (Dominica), 247
Cache River (Ill., U.S.A.), 71, 106, 112
Cairns (Qld., Australia), 392
calcium (*see also nutrients, elements*), 56, 113, 162–164, 230, 439, 443, 445
California (U.S.A.), 3, 6, 8, 57, 78, 96, 124
California Central Valley (U.S.A.), 123
Caltho-Alnetum, 441, 442
Caltion, 442
cambium, 72, 73
Cambridge (Cambridgeshire, U.K.), 409
Cambridgeshire (U.K.), 71
Campeche (Mexico), 287, 290–295
campos, 299
canacoital, 290
Canada, 3, 7, 26, 78, 88, 91, 114, 115, 124, 165, 171, 178
Canadian River (U.S.A.), 120
canals, 60, 61
Cananeia (São Paulo, Brazil), 168
Canberra (A.C.T., Australia), 355
Candelaria River (Mexico), 291
cane, 201
Caño Mora (Costa Rica), 274, 276, 277, 280, 282, 283, 285
Caño Palma (Costa Rica), 275–277, 281, 282
Caño Penitencia (Costa Rica), 275
Caño Servulo (Costa Rica), 274, 277, 278, 282
canoes, dugout, 133
canopy, 88, 99, 101, 102, 109, 118, 121, 267, 269, 280, 281, 283, 337, 338, 342–347, 349, 417, 419
–, closed, 269
– height, 123, 277, 282–285, 311–313, 315, 316, 318, 329
– leaves, 109
–, open, 269, 279, 280
– trees, 102
Cape Fear River (N.C., U.S.A.), 116
Cape York Peninsula (Qld., Australia), 399
capillary rise of water, 63
carbohydrates, 60, 75, 270
carbon, 70, 72, 74, 75, 79, 108, 161
– allocation, 104
–, atmospheric, 54, 55

GENERAL INDEX

carbon (*continued*)
– cycle, 53–55, 66, 70
– dioxide (CO_2), 10, 55, 70, 72, 75, 104, 159, 165
–, organic, 54, 57, 70, 72, 75
– sink, 54, 55, 70
– sources, 55
carbonate, 419–422
Carboniferous, 27, 28
Caribbean (*see also particular countries*), 4, 58, 89, 143, 148, 160, 213, 215, 236, 241, 247, 256, 259, 260, 269, 271, 273, 275, 282
Caribbean Basin, 54
Cariceto elongatae–Alnetum, 439
Caricetum elatae, 442
– *gracilis*, 442
– *vulpinae*, 442
Carici elongatae–Alnetum, 442–445
Carici–Alnetum, 442
Carici remotae-Fraxinetum, 442
carnivores, 428
Carolina Bay (U.S.A.), 28, 59, 65
carp(s), major, 372, 378
Carpathian Mountains (Poland), 441
Carpinion, 441, 443
carr(s), 440
–, alder, 440, 445
Casuarina kerapah, 316, 317
cat, fishing (*Prionailurus viverrinus*), 371
–, leopard (*P. bengalensis*), 371
catastrophic (periodic) events, 145, 147, 152
– floods, 8, 87
catchment, 90, 92, 93, 109, 113
– area, 59, 77, 79
catena, 304, 309, 318, 322, 326
catenary sequence, 309
catfish (*Clarias batrachus*), 378
–, stinging (*Heteropneustes fossilis*), 378
cation(s), 123
– exchange, 113
cativales, 58
cattail (*Typha*), 30, 34, 240, 289
cattle, 133
– grazing, 441
Cayenne (French Guiana), 217, 232, 234
Cayo Enrique (La Parquera, Puerto Rico), 146, 152
Cayos Caribe (Puerto Rico), 152
cedar, Atlantic white (*Chamaecyparis thyoides*), 5, 37, 176, 178, 180, 181, 183, 184, 188, 194, 201, 203, 205
–, northern white (*Thuja occidentalis*), 30, 120, 166, 175, 177–179, 183, 184, 186–189, 193, 194
Cedar Creek Bog (Minn., U.S.A.), 30
Cedar Creek Natural History Area (Minn., U.S.A.), 174, 179
cedar swamp, 179
Ceiba (Puerto Rico), 152
cell turgidity, 60
cellulose, 75
Cendrawasik (West Irian, Indonesia), 336
Central America, 4, 58, 213, 267, 269, 271
Central Europe, 7, 438, 440
Central Forest Region (U.S.A.), 118, 119
Cerro Tortuguero (Costa Rica), 277, 281–283, 285

Cerros de Coronel (Costa Rica), 275, 281–283
cesium, 79
chaffinch (*Fringilla coelebs*), 430
Chalabogi River (Nalianala Range, Bangladesh), 365, 368
chamaephytes, 418, 422, 424, 427
Champotón (Mexico), 291
channel, 87, 90–92, 102, 113, 120, 129, 131, 274, 275, 277, 282, 391, 396
–, braided, 89
– constriction, 130
– degradation, 132
– depth, 130
– discharge, 91
– drainage, 396
– fill, 89
– gradient, 129, 130
– meandering, 90
– morphology, 131
– overflow, 89
– scouring, 132
– shelf, 120
– slope, 59, 89, 92, 134
channelization, 57, 68, 69, 75, 412
–, stream, 129
chaparral, 121
Chapelle-lez-Herlaimont (Hainaut, Belgium), 409
Charandwip (Bangladesh), 364
charcoal, 245
chat, gray bush (*Saxicola ferrea*), 377
–, pied stone (*S. caprata*), 377
Chattahoochee River (U.S.A.), 116
chechenal (*Metopium brownei*), 291
check, 133
– dams, 133
chelation, 60, 75
chemical energy, 60, 70
chemoautotrophic, 75
Chemung River (N.Y., U.S.A.), 120
chenier complexes, 222
cherry, black (*Prunus serotina*), 118–180, 180, 432
Chesapeake Bay (U.S.A.), 116, 203
Chiapas (Mexico), 288
Chicago (Ill., U.S.A.), 3
Chichancanab (Mexico), 294
Chiffchaff (*Phylloscopus collybita*), 430
China, 163
Chippewa River (Wis., U.S.A.), 99, 120
Chittagong (Bangladesh), 358, 359, 361, 365, 368, 369, 384
Chittagong Hill Tract (Bangladesh), 358
chloride, 231, 368, 443
chlorinity, 158–160
chlorophyll, 22, 62
Chokolsee Bay (Fla., U.S.A.), 106
Chokoria (Sundarbans, Bangladesh), 359, 361–366, 368, 369, 384
chronic flooding, 158, 165, 166
chronology, 27
Cienaga Grande (Colombia), 234
Cienaga Grande De Santa Marta (Colombia), 106
Cimarron River (Kansas, U.S.A.), 71, 90
Circaeo-Alnetum, 441–445

498   GENERAL INDEX

civet, palm (*Paradoxurus hermaphroditus*), 371
–, small (*Viverra zibetha*), 371
clam, bay (*Meretmix meretmix*), 382
–, cockle (*Gafrarium tuimidum*), 382
–, false (*Paphia malabarica*), 382
–, inflated (*Katelysia opima*), 382
–, wedge (*Donax cuneatus*), 382
classification, 449
–, land, 15, 17
–, landscape, 17
–, wetlands, 17
clay, 93, 120, 222, 228, 231, 232, 420, 443
claypan, 27
clear-cutting, 133
clear-water rivers, 127
clearing, 58, 69, 76, 77
cliff(s), 21
– coasts, 430
climate, 59, 75, 89, 92, 113, 114, 134, 301, 302, 335, 341
–: air temperature, 360–362
–, arid, 145, 147, 150, 151, 387–391, 395, 400, 404
–: atmospheric pressure, 360
– change, 359
–: cool temperate, 395, 398, 403
– data, 409, 412
– diagrams, 302, 407, 408, 410
–, equatorial, 302, 303
– fronts, 360
–, humid, 387, 388, 392, 395, 401, 404
–: humidity, 302, 363
–, Mediterranean, 387
–, moist, 147, 149, 150
–, oceanic, 407
–, perhumid, 302
–: rainfall, 302, 305, 357, 360–363
–, semi-arid, 387, 390, 391, 395
–, subhumid, 396
–: temperature, 302
–: Thornthwaite's classification, 302
–: thunderstorms, 362
–, tropical, 389
–, tropical maritime, 362
–: wind, 302, 357, 360
climatograms, 411
climax community, 354
– vegetation,
climbers, 291, 338, 341, 346, 347, 417
cloud forest, 66
clumped distribution, 102
clumps, palm, 269, 277–281, 285
coal deposits, 27
coarse-grained deposits, 87, 90
coastal areas, 115, 124, 387, 389, 391, 393, 400
– basin, 252, 264
– forests, 353
– plain, 92, 270, 287, 288
– – rivers, 93
– riverine, 252, 256, 264
coastline, 293
–, subtropical, 171, 172, 181

–, tropical, 171, 172, 182
cobra (*Nana naja*), 373
codominants, 102
coefficient of similarity, 365
coffee crop, 78
colluvium, 89
Colombia, 4, 106, 127, 234, 237, 245, 267, 269, 272, 325
colonization, 123
Colorado (U.S.A.), 57, 94, 95, 102, 121, 131, 133
Colorado River (U.S.A.), 132
Columbia Forest Region (U.S.A.), 115
Combretocarpus–Dactylocladus association, 315, 316
Comilla (Bangladesh), 358, 360
community(ies), 87, 91, 93, 118, 387
–: closed-forest, 387, 392, 395–398, 400, 402, 404
–: dryland savannah woodlands, 297
–, dynamics of, 404
–: open-forest, 387, 392, 395, 397, 400, 402, 404
–: open-woodland, 387, 395, 400, 404
– patterns, 94
–: pure forest, 394, 395
– respiration, 104
– similarity, 277, 279, 285
–: woodland, 387, 395, 396, 399, 400, 404
community vegetation structure: basal area, 148–150, 152
– – –: complexity index, 147, 150, 152
– – –: effect of latitude, 148, 149
– – –: species richness, 146–149, 151, 152
– – –: stem density, 148, 149, 151, 152
– – –: tree height, 146–152, 162
competition, 99, 119
competitors, 101
complexity, 293, 447, 451
– index, 100, 282, 238, 284, 456
conch, 241, 243
conductivity, 175
Congaree Swamp (S.C., U.S.A.), 117
conifer(s), 171, 172, 412
coniferous swamp forest, 336, 344, 345
Connecticut River (New England, U.S.A.), 71, 119, 120
conservation, elements, 109
construction, 69, 75
consumers, 56, 57, 414
convection, 62
conversion of wetlands, 58
coolibah (*Eucalyptus microtheca*), 391, 403, 404
Copenhagen (Denmark), 411
coppice, 420, 421
coppicing, 237, 238, 245, 246, 255
coral reefs, 19
Cordillera Central (Costa Rica), 274, 282
Cork (Ireland), 407, 409
Corkscrew Swamp (U.S.A.), 27, 29, 98
cormorant, great (*Phalacrocorax carbo*), 377, 430
Coronie (Surinam), 228
corridors, 68
Corrouaie River (French Guiana), 231
Costa Rica (*see also particular localities*), 4, 98, 99, 100, 128, 152, 238, 261, 267, 269, 271–273, 275, 282, 283, 284, 447
cottonwood (*Populus deltoides, P. heterophylla*), 32, 94, 95, 116

GENERAL INDEX 499

cotyledon scars, 7
cover, percent, 124
coves, 117
Cox's Bazaar (Bangladesh), 358, 361
Cox's Bazaar Forest Division (Bangladesh), 361
crabs, 57, 245
Crassulacean Acid Metabolism (CAM), 159
crayfish, 74
creeks, 20, 93
creeper, short-toed tree (*Certhia brachydactyla*), 430
Creeping Swamp (N.C., U.S.A.), 112
Cretaceous, 90
Crique Gabrielle (French Guiana), 234
croaker (*Pama pama*), 378
crocodile, marsh (*Crodocylus palustris*), 372, 373
–, saltwater (*C. porosus*), 372, 373
Crocodilians, 246
crops, 133, 412, 421
crow, carrion (*Corvus corone corone*), 430
–, hooded (*C. c. cornix*), 429
–, house (*C. splendens*), 376
–, jungle (*C. macrorhynchos*), 376
crowns, 338, 341, 344, 346, 347, 349
Crustacea, 242
Cuba (West Indies), 4
cuckoo, banded bay (*Cacomanthis sonneratii*), 375
–, common hawk (*Cuculus varius*), 375
–, gray-breasted brush (*Cacomanthis merulinus*), 385
–, short-winged (*Cuculus micropterus*), 376
–, sirkeer (*Phaenicophacus leschenaultii*), 376
cultivation, 70, 76, 78
cunuri (*Micrandra sprucei*), 305, 319, 326
current(s), 19, 59, 91
– velocity, 15
cuticle, 62
Cuttaburra Creek (Qld., Australia), 391
cyclical succession, 65
cycling, 446
–, elemental, 108
–, nutrient, 88
cyclones, 16, 302, 357, 359, 362
cypress (*Taxodium*), 21, 37, 38, 40, 95–98, 101, 103–105, 107, 109, 174, 176, 178, 180, 181, 183–185, 188, 190, 194, 195, 201, 203–205, 287, 447, 453, 455, 457
–, bald (*T. distichum*), 2, 5, 31, 61, 102, 114, 116, 117, 121, 180, 181, 201
–, dome 19, 20, 27, 40, 62, 172, 173, 175–177, 181–188, 190, 192, 193–196
– floodplain forest, 108
– forests, 110, 447, 453, 455, 457
– head, 181
– knees, 102, 184, 185,
– –: role in gas exchange, 185
–, pond, scrub or dwarf (*T. distichum* var. *nutans*), 61, 175, 176, 181–188, 190–193
–, strand, 101, 110, 111
– swamp(s), 22, 177, 183, 184, 186, 188, 190, 192–194
Czechoslovakia, 95, 103, 105

Dacca (Bangladesh), 358, 369

Dacrydium kerapah, 316, 317, 319
– padang, 332
Dacrydium–Casuarina peat-swamp forest, 323, 331
damming, 68, 75, 241
dams, 131, 412, 419
Darien (Panama), 4
Darss Peninsula (Rostock, G.D.R.), 409
Darss River (G.D.R.), 442
darter, African (*Anhinga rufa*), 377
dating, ^{14}C, 70
dead wood, 425, 426, 432
debris, 110
– dams, 22
– pile, 111
decay, 72, 452
– coefficient, 111
– rate, 111
deciduous, 72, 453
– riparian forest, 287
– – woodlands, 287
– trees, 340
decomposers, 414
decomposition, 26, 30, 31, 34, 70, 72, 109, 110, 134, 186–188, 412–416, 418, 421, 425, 431
decomposition, 437, 445
–, anaerobic, 72
– coefficients, 188, 189
– constant, 161
–, effects of hydroperiod on, 188, 189, 195
– effects of litter quality on, 189
– leaves, 188, 195
–: litter, 110, 187–189, 194, 195
–: litter bag, 72
deer, 244, 243
–, barking (*Muntiacus muntiak*), 371
–, hog (*Axis porcinus*), 371
–, spotted (*A. axix*), 370, 371
defoliation, 61, 108
deforestation, 90
degradation, channel, 132
degrading, 89
Dehra Dun (India), 4
Delaware River (N. Y., U.S.A.), 119, 120
delta, 69, 267, 269, 271, 275, 287, 293, 299
deltaic peats, 302, 303, 305, 306, 318
– plain, 110
Dendermonde (Oost-Vlaanderen, Belgium), 409
Deniliquin (N.S.W., Australia), 388
denitrification, 112
Denmark, 71, 407, 409, 411, 415
density, 87, 99, 94, 95–97, 100, 107, 128, 133, 272, 440
–, bird, 56
–, stem, 268, 277–285
–, tree, 440
deposition, 87, 90, 110, 117, 120
–, alluvial, 92
–, fluvial, 70
–, organic, 70
deposits, 89, 90
depression(s), 101, 117, 267, 277, 281, 282, 390, 404

depression(s) (*continued*)
– landforms, 25, 26
–, seepage, 437, 440
depth, 89
–, flooding, 93
desiccation, 243, 245
destructive floods, 102
detritivores, 75
detritus, 15, 72, 74, 75, 159, 404
– foodwebs, 459
development, 25, 58
–: hydrologic mechanisms, 25
–: wetlands, 25
Devonian, 27
diaspores, 236
Dicrano-Pinion, 445
die-back, 147
Digul River (West Irian, Indonesia), 336
Dijon (Cote d'Or, France), 409
dikes, 131, 441, 442
diking, 69, 75
Dinajpur (Bangladesh), 358
dipper (*Cinclus cinclus*), 429
discharge, 94
disease(s), 247
–, Dutch Elm, 118, 121
Dismal Swamp, *see* Great Dismal Swamp
Dismal Swamp Canal (Great Dismal Swamp), 204
dispersal, 102
dissolved organic carbon (DOC), 56, 72
– oxygen, 229, 230
distribution of wetland tree species, 389–395
disturbance(s), 30, 33, 36, 38, 39, 101, 118, 124, 126, 172, 178, 180, 196
–, animal, 38
–, anthropogenic, 30, 38, 201, 204, 205
–: change in hydroperiod, 196
– cycle, 35, 36
–: drainage, 36, 37, 38, 196
–: drought, 36
–: effects on wetlands, 196
–: fire, 36, 37
–: flooding, 36, 38
–: harvesting, 179, 196
–: human, 196
–: impounding water, 196
–: oil spills, 196
–: peatslip, 38
–: sedimentation, 196
–: thermal stress, 196
–: toxic substances, 196
–: windthrows, 38
diversions, 131
diversity, 196
–, animal, 56
–, beta, 172
–, landscape, 121, 172, 176, 178
Dobrzany (Poland), 437
DOC/POC ratio, 56
dogfish (*Scoliodon sorrakowah*), 378

dolphin, 371
–, Malay(*Stenella malayana*), 371
dominance, 89, 102, 106, 119
dominants, 277
Dominica, 221, 236, 238–240, 245, 247, 261
Domkhali Khal (Sundarbans, Bangladesh), 368
Doon Valley (India), 4
Dorado (Puerto Rico), 251–253, 255–263
dormancy, 59
dormant, 108, 110
dove, red turtle (*Streptopelia tranquebarica*), 375
–, rock (*Columbia livia*), 375
–, stock (*C. oenas*), 430
downcutting, 89, 412
downstream movement, 89
dragonflies, 27
drain, 444
drainage, 36, 37, 54, 55, 57, 65, 69, 70, 73, 76–79, 87, 104, 113, 117, 119, 124, 127, 133, 134, 218, 226, 241, 247, 269, 270, 272, 275, 277, 282, 291, 294, 302, 306, 307, 308, 329, 388, 392, 396, 411, 412, 420, 421, 427, 431, 443, 456
–, basin, 92, 113, 134
– canals, 60, 61
– conditions, 335–337, 342, 344
– patterns, 94
draining, 440
Draved Mose (Denmark), 71
Dromling (Magdeburg, G.D.R.), 409
drought, 36, 38, 61, 62, 70, 76, 92, 112, 119, 132, 160, 176, 302, 318, 320
– stress, 335, 337, 340, 349
– stressed semidesert, 121
– tolerance, 102
drumlins, 20
dry air, 15
dryland, 338, 397, 398, 400–402, 404, 405
Dryopteri thelypteris-Betuletum pubescentis, 440
Dublin (Ireland), 409
duck(s), fulvous tree (*Dendrocygna bicolor*), 374
–, lesser whisting (*D. javanica*), 374
–, spot-billed (*Anas poecilorhyncha*), 374
–, white-winged wood (*Carina scutulata*), 374
duckweed (*Lemna*), 26, 30, 31
dugout canoes, 133
Dula Hazara (Sundarbans, Bangladesh), 358, 364
dunes, 19
–, coastal, 444
dunnock (*Prunella modularis*), 430
duration, flooding, 93
Dutch elm disease, 118, 121
dwarf forests, 150, 151
– palms, 281
dy, 439
dy-gyttja, 439
dykes, 412, 420
dynamics, 108

eagle, black (*Ictinaetus malayensis*), 374
–, crested serpent (*Spilornis cheela*), 374
–, gray-headed fishing (*Icthyophaga icthyaetus*), 374

GENERAL INDEX

eagle (*continued*)
–, lesser-spotted (*Aquila pomarina*), 374
–, Pallas' sea (*Haliaeetus leucoryphus*), 374
–, tawny (*Aquila rapax*), 374
–, white-bellied sea (*Haliaeetus leucogaster*), 374
East Anglian Fenlands (England), 77, 78
East Germany (G.D.R.), 407, 409, 416
Eastern Deciduous Forest Region (U.S.A.), 118, 119
Echuca (Vic., Australia), 388
ecosystem, 388, 400, 404
– chemistry, 458–460
– complexity, maintenance of, 455, 456, 458
– dynamics, 326, 447
– forms, 17, 18
– function, 326
– metabolism, 108
– respiration, 158, 159
– structure, 15, 88, 92, 447, 448
– type, 88
ecotone, 240, 252, 262, 264, 291, 292, 294, 295, 316, 330
–, abrupt, 263, 264
–, blending, 263, 264
edaphic, 274
– conditions, 69, 114
– factors, 451
edificator, 439
eel(s), 355
–, spiny (*Macrognathus aculeatum*), 378
egret, cattle (*Bubulcus ibis*), 374
–, great (*Egretta alba*), 374
–, intermediate (*E. intermedia*), 374
–, little (*E. garzetta*), 374
Eilanden River (West Irian, Indonesia), 336
El Encanto River (Guapi, Columbia), 106
Elbe River (Germany), 407, 429
electron acceptors, 74
– microscopy, 57
element(s) (*see also* nutrients *and specific elements*), 109, 110
–: calcium (Ca), 56, 174, 175, 177, 189, 190, 192–196, 449, 452, 453, 457
–: carbon (C), 53–57, 66, 70, 72, 74, 75, 79, 177, 455, 458, 459
–: chloride (Cl), 79
– cycling, 108, 109, 132
–: dissolved, 109
– distribution, 108
– in rainfall, 453
– in rivers, 453
–: iron (Fe), 443
–: magnesium (Mg), 174, 175, 177, 189, 190, 192–195, 449
–: manganese (Mn), 443
–: nitrogen (N), 55, 63, 64, 75, 79, 175, 177, 189–196, 448, 452, 453, 456, 458, 459
– : ammonium, 175
– : denitrification, 459
– fixation, 459
–, inorganic, 459
–: nitrate, 175, 459
–, organic, 459
–: total N, 175, 177
– oxygen (O_2), 55, 56, 74, 171, 185

– phosphorus (P), 63, 79, 175, 177, 189, 190, 192–196, 448, 452, 453, 457–459
–: potassium (K), 175, 177, 189, 190, 192–195
– sinks and sources, 458, 459
– – –, atmospheric, 458, 459
– – –, terrestrial, 458, 459
– sodium (Na), 443
– sulphur (S), 55, 75, 459
elevation, 87, 93, 99, 101, 122, 123
elm, American (*Ulmus americana*), 106, 117–121, 126, 166, 179
embankments, 407, 420, 429, 431
emergents, 272, 278, 280
Ems River (Germany), 407
encinares, 296
endemics, 127
endemism, 316, 324
energy, 39
–, actual, 42
– analysis, 43, 63
– caloric value, 454
–: circuit language, 26, 39
–, embodied 8, 29, 40–44, 454, 457
– environment, high, 30, 31
– –, low, 30, 31
–, environmental, 15
– flows, 67
–, fossil fuel, 15
– front, 16, 17, 19–22
–: frontal sources, 16, 19, 20, 11
–, kinetic, 15
– language, 63
– lines, 16, 17, 20–22
–, methods for calculating, 40
–, parallel (front and line), 16, 17, 19, 22
– patterns, 16
–, perpendicular (sheet and point), 16, 17, 19, 22
– point, 16, 17, 19, 21
–, potential, 15
– quality, 16, 40, 42, 64, 454
– sheet, 16, 17, 19, 21, 22
– signature, 8, 15, 16, 17, 22, 41, 42, 64, 77, 171, 172, 454
– –: core factors, 456, 458
– –: modifying factors, 456, 458
– source(s), 64, 67, 68
–, auxiliary, 15, 22
–, spatial, 15–18
– subsidies, 15
– transformation ratios, 41–43
– types, 15, 21
– value, 447, 457
engineering structures, 79
England (*see also particular localities*), 3, 33, 34, 39–41, 71, 76–78, 412
Enseñada Honda (Puerto Rico), 152
environmental change, 354
– stress, 337
– factors: fire, 402
– –: soil fertility, 397
– –: temperature, 398
– –: water balance, 400

epicormic buds, 402
epiphyte, 101, 181, 221, 222, 234, 253–255, 291, 313, 332, 339, 346, 347, 420, 439
– humus, 255
episodic events, 143, 152, 154, 155
equator, 335
ericaceous plants, 178, 179
erodibility, 90
erosion, 70, 78, 87, 89, 90, 91, 108, 119, 123, 407, 418
– agents, 25, 30, 31
Espiritu Santo Bay (Mexico), 292
estuarine, 271
estuary, 57, 132, 134, 143, 335, 336, 338, 346, 357, 416, 418
ethanol, 73
eucalypt (*Eucalyptus*), 165, 389–391, 404
Europe (*see also particular countries*), 3, 126, 407, 422, 428–432
euryhaline, 144
euthrophication, 30, 31, 33
eutrophic, 60, 79, 126
eutrophication, 166
evaporation, 59, 62, 93, 172, 387, 409
evaporative coefficient, 400–402
evapotranspiration, 25, 26, 59, 62, 77, 92, 93, 106, 123, 133, 171–173, 395, 400
–, factors affecting, 26
even-age stands, 155
Everglades (Fla., U.S.A.), 19, 20, 37, 71, 77, 78, 452, 453
evergreen, 63, 72, 452
–, bogs, 20
– briar, 201
– riparian woodlands, 287
– seasonal forest, 217, 218, 224, 236, 239
– shrubs, 181
exotic species, 431
exponential decay, 111
export, 55–57, 66, 68, 72
extinction, 429

facies, 222, 239
– sec, 222
falcon, red-headed (*Falco chiquera*), 374
False River (La., U.S.A.), 91
Farakka (Bangladesh), 360
Faridpuir (Bangladesh), 358–360, 369
farmland, 133, 437
felling, 404, 412, 421
fen, 70, 71, 78, 126, 173–175, 177, 179, 183, 184, 186–190, 193, 194, 416, 423–427, 429, 432, 437, 439
–, marginal, 120
fencing, 388
fermentation, 75
ferns, 27, 178, 221, 254, 255, 279, 310, 311
fertility, 89, 109, 112
fetch, 65
fetterbush (*Lyonia lucida*), 180–182
Ficario-Ulmetum campestris, 441
field capacity, 106
filtration coefficient, 60
Finger Lakes, (U.S.A.), 3
Finland, 3, 7, 8

fir, balsam (*Abies balsamea*), 126, 178, 179
fire, 5, 36, 37, 65, 68, 69, 70, 75, 76, 78, 134, 171, 172, 176, 218, 226, 241, 267, 272, 341, 343, 354, 402, 404, 405
– effects on biomass, 176
– – on landscape diversity, 172
– – on peat soils, 176, 193
– – on species composition, 176, 178
– – on wetlands, 37
– – on soil, 37
– – on vegetation, 37
–: plant adaptations, 178
–, tolerant, 124
firewood, 388
fish, 55, 57, 58, 88, 131, 237, 241, 243, 246, 268, 355
– migrations, 372
– nurseries, 55
– ponds, 55
– yield, 132
fisheries, 57, 129, 133, 372
–, commercial, 55, 57, 68
– production, 53
– sport, 68
fishing, 130, 133, 274
fish-owl, brown (*Ketupa (Bubo) zelonensis*), 377
Fivemile Creek (Wy., U.S.A.), 90
flats, 117
floating ice, 449
– vegetation, 30
– mat, 30, 31
floatsam, 150
flood(s), 16, 20, 22, 123, 129, 389, 396, 400, 404, 405
–, catastrophic, 87
– control, 133
– damage, 120
– duration, 93
– episode, 90
– events, 59, 77
– flash, 93, 127
– frequency, 90
–, high-power, low frequency, 88
–, low-power, high frequency, 88
– outburst, 94
– peak, 93
– plain, 87
– power, 87
– recurrence, 99
– tolerance, 288, 291
– water: chemical characteristics, 228, 230
– –: physical characteristics, 228, 230
flood-adapted, 108
flood-tolerant, 108
– – assemblages, 102
– – species, 93, 108
flooded areas, 59, 60, 73, 74
– –: pastures, 295, 296
– –: savannas, 292, 293, 295, 296
– –: seasonal, 392, 394, 395
– –: summer, 396
– –: timing, 396
– –: winter, 396

GENERAL INDEX

flooding (*see also* inundation), 87–90, 92–94, 101, 146, 171, 173, 176, 180–182, 184, 195, 196, 267, 270–272, 274, 291, 296, 308, 418, 419, 444, 449, 454
–, backwater, 90
–, chronic, 69, 72, 76
–, constant, 90
–, continuous (or permanent), 268, 274, 277
– depth, 176, 182–184, 188, 194
–, flash, 127
– frequency 94, 101, 132, 134, 176, 180, 183, 184, 188, 194, 284
–, intermittent, 268
– interval, 59
–, permanent, 106, 290
– recurrence, 59
– regime, 335–338, 341, 343, 345–347, 352–354
–, seasonal, 272, 288, 290
– stress, 108
– tolerance, 174
floodplain, 29, 36, 37, 39–41, 43, 55, 58, 70, 73, 79, 87–94, 101, 109–112, 114, 116, 124, 125, 127, 222, 241, 270, 274, 281, 282, 287, 290, 292, 312, 347, 349, 387, 388, 390, 391, 395, 400, 401, 404, 457
– communities, 90, 449, 454
– deposits, 89
– development, 92
–, eroding, 453, 459
– erosion, 70
– forests, 89, 91, 110, 440, 441
– inundation, 90
– soils, 109
– storage, 130
– topography, 90, 128
– vegetation, 93
floodwater, 88, 113, 404
Florida (U.S.A.) (*see also* particular localities), 2, 3, 6, 10, 19, 20, 27, 29, 32, 37, 38, 40, 41, 61, 71, 77, 78, 97, 98, 100, 101, 103–106, 109–112, 143, 149, 152–154, 156–161, 163, 164, 166, 171, 172, 176, 177, 181–186, 188, 190, 192–195, 196, 272, 447, 452, 453
floristic composition, 172, 178, 414
floristics, 447
flow(s), 93, 132
– velocity, 120
flowerpicker, thick-billed (*Dicaeum agile*), 376
–, Tickell's (*D. erythrorhynchos*), 376
flowthrough, 88
fluvial, 92, 109, 117, 129
– deposition rate, 69, 70
– inputs, 110
– landforms, 102
Fluvisols, 288, 290, 296
flycatcher, Asian paradise (*Tersiphone paradisi*), 377
–, black-naped monarch (*Monarcha azurea*), 376
–, spotted (*Muscicapa striata*), 430
Fly River (Papua New Guinea), 336, 342, 343, 345, 346
foliage, 107, 398, 400
– projective cover (F.P.C.), 395, 396, 400–402, 405
– definition, 395
– shoots, 398
forest, 87, 89

–, alluvial, 413
– basin, 59
– bog, 56, 58, 62, 70
– clearing, 69
–, closed, 387, 392, 395–397
– communities, 93
– degradation, 369
–, dry, 56, 272
–, dwarf, 294
–, evergreen seasonal, 217, 218 224, 236, 239
– floodplain, 55, 58, 79
– floor, 99, 101, 104, 109, 111, 269, 453, 455
– gallery, 234, 246, 267, 272, 287
–, humid, 59, 269
–: importance value, 278, 279–281, 285
– levee, 218, 231, 235
–, lower montane rain, 221
–, managed, 443
– management, 133
– marsh, 213, 214, 216, 218, 219, 221, 222, 226, 228, 229, 231, 234–236, 239, 240
–, montane, 56
–, open, 387, 392, 395, 397
–, palm, 65, 66, 269, 293
–, Panamanian, 128
–, peatland, 412, 415, 416, 421, 428, 432
–, pine, 445
–, premontane wet, 128
–, rain, 56, 218, 219, 221, 223, 225, 234, 244
–, riparian, 221, 235
–, riverine, 56–59, 62, 271, 287, 413, 415, 416, 428
–, secondary, 56
–, slope, 278, 279, 281–283, 285
–, spring-fed, 414
–, streamside, 87
– structure, 269, 274, 277, 278, 280–284
–, swamp, 213, 214, 219, 220, 228, 229, 234, 235, 238–240, 245, 246, 268, 271, 274, 412, 416, 422, 425, 426, 441
–, tall evergreen, 288, 292, 293
–, tropical, 56, 57
–, tropical moist, 128
– types, 88
–, upland, 56, 57
–, uses of (*see also* utilization), 330
–, várzea, 58
–, wet alder, 443, 445, 446
–, wetland, 94, 407, 411, 412, 414–416, 418, 420–422, 425, 427–432
– woodlands, 56
forested wetland (*see also* wetlands; swamps), 387–389, 395, 397, 398, 400–405
– –: area, 1
– –, biotic characteristics of, 9
– –, classification of, 1, 2, 5, 6, 7, 8
– –, cultivation of, 1
– – definition of, 2
– –, drainage of, 1
– –: driving forces, 2, 5
– –, global: ecologic, and economic roles, 1
– –: literature reviews, 1

forested wetland (*continued*)
– –: marginal wetlands, 5, 8
– –: research history, 1
– – types (*see also* wetlands; swamps *and alphabetical entries*)
– – –: alluvial swamp, 4
– – –: banados con seibal, 4, 5
– – –: basin forest, 3
– – –: bog forests, 3
– – –: bottomlands, 3
– – –: caatinga, 9
– – –: cedar swamps, 3
– – –: cypress heads or domes, 3
– – –: – swamps, 3, 5
– – –: deep water swamps, 3
– – –: fan palm oasis, 318
– – –: floodplain forest, 3, 5
– – –: forested bog, 3
– – –: fresh or saltwater swamps, 4, 6–8
– – –: gallery forest, 4
– – –: guandal, 4, 5
– – –: hyperseasonal savanna, 4
– – –: ígapo, 4, 5
– – –: inundation forest, 4, 5
– – –: low floodplains, 3
– – –: mangroves, 1, 2, 4, 5
– – –: mires, 2, 3, 7
– – –: mixed hardwood swamps, 4
– – –: montane wetlands, 5, 8
– – –: moors, 3, 7
– – –: mpanga forest, 3
– – –: occasionally flooded woodlands, 4
– – –: palm forests, 4, 8
– – –: – marsh forest, 4
– – –: peat bogs, 2
– – –: swamp, 4, 5
– – –: peatlands, 2, 7
– – –: peaty freshwater swamps, 3
– – –: perched water table forest, 4
– – –: phreatophyte systems, 3, 8
– – –: pocosins, 2, 5, 6
– – –: prairie swamps, 3
– – –: riparian forest, 5, 6
– – –: riverine, 3, 4
– – –: riverside swampland, 3
– – –: sajal, 4, 5
– – –: seasonal swamps, 3
– – –: shallow water swamps, 3
– – –: sitka spruce swamp, 5
– – –: spring-fed swamp, 4
– – –: swamp forests, 3, 4, 5, 8
– – –: – gallery forests, 4
– – –: tidal swamp, 4
– – –: tropical swamps, 4
– – –: várzea forests, 4, 5
– – –: woodlands, 4
forestry, 414, 426, 428, 431
fossil fuel, 64
fox, Bengal (*Vulpes bengalensis*), 371
France, 407, 408, 422, 425, 429, 431
Fraser Island (Qld., Australia), 391

Fraxino-Ulmetum, 440
freezing, 94, 114
French Guiana, 224, 229, 231, 233, 234, 237, 238, 241, 246, 247
freshwater, 128, 448, 449, 453, 456–459
– fringe, 345
– mangrove, 338, 340
– riverine forests, 94, 95, 104
– runoff, 104
– swamps, 335, 336–338, 340, 352, 354
– wetland, 105, 296
fowl, red jungle (*Galllus gallus*), 376
fringe, 127, 269, 270, 275, 292–295, 392, 440, 447–450, 458–460
– forest, 171
–, lagoon, 287, 288, 292, 293
–, river, 288
–, wetlands, freshwater, 165, 166
– –, high-energy, 143, 147, 151, 154, 161, 165
– –, location, 143
– –, low-energy, 143, 151, 154, 160, 165
– –, saltwater, 143
fronds, 270, 280
Frontera (Mexico), 288
frost, 76, 144, 345, 407
– hollows, 392
– line, 456
frozen, 94
fruit, 57, 106, 109, 162, 268
– fall, 104, 110
fruit-eating fish, 37
fruiting, 236, 311
fuel, 409, 412
fuelwood, 245

Gaarde (Schleswig-Holstein, F.R.G.), 409
Galibi, 246
Galion (Martinique), 222
gallery forests, 87, 90, 267, 272
Gammelmose (Frederiksborg, Denmark), 409
Ganderkhali (Nalianala Range, Bangladesh), 365
Ganges River (Bangladesh), 357, 359–361
gap(s), 101, 311, 314–316, 323, 327, 328, 330, 331
– phase regeneration, 102
– – replacement, 102
gas, 60
– diffusion, 60
– exchange, 69, 73, 144, 157–159, 185, 196, 447
gauge(s), staff, 284
–: water level, 284, 285
gaur (*Bos gaurus*), 370, 371
gekko(s), 373
–, tokay (*Gecko gecko*), 373
gene banks, 68
generation time, 88
genetic plasticity, 63
geographic variation, 88, 113
geology, 59, 92
geometry, 93
geomorphic features, 88, 113
geomorphology, 22, 60, 89, 130
geophytes, 416, 417, 422

GENERAL INDEX

Georgia (U.S.A.) (*see also particular localities*), 3, 6, 20, 21, 29, 71, 97, 113, 177, 183, 184, 186–188, 190, 192–194, 196
Germany, 71, 415, 425, 429, 431
germination, 102, 106, 236, 237, 400, 404, 405, 427
Ghana, 3, 269
Gila River (Ariz., U.S.A.), 88, 93, 123, 131, 132
gilgai, 390
glacial outwash deposit, 125
glaciation, 26–28, 30
glacier-dammed lakes, 94
Glen Canyon Dam (U.S.A.), 132
gleysol, 53, 288, 296
Glycerietum maximae, 442
Golfo de Uraba (Colombia), 269
Golfo Dulce (Costa Rica), 269
Gonystylus bancanus forests, 305
Gonystylus bancanus–Dactylocladus stenostachys–Neoscortechinia kingii association, 307, 309
goose, cotton pygmy (*Nettapus coromandelianus*), 374
–, knob-billed (*Sarkidiornis melanostos*), 374
Gordon River (Fla., U.S.A.), 106
goshawk (*Accipiter gentilis*), 430
gradients, 89, 92
graminoid layer, 395
Grampian Region (Scotland), 71
Grand Canyon (U.S.A.), 132
grasse(s), 122, 178, 422
– swamp, 335, 342–344, 346
grassland, 405
gravel, 89, 90, 134, 441
– bar, 124
– lenses, 90
grazers, 428
grazing, 75, 133, 404, 405
– livestock, 123, 130
Great Dismal Swamp (Va.–N. C., U.S.A.), 3, 19, 27, 29, 35, 71, 172, 176, 180, 187, 195, 201, 203, 206, 207, 210, 449
Great Dividing Range (Australia), 390
Great Lakes (North America), 120
Great Lakes–Saint Lawrence (North America), 115
grebe, great-crested (*Podiceps cristatus*), 377
–, little (*P. ruficollis salvadori*), 377
Green Swamp (Fla., U.S.A.), 19, 78
greenhouse gases, 55
Grijalva River (Mexico), 287
ground cover, 279, 285
– ice, 91
– layer, 279, 281, 338, 340, 341, 346, 347, 349, 351
groundwater, 2, 19, 22, 60, 65, 93, 113, 159, 401–414, 416–418, 420, 421, 425–428, 432, 439, 440
– moisture, 389
– recharge, 93
– seepage, 87, 113
– supply, 92
– table, 123
groves, 274
growing season, 106, 113, 122, 414, 421
growth, 108
– conditions, 99
– rates, 121

Guadeloupe, 215, 220, 221, 223, 228, 236–238, 240, 244–247, 251, 259
Guatemala, 245
Guianas, 213, 221, 224, 228, 237, 241, 245, 259
Gulf Coast (U.S.A.), 57
Gulf Coastal Plain (U.S.A.), 89
Gulf of Carpentaria (Qld–N.T., Australia), 397
Gulf of Mexico, 9, 116, 287
gull, black-headed (*Larus rudibundus*), 375
gully erosion, 92
– wall collapse, 92
gullying, 129
gum, 107, 180, 183, 184, 188, 194, 201, 105
–, black (*Nyssa sylvatica* var. *biflora*), 98, 105, 111, 116, 117, 121, 180–182, 185, 190, 191, 201
–, forest red (*Eucalyptus tereticornis*), 391, 392, 403
–, ghost (*Eucalyptus papuana*), 390
–, river red (*Eucalyptus camaldulensis*), 388, 390–392, 395–400, 403, 404
–, swamp (*Eucalyptus ovata*), 391, 403
–, sweet (*Liquidambar styraciflua*), 73, 116, 117, 121, 180, 182, 201
–, tupelo (*Nyssa aquatica*), 73, 75, 96, 97, 102, 105, 109, 111, 116, 117, 180, 201
–, Western Australian flooded (*Eucalyptus rudis*), 392, 403
–, white (*Eucalyptus alba*), 389, 403
Gunung Panti (Malaysia), 325
Guyana, 127, 214–216, 218, 228, 234, 236, 241, 245, 246, 271, 172
gyttja (*see also* dy-gyttja), 30, 437, 439

habitat diversity, 131
Hainan (China), 302, 329
Haiti, 221
halophytes, 182, 294
hammocks, 452, 456, 457
Hardinge Bridge (Bangladesh), 359
hardpan, 27
Hardura Canal (Nalianala Range, Bangladesh), 365, 368
hardwoods, 117, 171, 172, 176, 178, 183, 184, 190, 268–270, 277–282
–, evergreen, 171, 172, 177, 181
–, mixed, 181, 183, 184
–, – deciduous, 171, 179, 180
hare(s), 125
–, hispid (*Caprolagus hispidus*), 371
–, Indian (*Lepus nigrocollis*), 371
harvesting, 55, 58, 68, 69, 75, 165
–, peat, 58
–, timber, 58, 133
–, wood, 58
Harz (Niedersachsen, F.R.G.), 409
Hautes Fagnes (Liège, Belgium), 409
hawk, sparrow (*Accipiter nisus*), 430
hawk-owl, brown (*Ninox scutulata*), 377
Hay River (Canada), 3
hay-making, 424
headwater(s), 89, 122
– catchments, 120
– streams, 92

heat equivalents, 64
– effluents, 64
heath (*Calluna*), 38
– forest, 299
heathland, 395, 405
heavy metals, 67–69, 79, 130, 132
Heidewald, 299
Helgoland Island (Niedersachsen, F.R.G.), 409
helophytes, 424, 439, 442
hemicryptophytes, 416–418, 420, 422
hen, purple swamp (*Porphyrio porphyrio*), 376
herb(s), 59, 66, 67, 78, 102, 121, 178, 179, 181, 182, 291, 293, 295, 310, 311, 313, 315, 439, 440
– associations, 442
– layer, 413, 417, 418, 420–423, 425, 427, 428, 432, 445
herbaceous cover, 279
– layer, 395
– marsh, 267, 270, 273, 274, 277
– swamp, 214, 216, 218, 226, 228, 229, 232, 234, 235, 240, 268, 287, 335, 338, 343, 347, 349, 352, 353
– vegetation, 101, 102, 109, 119, 128
– wetland, 87, 287, 296
herbicides, 69, 130, 132
herbivores, 68, 69, 134, 428
herbivory, 125
heron, black crowned night (*Nycticorax nycticorax*), 374, 430
–, gray (*Ardea cinerea*), 374, 430
–, pond (*Ardea grayii*), 374
–, purple (*Ardea purpurea*), 374, 429
–, striated (*Butorides striatus*), 374
herring, 378
hierarchical control, 88
high forested wetlands, 287
– riparian forest, 287
Hilden (Nordrhein-Westfalen, F.R.G.), 409
Himalayas, 358, 359
histosol, 53, 78, 305
–, Eutric, 288, 296
hobby (*Falco subbuteo*), 430
Hochmoor, 305
hog, pygmy (*Sus salvanius*), 371
hollies, 201
hollows, 443
holly, American (*Ilex opaca*), 117, 180, 201, 205
Holocene, 226, 228
– terrace, 312, 320
Holtum (Niedersachsen, F.R.G.), 409
Honduras, 4, 218, 224, 240
hophornbeam, American (*Ostrya virginiana*), 180
horizon, 444
hornbeam (*Carpinus*), 441, 445
hornbeam, American (*C. caroliniana*), 117, 119, 180
horsetails, 27
Hottonio-Alnetum, 442
housing, 388
Hoxter (Nordrhein-Westfalen, F.R.G.), 409
Hudson Bay (Canada), 124
Hudson River (New England, U.S.A.), 120
Humacao (Puerto Rico), 252, 253, 255–258, 259, 261–264
human activity, 58, 69, 268, 271, 271, 282

– density, 369, 372
– education, 383
– health, 383
– impacts, 165, 166, 291, 295
– influence, 240
– life expectancy, 383
– nutrition, 383
– use, 447
humic acids, 34, 62, 231, 307, 308
– compounds, 75
– nitrogen, 75
humid climates, 113
– environments, 92, 456
– tropics, 87, 94, 134
humidity, 407
hummock(s), 438, 439, 443
– grass, 395
humus, 303, 305, 412, 417, 421, 427
hunting, 130, 133, 240, 246, 274, 429
hurricanes, 31, 36, 69, 76, 155
–, damage of, 347
Hutan Melingtang (Malaysia), 324
Hyde County (N. C., U.S.A.), 78
hydrarch succession, 29, 30, 33, 36, 44
hydraulic conductivity, 58, 63
– gradient, 93
hydric, 90
– plant communities, 87
– system, 202
Hydrocharo-Stratiotetum, 442
hydroelectric development, 58
hydrogen sulfide (H_2S), 69
hydrogeomorphic conditions, 102
hydrograph, 59, 77, 79, 93, 359
hydrographic peaks, 92
hydrologic conditions, 411, 415
– energy, 60, 64, 65, 143, 144, 146, 148, 156, 161
– factors, 451, 452, 456, 458
– fluxes, 60, 65
– regime, 89, 94
– settings, 88
hydrology, 25, 29, 44, 58, 92, 109, 129, 201, 204, 205, 251, 253, 264, 411, 427, 447, 460
– balance model, 25, 26
– budgets, 172, 173
–, drainage, 25
–, groundwater, 207, 451, 459
–: hydraulic head, 25
– mechanisms for development, 25
– potentiometric surface, 207, 209
– regime, 172, 196
–, surface water, 207
–, water table, 207, 209
hydroperiod, 2, 5–8, 10, 22, 29, 56, 58, 68, 73, 76–79, 87, 90, 92–94, 102, 104, 106, 108–110, 112, 116, 117, 126–129, 134, 143, 147, 161, 165, 174, 176, 180, 182, 185, 196, 251, 253, 258, 264, 282–284, 288, 290, 294, 296, 447, 456, 458, 447, 455, 456
–, depth, 58, 456
– duration, 58, 456
– effects on species composition, 196

hydroperiod (*continued*)
– frequency, 58
– influence on wetlands, 456
– season, 58
hydrophytes, 108, 294, 414, 416, 418, 420, 422, 423, 428
hydrosere, 29, 33, 218, 232, 235, 305, 443
–: classical model, 29
–, Clementsian, 29
hypersaline lagoon, 146, 151, 154, 155
hypersalinity, 69, 75

ibis, black (*Pseudibis papillosa*), 374
–, glossy (*Plegadis falcinellus*), 374
–, oriental (*Threskiornis melanocephala*), 374
ice, 56, 87, 94, 108, 134, 418, 419
– age, 409
– breakup, 94
– cover, 94
– damage, 108
– floes, 94, 441
– jamming, 94
– sheets, 89
–, storm damage of, 103
– wedges, 91
ichthyochory, 237
Idenburg River (Irian Yaya, Indonesia), 346
ígapo, 108, 127, 235, 271, 453
– forest, 109, 110, 111
– seasonal, 128
ignition residue, 415
Illinois (U.S.A.), 3, 39–41, 71, 93, 95, 96, 99, 101, 103–105, 109–112, 117, 118
Illinois River (U.S.A.), 76
immobilization, 72, 75, 110, 112, 113
– of heavy metals, 132
impact(s), 134
– on wetland, 129
importance value, 179, 278, 279, 280, 281, 285
impounded, 335
impoundment, 106, 121, 131, 132, 166
incision, 89
India, 4, 358, 361
Indian River (Dominica), 247
Indiana (U.S.A.), 76, 95, 119
industries, 388
Inery Creek (French Guiana), 232
infiltration, 25, 26, 93
inflows, 93
insecticides, 130, 132
insect(s), 56, 57, 242, 244, 245, 247, 313
– outbreaks, 313, 327
interglacial, 89
intermittent streams, 113, 122, 133
internal drainage, 93
interstitial water, 113
intertidal, 55
inundation (*see also flooding*), 2, 5–7, 87, 88, 102, 108, 117, 127, 213, 216–219, 221–223, 226, 231, 234–237, 239, 241, 243–245, 294, 335, 337, 338, 341, 342, 344, 349, 389, 391, 395, 411, 413, 420

– forests, 111
–, seasonal, 391, 392
iora, common (*Aegithina tiphia*), 376
Iowa (U.S.A.), 36, 40, 118
Iran, 4
Ireland, 3, 407, 409, 416, 429
Irian Jaya (Indonesia), 335, 336
iron content, 229, 230
irrigation, 93, 132, 247, 427, 432
Isla La Palma (Mexico), 100
Isla Roscell (Mexico), 152
island(s), 65, 131, 132
–, barrier, 19
–, coassal, 19, 20
–, offshore, 145, 146
–, overwash, 145, 246
–, tree, 120
Itanhaem (Brazil), 167
Itasca State Park (Minn., U.S.A.), 71

jacana, bronze-winged (*Metopidius indicus*), 375
jackal, golden (*Canis aureus*), 371
jay (*Garrulus glandarius*), 430
Jalankebun (Selangor, Malaysia), 78
Jalisco (Mexico), 287
Jamaica, 221, 236, 246, 247
Jamalpur (Bangladesh), 358
James Bay (Ont., Canada), 126
Jayapura (Irian Jaya, Indonesia), 336
Jessore (Bangladesh), 358–360, 369
jewfish (*Pama pama*), 378
Jobos Bay (Puerto Rico), 153, 158, 162, 163
Johore (Malaysia), 324
Joyuda Lagoon (Puerto Rico), 152
julubal (*Bravaisia tubiflora*), 289, 295
jungle cat (*Felis chaus*), 371
Jurong River (Singapore), 362, 364

Kaira (Sundarbans, Bangladesh), 368
Kalimantan (Indonesia), 299, 301, 324
Kankakee River (Ill., U.S.A.), 112
Kansas (U.S.A.), 71, 96
Kansas River (Kansas, U.S.A.), 71
Katka (Bangladesh), 370
Kaw River (French Guiana), 231, 234
Kenya, 129, 133
kerangas, 299, 303, 305, 306, 311, 316, 319–321, 325, 326, 329
–, *Agathis*, 319
–, forest 299, 310, 312, 315, 322–324, 326, 327
kerapah, 299, 303, 305, 306, 308, 315–317, 319, 320, 325, 329, 332
– forest, 300, 302, 308, 312, 318, 320–324, 326, 327
kestrel (*Falco tinnunculus*), 430
Khulna (Sundarbans, Bangladesh), 358–366, 368–370, 383
Kiel (Schleswig-Holstein, F.R.G.), 407, 409
Kikori River (Papua New Guinea), 336
Killarney (Kerry, Ireland), 409
kingfisher, black-capped (*Halcyon pileata*), 375
–, brown-winged (*Pelargopsis amauroptera*), 375
–, common (*Alcedo atthis*), 375, 430
–, lesser pied (*Ceryle rudis*), 375

kingfisher (*continued*)
–, stork-billed (*Pelargopsis capensis*), 375
–, white-breasted (*H. smyrnensis*), 375
–, white-colored (*H. chloris*), 375
kite, black (*Milvus migrans*), 374, 429
–, black-winged (*Elanus caeruleus*), 374
–, Brahmini (*Haliaster indus*), 374
Kortenhoef (Utrecht, The Netherlands), 409
Kuching (Sarawak), 299, 302, 305
Kumsi River (Papua New Guinea), 336
Kushtia (Bangladesh), 358
Kuskokwim River (Alaska, U.S.A.), 126

La Mancha (Mexico), 288
Labrador (Canada), 115
Labrador tea (*Ledum groenlandicum*), 178, 179
lag deposits, 89
lagg-zone, 412, 427
lagoon, 112, 116, 117, 147, 221, 274, 287, 288, 292–294, 345, 391
lagoonal forests, 353
Laguna de Terminos (Mexico), 293, 295
Laguna Las Salinas (Puerto Rico), 152
Laguna Penitencia (Costa Rica), 274
Laguna Samay (Costa Rica), 275
Laguna Terminos (Mexico), 152
Laguna Tortuguero (Costa Rica), 274
lake(s), 29, 30, 33, 59, 71, 77, 78, 87, 90, 143, 165, 166, 388, 437, 439
– beds, 390
ox bow, 88, 89, 439
Lake Agassiz (Minn., U.S.A.), 7, 29
Lake Agassiz Peatlands Natural Area (Minn., U.S.A.), 174, 178
Lake Drummond (Great Dismal Swamp), 201, 203, 204, 209, 210
Lake Erie, (North America), 41
Lake Michigan (U.S.A.), 3
Lake Ochlawaha (Fla., U.S.A.), 38
Lake Victoria (Africa), 167, 269
Lakekuma River (Papua New Guinea), 336, 341, 343
land classification, 15, 20
– subsidence, 110
landform, 43, 87, 88, 352, 454, 455
– accretion, 354
landscape, 17, 19, 20, 28, 43, 44, 120, 134, 400, 447
–: background, 17–21
–: center, 17–19, 21
–: diversity, 121
– form, 17, 21
–: island, 17–19, 21
– patterns, 16
–: string, 17–21
–: strip, 17, 18, 20, 21
–: zone, 17–19, 21
landslides, 60
landuse, 89
Lanterne River (Haute Saône, France), 409
lapwing, red-wattled (*Vanellus indicus*), 375
–, spurwinged (*V. spinosus*), 375
lark, sand (*Calandrella raytal*), 376
Las Choapas (Mexico), 288
Lassa Forest Reserve (Sarawak), 306–308, 330

lateral accretion, 89
– channel, 91
– roots, 109
latitude, 95, 456, 458
latitudinal effects, 148, 149
– gradient, 69
– patterns, 104
Lawas (Sarawak), 323, 330, 332
leaching, 60, 72, 146, 147, 161, 413
leaf(ves), 57, 69, 70, 76, 103, 104, 109, 110, 445
–, alder, 445
– area, 22
– – index, 26, 65, 70, 150, 152, 186, 187
– biomass, 108
– bird, gold-fronted (*Chloropsis aurifrons*), 376
– canopy, 156–158, 262
– characteristics, 311, 313, 314, 316, 319, 320, 326, 327
– –: adaptions to water regime, 326
– –: secondary compounds, 327
– epidermis, 144
–, evergreen, 162
– fall, 61, 72, 160, 163, 164, 342, 445
– half-life, 160
– litter, 101, 110, 112
–, needle-shaped, 62
– respiration, 454
–, sclerophyllous, 61
–, shade, 157, 158
–: specific leaf area, 163
– stomata, 144
–, sun, 157, 158
– wilting, 61
–, xeromorphic, 144, 157, 158, 160, 161
leatherleaf (*Chamaedaphne calyculata*), 178–180
Ledo-Pinetalia, 443
legumes, 218, 232, 269
lenticels, 72, 73
leopard (*Panthera pardus*), 370, 371
levee(s), 19, 75, 87, 90, 93, 116, 126, 131, 219, 222, 223, 226, 228, 229, 234, 236, 274
– forest, 218, 231, 235, 458
lianas, 88, 119, 127, 417, 421
lichens, 165
life cycles, 431
– form, 395, 401
– –: graminoid layer, 395
– –: hummock grass, 395
– –: shrub, 395
– – spectra, 414, 418, 423, 427
– –: tree, 395
– life zones (Holdridge), 19, 273
– –: tropical moist forest, 274
– –: – wet forest, 273, 282, 283
light, 99, 152, 156, 157, 454, 457
– seeded hardwoods, 117
– limitation, 99, 101
lightning, 347, 402, 404
– gaps, 311, 313, 315, 327, 328
lignotubers, 402
lime, 35

GENERAL INDEX

limiting factors, 104
Limon (Costa Rica,) 283
Linde River (Friesland, The Netherlands), 409
Linsley Pond (Conn., U.S.A.), 30
lithology, 134
litter, 76, 109–111, 113, 128, 129, 176, 187, 249, 258, 260, 303, 305, 323, 327, 412, 415, 418, 420, 421, 425, 427, 431, 445, 454
– accumulation, 453, 458
–, coarse, 187
– decomposition, 156, 160–162, 164, 187, 445, 448, 453, 454
– export, 161
– fall, 72, 75, 104, 105, 106, 110, 112, 159, 187, 188, 193, 194, 196, 444, 448, 453–458
–, fine, 187, 188
– ground, 150, 153, 161
– half-life, 161
–, microbial degradation, 452–454
–, nutrient content of, 193, 194
– production, 104, 127, 156, 160, 187, 196, 247, 259, 260, 437, 444, 445
– quality, 195
–, residence time of, 453
–, standing stock of, 187, 188, 196, 258, 445
– turnover, 156, 160–162, 164, 453
Little Tennessee River (N. C., U.S.A.), 120
liverworts, 420, 425, 426
livestock grazing, 123, 132
lizards, 244, 245
llanos, 272
Llanuras de Tortuguero (Costa Rica), 281
load, 89
loam soils, 122
logging, 117, 118, 238
logs, 101, 102
–, decaying, 76
Loire River (France), 407
Lomas de Sierpe (Costa Rica), 274, 280
longitude, 121
Loosdrechtse Plassen (Utrecht, The Netherlands), 409
Los Blanquizales (Trinidad), 216
Los Petenes (Mexico), 294
loss on ignition, 306, 443
Louisiana (U.S.A.), 9, 55, 57, 89, 98, 103, 105, 106, 130
low forested wetland, 290
lower montane rain forest, 221
lowlands, 124, 273, 282, 407
Lüneburger Heide (Niedersachsen, F.R.G.), 409
Luquillo (Puerto Rico), 221
Luquillo Forest (Puerto Rico), 251–259, 261, 263
Luquillo Mountains (Puerto Rico), 65, 67, 251
Luxembourg, 407, 409

Mackenzie River (N.W.T., Canada), 125
Mackenzie River Delta (N.W.T., Canada), 71, 91
macrofauna, 241, 242
macrofossils, 415, 416
Macroom (Cork, Ireland), 409
macrophytes, 19, 30, 31
Magdalena River (Colombia), 234, 237
Magdalena Valley (Colombia), 4

Maglemose (Frederiksborg, Denmark), 409
magnesium (*see also* nutrients), 113, 230, 443, 445
magpie (*Pica pica*), 430
mahogany, swamp (*Eucalyptus robusta*), 391, 392, 403
Mahury River (French Guiana), 233, 234
malate, 73
Malaya (Malaysia), 4, 7
Malayan Archipelago, 365
Malayan Uniform System, 331
Malaysia, 56, 57, 62, 78, 105, 106, 111, 164, 299, 321, 324, 325, 326, 363, 453, 456, 457
Malesian region, 299, 301, 324–326, 329
mallard (*Anas platyrhynchos*), 430
Malumghat (Sundarbans, Bangladesh), 364
Mamantel River (Mexico), 291
Mambare River (Papua New Guinea), 336
Mamberamo River (West Irian, Indonesia), 336
mammals (Mammalia), 88, 243, 244, 246
management, 106, 129, 213, 245, 247, 248, 447, 449
Manaus (Amazonas, Brazil), 110
Mandan (N.D., U.S.A.), 71
mangal, 223, 224, 226, 228, 229, 234, 235, 241, 244, 245, 247
mangrove(s), 19, 30–32, 34, 36, 37, 40, 41, 55–59, 66, 68, 75, 76, 114, 126–128, 143-169, 171–173, 182, 183, 185–188, 193, 194, 196, 214, 218–223, 226, 230, 234, 235, 244, 245, 269, 270, 271, 274, 283, 287, 291–296, 303, 308, 311, 315, 336–338, 340, 346, 347, 350, 389, 447, 450–452
–, biotic composition of, 363–367
–, black (*Avicennia germinans*), 31, 37, 75, 127, 144, 151, 157–161, 164, 173, 176, 182, 185, 188, 194, 195, 294, 453, 457
–: endangered species, 357, 370
– islands, 452
– management, 365
–, red (*Rhizophora mangle*), 2, 5, 31, 32, 37, 38, 75, 144, 146, 151, 156–164, 182, 185, 195, 263, 271, 293, 294, 453, 457
–, scrub, 185, 187, 188
–, short (*also* dwarf), 294
– structure, 363–365
– uses aesthetics, 359
– –: fuel, 359, 365, 369, 383
– –: honey, 369
– –: molasses, 369
– –: palm juice, 369
– –: pulp, 365, 384
– –: raw materials, 369
– –: thatching, 365, 369
– –: timber, 359, 383
–, white (*Laguncularia racemosa*), 31, 127, 144, 157–159, 161, 182, 253, 255, 256, 263, 269, 294, 453, 457
Manitoba (Canada), 29, 41, 174, 178, 182, 183, 186
Mantang (Port Weld, Perak, Malaysia), 106
map, 17
–: landscape, 17
–: vegetation, 17, 19
maple (*Acer*), 5, 30, 95, 97, 103–105, 120, 180, 183, 188, 201, 203, 209
maple, red (*Acer rubrum*), 33, 38, 99, 101, 111, 116, 117, 119, 120, 179–184, 188, 194, 201
Maranhao (Brazil), 272
marginal fen, 120

marine, 74
– regression, 223, 226
– transgression, 223, 226
marl, 30, 35, 453
Maroni River (French Guiana), 241
marsh, 26, 31, 36, 39, 41, 56, 59, 64, 110, 125, 131, 218, 219, 234, 239, 240, 243, 246, 280, 282, 284, 419, 437, 439
– forest, 214, 217–219, 221, 222, 226, 228, 229, 231, 234–236, 239, 240
–, herbaceous, 222, 226, 232, 246, 247, 267, 269, 270, 273, 274, 277, 284
–, humans dwelling, 270
–, palm, 214, 217, 218, 219, 221, 235, 240, 241
–, sedge, 221
Martinique, 222, 240
Maryborough (Qld., Australia), 393
Maryland (U.S.A.), 113, 116
Massachusetts (U.S.A.), 174, 180
massive tree mortality, 154
Mathoura (N.S.W., Australia), 395
matorral espinoso inundable, 295
matorral inerme inundable, 295
Mattang, 299
Matto Grosso (Brazil), 4, 272
maturity, 438
Mayagüez (Puerto Rico), 251–253, 255–264
McIntosh Index, 305, 316, 319, 322
McKenzie River (Ore., U.S.A.), 124
meadows, 412, 418, 437, 441, 442
meander, 89, 91, 93
– curve, 91
meandering, channel, 87, 89, 90
meanwater level, 143, 153
Medio Mundo Bay (Puerto Rico), 150, 152
Mediterranean climate, 387
Mediterranean Region (U.S.A.), 121, 123, 128
medium riparian forest, 287
Meervlakte (Irian Jaya, Indonesia), 336, 345
Meghna Estuary (Bangladesh), 358, 360
Merauke (Irian Jaya, Indonesia), 336, 342, 345, 346
merchantable timber, 124
Merurong plateau (Sarawak), 316–319, 321, 325
mesic, 123
– uplands, 117, 119
meso-oligotrophic, 439
mesophytes, 109
mesotrophic, 439
metabolic adaptations, 73
– pathways, 109
metabolism, 57, 60, 76, 109
metals, 68, 69, 79
meteorologic inputs, 89
methane, 55, 74
– evolution, 104
methanogenesis, 74
Metroxylon swamps, 305
Meuse River (W. Europe), 407
Mexico, 100, 152, 161, 183, 287, 288
Michigan (U.S.A.), 3, 41, 62, 65, 111, 178, 179, 183, 184, 186
Michoacan (Mexico), 287

microbes, 72, 74, 75, 79, 452, 453, 454
microbial biomass, 75
– communities, 109
– immobilization, 113
– processes, 74–76
– respiration, 74–76
microenvironments, 129
microflora, 327, 238, 331
micro-organisms, 445
microtopographic features, 101, 102
Middachten (Gelderland, The Netherlands), 409
Middle Ages, 412
migration, stream, 91, 119
Mildura (Vic., Australia), 404
Millewa State Forest (N.S.W., Australia), 396, 400, 404
Mindanao (Philippines), 302
mineral(s), 60, 409, 413, 414
– cycling, 15
mineralization, 72, 443
–: nitrogen, 413, 416
mining, tin, 58
Minnesota (U.S.A.), 7, 35, 60, 71, 174, 177, 178, 183, 184, 186–190, 193, 194
minnow, 378
mires, 56, 65, 387, 409, 412, 442, 443
Mississippi (La., U.S.A.), 43, 89, 106
Mississippi River (La., U.S.A.), 76, 90, 91, 110, 116, 119, 120, 130
Mississippi Valley (U.S.A.), 118
Missouri River (U.S.A.), 71, 91, 95, 106, 120–122, 131,
Mitachuri Canal (Sundarbans, Bangladesh), 364
mixed dipterocarp forest, 305, 312, 316, 319, 321–324, 326
mixed floodplain forest, 111
– forest, 109
– hardwood forest, 95, 96, 98, 103, 105, 109
models, 29, 31, 32, 34, 36, 39–41, 43
–, succession, 29, 36, 39, 40, 41, 43
Moin (Costa Rica), 100
Moira State Forest (N.S.W., Australia), 396
moisture, 121
– availability, 87
– index, 447
– –; definition of, 395, 400
– –, mean annual, 395, 396, 397, 400, 401, 403
– –, monthly, 400, 401
–, litter, 110
–, regimes, 134
–, soil, 108
mongoose, 371
–, small (*Herpestes auropunctatus*), 371
monitor, grey Indian (*Varanus bengalensis*), 373
–, ruddy snub-nosed (*V. flavescens*), 373
monocarpic stems, 269
monospecific stands, 102
– swamps, 267, 269, 270, 345, 346, 351
– wetland, 287, 295
monsoon, 335, 337, 341, 350, 352, 353, 357, 359, 360
monsoonal rainfall, 390
Montana (U.S.A.), 95
montane riverine, 251, 256, 264
Montreal (Que., Canada), 120

moor, ombrogenous (*see also* raised bog), 305
–, topogenous (*see also* fen), 305
moorhen, gray (*Gallinula chloropus*), 376, 430
moose, 38
moraines, 19
morass, 221
morichales, 216, 219, 220, 225, 230, 234, 235, 240, 241, 244, 247
morphologic adaptations, 72, 73
– features, 89
morphometry, 87
Morrelgonj (Bangladesh), 359, 360
mortality, 59, 72, 76, 102, 104, 121
mosquito(es), 244, 245, 247
moss, 17, 181, 190, 420, 425, 426, 439, 441
–, club, 27
–, *Sphagnum*, 34
Mothronwala Swamp (India), 4
mounds, 9
Mount Crawford Forest Reserve (S.A., Australia), 396
Mount Lofty Ranges (S.A., Australia), 396, 399
mountain(s), 93, 123, 441
– swamps, 349
mouse, fawn field (*Mus cervicolor*), 372
–, house (*M. musculus*), 372
–, tree (*Vandeleuria oleracea*), 372
mouth, river, 89
mucal (*Dalbergia brownei*), 296, 289, 290, 296
mucaleria, 296
muck, 102
mud, 68, 74
– islands, 346
– volcano, 227
mullet, freshwater (*Mugil corsula*), 378, 380
Mullinger (Westmeath, Ireland), 409
Mulu National Park (Sarawak), 302, 305, 327
munia, black-headed (*Lonchura malacca*), 377
–, spotted (*L. punctulata*, *L. subundulata*), 377
–, white-rumped (*L. striata*), 377
Murray River (Australia), 387, 388, 395, 404
Musa River (Papau New Guinea), 350
muskeg, 19, 22, 178, 179, 183, 184, 186, 189
muskrat (*Ondatra zibethica*), 38, 429
– eat-out, 38
mussel(s), 241
–, green (*Mytilus viridus*), 382
Muyil (Mexico), 294
mycorrhizal, 425
Mymensinge (Bangladesh), 358
myna, bank (*Acridotheres ginginianus*), 377
–, common (*A. tristis*), 377
–, jungle (*A. fuscus*), 377
Myriophyllo-Nupharetum, 442

N/P, 162, 163
Naardermeer (Noord-Holland, The Netherlands), 409
Nalianala Range (Sundarbans, Bangladesh), 364, 365, 368
Namanve Swamp (Uganda), 167
Namanvel (Uganda), 3
nanophanerophytes, 418

napunyah (*Eucalyptus ochrophloia*), 391, 403
Narino Swamp (Colombia), 4
Nariva (Trinidad), 216, 217, 222–227, 230, 231, 234–238, 240–244, 246, 247
natural heritage, 126
Nayarit (Mexico), 287
Nebraska (U.S.A.), 131
Neolithic man, 70
Neotropical, 213, 232, 239, 241, 245, 247
nesting, 430
net energy yield, 160
– release, 112
Netherlands, The, 3, 71, 407, 209, 412, 415, 416, 422, 424, 425, 429–432
Neuse River (N. C., U.S.A.), 116
New Britain (Bismarck Archipelago), 336, 343
New Caledonia, 394
New England (U.S.A.), 120
New Guinea, 4, 38, 89, 269, 270, 302, 305, 335–337, 345, 346, 354, 355
New Ireland (Bismarck Archipelago), 336, 343
New Jersey (U.S.A.), 41, 94–96, 99, 103
New Mexico (U.S.A.), 3, 6, 71, 97, 122, 123
New Papua, 4
New South Wales (Australia), 387–390
New York (U.S.A.), 119, 120
Newfoundland (Canada), 115
newt(s) (*Triturus*), 431
Nicaragua, 4, 245, 246, 269
niche, 91, 458
–, hypervolume, 458
Nickerie (Surinam), 226, 228, 229, 234, 247
Nigeria, 128, 129, 269
nightingale (*Luscinia megarhynchos*), 430
nightjar, Franklin's (*Caprimulgus affines*), 375
–, Indian (*C. asiaticus*), 375
–, jungle (*C. indicus*), 375
nipa palm (*Nype fruticans*), 270, 340, 347, 353, 357, 363, 366, 384
nitrate, 74, 79, 108, 112
nitrification, 112, 113
nitrite, 79
nitrogen, 55, 63, 64, 75, 79, 108, 109, 110, 112, 163, 164, 231, 232, 305, 397–399, 443
–, amide, 306, 307
–, ammonium, 306, 307
– fixation, 415, 417, 437
–, hexosamine, 306, 307
–, hydrolysable, 306, 307
Noakhal River (Sundarbans, Bangladesh), 368
Noakhali (Bangladesh), 358
node sedges (Restionaceae), 395
North America (*see also particular localities*), 3, 30, 32, 35, 114, 115, 171, 172, 184, 429
North Carolina (U.S.A.), 2, 6, 29, 37, 59, 71, 78, 96, 103–105, 109–113, 120, 167, 172, 180, 181, 201, 203, 259
North Dakota (U.S.A.), 71, 94, 95, 121, 122
North Oropuche (Trinidad), 216, 235, 245
North Sea, 407
North Stradbroke Island (Qld., Australia), 404
Northern Forest Region (U.S.A.), 115, 120

northern latitudes, 94
Northern Territory (Australia), 388, 389, 397
Northwest Territory (Canada), 125
Norway, 407, 409
nursery habitat, 243
nutria (*Myocaster coypus*), 429
nutrient(s), 89, 104, 108, 109, 113, 130, 134, 171, 173, 175–177, 187, 283, 323, 327, 411–414, 417, 419, 421, 423–425, 427, 443, 445
–, adaptations for conservation of, 191, 193, 194
– availability, 186, 194, 302, 327, 452, 455, 458
– budgets, 195
–: calcium (Ca), 413, 416, 431
– capital, 89
– concentrations, 109
– – in foliage, 189–191
– – in soil, 176, 177
– – in surface waters, 174, 175
– – in vegetation, 189–193
– conservation, 110, 190, 320, 329
– content, 127
– – in litterfall, 193–195
– – in soil, 193
– – in vegetation, 192, 193, 195
– cycle(s), 58, 65, 79
– –: abiotic control, 453, 454
– –: biotic control, 452, 454
– –, model of, 454, 455
– cycling, 87, 88, 130, 172, 186, 193, 195, 196, 329
– – efficiency, 193, 194
– – – index, 193, 194
– dynamics, 109, 189
– enrichment, 104, 369
–, exogenous sources of, 453–455
– export, 130, 449, 454, 455
– – and transport, 159–161
– immobilization, 194, 195, 454, 455
– impoverished soils, 108
– in litter fall, 448, 452, 454–457
– in rainfall, 306, 323
– in runoff, 264
– inputs, 173, 306
– leaching, 452, 454, 455
– levels, 229, 247
– limitation, 108, 127, 134
– loading, 127
– loss, 331
–: magnesium (Mg), 413, 416
– mineralization, 176, 195
– mobility, 59, 60
– mobilization, 454
– need, 454, 455
–: nitrogen (N), 413, 415–417
– -poor wetlands, 173, 176, 179, 194, 196
–: phosphorus (P), 413, 416
–: potassium (K), 413, 416
– recycling, 144, 147, 156, 161, 452, 454, 455
– –, efficiency of, 160, 162, 163, 165, 453–455
– regime, 328
– resorption, 109
– retranslocation, 190, 194, 195, 452, 454, 455
– return to uptake ratio, 196
– re-use, 452, 455
– –rich wetlands, 173
–: soil fertility, 148, 162
–, standing stocks of, 189, 191, 192
–, – – –: soils, 193
–, – – –: vegetation, 192, 193
– storage, 164–166
– supply, 108
– turnover, 452, 453
– uptake, 108, 162, 164, 184, 193, 195, 451, 454, 455
– use, 451, 454, 455
– –, efficiency of, 191
–: within-stand recycling efficiency index, 452–457, 459

oak (*Quercus*), 30, 33, 43, 96, 104, 107, 116, 119, 120, 179, 201, 442, 445
–, laurel (*Q. laurifolia*), 116, 117, 180
–, northern pin (*Q. ellipsoidalis*), 179
–, river (*Casuarina cunninghamiana*), 270, 389, 403
–, swamp chestnut (*Quercus michauxii*), 116, 117, 180
–, white (*Q. alba*), 116–118, 180
Oceania, 269, 270
oceans, 74, 143, 145
Odra River (Poland), 437
Ogeechee River (Ga., U.S.A.), 116
Ohio River (U.S.A.), 71, 118
Ohio River Valley (U.S.A.), 119
oil, 69, 76, 158, 166
– palm, 268
Okefenokee Swamp (U.S.A.), 3, 19, 27, 29, 37, 40, 172, 176, 178, 181, 182
Oklahoma (U.S.A.), 3, 79, 120, 121
Oklawaha River (Fla., U.S.A.), 76
oleoresin, 10
oligohaline, 127, 134, 144, 147, 151
oligotrophic, 35, 444
– rain-flooded wetlands, 299, 320
– – –, area of, 302
– forested wetland, 299, 301, 302, 316, 321, 324, 329, 330
– peat swamps, 299, 300, 303, 316
oligotrophy, 148
olive, Russian (*Elaeagnus augustifolia*), 122, 123, 132
Olympic Mountains (Wash., U.S.A.), 124
ombrogenous, 72, 409
ombrophilous, 439
ombrotrophic bogs, 411, 412, 422, 425, 427
Ontario (Canada), 126, 165, 166
open woodland, 128
open-water fetch, 91
Orapu River (French Guiana), 232
ordination, 324
Oregon (U.S.A.), 124
organic, 444
– acids, 34
– carbon, 70, 72, 74, 75, 110, 127
– – content, 104
– – export, 55, 56, 72

organic (*continued*)
– matter, 15, 19, 22, 27, 28, 34, 70–72, 76, 77, 109, 112, 120, 143, 146, 165, 171, 188, 194, 219, 220, 231, 274, 275, 412, 413, 415, 417, 419–421, 428, 443
– –, accumulation of, 171, 189, 196
– –, dynamics of, 172, 185
– – export, 146, 165, 171
– – production, 186, 187
– – turnover, 60
– nitrogen, 445
– sediments, 33
– soils, 126
organisms, 88
Orinoco Delta (Venezuela), 216, 219, 220, 222–224, 228, 231, 235
Orinoco River (Venezuela), 133, 267, 271
oriole, black-headed (*Oriolus xanthornus*), 377
–, golden (*Oriolus oriolus*), 430
orographic rains, 92
Osa Peninsula (Costa Rica), 283
osmoregulation, 160
Ostende (West-Vlaanderen, Belgium), 409
Ottawa (Canada), 190
otter, small-clawed (*Amblonyx cincrea*), 371
–, smooth, common (*L. lutra*), 371, 429
–, smooth-coated (*L. perspicillata*), 371
Ouachita River (Ark., U.S.A.), 93, 116
Oude Maas River (Zuid-Holland, The Netherlands), 409
outburst floods, 94
Outer Banks of North Carolina (U.S.A.), 203
outflows, 93
outwash, 122, 125
– plain, 125, 399
Overijssel River (The Netherlands), 429
overstory, 99, 119, 122, 124, 129, 268, 270–272
oxbow, 116
– depressions, 90
– lakes, 88, 89, 117, 439, 442
oxidation, 70, 76
oxidation–reduction, 54
Oxisol, 318, 330
Oxycocco-Sphagnetea, 443
oxygen, 55, 56, 68, 73–75, 79, 109, 112, 158, 165, 413, 439
– availability, 325
– deficiency, 303, 310, 320, 329
– diffusion, 72, 74
oyster(s), 234
–, backwater (*Crassostrea madrasensis*), 382
–, rock (*C. gryphoides*), 382
oystercatcher (*Haematopus ostralegus*), 375
O'Meara (Trinidad), 217, 239, 243

P/R, 156, 157, 158, 159
Pabna (Bangladesh), 358–360
Pacific, 57
Pacific Coast, 287
Pacific Northwest (U.S.A.), 8, 188
Pacific Northwest Region (U.S.A.), 115, 123
padang, 302, 314–316, 320, 332
Pahang (Malaysia), 325
Pakistan, 4

palm, 65, 66, 127–129, 217, 218, 221, 224, 234, 235, 239, 244, 246, 267, 272, 277, 280, 292, 293, 311, 337, 338, 340, 341, 346, 349, 354, 447, 449
–, babassu (*Orbignya martiana*), 272
–, clump, 269, 277–281, 283, 285
–, dwarf, 281
– forest, 98, 99, 109, 110, 269
– leaves, 109
– marsh, 214, 217, 218, 219, 221, 235, 240, 241
–, nipa (*Nypa fruticans*), 270, 340, 347, 353, 357, 363, 366, 384
– oases, 3, 8
–, oil (*Raphia*; *Elaeis*), 4, 37, 255, 267–271, 274–277, 279, 285
–, pangano (*Raphia*), 269
–, real (*Manicaria*), 214, 217, 240, 271, 275–277, 279–283, 285
–, sago (*Metroxylon sagu*), 268–270, 283, 347
– seedlings, 99
– species, 128
– swamp(s), 102, 214, 216, 218, 219, 220, 227, 234, 239, 240, 246, 247, 267–274, 280, 283, 284
– –: aguajales, 268
– –, *Manicaria*, 271, 272, 275–283, 285
– –, *Mauritia*, 267, 268
– –, *Metroxylon*, 269, 270, 284
– –, *Mora*, 283
– –, *Nypa*, 270, 271
– –: palma real, 271
– –, *Pandanus*, 269, 270, 271
– –: panganales, 269
– –, *Prioria*, 283
– –, *Raphia*, 268, 269, 271, 274, 276–282, 284, 285
– –: temichales, 271
– –, truli bush, 271
– –: yolillales, 269, 271
–, temiche (*Manicaria*), 271
–, truli (*Manicaria*), 271
–, uses of, 268, 271, 272
–, yolillo (*Raphia*), 269
palma real, 271
palmar inundable, 292
Palmas del Mar (Humacao, Puerto Rico), 252
paludification, 33, 70, 226, 327, 440–443
– effects on nutrients, 34, 35
– – – on water table, 33, 34
palynology, 226, 234, 241, 428
pan evaporation, 61, 395, 400
Panama, 4, 79, 99, 100, 103, 105, 109, 128, 162, 163, 165, 215, 224, 237, 269, 272
pandan, 129, 337, 338, 342, 345, 350–352
Pandan Nature Reserve (Singapore), 361–364, 366
Paneh Peninsula (Sumatra), 325
panganales, 269
paperbark(s) (*Melaleuca* spp.), 4, 325, 327, 341–343, 349, 352–354, 389, 390, 392, 394, 397, 404
Papua New Guinea, 7, 268, 270, 283, 335, 336, 341–343, 346, 348, 350, 389, 393, 394
papyrus, 26, 269
Para River (Brazil), 269
paradigms, 447–460
parakeet, Alexandrine (*Psittacula eupatria*), 377
–, roseringed (*P. krameri*), 377

Parc Natural de Guadeloupe (Guadeloupe), 247
Parguera (Puerto Rico), 146, 147
parisitic fungus, 440
Paroo River (Qld., Australia), 391
parrots, 236
particle size, 89
partridge, Assam black (*Francolinus francolinus*), 376
particulate organic carbon (POC), 56
Pascagoula River (Miss., U.S.A.), 116
pastures, 291, 296
Patillas (Puerto Rico), 251, 253, 255–264
Patuakhali (Bangladesh), 358, 369
Pearl River (La., U.S.A.), 98, 116
peat, 19, 26, 30, 31, 34, 38, 41, 60, 62, 63, 65, 70, 71, 76–78, 89, 120, 165, 171, 174, 176, 179, 186, 222, 231–235, 238, 270, 272, 303, 325, 329, 338, 341, 344, 349, 412, 422, 426, 437, 439–443, 444
– accumulation, 69, 70, 110, 120, 171, 189, 303, 350, 328, 329, 459
– age, 29
– blanket, 78
– bog, 70, 120, 171, 180, 299, 317, 319, 321, 325, 415, 440
– carbon, 307
– cutting, 414
– deposits, 70, 76, 103, 442
– depth, 172, 174, 176, 180, 303, 306, 316, 328, 451, 453, 459
– development, 303, 316, 307
– domes, 305, 309, 311, 315, 326, 329
– effect on biogeochemical cycles, 171, 186, 189, 193
–, fen, 415
–, forest-reed, 437
–, forest-sedge, 437
– formation, 77, 303, 305, 318, 327, 415, 443
– harvest, 58
–: humic acid fraction, 307
– loss, 196
– mass, 189
– nitrogen, 307
–, nutrient (or chemical) concentration of, 308
– nutrient sink, 193
–: pH, 171
– raft, 414, 425
–, sedge, 414
–, shallow, 439
– slip, 38
– subsidence, 330
– swamp(s), 447
– – forest, 299, 302, 305–309, 311, 313, 316–318, 320–326, 330
– thickness, 176, 189
– wetlands, 124
peatland, 20, 60, 126, 409
Pee Dee River (S. C., U.S.A.), 116
pegasse, 218, 231, 232, 234, 235, 237, 247
Pegunungan Van Rees Mountains (Irian Jaya, Indonesia), 336
pelican, spot-billed (*Pelecanus philippensis*), 377
Pennsylvania (U.S.A), 119
perch, climbing (*Anabus testudineus*), 378
Perche (Orne, France), 409
perennial streams, 133
periphyton, 159
permafrost, 91, 125

Permian, 27
perturbations, 130
Peru, 127, 267, 268
petrel, black-bellied storm (*Fregetta tropica*), 377
pH, 79, 102, 229, 230, 440, 443
phanerophytes, 418, 422
phasic communities, 306–309, 311–316, 321, 325
pheasant (*Phasianus colchicus*), 430
phenolic compounds, 75
phenology, 259–262
–: flowering, 259, 260
–: fruit fall, 259
–: fruiting, 259, 260
–: leaf fall, 260
–: role of rainfall, 259, 260
–: role of temperature, 260
–: seeding, 261, 262
phloem necrosis, 118
phosphorus (*see also* nutrients; elements), 63, 79, 108–110, 112, 162–164, 288, 397–399, 443
–: phosphate, 113, 229, 230
– deposition, 110
–, total 109
– translocation, 108
– uptake, 113
photic zone, 22
photorespiration, 72
photosynthesis, 156, 257, 421
–, net, 157–159
– –, daytime, 60, 67, 70
Phragmitetea, 443
phreatophyte, 3, 8, 26, 93, 113, 133
phreatophytic habit 102
– vegetation, 133
physical damage, 103, 108, 119, 134
physiognomy, 22, 447, 451, 447
–, forest 113, 126, 127, 128
physiographic provinces, 93
physiological adaptations, 62, 73, 114
– drought, 62
– stress, 130, 131
phytocoenoses, 437
phytoplankton, 33–35
phytosociological relevees, 438
– surveys, 439
Piceo-Sphagnetum girgensohnii, 440
piedmont, 274, 281, 282
Piedmont (U.S.A.), 92, 93
piezometric pressure, 93
pig, feral, 38
pigeon, orange-breasted green (*Treron bicincta*), 375
–, pompadour green (*T. pompadora*), 375
–, wood (*Columba palumbus*), 430
–, yellow-footed green (*T. phoenicoptera*), 375
Pinaui (Brazil), 272
pine (*Pinus*), 120, 176, 178, 201, 203, 415, 440, 444, 445
– flatwoods, 19
–, forest, 445
–, pond (*P. serotina*), 178, 181, 201
–, slash (*P.elliottii*), 9, 10, 181, 182

GENERAL INDEX

pine (*continued*)
– swamps, 437
–, white (*P.strobus*), 180
Pine Barrens (N. J., U.S.A.), 29, 99
pineapple crop, 78
Pino-Quercetum, 443
pintail (*Anas acuta*), 374
pipit, Richard's (*Anthus novaeseelandiae*), 376
–, tree (*A. trivialis*), 430
pioneer association, 439
Pitahaya (Puerto Rico), 148
Placaquods, 305
Plains Grassland Region (U.S.A.), 118, 120, 121, 124, 132
plant(s), biomass, 130, 132
–, climbing, 441
– communities, 395, 401
– classification, 395
– fossils, 27, 28
–, free-floating aquatic, 269
–, herbaceous aquatic, 269
–, herbaceous semi-aquatic, 269
– succession, 439
plantation(s), 61, 414, 412
– forestry, 246, 248
planting in fringes, 165
plateaux, 392
Platte Rivers (Nebr.– Colo., U.S.A.), 120, 131
Playa de Humacao (Puerto Rico), 221
Pleistocene, 27, 35, 38, 217, 226
– terraces, 120, 303, 312, 322
Ploiarium padang, 332
plover, grey (*Pluvialis squatarola*), 375
–, little ringed (*Charadrius dubius*), 375
–, ringed (*C. hiaticula*), 375
pluviometer, 284
plywood, 355
pneumathodes, 72
pneumatophores (*see also* breathing roots), 127, 153, 157, 158, 182, 185, 218, 239, 243, 270, 279, 280, 313
pneumatorrhizae, 65, 72
pocosins, 59, 78, 171, 172, 175, 176, 178, 181, 205
–, short, 181
–, tall, 181
podzol, 303, 305, 321, 329
–, groundwater (Tropaquod), 330, 331
–, groundwater humus (Aquod), 308, 320
–, humus (Humod), 308
podzolization, 327, 330
poikilothermous animals, 431
point bars, 87, 90, 91, 102, 116, 117, 120
poisonous chemicals, 68
Poland 7, 89, 437–441, 459
polders, 247
pollen analysis, 309, 324
Pom Lagoon (Mexico), 293
ponds, 20, 21
Pontian (Malaysia), 324
Pontoon (Mayo, Ireland), 409
pony fish (*Equula edentula*), 378, 379
Poortugaal (Zuid-Holland, The Netherlands), 409

poplar (*Populus*), 442
– willow forest, 440
–, yellow (*Liriodendron tulipifera*), 43, 73, 102, 117, 119, 120, 180, 201
populations, 102
porcupine, crestless Himalaya (*Hystrix hodgsonii*), 372
porpoise, little (*Neomeris phocanoides*), 371
Port Moresby (Papua New Guinea), 335, 336, 349, 351
Portland (Ont., Canada), 166
Potametea, 443
Potamion, 442
potassium, 113, 162, 163, 443, 445
potentiometric surface, 16
power front, 143
– line, 87, 88
– plant, 69
– sheets, 17
prairie, 122
Prairie Creek (Fla., U.S.A.), 112
precipitation, 59, 61, 92–94, 101, 106, 121, 128, 171, 172, 407, 408, 409, 411, 414
–, annual, 400, 403
–, orographic, 274
predation, 130
Preuilly (Indre-et-Loire, France), 409
primary production, 15, 72
– productivity, 68, 70, 72, 87, 104, 108, 130, 133, 143, 159, 196, 329, 330, 437, 447, 448, 451, 459
– –, gross, 70, 72, 104, 158, 159, 160, 185, 186
– –, net, 22, 156, 159, 160, 185, 186, 196
– –, – daytime, 70
– –, – 24-hour, 72
primates, 133, 244, 247
production, 88, 102, 104, 106, 186, 247, 413, 414, 440
–, aboveground, 72
–, biomass, 104, 186, 187
– allocation, 187
–, belowground, 72
–, – biomass, 104
–, biomass, 88, 104, 105, 106, 118, 134, 186, 196, 329
– leaves, 186, 187
– litter, 186, 247, 437
–, net, 27, 31, 41, 415
– – biomass, 72
–: P/B ratio, 39, 41
–: P/R ratio, 33
–, primary, 15, 72
–, seed, 236
–, understory, 186, 187
–, wood, 186, 187, 196, 448
profile(s) diagrams, 254, 255, 263
–, forest, 309
–, peat, 309
prop roots (*see also* stilt roots), 144, 146, 158
propagules, 60, 133
protein, 75
Pterocarpus forests (Puerto Rico), 251
– –, climax conditions for, 256, 264
– –, community structure of, 253, 254, 255, 256
– –, – –: basal area of, 256, 257, 263, 264

Pterocarpus forests (*continued*)
– –, – –: complexity index of, 256, 257, 263, 264
– –, – –: forest strata, 253
– –, – –: species diversity, 256, 257, 263, 264
– –, – –: – importance values, 253, 256
– –, – –: tree density, 256, 257, 263, 264
– –, – –: tree height, 254, 256
– –, – –: understory, 254, 255
– –, distribution of, 251, 252
– –, floristic composition of, 253
– fruits, characteristics of, 260, 261
– seeds, 260, 261, 262
– –, germination of, 260
– –: hydrology (role in implantation), 260, 261, 262
– –: massive implantation of, 262
Pueh Forest Reserve (Sarawak), 316, 320, 325
Puerto Limon (Costa Rica), 274
Puerto Rico (*see also particular localities*), 8, 65, 67, 76, 98, 100, 102, 105, 106, 109–111, 143, 145, 150, 152, 153, 158, 161, 163, 164, 166, 183, 188, 196, 220, 221, 234, 236, 245, 251–253, 256, 259–263, 453, 458
Puerto Viejo (Costa Rica), 4
pufferfish (*Tetraodon patoca*), 379, 380
Puget Sound (Wash., U.S.A.), 71
puktal (*Bucida buceras*), 221, 253, 255, 256, 291
Pulau Dolak (Irian Jaya, Indonesia), 336, 354
Pulau River (Irian Jaya, Indonesia), 336
pulsed hydroperiods, 258
pumping, 77
Punta Gorda (Puerto Rico), 152
Punta Pitahaya (La Parguera, Puerto Rico), 152
Purari River (Papua New Guinea), 336, 340
Pyrenees, 408

quail, common (*Coturnix coturnix*), 376
Quebec (Canada), 78
Queensland (Australia), 156, 387–389, 391, 392, 394, 397
Querceto-Betuletum molinietosum, 442
Quintana Roo (Mexico), 290–292, 294

raccoon, dog (*Nyctereutes procynoides*), 429
–, North American (*Procyon lotor*), 429
radial wood growth, 106
radioactive Cesium, 79
– isotopes, 68
rainfall, 15, 16, 25, 60, 65, 79, 126, 127, 171, 271, 274, 282, 284, 287, 335, 341, 387, 389, 392, 395, 396, 409, 452, 454, 455, 457
–, monsoonal, 387
raised bogs, 437, 440
Rajang Delta (Sarawak), 299, 306–308, 323, 330
Rajshahi (Bangladesh), 358
rangeland, 133
Rangpur (Bangladesh), 358
Raritan River (N.J., U.S.A.), 119
rat(s), 245
–, bandicot (*Bandicota indica*), 372
–, mole (*B. bangalensis*), 372
–, house (*Rattus rattus*), 245, 372
rattan, 129, 354
reclamation, 1, 2, 247, 248, 407

recolonization, 133
recovery, 130
recruitment, 133
recycling, 110, 112
red mangrove, *see* mangrove, red
Red River (U.S.A.), 116, 121
redox, 112
– potential, 74, 79, 171
redstart (*Phoenicurus phoenicurus*), 430
reducing environment, 74
reduction of wetlands, 53, 54
reed(s), 439
– swamp, 387
reefs, 19, 22
regeneration, 118, 119, 121, 255, 261, 330, 331, 341, 396, 403, 404, 440
regional role, 447
reiteration, 237
relevees, 418, 417, 419, 423–425, 438
relic levees, 88
remote sensing, 201
–: aerial photographs, 201, 203, 207
– –: landsat, 201, 203, 207
Rennes (Ille-et-Vilaine, France), 409
reproduction, 15, 133, 134
reptiles (Reptilia), 88, 253
research, 68, 447, 449
reservoirs, 106, 131, 132
residence time, 60, 78, 79
resources, 66
respiration, 27, 67, 68, 70, 72, 75, 76, 104, 156, 157, 159, 160, 185, 186, 452, 454
–, daytime, 72, 157
–, leaf, 454
–, nighttime, 70, 72, 156, 158
–, root, 69, 72, 454,
–, soil, 69, 72, 74, 75
–, 24-hr plant, 72
resprouting, 102, 402
restoration, 133
retranslocation, 109, 162
revetments, 131
rheophytes, 8, 337, 349, 350, 392, 394
rhesus monkey (*Macaca mulatta*), 373
Rhine River, 407, 415
rhizosphere, 73, 76, 88
rice, 228, 247
– production, 58
Rideau Lakes Region (Ont., Canada), 165
ridge-and-swale, 87–90
ridges, 117, 119
Ringbong River (Sundarbans, Bangladesh), 368
Rio Anton Ruiz (Puerto Rico), 100
Rio Atrato (Colombia), 269
Rio Branco (Brazil), 267, 271
Rio Cocal (Puerto Rico), 100
Rio Colorado (Costa Rica), 281–283
Rio De Las Canas (Mexico), 100
Rio Espiritu Santo (Puerto Rico), 100
Rio Grande (U.S.A.), 3, 8, 71, 122

Rio Grande–Pecos Rivers (Texas, U.S.A.), 121
Rio Hondo (Mexico), 294
Rio La Plata (Argentina), 4
Rio Lara (Panama), 128
Rio Negro (Venezuela–Brazil), 9, 271
Rio Paraguay (Brazil), 272
Rio Penitencia (Costa Rica), 274, 276, 277, 278, 280–282, 285
Rio San Pedro (Mexico), 294
Rio Tonale (Mexico), 288
Rio Tortuguero (Costa Rica), 274
Rio Ucayali (Peru), 268
riparian forests, 57, 79, 113, 124, 218, 287
–, deciduous, 287
–, evergreen, 287
–, high, 287
–, medium, 287
–, semi-evergreen, 287
–, vegetation, 121
– woodland, 288
river, 16, 17, 19–21, 87, 88, 93, 97, 119, 267, 269, 270, 272, 274, 282, 287, 288, 293, 299, 314, 389, 390, 396, 407, 411, 412, 416, 418, 440, 441
– banks, 89, 347, 350, 387, 390, 394
– basins, 413
– bed, 441, 439
– bends, 90
–, blackwater, 127, 271, 274, 275
– bottom, 97
– channels, 91, 93
–, clear-water, 127
– discharge, 359, 360
– flats, 390, 392
– level, 94
– piracy, 20
– plains, 337
– runoff, 19, 20
– scrolls, 338, 346, 349, 354
–, sediment-laden, 282
– stage, 93, 94
– swamps, 20
– tracks, 335, 338, 343
– valleys, 437, 441
–, white water, 271
River Indus (Pakistan), 4
River Nabu Kongole (East Africa), 3
Riverina (N.S.W., Australia), 399
riverine forests, 87–90, 92, 94, 99, 101–106, 108, 109, 130, 353
– –: mangrove forest, 128
– –: processes in, 87, 88
– swamp, 222
road construction, 69, 75
Roanoke River (N. C., U.S.A.), 116
robin (*Erithacus rubecula*), 430
–, magpie (*Copsychus erimelas*), 376
Roblitos (Mexico), 100
rock, 16
Rocky Mountains (U.S.A.), 115, 123
rodents, 245, 247
roe deer (*Capreolus capreolus*), 482
rook (*Corvus frugilegus*), 430

Rookery Bay (Fla., U.S.A.), 152
root, 9, 60–63, 65, 69, 72, 73, 76, 79, 101, 113, 124, 128, 149, 153, 156, 158, 162, 164, 165, 171, 173, 184, 193, 237, 239, 258, 259, 264, 279, 280, 389, 404, 414, 419, 424, 425, 427, 438, 439
–, adventitious, 65, 72, 73, 338, 341, 347, 349
–, aerial, 184, 185, 320, 323, 396
– biomass, 184, 258, 264
– breathing, 338, 347, 349
– depth, 92, 258, 259
– development, 60
– growth, 108
– knee, 72
–: lateral knee-roots, 72
– mat, 60
– network, 236
– nodules, 415, 417
– peg, 72
– pneumathodes, 72
–, prop, 185, 239, 338, 347, 349
–: serial knee-roots, 72
– stilt, 221, 222, 318, 320, 338, 341, 347, 349
–: stilted peg, 72
–, succulent, 72
– systems, 310, 320
–, temporary, 72
root-to-shoot ratio, 73, 144
rooting zone, 401
Roseau (Dominica), 225
ruderals, 101, 102
run-on, 335, 400, 403
runoff, 17, 19, 21, 57, 60, 66, 75, 77–79, 104, 113, 132, 143, 147, 150, 160, 171, 357, 362, 387, 396, 400, 403, 421
Russia (U.S.S.R.), 2, 3, 7

Sabal Forest Reserve (Sarawak), 323
Sacramento River (Calif., U.S.A.), 121
Sacramento Slope (N. M., U.S.A.), 97
Sacramento Valley (Calif., U.S.A.), 6, 123, 124
sago, 268, 271
– flour, 354
– swamp, 337, 342, 347, 349, 353
Saint Lawrence River (Canada), 120
salamander, fire (*Salamandra salamandra*), 431
Salicetum albae, 418
Salicetum albae–S. fragilis, 439
S. pentandro–S. cinereae, 439
S. triandro–S. viminalis, 441, 443
Salici-Franguletum, 439
S.–Populetum, 440, 441, 443
Salicion albae, 441
S. eleagni, 441
saline, 389, 394
– flats, 349, 350
– intrusion, 251, 252
salinity, 76, 104, 113, 114, 126–128, 134, 143, 144, 146–151, 157, 158, 160, 222, 223, 228, 232, 274, 291, 337, 347, 352, 357, 363
Salix–Rhamnus, 442
salmonoid fish, 57
salt, 123
– flats, 146, 147, 149

salt (*continued*)
– marsh, 19, 20, 114, 389
– water, 103, 127
Salt-Gila River (Ariz., U.S.A.), 121
saltcedar (*Tamarix pentandra*), 101, 121–123, 131–133
saltpans, 388
saltwater wedges, 357
Salybia (Trinidad), 236–239
San Andreas Fault (Calif., U.S.A.), 3
San Carlos de Rio Negro (Venezuela), 305, 309, 325, 326, 331
San Cayetano (Mexico), 288
San Joaquin Delta (Calif., U.S.A.), 78
San Joaquin River (Calif., U.S.A.), 121
San Jose (Costa Rica), 285
Sand Hill (Trinidad), 226, 227, 244
sand(s), 89, 90, 93, 120
– bars, 19, 93, 123, 131
sandpiper, common (*Tringa hypoleucos*), 375
sandy soils, 122
Sangamon River (Ill., U.S.A.), 93, 112, 118
Santa Rosa (Costa Rica), 152
Santee River (S. C., U.S.A.), 116
sapling(s), 119, 121, 157, 400, 428, 429
– density, 99
– growth, 102
saprobiotic, 446
saprophages, 445
Sarawak, 4, 29, 299, 302, 303, 305, 306, 308, 309, 312, 313, 316–325, 327–329
Saskatchewan River (Canada), 126
Satkhira (Bangladesh), 360
saturated soil, 108
saturation, 102, 117
savanna, 88, 120, 214, 215–218, 234, 235, 267, 272, 399
–, seasonal, 267
Savanne Sarcelle (French Guiana), 234
sawgrass (*Cladium jamaicense*), 19, 38, 441
Scheldt River (Belgium), 407
Scheuchzerio-Caricetea fuscae, 443
schistosoma, 247
Schwarzwald (Baden-Württemberg, F.R.G.), 409, 416
Scirpo-Phragmitetum, 442
scleromorphic features, 326, 327
sclerophyll forest, 398, 399
sclerophyllous, 395
– leaves, 127
sclerophylly, 2, 9, 62, 63, 172, 314, 315, 318, 326, 327, 451
Scotland, 71, 407
scour, 94, 130, 134
scouring, 89, 102, 119
scrolls, 90
scrub, 418, 430
sea bass (*Lates calcarifer*), 378, 379
– level, 70, 76, 89, 221, 223, 226, 231, 287
– – rise, 110
– water, 150, 158, 407
seagrass, 32
seasonal flooding, 87, 88
– ígapo, 127, 128
– várzea forest, 99, 103, 127

secondary biochemical compounds, 327, 328
– – –: alkaloids, 327
– – –: aromatic oils, 327
– – –: polyphenols, 303, 323, 327, 328
– – –: tannins, 327
– – –: terpenes, 327
– succession, 117
sedge(s), 279, 281, 310, 311, 313, 315, 317, 318, 395, 422, 439
– grass tundra, 125
– peat, 414
Sedili (Malaysia), 324
sediment, 27, 29–31, 33, 36, 44, 71, 74–76, 87, 89, 102, 108–110, 112, 113, 129, 131, 132, 134, 143, 148, 158, 159, 228, 231, 237, 274, 282, 297, 308, 357, 451, 453
–, allogenic, 32
–, alluvial, 70
–, autogenic, 32
– deposition, 87, 88, 102
–, erosion of, 87
– exchange, 129
–, flooded, 59
–: grain size, 102, 418
–, inorganic, 33, 70
– load, 129
–, organic, 33, 42, 70
– profiles, 27
–, suspended, 90
– transport, 89
–, wetland, 59
sedimentary rocks, 90
sedimentation, 60, 70, 71, 109, 110, 112, 124, 346, 351, 354, 411, 412, 418, 420, 442, 458
– rate, 112
seed, 57, 58, 279, 403, 404, 405, 439
– dispersal, 31, 36, 37, 236, 237, 244
– germination, 87, 117, 236–239, 246
– production, 236
–, water dispersed, 236
seedbed, 121
seedling, 58, 65, 69, 108, 119, 241, 271–281, 344, 347, 387, 400, 402–405, 427
– biomass, 156
– colonization, 145, 154, 156
– density, 261
– dispersal, 144, 156
– distribution, 255
– establishment, 156, 236, 237, 363, 365
– growth, 150, 156, 157, 159
– height, 153, 157, 261
– mortality, 156
–, palm, 99
– photosynthesis, 156, 157, 159
– predation, 156
– respiration, 156, 157
–: size classes, 157
– survival, 88, 121
–, tidal sorting of, 150, 156
– vivipary, 144, 156, 162, 363, 385
seepage, 21, 87, 400, 402
seiches, 59, 143, 458

Seine River (France), 407
Sela River (Bangladesh), 368
Selangor (Malaysia), 78
selective cutting, 133
– removal, 133
self-thinning, 440
Selma (Ala., U.S.A.), 3
selva alta riparia, 287
– *baja inundable*, 290
– *mediana riparia*, 287
semi-evergreen riparian forest, 287
senescence, 72, 75, 452
Sennestadt (Nordrhein-Westfalen, F.R.G.), 409
Sepik River (Papua New Guinea), 336, 341, 345, 349
seral stages, 442
sere, 442
serotony, 178
sewage effluent, 110, 186, 189, 194, 196
– enrichment, 447
– enriched wetlands, 109
Sewanee River (Fla., U.S.A.), 116
shad, jewelled (*Illisha filigera*), 378
shade, 101, 102, 106
– intolerance, 125
– plants, 422
– tolerant, 102, 128
– – hardwoods, 117
shallow water, 111
Sharankhola Range (Sundarbans, Bangladesh), 364, 365, 368
shellfish, 57
shelter belts, 19
Shenandoah River (U.S.A.), 119
Shibsa River (Nalianala Range, Bangladesh), 365, 368
Shippea Hill (Cambridgeshire, England), 71
shipping, 418
shoot growth, 402
Shorea albida consociation, 307, 311
– – forests, 329
– – peatswamp consociation, 306
– – –*Gonystylus bancanus–Stemonurus secundiflorus* association (Alan forest), 307, 311
– – – *Litsea–Parastemon* association, 307, 314
shores, rocky intertidal, 19
shrike, ashy shallow (*Artamus fuscus*), 376
–, commonwood (*Tephrodornis pondicerianus*), 376
–, black-headed (*Lanius schach*), 376
–, large cuckoo (*Coracina novaehollandiae*), 376
shrimp, 55, 57
shrub(s), 94, 99, 101, 119, 120–123, 125, 127, 128, 178–182, 279, 338, 341, 347, 387, 393, 395, 400, 405, 439, 440
– biomass, 101
–, deciduous, 191
–, evergreen, 191, 201, 203
– layer, 417, 419–421, 428, 430, 432
– seedlings, 102
– thickets, 134
– wetlands, 295
shrublands, 441
Signy Island (Antarctic), 71
silica, 443

silicates, 229, 230, 245
silt, 93, 120, 222
siltation, 69, 110, 357, 359, 360, 362, 443
Silurian, 27
silviculture, 130, 133, 330–332
simulation models, 28, 30, 39
– – of succession, 39, 40
Singapore, 361–366, 369, 370
Sinnamary (French Guiana), 238
sinuosity, 129, 130
siriuba (black mangrove), 127
size, 88
– classes, 102
slopes, 65, 66
sloughing, 91
sloughs, 119
snags, 131
snail(s), 243
– hosts, 247
snake(s), 244
–, blind (*Typhlops diardi*), 373
–, tree (*Dryophis mycterizens*), 373
snipe (*Gallinago*), 375
snowmelt, 59, 60, 94, 357
Soajonya (Sundarbans, Bangladesh), 368
Sobger River (Irian Jaya, Indonesia), 336, 346
Socorro (N.M., U.S.A.), 71
Sogeri Plateau (Papua New Guinea), 351
soil, 77, 79, 87, 89, 90, 92, 104, 108, 109, 113, 117, 120, 122, 134, 171, 172, 176–178, 196, 218, 219, 222, 228, 231, 232, 234–237, 244, 245, 247, 251–253, 258–264, 271, 296, 337, 341, 390, 393, 404, 437–441, 443–454
–, acid, 303, 327
–, aerated, 261
– aeration, 2, 8–10, 94, 447, 449
–, alkaline, 338, 349, 391, 413, 421
–, alluvial, 270, 272, 274, 281, 287, 288, 290, 296, 391, 392, 397, 405, 413, 421
–, anaerobic, 2, 8, 9, 62, 65, 73, 261, 415
– area, 53, 54
–, black, 393
– C/N, 288
–, calcareous, 288, 291, 294
– carbon, 54, 55, 65
– – content, 363, 368
– : cation exchange capacity, 177
– characteristics, 108
–, chemical characteristics of, 177, 182, 303
– chemistry, 438, 451
– chloride, 368
–, clay 272, 274, 303, 308, 317, 319, 338, 341, 363, 368, 390, 391
– clay content, 296
– clay loam, 390
– color, 368
– depth, 182
–: Distric Histosol, 296
–, dry, 262
–, erosion, 60
–: Eutric Histosol, 288, 296
–: eutrophic muck, 303

soil (*continued*)
– fauna, 431
– fertility, 2, 5, 9, 58, 69, 187, 190, 193, 322
–, flooded, 261
– flow, 91
–: Fluvisol, 288, 290, 296
– formation, 420
–, gley, 303, 305, 317, 321
–: gleysol, 53, 54, 288, 296
–, halomorphic, 288, 296
–, heavy clay, 412
–: Histosols, 53, 54
– horizons, 60
–, humus rich, 421
– hydraulic conductivity, 58, 63
–, hydromorphic, 275, 288, 296, 413
–, infertile, 395
–, interstitial salinity of, 146–151, 158
–, loam, 63
–, mesotrophic muck, 303
– mineral(s), 27, 282, 299, 417, 425
–, mineral-rich, 120
– moisture, 61, 63, 65, 94, 106, 108, 120–122, 134, 260, 389, 392, 400, 403, 414
–, mountain, 305
– nitrogen content, 306, 327
– nutrient(s), 63, 296, 397, 398, 399, 405
– – concentrations, 176, 177
– –, standing stocks of, 193
–, nutrient-poor, 422
–, nutrient-rich, 426
–, oligotrophic, 308, 327
– oligotrophy, 5
–, organic, 70, 126, 201, 274, 275, 282, 338, 341, 414
– – alluvial, 296
– – carbon, 288
– – matter, 176, 182, 187, 193, 219, 220, 231, 258, 259, 264, 287, 288, 290, 291, 294, 296, 297
– oxygen, 8, 207
– particle distribution, 444
–, peat, 413, 414
–, peaty, 270, 272, 299, 303, 306, 317
– pH, 177, 288, 419–426, 428, 439
– phosphorus, 288
– – content, 306
– physical characteristics, 176, 182
– physio-chemical characteristics, 288
–, podzolic, 221
–, poorly drained, 392
– porosity, 63
–, redox, 104
–: Regosols, 275
–, ripening, 222, 226, 228, 231
–, rocky, 393
–, saline, 292, 294
– salinity, 2, 5, 7, 75, 171, 176, 187, 196, 251–253, 263, 264, 450, 458
–, sand, 60, 63, 368, 441, 443, 444
–, sandy, 176, 187, 193, 310, 317, 338, 343, 349, 390, 394, 399
–, sandy clay, 412, 428

–: sandy Tropaquod, 303
–, saturated, 267, 270, 277
– saturation, 2, 5, 8, 10, 60, 65
–, shallow, 416
–, silt, 404
–: sodium content, 296
–: Spodosols, 9
–, stony, 343, 349, 411, 417
– sulfide, 294
– temperature, 62, 63
– texture, 120, 134, 288, 290, 294, 363, 368
–: total nitrogen, 288
– types, 223, 232, 305
–, upland, 89
– water, 108
– – storage, 92
–, waterlogged, 73, 267, 270, 393, 394
–: white sands, 299, 332
–, zonal, 274
Solomon Islands, 336, 343
Sontecomapan Lagoon (Mexico), 291
sorption, 113
sourwood (*Oxydendrum arboreum*), 180
South America, 4, 54, 133, 234, 241, 244, 246, 247, 267–269, 272, 299, 429
South Australia (Australia), 387, 388, 396
– –, southeast, 398, 399, 403
South Carolina (U.S.A.), 29, 57, 97, 102, 103, 105, 117
South Dakota (U.S.A.), 95, 102
South Florida (U.S.A.), 182, 259
South Pacific Islands, 267, 270
South Platte River (Colo., U.S.A.), 95, 101, 121
Southeast Asia, 269, 271, 305
Southeastern Atlantic States (U.S.A.), 91
Southern Forest Region (U.S.A.), 94, 109, 116, 118
Spain, 430
Spanish moss (*Tillandsia usneoides*), 181
sparrow, house (*Passer domesticus*), 377
spatial patterns, 102
speciation, 89
species, 95
– abundance, 101
– composition, 22, 88, 89, 94, 113, 117, 118, 120, 127, 134, 178–182, 196, 278, 280, 281, 283, 412, 417, 419, 423, 424
– diversity, 127, 172, 178, 267, 277, 281, 282, 284, 305, 316–318, 320, 325, 328
– dominance, 119
– extinction, 370
– number, 94, 96, 97, 99, 100, 183
– response to hydrology, 447, 449
– richness, 58–60, 65, 119–121, 126, 131, 176, 181, 182, 290, 299, 305, 309, 316–318, 320, 322, 325, 328, 448, 456, 448, 456
specific conductivity, 230
– yield, 63
Sphagnetea, 443
Sphagno girgensohnii–Piceetum, 440
Sphagno–Piceetum, 440
Sphagnum, 67–71
– bog, 120
– peat, 305

GENERAL INDEX

spills, accidental, 132
splays, 349, 350
Spodosol, 303, 305, 308, 318, 319, 321, 324, 326, 329
–, Orthod, 308
spoonbill, white (*Platalea leucorodia*), 374, 430
springs, 21, 413, 414, 441
sprout, 134, 285
spruce (*Picea*), 439, 440
–, black (*P. mariana*), 37, 125, 175, 177–180, 183, 189, 190, 193
– squirrel(s), 372
–, beautiful (*Calloscrinus pygerythus*), 372
St. Francis River (Ark., U.S.A.), 116
St. Lucia, 222
St. Martin Island (Bangladesh), 358
stability, 439
stagnant swamp, 335, 337
– water, 108
stand age, 104
– density, 133
– height, 100
– topography, 108
standing stocks, 90, 102, 109
starling, pied (*Sturnus contra*), 377
steady state, 89, 104
stem, 73, 157, 158, 269, 270, 277, 278
– density, 65, 87, 94, 104, 128, 182, 277–285, 414
–, monocarpic, 269
– wood, 110
– – volume, 104, 107
stenohaline, 144
stilt roots, 221, 222, 338, 341, 347, 349
stint, little (*Calidris minutus*), 375
stochastic events, 117
stomata, 60, 62, 72
stomatal resistance, 72
stone-curlew, beach (*Esacus magnirostris*), 375
storage capacity, 92, 93
stork, Asian openbill (*Anastomus oscitans*), 374
–, black-nosed (*Xenorhynchus asiaticus*), 374
–, greater adjutant (*Leptoptilos dubius*), 374
–, lesser adjutant (*L. javanicus*), 374
–, painted (*Ibis leucocephalus*), 374
–, white (*Ciconia ciconia*), 374
–, white-necked (*C. episcopus*), 374
storm(s), 145, 147, 155
– hydrograph, 77, 78
– periods, 92
Strasbourg (Bas-Rhin, France), 409
stratum, 395
stream, 89, 90, 92, 93, 102, 109, 116, 119–121, 390, 392, 411, 416, 417
– bank, 130, 392
– beds, 387
–, braided, 125, 131
– channel, 87, 93, 116, 129, 412
– channelization, 129, 130
– discharge, 93
–, ephemeral, 92
– flow, intermittent, 391, 392
– –, perennial, 387, 391

– – regulation, 123
–, intermittent, 91, 133
– levees, 112
– migration, 119
– morphology, 91
–, perennial, 92
streamside elevation, 99
– forests, 87
– vegetation, 87, 88, 92, 94
stress (*see also* subsidy), 58, 59, 64, 67, 74–76, 101, 149, 156, 158–160
– tolerance, 102
– tolerators, 101, 102
stressors (*see also* disturbances), 64, 66, 67, 68, 75, 76, 104, 106, 134, 196
Strickland River (Papua New Guinea), 336, 346
string bogs (strangmoor), 19
stringers, 90
structural, 100
– characteristics, 94, 95
– complexity, 69
– indices, 182, 183, 196
– –: basal area, 182, 183, 196
– –: canopy height, 182, 183, 196
– –: stem density, 182, 183, 187
– –: tree species, 182, 183
structure, effect of climatic zone on, 395, 396, 397
–, forest 88, 94, 269, 274, 277, 278, 280–284
– of wetlands, 395
–, vegetation, 419, 425, 426, 431
stumps, 101
stunted, 127
– vegetation, 397, 404, 405
Stuttgart (Baden-Württemberg, F.R.G.), 407, 409
subalpine, 115
– region, 416, 428
subboreal region, 415
subcanopy, 99
subsidence, 76
subsidy stress, 59, 449, 455, 456
– –: concept, 49
– –: factors : aeration, 449
– – –: deep water, 456
– – –: dispersal, 449
– – –: drought, 456
– – –: fire, 456
– – –: frost, 456
– – –: growing season, 456
– – –: hurricanes, 456
– – –: hydroperiod, 456, 458
– – –: nutrients, 449
– – –: salinity, 456
– – –: storms, 456
– – –: ventilation of roots, 449
subsoil, 400, 404, 414
substrate, 89, 102, 414
subtropics, 88, 114, 115, 126, 128, 134, 395, 398, 403
succession, 28, 29, 32, 35, 38–40, 42, 59, 64, 65, 91, 101, 117–119, 152, 165, 218, 220–222, 232, 235, 236, 239, 240, 269, 282–284, 311, 351, 354, 416, 429, 430, 439, 441, 442

succession (*continued*)
–, allogenic, 65, 145, 165
–, autogenic, 65, 165
–: autosuccession, 154
–, Cleméntsian, 29, 35
–, cyclic, 35, 36, 38, 44, 65, 145
–, energy control of, 30
–, freshwater, 165, 166
–, hydrarch, 29, 30, 33, 36, 44
– models, 29, 36, 39, 40
– reversal, 65
–, reverse, 33, 35
–, saltwater, 145, 154
– stages, 30
successional development, 91
– stage (*see also seral stage*), 439
succulent plants, 159, 347
suckers, 349
Sudan, 26
sulphate reduction, 68, 74, 75
sulphide, 74, 75
sulphur, 55, 75, 231, 303
Sumatra, 299, 302, 305, 325
summer storms, 94
Sunati Khal Canal (Sundarbans, Bangladesh), 364, 368
sunbird, purple rumped (*Nectarinia zeylonica*), 377
Sundarbans (Bangladesh), 357–365, 368–373, 381–384
–, area of, 357, 358, 361
–, environment of, 357, 361, 362
–, fauna of, 359, 364, 365, 370, 371
–, –: birds, 372, 374–376, 377
–, – –: decapods, 373, 381
–, – –: fish, 372–380
–, – –: mammals, 370, 371, 372
–, – –: molluscs, 382
–, – –: reptiles, 370, 373
–, flora of, 359, 366, 367, 383, 384
–, location of, 357, 358
Sungai Dalam Forest Reserve (Sarawak), 312, 322, 323
Sungai Merbok Estuary (Kedah, Malaysia), 106, 168
sunlight, 15, 16, 64, 129
surface water, 109
Surinam, 214, 218, 222, 224, 228, 229, 234, 244, 246, 247
survivorship, 76
suspended loads, 131
– sediments, 90, 127
Susquehanna River (N. Y., U.S.A.), 119, 120
Suwalski Lake District (Poland), 439
swales, 93, 117
swamp, 3–5, 7, 8, 26, 27, 30, 31, 97, 105, 106, 110, 171, 173, 175, 177, 183, 187, 190, 217–219, 222, 226, 231, 239, 241, 243, 246, 267, 270, 272, 279, 280, 387
–, alluvial, 311
–, back, 116, 221–223, 231, 234, 235, 239, 240, 241
–, brackish, 270
– carr, 422
– depressions, 93
– forest, 120, 214, 215, 216, 219, 220, 223, 224, 228, 229, 234, 235, 238–240, 245, 246, 268, 271, 272, 274
– –, 336–338, 441

– –, alder, 437
– –, coniferous, 337, 344, 345
– –, mixed, 337–340, 352, 354
– –, pine, 437
– –, scrub, 336, 337, 349
– –, –, mangrove, 337, 350
– –, –, mixed, 337, 349
– –, woodland, 336, 337, 347, 348
– –, –, mixed, 337, 347, 348, 352, 353
– –, –, pandan, 337, 349
– –, –, sago, 337, 347
– herbaceous, 214, 216, 218, 226, 228, 229, 232, 234, 235, 240, 247, 268, 302, 387
–, history of, 27
–, intermittently flooded, 268
–, littoral, 311
–, mangrove, 214
–, mixed hardwood forest, 270, 274, 275, 282
–: mixed woods, 214, 218, 223, 224, 228, 230, 231, 234, 235, 239, 240
–, monospecific, 267, 269, 270
–, oligotrophic peaty, 299
–, palm, 214, 216, 218–220, 227, 234, 239, 240, 246, 247, 267–274, 280, 283, 284
–, papyrus, 269
–, peat, 271, 305
–, permanently flooded, 268
–, *Pterocarpus*, 67
–, rain-fed, 270, 274–284
–, reed 221, 241, 387
–, riparian, 218, 222, 235
–, river-fed, 274, 276–284
–, riverine, 222, 245
–, sago palm, 269, 270, 302
–, seasonal, 215, 267, 268, 272
–, – woodland, 268
–, sheoak (*Casuarina glauca*), 389, 391, 403
–, species-rich, 267, 270
–, thermally stressed, 101
– values, 246
–, várzea, 269, 270
–, wooded, 291
– woods, 218, 219, 221, 224, 225, 227, 228, 240
sweet pepperbrush (*Clethra alnifolia*), 180, 181
swift (*Apus*), 374
–, crested tree (*Hemiprocne longipennis*), 375
–, palm (*Cypsiurus parvus*), 374
Switzerland, 40
Sydney (N.S.W., Australia), 394
Syke (Niedersachsen, F.R.G.), 409
Sylhat (Bangladesh), 358
symbionts, nitrogen-fixing, 425
synchronous tree maturation, 154, 155
synergy, 456
synthetic chemicals, 132
Szczecin Lagoon (Poland), 437

Tabasco (Mexico), 287–290, 293, 294
taiga, 56
tailorbird, common (*Orthotomus sutorius*), 376

tamarack (*Larix laricina*), 5, 26, 30, 37, 120, 166, 175, 178–180, 190
Tamaulipas (Mexico), 287
Tampico (Mexico), 287
Tana River (Kenya), 133
Tangail (Bangladesh), 358
Tanjong Penjuru Road (Singapore), 362
tannins, 165
Tar River (N.C., U.S.A.), 116
Tasbapauni, 246
Tasek Bera (Malaysia), 57
tasistal(es) (*Acoelorrhaphe wrightii*), 290, 292
Tasmania (Australia), 391
Tatau (Sarawak), 313
Tauri River (Papua New Guinea), 336, 342, 343
teal (*Anas crecca*), 430
Teapa (Mexico), 288
Tecapan Lagoon (Mexico), 293
tectonic activity, 89
Temesh River (Belize), 271
temichales, 271
temperate, 421
– region, 54–56
– zone, 91, 109, 114
temperature, 2, 5, 94, 114, 123, 146, 157, 160, 165, 171, 172, 287, 404, 407, 408, 414
– effects on species distribution, 172
–, frost-free, 172
– zones, 87
temporal patterns of function, 447
Ten Thousand Islands (Fla., U.S.A.), 100, 106, 152
Tennessee Valley (Tenn., U.S.A.), 165
tern, greater crested (*Sterna bergii*), 375
–, gull-billed (*S. nilotica*), 375
–, least (*S. albifrons*), 375
–, lesser crested (*S. bengalensis*), 375
–, river (*S. aurantia*), 375
–, sooty (*S. fuscata*), 375
–, spotted (*Streptopelia chinensis*), 375
terra firma (see also tierra firme), 127
terrace(s), 19, 90, 95, 124, 125
– cultivation, 359
Tertiary, 90, 303
Texas (U.S.A.), 9, 43, 57, 98, 168, 171
Thailand, 149, 153
Thames River (U.K.), 407
thatching, 354
thermal stress, 58, 64
thermo-erosional niches, 91
– processes, 91
therophytes, 416, 420, 422, 418
thicket, 119, 124, 416–418, 420
Thoreau's Bog (Concord, Mass., U.S.A.), 174
thorn woodlands, 129
thorn-shrub wetland, 295
thornless-shrub wetland, 295
Thornwaite's classification, 302
thrush(es), mistle (*Turdus viscivorus*), 430
–, song (*T. philomelos*), 430
Thüringer Wald (G.D.R.), 409

tidal, 412, 416, 418, 419
– amplitude, 144, 152
– bores, 359
– creeks, 270
– energy, 64, 143, 145, 154, 161
– exchange, 171–173, 176
– factors, 144, 147, 150, 159
– flats, 270, 347, 349
– flooding, 346, 347, 349, 351
– frequency, 176
– influence, 269–271, 274
– inundation, 158, 159
– seasonality, 176
– swamp, 59
– várzea, 127
– zone, 270
tides, 20, 22, 59, 64, 127, 277, 335, 338, 347, 349, 375, 359, 362, 364, 365, 450, 458
tierra firme (see also terra firma), 305, 326
tigasso oil, 354
tiger, royal Bengal (*Panthera tigris*), 370, 371
Tilio-Carpinetum, 442
tilling, 76
timber, 117, 238, 246, 247, 324, 388, 437, 440
– characteristics, 311
– harvest, 130, 133
–, merchantable, 124
– removal, 55, 58
Timor (Indonesia), 389
tintal(es) (*Haematoxylum campechianum*), 221, 246, 289–292, 296
Tippecanoe River (Ind., U.S.A.), 96, 119
tissue, 446
tit, great (*Parus major*), 377, 430
–, grey (*P. major*), 377, 430
–, long-tailed (*Aegithalos caudatus*), 430
–, willow (*P. montanus*), 430
Tocumwal (N.S.W., Australia), 388
Tombigbee River (Ala., U.S.A.), 116
Tonnère (Yonne, France), 409
topographic features, 93
– relief, 129
– variation, 87
topography, 58, 59, 87, 92, 93, 110, 117, 134, 146, 147, 220, 222, 226, 232
–: depressions, 385
–: floodplain, 90, 128
–: mounds, 363
–: ridge-and-swale, 88
Tortuguero (Costa Rica), 274, 275, 282, 284
total hardness, 230
– organic carbon (TOC), 55, 56, 57, 79
toxic compounds, 130, 132
– metals, 132
– substances, 171
toxins, 59, 60, 68, 73, 74
tracheophytes, 165
trade winds, 65, 302, 335
transects, 87, 277, 279–281, 285
transgressions, 407, 415

transition, 120
– bog, 439
translocation, 60
transmissivity, 93
transpiration, 26, 60–63, 73, 156–160, 327, 454, 455, 457
transpiration-to-pan ratio, 62
Travancore (India), 4
tree(s), 59, 79, 87, 92, 100, 121, 387, 390, 395, 400, 404, 405, 440
– age, 91
–, alder, 439
– bases, 101
– death, 405
– density, 148, 149, 151, 152, 343, 396, 447
– diameter, 61, 102, 148, 149, 151, 152, 159, 389–392
– die-out, 252
– dieback, 61, 76
–, emergent, 336, 338, 340
– form, 311
– gaps, 119
– growth, 106, 149, 156, 159, 160, 182, 451
– – form, 145
– height, 128, 146–152, 155, 182, 183, 196, 288, 290–293, 295, 343, 344, 347, 389–395
– – factors affecting, 2, 5
– height/diameter ratio, 305, 311, 313, 327, 331
–, mature, 396, 405
– mortality, 101, 114, 146–149, 151, 154, 155, 196, 447
– palms, 337, 338
– phytomass, 329
– reproduction, 133, 134
– species, 88
– stratum, 101, 400, 404, 405
–, stunted, 391, 397, 404, 405
– trunk, 402
– zonation, 450, 451
Tremblois (Ardennes, France), 409
trenching, 10
tributaries, 89
Trinidad (*see also particular localities*), 4, 166, 213, 214, 217, 221–226, 228–231, 235–247, 262
Trinidad and Tobago, 271, 272
Trinité (Martinique), 222
Tristania–Parastemon–Palaquium association, 315
trophic equilibrium, 33
– levels, 75
– structure, 130
Tropic of Capricorn, 388, 389
tropical, 389, 398, 403
– coastlines, 145
– forests, 159
– rain forest, 7, 8, 302
– region, 53, 54, 55
– rivers, 133
– shelterwood system (TSS), 331
tropical America, 172, 182
tropics 53–56, 60, 62, 88, 110, 114, 115, 126, 128, 134, 287, 395
–, humid, 87, 134
–, wet, 392
tropic bird, long-tailed (*Phaethon lepturus*), 377
– –, red-tailed (*P. rubricauda*), 377

truli bush, 271
Tuil (Gelderland, The Netherlands), 409
tundra, 56
–, Arctic, 134
– climates, 114
–, sedge-grass, 124
–, tussock-heath, 125
Turama River (Papua New Guinea), 336
turnover, 44
– rate, 39
– time, 39, 41, 44, 187
–, – biomass, 187
turtle(s), 241, 246
–, marine, 274
–, black mud (*Trionyx hurum*), 373
–, green (*Chelonia mydas*), 373
–, hawksbill (*Eretmochelys inbricata*), 373
–, marine leatherneck (*Dermochelys coriacea*), 373
–, roof (*Kachuga tectum*), 373
–, sea (*Chelonia amboinensis, C. emys*), 373
–, soft-shelled (*Trionyx gangeticus*), 373
tussocks, 411, 427
twigs, 72, 107

Uganda, 3, 129, 165
ulat bulu (moth caterpillar), 311, 315, 327, 331
Ulta Khal Canal (Sharankhola Range, Bangladesh), 365, 368
Ultisol, 305, 318, 319, 323, 324, 330
undercutting, river, 91, 124
understory, 94, 99, 101, 128, 129, 178–182, 184, 186, 187, 190, 268, 279–281, 283, 285
– species, 99
– trees, 108
undrained, 104
United Kingdom, 7, 407, 425, 430
United States (of America) (*see also particular localities*), 1, 2, 6–9, 55, 57, 58, 61, 62, 65, 76–79, 88–92, 107, 113–115, 128, 131, 157, 165, 171, 174, 259, 449
– –, eastern, 171
– –, glaciated region of, 178
– –, north central, 172, 180
– –, northeastern, 180
– –, northern, 26
– –, southeastern, 19, 22, 88, 94, 102, 104, 116, 171, 172, 195
– –: southeastern coastal plain, 171
– –, southern, 171, 237
– –, southwestern, 26, 88, 94, 108, 133
upland, 73, 77, 79, 87, 92–94, 109, 116, 119, 121
– communities, 92, 128
– forests, 87, 94, 99, 101, 118, 120, 129
–, mesic, 87
–, species 119
– vegetation, 122
uplift, 16, 89
urea, 108
Urtico-Alnetum, 442
Uruguay, 272
uses (*see also utilization*): crabs, 355
–: exploitation, 354
–: food, 354, 355

GENERAL INDEX

uses (*continued*)
–: oil, 354
–: packing cases, 354
–: plywood, 354
–: starch, 354
–: wood, 354
Usumacinta River (Mexico), 287, 293
utilization, 330, 331, 332
–: conversion to agriculture, 330
–: fibre, 332
–: firewood, 332
–: logging, 330, 331
–: management, 331
–: timber exploitation, 330
Utuado (Puerto Rico), 251

Vaccinio myrtilli–Pinetum, 445
Vaccinio–Piceetea, 443
Vaccinio uliginosi–Pinetum, 440, 445
V. uliginosi–Pinetum ericetosum tetralicis, 444
Vacia Talega (Puerto Rico), 100, 106, 163
valley, 89, 90, 92, 93
– cutting, 89
value, 68
–, economic, 246, 248
–, non-forestry, 246
–, swamp, 246, 247
vampire, Indian false (*Megaderma lyra*), 371
vapor pressure deficit, 61
várzea, 127, 235, 237, 244, 269, 270
– forest, 58, 111
–, seasonal 127
–, tidal 127
vegetation, 92–94, 102, 112, 116, 129, 388
–, anthropogenic, 443
– map, 225, 227–229, 233, 311, 388
–, measurement of, 285
– ordination, 117
– profiles, 287, 289
– structure, 144, 146–148, 150–152, 309, 318, 325, 326
– transition, 118
– zonation, 144, 146–151, 154, 155, 165
vegetative propagation, 369
Venezuela, 9, 219, 220, 241, 267, 268, 271, 272
ventilation, 60
Veracruz (Mexico), 287, 290–292
Verlandung, 213, 226, 228
vertebrates (Vertebrata), 56, 242, 244
vertical accretion, 89, 90, 110
Vesbeck (Niedersachsen, F.R.G.), 409
Victoria (Australia), 387, 388
Viet Nam, 325
vines, 121, 128, 349
viper, carpet (*Echis carinata*), 373
–, Russell's (*Vipera russelli*), 373
Virginia (U.S.A.), 3, 71, 96, 116, 117, 119, 171, 180, 181, 183, 184, 188, 194, 195, 201, 203
viruses, 247
Vishwamitri River (India), 4
Vogelkop (Irian Jaya, Indonesia), 336

volcanic activity, 354
vole, bank (*Clethrionomys glareolus*), 429
–, tundra (*Microtus oeconomus*), 429
–, water (*Arvicola terrestris*), 374
vulture, griffon (*Gyps fulvus*), 374
–, white-backed (*G. benghalensis*), 374

Wabash River (Ind., U.S.A.), 95, 118, 119
Wageningen (Surinam), 228, 247
Wales, 71
Wanssum (Limburg, The Netherlands), 409
war, 69
Warao, 268, 271
warbler, bristled grass (*Chaetornis striatus*), 376
–, golden (*S. borin*), 430
–, grasshopper (*Locustella naevia*), 430
–, icterine (*Hippolais icterina*), 430
–, marsh (*Acrocephalus palustris*), 430
–, reed (*A. scirpaeus*), 430
–, sedge (*A. schoenobaenus*), 430
–, striated marsh (*Megalurus palustris*), 376
–, willow (*Phylloscopus trochilus*), 430
warm temperate, 398, 403
Warragul (Vic., Australia), 392
Warszawa (Poland), 438
Warta River (Poland), 439
Washington (U.S.A.), 43, 72, 95–124
Washington County (N.C., U.S.A.), 78
wastewater, 108
water, 15, 87, 104, 134
– availability, 457
– balance, 447
–, black, 231, 299, 301, 306, 320
– -borne diseases, 383
–, brackish, 219, 223, 228, 229, 235, 240, 241, 243, 245, 269–271
– budget, 8, 26, 282
– chemistry, 131, 159, 166, 175, 229, 230, 451
– conservation, 61, 400
–, consumptive use of, 133
– courses, 389, 390, 391, 400, 401, 404
– currents, 15, 19, 106, 108, 110
– deficit, 61–63, 88, 400, 404
– demand, 388
– depth, 216, 292, 297
– drainage, 36, 196
– flood, 228, 245
– flooding, 31, 36, 37, 38
–: floodwater, 171, 176
– flow, 60, 73, 104, 129, 146, 155, 156, 165, 447, 449, 453
– – direction, 447, 449
– –: kinetic energy, 447, 449
– –: physical damage, 449
– –: subsidy-stress, 449
– flowing, 32, 228, 229
– flushing, 60
–, fresh, 217, 219, 222, 226, 229, 234, 235, 240, 241, 243, 244, 245, 247, 292, 294
–, –: seeps 292
–, ground, 25, 33, 171, 173, 174, 183

water (*continued*)
– hyacinth (*Eichhornia crassipes*), 26, 31, 383
– infiltration, 173
– interception, 172
– level, 27, 36, 87, 90, 108, 171, 174, 182, 185, 219, 247, 274, 276, 277, 284, 441, 442
– – fluctuations, 276, 284
– motion, 288
– movement, 438
–, nutrient-poor, 270
–, oligotrophic, 308
–: overland flow, 171, 173
–: pH, 173, 175
– potential, 106, 159
– quality, 60, 65, 79, 126, 134, 175
– recharge, 172
– regime, 326–328
–, running, 440
–, salt, 271
–: seawater, 171, 176
– seepage, 25, 33, 171, 173
– seeps, 172
– shed(s), 17, 113, 451
– sheer flow, 171
shrew (*Neomys fodiens*), 429
–, stagnant, 222, 228, 243, 270, 437, 440
–: stillwater, 32
– storage, 25
– stress, 61–63, 108, 113, 114, 119, 127, 128
–: subsurface flow, 173
–, surface, 176, 177
–: surface flow, 172, 173
–: surface runoff, 25, 26
– table, 5, 8, 25, 33, 37, 60, 61, 76, 77, 93, 110, 113, 129, 171, 174, 326, 337, 340, 347, 349, 392, 395, 396, 404
– tolerance of plants, 174
– tracks, 60, 183, 449
– turbulence, 144
– turnover, 60, 79
– usage, 388, 400
– vapor pressure deficit, 61
waterfowl, 56
water-hen, white-breasted (*Amaurornis phoenicurus*), 376
waterlogged, 392–395, 404
waterlogging, 217, 219, 221, 222, 235, 302, 303, 305, 308, 318, 320, 321, 325, 329, 332, 338, 413, 414, 419
waves, 15, 64, 69, 91, 143, 144–147, 150, 154, 161, 357
Wavreille (Namur, Belgium), 409
weather front, 16
weathering, 89, 420
weaver bird, black-throated (*Ploceus bengalensis*), 377
–, streaked (*P. manyar*), 377
Weerribben (Overijssel, The Netherlands), 409, 426
Weser River (Germany), 407
West Bengal (India), 358
West Germany, 407, 409, 416
West Indies, 4, 213
West Irian (*see* Irian Jaya), 335
West New Britain (Papua), 4
Western Arid Forest Region (U.S.A.), 121, 124

Western Australia (Australia), 388, 392
Western Europe, 89
Western Hemisphere, 127
wet alder forests, 442, 443
– – woods, 438
– birch forests, 439
wet heathland, 387, 395
wetland, 108, 268, 269, 273, 275, 400, 402, 404
–, animals in, role of, 452
– area, 53, 54
–, basin, 59, 62, 70, 73, 77, 171, 177, 182–186, 188, 192, 193, 196, 271, 447–449, 453, 456–459
– bog, 172–175, 178, 180, 183, 184, 190, 191, 193, 196
– boundaries, 458
–, brushy bog, 174
–, brushy fen, 174
– classification, 449
–, coastal, 25, 26, 29, 30, 32, 33, 44, 55, 58, 389
–, – basin, 458
–, – riverine, 458, 459
–, coniferous forested, 171, 172, 181, 190
–: detritus foodwebs, 459
–, dwarfed, 456, 457
–, eutrophic, 451
–, experimental research on, 447
– floodplain, 457
– –, eroding, 453, 459
– floristics, 447
–, forested, 54, 58, 267, 273, 387–389, 395, 448
–, –: acalches, 291
–, –: chechenal, 291
–, –, evergreen, 288
–, –: puktal, 291
–, –, species-poor, 296
–, –, species-rich, 296
– formation, 28
–, freshwater, 54, 56, 58, 74, 172, 177, 182–186, 188, 192, 267, 296, 448, 449, 453, 456–459
– fringe, 59, 75, 171, 172, 270, 447–450, 458–460
– geomorphology, 28
–, global role of, 53–55
–, graminoid, 395
–, grass–sedge, 404
–, herbaceous, 287, 296
–, high forested, 287, 291, 296
–, human use of, 447
– hydroperiod, 447, 455, 456
– levee forests, 458
–, literature on, 447
–, low forested, 290
– management, 447, 449
–, mangrove, 55–58, 66, 68, 75, 76
–: – islands, 452
–: marginal fens, 173–175, 177, 179, 183, 184, 186–190, 193, 194
–, marsh, 174
–, minerotrophic, 173, 174, 178, 451
–, mixed types of, 458
–, monospecific, 287, 295, 458
–, montane, 65, 66
–, muskeg, 178, 179, 183, 184, 186, 189

GENERAL INDEX 527

wetland (*continued*)
–, needle-leaved deciduous, 181
–, – –, evergreen, 181
– non-forested, 458
–, nutrient-poor, 451, 453
–, nutrient-rich, 451
–, oligotrophic, 451
–, ombrotrophic, 173, 174, 176, 178, 451, 459
–, palm, 269, 272
–: – thickets, 292
– paradigms, 447–460
–, peat, accumulation of, 459
–, –, depth of, 451, 453, 459
– physiognomy, 447
–, regional role of, 447
–, riverine, 56–59, 62, 172, 271, 447–450, 453, 456–458
–, saltwater, 172, 176, 182–188, 196, 447–451, 453, 456–459
– – forested, 53–56, 74
– , sedge meadow, 174
–, sewage enrichment of, 447
–, shrub, 295
– soils, 53, 54
–, study of: holistic approach, 447, 448, 459
–, – –: multi factorial approach, 455, 456
–, – –: single factor approach, 455, 456
–, sub-optimal function of, 449
– swamps, 171, 173, 175, 177, 183, 187, 190
– –, species-poor, 178, 179
– –, species-rich, 178, 179
–, temperate freshwater, 54–56
–, thorn-shrub, 295
–, thornless-shrub, 295
–, treeless, 400, 404
–, tropical, 29, 267
–, uses of, 388
–, – –: fencing, 388
–, – –: firewood, 388
–, – –: housing, 388
–, – –: timber, 388
–, wood biomass production in, 448
whistler, mangrove (*Pachycephala cinerea*), 376
white mangrove, (*see* mangrove, white)
White River Basin (Ark., U.S.A.), 90
white sands, 299, 332
whitethroat (*Sylvis communis*), 430
–, lesser (*S. curruca*), 430
white-water rivers, 127
Wierden (Overijssel, The Netherlands), 409
Wierzbiczanskie Lake (Poland), 437
Wijk bij Duurstede (Utrecht, The Netherlands), 409
wildlife, 57, 58
– management, 133
willow (*Salix*), 30, 32, 71, 78, 102, 111, 117, 120, 121, 123–125, 287, 412, 415, 418, 419–421, 425, 440, 441
wilting, 61, 63
wind, 15, 59, 61, 64, 65, 69, 76, 78, 91, 143, 144, 148, 156, 165, 335
– breaks, 19
– damage, 313, 331

– erosion, 76, 78
windthrows, 38, 69, 311, 313, 315, 317, 327, 328
Wisconsin (U.S.A.), 3, 6, 30, 37, 71, 77, 78, 99, 112, 174, 178
Wisla river (Poland), 442
Wissahickon Creek (Penn. U.S.A.), 119
wood, 55, 72, 109, 110, 415–417, 420, 425, 426, 432
– biomass production, 104–106, 448
– cutting, 274
–, dead, 188
–, fuel, 245
– growth, 104, 160, 165
– harvesting, 441
– harvests, 58
– production, 104, 186, 187
– products, 133
– proporties, 245, 246
–, standing dead, 188
– uses, 245, 246
– volume, 61
woodcock (*Scolopax rusticola*), 430
woodland(s), 411–415, 417, 419–423, 425, 428–430
– swamp, 268
woodowl, brown (*Strix leptogammica*), 377
woodpecker(s), 58
–, greater spotted (*Dendrocopos major*), 430
–, lesser golden-backed (*Dinopium bengalensis*), 377
–, lesser spotted (*Dendrocopos minor*), 430
–, pygmy (*Picoides manus*), 377
–, rufous (*Micropterus brachyarus*), 377
– vegetation, 94, 108, 123
– vines, 254, 255
woody debris, 188
– vegetation, 94, 108, 123
– vines, 254, 255
wren (*Troglodytes troglodytes*), 430
wren-warbler, Franklin's (*Prinia hodgsoni*), 376
– –, yellow-bellied (*P. flaviventris*), 376
Wybunbury Moss (Cheshire, England), 3

xeric, 122
xeromorphic, 61–63
– features, 304, 316, 318–320, 326, 327
xeromorphism : bog xerophylly, characteristsics of leaves, 27
xeromorphy, 451
xerophytes, 62

Yazoo River (U.S.A.), 116
yellow fever, 247
yevaro (*Eperua purpurea*), 305, 319, 326
yolillales, 269, 271
Yucatan (Mexico), 291–294

zarzal (*Mimosa pigra*), 289, 295, 296
zonation, 29, 37, 58–60, 64, 65, 143, 144, 146, 147, 149–151, 154, 155, 165, 218, 222, 223, 232, 234, 267, 269, 270, 282, 307, 318
–, vegetation, 88, 93, 124
zone, 19

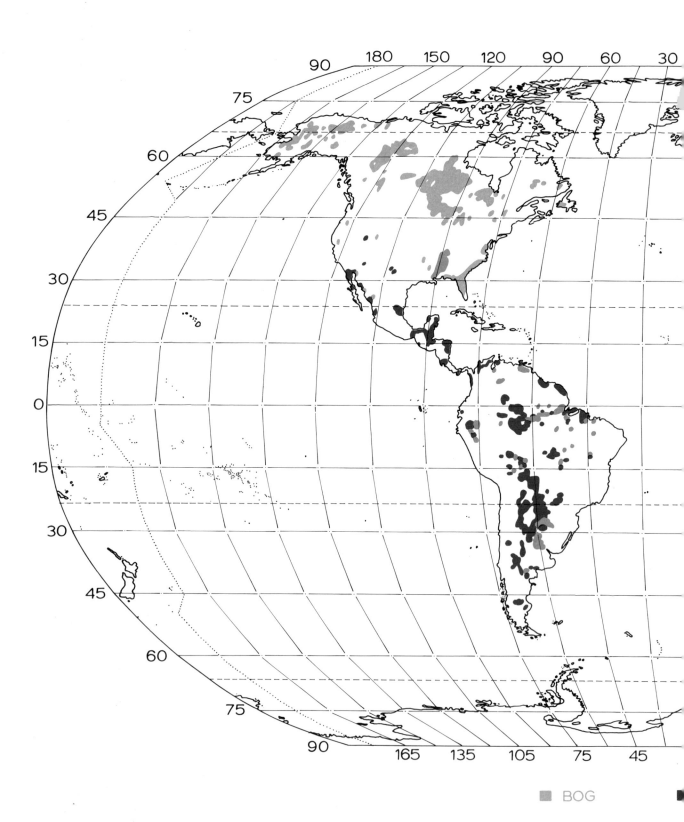
BOG